The Handbook of Microbial Metabolism of Amino Acids

Edited by

J.P.F. D'Mello

FSC
www.fsc.org
MIX
Paper from responsible sources
FSC® C013604

This volume is dedicated to my late daughter,
Deborah Claire D'Mello.

The Handbook of Microbial Metabolism of Amino Acids

Edited by

J.P.F. D'Mello

Formerly of SAC, University of Edinburgh King's Buildings Campus, Edinburgh, UK

CABI

CABI is a trading name of CAB International

CABI
Nosworthy Way
Wallingford
Oxfordshire OX10 8DE
UK

Tel: +44 (0)1491 832111
Fax: +44 (0)1491 833508
E-mail: info@cabi.org
Website: www.cabi.org

CABI
745 Atlantic Avenue
8th Floor
Boston, MA 02111
USA

Tel: +1 (617)682-9015)

E-mail: cabi-nao@cabi.org

A catalogue record for this book is available from the British Library, London, UK.

Library of Congress Cataloging-in-Publication Data

Names: D'Mello, J. P. Felix, editor. | C.A.B. International, issuing body.
Title: The handbook of microbial metabolism of amino acids / edited by J.P.F. D'Mello.
Other titles: Microbial metabolism of amino acids
Description: Oxfordshire, UK ; Boston, MA : CAB International, [2017] | Includes bibliographical references and index.
Identifiers: LCCN 2016028986 (print) | LCCN 2016029640 (ebook) | ISBN 9781780647234 (alk. paper) | ISBN 9781780647241 (pdf) | ISBN 9781780647258 (ePub)
Subjects: | MESH: Amino Acids--metabolism | Microbiological Phenomena | Amino Acids--biosynthesis | Biosynthetic Pathways--physiology
Classification: LCC QP562.B73 (print) | LCC QP562.B73 (ebook) | NLM QU 60 | DDC 572/.6519--dc23
LC record available at https://lccn.loc.gov/2016028986

ISBN-13: 978 1 78064 723 4

Commissioning editor: Rachael Russell
Editorial assistant: Alexandra Lainsbury
Production editor: James Bishop

Typeset by SPi, Pondicherry, India
Printed and bound by CPI Group (UK) Ltd, Croydon, CR0 4YY

Contents

Contributors

Abe, K., Department of Bioengineering, Nagaoka University of Technology, 1603-1 Kamitomioka, Nagaoka, Niigata 940-2188, Japan. E-mail: katsuabe@vos.nagaokaut.ac.jp

Afzal, M.I., Department of Biosciences, COMSATS Institute of Information Technology, Park Road, Tarlai Kalan, Islamabad 45550, Pakistan. E-mail: inam.afzal@comsats.edu.pk/dr-miafzal@hotmail.com

Akhtar, M., School of Biological Sciences, University of Southampton, University Road, Southampton, SO17 1BJ, UK. E-mail: M.Akhtar@soton.ac.uk

Alves, J.M.P., Department of Parasitology, Institute of Biomedical Sciences, Universidade de São Paulo, Av. Prof. Lineu Prestes, 1374, Cidade Universitária, SP, Brazil. E-mail: jotajj@usp.br

Balaña Fouce, R., Departamento de Ciencias Biomédicas, Universidad de Léon, Campus de Vegazana, s/n 24071 León, Spain and Instituto de Biotecnología de León (INBIOTEC), Avda Real no. 1, Parque Cientifico de León, 24006, Spain. E-mail: rbalf@unileon.es

Bashir, Q., School of Biological Sciences, University of the Punjab, Quaid-e-Azam Campus, Lahore 54590, Pakistan. E-mail: qbashir@gmail.com

Bender, R.A., Molecular, Cellular and Developmental Biology, University of Michigan, 830 N. University Ave., Ann Arbor, MI 48109-10888 , USA. E-mail: rbender@umich.edu

Brown, S., The Deviot Institute, Deviot, TAS 7275, Australia. E-mail: Simon.Brown@deviotinstitute.org

Cailliez-Grimal, C., Laboratoire d'Ingénierie des Biomolécules, Université de Lorraine, 2 avenue de la Forêt de Haye, BP 172, Vandoeuvre-lès-Nancy, F-54505, France. E-mail: catherine.cailliez@univ-lorraine.fr

Caldwell, R.W., Department of Pharmacology and Toxicology, Medical College of Georgia, Augusta University, 1120 15th Street, Augusta, GA 30912-2500, USA. E-mail: WCALDWEL@augusta.edu

Campanini, B., Dipartimento de Farmacia, Università degli Studi de Parma, Parco Area delle Scienze 23/A, 43124 Parma, Italy. E-mail: barbara.campanini@unipr.it

Cao, Y., School of Pharmacy, Second Military Medical University, Shanghai, China. E-mail: caoyingying608@163.com

Chakraborty, T., Institut für Medizinische Mikrobiologie, Justus-Liebig University Giessen, Justus-Liebig-Universität Gießen, Schubertstr. 81, 35392 Giessen, Germany. E-mail: trinad.chakraborty@mikrobio.med.uni-giessen.de

Charlier, D., Research Group of Microbiology, Faculty of Sciences and Bioengineering Sciences, Vrije Universiteit Brussel, Pleinlaan 2, 1050 Elsene, Brussels, Belgium. E-mail: dcharlie@vub.ac.be

Chowdhury, N.P., Max Planck Institute for Terrestrial Microbiology, Karl-von-Frisch-Straße 10, D-35043 Marburg, Germany. E-mail: nilanjanchaudhry@gmail.com

Coker, O.O., Department of Medicine and Therapeutics, Faculty of Medicine, Chinese University of Hong Kong, Hong Kong. E-mail: bisistill@yahoo.com

Dawes, I.W., School of Biotechnology and Biomolecular Sciences, The University of New South Wales, Sydney, NSW 2052, Australia. E-mail: i.dawes@unsw.edu.au

De Biase, D., Istituto Pasteur Italia - Fondazione Cenci-Bolognetti, Dipartimento di Scienze e Biotecnologie Medico-Chirurgiche, Sapienza Università di Roma, Corso della Repubblica 79, 04100 Latina, Italy. E-mail: daniela.debiase@uniroma1.it

Delaunay, S., Laboratoire Réactions et Génie des Procédés (LRGP), UMR 7274 du CNRS (Centre National de la Recherche Scientifique), 2 avenue de la Forêt de Haye, TSA 40602, Vandoeuvre-lès-Nancy, F-54518, France and Université de Lorraine, 2 avenue de la Forêt de Haye, TSA 40602, Vandoeuvre-lès-Nancy, F-54518, France. E-mail: stephane.delaunay@univ-lorraine.fr

D'Mello, J.P.F., Formerly of SAC, University of Edinburgh Campus, West Mains Road, Edinburgh EH9 3JG, UK. E-mail: jpfdmello@hotmail.co.uk

Dobson, R.C.J., Biomolecular Interaction Centre, School of Biological Sciences, University of Canterbury, Private Bag 4800, Christchurch, New Zealand and Department of Biochemistry and Molecular Biology, Bio21 Institute, University of Melbourne, Parkville, Australia. E-mail: renwick.dobson@canterbury.ac.nz

Dunn, M.F., Programa de Genómica Functional de Procariotes, Centro de Ciencias de Genómicas, Universidad Nacional Autónoma de México, Cuernavaca, Morelos, CP 62210, Mexico. E-mail: mike@ccg.unam.mx

Fulton, D., Vascular Biology Center and Department of Pharmacology and Toxicology, Medical College of Georgia, Augusta University, 1120 15th Street, Augusta, GA 30912-2500, USA. E-mail: dfulton@augusta.edu

Giovannercole, F., Istituto Pasteur Italia – Fondazione Cenci-Bolognetti, Dipartimento di Scienze e Biotecnologie Medico-Chirurgiche, Sapienza Università di Roma, Corso della Repubblica 79, 04100 Latina, Italy. E-mail: fabio.giovannercole@gmail.com

Holmes, D.E., Department of Physical and Biological Sciences, Western New England University, 1215 Wilbraham Rd, Springfield, MA 01119, USA. E-mail address: dholmes@bio.umass.edu

Hudson, A.O., Thomas H. Gosnell School of Life Sciences, Rochester Institute of Technology, 85 Lomb Memorial Drive, Rochester, NY 14623, USA. E-mail: aohsbi@rit.edu

Kera, Y., Department of Bioengineering, Nagaoka University of Technology, 1603-1 Kamitomioka, Nagaoka, Niigata 940-2188, Japan. E-mail: yoshkera@vos.nagaokaut.ac.jp

Kornfeld, G.D., School of Biotechnology and Biomolecular Sciences, The University of New South Wales, Sydney, NSW 2052, Australia. E-mail: g.kornfeld@unsw.edu.au

Leroy, F., Research Group of Industrial Microbiology and Food Biotechnology, Faculty of Sciences and Bioengineering Sciences, Vrije Universiteit Brussel, Pleinlaan 2, 1050 Elsene, Brussels, Belgium. E-mail: fleroy@vub.ac.be

Liao, Z., School of Pharmacy, Second Military Medical University, Shanghai, China. E-mail: 18701832213@163.com.

Locy, R.D., Department of Biological Sciences, 101 Rouse Life Sciences Building, Auburn University, Auburn, AL 36849, USA. E-mail: locyrob@auburn.edu

Lott, J.S., School of Biological Sciences, University of Auckland, 3a Symonds Street, Auckland 1142, New Zealand. E-mail: s.lott@auckland.ac.nz

Lucas, P.M., Institut des Sciences de la Vigne et du Vin (ISVV), Université de Bordeaux, 210 Chemin de Leysotte, 33882 Villenave d'Ornon, France. E-mail: patrick.lucas@u-bordeaux.fr

Lucas, R., Vascular Biology Center, Department of Pharmacology and Toxicology, Medical College of Georgia, Augusta University, 1120 15th Street, Augusta, GA 30912-2500, USA. E-mail: rlucas@augusta.edu

Ludovico, P., Life and Health Sciences Research Institute (ICVS), School of Health Sciences, Universidade do Minho, Campus de Gualtar, 4710-057 Braga, Portugal and ICVS/3B's-PT Government Associate Laboratory, Guimarães, Braga, Portugal. E-mail: pludovico@ecsaude.uminho.pt

Morán, J.M., Departamento de Ciencias Biomédicas, Universidad de Léon, Campus de Vegazana, s/n 24071 León, Spain. E-mail: mmorac00@estudiantes.unileon.es

Mozzarelli, A., Dipartimento de Farmacia, Università degli Studi de Parma, Parco Area delle Scienze 23/A, 43124 Parma, Italy. E-mail: andrea.mozzarelli@unipr.it

Nieuwkoop, A.J., Biomedical Sciences Department, Grand Valley State University, 212 Henry Hall, 1 Campus Drive, Allendale, MI 49401, USA. E-mail: nieuwkot@gvsu.edu

Osório, N.S., Life and Health Sciences Research Institute (ICVS), School of Health Sciences, Universidade do Minho, Campus de Gualtar, 4710-057 Braga, Portugal and ICVS/3B's-PT Government Associate Laboratory, Guimarães, Braga, Portugal. E-mail: nosorio@ecsaude.uminho.pt

Paietta, J.V., Department of Biochemistry and Molecular Biology, Wright State University, Dayton OH 45435, USA. E-mail: john.paietta@wright.edu

Palittapongarnpim, P., Department of Microbiology, Faculty of Science, Mahidol University, Thailand. E-mail: prasit.pal@mahidol.ac.th

Parker, E.J., Maurice Wilkins Centre for Molecular Biodiscovery, Biomolecular Interaction Centre and Department of Chemistry, University of Canterbury, Private Bag 4800, Christchurch, New Zealand. E-mail: emily.parker@canterbury.ac.nz

Parthasarathy, A., Formerly of Department of Civil and Environmental Engineering, Stanford University, Stanford, CA 94305, USA; currently at Science Academy of India, Chennai 600114, India. E-mail: harikuttanp@gmail.com

Pearce, F.G., Biomolecular Interaction Centre, School of Biological Sciences, University of Canterbury, Private Bag 4800, Christchurch, New Zealand. E-mail: grant.pearce@canterbury.ac.nz

Pennacchietti, E., Istituto Pasteur Italia – Fondazione Cenci-Bolognetti, Dipartimento di Scienze e Biotecnologie Medico-Chirurgiche, Sapienza Università di Roma, Corso della Repubblica 79, 04100 Latina, Italy. E-mail: eugenia.pennacchietti@uniroma1.it

Pérez-Pertejo, Y., Departamento de Ciencias Biomédicas, Universidad de León, Campus de Vegazana, s/n 24071 León, Spain. E-mail: myperp@unileon.es

Phillips, R.S., Departments of Chemistry and of Biochemistry and Molecular Biology, University of Georgia, Athens, GA 30602, USA. E-mail: plp@uga.edu

Quinn, P.J., Department of Biochemistry, King's College London, 150 Stamford Street, London SE1 9NH, UK. E-mail: p.quinn@kcl.ac.uk

Rashid, N., School of Biological Sciences, University of the Punjab, Quaid-e-Azam Campus, Lahore 54590, Pakistan. E-mail: neem.ff.sbs@pu.edu.pk/naeemrashid37@hotmail.com

Rodrigues, F., Life and Health Sciences Research Institute (ICVS), School of Health Sciences, Universidade do Minho, Campus de Gualtar, 4710-057 Braga, Portugal and ICVS/3B's-PT Government Associate Laboratory, Guimarães, Braga, Portugal. E-mail: frodrigues@ecsaude.uminho.pt

Romero Lucas, M.J., Vascular Biology Center and Department of Pharmacology and Toxicology, Medical College of Georgia, Augusta University, 1120 15th Street, Augusta, GA 30912-2500, USA. E-mail: mromerolucas@augusta.edu

Sampaio-Marques, B. Life and Health Sciences Research Institute (ICVS), School of Health Sciences, Universidade do Minho, Campus de Gualtar, 4710-057 Braga, Portugal and ICVS/3B's-PT Government Associate Laboratory, Guimarães, Braga, Portugal. Email: mbmarques@ecsaude.uminho.pt

Savka, M.A., Thomas H. Gosnell School of Life Sciences, Rochester Institute of Technology, 85 Lomb Memorial Drive, Rochester, NY 14623, USA. E-mail: massbi@rit.edu

Shibata, K., Department of Chemistry and Biochemistry, Fukushima National College of Technology, 30 Nagao, Iwaki, Fukushima 970-8034, Japan. E-mail: shibata@fukushima-nct.ac.jp

Simcock, D.C., The Deviot Institute, Deviot, Australia; Institute of Food, Nutrition and Human Health, Massey University, Palmerston North, New Zealand; and Faculty of Medicine and Bioscience, James Cook University, Cairns, QLD 4870, Australia. E-mail: david.simcock@jcu.edu.au

Simpson, H.V., Institute of Veterinary, Animal and Biomedical Sciences, Private Bag 11-222, Massey University, Palmerston North, New Zealand. E-mail: H.V.Simpson@massey.ac.nz

Smith, J.A., Department of Biology, American International College, 1000 State St., Springfield, MA 01109, USA. E-mail: jessica.smith@aic.edu

Takahashi, S., Department of Bioengineering, Nagaoka University of Technology, 1603-1 Kamitomioka, Nagaoka, Niigata 940-2188, Japan. E-mail: shoutaka@vos.nagaokaut.ac.jp

Umair, S., AgResearch Ltd, Private Bag 11-008, Palmerston North, New Zealand. E-mail: saleh.umair@agresearch.co.nz

Wang, X., State Key Laboratory of Food Science and Technology and Synergetic Innovation Center of Food Safety and Nutrition, Jiangnan University, Wuxi 21422, Jiangsu Province, China. E-mail: xwang@jiangnan.edu.cn

Wargo, M.J., Department of Microbiology and Molecular Genetics, University of Vermont College of Medicine, 201 Stafford Hall, 95 Carrigan Drive, Burlington, VT 05405, USA. E-mail: matthew.wargo@med.uvm.edu

Watanabe, S., Department of Bioscience, Graduate School of Agriculture, Ehime University, 3-5-7 Tarumi, Matsuyama 790-8566, Japan. E-mail: irab@agr.ehime-u.ac.jp

Wencker, F., Institut für Molekulare Infektionsbiologie, Universität Würzburg, Josef-Schneider-Str. 2/Bau D15, 97080 Würzburg, Germany. E-mail: freya.wencker@uni-wuerzburg.de

Ziebuhr, W., Institut für Molekulare Infektionsbiologie, Universität Würzburg, Josef-Schneider-Str. 2/Bau D15, 97080 Würzburg, Germany. E-mail: w.ziebuhr@mail.uni-wuerzburg.de

Preface: Concluding the Series

This edition of *The Handbook of Microbial Metabolism of Amino Acids* completes my specialist CAB International monographs on the metabolism of amino acids in living organisms. The series commenced with the publication of *Amino Acids in Farm Animal Nutrition* (D'Mello, 1994), followed by an updated edition entitled *Amino Acids in Animal Nutrition* (D'Mello, 2003). *Amino Acids in Human Nutrition and Health* appeared 9 years later (D'Mello, 2012), while *Amino Acids in Higher Plants* represents my most recent addition in this sequence (D'Mello, 2015). I have always strived to reflect the rich diversity of unique and interspecies issues in pathways of amino acid utilization and this approach has been adopted again in the present handbook.

As with the preceding titles, *The Handbook of Microbial Metabolism of Amino Acids* has been designed specifically to promote research in the academic and corporate sectors in Europe, the USA, Canada, Japan, Australia and New Zealand, but generally in all countries where English is a primary medium of communication. The handbook should appeal to a wide readership, including advanced undergraduates and graduate students, as well as teaching and research staff in colleges, universities and state-funded institutes. It is anticipated that *The Handbook of Microbial Metabolism of Amino Acids* will be recommended reading for courses in the biological, medical and veterinary sciences, including microbiology, biochemistry, genetics and pathology. This volume should also be of interest to the commercial sector, particularly corporate organizations with active R&D programmes, as in the pharmaceutical industry.

The Handbook of Microbial Metabolism of Amino Acids: An Overview

The chapters in this volume are arranged in nine sections, each representing a particular theme and demonstrating the diversity of amino acid metabolism in disparate types of microorganisms. The nature of the subjects under review and the need for continuity necessarily involves a certain degree of overlap. This is not envisaged as a detraction, because as a consequence, individual chapters are self-contained, thereby reducing the need for cross-referencing to other parts of the book. More importantly, this approach has also allowed authors increased flexibility in terms of emphasis and interpretation. Indeed, the diversity of microorganisms under review means that the metabolism of key amino acids should be considered for bacteria, fungi and apicomplexan parasites in order to discern comparative issues.

In order to maximize benefits, readers are recommended to consult *Amino Acids in Higher Plants* as a companion text to *The Handbook of Microbial Metabolism of Amino Acids*. There are numerous instances of comparisons, particularly in biosynthetic pathways. Equally, there are examples of contrasting issues now emerging across several chapters which demonstrate the unique biochemical diversity of microorganisms.

Rationale

The conceptual basis of each section and its constituent chapters is presented below. Although this volume is structured along conventional lines with chapters on the metabolism of individual amino acids, it was considered important that recent developments should be viewed in terms of the practical implications for food microbiology and pathogen characterization. An understanding of the metabolic pathways underlying the design of novel drugs has been a recurring theme in the scientific literature. Furthermore, the ecological and environmental interactions influencing the microbial metabolism of amino acids deserve detailed attention. Here, the emphasis is currently directed at aspects relating to microniches, the mammalian/avian microbiome and the impact on amino acid metabolism in host–microbe relationships.

Part I Glutamate

Glutamate occupies a key position in the biochemistry of all living organisms, to the extent that Young and Ajami (2000) identified it as an amino acid of 'particular distinction', albeit in the context of mammalian metabolism. As in other living things, the fate of glutamate in microorganisms is defined by specific enzymes with dehydrogenase, transaminase and decarboxylase activities. It is appropriate, therefore, that chapters on glutamate dehydrogenase and glutamate decarboxylase should form the basis of this section. In conjunction with similar chapters in *Amino Acids in Higher Plants* (D'Mello, 2015) and *Amino Acids in Human Nutrition and Health* (D'Mello, 2012), readers will be provided with the opportunity to become familiar with important issues in the comparative biochemistry of the glutamate family.

Part II Lysine, arginine and hydroxyproline

Physical characteristics, including structural configuration and chromatographic elution sequence, unite the basic amino acids, but there is ample evidence of shared biochemistry in the metabolic fate of lysine, arginine and ornithine in microorganisms. Thus, all three amino acids undergo decarboxylation reactions, although catalysis is effected by enzymes with differing specificities. Ornithine is derived from arginine through the action of arginase. In comparison with arginine, however, lysine is still regarded as an inert amino acid. Arginine metabolism ultimately leads to the synthesis of polyamines via ornithine and is also the precursor of the signalling molecule, nitric oxide, through the action of nitric oxide synthase. The universal synthesis of nitric oxide in living organisms is now emerging as significant function of arginine. The clinical implications of arginine metabolism are currently under active consideration regarding microbial pathogenicity and the identification of potential therapeutic targets. Comparative issues are highlighted in the metabolism of several amino acids in this section, including hydroxyproline, with differences and similarities between microorganisms and higher animals noted. There are opportunities in this section to evaluate differences between yeast, bacteria and trypanosomatids in the metabolism of basic amino acids.

Part III Serine and threonine

Although serine and threonine follow distinct pathways in the metabolism of living organisms, there are common features which are at an appropriate stage for review. Hence, there is continuing interest in the classical serine–glycine interconversion and the mediation of the enzyme serine hydroxymethyltransferase in this reaction. The degradation of threonine is also under regular review as part of an effort to compare pathways in hyperthermophilic and other microorganisms. It is known that organisms that thrive in harsh environments display unusual biological features, and amino acid metabolism under these conditions is perceived as an attractive research topic. At the post-translational level, several recent studies demonstrate the role of serine/threonine kinases in bacterial signalling and regulation. Other investigations highlight the modulation of host–pathogen interactions by bacterial serine/threonine protein kinases. The role of these protein kinases in the physiology and pathogenicity of important disease organisms is becoming ever more apparent.

Part IV Sulfur amino acids

Research on the sulfur amino acids inevitably gravitates towards methionine, not least due to its multiple roles in diverse pathways of amino acid metabolism of microorganisms and its vital role in the initiation of protein synthesis. Its functions as a source of sulfur and methyl groups confer additional significance, resulting in the synthesis of critical molecules, not just in microorganisms. In addition, methionine is widely associated with metabolic interactions, providing links with disparate amino acid families. Methionine is an essential amino acid for animals, serving as a precursor of cysteine via an irreversible pathway. Results at the gene level reveals novel aspects of methionine biosynthesis in bacteria and pathogenic fungi. RNA-based regulatory mechanisms in methionine biosynthesis may present opportunities for the development of novel antibacterial drugs. The characterization of key enzymes and the potential biopharmaceutical implications of recent results point to a favourable research scenario for sulfur amino acids. The catabolism of methionine proceeds along established pathways, but of particular relevance is its degradation to those volatile sulfur compounds associated with the development of flavour in cheese and the microbial interactions occurring within this process. The scope for extracting practical value should be a powerful driving force in such investigations.

Part V Branched-chain amino acids

The branched-chain amino acids have long been associated with profound metabolic and regulatory roles in living organisms, characterized by complex structurally related interactions not only within the group but also with other amino acids, particularly methionine. The association between the branched amino acids and methionine occurs in plants as well as microorganisms, while the mechanisms underlying this interaction appear to be linked with the activity of a branched-chain amino acid aminotransferase. The expression in certain bacteria of functional genes linked to amino acid breakdown and cheese flavour appears to be determined by branched-chain amino acids. The role of transamination reactions as an initial step in the pathway to *S*-methyl thioesters has been proposed. In addition, there is an emerging consensus that branched-chain amino acid biosynthesis is essential for effective symbiosis in rhizobia–legume relationships. Furthermore, transport and availability may be important for pathogenicity and virulence in disease-causing microorganisms. There is also interest in the putative role of branched-chain amino acid aminotransferase in conferring acid tolerance to bacteria involved in the development of carious lesions. As the biosynthesis of the branched-chain amino acids is absent in humans, it has been suggested that key enzymes involved in *de novo* synthesis of these amino acids in bacterial pathogens may offer potential for the development of new drugs.

Part VI Aromatic amino acids and histidine

In this section, the microbial metabolism of the aromatic amino acids justifies enhanced significance due to its impacts on food quality and safety, human pathology and the synthesis of bioactive compounds. In general, the degradation of certain aromatic amino acids is associated with the synthesis of undesirable floral aromas in cheese. Aromatic amino acids also directly or indirectly contribute to the structures or provide precursors or act as donors of active groups in the biosynthesis of mycotoxins, the secondary compounds produced by a wide range of plant pathogenic or storage fungi. The aromatic amino acids are well recognized as sources of important secondary metabolites with marked bioactive properties. In at least one case, signalling compounds may be generated to affect pathogenic virulence and biofilm formation. Increasingly, the experimental evidence points to these metabolites serving as scaffolds for drug design and development. Specific chapters address biosynthesis as well as degradation, but aspects of host–pathogen interactions are also considered in the context of drug targets and immunology. The molecular characterization of specific pathogens in relation to the metabolism and post-translational modifications of amino acids is an important development worth exploring in this section.

Part VII D-Amino acids

The occurrence of D-amino acids in the structure of peptidoglycan has long been recognized as a characteristic feature of bacteria. However, there is a need to update traditional concepts of the functions of particular D-amino acids and the characterization of specific and broad-spectrum racemases as well as of those enzymes with different cofactor requirements. Fermentation reactions in the rumen mean that substantial quantities of bacterial protein become available to the ruminant animal. The fate and dynamics of the D-amino acids derived from the peptidoglycan components will be considered in this section. In addition, it is important to recall that D-serine serves as a central nervous system co-agonist in higher animals, including humans. It has been suggested that in plants, D-amino acids taken up from the soil may be utilized via central metabolic hubs.

Part VIII Ecology

The ubiquitous distribution of microorganisms in diverse ecosystems justifies an entire section of chapters to address the adaptive complexities in amino acid metabolism and the implications for plant growth and human health. The bacteroids residing in the root nodules of leguminous plants utilize host dicarboxylic acids as sources of energy to reduce atmospheric nitrogen to ammonia which is exported to the plant for the synthesis of amino acids. Another example of considerable interest in microecology is amino acid biosynthesis in endosymbiont-harbouring parasites of vertebrates and plants. Nevertheless, the overriding issue in parasitism is the acquisition of nutrients, especially amino acids, at the expense of host metabolism and survival. Intracellular pathogens consistently rely on their respective hosts to provide essential nutrients, particularly amino acids, to satisfy their requirements for protein biosynthesis, energy, one-carbon units and sulfur, usually under limiting physical conditions such as local hypoxia, salinity and adverse pH. The concept of 'nutritional virulence' has recently been advanced to exemplify the changing emphasis in studies on host–pathogen interactions. Thus, it is essential to identify the amino acid requirements of parasites and other pathogens as a prerequisite for the characterization of immune responses and potential treatments to mitigate adverse effects. While the issues of host–microbe relationships form an underlying theme throughout *The Handbook of Microbial Metabolism of Amino Acids*, this section of the volume provides the opportunity to explore further the molecular dynamics of such interactions. Anoxic microniches occur widely in soils, sewage and sediments, and the microbial degradation of amino acids under anaerobic conditions is highly relevant in this section. Other environmental bacteria utilize the derivatives of several amino acids as sources of nutrients and for osmoprotection. Hence, the role of amino acids in microbial stress responses is an emerging issue, which in the long term, may have implications for biotechnology and environmental remediation. Future advances in biofilm and gut microbiome research should highlight the impact of interspecific and intraspecific interactions in microbial amino acid metabolism.

Part IX Conclusions

The underlying aim in this section is to unlock the practical implications of recent and ongoing research on the amino acid metabolism of diverse types of microorganisms. Even in a volume as comprehensive as this handbook, there is scope to integrate developments from a wide and fast-moving field of research. Emerging advances centre on food quality and safety, the biochemical characterization of pathogenicity and virulence in disease-causing microorganisms as exemplified in pathways of amino acid biosynthesis and degradation, and the application of the results to identify potential targets for the design of more effective drugs against a wide range of pathogens. The transport and metabolism of amino acids in microorganisms, particularly involving pathways absent in the mammalian host, provide multiple sites for testing a wide range of inhibitory compounds. However, further developments will depend on an improved understanding of amino acid metabolism in host–pathogen interactions. In addition, the safety of any new drugs will have to be established in rigorous clinical trials.

Acknowledgements

As in the preceding volumes of this series, I have assembled an international team of authors with an established reputation in publishing original papers in peer-reviewed research journals. I am indebted to them for devoting valuable time to preparing authoritative and comprehensive chapters. I am convinced that their efforts will serve our readers well for many years to come.

In planning this volume, I decided that authors should be given freedom of expression in their unique specialisms, while recognizing that the terminology and technology becomes increasingly more complex with successive advances in the amino acid metabolism of living organisms.

Disclaimer

The Handbook of Microbial Metabolism of Amino Acids inevitably contains references to and descriptors of experimental and commercial products. Authors were asked to refrain from excessive usage of trade names unless there were compelling reasons for doing so. It is emphasized that no endorsement of these products is implied or should be attributed to the Editor or to CAB International, the publisher of this volume. We confirm our absolute impartiality in the selection of chapter titles and in the appointment of authors and that there has been no conflict of interests during this process.

The information set out in this volume is presented in good faith and in accordance with the principles of 'best practice' and 'due diligence'. Although every effort has been made to verify the facts and figures, neither the Editor nor CAB International can accept responsibility for the data presented in individual chapters or for any consequences of their use.

The popular press and particularly Internet sites contain numerous articles extolling the virtues of a variety of amino acids. A number of amino acids are also available in health food stores in the high street. However, publication of this volume should not be interpreted as a recommendation for readers to use these or any other amino acid or any microbial or antimicrobial preparation for whatever purpose. This handbook is intended exclusively for use as a text in education and research.

References

D'Mello, J.P.F. (ed.) (1994) *Amino Acids in Farm Animal Nutrition*. CAB International, Wallingford, UK.
D'Mello, J.P.F. (ed.) (2003) *Amino Acids in Animal Nutrition*. CAB International, Wallingford, UK.
D'Mello, J.P.F. (ed.) (2012) *Amino Acids in Human Nutrition and Health*. CAB International, Wallingford, UK.
D'Mello, J.P.F. (ed.) (2015) *Amino Acids in Higher Plants*. CAB International, Wallingford, UK.
Young, V.R. and Ajami, A.M. (2000) Glutamate: an amino acid of particular distinction. *Journal of Nutrition* 130, 892S–900S.

Glossary

Introduction

Specific nomenclature and technical descriptors are now firmly embedded in the literature associated with amino acid metabolism. Although many of the terms and acronyms appearing in this handbook are already in common use, it is important to provide a comprehensive glossary to assist those readers who are new to this field. Additional definitions are available in a wide range of scientific dictionaries including, for example, the compilations of Parish *et al.* (2006), Singleton and Sainsbury (2006) and Lackie (2013). Readers will be aware that several free dictionaries are available and readily accessible online. Glossaries and other relevant information have also been provided in specialist monographs and handbooks such as those edited by D'Mello *et al.* (1991), D'Mello (1997), D'Mello (2012) and D'Mello (2015). These sources are recommended on the basis of a common vocabulary in the different disciplines of the life sciences. In addition, standard titles, including works by Alberts *et al.* (2008), Lodish *et al.* (2013), Nelson and Cox (2013), Klug *et al.* (2014), Madigan *et al.* (2015) and Strelkauskas *et al.* (2016), are recommended as sources of in-depth information on different aspects of the biological sciences. It is assumed that readers will be conversant with general biochemistry, microbiology and molecular genetics to levels covered by these textbooks. However, the language associated with microbial metabolism is continually evolving and unremittingly complex, even to experienced researchers, though a highly effective way of updating information and the relevant terminology is via the various research articles cited at the end of each chapter of this volume.

Definition of Terms and Acronyms

The important terms and acronyms appearing in *The Handbook of Microbial Metabolism of Amino Acids* are listed and explained in Table 1. This compilation includes definitions of standard as well as unique chapter-specific abbreviations or terms. Where appropriate, cross-referencing to individual chapters in this volume is included in order to facilitate a greater appreciation of the context of usage of particular terms.

Table 1. Definitions of terms and acronyms used in *The Handbook of Microbial Metabolism of Amino Acids*.

Abbreviation or Term	Definition
AA	Amino acid
AAA	α-Aminoadipic acid (Chapter 4)
AAAH	Aromatic amino acid hydroxylase (Chapter 28)
AAAP	Amino acid/auxin permease (Chapter 7)
AbuCad	Aminobutylcadaverine (Chapter 26)
AbuHspd	Aminobutyrylhomospermidine (Chapter 26)
Acetyl-CoA	Acetyl coenzyme A (Chapters 3, 4, 12, 13, 16, 18, 28, 29, 30)
ACME	Arginine catabolic mobile elements (Chapter 5)
Acute toxicity	Severe adverse effects occurring within a relatively short period of time after exposure to a harmful compound
AD	Arginine deiminase (Chapter 6)
ADC	Arginine decarboxylase (Chapters 7, 8, 26 and 28)
ADI	Arginine deiminase (Chapters 5 and 8)
ADIC	2-Amino-2-deoxyisochorismate (Chapter 19)
AdoHcy	*S*-Adenosylhomocysteine (Chapter 7)
AdoHcyH	*S*-Adenosylhomocysteine hydrolase (Chapter 7)
AdoMet	*S*-Adenosyl-L-methionine (Chapters 7, 9 and 28)
AdoMetDC	*S*-Adenosylmethionine decarboxylase (Chapter 7)
ADP	Adenosine diphosphate (Chapters 4, 28, 29)
ADS	Arginine deiminase system (Chapter 6)
AgDI	Agmatine deiminase (Chapter 8)
AGM	Agmatinase (Chapter 7)
AGX1	Gene encoding alanine:glyoxylate aminotransferase 1 (Chapter 11)
AHAIR	Acetohydroxyacid isomeroreductase (Chapter 16)
AHAS	Acetohydroxyacid synthase (Chapters 16 and 32)
AICAR	5-Amino-1-(5'-phosphoribosyl)-imidazole-4-carboxamide (Chapter 11)
AIR	1-(5'-Phosphoribosyl)-5-aminoimidazole (Chapter 11)
AJ	Adherens junctions (Chapter 6)
AK	Aspartate kinase (Chapter 4)
β-Ala	β-Alanine (Chapters 3 and 8)
AlaAT	Alanine aminotransferase (Chapter 28)
ALI	Acute lung injury (Chapter 6)
Alr	Alanine racemase (Chapter 24)
ALDH4A1	Aldehyde dehydrogenase 4 (Chapter 7)
ALDH-SF	Aldehyde dehydrogenase superfamily (Chapter 3)
Allosteric	Multisite enzyme modulation of structure and activity
AMP	Adenosine monophosphate (Chapter 5)
AnPRT (TrpD)	Anthranilate phosphoribosyltransferase (Chapters 19 and 20)
Antagonist	A compound acting as a metabolic inhibitor by virtue of structural analogy or functional competition
AOR	Aldehyde:ferredoxin oxidoreductase (Chapter 18)
APAO	N^1-acetyl polyamine oxidase (Chapter 7)
APS	Adenosine-5′-phosphosulfate (Chapter 14)
AR	Acid resistance (Chapter 2)
arcA	Gene coding for ADI (Chapter 5)
arcB	Gene coding for OTC (Chapter 5)
arcC	Gene coding for CK (Chapter 5)
arcD	Gene coding for Arg/Orn antiporter (Chapters 5 and 6)
ARDS	Acute respiratory distress syndrome (Chapter 6)
Arg	Arginine
ARG	Arginase (Chapter 7)

Continued

Table 1. Continued.

Abbreviation or Term	Definition
ArgD	Bifunctional acetylornithine/succinyldiaminopimelate aminotransferase (Chapter 26)
ArgE	*N*-Acetylornithine deacetylase (Chapter 26)
ArgJ	Glutamate *N*-acetyltransferase (Chapter 26)
ArgR	Arginine racemase (Chapter 24)
AS	Anthranilate synthase (Chapters 19 and 20)
ASA	Aspartate β-semialdehyde (Chapter 4)
ASCT2	Alanine-serine-cysteine transporter 2 (Chapter 25)
Asd, ASD	Aspartate β-semialdehyde dehydrogenase (Chapter 32)
ASL	Argininosuccinate lyase (Chapters 7 and 28)
Asn	Asparagine
Asp	Aspartate
AspAT	Aspartate aminotransferase (Chapter 28)
AspR	Aspartate racemase (Chapter 24)
ASS, ASS1	Argininosuccinate synthase, argininosuccinate synthase1 (Chapters 7 and 28)
AT	Unspecific aminotransferase (Chapter 7)
ATP	Adenosine triphosphate (Chapters 2, 5, 7 and 8)
ATR	Acid tolerance response (Chapter 2)
BCAA	Branched-chain amino acid (Chapters 10, 28 and 32)
BCAT	Branched-chain amino acid aminotransferase (Chapter 28)
Bcd	Butyryl-CoA dehydrogenase (Chapter 29)
BCKA	Branched-chain keto acid
BCR	Benzoyl-CoA reductase (Chapter 18)
BFT	Biopterin folate transporter (Chapter7)
BH_4	5,6,7,8-Tetrahydrobiopterin (Chapter 9)
BsPhoE	Acid phosphatase (PhoE) from *Bacillus stearothermophilus* (Chapter 23)
Cad	Cadaverine (Chapter 26)
CAIR	1-(5′-Phosphoribosyl)-5-aminoimidazole-4-carboxylate (Chapter 11)
CAP	Community-acquired pneumonia (Chapter 6)
CAT1	Cationic amino acid transporter 1 (Chapter7)
CBL	Cystathionine-beta (β-) lyases (Chapter 13)
CBS	Cystathionine beta (β-) synthase (Chapter 7)
CCR	Carbon catabolite repression (Chapter 5)
C4DHyp	*Cis*-4-Hydroxy-D-proline (Chapter 10)
CdRP	1-(2-Carboxyphenylamino)-1-deoxy-ribulose 5-phosphate (Chapter 19)
CGL	Cystathionine gamma (γ-) lyase (Chapter 7)
CGS	Cystathionine gamma (γ-) synthase (Chapters 4 and 13)
CH_3-THF	N^5-Methyl-tetrahydrofolate (Chapter 13)
CH_3-THPTG	N^5-Methyl-tetrahydropteroyl-(tri) glutamate (Chapter 13)
Cit	Citrulline
CK	Carbamate kinase (Chapters 5, 6 and 8)
CL	Cutaneous leishmaniasis (Chapter 7)
C3LHyp	*Cis*-3-Hydroxy-L-proline (Chapter 10)
CLS	Chronological lifespan (Chapter 9)
CM	Chorismate mutase (Chapter 28)
CNS	Central nervous system (of animals) (Chapters 7, 9 and 25)
CNSpd	Carboxynorspermidine (Chapter 26)
CoA	Coenzyme A
cOCT	Catabolic ornithine carbamoyltransferase (Chapter 6)
CodY	Global transcription factor repressing many amino acid biosynthetic pathways (Chapter 13)
COX	Cytochrome c oxidase (Chapter 9)
CPS	Carbamoylphosphate synthetase (Chapter 7)
CR	Caloric restriction (Chapter 9)
CSC	Commercial starter cultures (Chapter 17)

Continued

Table 1. Continued.

Abbreviation or Term	Definition
CSF	Cerebrospinal fluid (Chapter 6)
Cys	Cysteine
Cysth	Cystathionine (Chapter 7)
DAA	Dispensable amino acid(s) (for animals; see also NEAA)
DAAO	D-Amino acid oxidase (Chapter 25)
D-AAT	D-amino acid transaminase (Chapter 24)
DABA	α,γ-Diaminobutyric acid (Chapter 26)
DAD	D-amino acid dehydrogenase (Chapter 24)
DALY	Annual disability-adjusted life years (Chapter 7)
DAO	D-amino acid oxidase (Chapter 24)
DAP	Diaminopimelate (Chapters 4 and 27)
dcAdoMet	Decarboxylated *S*-adenosylmethionine (Chapter 7)
Ddh	Diaminopimelate dehydrogenase (Chapter 4)
DDO	D-Aspartate oxidase (Chapter 24)
DDR	Dihydrodipicolinate reductase (Chapter 32)
DFMO	Difluoromethylornithine (Chapter 26)
DHAD	Dihydroxyacid dehydratase (Chapter 16)
DHDP	Dihydrodipicolinate (Chapter 4)
DHDPS/NAL	Dihydrodipicolinate synthase/*N*-acetylneuraminate lyase (Chapter 10)
DHK-MTPene	1,2-Dihydroxy-3-keto-5-methylthiopentane (Chapter 7)
3,4-DHP	3-Hydroxy-4(1H)-pyridone (Chapter 32)
DHPS	Dihydrodipicolinate synthase (Chapter 4)
DMGDM	Dimethylglycine demethylase (Chapter 30)
DNA	Deoxyribonucleic acid (Chapters 11 and 13)
DOPA	Dihydroxyphenylalanine (Chapter 28)
DPDC	Diaminopimelate decarboxylase (Chapter 32)
dPGM	(Cofactor)-dependent phosphoglycerate mutases (Chapter 23)
DSD	D-Serine dehydratase (Chapter 24)
DTA	D-Threonine aldolase (Chapter 24)
EAA	Essential amino acid(s) (for animals)
EcolidPGM	(Cofactor)-dependent phosphoglycerate mutase from *Escherichia coli* (Chapter 23)
EcPhytase	Phytase from *Escherichia coli* (Chapter 23)
eNOS	Endothelial nitric oxide synthase (Chapters 6 and 9)
E1Phos	Enolase phosphatase E1 (Chapter 7)
EPS	Exopolysaccharide (Chapter 26)
ERG	Ergothioneine (Chapter 30)
ERGase	Ergothionase (Chapter 30)
ET	Ethylene (Chapter 32)
Etf	Electron-transferring flavoprotein (Chapter 29)
EU	European Union
FAD	Flavin adenine dinucleotide (Chapters 16, 24 and 29)
FAICAR	5-Formamido-1-(5′-phosphoribosyl)-imidazole 4-carboxamide (Chapter 11)
FBEB	Flavin-based electron bifurcation (Chapter 29)
FG	Formylglutamate (Chapter 22)
FGAR	5′-Phosphoribosyl-*N*′-formylglycinamide (Chapter 11)
FGase	Formylglutamate hydrolase (Chapter 22)
FgMETB	*Fusarium graminearum* cystathionine γ-synthase gene (Chapter 32)
FIG	Formiminoglutamate (Chapter 22)
FIGase	Formiminoglutamate hydrolase (Chapter 22)
FMN	Flavin mononucleotide (Chapters 4 and 30)
FtHap	Histidine acid phosphatase from *Francisella tularensis* (Chapter 23)
GABA	Gamma (γ-) aminobutyrate (Chapters 2, 3, 28 and 29)
GabP	GABA permease (Chapter 32)

Continued

Table 1. Continued.

Abbreviation or Term	Definition
GABA-T	GABA transaminase (Chapters 3 and 28)
GAD/Gad	Glutamic acid (glutamate) decarboxylase (Chapters 2, 3 and 32)
GAR	Glycinamide ribonucleotide (5′-phosphoribosylglycinamide (Chapter 11)
GAT	Glutamine amidotransferase (Chapter 20)
GB	Glycine betaine (Chapter 30)
GBDM	Glycine betaine demethylase (Chapter 30)
GBMT	Glycine betaine methyltransferase (Chapter 30)
γ-GC	γ-Glutamylcysteine (Chapter 7)
GCD	Glycine decarboxylase
GC-MS	Gas chromatography-mass spectroscopy
γ-GCS	γ-Glutamylcysteine synthetase (Chapter 7)
GCV	Glycine cleavage complex (Chapter 11)
GDAR	Glutamate-dependent acid resistance (Chapter 2)
GDH	Glutamate dehydrogenase (Chapters 1 and 28)
GIT	Gastrointestinal tract (Chapter 2)
Gln	Glutamine
Glu	Glutamate
Gly	Glycine
GLR	Glutamate-like receptor
GMO	Genetically modified organism
GMP	Guanosine monophosphate (Chapter 26)
GOGAT	Glutamate synthase/Glutamine oxoglutarate aminotransferase (Chapters 1 and 28)
GS	Glutamine synthetase (Chapter 28)
GSH	Reduced glutathione (Chapters 7 and 28)
Gsp	Glutathionylspermidine (Chapter 7)
GSS	Glutathionylspermidine synthetase (Chapter 7)
GST	Glutathione *S*-transferase(s) (Chapter 28)
GTP	Guanosine triphosphate (Chapter 6)
2-HADH	2-Hydroxyacyl-CoA dehydratase (Chapter 29)
HAP	Histidine acid phosphatases (Chapter 23)
HAT	Homoserine *O*-acetyltransferases (Chapter 13)
4-HBDH	4-Hydroxybutyryl-CoA dehydratase (Chapter 29)
Hcy	Homocysteine (Chapter 7)
Hepatotoxic	Toxic to the liver
HGT	Horizontal gene transfer (Chapter 27)
HiOASS	*O*-Acetylserine sulfhydrylase from *Haemophilus influenzae* (Chapter 15)
His	Histidine
His-B	Histidine betaine (Chapter 30)
HIV	Human immunodeficiency virus (Chapter 20)
HPLC	High-performance liquid chromatography (Chapter 26)
HPP	Histidinol phosphate phosphatase (Chapter 27)
HSD	Homoserine dehydrogenase (Chapter 4)
HSpd	Homospermidine (Chapter 26)
HST	Homoserine *O*-succinyltransferases (Chapter 13)
5-HT	5-Hydroxytryptamine
HTA	Homoserine transacetylase (Chapter 13)
5-HTP	5-Hydroxytryptophan
Hut	Histidine utilization (Chapter 22)
L-Hyp	L-Hydroxyproline (Chapter 10)
HypE	Hydroxyproline 2-epimerase (Chapter 10)
D-HypDH	D-Hydroxyproline dehydrogenase (Chapter 10)
5-HVDH	5-Hydroxyvaleryl-CoA dehydratase (Chapter 29)
IAA	Indole acetic acid
IAV	Influenza A virus (Chapter 6)

Continued

Table 1. Continued.

Abbreviation or Term	Definition
ICL1	Gene encoding isocitrate lyase 1 (Chapter 11)
ICS	Isochorismate synthase (Chapter 19)
IDH	Isocitrate dehydrogenase (Chapter 1)
IDO	Indoleamine 2,3-dioxygenase (Chapters 20, 21 and 32)
IFN-γ	Interferon-γ/Interferon gamma (Chapters 20 and 32)
iGluR	Ionotropic glutamate receptor
IGP	Indole-3-glycerol phosphate (Chapter 19)
IGPS	Indole-3-glycerol phosphate synthase (Chapter 20)
Ile	Isoleucine
IMP	Inosine monophosphate (inosinate)
iNOS	Inducible NOS (Chapters 6, 7 and 9)
IP	Imidazolone propionate (Chapter 22)
IPA	Indole-3-pyruvic acid
IPase	Imidazolone propionate hydrolase (Chapter 22)
ITC	Isothermal titration microcalorimetry (Chapter 4)
JA	Jasmonic acid (Chapter 32)
KADC	α-Keto-acid decarboxylase (Chapter 17)
KADH	α-Keto-acid dehydrogenase (Chapter 17)
KBL	2-Amino-3-ketobutyrate CoA ligase (Chapter 12)
L-KDA	L-2-Keto-deoxyarabinonate (Chapter 10)
D-KDG	D-5-Keto-4-deoxyglucarate (Chapter 10)
α-KGSA	α-Ketoglutaric semialdehyde (Chapter 10)
KMTB	α-Ketomethiobutyrate (Chapter 7)
*kyn*A	Tryptophan dioxygenase Chapter 21)
*kyn*B	Kynurenine formamidase (Chapter 21)
*kyn*R	Transcriptional regulator of the Lrp (leucine-responsive regulatory protein)/AsnC (leucine-responsive regulatory protein) family
*kyn*U	Kynureninase (Chapter 21)
LAB	Lactic acid bacteria (Chapters 8 and 17)
LD_{50}	The dose that causes lethality in 50% of a population of test organisms
LDC	Lysine decarboxylase (Chapter 8)
LDH	Lactate dehydrogenase
Leu	Leucine
LKR	Lysine-ketoglutarate reductase (Chapter 28)
LNAA	Large neutral amino acid
Lys	Lysine
MAO	Monoamine oxidase
MAT	Methionine adenosyltransferase (Chapter 7)
McbR	Transcriptional repressor belonging to the TetR (Tet repressor protein which regulates tetracycline resistance) family of transcription factors (Chapter 13)
MDH	Malate dehydrogenase
ME	Malic enzyme
MeCbl	Methylcobalamin (Chapter 13)
Met	Methionine
MetJ	Transcriptional repressor (Chapter 13)
MetR	Transcriptional regulator of methionine biosynthesis (Chapter 13)
MetS	Methionine synthase (Chapter 7)
MGLP	Methylglucose lipopolysaccharides (Chapter 23)
MIO	4-Methylidene-imidazole-5-one (Chapter 22)
MLC	Myosin light chain (Chapter 6)
mRNA	Messenger RNA
MRSA	Methicillin-resistant *Staphylococcus aureus* (Chapter 32)
MSOX	Monomeric sarcosine oxidase (Chapter 30)
MTA	Methylthioadenosine (Chapter 7)

Continued

Table 1. Continued.

Abbreviation or Term	Definition
MTAN	Methylthioadenosine nucleosidase (Chapter 7)
MTA-Pase	Methylthioadenosine phosphorylase (Chapter 7)
MtaR	Member of the LysR type of transcription factors (Chapter 13)
MtbdPGM	(Cofactor)-dependent phosphoglycerate mutase (Rv0489) from *Mycobacterium tuberculosis* (Chapter 23)
MtbRv2135c	Acid phosphatase Rv2135c from *Mycobacterium tuberculosis* (Chapter 23)
MtbRv2419c	Glucosyl-3-phosphoglycerate phosphatase (Rv2419c) from *Mycobacterium tuberculosis* (Chapter 23)
MtbRv3214	Acid phosphatase (Rv3214) from *Mycobacterium tuberculosis* (Chapter 23)
MTHRR	Methylene tetrahydrofolate reductase (Chapter 7)
MtOASS	*O*-Acetylserine sulfhydrylase from *Mycobacterium tuberculosis* (Chapter 15)
MTRK	5′-Methylthioribose kinase (Chapter 7)
MTRP	5′-Methylthioribose-1-phosphate (Chapter 7)
MTRu-1-P	5′-Methylthioribulose-1-phosphate (Chapter 7)
Mutagenic	Causing mutations
N	Nitrogen
NAA	Neutral amino acids
NAD^+	Nicotinamide adenine dinucleotide (oxidized)
NADH	Nicotinamide adenine dinucleotide (reduced)
$NADP^+$	Nicotinamide adenine dinucleotide phosphate (oxidized)
NADPH	Nicotinamide adenine dinucleotide phosphate (reduced)
NAME	N^{ω}-Nitro-L-arginine methyl ester (Chapter 9)
NAO	*N*-Acetylornithine (Chapter 26)
NAS	*N*-Acetylserine (Chapter 15)
NEAA	Non-essential amino acid(s) (Chapter 25)
Nephrotoxic	Toxic to the kidney
NMDA	*N*-Methyl-D-aspartate (Chapter 25)
N-mePro-DM	*N*-methylproline demethylase (Chapter 30)
NMR	Nuclear magnetic resonance (Chapters 4 and 11)
NO	Nitric oxide (Chapters 6 and 9)
NOHA	N^{ω}-hydroxy-nor-L-arginine (Chapter 7)
NORE	NO-responsive element (Chapter 9)
NOS	Nitric oxide synthase (Chapters 9 and 28)
nNOS	Neuronal NOS (Chapter 9)
NSpd	Norspermidine (Chapter 26)
NTD	Neglected tropical diseases (Chapter 7)
OAA	Oxaloacetate
OAS	*O*-Acetylserine (Chapter 15)
OASS	*O*-Acetylserine sulfhydrylase (Chapter 15)
OASTL	*O*-Acetylserine (thiol) lyase
OAT	Ornithine aminotransferase (Chapter 7)
OCD	Ornithine cyclodeaminase (Chapter 10)
ODC	Ornithine decarboxylase (Chapters 6, 7, 8, 26 and 28)
ODH	2-Oxoglutarate dehydrogenase (Chapter 1)
OGC	2-Oxoglutarate decarboxylase (Chapter 1)
$OONO^-$	Peroxynitrite (Chapter 9)
OPH	*O*-Phosphohomoserine
ORF	Open reading frame(s) (Chapter 20)
Orn	Ornithine
OTC	Ornithine transcarbamoylase/Ornithine carbamoyl-transferase (Chapters 5, 7 and 8)
PA	Polyamine(s)
PAGE	Polyacrylamide gel electrophoresis
PAH	Phenylalanine hydroxylase (Chapter 28)
PAL	Phenylalanine ammonia-lyase (Chapter 30)

Continued

Table 1. Continued.

Abbreviation or Term	Definition
PAPS	3′-Phosphoadenosine-5′-phosphosulfate (Chapter 14)
P5C	Δ^1-Pyrroline-5-carboxylate
P5CDH	Δ^1-Pyrroline-5-carboxylate dehydrogenase
PCET	Proton-coupled electron transfer (Chapter 29)
PCP	Peptidyl-carrier protein (Chapter 24)
PCR	Polymerase chain reaction (Chapter 8)
PDC	Pyruvate decarboxylase
PDHC	Pyruvate dehydrogenase complex (Chapter 16)
PEP	Phosphoenolpyruvate (Chapter 27)
PG	Peptidoglycan (Chapter 4)
PGA	Poly-γ-glutamic acid (Chapter 24)
3-PGA	3-Phosphoglycerate (Chapters 11 and 15)
PGDH	3-Phosphoglycerate dehydrogenase (Chapter 15)
Phe	Phenylalanine
PKC	Protein kinase C (Chapter 6)
PLP	Pyridoxal phosphate (Chapters 2, 4, 21 and 29)
PMF	Proton motive force (Chapter 2)
PMP	Pyridoxamine phosphate
PQS	*Pseudomonas* quinolone signalling molecule (Chapter 21)
PR	D-Proline reductase (Chapter 24)
PRA	5-Phosphoribosylamine (Chapter 11)
Pro	Proline
L-ProDH	L-Proline dehydrogenase (Chapter 10)
Pro-B	Proline betaine (Chapter 30)
Pro-B-DM	Proline betaine demethylase (Chapter 30)
ProR	Proline racemase (Chapter 24)
PRPP	5-Phosphoribosyl diphosphate (Chapters 11, 19 and 27)
PRT	Phosphoribosyltransferase (Chapter 19)
PSAT	3-Phosphoserine aminotransferase (Chapter 15)
PSP	Phosphoserine phosphatase(s) (Chapters 15 and 23)
PTC	Putrescine transcarbamylase (Chapter 8)
PTP	Gene coding for a putative Cit/Orn transporter (present in some bacteria) (Chapter 5)
Put	Putrescine (Chapters 7 and 26)
PutA	Proline dehydrogenase (Chapter 30)
PYCR2	Pyrroline-5-carboxylate reductase (Chapter7)
Pyr	Pyruvate
Pyr2C	Δ^1-Pyrroline-2-carboxylate (Chapter 10)
Pyr4H2C	Δ^1-Pyrroline-4-hydroxy-2-carboxylate (Chapter 10)
QS	Quorum sensing (Chapter 31)
R&D	Research and development
RDP	Rumen degradable protein (Chapter 25)
RNA	Ribonucleic acid (Chapter 13)
RNS	Reactive nitrogen species (Chapter 9)
ROS	Reactive oxygen species (Chapters 3, 6, 7, 9, 29 and 32)
RT	Regular trypanosomatid (Chapter 27)
SA	Salicylic acid (Chapter 32)
SacDH	Saccharopine dehydrogenase (Chapter 28)
SAH	*S*-Adenosylhomocysteine (Chapters 13 and 14)
SAICAR	1-(5′-Phosphoribosyl)-4-(*N*-succinocarboxamido)-5-aminoimidazole (Chapter 11)
SAM	*S*-Adenosylmethionine (Chapters 13, 14, 27 and 29)
SarDH	Sarcosine dehydrogenase (Chapter 28)
SAT	Serine acetyltransferase/serine *O*-acetytransferase (Chapters 15 and 32)
SCF	Skp/Cullin/F-box (Chapter 14)
SCFA	Short-chain fatty acid(s) (Chapters 2 and 29)

Continued

Table 1. Continued.

Abbreviation or Term	Definition
SCID	Severe combined immune-deficient (Chapter 20)
SCoAS	Succinyl CoA synthetase (Chapter 1)
SCON2	Sulfur controller 2 (Chapter 14)
SCV	Small colony variants (Chapter 5)
SeMet	Selenomethionine (Chapter 32)
Ser	Serine
SERAT	Serine acetyltransferase
SfNeo13	Histidine phosphatase Neo13 from *Streptomyces fradiae* (Chapter 23)
SGD	Saccharomyces Genome Database (Chapter 3)
SHM2	Gene encoding hydroxymethyl transferase (Chapter 11)
SHT	Symbiont-harbouring trypanosomatids (Chapter 27)
SLP	Substrate-level phosphorylation (Chapter 29)
SMCO	*S*-Methylcysteine sulfoxide (Chapter 32)
SOD	Superoxide dismutase (Chapters 3, 9 and 26)
SOX	Sarcosine oxidase (Chapter 28)
Spd	Spermidine (Chapters 7 and 26)
SpdS	Spermidine synthase (Chapter 7)
Spm	Spermine (Chapter 7)
SpmS	Spermine synthase (Chapter 7)
Srr	Serine racemase (Chapter 24)
SSA	Succinate semialdehyde (Chapter 3)
SSADH	Succinate semialdehyde dehydrogenase (Chapters 1 and 3)
SSAT	Spermidine/spermine N^1-acetyltransferase (Chapter 7)
STCobe	Histidine phosphatase cobC from *Salmonella enterica* serotype Typhimurium (Chapter 23)
StOASS	*O*-Acetylserine sulfhydrylase from *Salmonella enterica* serovar Typhimurium (Chapter 15)
TA	Transaminase (Chapter 16)
TB	Tuberculosis (Chapter 20)
TCA	Tricarboxylic acid (with reference to the TCA cycle) (Chapters 1, 2, 3, 16, 27, 31 and 32)
TD	Threonine deaminase
TDO	Tryptophan-2,3-dioxygenase (Chapter 32)
Teratogenic	Causing birth defects
TF	Transcription factors (Chapter 13)
TH	Tyrosine hydroxylase
THDP	Tetrahydrodipicolinate (Chapter 4)
THF	Tetrahydrofolate (Chapters 7, 13, 27 and 28)
ThPP	Thiamine diphosphate (Chapter 27)
Thr	Threonine
TLC	Thin layer chromatography (Chapter 26)
T3LHyp	*trans*-3-Hydroxy-L-proline (Chapter 10)
T4LHyp	*trans*-4-Hydroxy-L-proline (Chapter 10)
T4LHyp-B	*trans*-4-Hydroxy-L-proline betaine (Chapter 10)
TMA	Trimethylamine (Chapter 30)
TNF	Tumour necrosis factor (Chapter 6)
TPE	Trypanosomatid proteobacterial endosymbiont (Chapter 27)
TPP	Thiamine pyrophosphate (Chapter 12)
TpS	Trypanothione synthase (Chapter 7)
TR	Trypanothione reductase (Chapter 7)
tRNA	Transfer RNA
Trp	Tryptophan
Trp-B	Tryptophan betaine (Chapter 30)

Continued

Table 1. Continued.

Abbreviation or Term	Definition
TryR	Trypanothione reductase (Chapter 7)
TryS	Trypanothione synthetase (Chapter 7)
TS_2	Oxidized trypanothione (Chapter 7)
$T(SH)_2$	Reduced trypanothione (Chapter 7)
TSOX	Tetrameric sarcosine oxidase (Chapter 30)
Tyr	Tyrosine
Tyr-B	Tyrosine betaine (Chapter 30)
UDP	Uridine diphosphate (Chapter 12); undegradable dietary protein (Chapter 25)
5′-UTRs	5′-Untranslated regions (Chapter 13)
Val	Valine
VL	Visceral leishmaniasis (Chapter 7)
VSC	Volatile sulfur compounds (Chapter 32)
WHO	World Health Organization (United Nations)
XRD	X-ray diffraction (Chapter 29)
Yct1p	Cysteine-specific transporter (Chapter 14)

Microorganisms

A diverse number of microorganisms are cited in this volume, ranging from bacteria to apicomplexan parasites. It is standard convention to use the abbreviated nomenclature on second and subsequent citations in the text. The list presented in Table 2 shows the major species in abbreviated and complete forms to facilitate the rapid and positive identification of the genus.

Table 2. Abbreviated forms of names of major species of microorganisms used in *The Handbook of Microbial Metabolism of Amino Acids*.

Abbreviated form	Full name
A. aeolicus	*Aquifex aeolicus*
A. brasilense	*Azospirillum brasilense*
A. caviae	*Aeromonas caviae*
A. fermentans	*Acidaminococcus fermentans*
A. fulgidus	*Arhaeoglobus fulgidus*
A. galli	*Ascaridia galli*
A. nidulans	*Aspergillus nidulans*
A. pleuropneumoniae	*Actinobacillus pleuropneumoniae*
A. tumefaciens	*Agrobacterium tumefaciens*
A. xylosoxidans	*Alcaligenes xylosoxidans*
B. bifidum	*Bifidobacterium bifidum*
B. brevis	*Bacillus brevis*
B. burgdorferi	*Borrelia burgdorferi*
B. cepacia	*Burkholderia cepacia*
B. japonicum	*Bradyrhizobium japonicum*
B. licheniformis	*Bacillus licheniformis*
B. linens	*Brevibacterium linens*
B. microti	*Brucella microti*
B. pahangi	*Brugia pahangi*
B. pseudomallei	*Burkholderia pseudomallei*
B. sphaericus	*Bacillus sphaericus*

Continued

Table 2. Continued.

Abbreviated form	Full name
B. subtilis	*Bacillus subtilis*
B. thailandensis	*Burkholderia thailandensis*
C. albicans	*Candida albicans*
C. boidinii	*Candida boidinii*
C. carbonum	*Cochliobolus carbonum*
C. crescentus	*Caulobacter crescentus*
C. difficile	*Clostridium difficile*
C. elegans	*Caenorhabditis elegans*
C. glutamicum	*Corynebacterium glutamicum*
C. humicola	*Cryptococcus humicola*
C. hutchisonii	*Cytophaga hutchisonii*
C. jejuni	*Campylobacter jejuni*
C. maltaromaticum	*Carnobacterium maltaromaticum*
C. neoformans	*Cryptococcus neoformans*
C. psychreythraea	*Colwellia psychrerythraea*
C. sporogenes	*Clostridium sporogenes*
C. sticklandii	*Clostridium sticklandii*
C. symbiosum	*Clostridium symbiosum*
C. trachomatis	*Chlamydia trachomatis*
C. trifolii	*Colletotrichum trifolii*
D. hafniense	*Desulfitobacterium hafniense*
D. immitis	*Dirofilaria immitis*
E. cloacae	*Enterobacter cloacae*
E. coli	*Escherichia coli*
E. faecalis	*Enterococcus faecalis*
E. histolytica	*Entamoeba histolytica*
F. graminearum	*Fusarium graminearum*
F. hepatica	*Fasciola hepatica*
F. moniliforme	*Fusarium moniliforme*
F. nucleatum	*Fusobacterium nucleatum*
F. placidus	*Ferroglobus placidus*
F. tularensis	*Francisella tularensis*
G. acetivorans	*Geoglobus acetivorans*
G. candidum	*Geotrichum candidum*
G. intestinalis	*Giardia intestinalis*
H. arabidopsidis	*Hyaloperonospora arabidopsidis*
H. contortus	*Haemonchus contortus*
H. diminuta	*Hymenolepis diminuta*
H. influenzae	*Haemophilus influenzae*
H. polygyrus	*Heligmosomoides polygyrus*
H. pylori	*Helicobacter pylori*
H. salinarum	*Halobacterium salinarum*
K. pneumoniae	*Klebsiella pneumoniae*
L. acidophilus	*Lactobacillus acidophilus*
L. aggregata	*Labrenzia aggregata*
L. amazonensis	*Leishmania amazonensis*
L. braziliensis	*Leishmania braziliensis*
L. brevis	*Lactobacillus brevis*
L. casei	*Lactobacillus casei*
L. curvatus	*Lactobacillus curvatus*
L. delbrueckii	*Lactobacillus delbrueckii*
L. donovani	*Leishmania donovani*
L. hilgardii	*Lactobacillus hilgardii*

Continued

Table 2. Continued.

Abbreviated form	Full name
L. infantum	*Leishmania infantum*
L. lactis	*Lactococcus lactis*
L. major	*Leishmania major*
L. mexicana	*Leishmania mexicana*
L. monocytogenes	*Listeria monocytogenes*
L. panamensis	*Leishmania panamensis*
L. paracasei	*Lactobacillus paracasei*
L. plantarum	*Lactobacillus plantarum*
L. sacchari	*Laceyella sacchari*
L. saerimneri	*Lactobacillus saerimneri*
L. sakei	*Lactobacillus sakei*
M. avium	*Mycobacterium avium*
M. oryzae	*Magnaporthe oryzae*
M. phenylpyruvica	*Moraxella phenylpyruvica*
M. pneumoniae	*Mycoplasma pneumoniae*
M. smegmatis	*Mycobacterium smegmatis*
M. thermoautotrophicum	*Methanobacterium thermoautotrophicum*
M. tuberculosis	*Mycobacterium tuberculosis*
N. brasiliensis	*Nippostrongylus brasiliensis*
N. crassa	*Neurospora crassa*
P. aeruginosa	*Pseudomonas aeruginosa*
P. berghei	*Plasmodium berghei*
P. camembertii	*Penicillium camembertii*
P. chrysogenum	*Penicillium chrysogenum*
P. denitrificans	*Paracoccus denitrificans*
P. falciparum	*Plasmodium falciparum*
P. fluorescens	*Pseudomonas fluorescens*
P. horikoshii	*Pyrococcus horikoshii*
P. islandicum	*Pyrobaculum islandicum*
P. putida	*Pseudomonas putida*
P. redivivus	*Panagrellus redivivus*
P. syringae	*Pseudomonas syringae*
R. capsulatus	*Rhodobacter capsulatus*
R. eutropha	*Ralstonia eutropha*
R. leguminosarum	*Rhizobium leguminosarum*
R. sphaeroides	*Rhodobacter sphaeroides*
R. tropici	*Rhizobium tropici*
S. aureus	*Staphylococcus aureus*
S. cerevisiae	*Saccharomyces cerevisiae*
S. coelicolor	*Streptomyces coelicolor*
S. enterica	*Salmonella enterica*
S. epidermidis	*Staphylococcus epidermidis*
S. faecalis	*Streptococcus faecalis*
S. flexneri	*Shigella flexneri*
S. lactis	*Streptococcus lactis*
S. lugdunensis	*Staphylococcus lugdunensis*
S. mansoni	*Schistosoma mansoni*
S. marcescens	*Serratia marcescens*
S. meliloti	*Sinorhizobium meliloti*
S. mutans	*Streptococcus mutans*
S. oneidensis	*Shewanella oneidensis*
S. pneumoniae	*Streptococcus pneumoniae*
S. pombe	*Schizosaccharomyces pombe*
S. pyogenes	*Streptococcus pyogenes*

Continued

Table 2. Continued.

Abbreviated form	Full name
S. ruminantium	*Selenomonas ruminantium*
S. salivarius	*Streptococcus salivarius*
S. sibiricum	*Streptosporangium sibiricum*
S. solfataricus	*Sulfolobus solfataricus*
S. thermophiles	*Streptococcus thermophiles*
S. thermophilum	*Symbiobacterium thermophilum*
S. typhimurium	*Salmonella typhimurium*
S. xylosus	*Staphylococcus xylosus*
T. aromatica	*Thauera aromatica*
T. brucei brucei	*Trypanosoma brucei brucei*
T. brucei gambiense	*Trypanosoma brucei gambiense*
T. brucei rhodesiense	*Trypanosoma brucei rhodesiense*
T. cruzi	*Trypanosoma cruzi*
T. circumcincta	*Teladorsagia circumcincta*
T. gondii	*Toxoplasma gondii*
T. harzianum	*Trichoderma harzianum*
T. litoralis	*Thermococcus litoralis*
T. maritima	*Thermotoga maritima*
T. niveum	*Tolypocladium niveum*
T. thermophilus	*Thermus thermophilus*
V. cholerae	*Vibrio cholerae*
V. vulnificus	*Vibrio vulnificus*

References

Alberts, B., Bray, D., Hopkin, K., Johnson, A., Lewis, J., Raff, M., Roberts, K. and Walter, P. (2014) *Essential Cell Biology,* 4th edn. Garland Science (Taylor & Francis), New York and Abingdon, UK.

D'Mello, J.P.F. (ed.) (1997) *Handbook of Plant and Fungal Toxicants,* 1st edn. CRC Press, Boca Raton, Florida.

D'Mello, J.P.F. (ed.) (2012) *Amino Acids in Human Nutrition and Health*, 1st edn. CAB International, Wallingford, UK.

D'Mello, J.P.F. (ed.) (2015) *Amino Acids in Higher Plants*, 1st edn. CAB International, Wallingford, UK.

D'Mello, J.P.F., Duffus, C.M. and Duffus, J.H. (eds) (1991) *Toxic Substances in Crop Plants*, 1st edn. The Royal Society of Chemistry, Cambridge, UK.

Klug, W.S., Cummings, M.R., Spencer, C. and Palladino, M. (2014) *Concepts of Genetics*, 10th edn. Pearson Education, Harlow, UK.

Lackie, J. (ed.) (2013) *The Dictionary of Cell and Molecular Biology*, 5th edn. Academic Press, New York and London.

Lodish, H., Berk, A., Kaiser, C.A., Krieger, M., Bretscher, A., Ploegh, H., Amon, A. and Scott, M.P. (2013) *Molecular Cell Biology*, 7th edn. W.H. Freeman, New York.

Madigan, M., Martinko, J., Bender, K., Buckley, D. and Stahl, D. (2015) *Brock Biology of Microorganisms*, 14th edn. Pearson Education, Harlow, UK.

Nelson, D.L. and Cox, M.M. (2013) *Lehninger: Principles of Biochemistry*, 6th edn. Macmillan Higher Education, Basingstoke, UK.

Parish, H., Smith, T., Stirling, J. and Vella, F. (eds) (2006) *Oxford Dictionary of Biochemistry and Molecular Biology*, 2nd edn. Oxford University Press, Oxford, UK and New York.

Singleton, P. and Sainsbury, D. (2006) *Dictionary of Microbiology and Molecular Biology*, 3rd edn. Wiley-Blackwell, Hoboken, New Jersey.

Strelkauskas, A., Fahnert, B., Pryor, G. and Strelkauskas, J. (2016) *Microbiology: A Clinical Approach*, 2nd edn. Garland Science (Taylor & Francis) New York and Abingdon, UK.

1 Structural and Functional Properties of Glutamate Dehydrogenases

S. Brown[1]* and D.C. Simcock[1,2,3]

[1]*The Deviot Institute, Deviot, Tasmania, Australia;* [2]*Institute of Food, Nutrition and Human Health, Massey University, Palmerston North, New Zealand;* [3]*Faculty of Medicine and Bioscience, James Cook University, Cairns, Australia*

1.1 Abstract

Glutamate dehydrogenases are homooligomeric enzymes that catalyse the reversible oxidative deamination of glutamate to 2-oxoglutarate, thereby linking nitrogen metabolism and the tricarboxylic acid cycle. Four different classes of the enzyme are known in microbes that differ in size and cofactor specificity. Despite some common features, the kinetic properties of the enzymes from different species vary considerably. We consider the relationship between the sequences, those structures that are available and the enzyme kinetics in the context of the metabolic role of the enzyme.

1.2 Introduction

Glutamate dehydrogenase (GDH) catalyses the reversible reductive amination of 2-oxoglutarate to glutamate:

$$\begin{aligned} &\text{2-Oxoglutarate} + \text{NAD(P)H} + \text{NH}_4^+ \rightleftharpoons \\ &\quad \text{Glutamate} + \text{NAD(P)}^+ + \text{H}_2\text{O} + \text{H}^+. \end{aligned} \quad (1.1)$$

The enzyme may be NAD^+ specific (EC 1.4.1.2), $NADP^+$ specific (EC 1.4.1.4) or able to use either cofactor (EC 1.4.1.3). Irrespective of the cofactor, the standard transformed Gibbs free energy of reaction, $\Delta_r G'^0 \approx -40$ kJ mol^{-1} (Olson and Anfinsen, 1953; Strecker, 1953; Engel and Dalziel, 1967; Subramanian, 1978; Cook *et al.*, 1980) and the reaction operates in either direction depending on the relative magnitudes of

$$\begin{aligned} Q = &[\text{Glutamate}][\text{NAD(P)}^+]/ \\ &[\text{2-Oxoglutarate}][\text{NAD(P)H}][\text{NH}_4^+], \end{aligned}$$

where Q is the reaction quotient and

$$K' = \exp\left(-\Delta G'^0/RT\right) \approx 10^{-7},$$

where K' is the apparent equilibrium constant, R is the gas constant and T is the absolute temperature. If $Q < K'$, the reaction will tend to form glutamate.

In bacteria, GDH can be involved in the assimilation of NH_4^+ (Kanamori *et al.*, 1987; Bruggeman *et al.*, 2005), in glutamate synthesis (Helling, 1994) and in controlling the intracellular concentration of 2-oxoglutarate to ensure the operation of 2-oxoglutarate dehydrogenase (ODH) and the whole tricarboxylic acid (TCA) cycle (Struhl and Maganasik, 1976; Janes *et al.*, 2001; Doucette *et al.*, 2011). We consider the structural and kinetic properties of the different forms of GDH, their taxonomic distribution and the roles of the enzyme in the metabolism of bacteria.

*Corresponding author: E-mail address: Simon.Brown@deviotinstitute.org

1.3 Enzyme Structure

Four broad classes of GDH, designated GDH1–4 (Andersson and Roger, 2003), have been distinguished using structural and functional criteria. Both GDH1 and GDH2 have monomers of about 50 kDa and are distinguished by their cofactor specificity: GDH1 is NAD^+ specific, whereas GDH2 is $NADP^+$-specific. In contrast, the monomers of GDH3 and GDH4 are of about 115 and 180 kDa, respectively (Fig. 1.1), but are not as well characterized.

Both GDH1 and GDH2 are about 450 residues in length, are highly conserved and are much more similar to one another than they are to GDH3 and GDH4 (Fig. 1.2). In the alignment on which Fig. 1.2 is based, the identity among the 14 GDH1 and GDH2 sequences shown is 21–89% (40–96% similarity; average overall distance = 0.80 substitutions per site), whereas the seven GDH3 sequences have 17–65% identity (32–79% similarity; mean overall distance = 1.22 substitutions per site) and the seven GDH4 sequences are more highly conserved (31–65% identity; 47–77% similarity; overall mean distance = 0.88 substitutions per site).

1.3.1 Substrate and cofactor binding

The residues involved in binding the substrates or the cofactor are mostly completely conserved among GDH1 and GDH2 sequences (Fig. 1.3), and some of these are also conserved in GDH3 and GDH4 sequences. Of particular note among the latter are the lysine residue K126 and the aspartate residue D166, which, in concert with the α-amino group of the substrate glutamate and a water oxygen, form a tetrad capable of a wide range of hydrogen bonding patterns that may be catalytically significant (Fisher and Maniscalco, 2002).

The determinants of the cofactor specificity of GDH have not yet been identified unambiguously. It has been proposed that the Gly-X-Gly-X-X-Gly/Ala motif in the P-loop of the Rossmann fold is involved in determining the cofactor specificity of dehydrogenases (Bellamacina, 1996). An alanine in the final position (P6 or the alanine A244 in *Escherichia coli*) is supposed to favour $NADP^+$ binding, whereas a glycine favours NAD^+ binding, but the *Clostridium symbiosum* GDH1 has an alanine and the *Psychrobacter* sp. TAD1 GDH2 has a serine, so neither of these conform to this model. No consensus about P6 is apparent from the GDH3 and GDH4 sequences. Similarly, the presence of a negatively charged group near the 2′-OH and 3′-OH of adenosine (P7 or D263 in *E. coli*) has been suggested to favour NAD^+ rather than $NADP^+$ binding. However, as Sharkey *et al.* (2013) observed, there is an aspartate (or a glutamate in *Peptinophilus asaccharolyticus*) in this position in most GDH sequences, with the exceptions of the *C. symbiosum* GDH1 and *Pseudomonas aeruginosa* GDH4. Based on the modelling of $NADP^+$ binding to GDH2, it has been suggested that $NADP^+$

Fig. 1.1. The domains of representative bacterial glutamate dehydrogenases (GDHs). The *Peptoniphilus asaccharolyticus* (AAA25611.1), *Escherichia coli* (P00370.1), *Desulfovibrio vulgaris* (ABM29040.1) and *Pseudomonas aeruginosa* (WP_053816968.1) enzymes are examples of GDH1, GDH2, GDH3 and GDH4, respectively. The domains identified using Pfam (the protein families database; Finn *et al.*, 2014) and CATH (class, architecture, topology, homology – a hierarchical protein domain classification system; Cuff *et al.*, 2011) are an NAD-binding Rossmann-like domain (use of 'NAD' intends no implication on cofactor specificity), a bacterial GDH domain (BAC) and a domain common to glu/leu/phe/val dehydrogenases (the substrate-binding domain, ELFV). For the *P. aeruginosa* sequence, Pfam identified a more extensive BAC domain (the dashed rectangle) than did CATH (the black rectangle).

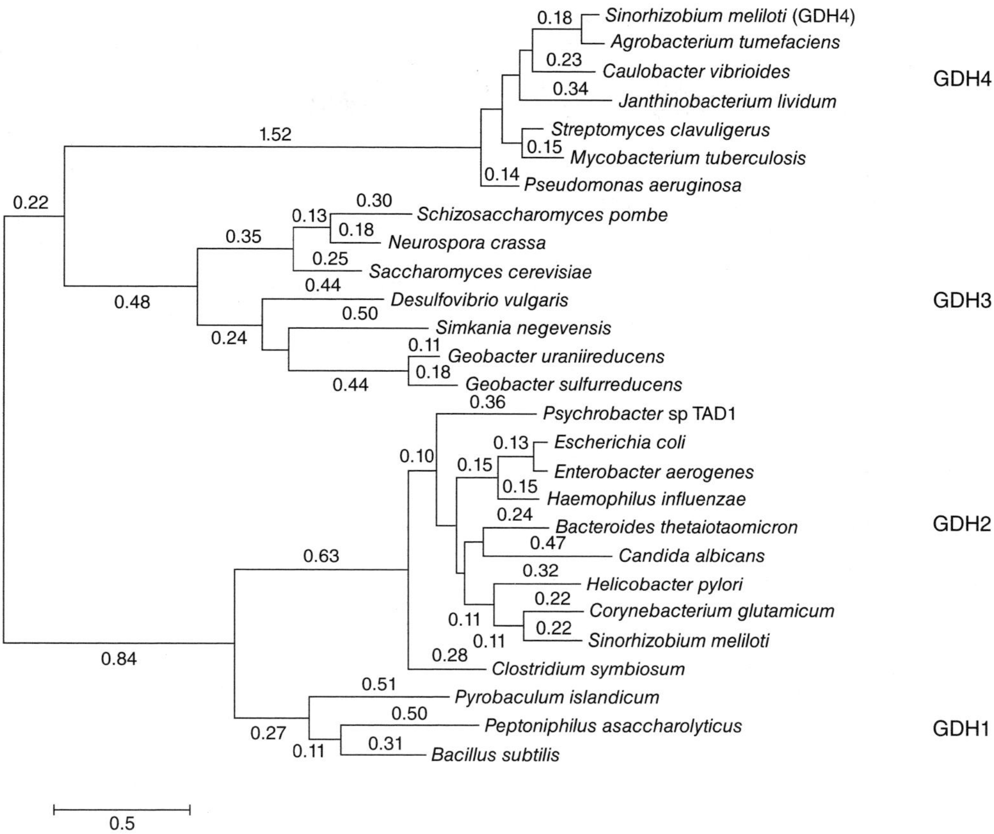

Fig. 1.2. The bootstrap consensus phylogram derived by maximum likelihood from an alignment of 27 glutamate dehydrogenase (GDH) amino acid sequences (375 gap-free positions) using molecular evolutionary genetics analysis (MEGA5; Tamura *et al.*, 2011). The maximum likelihood was based on the JTT (Jones, Taylor, Thornton) matrix-based model (Jones *et al.*, 1992) using 500 bootstrap replicates (Felsenstein, 1985). The branch lengths (number of substitutions per site) are indicated below the branches unless the value was less than 0.1. The shape parameter of the discrete gamma distribution was 1.637.

specificity may be conferred by: (i) three positively charged residues (lysine K286, and arginines R289, R292 in the *E. coli* enzyme); and (ii) a serine (S264 in *E. coli*). In this model, these four residues are adjacent to the 2′-phosphate of $NADP^+$ and stabilize the negative charge (Sharkey *et al.*, 2013). While K286 is almost completely conserved among GDH1 and GDH2, and S264 is also highly conserved (although neither is present in the *Burkholderia thailandensis* sequence), R289 and R292 are both absent from the $NADP^+$-specific *Psychrobacter* enzyme (Di Fraia *et al.*, 2000). Overall, as Sharkey *et al.* (2013) suggest, cofactor specificity cannot be reliably inferred from primary sequence.

1.3.2 Tertiary and quaternary structure

As with other dehydrogenases, GDH has a substrate-binding domain (ELFV), which is similar to those of other amino acid dehydrogenases, and an $NAD(P)^+$-binding domain (NAD), both of which can be identified from sequence homology alone (Fig. 1.2). Crystal structures are available for only GDH1 and GDH2, but these are similar to those of the human and bovine enzymes, except for the absence of the antenna domain. In each case, the ELFV domain consists of a series of β-strands sandwiched between α-helices and the NAD domain is a modified Rossmann fold. The nucleotide binds

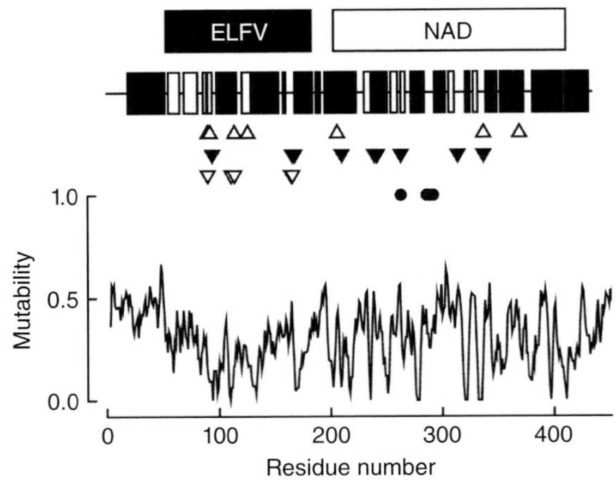

Fig. 1.3. Summary of the structural properties of the NAD$^+$-specific glutamate dehydrogenase (GDH) from *Burkholderia thailandensis* and the mutability of GDH1 and GDH2 sequences. The triangles (▲, ▼, △) indicate residues that are hydrogen bonded to glutamate, 2-oxoglutarate or NAD$^+$ based on the crystal structures of the *B. thailandensis* (4XGI) and *Clostridium symbiosum* (1BGV) enzymes. The circles (●) indicate the approximate position of the residues involved in determining the NADP$^+$-specificity of the *Escherichia coli* enzyme (Sharkey *et al.*, 2013). The secondary structure (4XGI) indicates only the positions of α-helices and β-strands (black and white rectangles, respectively). The mutability was calculated as described previously (Brown *et al.*, 2014) from the sequence alignment of the GDH1 and GDH2 sequences identified in Fig. 1.2. Abbreviations: ELFV, a domain common to glu/leu/phe/val dehydrogenases (the substrate-binding domain); NAD, a NAD-binding Rossmann-like domain (use of 'NAD' intends no implication on cofactor specificity).

in a cleft between the domains and is ligated by residues from each (Fig. 1.3).

The substrate and cofactor are necessarily close together. In both the bovine enzyme and in *B. thailandensis* GDH1, the substrate binds towards the narrow end of the cleft and NAD$^+$ binds further towards the wider end of the cleft with the nicotinamide moiety oriented towards the substrate. In each case, they are so close that the 3′-OH of NAD$^+$ and the carboxyl group of the substrate are both H bonded to a single asparagine residue.

Most microbial GDHs are homohexamers, although the GDH2 of *Streptomyces fradiae* (Vančurová *et al.*, 1989), the GDH3s of *Agaricus bisphorus* and *Neurospora crassa* (Smith *et al.*, 1975) and the GDH4 of *P. aeruginosa* (Lu and Abdelal, 2001) are reportedly tetramers. Unusually, the GDH4 of *Psychrobacter* is dimeric and while some kinetic data have been reported (Camardella *et al.*, 2002), the primary sequence is not yet available. The crystal structures that are available exhibit D_3 symmetry in which monomers are located at the vertices of triangles and these are stacked in pairs with the monomers approximately aligned. In each monomer the cleft between the domains opens away from, and the nicotinamide moiety of NAD$^+$ is oriented towards, the threefold axis. It is inevitable that the tetrameric GDHs will have very different quaternary structures.

1.4 Enzyme Mechanism and Kinetics

The kinetics of GDH are complex, both because of the number of reactants and products (Eqn 1.1) and because of the variety of behaviours observed (Barton and Fisher, 1971; Bailey *et al.*, 1982; Syed and Engel, 1987). For some forms of the enzyme the usual rectangular hyperbolic relationship between initial rate (v_0) and substrate concentration (s) is observed, but cooperative kinetics have also been reported (Table 1.1). Reports of non-linear double reciprocal, Eadie–Hofstee or other transformed plots are common for both the mammalian (Barton and Fisher, 1971; Bailey *et al.*, 1982) and microbial enzymes (Table 1.1).

Table 1.1. Kinetic characteristics of some microbial glutamate dehydrogenases (GDHs). Where cooperativity is reported, the Hill coefficient (n_H) is given below the K_m or $s_{0.5}$ in parentheses.

		K_m or $s_{0.5}$ (n_H)				
Species	Cofactor	2-OG[a] (mM)	Glu[b] (mM)	NH_4^+ (mM)	NAD(P)[+] (μM)	NAD(P)H (μM)
GDH1						
Bacillus cereus DSM 31 (Jahns and Kaltwasser, 1993)	NAD[+]	[c]	7.4	96	560	33
Bacillus subtilis ISW1214 (Khan *et al.*, 2005)	NAD[+]	0.65	0.34	55.6	80	70
Bacteroides thetaiotaomicron	NAD[+]		12.3		610	
(Glass and Hylemon, 1980)	NADP[+]	0.24	3.7	5.0	40	17
Peptostrepococcus asaccharolyticus (Hornby and Engel, 1984)	NAD[+]	0.82	6.6	18.4	31	66
Psychrobacter sp. TAD1 (Camardella *et al.*, 2002)	NAD[+]	2.36	28.6	24.6	500	40 (1.8)
Pyrobaculum islandicum	NAD[+]	0.066	0.17	9.7	25	5.0
(Kujo and Ohshima, 1998)	NADP[+]				240	270
GDH2						
Escherichia coli B/r (Sakamoto *et al.*, 1975)	NADP[+]	0.64	1.3	1.1	42	40
Psychrobacter sp. TAD1 (Di Fraia *et al.*, 2000)	NADP[+]	6.8 (3.6)	67.9 (1.2)	4.5 (1.1)	52 (0.8)	23 (2.3)
Ruminococcus flavefaciens (Duncan *et al.*, 1992)	NADP[+]	0.41	62.19	19.22	–	34
Saccharomyces cerevisiae (DeLuna *et al.*, 2001)	NADP[+]	1.27 (1.5)	6.36	5.00	10.5	331
Saccharomyces cerevisiae (DeLuna *et al.*, 2001)	NADP[+]	0.29 (1.3)	9.79	5.96	14.1	113
GDH3						
Benjaminiella poitrassi (Joshi *et al.*, 2013)	NAD[+]	3.12	33.3	27.7	170	60
Agaricus bisporus (Kersten *et al.*, 1999)	NAD[+]	3.5	37.1	6.5	46	60
Neurospora crassa (Smith *et al.*, 1975)	NAD[+]	4.6	5.5	17	330	550
GDH4						
Janthinobacterium lividum (Kawakami *et al.*, 2007, 2010)	NAD[+]	9.1 (1.7)	7.1	20.9	2100	68 (1.8)
Pseudomonas aeruginosa (Lu and Abdelal, 2001)	NAD[+]	–	4.4 (1.7)	–	130	–
Streptomyces clavuligerus (Miñambres *et al.*, 2000)	NAD[+]	1.33 (1.5)	2.27 (1.39)	33.9	51.16 (1.48)	19.87 (1.52)

[a]2-OG, 2-oxoglutarate.
[b]Glu, glutamate.
[c]Response is sigmoidal, but no data are reported.

Rife and Cleland (1980b) proposed a model of the catalytic mechanism of GDH which is largely consistent with the crystal structures that have been found (Stillman *et al.*, 1993) and subsequent analyses of the kinetics (Fisher, 2005). The model shown in Fig. 1.4 is essentially that of Rife and Cleland (1980b) for bovine liver GDH. The enzyme-bound 2-oxoglutarate is aminated, transiently forming a quaternary complex (ERNK in Fig. 1.4), to a 2-carbinolamine (2-hydroxy-2-iminoglutarate). This is dehydrated to 2-iminoglutarate which is then reduced to glutamate. At this point, the complex is protonated and then water is released

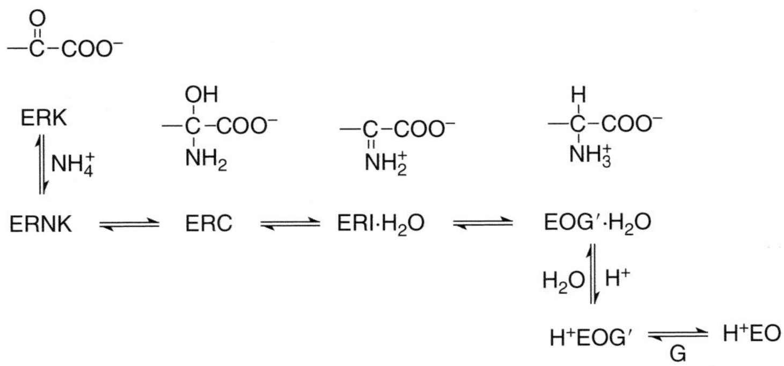

Fig. 1.4. Catalytic mechanism of bovine glutamate dehydrogenase (GDH) adapted from Rife and Cleland (1980b), Maniscalco *et al.* (1996) and Fisher (2005). The sequence of binding of NADH and 2-oxoglutarate is not shown. The structural fragments above each intermediate represent C1 and C2 of the ligand starting with 2-oxoglutarate (left) and ending with glutamate (right). The standard abbreviations used below the fragments are: C, 2-carbinolamine; E, enzyme; G, glutamate; I, 2-iminoglutarate; K, 2-oxoglutarate; N, NH_4^+; O, NAD^+; R, NADH.

from bovine liver GDH, although these steps are reversed in the *C. symbiosum* enzyme (Maniscalco *et al.*, 1996), before glutamate is released (Fig. 1.4).

The most commonly used initial rate (v_0) equation for GDH is the empirical expression

$$\frac{e}{v_0} = \phi_0 + \frac{\phi_1}{s_1} + \frac{\phi_2}{s_2} + \frac{\phi_3}{s_3} + \frac{\phi_{12}}{s_1 s_2} + \frac{\phi_{13}}{s_1 s_3} + \frac{\phi_{23}}{s_2 s_3} + \frac{\phi_{123}}{s_1 s_2 s_3} \quad (1.2)$$

or (see Eqn 1.3 at bottom of the page):

where e is the concentration of GDH, the s_i (s_1, s_2, s_3) are the concentrations of NAD(P)H, NH_4^+ and 2-oxoglutarate, and different combinations of rate and dissociation constants are distinguished by particular subscripts to ϕ. In Eqns 1.2 and 1.3, e/ϕ_0 is the maximum velocity (V_{max}), the K_m for s_i is ϕ_i/ϕ_0 and all substrates are independent and have rectangular hyperbolic kinetics. This expression is based on a general random binding order model involving the formation of a quaternary complex, but it includes, as special cases, other models, including those with specific binding sequences (Dalziel, 1969). For example, some authors have concluded that there was, in at least one direction, an ordered sequence of binding to the bovine enzyme (Frieden, 1959; Rife and Cleland, 1980a), whereas others have argued that the data are consistent with the random binding order and have obtained estimates of all eight parameters (Engel and Dalziel, 1970). The implication of the mechanism shown in Fig. 1.4 is that binding is at least partially ordered: NH_4^+ binds after NAD(P)H and 2-oxoglutarate and, in the opposite direction, glutamate binds after $NAD(P)^+$.

The parameters of Eqn 1.2 are often obtained sequentially from linearized plots (Dalziel, 1969) which necessarily introduces bias into the estimates and may explain, in part, why it is sometimes difficult to obtain values for some parameters (Engel and Dalziel, 1970). Unsurprisingly, it is also clear that Eqn 1.2 cannot account for the reported cooperative kinetics (Di Fraia *et al.*, 2000; Miñambres *et al.*, 2000) of GDH and does not account for substrate inhibition (Sharkey and Engel, 2008) or the activation by high NAD^+ concentrations (Olson and Anfinsen, 1953; Engel and Dalziel, 1970). Other empirical expressions have been devised to account for some of these phenomena, and of these three

$$v_0 = \frac{s_1 s_2 s_3 e}{\phi_0 s_1 s_2 s_3 + \phi_1 s_2 s_3 + \phi_2 s_1 s_3 + \phi_3 s_1 s_2 + \phi_{12} s_3 + \phi_{13} s_2 + \phi_{23} s_1 + \phi_{123}}, \quad (1.3)$$

are of particular interest. First, Engel and Ferdinand (1973) proposed a rational rate equation:

$$v_0 = \frac{\sum_{j=1}^{q} D_j s^j e}{N_0 + \sum_{i=1}^{q} N_i s^i}, \quad (1.4)$$

which can describe an almost unlimited range of behaviour, as has been considered in some detail (Bardsley and Childs, 1975; Bardsley *et al.*, 1980; Brown and Simcock, 2013). Second, Kurganov (2000) proposed that the K_m varies continuously with v_0/V_{max} and modified the standard Michaelis–Menten expression accordingly:

$$v_0 = \frac{V_{max} s}{\left(K_0 + (K_{lim} - K_0)\frac{v_0}{V_{max}}\right) + s}, \quad (1.5)$$

where K_0 and K_{lim} are the effective values of K_m as v_0/V_{max} approaches 0 and 1, respectively. This can be written as

$$s = (K_0 - K_{lim})\frac{v_0}{V_{max}} + K_{lim}\frac{v_0}{V_{max} - v_0}, \quad (1.6)$$

so if $K_{lim} < K_0$, the linear v_0–s regime dominates and the transition to the saturated regime occurs over a small range of s. If $K_{lim} > K_0$, a biphasic double reciprocal plot is generated which is reminiscent of those often reported for GDH (Barton and Fisher, 1971; Bailey *et al.*, 1982; Syed and Engel, 1987). Both Eqns 1.4 and 1.5 account for only one of the substrates and, like Eqn 2, their main purpose is to provide a means of summarizing particular aspects of the kinetics of GDH. This weakness is overcome, to some extent, by the third example, Eqn 1.6 – the expression devised by Barton and Fisher (1971) to account for the kinetics of the oxidative deamination reaction. Then again, this involves 13 parameters, few of which have any obvious mechanistic significance. A complete kinetic model is yet to be devised for the enzyme.

Despite rate equations such as shown in Eqns 1.2–1.5, the usual practice has been to estimate the K_m (or $s_{0.5}$ where there is cooperativity) and V_{max} by independently varying the concentration of each reactant. The small selection of data in Table 1.1 prompts hypothesis that there might be some differences between GDHs. For example, it appears that the K_m values for both NH_4^+ and $NAD(P)^+$ may tend to be larger in GDH1 than in GDH2 (Table 1.1). While this is merely speculation, careful analysis of the available data is warranted.

1.5 Genomics

The taxonomic distribution of the forms of GDH is uneven. Both GDH1 and GDH2 are widespread among bacteria and fungi, but GDH3 and GDH4 have more restricted distributions (Andersson and Roger, 2003). Based on the data available in Pfam (the protein families database; Finn *et al.*, 2014), homologues of GDH4 (based on the presence of the BAC domain shown in Fig. 1.1, Pfam accession code PF05088) can be found among the α-, β- and γ-proteobacteria (149, 43 and 146 species, respectively), and in the actinobacteria (84 species), as well as in a small number of fungi (11 species) and metazoa (five species). Among the bacteria that have GDH4 homologues are a number of *Mesorhizobium* spp., *Sinorhizobium* spp. and *Bradyrhizobium* spp., and some, such as *Sinorhizobium meliloti* (Fig. 1.2), also have other forms of GDH. However, rhizobia generally lack significant GDH activity (Ali *et al.*, 1981; Bravo and Mora, 1988; Rossi *et al.*, 1989), and several species of bacteria lack any GDH sequence (Andersson and Roger, 2003).

One of the many challenges of genomic analysis is the determination of the presence or absence of a gene (Cordwell, 1999; Osterman and Overbeek, 2003), but it is just one of many. Even if it can be confirmed that a gene occurs in a genome, that need not mean that it is expressed or, if it is, that the product has only the anticipated consequence. The identification of any homologue of a gene can be an even greater challenge if only because of insufficient data. For example, GDH4 is quite different from GDH1 (see Fig. 1.1) and so a database search may not identify all of the four forms of the enzyme. Depending on the circumstances, even the detection of a particular enzyme activity need not be unambiguous. For example, ODH was reported to be present in *Mycobacterium tuberculosis* (Cordwell, 1999) based on the presence of the genes and the oxidation of the substrate, but even if the genes are expressed they need not be incorporated into ODH (Tian *et al.*, 2005b) and the apparent ODH activity can be due to other enzymes entirely

(Tian *et al.*, 2005a). Further, there is the growing realization that an enzyme might have more than one function. For example, mammalian GDH may link lysosomes to microtubules (Rajas *et al.*, 1996) and bind mRNA (Preiss *et al.*, 1995), perhaps regulating the translation of proteins targeted to the mitochondria, while the aconitase apoprotein is involved in the regulation of iron metabolism (Beinert *et al.*, 1997; Brzóska *et al.*, 2006).

Nevertheless, the presence or absence of an enzyme such as GDH is biochemically and physiologically significant because it links carbon and nitrogen metabolism (Fig. 1.5). Notwithstanding, the TCA cycle has many variants (Weitzman, 1981; Huynen *et al.*, 1999) and so this metabolic intersection can be especially complex. Bearing in mind the intrinsic difficulties already described, several species do not have genes encoding any of the GDHs (Andersson and Roger, 2003), including *Aquifex aeolicus*, *Archaeoglobus fulgidus*, *Borrelia burgdorferi*, *Methanobacterium thermoautotrophicum* and *Mycoplasma pneumoniae* (Cordwell, 1999). All five of the named species just listed also lack ODH activity, and two of them, *B. burgdorferi* and *M. pneumoniae*, do not have isocitrate dehydrogenase (IDH) activity either (Cordwell, 1999). This type of listing can be extended indefinitely (see Table 1.2). While such a list does not show the extent to which alternative reactions can substitute for any of these enzymes, it does prompt consideration of the possible metabolic variants.

1.6 Metabolic Context

GDH is an important link between amino acid and carbon metabolism through the TCA cycle (Fig. 1.5). Specifically, the 2-oxoglutarate produced by IDH (EC 1.1.1.42), is a substrate for

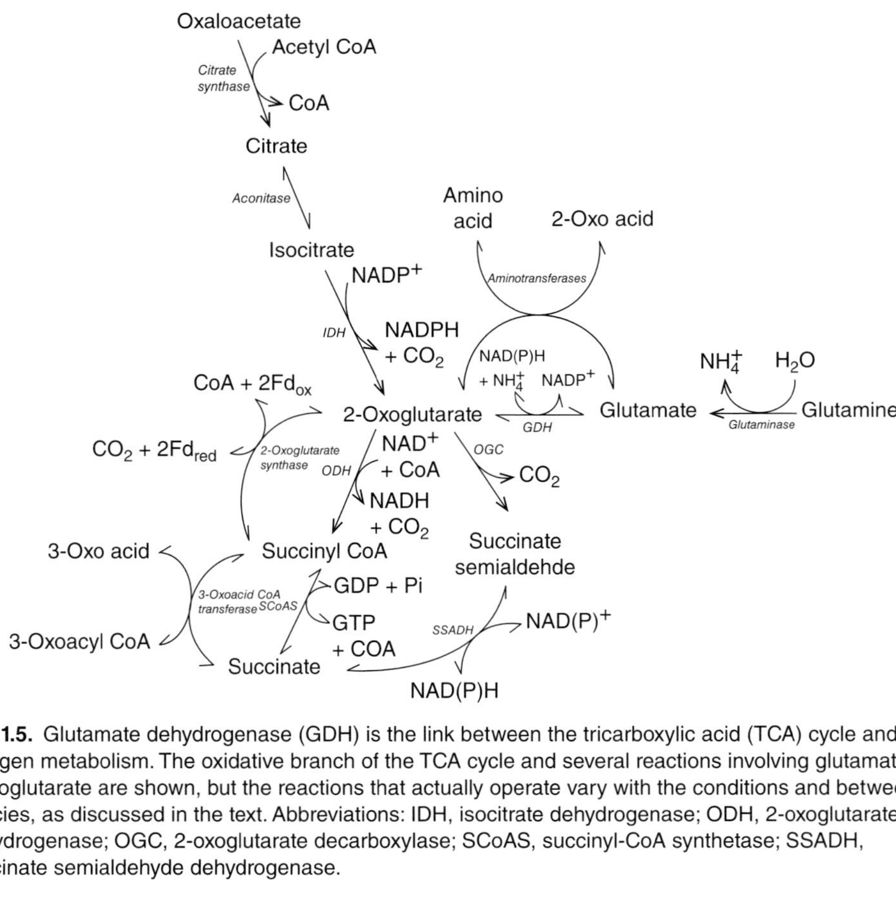

Fig. 1.5. Glutamate dehydrogenase (GDH) is the link between the tricarboxylic acid (TCA) cycle and nitrogen metabolism. The oxidative branch of the TCA cycle and several reactions involving glutamate and 2-oxoglutarate are shown, but the reactions that actually operate vary with the conditions and between species, as discussed in the text. Abbreviations: IDH, isocitrate dehydrogenase; ODH, 2-oxoglutarate dehydrogenase; OGC, 2-oxoglutarate decarboxylase; SCoAS, succinyl-CoA synthetase; SSADH, succinate semialdehyde dehydrogenase.

Table 1.2. The presence (+) or absence (–) of glutamate dehydrogenase (GDH), isocitrate dehydrogenase (IDH), 2-oxoglutarate dehydrogenase (ODH) and/or succinyl-CoA synthetase (SCoAS) in several species for which a complete genome sequence is available. The data for IDH, OGDH and SCoAS were adapted from Cordwell (1999).

Species	GDH	IDH	ODH	SCoAS
Aquifex aeolicus	–	+	–	+
Archaeoglobus fulgidus	–	+	–	+
Bacillus subtilis	+	+	+	+
Borrelia burgdorferi	–	–	–	–
Chlamydia trachomatis	–	–	+	+
Corynebacterium glutamicum	+	+	–	+
Escherichia coli	+	+	+	+
Haemophilus influenzae	+	–	+	+
Helicobacter pylori	+	+	–	–
Methanobacterium thermoautotrophicum	–	+	–	+
Methanococcus jannaschii	?	+	–	+
Mycobacterium tuberculosis	+	+	–	?
Mycoplasma genitalium	?	–	–	–
Mycoplasma pneumoniae	–	–	–	–
Pyrococcus horikoshii	+	–	–	–
Synechocystis sp. PCC6803	+	+	–	+
Treponema pallidum	?	–	–	–

both GDH and the ODH complex, which consists of multiple copies of a thiamine pyrophosphate-dependent dehydrogenase (E1; EC 1.2.4.2), dihydrolipoamide succinyltransferase (E2; EC 2.3.1.61) and dihydrolipoamide dehydrogenase (E3; EC 1.8.1.4). While GDH catalyses a reversible reaction, the citrate synthase, IDH and ODH reactions are irreversible. In fact, 2-oxoglutarate is among the key metabolites in the metabolic network of *E. coli* and glutamate is the most highly connected metabolite in this network (Wagner and Fell, 2001).

The interaction of the great variety of enzymes and reactions of the TCA cycle observed among microbes (Weitzman, 1981; Huynen *et al.*, 1999) with GDH has profound metabolic implications. Considered simply, the three enzymes directly involved in 2-oxoglutarate metabolism are IDH, ODH and GDH. In *E. coli*, IDH and GDH2 are $NADP^+$ dependent with K_m = 17 μM $NADP^+$ (Hurley *et al.*, 1996) and 42 μM NADPH (Table 1.1), respectively, while ODH is NAD^+ dependent with K_m = 400 μM NAD^+, and is also subject to product inhibition (K_i = 1 mM) (Schmincke-Ott and Bisswanger, 1981). However, this difference in cofactor specificity is less significant than it might seem because *E. coli* has two transhydrogenases, one soluble (Cao *et al.*, 2011), and the other in the plasma membrane (Bizouarn *et al.*, 2000), that keep $[NAD^+]/[NADH] \approx 19$ and $[NADP^+]/[NADPH] \approx 1.2$ (Andersen and von Meyenburg, 1977). The K_m for 2-oxoglutarate is similar for GDH2 (0.64 mM; Table 1.1) and ODH (0.1 mM; Gupta and Dekker, 1980), but GDH2 also requires NH_4^+ (K_m = 1.1 mM; Table 1.1), so the flux through ODH would tend to be greater than that through GDH2.

Under aerobic conditions, *E. coli* operates a complete TCA cycle, but in anaerobic conditions it lacks ODH activity and the pathway operates as two branches: an oxidative branch leading to the production of 2-oxoglutarate and a reductive branch yielding succinate (Hager and Kornberg, 1961; Amarasingham and Davis, 1965). *E. coli* only assimilates NH_4^+ under aerobic conditions and GDH is involved in this process in energy- or glucose-limited conditions, when the concentration of NH_4^+ is sufficiently high, whereas glutamine synthetase (GS; EC 6.3.1.2) and glutamate synthase (GOGAT, E. C. 1.4.1.13) are responsible in energy-rich conditions (Helling, 1994, 1998). In glucose-limited conditions, GDH and ODH share the available 2-oxoglutarate.

However, when ODH activity is low and 2-oxoglutarate accumulates, GDH2 can utilize the available NH_4^+ and the NADPH generated by

IDH to produce glutamate (if $Q < K'$). For example, the *E. coli* mutant 309-1, which constitutively lacks the capacity to oxidize 2-oxoglutarate, excretes glutamate when grown aerobically on glucose in the presence of succinate (Davis *et al.*, 1959; Shiio *et al.*, 1961). A similar *Enterobacter aerogenes* mutant produces 2-oxoglutarate from glucose in anaerobic conditions (Stouthamer, 1966). Conversely, if the 2-oxoglutarate supply is low, then GDH2 can oxidatively deaminate glutamate (using $NADP^+$; $K_m = 40\ \mu M$) to produce more, providing that $Q > K'$. It should be noted though that while glutamate is an effective inducer of ODH activity (Amarasingham and Davis, 1965) and *E. coli* can use glutamate as a sole nitrogen source, the maximum growth rate is low whether or not glucose is limiting (Halpern and Umbarger, 1961; Senior, 1975).

Corynebacterium glutamicum (formerly *Brevibacterium flavum* and *B. lactofermentum*) has significant $NADP^+$-dependent GDH activity (Shiio *et al.*, 1959), which is presumably GDH2 (Shiio and Ozaki, 1970), along with appreciable IDH activity (Shiio *et al.*, 1961), but its ODH activity declines during growth (Kawahara *et al.*, 1997) and, in at least some circumstances, it can be undetectable (Shiio *et al.*, 1961). During aerobic growth on glucose, 2-oxoglutarate, or in the presence of NH_4^+, glutamate accumulates (Shiio *et al.*, 1959), and this can be substantially reduced by biotin supplementation (Shiio *et al.*, 1962). Deleting the gene for the E1 enzyme of ODH completely inactivates ODH and has little effect on the accumulation of glutamate (Asakura *et al.*, 2007), whereas deletion of the GDH gene reduces the intracellular concentration of glutamate and increases those of NH_4^+ and 2-oxoglutarate (Müller *et al.*, 2006).

The oxidation of citrate and isocitrate by *M. tuberculosis* can be estimated from the rate of O_2 uptake, but this requires $NADP^+$ and is independent of CoA (Murthy *et al.*, 1962). This is inconsistent with the operation of ODH, which requires NAD^+ and CoA, and arises from the replacement of ODH and succinyl CoA synthetase (SCoAS; EC 6.2.1.4) by a combination of reactions catalysed by 2-oxoglutarate decarboxylase (OGC; EC 4.1.1.71)

$$\text{2-Oxoglutarate} \rightleftharpoons \text{Succinate semialdehyde} + CO_2$$

and succinate semialdehyde dehydrogenase (SSADH; EC 1.2.1.16)

$$\text{Succinate semialdehyde} + \text{NAD(P)}^+ + \text{H}_2\text{O} \rightleftharpoons \text{Succinate} + \text{NAD(P)H} + 2\text{H}^+$$

(Tian *et al.*, 2005a). The K_m for 2-oxoglutarate of OGC is 0.48 mM (Tian *et al.*, 2005a) is much lower than that of *M. smegmatis* (5 mM; Sarada *et al.*, 1980), the only GDH2 that has been characterized from *Mycobacterium* spp., which, if a similar situation pertains in *M. tuberculosis*, would tend to favour the operation of OGC/SSADH rather than GDH. While these, and other (Piston *et al.*, 1999), alternative reactions are interesting, what is remarkable is that *M. tuberculosis* has genes encoding the enzymes of ODH, but that they are used for other purposes (Maksymiuk *et al.*, 2015). This reinforces a point made previously: it is unwise to infer the presence of a gene product or an expected biological function from the presence of a gene.

1.7 Conclusions

Of the four forms of GDH in microbes, GDH1 and GDH2 have been well characterized, but GDH3 and GDH4 are less well understood. Crystal structures are available for GDH1 and GDH2 and much work has been done to analyse the kinetics of these enzymes, but there is little information on GDH3 or GDH4, and much remains to be done to generate a coherent whole from the rapidly growing volume of data. Some of the difficulty arises from the extraordinary metabolic and physiological flexibility of bacteria and fungi, the consequence of which is that genera (and sometimes species) have less in common than might be anticipated. This leads to the conclusion that more data are only part of the answer, and that it is also necessary to develop the means to integrate the data in meaningful ways.

References

Ali, H., Niel, C. and Guillaume, J.B. (1981) The pathways of ammonium assimilation in *Rhizobium meliloti*. *Archives of Microbiology* 129, 391–394.

Amarasingham, C.R. and Davis, B.D. (1965) Regulation of α-ketoglutarate dehydrogenase formation in *Escherichia coli*. *The Journal of Biological Chemistry* 240, 3664–3668.

Andersen, K.B. and von Meyenburg, K. (1977) Charges of nicotinamide adenine nucleotides and adenylate energy charge as regulatory parameters of the metabolism in *Escherichia coli*. *The Journal of Biological Chemistry* 252, 4151–4156.

Andersson, J.O. and Roger, A.J. (2003) Evolution of glutamate dehydrogenase genes: evidence for lateral gene transfer within and between prokaryotes and eukaryotes. *BMC Evolutionary Biology* 3:14.

Asakura, Y., Kimura, E., Usuda, Y., Kawahara, Y., Matsui, K., Osumi, T. and Nakamatsu, T. (2007) Altered metabolic flux due to deletion of *odhA* causes L-glutamate overproduction in *Cornyebacterium glutamicum*. *Applied and Environmental Microbiology* 73, 1308–1319.

Bailey, J., Bell, E.T. and Bell, J.E. (1982) Regulation of bovine glutamate dehydrogenase. The effects of pH and ADP. *The Journal of Biological Chemistry* 257, 5579–5583.

Bardsley, W.G. and Childs, R.E. (1975) Sigmoid curves, non-linear double-reciprocal plots and allosterism. *Biochemical Journal* 149, 313–328.

Bardsley, W.G., Leff, P., Kavanagh, J. and Waight, R.D. (1980) Deviations from Michaelis–Menten kinetics. The possibility of complicated curves for simple kinetic schemes and the computer fitting of experimental data for acetylcholinesterase, acid phosphatase, adenosine deaminase, arylsulphatase, benzylamine oxidase, chymotrypsin, fumarase, galactose dehydrogenase, β-galactosidase, lactate dehydrogenase, peroxidase and xanthine oxidase. *Biochemical Journal* 187, 739–765.

Barton, J.S. and Fisher, J.R. (1971) Nonlinear kinetics of glutamate dehydrogenase. Studies with substrates – glutamate and nicotinamide-adenine dinucleotide. *Biochemistry* 10, 577–585.

Beinert, H., Holm, R.H. and Münck, E. (1997) Iron–sulfur clusters: nature's modular, multipurpose structures. *Science* 277, 653–659.

Bellamacina, C.R. (1996) The nicotinamide dinucleotide binding motif: a comparison of nucleotide binding proteins. *FASEB Journal* 10, 1257–1269.

Bizouarn, T., Fjellström, O., Meuller, J., Axelsson, M., Bergkvist, A., Johansson, C., Karlsson, B.G. and Rydström, J. (2000) Proton translocating nicotinamide nucleotide transhydrogenase from *Escherichia coli*. Mechanism of action deduced from its structural and catalytic properties. *Biochimica et Biophysica Acta* 1457, 211–228.

Bravo, A. and Mora, J. (1988) Ammonium assimilation in *Rhizobium phaseoli* by the glutamine synthetase–glutamate synthase pathway. *Journal of Bacteriology* 170, 980–984.

Brown, S., Muhamad, N., Walker, L.R., Pedley, K.C. and Simcock, D.C. (2014) An *in silico* analysis of the glutamate dehydrogenases of *Teladorsagia circumcincta* and *Haemonchus contortus*. *Journal of Bioscience and Biotechnology* 3, 49–60.

Brown, S. and Simcock, D.C. (2013) Confidence bands for rational rate equations. *International Journal of Emerging Sciences* 3, 15–27.

Bruggeman, F.J., Boogerd, F.C. and Westerhoff, H.V. (2005) The multifarious short-term regulation of ammonium assimilation of *Escherichia coli*: dissection using an *in silico* replica. *FEBS Journal* 272, 1965–1985.

Brzóska, K., Meczynska, S. and Kruszewski, M. (2006) Iron–sulfur cluster proteins: electron transfer and beyond. *Acta Biochimica Polonica* 53, 685–691.

Camardella, L., Di Fraia, R., Antignani, A., Ciardiello, M.A., di Prisco, G., Coleman, J.K., Buchon, L. and Russell, N.J. (2002) The Antarctic *Psychrobacter* sp. TAD1 has two cold-active glutamate dehydrogenases with different cofactor specificities. Characterisation of the NAD^+-dependent enzyme. *Comparative Biochemistry and Physiology* 131A, 559–567.

Cao, Z., Song, P., Xu, Q., Su, R. and Zhu, G. (2011) Overexpression and biochemical characterization of soluble pyridine nucleotide transhydrogenase from *Escherichia coli*. *FEMS Microbiology Letters* 320, 9–14.

Cook, P.F., Blanchard, J.S. and Cleland, W.W. (1980) Primary and secondary deuterium isotope effects on equilibrium constants for enzyme-catalyzed reactions. *Biochemistry* 19, 4853–4858.

Cordwell, S.J. (1999) Microbial genomes and “missing” enzymes: redefining biochemical pathways. *Archives of Microbiology* 172, 269–279.

Cuff, A.L., Sillitoe, I., Lewis, T., Clegg, A.B., Rentzsch, R., Furnham, N., Pellegrini-Calance, M., Jones, D., Thornton, J. and Orengo, C.A. (2011) Extending CATH: increasing coverage of the protein structure universe and linking structure with function. *Nucleic Acids Research* 39(Suppl. 1), D420–D426.

Dalziel, K. (1969) The interpretation of kinetic data for enzyme-catalysed reactions involving three substrates. *Biochemical Journal* 114, 547–556.

Davis, B.D., Kornberg, H.L., Nagler, A., Miller, P. and Mingioli, E. (1959) Formation and functions of succinate in *Escherichia coli*. *Federation Proceedings* 18: 211.

DeLuna, A., Avendaño, A., Riego, L. and González, A. (2001) NADP-glutamate dehydrogenase isoenzymes of *Saccharomyces cerevisiae*. Purification, kinetic properties, and physiological roles. *The Journal of Biological Chemistry* 276, 43775–43783.

Di Fraia, R., Wilquet, V., Ciardiello, M.A., Carratorre, V., Antignane, A., Camardella, L., Glansdorff, N. and di Prisco, G. (2000) $NADP^+$-dependent glutamate dehydrogenase in the Antarctic psychrotolerant bacterium *Psychrobacter* sp. TAD1. Characterization, protein and DNA sequence, and relationship to other glutamate dehydrogenases. *European Journal of Biochemistry* 267, 121–131.

Doucette, C.D., Schawb, D.J., Wingreen, N.S. and Rabinowitz, J.D. (2011) α-Ketoglutarate coordinates carbon and nitrogen utilization via enzyme I inhibition. *Nature Chemical Biology* 7, 894–901.

Duncan, P.A., White, B.A. and Mackie, R.I. (1992) Purification and properties of NADP-dependent glutamate dehydrogenase from *Ruminococcus flavefaciens* FD-1. *Applied and Environmental Microbiology* 58, 4032–4037.

Engel, P.C. and Dalziel, K. (1967) The equilibrium constants of the glutamate dehydrogenase systems. *Biochemical Journal* 105, 691–695.

Engel, P.C. and Dalziel, K. (1970) Kinetics and mechanism of glutamate dehydrogenase. In: Sund, H. (ed) *Pyridine Nucleotide-dependent Dehydrogenases*. Springer, Heidelberg, Germany, pp. 245–256.

Engel, P.C. and Ferdinand, W. (1973) The significance of abrupt transitions in Lineweaver–Burk plots with particular reference to glutamate dehydrogenase. Negative and positive co-operativity in catalytic rate constants. *Biochemical Journal* 131, 97–105.

Felsenstein, J. (1985) Confidence limits on phylogenies: an approach using the bootstrap. *Evolution* 39, 783–793.

Finn, R.D., Bateman, A., Clements, J., Coggill, P., Eberhardt, R.Y., Eddy, S.R., Heger, A., Hetherington, K., Holm, L., Mistry, J. *et al.* (2014) Pfam: the protein families database. *Nucleic Acids Research* 42, D222–D230.

Fisher, H.F. (2005) Transient-state kinetic approach to mechanisms of enzymatic catalysis. *Accounts of Chemical Research* 38, 157–166.

Fisher, H.F. and Maniscalco, S.J. (2002) A close-packed planar 4-atom motif serves as a variable pathway mechanistic switching device in enzymatic catalysis. *Bioorganic Chemistry* 30, 199–210.

Frieden, C. (1959) Glutamic dehydrogenase. III. The order of substrate addition in the enzymatic reaction. *The Journal of Biological Chemistry* 234, 2891–2896.

Glass, T.L. and Hylemon, P.B. (1980) Characterization of a pyridine nucleotide-nonspecific glutamate dehydrogenase from *Bacteroides thetaiotaomicron*. *Journal of Bacteriology* 141, 1320–1330.

Gupta, S.C. and Dekker, E.E. (1980) Evidence for the identity and some comparative properties of α-ketoglutarate and 2-keto-4-hydroxyglutarate dehydrogenase activity. *The Journal of Biological Chemistry* 255, 1107–1112.

Hager, L.P. and Kornberg, H.L. (1961) On the mechanism of α-oxoglutarate oxidation in *Escherichia coli*. *Biochemical Journal* 78, 194–198.

Halpern, Y.S. and Umbarger, H.E. (1961) Utilization of L-glutamic and 2-oxoglutaric acid as sole sources of carbon by *Escherichia coli*. *Journal of General Microbiology* 26, 175–188.

Helling, R.B. (1994) Why does *Escherichia coli* have two primary pathways for synthesis of glutamate? *Journal of Bacteriology* 176, 4664–4668.

Helling, R.B. (1998) Pathway choice in glutamate synthesis in *Escherichia coli*. *Journal of Bacteriology* 180, 4571–4575.

Hornby, D.P. and Engel, P.C. (1984) Characterization of *Peptostreptococcus asaccharolyticus* glutamate dehydrogenase purified by dye–ligand chromatography. *Journal of General Microbiology* 130, 2385–2394.

Hurley, J.H., Chen, R. and Dean, A.M. (1996) Determinants of cofactor specificity in isocitrate dehydrogenase: structure of an engineered $NADP^+ \rightarrow NAD^+$ specificity-reversal mutant. *Biochemistry* 35, 5670–5678.

Huynen, M.A., Dandekar, T. and Bork, P. (1999) Variation and evolution of the citric-acid cycle: a genomic perspective. *Trends in Microbiology* 7, 281–291.

Jahns, T. and Kaltwasser, H. (1993) Properties of the cold-labile NAD^+-specific glutamate dehydrogenase from *Bacillus cereus* DSM 31. *Journal of General Microbiology* 139, 775–780.

Janes, B.K., Pomposiello, P.J., Perez-Matos, A., Najarian, D.J., Goss, T.J. and Bender, R.A. (2001) Growth inhibition caused by overexpression of the structural genes for glutamate dehydrogenase (*gdhA*) from *Klebsiella aerogenes*. *Journal of Bacteriology* 183, 2709–2714.

Jones, D.T., Taylor, W.R. and Thornton, J.M. (1992) The rapid generation of mutation data matrices from protein sequences. *Computer Applications in the Biosciences* 8, 275–282.

Joshi, C.V., Pathan, E.K., Punekar, N.S., Tupe, S.G., Kapadnis, B.P. and Deshpande, M.V. (2013) A biochemical correlate of dimorphism in a zygomycete *Benjaminiella poitrasii*: characterization of purified NAD-dependent glutamate dehydrogenase, a target for antifungal agents. *Antonie van Leeuwenhoek* 104, 25–36.

Kanamori, K., Weiss, R.L. and Roberts, J.D. (1987) Role of glutamate dehydrogenase in ammonia assimilation in nitrogen-fixing *Bacillus macerans*. *Journal of Bacteriology* 169, 4692–4695.

Kawahara, Y., Takahashi-Fuke, K., Shimizu, E., Nakamatsu, T. and Nakamori, S. (1997) Relationship between the glutamate production and the activity of 2-oxoglutarate dehydrogenase in *Brevibacterium lactofermentum*. *Bioscience, Biotechnology and Biochemistry* 61, 1109–1112.

Kawakami, R., Sakuraba, H. and Ohshima, T. (2007) Gene cloning and characterization of the very large NAD-dependent L-glutamate dehydrogenase from the psychrophile *Janthinobacterium lividum*, isolated from cold soil. *Journal of Bacteriology* 189, 5626–5633.

Kawakami, R., Oyama, M., Sakuraba, H. and Ohshima, T. (2010) The unique kinetic behavior of the very large NAD-dependent L-glutamate dehydrogenase from *Janthinobacterium lividum*. *Bioscience, Biotechnology and Biochemistry* 74, 884–887.

Kersten, M.A.S.H., Müller, Y., Baars, J.J.P., Op den Camp, H.J.M., van der Drift, C., Van Griensven, L.J.L.D., Visser, J. and Schaap, P.J. (1999) NAD^+-dependent glutamate dehydrogenase of the edible mushroom *Agaricus bisporus*: biochemical and molecular characterization. *Molecular and General Genetics* 261, 452–462.

Khan, M.I.H., Ito, K., Kim, H., Ashida, H., Ishikawa, T., Shibata, H. and Sawa, Y. (2005) Molecular properties and enhancement of thermostability by random mutagenesis of glutamate dehydrogenase from *Bacillus subtilis*. *Bioscience, Biotechnology and Biochemistry* 69, 1861–1870.

Kujo, C. and Ohshima, T. (1998) Enzymological characteristics of the hyperthermostable NAD-dependent glutamate dehydrogenase from the archaeon *Pyrobaculum islandicum* and effects of denaturants and organic solvents. *Applied and Environmental Microbiology* 64, 2152–2157.

Kurganov, B.I. (2000) New approach to analysis of deviations from hyperbolic law in enzyme kinetics. *Biokhimiya* 65, 898–909.

Lu, C.-D. and Abdelal, A.T. (2001) The *gdhB* gene of *Pseudomonas aeruginosa* encodes an arginine-inducible NAD^+-dependent glutamate dehydrogenase which is subject to allosteric regulation. *Journal of Bacteriology* 183, 490–499.

Maksymiuk, C., Balakrishnan, A., Bryk, R., Rhee, K.Y. and Nathan, C.F. (2015) E1 of α-ketoglutarate dehydrogenase defends *Mycobacterium tuberculosis* against glutamate anaplerosis and nitroxidative stress. *Proceedings of the National Academy of Sciences of the United States of America* 112(43), E5834–E5843.

Maniscalco, S.J., Saha, S.K., Vicedomine, P. and Fisher, H.F. (1996) A difference in the sequence of steps in the reaction catalyzed by two closely homologous forms of glutamate dehydrogenase. *Biochemistry* 35, 89–94.

Miñambres, B., Olivera, E.R., Jensen, R.A. and Luengo, J.M. (2000) A new class of glutamate dehydrogenases (GDH). Biochemical and genetic characterization of the first member, the AMP-requiring NAD-specific GDH of *Streptomyces clavuligerus*. *The Journal of Biological Chemistry* 275, 39529–39542.

Müller, T., Strösser, J., Buchinger, S., Nolden, L., Wirtz, A., Krämer, R. and Burkovski, A. (2006) Mutation-induced metabolite pool alterations in *Corynebacterium glutamicum*: towards the identification of nitrogen control signals. *Journal of Biotechnology* 126, 440–453.

Murthy, P.S., Sirsi, M. and Ramakrishnan, T. (1962) Tricarboxylic acid cycle and related enzymes in cell-free extracts of *Mycobacterium tuberculosis* $H_{37}R_V$. *Biochemical Journal* 84, 263–269.

Olson, J.A. and Anfinsen, C.B. (1953) Kinetic and equilibrium studies on crystalline L-glutamic acid dehydrogenase. *The Journal of Biological Chemistry* 202, 841–856.

Osterman, A. and Overbeek, R. (2003) Missing genes in metabolic pathways: a comparative genomics approach. *Current Opinion in Chemical Biology* 7, 238–251.

Piston, S.M., Mendz, G.L., Srinivasan, S. and Hazell, S.L. (1999) The tricarboxylic acid cycle of *Helicobacter pylori*. *European Journal of Biochemistry* 260, 258–267.

Preiss, T., Sang, A.E., Chrzanowska-Lightowlers, Z.M.A. and Lightowlers, R.N. (1995) The mRNA-binding protein COLBP is glutamate dehydrogenase. *FEBS Letters* 367, 291–296.

Rajas, F., Gire, V. and Rousse, B. (1996) Involvement of a membrane-bound form of glutamate dehydrogenase in the association of lysosomes to microtubules. *The Journal of Biological Chemistry* 271, 29882–29890.

Rife, J.E. and Cleland, W.W. (1980a) Kinetic mechanism of glutamate dehydrogenase. *Biochemistry* 19, 2321–2328.

Rife, J.E. and Cleland, W.W. (1980b) Determination of the chemical mechanism of glutamate dehydrogenase from pH studies. *Biochemistry* 19, 2328–2333.

Rossi, M., Defez, R., Chiurazzi, M., Lamberti, A., Fuggi, A. and Iaccarino, M. (1989) Regulation of glutamine synthetase isoenzymes in *Rhizobium leguminosarum* biovar *viceae*. *Journal of General Microbiology* 135, 629–637.

Sakamoto, N., Kotre, A.M. and Savageau, M.A. (1975) Glutamate dehydrogenase from *Escherichia coli*: purification and properties. *Journal of Bacteriology* 124, 775–783.

Sarada, K.V., Rao, N.A. and Venkitasubramanian, T.A. (1980) Isolation and characterisation of glutamate dehydrogenase from *Mycobacteriun smegmatis* CDC 46. *Biochimica et Biophysica Acta* 615, 299–308.

Schmincke-Ott, E. and Bisswanger, H. (1981) Dihydrolipoamide dehydrogenase component of the pyruvate dehydrogenase complex from *Escherichia coli* K12. Comparative characterization of the free and the complex-bound component. *European Journal of Biochemistry* 114, 413–420.

Senior, P.J. (1975) Regulation of nitrogen metabolism in *Escherichia coli* and *Klebsiella aerogenes*: studies with the continuous-culture technique. *Journal of Bacteriology* 123, 407–418.

Sharkey, M.A. and Engel, P.C. (2008) Apparent negative co-operativity and substrate inhibition in overexpressed glutamate dehydrogenase from *Escherichia coli*. *FEMS Microbiology Letters* 281, 132–139.

Sharkey, M.A., Oliveira, T.F., Engel, P.C. and Khan, A.R. (2013) Structure of $NADP^+$-dependent glutamate dehydrogenase from *Escherichia coli* – reflections on the basis of coenzyme specificity in the family of glutamate dehydrogenases. *FEBS Journal* 280, 4681–4692.

Shiio, I. and Ozaki, H. (1970) Regulation of nicotinamide adenine dinucleotide phosphate-specific glutamate dehydrogenase from *Brevibacterium flavum*, a glutamate-producing bacterium. *The Journal of Biochemistry* 68, 633–647.

Shiio, I., Ôsuka, S.-I. and Tsunoda, T. (1959) Glutamic acid formation from glucose by bacteria. II. Glutamic acid and α-ketoglutaric acid formation by *Brevibacterium flavum*, No. 2247. *The Journal of Biochemistry* 46, 1597–1605.

Shiio, I., Otsuka, S. and Takahashi, M. (1961) Significance of α-ketoglutarate dehydrogenase on the glutamic acid formation in *Brevibacterium flavum*. *The Journal of Biochemistry* 50, 164–165.

Shiio, I., Nariui, K., Yahaba, N. and Takahashi, M. (1962) Free intracellular amino acid of a glutamate-forming bacterium, *Brevibacterium flavum* No. 2247. *The Journal of Biochemistry* 51, 109–111.

Smith, E.L., Austen, B.M., Blumenthal, K.M. and Nyc, J.F. (1975) Glutamate dehydrogenases. In P. Boyer (ed) *The enzymes*, vol. XI. Academic Press, New York pp. 293–367.

Stillman, T.J., Baker, P.J., Britton, K.L. and Rice, D.W. (1993) Conformational flexibility in glutamate dehydrogenase. Role of water in substrate recognition and catalysis. *Journal of Molecular Biology* 234, 1131–1139.

Stouthamer, A.H. (1966) Some properties of mutants of *Aerobacter aerogenes* blocked in the formation of succinyl-Co A. *Antonie van Leeuwenhoek* 32, 452.

Strecker, H.J. (1953) Glutamic dehydrogenase. *Archives of Biochemistry and Biophysics* 46, 128–140.

Struhl, K. and Maganasik, B. (1976) Ammonia-sensitive mutant of *Klebsiella aerogenes*. *Journal of Bacteriology* 126, 739–742.

Subramanian, S. (1978) Thermodynamics of the glutamate dehydrogenase catalytic reaction. *Biophysical Chemistry* 7, 375–378.

Syed, S.-E.H. and Engel, P.C. (1987) Glutamate dehydrogenase of *Clostridium symbiosum*: another example of an abrupt transition in the kinetic plots. *Biochemical Society Transactions* 15, 237–238.

Tamura, K., Peterson, D., Peterson, N., Stecher, G., Nei, M. and Kumar, S. (2011) MEGA5: molecular evolutionary genetics analysis using maximum likelihood, evolutionary distance, and maximum parsimony methods. *Molecular Biology and Evolution* 28, 2731–2739.

Tian, J., Bryk, R., Itoh, M., Suematsu, M. and Nathan, C. (2005a) Variant tricarboxylic acid cycle in *Mycobacterium tuberculosis*: identification of α-ketoglutarate decarboxylase. *Proceedings of the National Academy of Sciences of the United States of America* 102, 10670–10675.

Tian, J., Bryk, R., Shi, S., Erdjument-Bromage, H., Tempst, P. and Nathan, C. (2005b) *Mycobacterium tuberculosis* appears to lack α-ketoglutarate dehydrogenase and encodes pyruvate dehydrogenase in widely separated genes. *Molecular Microbiology* 57, 859–868.

Vančurová, I., Vančura, A., Volc, J., Kopecý, J., Neužil, J., Basařová, G. and Běhal, V. (1989) Purification and properties of NADP-dependent glutamate dehydrogenase from *Streptomyces fradiae*. *Journal of General Microbiology* 135, 3311–3318.

Wagner, A. and Fell, D.A. (2001) The small world inside large metabolic networks. *Proceedings of the Royal Society of London* 268B, 1803–1810.

Weitzman, P.D.J. (1981) Unity and diversity in some bacterial citric acid-cycle enzymes. *Advances in Microbial Physiology* 22, 185–244.

2 Glutamate Decarboxylase in Bacteria

F. Giovannercole, E. Pennacchietti and D. De Biase*
Istituto Pasteur Italia – Fondazione Cenci-Bolognetti, Sapienza Università di Roma, Latina, Italy

2.1 Abstract

γ-Aminobutyrate (GABA) is a non-proteinaceous amino acid which plays different roles in the living systems. GABA is biochemically produced by the irreversible α-decarboxylation of L-glutamate catalysed by glutamate decarboxylase (Gad), a widespread enzyme distributed among eukaryotes and prokaryotes. Structural features of the plant, mammalian and bacterial Gads are reported in the literature. In pathogenic bacteria, Gad activity has been linked to protection from acid stress, but the possibility cannot be excluded that it might perform other physiological roles. Indeed, the screening of lactic acid bacteria (LAB) based on their capacity to synthesize GABA opens a new perspective for the production of GABA-containing dairy foods. The aim of this chapter is to provide an overview of the best characterized Gad enzymes in pathogenic and non-pathogenic bacteria, along with a description of the potential use of Gad as a source of GABA in functional foods and in biotechnological products.

2.2 Introduction

2.2.1 Centrality of L-glutamate in bacterial metabolism

The natural substrate of the Gad enzyme is L-glutamate, one of the 20 amino acids that are commonly used to make proteins in all living systems. L-Glutamate (henceforth L-glutamate or glutamate) represents one of 11 non-essential amino acids in humans, that is, our cells can synthetize it *de novo* without the need for it being supplied through the diet.

As depicted in Fig. 2.1, in bacteria, L-glutamate, in addition to being incorporated into proteins, is a key metabolite that plays a central role in a wide range of metabolic processes. Indeed, its intracellular fate is strongly linked to many essential biosynthetic pathways, some of which are vital. Intriguingly, in *Escherichia coli* glutamate has been shown to be the most abundant intracellular metabolite, comprising more than 40% (i.e. 120 mM) of the total metabolome (Bennett *et al.*, 2009) and acting as the major cytoplasmic counter-ion of K^+ (McLaggan *et al.*,

*Corresponding author. E-mail address: daniela.debiase@uniroma1.it

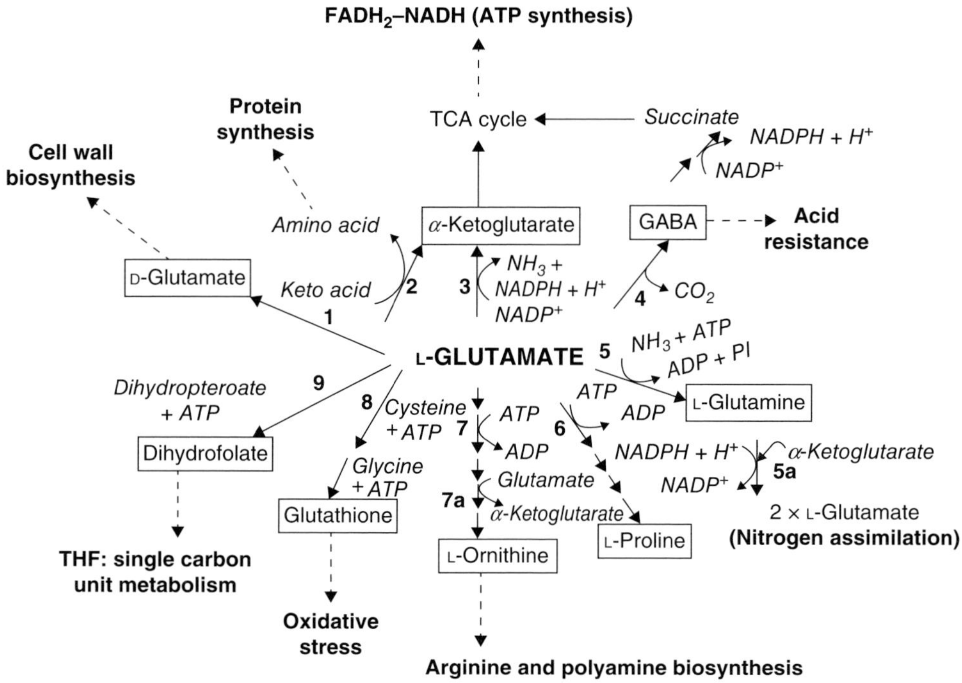

Fig. 2.1. Centrality of L-glutamate in metabolism. The bold numbers shown abutting the reaction arrows refer to the most well-known enzymatic transformations of L-glutamate: racemization (**1**); transaminations (**2**) with L-glutamate as amino donor; oxidative deamination (**3**); decarboxylation (**4**); and ATP-dependent ligations (**5–9**). Reactions (**1–3**) can be reversible. The metabolites in the boxes will then enter into metabolic pathways (dashes and arrows) affecting many biological processes (in bold). Abbreviations: $FADH_2$, reduced form of flavin adenine nucleotide; GABA, γ-aminobutyrate; TCA cycle, tricarboxylic acid (or Krebs) cycle.

1994). Furthermore, glutamate is a key molecule required for the survival of many enteric pathogens exposed to an extreme acid stress (pH < 2.5), such as that of the gastric compartment of the animal host (cf. Section 2.2.2) (De Biase and Pennacchietti, 2012; Lund *et al.*, 2014).

The most well-known transformations that L-glutamate can undergo in bacteria (Fig. 2.1) are the following:

1. Racemization from L-glutamate into D-glutamate by glutamate racemase (the product of the *murI* gene in *E. coli*): this enzyme is required to generate D-glutamate, an essential component of peptidoglycans in the bacterial cell wall.
2. Transamination, in which glutamate acts as a donor of the α-amino group, which is transferred to different α-keto acid acceptors via specific amino acid transaminases, thereby leading to the synthesis of new amino acids and α-ketoglutarate (Bennett *et al.*, 2009).
3. Oxidative deamination, catalysed by glutamate dehydrogenase, leads to the production of ammonia, NADPH and a molecule of α-ketoglutarate (the same product as that of the transamination reactions), and this latter molecule directly enters the Krebs cycle or tricarboxlic acid (TCA cycle in Fig. 2.1) to yield energy but can also act as a substrate in reactions **2** (in the reverse direction) and **5a** in Fig. 2.1.
4. Decarboxylation into GABA by glutamate decarboxylase (Gad) via a proton-consuming reaction, employed to cope with acid stress but also providing the starting molecule, i.e. GABA, for the GABA-shunt, a short metabolic pathway which leads to succinate and NADPH production and is useful for bypassing the TCA step catalysed by α-ketoglutarate dehydrogenase (this is notably missing in *Mycobacterium tuberculosis* and *Listeria monocytogenes*, and in cyanobacteria) (Tian *et al.* 2005; Steinhauser *et al.*, 2012; Feehily *et al.*, 2013).

5. Synthesis of L-glutamine via glutamine synthetase, which catalyses an ATP-dependent reaction condensing free ammonia (NH_3) to glutamate to yield L-glutamine, a key molecule for ammonia assimilation (path **5a** in Fig. 2.1).
6. Proline biosynthesis, consisting of four reactions, the first of which involves the activation of glutamate to γ-glutamyl phosphate by γ-glutamyl kinase.
7. The ornithine biosynthetic pathway also uses glutamate as a starting molecule; this is converted into L-ornithine via *N*-acetylated intermediates.
8. Biosynthesis of glutathione catalysed by the consecutive action of two ATP-dependent enzymes: glutamate-cysteine ligase and glutathione synthetase.
9. Biosynthesis of dihydrofolate catalysed by dihydrofolate synthetase in an ATP-dependent manner.

2.2.2 Importance of L-glutamate in the acid stress response: the GDAR system

Commensal and pathogenic enteric bacteria, both Gram positive and Gram negative, are neutralophiles because they grow best when the intracellular pH (pH_i) is close to neutrality and are capable of keeping intracellular pH in the range 7.0–7.8 when the extracellular pH (pH_o) ranges from 5.5 to 9.0 (Slonczewski *et al.*, 2009). In the absence of any protective mechanism, a significant fall in pH_o would lead to growth arrest or even be life-threatening because acid pH_i impairs (i.e. drastically slows or, even worse, halts) the activity of biological macromolecules, which carry out their functions only within a limited pH range. Furthermore, an intracellular acidification would alter the ΔpH across the bacterial membranes, which is essential for keeping the electrochemical gradient as the proton motive force (PMF) required for ATP synthesis (Foster, 2004; Lund *et al.*, 2014). Thus, any fluctuation of pH_i is clearly undesirable.

Fermented foods, the gastrointestinal tract (GIT) of the host, dental plaque, food preservatives, plant roots, soil and acid mine drainage are the environments in which neutralophiles encounter an acid stress (Lund *et al.*, 2014 and references therein). In particular, in many species of bacteria acquired through the oral route, L-glutamate is required for the functioning of the glutamate-dependent acid resistance (GDAR) system, the most efficient molecular system in conferring protection from the extreme acid stress encountered in the gastric compartment (pH 1.5–3.0) before the bacteria reach the host gut. Even though the stomach represents the major bactericidal barrier (Giannella *et al.*, 1972; Martinsen *et al.*, 2005) of the GIT, the mild acidic environments of the right colon (pH 5.5–7.5), where the fermentation of sugars by resident anaerobic bacteria produces short-chain fatty acids (SCFAs), and that of the macrophagic phagolysosome (pH 4.5–5.5) are additional sources of acid stress. Clearly, in food-borne pathogens the ability to respond to a mild-to-harsh acid stress is an essential prerequisite to succeeding in host colonization. Thus *E. coli*, *Shigella flexneri*, *Salmonella typhimurium*, *L. monocytogenes* and *Vibrio cholerae* have evolved different strategies to overcome this stress (reviewed by: Bearson *et al.*, 1997; Cotter *et al.*, 2001; Merrell and Camilli, 2002; De Biase and Pennacchietti, 2012; Lund *et al.*, 2014; De Biase and Lund, 2015).

Bacterial acid survival strategies are typically classified either as an ATR (acid tolerance response) or as AR (acid resistance). They differ in that the ATR (generally in the log or stationary phases of growth) requires a pre-exposure to a mild acidic environment, i.e. pH 4.5–5.0, and provides protection down to a pH_o of 3, whereas the AR is typical, though not exclusive to stationary-phase unadapted cells and protects them from a more severe acid stress (pH < 3.0) (Lin *et al.*, 1996; Bearson *et al.*, 1997; Foster, 2001). Among the mechanisms employed by different bacteria to counteract acid stress, the most effective is the removal of cytoplasmic protons by their irreversible incorporation into small molecules which are then exported outside the cell; this is exactly what the pH-inducible amino acid decarboxylases carry out (Kanjee and Houry, 2013; De Biase and Lund, 2015).

Acid pH-inducible bacterial amino acid decarboxylases are cytoplasmic pyridoxal 5′-phosphate (PLP)-dependent enzymes that catalyse an irreversible decarboxylation on a specific amino acid substrate (i.e. glutamate, arginine, lysine, ornithine) and act in concert with a cognate antiporter, which in turn expels the decarboxylation product while importing the amino acid substrate (for a review, see Lund *et al.*, 2014).

The GDAR system (Fig. 2.2A) is an extremely efficient AR system that is found in commensal and pathogenic *E. coli*, *L. monocytogenes*, *Lactococcus lactis*, *S. flexneri*, various newly discovered species belonging to the *Brucella* genus and many orally acquired bacteria (De Biase and Pennacchietti, 2012; Damiano *et al.*, 2015). In order to work, the system requires only glutamate (≤1.5 mM) in the challenge minimal medium at pH 1.5–2.5 for a period of 2 h or more (Lin *et al.*, 1996). There are three key components of this system: two homologous α-decarboxylases, GadA and GadB, and the cognate Glu_{in} (glutamate in)/$GABA_{out}$ antiporter, GadC. According to the proposed model for GDAR in *E. coli* (De Biase and Pennacchietti, 2012; Fig. 2.2A), when the cell is exposed to an extreme acid challenge (pH ≤ 2.5), it experiences an uncontrolled entry of protons intracellularly and the pH_i drops below 5.0 (Richard and Foster, 2004). This first event triggers the $Glu_{in}/GABA_{out}$ exchange by GadC and the activity of GadA/B, which are maximally operative in the pH range 4.0–5.0 (Shukuya and Schwert, 1960; Pennacchietti *et al.*, 2009; Grassini *et al.*, 2015). Hence, the harmful drop in pH_i is quickly contrasted by the efficient enzymatic activity of the decarboxylase, which replaces the leaving α-carboxyl group of glutamate with a cytoplasmic proton, yielding GABA, which in turn is exported via GadC in exchange for a new glutamate molecule. GadC acts via a 'logic-gate' mechanism, i.e. it recognizes and discriminates the two substrates by their net charge, so it guarantees the proper export of positive charges (0.9 H^+/cycle) (Ma *et al.*, 2013; Tsai *et al.*, 2013). Moreover, it has been suggested that the electrogenic export of GABA helps the maintenance of the membrane potential, hence supporting the synthesis of ATP, even when the electron transport chain is not working (Small and Waterman, 1998; Shepherd *et al.*, 2010).

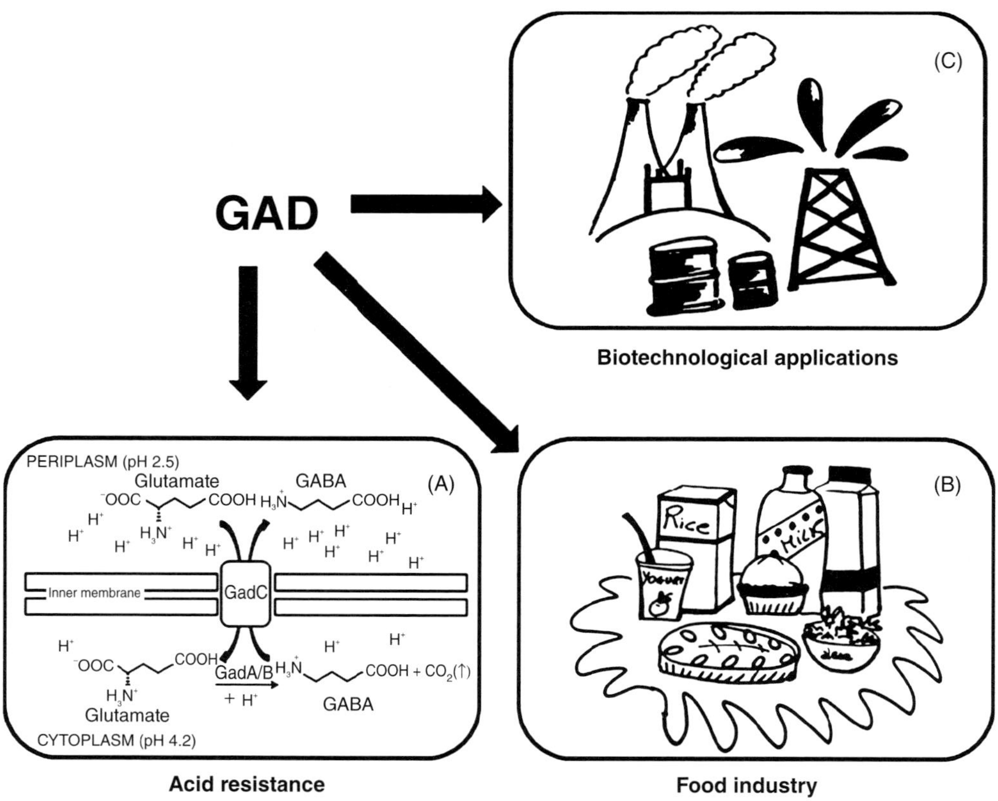

Fig. 2.2. Bacterial glutamate decarboxylase (Gad) in (A) acid resistance, and as a source of γ-aminobutyrate (GABA) for (B) the food industry and (C) biotechnological applications.

An interesting and unique feature of GDAR is that glutamate and GABA possess a γ-carboxyl group whose acid dissociation constant, pK_a (4.1 for Glu and 4.0 for GABA), is close to the pH_i under acid stress, thus making these molecules able to act as intracellular buffers better than the amino acids and decarboxylation products from the other amino acid-dependent AR systems, which in fact rely on positively charged amino acids (Lund *et al.*, 2014).

2.3 Glutamate Decarboxylase (Gad) in Bacteria

Glutamate decarboxylase (L-glutamate 1-carboxy-lyase, Gad; EC 4.1.1.15) is largely distributed in higher organisms, such as plants and animals, where it takes part in important physiological activities (Ueno, 2000). As summarized in Table 2.1, Gad has been biochemically characterized from *E. coli*, *Clostridium perfringens*, *L. lactis*, *Lactobacillus* spp. and *Brucella microti*, and also from hypertermophilic archea, i.e. *Pyrococcus* spp.; the number of species is certainly bound to increase.

2.3.1 *Escherichia coli* Gad

As mentioned before, Gad is a PLP-dependent enzyme that is widely distributed among living organisms. It catalyses the irreversible α-decarboxylation reaction of L-glutamate to yield CO_2 and GABA, with consumption of one proton at each catalytic cycle (Fig. 2.2A). *E. coli* possesses two cytoplasmic Gad isoforms, namely GadA and GadB, which share 99% sequence identity and display an acidic pH optimum of activity (pH 3.8–4.6). The relevant genes, i.e. *gadA* and *gadB* (the latter is in an operon arrangement with the *gadC* gene that codes for GadC), are 2.1 megabases from each other on the chromosome of the *E. coli* K12 reference strain MG1655 and code for proteins of the same length, i.e. 466 amino acids. The proteins differ in five amino acid residues, four of which are located in the *N*-terminal region (i.e. within the first 22 amino acids) while the last is at position 153 (De Biase *et al.*, 1996). Interestingly, the *gadA* gene belongs to the acid fitness island, a 14 kb genome region comprising 13 additional genes which contribute at various levels to AR (De Biase and Pennacchietti, 2012). The separate expression of the two isoforms has provided evidence that they are biochemically and functionally/structurally very similar (De Biase *et al.*, 1996; Capitani *et al.*, 2003; Dutyshev *et al.*, 2005). Because recombinant GadB was overexpressed at higher levels than GadA (De Biase *et al.*, 1996), it was chosen for further structural and biochemical studies, which are described below.

2.3.1.1 *Structural studies*

In 2003, the crystal structure of *E. coli* GadB was reported at both an acidic pH (4.6) and a neutral pH (7.6) (Capitani *et al.*, 2003). This was the first structure of a glutamate decarboxylase to be released and has provided insights into the molecular basis of its intracellular activation. Subsequently, the structures of *E. coli* GadA (Dutyshev *et al.*, 2005), human Gad67 and Gad65 (Fenalti *et al.*, 2007) and plant Gad1 (Gut *et al.*, 2009) were published too, thus providing further information of the mechanisms controlling Gad activity in different organisms. *E. coli* GadB (*Ec*GadB) is a homohexamer of approximately 320 kDa which can be described as a trimer of dimers, with a PLP molecule in each subunit. The GadB structure is arranged on two layers with each dimer providing a monomer to each layer. The entangled hexameric structure of *Ec*GadB is rather different from that of the other pH-inducible decarboxylases, which have higher structural arrangements, though they are less intertwined, e.g. a dodecamer for ornithine decarboxylase (Momany *et al.*, 1995), and a decamer for arginine and lysine decarboxylases (Andréll *et al.*, 2009; Kanjee *et al.*, 2011a).

Like other PLP-dependent enzymes, in GadB the dimer is also the functional unit as each active site is made up of amino acid residues that are contributed by both monomers in the dimer. The monomer is organized into three distinct domains: (i) *N*-terminal (residues 1–57); (ii) central, with PLP bound (residues 58–346); and (iii) *C*-terminal (residues 347–466). The *N*-terminal domain acts as a hook, departing perpendicularly from the subunit surface and reaching the opposite layer where it makes interactions both with the counterpart in the functional dimer and with a subunit belonging to a neighbouring dimer. In this way, it confers stability on the entire structure.

Table 2.1. *Escherichia coli* glutamate decarboxylase (GadB) compared with other bacterial Gads.

	Subunit features			Optimum				
Species	N°	MW (kDa)	No. amino acids	pH	Temp. (°C)	Sequence identity (%)	Accession code (Uniprot)	Reference(s)
Escherichia coli K12	6	53	466	3.8–4.6	37	–	P69910	Shukuya and Schwert, 1960; De Biase *et al.*, 1996; Capitani *et al.*, 2003; Pennacchietti *et al.*, 2009
Brucella microti	6	52	464	3.8–4.5	–	73.0	C7LHI0	Grassini *et al.*, 2015
Clostridium perfringens	6	53	464	4.7	–	48.0	Q8XIQ6	Cozzani *et al.*, 1970, 1975
Lactobacillus brevis IFO12005	2–4	54	480	4.2–5.0	30	38.1	A9ZM78	Ueno *et al.*, 1997; Hiraga *et al.*, 2008
Lactobacillus paracasei	2	57	481	5.0	50	38.0	B1B389	Komatsuzaki *et al.*, 2008
Lactobacillus plantarum	2	53	468	5.0	40	42.6	D7VEX3	Shin *et al.*, 2014
Lactobacillus sakei	–	54	479	5.0	55	38.0	A0A0D3R1A0	Sa *et al.*, 2015
Lactococcus lactis	–	54	466	4.7	30	45.0	T0V974	Nomura *et al.*, 1999
Pyrococcus furiosus	–	41	371	6.0	75	27.0	Q8U1P6	Lee *et al.*, 2013
Pyrococcus horikoshii	1	42–45	358	8.0	>97	21.0	PH0973	Kim *et al.*, 2009a
Thermococcus kodakaraensis KOD1	1	42	384	8.0	90	27.0	Q5JJ82	Hong *et al.*, 2012

Unlike the other pH-inducible decarboxylases (Kanjee *et al.*, 2011b), the acidic ↔ neutral pH switch apparently does not affect the stability of the hexamer but rather triggers three important structural reorganizations involving the first and the last 15 amino acid residues in the *N*- and *C*-terminal domains (residues 1–14 and 452–466, respectively), and 14 amino acid residues in the central domain which make up a β-hairpin (residues 300–313). Briefly, at acidic pH, the *N*-terminal tails assume α-helical conformations, which close the hexamer central channel and give rise to two triple helical bundles, parallel to the threefold axis of the protein, each protruding on each (top and bottom) side of the hexamer. Because these latter regions are rich in acidic residues, as soon as they become protonated, as a consequence of acidification, repulsion is abolished and bundle formation is allowed to occur. Interestingly, the interaction with halides, especially Cl^- ions (biologically relevant as they are present in the gastric juice) confers an additional structural stability to the triple helical bundles, thereby acting as positive allosteric modulators (Gut *et al.*, 2006; Pennacchietti *et al.*, 2009).

In contrast, the *C*-terminal residues 452–466, which are likely to be in a flexible conformation at acidic pH, are well structured at neutral pH and protrude into the active site of each monomer, with residues His465 and Thr466 precluding the access of glutamate into the enzyme active site. Biochemical and structural studies have clarified the role of *C*-terminal tail in the active site at neutral pH, providing evidence for the existence of a ternary complex, i.e. a substituted aldamine (arising from the reaction of the distal nitrogen of the imidazole ring of His465 with the PLP-Lys276 Schiff base), which constitutes the catalytically inactive form of GadB. As confirmed by the biochemical and spectroscopic properties of the mutants GadB-H465A and GadB-ΔHT (which lack both His465 and Thr466), the locking of the active site represents an elegant auto-inhibition mechanism, which ensures that the enzyme autonomously controls its catalytic activity in an acidic pH-dependent fashion (Gut *et al.*, 2006; Pennacchietti *et al.*, 2009).

2.3.1.2 Spectroscopic properties

Spectroscopic studies have provided evidence that the absorption spectrum of GadB changes with pH (Gut *et al.*, 2006; Pennacchietti *et al.*, 2009); at pH 4.5–5.0 the enzyme displays its maximal activity and is yellow in colour because the coenzyme is in the ketoenamine form with maximum absorbance at 420 nm, whereas at pH ≥ 6.0 GadB is inactive and colourless because the cofactor is in the substituted aldamine form, which absorbs maximally at 340 nm. The pH-dependent UV-visible, circular dichroism and fluorescence spectroscopic changes are influenced by the different positioning of a proton on the cofactor, as a consequence of the reduced hydration state of the active site (more apolar), and by the formation of the substituted aldamine as the pH increases (Pennacchietti *et al.*, 2009).

Besides the crystallographic studies, steady-state spectroscopic studies show that the interconversion of the PLP coenzyme between the 340 nm-absorbing inactive form and the 420 nm-absorbing active form is a cooperative process that occurs over a very narrow range of pH (5.25–5.4) and is also influenced by interaction with halides in the order $I^- > Br^- > Cl^-$ (Gut *et al.*, 2006). The midpoint of the 420 ↔ 340 nm spectroscopic transition is shifted to pH 5.8 in the presence of chloride ions, which therefore act as positive allosteric modulators of the enzyme, i.e. they stabilize GadB in the form competent for catalysis (the ketoenamine form). Markedly important in this competence is the biological connection between the presence of Cl^- ions in gastric secretions and their involvement in the structural stability of GadB.

2.3.1.3 Gad signature

As a member of the group II PLP-dependent decarboxylases, which belong to the aspartate aminotransferase family, GadB shows a strong conservation of eight residues in the active site: Lys276, forming the Schiff base with the C4′ of PLP; Asp243, which contacts the protonated pyridine nitrogen of PLP; Ala245, whose methyl group is in hydrophobic contact with the pyridine ring; His241 and Ser269, which play a critical structural role in stabilizing the interaction between two β-filaments and an α-helix in the central domain; Thr212, pointing towards the hydroxyl group of PLP and possibly providing a proton during catalysis; and Gly120 and Gly210, which are speculated to play an important role for the stability of the protein (De Biase and Pennacchietti, 2012 and references therein).

In addition, by a Clustal alignment that was generated using 20 bacterial Gads, it was found that 84 amino acids are strictly conserved (De Biase and Pennacchietti, 2012), among which some were added to those listed above to propose a Gad 'signature'. These include: Asp86* (where * indicates that the amino acid residue is from the neighbouring subunit in the functional dimer), which at a permissive pH establishes a contact with its side chain and the γ-carboxyl of L-glutamate in the active form; Glu89, close to Asp86 only in the active form; Gln163, which stacks the pyridine ring of PLP; Thr62 and Phe63, which provide additional hydrogen bonds to the γ-carboxylate of L-glutamate with the side chain hydroxyl and amide nitrogen, respectively; His275, which guarantees the correct positioning of the cofactor by hydrogen bonding its side chain τ-nitrogen to the phosphate group of PLP; Tyr305* and Leu306*, at the tip of the β-hairpin (amino acid residues 300–313), which interact with residues 461–463 of the *C*-terminal tail in the inactive form; Tyr305*, which in addition may be involved in protonating the substrate during the decarboxylation reaction; and His465, which is responsible for GadB auto-inhibition via formation of the substituted aldamine at neutral pH.

2.3.2 Other bacterial Gads

It has been hypothesized (De Biase and Pennacchietti, 2012) that in the genomes of several bacteria, mostly orally acquired, the adjacency of the gene loci of the functionally linked *gadB* and *gadC* genes would favour their co-expression and thus improve the ability of the bacteria to colonize or transit through the host GIT. This, for example, has been shown in *Brucella microti* (Occhialini *et al.*, 2012). Indeed, the *gadA and gadB* genes have been found in many microorganisms, although a full biochemical characterization of the protein products is still restricted to a few of them. In Table 2.1, only those GadBs that have been more thoroughly characterized at the biochemical level are compared with the *E. coli* counterpart (*Ec*GadB). Overall, the Gads from *Pyrococcus* spp. show rather different biochemical properties. The relevant references are provided in Table 2.1.

2.3.2.1 *GadB from* Clostridium perfringens

This microorganism is a major cause of food poisoning in developed countries. GadB, purified to homogeneity, exhibits an oligomeric assembly, and spectroscopic and kinetic properties similar to those of *Ec*GadB. Notably, the *C. perfringens gadB* and *gadC* genes are separated by a gene (CPE2059) that is likely to be implicated in protection from acid stress (De Biase and Pennacchietti, 2012).

2.3.2.2 *GadB from* Brucella microti

This bacterium belongs to a group of recently-described, fast-growing *Brucella* spp. that possess both *gadB* and *gadC* genes intact (unlike the classical species, which are well known for their pathogenic potential; Damiano *et al.*, 2015). The role of the GDAR system in *B. microti* was investigated by Occhialini *et al.* (2012). Moreover, biochemical properties of GadB from *B. microti* were compared with those of *Ec*GadB (Grassini *et al.*, 2015). Apart from displaying a pH-dependent spectroscopic transition shifted towards acidic pH values, this enzyme shares with *Ec*GadB many properties, including the ability to undergo auto-inactivation via the formation of a substitute aldamine in the active site. Unlike *Ec*GadB, *B. microti* GadB does not form triple helical bundles at acidic pH (cf. 2.3.1.1).

2.3.2.3 *GadB from lactic acid bacteria*

An interesting group of bacteria in which Gad activity is of importance is that of the LAB (Li and Cao, 2010). These include several species and strains belonging to the genera *Lactobacillus* and *Lactococcus*, isolated from a variety of fermented foods, such as yogurt, milk and cheese, and naturally associated with the mucosae of animals (vagina, small intestine and colon). LAB ferment hexose sugars to produce mainly lactic acid. Often exploited for the preservation of foods and beverage, they are undergoing extensive genomic and proteomic studies (Makarova *et al.*, 2006).

The GadB from *Lactobacillus brevis* IFO12005, isolated from L-glutamate-enriched foods as *kimchi* (a traditional Korean food), shares with *Ec*GadB the acidic pH optimum (pH 4.2–5.0) and the ability to decarboxylate only

L-glutamate. However, it appears to be very different from *Ec*GadB in other respects: it is either dimeric or tetrameric, shares only 38% of sequence identity with *Ec*GadB and NaCl addition has no effect on the enzymatic activity. This latter might be linked to the different oligomeric assembly. Notably, the residue corresponding to *Ec*GadB His465 is missing in GadB from *L. brevis*. Hiraga and co-workers (Hiraga *et al.*, 2008) showed GadB from *L. brevis* IFO12005 to be dimeric in the inactive form (at pH 7.0) and tetrameric in the active form (at pH < 5.5). Treatment of Gad with high concentrations of ammonium sulfate and subsequent dilution with sodium glutamate were essential for tetrameric assembly and activation.

The GadB from *Lactobacillus paracasei* strain NFRI, isolated from a Japanese traditional fermented fish (funa-sushi), has also been characterized biochemically. In addition to the properties listed in Table 2.1, it is worth noticing that this enzyme is dimeric and reported to be positively influenced by many ions (Ca^{2+} or $(NH_4)_2SO_4$ stimulate the enzymatic activity); in contrast, Cl^- has a negative effect on its activity. This strain might have a strong impact in the food industry because it produces large amounts of GABA (Komatsuzaki *et al.*, 2008).

The GadB from *Lactobacillus plantarum*, in addition to sharing a dimeric assembly with the enzyme from *L. paracasei*, has the properties listed in Table 2.1. A notable finding is that in this GadB also, the C-terminal tail participates in the auto-inactivation of the enzyme at neutral pH; the mutants for the segment Ile454–Thr468, corresponding to Pro452–Thr466 in *Ec*GadB, were in fact shown to be still active at neutral pH (Shin *et al.*, 2014). This finding underlines that the control of the enzymatic activity via the C-terminal tail represents a common motif among evolutionarily distant GadBs.

Lactococcus lactis is a microorganism that is essential for milk fermentation, especially in the cheesemaking process. It spends a large part of its growth cycle at low pH and exhibits a higher survival ability when preadapted to mild acidity or in starvation. Sanders *et al.* (1998) found that the genes coding for GadB and GadC (*gadB* and *gadC*) are organized in an operon, whose expression is positively influenced by NaCl and L-glutamate in a medium at low pH in cells at the stationary phase of growth.

2.4 Use of Gad for γ-Aminobutyrate (GABA) Production: a Beneficial Molecule for our Society

In recent years, Gad has received attention as a tool for GABA synthesis for biotechnological applications and in the food industry (Figs 2.2B and C). GABA is regarded as a lead molecule for the production of eco-friendly goods as well as healthy foods with beneficial effects on our well-being. In this respect, Gad plays a pivotal role as a source of GABA, and therefore work is in progress to improve the yield and the efficiency of the decarboxylation reaction, in both the isolated enzyme and in whole cell systems.

2.4.1 GABA in health and disease

GABA is a four carbon non-proteinaceous amino acid that is widely distributed in all living organisms. The interest in GABA-enriched foods has been growing in the last decade as a consequence of their numerous healthy effects on humans (Li and Cao, 2010 and references therein; De Biase and Pennacchietti, 2012; Dhakal *et al.*, 2012). Briefly, these include: (i) hypotensive, tranquillizing and diuretic effects; (ii) the lowering of systolic blood pressure in animals and human subjects; (iii) acting as a strong secretagogue of insulin from the pancreas, thereby preventing diabetes; (iv) promoting the healing process of cutaneous wounds; (v) exhibiting positive effects on alcohol-related symptoms; (vi) interfering with the synthesis of hyaluronic acid and enhancing the survival rate of the dermal fibroblasts when they are exposed to H_2O_2 as a reactive species; (vii) preventing cancer cell proliferation (Oh and Oh, 2004; Schuller *et al.*, 2008); and (viii) stimulating immune cells (Oh and Oh, 2003).

In the last few decades, the hypotensive property of GABA has been reported in different foods: green tea (Abe *et al.*, 1995), red mould rice (prepared using *Monascus* spp.; Kohama *et al.*, 1987) and the single-celled green alga *Chlorella* (Nakamura *et al.*, 2000). Moreover, the accumulation of GABA in foods has been studied in many dietary products, such as Japanese rice germ during water soaking (Saikusa *et al.*, 1994), soy products (Aoki *et al.*, 2003; Shizuka *et al.*, 2004;

Tsai *et al.*, 2006) and fermented milk (Inoue *et al.*, 2003; Hayakawa *et al.*, 2004). Notably, the frequent consumption of germinated brown rice also has healthy effects, including boosting of the immune system, reducing lipids in the blood, combating cancer development (by preventing cancer cell proliferation), preventing heart diseases, and protecting from Alzheimer's disease and dementia, and from hypertension, which are all associated with its high GABA content (Oh, 2003; Patil and Khan, 2011).

LAB seem to show prospects for application in the production of diary and other healthy products as a starter culture for GABA-enriched foods, such as yogurt (Park and Oh, 2007), cheese starters (Nomura *et al.*, 1998; Siragusa *et al.*, 2007), fermented fish and pepino milk (*Solanum muricatum*; Komatsuzaki *et al.*, 2005; Chiu *et al.*, 2012), kimchi (Kim *et al.*, 2007; Seok *et al.*, 2008), black raspberry juice (Kim *et al.*, 2009b) and soybean (Tsai *et al.*, 2006).

2.4.2 GABA as an alternative promising molecule for sustainable resources

The global economy relies on fossil fuels made from non-renewable carbon deposits, such as petroleum, oil and natural gas, which are limited resources, although they are required for many purposes. Thus, there is a need for a sustainable chemical industry (Nikolau *et al.*, 2008).

Glutamate, a non-essential amino acid, can be regarded as an alternative resource for the production, by microbial fermentation, of a number of useful chemicals as building blocks for five carbon polymers, mostly polyesters, polyamides and lactams, such as glutaminol, norvoline, *N*-methylpyrrolidone, *N*-vinylpyrrolidone and many others (Werpy and Peterson, 2004). The microbial production of Nylon-4, PGA (poly-γ-glutamic acid) and ALA (5-aminolevulinic acid) are by far the most investigated. Nylon-4 is both an example of an eco-friendly heat-stable polyamide and a new promising biocompatible and biodegradable polymer for industrial applications (Kawasaki *et al.*, 2005). For this purpose, a promising research area involves the improvement of the biochemical production of GABA by using either immobilized GadB enzyme (Lammens *et al.*, 2009) or immobilized whole cells. Recently, *Corynebacterium glutamicum*, a non-pathogenic microorganism, has been proposed for industrial use, owing to its high production of amino acids (mostly glutamate and lysine) (Choi *et al.*, 2015). As *C. glutamicum* does not seem to possess *gad* genes, the production of GABA in the species was implemented by the recombinant expression in this bacterium of *gad* genes from *E. coli* (Takahashi *et al.*, 2012) or *L. brevis* (Shi *et al.*, 2013); also, in order to increase the yield, strategies based on deletion mutants were provided (Okai *et al.*, 2014; Wang *et al.*, 2015). Even fungal species can be exploited for this purpose (Jiang *et al.*, 2011), and this is a clear indication that an improvement in GABA production and yield is currently in progress in these living systems. In addition, by using *Ec*GadB (both the wild type and mutant variants), it has been shown that the reaction can be optimized for the downstream isolation and utilization of GABA (Pennacchietti *et al.*, 2009; Lammens *et al.*, 2012; Choi *et al.*, 2015).

The characterization of archaeal Gad might turn out useful for biotechnological applications because of their high stability under different environmental conditions, such as different temperatures and pHs. Thus, Gad from the archaeal microorganisms *Pyrococcus* spp. might be interesting in this respect (Table 2.1).

2.5 Conclusions

Bacterial Gad is an interesting object of study. First, within L-glutamate metabolism (Fig. 2.1), the enzyme catalyses a unique irreversible ATP-independent, proton-consuming reaction which leads to the production of GABA, a molecule whose role in microbe–microbe and microbe–host interactions still needs to be fully elucidated. Second, GABA production from purified Gad (or whole cells overexpressing Gad) is currently exploited both in the food industry and for biotechnological applications. Indeed, GABA is regarded as an interesting ingredient for functional food and also as the precursor of pyrrolidones for polymer biosynthesis (Zhang *et al.*, 2016). In addition, the occurrence in some bacteria of two or more isoforms of Gad suggests that their biological roles might overlap only in part (De Biase and Pennacchietti, 2012; Feehily *et al.*, 2013). Ongoing and future work will certainly shed light on these biological cues.

References

Abe, Y., Umemura, S., Sugimoto, K., Hirawa, N., Kato, Y., Yokoyama, N., Yokoyama, T., Iwai, J. and Ishii, M. (1995) Effect of green tea rich in γ-aminobutyric acid on blood pressure of Dahl salt-sensitive rats. *American Journal of Hypertension* 8, 74–79.

Andréll, J., Hicks, M.G., Palmer, T., Carpenter, E.P., Iwata, S. and Maher, M.J. (2009) Crystal structure of the acid-induced arginine decarboxylase from *Escherichia coli*: reversible decamer assembly controls enzyme activity. *Biochemistry* 48, 3915–3927.

Aoki, H., Furuya, Y., Endo, Y. and Fujimoto, K. (2003) Effect of gamma-aminobutyric acid-enriched tempeh-like fermented soybean (GABA-tempeh) on the blood pressure of spontaneously hypertensive rats. *Bioscience, Biotechnology, and Biochemistry* 67, 1806–1808.

Bearson, S., Bearson, B. and Foster, J.W. (1997) Acid stress responses in enterobacteria. *FEMS Microbiology Letters* 147, 173–180.

Bennett, B.D., Kimball, E.H., Gao, M., Osterhout, R., Van Dien, S.J. and Rabinowitz, J.D. (2009) Absolute metabolite concentrations and implied enzyme active site occupancy in *Escherichia coli. Nature Chemical Biology* 5, 593–599.

Capitani, G., De Biase, D., Aurizi, C., Gut, H., Bossa, F. and Grütter, M.G. (2003) Crystal structure and functional analysis of *Escherichia coli* glutamate decarboxylase. *The EMBO Journal* 22, 4027–4037.

Chiu, T.-H., Tsai, S.-J., Wu, T.-Y., Fu, S.-C. and Hwang, Y.-T. (2012) Improvement in antioxidant activity, angiotensin-converting enzyme inhibitory activity and *in vitro* cellular properties of fermented pepino milk by *Lactobacillus* strains containing the glutamate decarboxylase gene. *Journal of the Science of Food and Agriculture* 93, 859–866.

Choi, J.W., Yim, S.S., Lee, S.H., Kang, T.J., Park, S.J. and Jeong, K.J. (2015) Enhanced production of gamma-aminobutyrate (GABA) in recombinant *Corynebacterium glutamicum* by expressing glutamate decarboxylase active in expanded pH range. *Microbial Cell Factories* 14:21.

Cotter, P.D., Gahan, C.G. and Hill, C. (2001) A glutamate decarboxylase system protects *Listeria monocytogenes* in gastric fluid. *Molecular Microbiology* 40, 465–475.

Cozzani, I., Misuri, A. and Santoni, C. (1970) Purification and general properties of glutamate decarboxylase from *Clostridium perfringens*. *Biochemical Journal* 118, 135–141.

Cozzani, I., Barsacchi, R., Dibenedetto, G., Saracchi, L. and Falcone, G. (1975) Regulation of breakdown and synthesis of L-glutamate decarboxylase in *Clostridium perfringens. Journal of Bacteriology* 123, 1115–1123.

Damiano, M.A, Bastianelli, D., Al Dahouk, S., Kohler, S., Cloeckaert, A., De Biase, D. and Occhialini, A. (2015) Glutamate decarboxylase-dependent acid resistance in *Brucella* spp.: distribution and contribution to fitness under extremely acidic conditions. *Applied and Environmental Microbiology* 81, 578–586.

De Biase, D. and Lund, P.A. (2015) The *Escherichia coli* acid stress response and its significance for pathogenesis. *Advances in Applied Microbiology* 92, 49–88.

De Biase, D. and Pennacchietti, E. (2012) Glutamate decarboxylase-dependent acid resistance in orally acquired bacteria: function, distribution and biomedical implications of the *gadBC* operon. *Molecular Microbiology* 86, 770–768.

De Biase, D., Tramonti, A., John, R.A. and Bossa, F. (1996) Isolation, overexpression, and biochemical characterization of the two isoforms of glutamic acid decarboxylase from *Escherichia coli. Protein Expression and Purification* 8, 430–415.

Dhakal, R., Bajpai, V.K. and Baek, K.-H. (2012) Production of GABA (γ-aminobutyric acid) by microorganisms: a review. *Brazilian Journal of Microbiology* 43, 1230–1241.

Dutyshev, D.I., Darii, E.L., Fomenkova, N.P., Pechik, I.V., Polyakov, K.M. Nikonov, S.V., Andreeva, N.S. and Sukhareva, B.S. (2005) Structure of *Escherichia coli* glutamate decarboxylase (GADalpha) in complex with glutarate at 2.05 angstroms resolution. *Acta Crystallographica* 61, 230–235.

Feehily, C., O'Byrne, C.P. and Karatzas, K.A. (2013) Functional gamma-aminobutyrate shunt in *Listeria monocytogenes*: role in acid tolerance and succinate biosynthesis. *Applied and Environmental Microbiology* 79, 74–80.

Fenalti, G., Law, R.H., Buckle, A.M., Langendorf, C., Tuck, K., Rosado, C.J., Faux, N.G., Mahmood, K., Hampe, C.S., Banga, J.P. *et al.* (2007) GABA production by glutamic acid decarboxylase is regulated by a dynamic catalytic loop. *Nature Structural and Molecular Biology* 14, 280–286.

Foster, J.W. (2001) Acid stress responses of *Salmonella* and *E. coli*: survival mechanisms, regulation, and implications for pathogenesis. *Journal of Microbiology-Seoul* 39, 89–94.

Foster, J.W. (2004) *Escherichia coli* acid resistance: tales of an amateur acidophile. *Nature Reviews* 2, 898–907.

Giannella, R.A., Broitman, S.A and Zamcheck, N. (1972) Gastric acid barrier to ingested microorganisms in man: studies *in vivo* and *in vitro*. *Gut* 13, 251–256.

Grassini, G., Pennacchietti, E., Cappadocio, F., Occhialini, A. and De Biase, D. (2015) Biochemical and spectroscopic properties of *Brucella microti* glutamate decarboxylase, a key component of the glutamate-dependent acid resistance system. *FEBS Open Bio* 5, 209–218.

Gut, H., Pennacchietti, E., John, R.A., Bossa, F., Capitani G., De Biase, D. and Grütter, M.G. (2006) *Escherichia coli* acid resistance: pH-sensing, activation by chloride and autoinhibition in GadB. *The EMBO Journal* 25, 2643–2651.

Gut, H., Dominici, P., Pilati, S., Astegno, A., Petoukhov, M.V., Svergun, D.I., Grütter, M.G. and Capitani, G. (2009) A common structural basis for pH- and calmodulin-mediated regulation in plant glutamate decarboxylase. *Journal of Molecular Biology* 392, 334–351.

Hayakawa, K., Kimura, M., Kasaha, K., Matsumoto, K., Sansawa, H. and Yamori, Y. (2004) Effect of a gamma-aminobutyric acid-enriched dairy product on the blood pressure of spontaneously hypertensive and normotensive Wistar-Kyoto rats. *British Journal of Nutrition* 92, 411–417.

Hiraga, K., Ueno, Y. and Oda, K. (2008) Glutamate decarboxylase from *Lactobacillus brevis*: activation by ammonium sulfate. *Bioscience, Biotechnology, and Biochemistry* 72, 1299–1306.

Hong, S.-J., Ullah, I., Park, G.-S., Lee, C. and Shin, J.-H. (2012) Overexpression and characterization of recombinant glutamate decarboxylase from *Thermococcus kodakaraensis* KOD1. *Journal of Korean Society for Applied Biological Chemistry* 55, 213–218.

Inoue, K., Shirai, T., Ochiai, H., Kasao, M., Hayakawa, K., Kimura, M. and Sansawa, H. (2003) Blood-pressure-lowering effect of a novel fermented milk containing gamma-aminobutyric acid (GABA) in mild hypertensives. *European Journal of Clinical Nutrition* 57, 490–495.

Jiang, D., Ji, H., Ye, Y. and Hou, J. (2011) Studies on screening of higher γ-aminobutyric acid-producing *Monascus* and optimization of fermentative parameters. *European Food Research and Technology* 232, 541–547.

Kanjee, U. and Houry, W.A. (2013) Mechanisms of acid resistance in *Escherichia coli*. *Annual Review of Microbiology* 67, 65–81.

Kanjee, U., Gutsche, I., Alexopolous, E., Zhao, B., El Bakkouri, M., Thibault, G., Liu, K., Ramachandran, S., Snider, J., Pai, E.F. and Houry, W.A. (2011a) Linkage between the bacterial acid stress and stringent responses: the structure of the inducible lysine decarboxylase. *The EMBO Journal* 30, 931–944.

Kanjee, U., Gutsche, I., Ramachandran, S. and Houry, W.A. (2011b) The enzymatic activities of the *Escherichia coli* basic aliphatic amino acid decarboxylases exhibit a pH zone of inhibition. *Biochemistry* 50, 9388–9398.

Kawasaki, N., Nakayama, A., Yamano, N., Takeda, S., Kawata, Y., Yamamoto, N. and Aiba, S. (2005) Synthesis, thermal and mechanical properties and biodegradation of branched polyamide 4. *Polymer* 46, 9987–9993.

Kim, H.-W., Kashima, Y., Ishikawa, K. and Yamano, N. (2009a) Purification and characterization of the first archaeal glutamate decarboxylase from *Pyrococcus horikoshii*. *Bioscience, Biotechnology, and Biochemistry* 73, 224–227.

Kim, J.Y., Lee, M.Y., Ji, G.E., Lee, Y.S. and Hwang, K.T. (2009b) Production of gamma-aminobutyric acid in black raspberry juice during fermentation by *Lactobacillus brevis* GABA100. *International Journal of Food Microbiology* 130, 12–16.

Kim, S.-H., Shin, B.-H., Kim, Y.-H., Nam, S.-W. and Jeon, S.-J. (2007) Cloning and expression of a full-length glutamate decarboxylase gene from *Lactobacillus brevis* BH2. *Biotechnology and Bioprocess Engineering* 12, 707–712.

Kohama, Y., Matsumodo, S., Mimura, T., Tanabe, N., Inada, A. and Nakanishi, T. (1987) Isolation and identification of hypotensive principles in red-mold rice. *Chemical and Pharmaceutical Bulletin* 35, 2484–2489.

Komatsuzaki, N., Shima, J., Kawamoto, S., Momose, H. and Kimura, T. (2005) Production of γ-aminobutyric acid (GABA) by *Lactobacillus paracasei* isolated from traditional fermented foods. *Food Microbiology* 22, 497–504.

Komatsuzaki, N., Nakamura, T., Kimura, T. and Shima, J. (2008) Characterization of glutamate decarboxylase from a high gamma-aminobutyric acid (GABA)-producer, *Lactobacillus paracasei*. *Bioscience, Biotechnology, and Biochemistry* 72, 278–285.

Lammens, T.M., De Biase, D., Franssen, M.C.R., Scott, E.L. and Sanders, J.P.M. (2009) The application of glutamic acid a-decarboxylase for the valorization of glutamic acid. *Green Chemistry* 11, 1562–1567.

Lammens, T.M., Gangarapu, S., Franssen, M.C.R., Scott, E.L. and Sanders, J.P.M. (2012) Techno-economic assessment of the production of biobased chemicals from glutamic acid. *Biofuels, Bioproducts and Biorefining* 6, 177–187.

Lee, E.-S., Kim, H.-W., Kim, D.-E., Kim, Y.-H., Nam, S.-W., Kim, B.-W. and Jeon, S.-J. (2013) Gene expression and characterization of thermostable glutamate decarboxylase from *Pyrococcus furiosus*. *Biotechnology and Bioprocess Engineering* 18, 375–381.

Li, H. and Cao, Y. (2010) Lactic acid bacterial cell factories for gamma-aminobutyric acid. *Amino Acids* 39, 1107–1116.

Lin, J., Smith, M.P., Chapin, K.C., Baik, H.S., Bennett, G.N. and Foster, J.W. (1996) Mechanisms of acid resistance in enterohemorrhagic *Escherichia coli*. *Applied and Environmental Microbiology* 62, 3094–3100.

Lund, P., Tramonti, A. and De Biase, D. (2014) Coping with low pH: molecular strategies in neutralophilic bacteria. *FEMS Microbiology Reviews* 38, 1091–1125.

Ma, D., Lu, P. and Shi, Y. (2013) Substrate selectivity of the acid-activated glutamate/gamma aminobutyric acid (GABA) antiporter GadC from *Escherichia coli*. *The Journal of Biological Chemistry* 288, 15148–15153.

Makarova, K., Slesarev, A., Wolf, Y., Sorokin, A., Mirkin, B., Koonin, E., Pavlov, A., Pavlova, N., Karamychev, V., Polouchine, N. *et al*. (2006) Comparative genomics of the lactic acid bacteria. *Proceedings of the National Academy of Sciences of the United States of America* 103, 15611–15616.

Martinsen, T.C., Bergh, K. and Waldum, H.L. (2005) Gastric juice: a barrier against infectious diseases. *Basic and Clinical Pharmacology and Toxicology* 96, 94–102.

McLaggan, D., Naprstek, J., Buurman, E.T. and Epstein, W. (1994) Interdependence of K^+ and glutamate accumulation during osmotic adaptation of *Escherichia coli*. *The Journal of Biological Chemistry* 269, 1911–1917.

Merrell, D.S. and Camilli, A. (2002) Acid tolerance of gastrointestinal pathogens. *Current Opinion in Microbiology* 5, 51–55.

Momany, C., Ernst, S., Ghost, R., Chang, N-L. and Hackert, M.L. (1995) Crystallographic structure of a PLP-dependent ornithine decarboxylase from *Lactobacillus* 30a to 3.0 A resolution. *Journal of Molecular Biology* 252, 643–655.

Nakamura, T., Matsubayashi, T., Kamachi, K., Hasegawa, T., Ando, Y. and Omori, M. (2000) γ-Aminobutyric acid (GABA)-rich *Chlorella* depresses the elevation of blood pressure in spontaneously hypertensive rats (SHR). *Nippon Nogeikagaku Kaishi* 74, 907–909. [In Japanese].

Nikolau, B.J., Perera, M.A.D.N., Brachova, L. and Shanks, B. (2008) Platform biochemicals for a biorenewable chemical industry. *The Plant Journal* 54, 536–545.

Nomura, M., Kimoto, H., Someya, Y., Furukawa, S. and Suzuki, I. (1998) Production of γ-aminobutyric acid by cheese starters during cheese ripening. *Dairy Foods* 81, 1486–1491.

Nomura, M., Nakajima, I., Fujita, Y., Kobayashi, M., Kimoto, H., Suzuki, I. and Aso, H. (1999) *Lactococcus lactis* contains only one glutamate decarboxylase gene. *Microbiology* 145, 1375–1380.

Occhialini, A., Jiménez de Bagüés, M.P., Saadeh, B., Bastianelli, D., Hanna, N., De Biase, D. and Köhler, S. (2012) The glutamic acid decarboxylase system of the new species *Brucella microti* contributes to its acid resistance and to oral infection of mice. *The Journal of Infectious Diseases* 206, 1424–1432.

Oh, C.-H. and Oh, S.-H. (2004) Effects of germinated brown rice extracts with enhanced levels of GABA on cancer cell proliferation and apoptosis. *Journal of Medicinal Food* 7, 19–23.

Oh, S.H. (2003) Stimulation of γ-aminobutyric acid synthesis activity in brown rice by a chitosan/glutamic acid germination solution and calcium/calmodulin. *Journal of Biochemistry and Molecular Biology* 36, 319–325.

Oh, S.-H. and Oh, C.-H. (2003) Brown rice extracts with enhanced levels of GABA stimulate immune cells. *Food Science and Biotechnology* 12, 248–252.

Okai, N., Takahashi, C., Hatada, K., Ogino, C. and Kondo, A. (2014) Disruption of *pknG* enhances production of gamma-aminobutyric acid by *Corynebacterium glutamicum* expressing glutamate decarboxylase. *AMB Express* 4: 20.

Park, K.-B. and Oh, S.-H. (2007) Production of yogurt with enhanced levels of gamma-aminobutyric acid and valuable nutrients using lactic acid bacteria and germinated soybean extract. *Bioresource Technology* 98, 1675–1679.

Patil, S.B. and Khan, Md. K. (2011) Germinated brown rice as a value added rice product: a review. *Journal of Food Science and Technology* 48, 661–667.

Pennacchietti, E., Lammens, T.M., Capitani, G., Franssen, M.C.R., John, R.A., Bossa, F. and De Biase, D. (2009) Mutation of His465 alters the pH-dependent spectroscopic properties of *Escherichia coli* glutamate decarboxylase and broadens the range of its activity toward more alkaline pH. *The Journal of Biological Chemistry* 284, 31587–31596.

Richard, H. and Foster, J.W. (2004) *Escherichia coli* glutamate- and arginine-dependent acid resistance systems increase internal pH and reverse transmembrane potential. *Journal of Bacteriology* 186, 6032–6041.

Sa, H.D., Park, J.Y., Jeong, S.-J., Lee, K.W. and Kim, J. H. (2015) Characterization of glutamate decarboxylase (GAD) from *Lactobacillus sakei* A156 isolated from *jeot-gal*. *Journal of Microobiology and Biotechnology* 25, 696–703.

Saikusa, T., Horino, T. and Mori Y. (1994) Accumulation of γ-aminobutyric acid (GABA) in the rice germ during water soaking. *Bioscience, Biotechnology, and Biochemistry* 58, 2291–2292.

Sanders, J.W., Leenhouts, K., Burghoorn, J., Brands, J.R., Venema, G. and Kok, J. (1998) A chloride-inducible acid resistance mechanism in *Lactococcus lactis* and its regulation. *Molecular Microbiology* 27, 299–310.

Schuller, H.M., Al-Wadei, H.A. and Majidi, M. (2008) Gamma-aminobutyric acid, a potential tumor suppressor for small airway-derived lung adenocarcinoma. *Carcinogenesis* 29, 1979–1985.

Seok, J.-H., Park, K.-B., Kim, Y.-H., Bae, M.-O., Lee, M.-K. and Oh, S.H. (2008) Production and characterization of *kimchi* with enhanced levels of γ-aminobutyric acid. *Food Science and Biotechnology* 17, 940–946.

Shepherd, M., Sanguinetti, G., Cook, G.M. and Poole, R.K. (2010) Compensations for diminished terminal oxidase activity in *Escherichia coli*: cytochrome bd-II-mediated respiration and glutamate metabolism. *The Journal of Biological Chemistry* 285, 997–1002.

Shi, F., Jiang, J., Li, Y., Li, Y. and Xie, Y. (2013) Enhancement of γ-aminobutyric acid production in recombinant *Corynebacterium glutamicum* by co-expressing two glutamate decarboxylase genes from *Lactobacillus brevis*. *Journal of Industrial Microbiology and Biotechnology* 40, 1285–1296.

Shin, S.-M., Kim, H., Joo, Y., Lee, S.-J., Lee, Y.-J. Lee, S.J. and Lee, D.W. (2014) Characterization of glutamate decarbozylase from *Lactobacillus plantarum* and its *C*-terminal function for the pH dependence of activity. *Journal of Agricultural and Food Chemistry* 62, 12186–12193.

Shizuka, F., Kido, Y., Nakazawa, T., Kitajima, H., Aizawa, C., Kayamura, H. and Ichijo, N. (2004) Antihypertensive effect of γ-amino butyric acid enriched soy products in spontaneously hypertensive rats. *BioFactors* 22, 165–167.

Shukuya, R. and Schwert, G.W. (1960) Glutamic acid decarboxylase. I. Isolation procedures and properties of the enzyme. *The Journal of Biological Chemistry* 235, 1649–1652.

Siragusa, S., De Angelis, M., Di Cagno, R., Rizzello, C.G, Coda, R. and Gobbetti, M. (2007) Synthesis of γ-aminobutyric acid by lactic acid bacteria isolated from a variety of Italian cheeses. *Applied and Environmental Microbiology* 73, 7283–7290.

Slonczewski, J.L., Fujisawa, M., Dopson, M. and Krulwich, T.A. (2009) Cytoplasmic pH measurement and homeostasis in bacteria and archaea. *Advances in Microbial Physiology* 55, 1–79.

Small, P.L. and Waterman, S.R. (1998) Acid stress, anaerobiosis and *gadCB*: lessons from *Lactococcus lactis* and *Escherichia coli*. *Trends in Microbiology* 6, 214–216.

Steinhauser, D., Fernie, A.R. and Araújo, W.L. (2012) Unusual cyanobacterial TCA cycles: not broken just different. *Trends in Plant Science* 17, 503–509.

Takahashi, C., Shirakawa, J., Tsuchidate, T., Okai, N., Hatada K., Nakayama, H., Tateno, T., Ogino, C. and Kondo A. (2012) Robust production of gamma-amino butyric acid using recombinant *Corynebacterium glutamicum* expressing glutamate decarboxylase from *Escherichia coli*. *Enzyme and Microbial Technology* 51, 171–176.

Tian, J., Bryk R., Itoh, M., Suematsu, M. and Nathan, C. (2005) Variant tricarboxylic acid cycle in *Mycobacterium tuberculosis*: identification of alpha-ketoglutarate decarboxylase. *Proceedings of the National Academy of Sciences of the United States of America* 102, 10670–10675.

Tsai, J.S., Lin, Y.S., Pan, B.S. and Chen, T.J. (2006) Antihypertensive peptides and γ-aminobutyric acid from prozyme 6 facilitated lactic acid bacteria fermentation of soymilk. *Process Biochemistry* 41, 1282–1288.

Tsai, M.F., McCarthy, P. and Miller, C. (2013) Substrate selectivity in glutamate-dependent acid resistance in enteric bacteria. *Proceedings of the National Academy of Sciences of the United States of America* 110, 5898–5902.

Ueno, H. (2000) Enzymatic and structural aspects on glutamate decarboxylase. *Journal of Molecular Catalysis* 10, 67–79.

Ueno, Y., Hayakawa, K., Takahashi, S. and Oda, K. (1997) Purification and characterization of glutamate decarboxylase from *Lactobacillus brevis* IFO 12005. *Bioscience, Biotechnology, and Biochemistry* 61, 1168–1171.

Wang, N., Ni, Y. and Shi, F. (2015) Deletion of *odhA* or *pyc* improves production of γ-aminobutyric acid and its precursor L-glutamate in recombinant *Corynebacterium glutamicum*. *Biotechnology Letters* 37, 1473–1481.

Werpy, T. and Peterson, G. (eds) (2004) *Top Value Added Chemicals from Biomass. Volume I – Results of Screening for Potential Candidates from Sugars and Synthesis Gas*. Office of the Biomass Program, Energy Efficiency and Renewable Energy, US Department of Energy, Washington, DC.

Zhang, J., Kao, E., Wang, G., Baidoo, E.E.K., Chen, M. and Keasling, J.D. (2016) Metabolic engineering of *Escherichia coli* for the biosynthesis of 2-pyrrolidone. *Metabolic Engineering Communications* 3, 1–7.

3 The Yeast γ-Aminobutyrate (GABA) Shunt

Robert D. Locy*
Department of Biological Sciences, Auburn University, Auburn, Alabama, USA

3.1 Abstract

γ-Aminobutyrate acid (GABA) is a 4-carbon non-protein amino acid that is ubiquitously found in biological systems. It is both produced and degraded by a simple three-enzyme metabolic pathway consisting of glutamate decarboxylase (GAD; EC 4.1.1.15), 4-aminobutyrate aminotransferase (GABAT; EC 2.6.1.19), and succinate semialdehyde dehydrogenase (SSADH; EC 1.2.1.16). The pathway is named the GABA shunt because it moves carbon from 2-oxoglutarate to succinate by bypassing the 2-oxoglutarate decarboxylase and succinate thiokinase reactions of the tricarboxylic acid (TCA) cycle. The pathway serves to catabolize both L-glutamate and GABA. Enzymes and metabolites of the GABA shunt play a variety of roles in biological systems, including N-catabolism/assimilation, cytosolic pH regulation, stress metabolite production, environmental stress tolerance, neurotransmission inhibition and autoimmune responses production. The fungi appear to have provided the opportunity for all of the varied eukaryotic functions of the GABA shunt to have arisen. This is demonstrated by looking at the properties of the individual enzymes of the pathway in terms of the number of isoforms present, intracellular location, substrate specificities, inhibitor preferences, and the transcriptional and post-translational regulatory properties of the enzymes of the GABA shunt. In yeast, these properties link closely to known mechanisms of the general stress response and the oxidative stress response in particular. This positions the GABA shunt pathway at the nexus of carbon metabolism, nitrogen metabolism, redox regulation and reactive oxygen scavenging in *Saccharomyces cerevisiae* (Baker's yeast).

3.2 Introduction

The GABA shunt is a simple metabolic pathway that involves three enzymes: glutamate decarboxylase, 4-aminobutyrate aminotransferase and succinate semialdehyde dehydrogenase (Fig. 3.1). It is called the GABA shunt because it serves to move carbon from 2-oxoglutarate to succinate, both metabolites of the tricarboxylic acid (TCA) cycle, by bypassing two reactions of this cycle. These reactions are virtually ubiquitous in all cellular life forms, but across the diversity of life the pathway has evolved varying functions, including roles in the assimilation of L-glutamate and GABA as nitrogen sources (Ramos *et al.*, 1985; Schneider *et al.*, 2002; Kurihara *et al.*, 2005),

*E-mail address: locyrob@auburn.edu

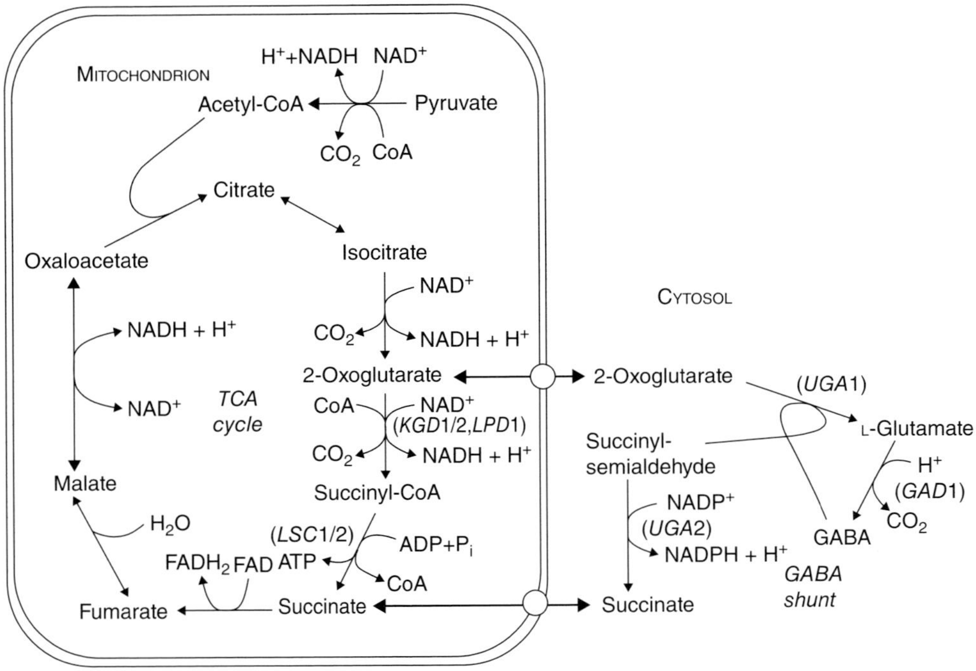

Fig. 3.1. The GABA shunt in *Saccharomyces cerevisiae* is a pathway consisting of three enzymes: L-glutamate decarboxylase, the product of the *GAD*1 gene; 4-aminobutyrate (γ-aminobutyrate) aminotransferase (GABAT), the product of the *UGA*1 gene; and succinate semialdehyde dehydrogenase (SSADH), product of the *UGA*2 gene. The GABA shunt is so named because it serves to move 2-oxoglutarate to succinate in the cytosol, bypassing the 2-oxoglutarate dehydrogenase (product of the *KGD*1, *KGD*2 and *LPD*1 genes) and succinyl-CoA ligase (product of the *LSC*1 and *LSC*2 genes) reactions in the TCA (tricarboxylic acid) cycle in the mitochondria.

the remediation of cytosolic acidification (Christensen and Schmit, 1980: Davies, 1980; Raven, 1985; Castanie-Cornet *et al.*, 1999; Karatzas *et al.*, 2012; Feehily and Karatzas, 2013), cellular redox regulation and the oxidative stress responses (Coleman *et al.*, 2001; Bouché *et al.*, 2003; Ludewig *et al.*, 2008; Cao *et al.*, 2013a), tolerance to various environmental stresses (Locy *et al.*, 2000; Renault *et al.*, 2010; Al-Quraan *et al.*, 2013; Yang *et al.*, 2015), nerve transmission inhibition (Erdö *et al.*, 1986) and a role in autoimmunity in the mammalian brain and pancreas (Reetz *et al.*, 1991; Maclaren, 1992; Winter, 1996; Fiorina, 2013).

Because the GABA shunt is ubiquitous and has such varied functions, remarkably little is understood about the mechanistic details of its involvement in these various biological phenomena. In comparison, the GABA shunt in the yeast, *Saccharomyces cerevisiae* Hansen, 1983, has been well studied and is well documented (Coleman *et al.*, 2001; Cao *et al.*, 2013a). *S. cerevisiae* was the first eukaryote to have its genome fully sequenced (Goffeau *et al.*, 1996), and it has served as a genetic model organism for virtually all of the Eukaryota. A key resource for yeast genomics and genetics community is the Saccharomyces Genome Database (SGD; http://www.yeastgenome.org) where comprehensive information about all yeast genes and proteins is housed, including the genes and proteins of the GABA shunt and related genes (Cherry *et al.*, 2012).

This chapter provides a comprehensive review of the GABA shunt genes and enzymes, their catalytic functions and their role in general metabolism, as well as a summary of the regulation of these enzymes, and an account of their function in the general stress response as exemplified by the role of the GABA shunt in the oxidative stress response and the heat stress response of yeast cells.

3.3 Metabolism of GABA – the GABA Shunt

GABA is the four-carbon straight chain, non-protein amino acid in the series of amino acids having amino groups on their terminal carbon (ω-amino acids). In this series, the biologically significant orthologues of GABA are: glycine, the two-carbon protein amino acid; β-alanine, the three-carbon analogue that is involved in CoA biosynthesis (White *et al.*, 2001); γ-aminobutyric acid, the four-carbon ω-amino acid (GABA); and δ-aminovalerate, the five-carbon analogue that is made during lysine degradation (Guengerich and Broquist, 1976).

In virtually all organisms, GABA is metabolized by a pathway referred to as the GABA shunt (Fig. 3.1). The main reaction producing GABA (when it is not found in the growth substrate) is the decarboxylation of L-glutamate, a proton-consuming reaction that is catalysed by the enzyme glutamic acid (glutamate) decarboxylase (GAD; EC 4.1.1.15). This is a biologically ubiquitous enzyme, and the products of L-glutamate decarboxylation are GABA and CO_2. The carbon portion of GABA is then converted to succinate semialdehyde (SSA) using 2-oxoglutarate as the amino acceptor and producing L-glutamate (Cao *et al.*, 2013a) in *S. cerevisiae* and all organisms other than green plants. In plants, the amino group is transaminated to pyruvate producing L-alanine. The yeast reaction is catalysed by 4-aminobutyrate aminotransferase, also called GABA transaminase (GABAT; EC 2.6.1.19). SSA is then oxidized to succinate by the $NAD(P)^+$-utilizing succinate semialdehyde dehydrogenase (SSADH; EC 1.2.1.16). These three reactions then provide an alternative to the conversion of 2-oxoglutarate into succinate in the TCA cycle, thus shunting carbon around the 2-oxoglutarate dehydrogenase and the succinyl-CoA ligase reactions while involving amino acid metabolism via L-glutamate in the process.

3.3.1 Glutamate decarboxylase

Although L-glutamate is normally degraded using the NAD^+-dependent glutamate dehydrogenase encoded by *GDH2* (Systematic name: YDL215C) in *S. cerevisiae* (Miller and Magasanik, 1990), under certain conditions L-glutamate is also degraded by the GABA shunt pathway (Fig 3.1) using glutamate decarboxylase (Gad1p; YMR250W). *S. cerevisiae* Gad1p contains 585 amino acids and has a molecular mass of 64,979.6 kDa (Coleman *et al.*, 2001). This enzyme catalyses the conversion of L-glutamate into GABA and CO_2 while consuming a proton (Fig. 3.1). In many organisms, particularly bacteria (Wang *et al.*, 2011; De Biase and Pennacchietti, 2012), fungi (Hao and Schmit, 1991; Strigacova *et al.*, 2001; Tsuchiya *et al.*, 2003) and plants (Baum *et al.*, 1993; Chen *et al.*, 1994; Gut *et al.*, 2009), GADs are known to have acidic pH optima in the range of pH 2–6, while they are relatively inactive at more typical physiological (neutral) pHs. This supports the idea that GAD acts to metabolically remediate cytoplasmic acidosis, i.e. it acts as a metabolic pH stat (Christensen and Schmit, 1980; Davies, 1980; Raven, 1985; Castanie-Cornet *et al.*, 1999; Karatzas *et al.*, 2012; Feehily and Karatzas, 2013) when the cytoplasmic pH falls to lower levels in organisms that have GADs with relatively acidic pH optima. However, GADs with neutral to mildly alkaline pH optima are found in most animals and other metazoans (Wu *et al.*, 1973; Denner *et al.*, 1987; Ilg *et al.*, 2013), and there is no information supporting a role for these GADs in pH regulation.

The *S. cerevisiae GAD*1 gene appears to be the only gene in the *S. cerevisiae* genome coding for a glutamate decarboxylase enzyme activity (SGD, Cherry *et al.*, 2012). There are 234 fungal reference sequences annotated as GADs found in the NCBI (US National Center for Biotechnology Information) Reference Sequence Database (Tatusova *et al.*, 2014) that have been queried using the NCBI Taxonomy Browser or BLAST searches. These sequences derive from 154 fungal genomes that have from one to three GAD genes per genome with an average of 1.52 GADs per genome (Table 3.1). In the Ascomycota, 38/120 sequenced genomes have a single GAD gene, including *S. cerevisiae*, while in the Basidiomycota 15/34 sequenced genomes show only a single GAD gene. In the Saccharomycetales, most of the sequenced genomes (21/23) show only a single GAD gene. The Archaea and Bacteria average 1.33 and 1.61 GAD genes per genome, respectively (Table 3.1), e.g. *Escherichia coli* has two GAD genes (De Biase *et al.*, 1996). Thus, most archaeal, bacterial and fungal genomes

Table 3.1. Taxonomic distribution and information about annotated glutamate decarboxylase (GAD) genes in the Archaea, Bacteria and Eukaryota (divided into Fungi, Metazoa and Viridiplantae).[a]

Taxon	Total sequenced genomes	Annotated sequenced genomes	GAD sequences	Average per genome	Range	Calmodulin-binding domain
Archaea	32	21	28	1.33	1–2	No
Bacteria	2231	1239	2003	1.61	1–5	No
Eukaryota	7875	568	1259	2.22	1–9	Yes/No
Fungi	720	154	234	1.52	1–3	Yes/No
Ascomycota	493	120	186	1.55	1–3	Yes/No
Saccharomycetales	48	23	25	1.08	1–2	Yes/No
Basidiomycota	153	34	48	1.41	1–3	Yes/No
Metazoa (metazoans)	5794	341	766	2.25	1–5	No
Viridiplantae	1002	73	259	3.55	1–9	Yes

[a]Reference sequences found in the NCBI Protein database originating from sequenced genomes that are annotated as glutamate decarboxylases (or a synonym) or that have an E-value below 1 e-50 in a BLAST search using known verified GAD genes in each respective taxon.

contain only one or two GAD genes. Metazoans appear to have two GADs (Erlander *et al.*, 1991; Langendorf *et al.*, 2013), and a number of green plants (Viridiplantae in Table 3.1) with sequenced genomes have 3–5 GAD genes in their genomes (average 3.6 GADs per genome). Hence, a characteristic of the fungi, and of the Saccharomycetales in particular, is that they demonstrate a single GAD gene.

Further comparison of the complexity of GAD sequences reveals that in bacteria, the species with two GADs have two nearly identical copies of the gene, and the major role of these GADs involves the catabolism of L-glutamate as a nitrogen source and pH regulation under stress conditions. The only regulation of bacterial GADs tends to be transcriptional regulation (De Biase *et al.*, 1996), and the pH regulation described above. More complex eukaryotes have increasing numbers of GAD genes with more specialized roles relating to the apparent functional significance of their GABA shunt pathway (Erlander *et al.*, 1991; Zik *et al.*, 1998; Akama *et al.*, 2001). Specifically, the carboxy-terminal 30–40 amino acids have emerged as a critical domain involved in pH regulation (Yu *et al.*, 2012; Langendorf *et al.*, 2013), calmodulin binding and activation (Coleman *et al.*, 2001; Gut *et al.*, 2009) and autoimmune function (Arafat *et al.*, 2009; Kass *et al.*, 2014; Mackay *et al.*, 2014) in various eukaryotes. Because the fungi have numerous species with a single GAD gene, and yet some fungal species have multiple GAD genes that demonstrate properties related to the more complex eukaryotes (discussed further below), it appears that the complex functions of GADs associated with the higher eukaryotes evolved first in the fungi.

3.3.2 GABA aminotransferase

GABA is metabolized to succinate semialdehyde in *S. cerevisiae* by a 4-aminobutyrate:2-oxoglutarate transaminase (EC 2.6.1.19; also called GABA aminotransferase, GABA transaminase and Uga1p). The protein, which is also referred to as GABAT or UGP1p, is the gene product of the *UGA1* gene of *S. cerevisiae*, and it catalyses the transamination of the ω-amino group of GABA to 2-oxoglutarate, producing L-glutamate and succinate semialdehyde. This appears to be the only gene in the *S. cerevisiae* genome that utilizes GABA for transamination (Andersen *et al.*, 2007), and it is localized in the cytosol (Cao *et al.*, 2013b) of that organism. Additionally, *Saccharomyces* (*Lachancea*) *kluyveri* contains two transaminases that work on GABA, one UGA1p-like enzyme, and a second enzyme that appears to prefer β-alanine rather than GABA (Andersen *et al.*, 2007). Whether both of these are utilized for GABA transamination *in situ* is unclear, but clearly the Uga1p-like activity is used for that function.

Based on reference sequences found in the NCBI Gene database, it appears that 16 fungal species in the Saccharomycetales harbour only a single *UGA1*-like gene similar to *UGA1*, while 12 additional fungal species harbour both the *UGA1*-like activity and a second GABAT that prefers β-alanine to GABA as an amino donor (Table 3.2). Both of these activities transaminate 2-oxoglutarate and do not utilize pyruvate as do plant GABATs (Andersen *et al.*, 2007).

In prokaryotes, two GABA aminotransferases are generally found. In *E. coli*, for example, these are the products of the *gab*T (Metzer *et al.*, 1979; Niegemann *et al.*, 1993; Schneider *et al.*, 2002) and *puu*E (Kurihara *et al.*, 2005) genes. The former is involved in L-glutamate catabolism (Mehta *et al.*, 1993), while the latter is associated with polyamine (putrescine) degradation (Kurihara *et al.*, 2005). The two GABA aminotransferases both utilize GABA but utilize β-alanine (poorly) to transaminate 2-oxoglutarate (Mehta *et al.*, 1993). There is also a separate β-alanine:pyruvate aminotransferase in prokaryotes (Yun *et al.*, 2004).

In comparison with the yeasts, metazoans have a single mitochondrial GABA aminotransferase (Cooper, 1985). This enzyme uses GABA (or β-alanine) to transaminate 2-oxobutyrate, producing succinate semialdehyde (or malonyl semialdehyde) and L-glutamate. Green plants have been reported to contain a similar enzyme activity that utilizes 2-oxoglutarate as an amino acceptor (Van Cauwenberghe and Shelp, 1999; Akihiro *et al.*, 2008), although this activity has not been conclusively demonstrated. However, plants do have a GABA transaminase that uses either pyruvate or glyoxylate as the amino acceptor (Clark *et al.*, 2009a,b; Shimajiri *et al.*, 2013). These are localized in the cytosol, mitochondria, chloroplasts and peroxisomes, reflecting a more complex pattern of GABA metabolism in green plants.

3.3.3 Succinate semialdehyde dehydrogenase

The *UGA2* gene in *S. cerevisiae* produces a GABA shunt enzyme called succinate semialdehyde dehydrogenase (SSADH). The UGA2p protein catalyses the production of succinate from succinate semialdehyde (SSA) while reducing NAD^+ or $NADP^+$ (Ramos *et al.*, 1985; Vissers *et al.*, 1990; Cao *et al.*, 2014). The ability of SSADH from *S. cerevisiae* (EC 1.2.1.16) to utilize NAD^+ or $NADP^+$ makes it a member of the NAD(P)+-dependent aldehyde dehydrogenase superfamily (ALDH-SF), which contains an *ALDH_F5_SSADH_GabD* domain (Marchler-Bauer *et al.*, 2014). The EC1.2.1.16 family of SSADHs is found primarily

Table 3.2. Taxonomic distribution of and information about annotated 4-aminobutyrate aminotransferase (GABAT) genes in the Archaea, Bacteria and Eukaryota (divided into Fungi, Metazoa and Viridiplantae).[a]

Taxon	Sequenced genomes	GABAT sequences	Average per genome	Substrate used[b]		Mitochondrial localization	
				2-KG	Pyr/Glyox	No. species[c]	%
Archaea	138	243	1.76	243	0		
Bacteria	3088	7481	1.91	7481	0		
Eukaryota	422	796	1.89	525	271	679	85.3
Fungi	139	203	1.46	203	0	129	63.5
Ascomycota	123	181	1.47	181	0	107	59.1
Saccharo-mycetales	28	44	1.57	44	0	14	31.8
Basidiomycota	16	22	1.38	22	0	21	95.5
Metazoa	236	322	1.36	322	0	318	98.8
Viridiplantae	47	271	5.77	0	271	232	85.6

[a]Reference sequences found in the NCBI Protein database originating from sequenced genomes that are annotated as 4-aminobutyrate aminotransferases (or a synonym) or that have an E-value, below 1 e-50 in a BLAST search using known verified 4-aminobutyrate aminotransferases protein sequences in each respective taxon.
[b]2-KG, 2-oxoglutarate (also known as 2-ketoglutarate); Pyr/Glyox, pyruvate or glyoxylate.
[c]No. of species within the taxon in col. 1 where mitochondrial location is indicated in the sequence annotation in the NCBI.

in a number of fungi. The enzyme is found either as the sole SSADH activity, or in some fungi, the dual-reductant enzyme is present with an NAD^+-specific SSADH (EC 1.2.1.24), as found in most prokaryotes (Donnelly and Cooper, 1981) and most other eukaryotes (Bouché *et al.*, 2003; Leone *et al.*, 2007; Rothacker and Ilg 2008). Additionally, there are a handful of prokaryotes that appear to contain an $NADP^+$-specific (EC 1.2.1.79) SSADH (Sanchez *et al.*, 1989; Lutke-Eversloh and Steinbuchel, 1999; Langendorf *et al.*, 2010; Park and Rhee, 2013), and X-ray crystallographic data have shown that as few as three amino acid residues control the pyridine nucleotide specificity of SSADHs (Langendorf *et al.*, 2010; Park and Rhee, 2013). Thus, the enzymes of this family are all rather closely related.

Further structural and biochemical investigations of the substrate specificity of SSADHs have suggested a role for the NAD(P)-binding pocket in the redox regulation of SSADH activity (Kim *et al.*, 2011; Yuan *et al.*, 2013; Jang *et al.*, 2015). Consistent with this demonstration is the fact that nearly all SSADHs are inhibited by AMP and by NADH (Blaner and Chruchich, 1980; Rivett and Tipton, 1981; Busch and Fromm, 1999; Jeong *et al.*, 2005), including SSADHs from the fungi *S. cerevisiae* (Cao *et al.*, 2014) and *Aspergillus niger* (Kumar *et al.*, 2015). Given the mitochondrial location of most eukaryotic SSADHs (Ryzlak and Pietruszko, 1988; Busch and Fromm, 1999; Blasi *et al.*, 2006), inhibition by NADH would seem to involve a significant component of the redox and energy regulation of SSADH activity in most eukaryotes, and in *S. cerevisiae* specifically.

However, in many fungi, a cytosolic SSADH (Table 3.3) can be found. Table 3.3 shows that in the Saccharomycetaceae most species (with a sequenced genome) have only a single cytosolic NAD(P)-dependent SSADH. In the remainder of the Ascomycota, a significant number of species have both the cytosolic orthologues and mitochondrial SSADHs (Table 3.3). In the Basidiomycota, many more species have both the cytosolic and a mitochondrial SSADHs (Table 3.3). This trend is not apparent in the remainder of the Eukaryota, as most metazoans and plants have only mitochondrial SSADHs (Table 3.3). The uniqueness of cytosolic SSADHs in the fungi leads to the hypothesis that the functional roles for this enzyme may also be unique in the fungi. This will be further discussed in Section 3.4.

3.4 Regulation of the Yeast GABA Shunt

The enzymes of the GABA shunt appear to be regulated both transcriptionally (Ramos *et al.*, 1985; Vissers *et al.*, 1989; Talibi *et al.*, 1995; Coleman *et al.*, 2001) and post-translationally (Coleman *et al.*, 2001; Cao *et al.*, 2014). Transcript levels of *GAD1*, *UGA1*, and *UGA2*, the genes for GAD and GABAT, are mediated by the GABA-specific transcription activator Uga3p, along with the Dal81p transcription activator (Ramos *et al.*, 1985; Vissers *et al.*, 1989; Talibi

Table 3.3. Taxonomic distribution and information about annotated succinate semialdehyde dehydrogenase (SSADH) genes in the Archaea, Bacteria and Eukaryota (divided into Fungi, Metazoa, and Viridiplantae).[a]

Taxon	Sequenced genomes	SSADH sequences	Average per species	Mitochondrial SSADHs	Cytosolic SSADHs	% Mitochondrial
Archaea	127	195	1.54			
Bacteria	3744	11143	2.98			
Eukaryota	469	768	1.64	420	348	54.7
Fungi	162	414	2.56	69	345	16.7
Ascomycota	127	356	2.80	48	308	13.5
Saccharomycetales	13	20	1.54	2	18	10.0
Basidiomycota	35	46	1.31	21	25	45.7
Metazoa	251	264	1.05	262	2	99.2
Viridiplante	56	90	1.61	89	1	98.9

[a]Reference sequences found in the NCBI Protein database originating from sequenced genomes that are annotated as succinate semialdehyde dehydrogenases (or a synonym) or that have an E-value, below 1 e-50 in a BLAST search using known verified succinate semialdehyde dehydrogenase protein sequences in each respective taxon.

et al., 1995; Coleman *et al.*, 2001). These factors appear to bind to a UAS-GABA element found in the promoters of these genes (Vissers *et al.*, 1989; Talibi *et al.*, 1995). The nitrogen catabolite repression activator Gln3p also appears to be implicated in the regulation of the UGA regulon and the *GAD*1 gene by competing with Dal80p (Coleman *et al.*, 2001), a transcriptional repressor for the UAS-GATA promoter elements found in conjunction with the UAS-GABA element (Talibi *et al.*, 1995; Idicula *et al.*, 2002).

While the transcriptional activators and enhancer sequence elements regulating the expression of the GABA shunt are known in *Saccharomyces*, the nature of the signalling pathways producing these effects is less clear. It is known that the GABA shunt enzymes Uga1p (GABAT) and Uga2p (SSADH), along with the Uga4p GABA vacuolar transporter (Vissers *et al.*, 1989), make up the UGA regulon, and that UGA regulon genes are induced by GABA (Ramos *et al.*, 1985; Vissers *et al.*, 1989) in a process that is mediated by the Uga3p transcriptional activator. However, the UGA regulon and Gad1p are also induced during the oxidative stress response, e.g. by H_2O_2 (Coleman *et al.*, 2001). Thus, in *S. cerevisiae*, an increase in cytosolic GABA, either because of GABA supplied in the medium as a nitrogen source, or as a result of activation of the Gad1p reaction, will promote the catabolism of GABA by Uga1p and Uga2p (Fig. 3.2), and/or GABA uptake into vacuoles by Uga4p. This accounts for the frequently observed accumulation of GABA in response to environmental stresses, including oxidative stresses (Cao *et al.*, 2013a).

Post-translational regulation of the GABA shunt has focused on two areas; the regulation of Gad1p activity, and product inhibition of Uga2p by NADH. In the absence of a GABA supply as an N source, yeast has the capacity to make GABA using Gad1p by the decarboxylation of L-glutamate (see Section 3.3.1. above). This is a proton-consuming decarboxylation reaction that is activated either by low cytosolic pH or via the calcium-binding protein, calmodulin (Coleman *et al.*, 2001). Because abiotic stresses, including oxidative stresses (Popa *et al.*, 2010), are known to increase cytosolic calcium levels, it is clear that the stress-induced accumulation of GABA can be mediated by calcium/calmodulin and/or pH regulation of Gad1p (see Fig. 3.2).

The second aspect of post-transcriptional regulation of the GABA shunt involves product inhibition of Uga2p by NADH. Note that Uga2p can utilize NAD^+ or $NADP^+$ and has only a slightly lower affinity for $NADP^+$ than NAD^+ (Cao *et al.*, 2014); the measured ratio of $(NADP^+ + NADPH)/(NAD^+ + NADH)$ is *c*.35:1 in *S. cerevisiae* (Anderlund *et al.*, 1999). Thus, it seems likely that the critical pyridine dinucleotide used to oxidize succinate semialdehyde to succinate via Uga2p is $NADP^+$ rather than NAD^+, and that conditions favouring the reduction the pyridine nucleotide pool will consequently increase the accumulation of succinate semialdehyde. Subsequently, GABA will also accumulate as the Uga1p-catalysed reaction reaches equilibrium.

Figure 3.2 shows a summary of the processes described above. In summary, a condition that produces an increase in cytosolic calcium will lead to the activation of GAD via calmodulin, or a condition that produces a lower cytosolic pH will lead to the activation of GAD. The production of GABA from L-glutamate with the stoichiometric consumption of a proton results from GAD activation. L-Glutamate is restored via 4-aminobutryate aminotransferase (GABAT), with the production of SSA from GABA. Additionally, accumulated GABA activates the transcription of the Uga4p GABA vacuolar transporter and *UGA*1 and *UGA*2 genes mediated by Uga3p binding to UAS-GABA sequence elements in the *UGA*1, *UGA*2 and *UGA*4 genes. SSA is converted to succinate with the production of NADPH and a proton in the cytosol, until the NADP pool is fully reduced and/or the level of NADH rises sufficiently to slow or stop succinate production. At that point, SSA and GABA accumulation result in fully reduced NADPH and NADH pools.

Note that when yeast is grown on GABA as an N source, the Gad1p reaction is not used, but the remainder of the pathway and regulation described above occurs, i.e. GABA is accumulated in vacuoles, and SSA levels rise as the pyridine nucleotide pools become more reduced. As succinate is oxidized in the mitochondria, and/or reduced pyridine nucleotides are consumed, further GABA metabolism occurs until the GABA is consumed.

In summary then, the redox status of the NAD/NADH and NADP/NADPH pools appears

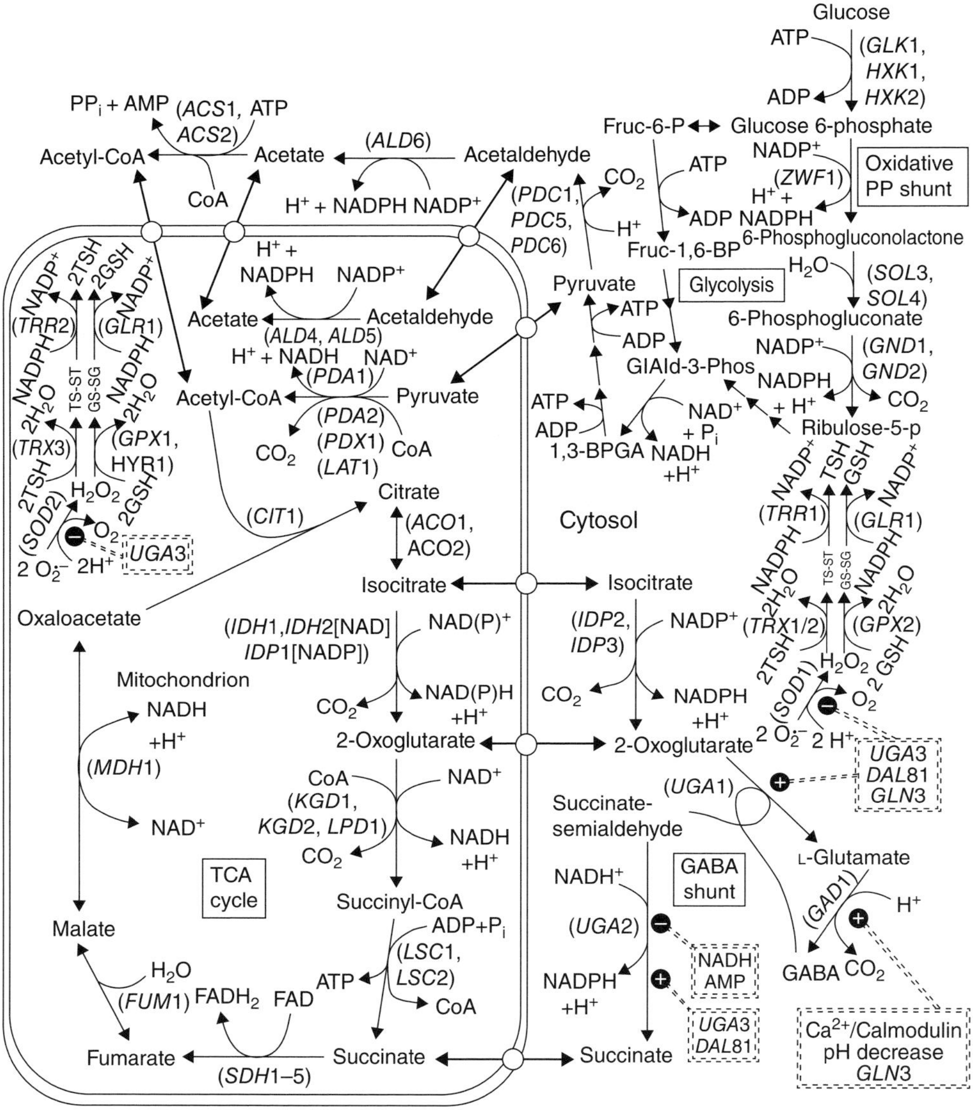

Fig. 3.2. Integration of the GABA shunt with cellular metabolism showing reactions that influence the oxidative stress response and the heat stress response in *Saccharomyces cerevisiae*. Reactions are labelled by their gene symbols, and the proteins (enzymes) corresponding to the gene symbols are given in Table 3.4. Dashed lines show regulatory interactions mediating regulation during the oxidative stress response and the heat stress response. The double-dashed boxes in the figure indicate regulator genes of the reactions concerned. Further information about key steps in the TCA (tricarboxylic acid) cycle, glycolysis, the oxidative PP (pentose phosphate) pathway (shunt), the cytosolic TCA bypass, and Redox/ROS (reactive oxygen species) regulation can be found in Table 3.4.

to play a critical role in regulating the flow of carbon through the GABA shunt. Further, the cytosolic location of the GABA shunt enzymes, and an apparent likelihood to utilize NADP rather than NAD, suggest that GABA shunt regulation serves to reduce NADH production in the mitochondria while increasing NADPH production in the cytosol.

Table 3.4. Key to the (majority of) symbols used in Fig. 3.2, which shows the integration of the GABA shunt with cellular reactions influencing the oxidative and heat stress responses in *Saccharomyces cerevisiae*. Gene locations are shown, followed by the gene systematic name, the gene symbol used in Fig. 3.2 and the EC number for each protein/enzyme. The description (full name) of the enzyme is then given along with information on the phenotype of the null mutant of each protein for the oxidative stress response (OX) or the heat stress response (HS); a (+) in the OX and HS columns indicates that the null mutant shows increased sensitivity to the stress, while a (–) indicates that the null mutant is less sensitive to the stress. Information for this table was mined from the Saccharomyces Genome Database (SGD; see http://www.yeastgenome.org).

Location/ System	Systematic name	Symbol	EC No.	Description	OX	HS
GABA shunt –Cytosolic						
	YMR250W	*GAD*1	EC 4.1.1.15	Glutamate decarboxylase	+	+
	YGR019W	*UGA*1	EC 2.6.1.19	4-aminobutyrate aminotransferase	+	+
	YBR006W	*UGA*1	EC 1.2.1.16	Succinate semialdehyde dehydrogenase	+	+
GABA transporters/transcription factors						
	YDL170W	*UGA*4		GABA-specific transcriptional activator		
	YDL210W	*UGA*4		GABA-specific transporter, vacuolar		
	YKR039W	*GAP*1		General amino acid permease, plasma membrane		
	YOR348C	*PUT*4		Proline, amino acid permease, plasma membrane		
Mitochondrial enzymes/systems						
Pyruvate dehydrogenase complex						
	YER178W	*PDA*1	EC 1.2.4.1	Pyruvate dehydrogenase, subunit E1 alpha	+	+
	YBR221C	*PDB*1	EC 1.2.4.1	Pyruvate dehydrogenase, subunit E1 beta		+
	YNL071W	*LAT*1	EC 1.2.4.1	Dihydrolipoamide acetyltransferase (E2)	+	+
	YGR193C	*PDX*1	EC 1.2.4.1	E3-binding protein		+
TCA cycle						
	YNR001C	*CIT*1	EC 2.3.3.16	Citrate synthase, mitochondrial		+
	YLR304C	*AC01*	EC 4.2.1.3	Aconitase, mitochondrial		
	YJL200C	*ACO*2	EC 4.2.1.3	Aconitase, mitochondrial		
	YDL066W	*IDP*1	EC 1.1.1.42	Isocitrate dehydrogenase, mitochondrial (NADP)	+	
	YNL037C	*IDH*1	EC 1.1.1.41	Isocitrate dehydrogenase, mitochondrial (NAD)	+	–
	YOR136W	*IDH*2	EC 1.1.1.41	Isocitrate dehydrogenase, mitochondrial (NAD)	+	+
	YIL125W	*KGD*1	EC 1.2.4.2	2-oxoglutarate dehydrogenase, mitochondrial	+	
	YDR148C	*KGD*2	EC 1.2.4.2	2-oxoglutarate dehydrogenase, mitochondrial	+	
	YFL018C	*LPD*1	EC 1.2.4.2	2-oxoglutarate dehydrogenase, mitochondrial	+	+
	YOR142W	*LSC*1	EC 6.2.1.5	Succinyl-CoA ligase, alpha subunit		
	YGR244C	*LSC*2	EC 6.2.1.5	Succinyl-CoA ligase, beta subunit		
	YKL148C	*SDH*1	EC 1.3.5.1	Succinate dehydrogenase, flavoprotein subunit		

Continued

Table 3.4. Continued.

Location/ System	Systematic name	Symbol	EC No.	Description	OX	HS
	YLL041C	*SDH*2	EC 1.3.5.1	Succinate dehydrogenase Iron-S subunit		
	YKL141W	*SDH*3	EC 1.3.5.1	Succinate dehydrogenase, cyt b subunit		
	YDR178W	*SDH*4	EC 1.3.5.1	Succinate dehydrogenase, membrane anchor subunit		
	YOL071W	*SDH*5	EC 1.3.5.1	Succinate dehydrogenase, assembly factor	+	
	YPL262W	*FUM*1	EC 4.2.1.2	Fumarase	+	–
	YKL085W	*MDH*1	EC 1.1.1.37	Malate dehydrogenase, mitochondrial		
Acetate production						
	YOR374W	*ALD*4	EC 1.2.1.3	Aldehyde dehydrogenase ($NADP^+$), mitochondrial		
	YER073W	*ALD*5	EC 1.2.1.3	Aldehyde dehydrogenase ($NAD(P)^+$),mitochondrial		
Redox system – detoxification of reactive oxygen species (ROS)						
	YHR008C	*SOD*2	EC 1.15.1.1	Manganese superoxide dismutase, mitochondrial	+	+
	YPL091W	*GLR*1	EC 1.8.1.7	Glutathione oxidoreductase, mitochondria	+	+
	YKL026C	*GPX*1	EC 1.11.1.12	Glutathione peroxidase, mitochondrial OM	+	+
	YIR037W	*HYR*1	EC 1.11.1.7	Thiol peroxidase, paralog of GPX1, mitochondrion	+	
	YHR106W	*TRR*2	EC 1.8.1.9	Thioredoxin reductase, mitochondrial	+	+
	YCR083W	*TRX*3	EC 1.11.1.15	Thioredoxin, mitochondrial		
Cytosolic enzymes/systems						
TCA bypass						
	YLR044C	*PDC*1	EC 4.1.1.1	Pyruvate decarboxylase	+	+
	YLR134W	*PDC*5	EC 4.1.1.1	Pyruvate decarboxylase	+	
	YGR087C	*PDC6*	EC 4.1.1.1	Pyruvate decarboxylase		
	YPL061W	*ALD6*	EC 1.2.1.4	Aldehyde dehydrogenase (cytosol)	+	
	YAL054C	*ACS*1	EC 6.2.1.1	Acetyl-CoA synthetase, cytosolic	+	
	YLR153C	*ACS*2	EC 6.2.1.1	Acetyl-CoA synthetase, cytosolic		
	YCR005C	*CIT*2	EC 2.3.3.16	Citrate synthase, peroxisomal		–
	YLR304C	*ACO*1	EC 4.2.1.3	Aconitase, cytosolic		
	YJL200C	*ACO*2	EC 4.2.1.3	Aconitase, cytosolic		
	YLR174W	*IDP*2	EC 1.1.1.42	Isocitrate dehydrogenase, cytosol (NADP)	+	
Glycolysis						
	YCL040W	*GLK*1	EC 2.7.1.2.	Glucokinase	+	
	YFR053C	*HXK*1	EC 2.7.1.1.	Hexokinase, fermentation		
	YGL253W	*HXK*2	EC 2.7.1.1.	Hexokinase, aerobic, signalling	+	–
	YJL052W	*TDH*1	EC 1.2.1.12	Glyceraldehyde-3-P dehydrogenase (NAD), cytosol/cell wall	–	
	YJR009C	*TDH*2	EC 1.2.1.12	Glyceraldehyde-3-P dehydrogenase (NAD), cytosol/cell wall		+

Continued

Table 3.4. Continued.

Location/ System	Systematic name	Symbol	EC No.	Description	OX	HS
	YGR192C	*TDH*3	EC 1.2.1.12	Glyceraldehyde-3-P dehydrogenase (NAD), cytosol/cell wall	–	
	YNL241C	*ZWF*1	EC 1.1.1.79	Glucose-6-phosphate dehydrogenase	+	
Oxidative pentose phosphate shunt						
	YGR248W	*SOL*3	EC 3.1.1.31	6-Phosphogluconolactonase	+	+
	YGR248W	*SOL*4	EC 3.1.1.31	6-Phosphogluconolactonase		
	YHR183W	*GND*1	EC 1.1.1.44	6-phosphogluconate dehydrogenase	+	
	YGR256W	*GND*2	EC 1.1.1.44	6-Phosphogluconate dehydrogenase	+	
Cytosolic redox system						
	YJR104C	*SOD*1	EC 1.15.1.1	Copper-zinc superoxide dismutase, cytosol	+	+
	YBR244W	*GPX*2	EC 1.11.1.12	Glutathione peroxidase, cytosol	+	
	YDR353W	*TRR*1	EC 1.8.1.9	Thioredoxin reductase, cytosol	+	
	YLR043C	*TRX*1	EC 1.11.1.15	Thioredoxin, cytosol	+	+
	YGR209C	*TRX*2	EC 1.11.1.15	Thioredoxin, cytosol	+	

3.5 The Role of the Yeast GABA Shunt in Environmental Stress Responses

Based on genome-wide expression analysis and analysis of null mutations of specific genes in *S. cerevisiae*, the response of yeast cells to environmental stresses has been characterized very broadly (Gasch *et al.*, 2000). What emerges from these studies and from additional specific studies (Howlett and Avery, 1997; Morano *et al.*, 2012; Michán *et al.*, 2013; Bubnová *et al.*, 2014) is the notion that an oxidative stress is a secondary component of virtually all environmental stresses in *Saccharomyces*. This derives from the fact that virtually all organisms growing aerobically produce reactive oxygen species (ROS) as a result of the flow of electrons through their respiratory electron transport chain (Thorpe *et al.*, 2013, and references cited therein). In yeast, superoxide is produced as a result of mitochondrial dysfunction primarily at Complex I and/or Complex III in the electron transport chain (Murphy, 2009; Morano *et al.*, 2012), and environmental stresses typically produce such mitochondrial dysfunction.

Superoxide (O_2^-) produced in the mitochondria is detoxified by conversion to H_2O_2 via a cytosolic (*SOD*1; Bermingham-McDonogh *et al.*, 1988) or a mitochondrial (*SOD*2; van Loon *et al.*, 1986) superoxide dismutase (see Fig. 3.2). A number of enzyme systems are known to reduce H_2O_2, including a peroxisomal catalase (Cta1p) and a cytosolic catalase (Ctt1p). However, double mutants lacking both of these enzymes are not more sensitive to H_2O_2 stress than wild types except in the stationary phase (Izawa *et al.*, 1996). The main mechanism for mediating oxidative damage during oxidative stress appears to be the glutathione/glutaredoxin and thioredoxin systems (Grant, 2001). Both of these systems consist of cytosolic (Glr1p, Trr1p) and mitochondrial (Glr1p, Trr2p) glutathione and thioredoxin reductases that acquire electrons from NADPH (Fig. 3.2), and keep protein sulfhydryl groups in the reduced state in the face of oxidative stresses (Izawa *et al.*, 1998; Outten and Culotta, 2004; Kojer *et al.*, 2012). A number of additional components, including low molecular weight thiol reagents, glutathione (GSH) and thioredoxin (TSH), and glutathione and thioredoxin thiol peroxidases found in either the mitochondria (Grx1p, Hyr1p, Trx3p) or in the cytosol (Grx2p, Trx1p, Trx2p) are also implicated in this process (Fig. 3.2).

*HYR*1p, *GRX*1p and *GRX*2p are the glutathione peroxidase homologues (Inoue *et al.*, 1999) identified in yeast cytosol and mitochondria, respectively (Avery and Avery, 2001; Avery *et al.*, 2004). Hyr1p is able to act on glutathione, phospholipid hydroperoxides and thioredoxin as substrates (Inoue *et al.*, 1999; Avery and Avery, 2001; Delaunay *et al.*, 2002), and it can also act as a hydroperoxide sensor for the oxidative stress transcription factor Yap1p (Izawa *et al.*, 1999; Delaunay *et al.*, 2002; Avery *et al.*, 2004; Ma *et al.*, 2007). The Hyr1p-dependent activation of Yap1p is also dependent on the protein Ybp1p (Okazaki *et al.*, 2005). Via this pathway, oxidative damage is minimized in mitochondria and other intracellular compartments by using NADPH to reduce the ROS level that accumulates in cells (Fig. 3.2.).

It is well documented that a source of NADPH required for oxidative stress tolerance arises in the cytosol from the oxidative branch of the pentose phosphate shunt (Juhnke *et al.*, 1996; Slekar *et al.*, 1996; Lee *et al.*, 1999), specifically from the action of glucose-6-phosphate dehydrogenase (Zwf1p; Jarori and Maitra, 1991; Izawa *et al.*, 1998) and 6-phosphogluconate dehydrogenase (Gnd1p, Gnd2p; Sinha and Maitra, 1992). Zwf1p catalyses the important rate-controlling step of the pentose phosphate shunt, and it is essentially constitutively expressed (Minard and McAlister-Henn, 2005), although it is inhibited by NADPH (Llobell *et al.*, 1988). This allows the consumption of NADPH to control the activity of the NADPH-producing reactions of the oxidative pentose phosphate shunt.

At the same time, the *S*-thiolation of enzymes in lower glycolysis, particularly glyceraldehyde-3-phosphate dehydrogenase, protects them from irreversible damage by H_2O_2 during oxidative stress (Shenton and Grant, 2003; Grant, 2008), while inhibiting the activity of these enzymes. Thiolation by glutathione (for example) during oxidative stress reconfigures carbohydrate metabolism from glycolysis in favour of the pentose phosphate shunt (Ralser *et al.*, 2007, 2009), and is a critical component of the stress tolerance mechanism as it limits carbon flux into the NADH-generating TCA cycle in mitochondria, while increasing the supply of cytosolic NADPH required to moderate oxidative damage.

The NADPH required in mitochondria to reduce the oxidative stress response appears to arise from three sources: by phosphorylation of NADH by NADH kinase (*POS5*; Outten and Culotta, 2003; Strand *et al.*, 2003), from mitochondrial NADP-specific isocitrate dehydrogenase (IDP1; Minard *et al.*, 1998), and from the oxidation of acetaldehyde to acetate by mitochondrial NADPH-specific aldehyde dehydrogenases (Ald4p, Ald5p; Boubekeur *et al.*, 2001). This latter reaction is part of a bypass of the NADH-generating pyruvate dehydrogenase (PDH) complex in the mitochondria (Boubekeur *et al.*, 1999) that involves cytosolic pyruvate decarboxylase (*PDC1*, *PDC5*) and acetyl CoA synthetase (*ACS1*, *ACS2*). If $NADP^+$ is available in the cytosol, the cytosolic Ald6p dehydrogenase (Aranda and del Olmo, 2003) could also produce acetate in the cytosol (see Fig. 3.2). During episodic oxidative stresses, Ald6p translocates to the outer surface of mitochondria although it still remains functionally cytosolic (Saint-Prix *et al.*, 2004; Rinnerthaler *et al.*, 2006). Regulation of the PDH bypass (see Fig. 3.2) appears to involve general stress response regulators Msn2/4p and Hsf1p (Aranda and del Olmo, 2003). Thus, regulation of carbohydrate metabolism and mitochondrial energy production during general stress responses involving oxidative stress responses works to minimize NADH generation in the mitochondria which would subsequently produce increasing amounts of ROS from mitochondrial electron transport, and promotes cytosolic and mitochondrial NADPH production. In this way, the redox status of the cell is maintained while supporting the elimination of ROS produced by dysfunction of the mitochondrial electron transport chain (ETC).

The regulatory properties of the GABA shunt described in Section 3.4 that were described by Coleman *et al.* (2001) are consistent with the cellular strategy for maintaining redox balance during the oxidative and general stress responses described above (Coleman *et al.*, 2001). 2-Oxoglutarate (2-KG) and/or isocitrate (which can be converted to 2-KG in the cytosol) are shunted to the cytosol where 2-oxoglutarate is transaminated by the Uga1p, 4-aminobutyrate aminotransferase, producing SSA and L-glutamate. The L-glutamate can be decarboxylated by Gad1p to re-form GABA, which can either be stored in the vacuole when NADH is abundant, or used for further transamination. SSA can be oxidized to succinate, producing cytosolic NADPH, and so NADH-producing reactions in the mitochondria are bypassed in favour of NADPH-producing reactions in the cytosol. Hence, the actions of the GABA shunt utilize normal L-glutamate and

GABA catabolic processes to maintain redox status and provide N storage and utilization needs during the oxidative stress response.

During the heat stress response, Cao *et al.* (2013a) have shown that the *S. cerevisiae* deletion mutants *Δgad*1, *Δuga*1 and *Δuga*2 have increased sensitivity to heat stress, and produce elevated levels of ROS in comparison with their respective wild-type strains. These studies confirm that for the heat stress response, as for the oxidative stress response, carbon flow through the GABA shunt is necessary for heat stress tolerance. Clearly, an oxidative stress response is a component of the heat stress response. The relationship between the heat stress and oxidative stress responses has also been intimated because the regulation of both *SOD*1 and *SOD*2 is altered so that in the *Δuga*3 mutants, SOD expression is constitutive even in the absence of heat stress, while in wild types, *UGA*3 expression of both SOD genes is upregulated during heat stress (Cao, 2008).

3.6 Conclusions

In most microorganisms, the GABA shunt plays a role in the catabolism of L-glutamate and GABA as growth-supporting nitrogen sources. *S. cerevisiae* contains three enzymes (Gad1p, Uga1p, and Uga2p) and a vacuolar GABA transporter (Uga4p) that are responsible for the storage of GABA and the catabolic utilization of both glutamate and GABA as N sources. The carbon skeletons of both glutamate and GABA are converted into succinate, which is catabolized in the TCA cycle in mitochondria.

Unlike most other eukaryotes, the yeast GABA shunt consists of a single glutamate decarboxylase gene (*GAD*1), a single 4-aminobutyrate aminotransferase gene (*UGA*1), and a single succinate semialdehyde dehydrogenase gene (*UGA*2). Most other eukaryotes have multiple copies of one or more of these enzymes and these multiple forms have different properties suited to their multicellular, differentiated nature.

The *Saccharomyces GAD*1 gene codes for the single protein that contains a regulatory carboxy-terminal peptide that binds the calcium sensor, calmodulin (Coleman *et al.*, 2001). Carboxy-terminal regulatory sequences can be bioinformatically identified in most, if not all, eukaryotic glutamate decarboxylase genes, but they appear not to be found in prokaryotic genes. However, control of the carbon flow through the GABA shunt appears to involve the post-translational regulation of GAD in virtually all domains of cellular life forms. In prokaryotes, fungi and plants in particular, GADs have low pH optima, and intracellular pH has a role in this post-translational regulation as a means of remediating acid-related stress, while in fungi, metazoa and plants, the *C*-terminal regulatory sequence can also activate a relatively constitutively produced, but inactive GAD at normal, neutral intracellular pHs. This lack of involvement of GADs in pH control is most likely because eukaryotic cells have multiple mechanisms of pH regulation compared with prokaryotes in general. It appears that in the fungi, the utilization of GAD in pH regulation evolved to more complex GAD regulatory sequences involving the *C*-terminal amino acid sequences.

In the Viridiplantae, *C*-terminal calmodulin regulatory genes like the *S. cerevisiae* gene are common, while in the Metazoa, particularly mammals, the GAD sequences appear to have a *C*-terminal regulatory sequence, though this sequence does not necessarily bind to calmodulin (Arafat *et al.*, 2009; Kass *et al.*, 2014). In the fungi, there are a number of species, particularly in the Saccharomycetales, with a single GAD gene, but there are also a number of species containing two GAD genes, and a handful with multiple GAD genes. This pattern of GAD gene regulatory sequences suggests that the multiple GAD genes with differentiated functions arose in the fungi. Subsequently, the use of these multiple GAD genes evolved into specialized functions for this enzyme as the GABA shunt pathway took on multiple different eukaryotic functions.

Saccharomyces apparently retained a simpler role for GAD in cellular regulation similar to that in prokaryotes, but GAD regulatory sophistication that became the hallmark of multicellular eukaryotes is also demonstrated in the yeasts. In the genus *Saccharomyces*, GAD evolved a function as a regulatory gene mediating the general stress response via the oxidative stress response, and this regulatory behaviour is also demonstrated in multicellular eukaryotes, particularly in the Viridiplantae.

In *S. cerevisiae*, a single cytosolic 4-aminobutyrate aminotransferase (GABAT) gene, *UGA*1, is combinatorially regulated along with succinate semialdehyde dehydrogenase, Uga2p, and vacuolar GABA-permease (Uga4p) in the Uga regulon. The expression of Uga1p in yeast and in other microorganisms is upregulated under any condition that favours GABA synthesis and accumulation. This appears to be mediated primarily by Uga3p, and the balance of GABA synthesis and degradation by this regulon leads to the accumulation of GABA resulting from the GABA shunt activation.

In many bacteria, as exemplified by *E. coli*, there are at least two GABAT genes. One gene (*gab*T) appears to be part of the *gab*DTPC operon that is involved in GABA utilization as an N source (Metzer *et al.*, 1979; Niegemann *et al.*, 1993; Schneider *et al.*, 2002), while the other *puu*E appears to be involved in a putrescine catabolic operon (Kurihara *et al.*, 2005). Thus, in prokaryotes it appears that GABAT is regulated by the metabolic need to catabolize GABA using an appropriate operon associated with a specific catabolic pathway. As genes can be individually and coordinately regulated in eukaryotes, such as yeast and multicellular eukaryotes, the ability to regulate the Uga regulon and the GABA shunt pathway other than by nitrogen metabolite catabolism has allowed evolution of the functions that can be assigned to this pathway in yeast and multicellular eukaryotes.

In yeast, a role for the GABA shunt involving redox regulation (Coleman *et al.*, 2001), particularly during episodic environmental stresses such as heat stress that are known to produce oxidative stress responses (Cao *et al.*, 2013a), has evolved. The GABA shunt sits at the nexus of carbohydrate and amino acid metabolism, circumvents reactions producing NADH in mitochondrial metabolism leading to ROS production, and promotes NADPH production, which is critical to ROS remediation in the cytosol and mitochondria. This role is consistent with our understanding of the oxidative stress response in yeast, but it is equally consistent with the roles of the GABA shunt in more complex eukaryotes. It is perhaps easier to conduct genetic analysis of these complex phenomena where single genes are involved in the GABA shunt pathway, and so these studies provide valuable information on the function of the GABA shunt pathway in all organisms.

Acknowledgements

The author is grateful to colleagues and students Narendra K. Singh, Juxiang Cao and Jose Barbosa for their generous collaboration on work on the GABA shunt and heat stress tolerance, and especially acknowledges the late Professor Joe H. Cherry, the author's mentor, both for introducing him to GABA metabolism and for his support and encouragement throughout the author's career, which made this contribution possible.

References

Akama, K., Akihiro, T., Kitagawa, M. and Takaiwa, F. (2001) Rice (*Oryza sativa*) contains a novel isoform of glutamate decarboxylase that lacks an authentic calmodulin-binding domain at the C-terminus. *Biochimica et Biophysica Acta* 1522, 143–150.

Akihiro, T., Koike, S., Tani, R., Tominaga, T., Watanabe, S., Iijima, Y., Aoki, K., Shibata, D., Ashihara, H., Matsukura, C. *et al.* (2008) Biochemical mechanism on GABA accumulation during fruit development in tomato. *Plant and Cell Physiology* 49, 1378–1389.

Al-Quraan, N.A., Sartawe, F.A. and Qaryouti, M.M. (2013) Characterization of γ-aminobutyric acid metabolism and oxidative damage in wheat (*Triticum aestivum* L.) seedlings under salt and osmotic stress. *Journal of Plant Physiol*ogy 170, 1003–1009.

Anderlund, M., Nissen, T.L., Nielsen, J., Villadsen, J., Rydstrom, J., Hahn-Hagerdal, B. and Kielland-Brandt, M.C. (1999) Expression of the *Escherichia coli pntA* and *pntB* genes, encoding nicotinamide nucleotide transhydrogenase, in *Saccharomyces cerevisiae* and its effect on product formation during anaerobic glucose fermentation. *Applied and Environmental Microbiol*ogy 65, 2333–2340.

Andersen, G., Andersen, B., Dobritzsch, D., Schnackerz, K.D. and Piskur, J. (2007) A gene duplication led to specialized gamma-aminobutyrate and beta-alanine aminotransferase in yeast. *The FEBS Journal* 274, 1804–1817.

Arafat, Y., Fenalti, G., Whisstock, J.C., Mackay, I.R., Garcia de la Banda, M., Rowley, M.J. and Buckle, A.M. (2009) Structural determinants of GAD antigenicity. *Molecular Immunol*ogy 47, 493–505.

Aranda, A. and del Olmo, M.M. (2003) Response to acetaldehyde stress in the yeast *Saccharomyces cerevisiae* involves a strain-dependent regulation of several *ALD* genes and is mediated by the general stress response pathway. *Yeast* 20, 747–759.

Avery, A.M. and Avery, S.V. (2001) *Saccharomyces cerevisiae* expresses three phospholipid hydroperoxide glutathione peroxidases. *The Journal of Biological Chemistry* 276, 33730–33735.

Avery, A.M., Willetts, S.A. and Avery, S.V. (2004) Genetic dissection of the phospholipid hydroperoxidase activity of yeast Gpx3 reveals its functional importance. *The Journal of Biological Chemistry* 279, 46652–46652.

Baum, G., Chen, Y., Arazi, T., Takatsuji, H. and Fromm, H. (1993) A plant glutamate decarboxylase containing a calmodulin binding domain. Cloning, sequence, and functional analysis. *The Journal of Biological Chemistry* 268, 19610–19617.

Bermingham-McDonogh, O., Gralla, E.B. and Valentine, J.S. (1988) The copper, zinc-superoxide dismutase gene of *Saccharomyces cerevisiae*: cloning, sequencing, and biological activity. *Proceedings of the National Academy of Sciences of the United States of America* 85, 4789–4793.

Blaner, W.S. and Churchich, J.E. (1980) The binding of NADH to succinic semialdehyde dehydrogenase. *European Journal of Biochem*istry 109, 431–437.

Blasi, P., Palmerio, F., Aiello, A., Rocchi, M., Malaspina, P. and Novelletto, A. (2006) SSADH variation in primates: intra- and interspecific data on a gene with a potential role in human cognitive functions. *Journal of Molecular Evol*ution 63, 54–68.

Boubekeur, S., Bunoust, O., Camougrand, N., Castroviejo, M., Rigoulet, M. and Guerin, B. (1999) A mitochondrial pyruvate dehydrogenase bypass in the yeast *Saccharomyces cerevisiae*. *The Journal of Biological Chemistry* 274, 21044–21048.

Boubekeur, S., Camougrand, N., Bunoust, O., Rigoulet, M. and Guerin, B. (2001) Participation of acetaldehyde dehydrogenases in ethanol and pyruvate metabolism of the yeast *Saccharomyces cerevisiae*. *European Journal of Biochem*istry 268, 5057–5065.

Bouché, N., Fait, A., Bouchez, D., Møller, S.G. and Fromm, H. (2003) Mitochondrial succinic-semialdehyde dehydrogenase of the gamma-aminobutyrate shunt is required to restrict levels of reactive oxygen intermediates in plants. *Proceedings of the National Academy of Sciences of the United States of America* 100, 6843–6848.

Bubnová, M., Zemančíková, J. and Sychrová, H. (2014) Osmotolerant yeast species differ in basic physiological parameters and in tolerance of non-osmotic stresses. *Yeast* 31, 309–321.

Busch, K.B. and Fromm, H. (1999) Plant succinic semialdehyde dehydrogenase. Cloning, purification, localization in mitochondria, and regulation by adenine nucleotides. *Plant Physiol*ogy 121, 589–597.

Cao, J. (2008) Functional genomics of GABA metabolism in yeast thermotolerance. PhD Dissertation, Auburn University, Alabama.

Cao J., Barbosa, J.M., Singh, N. and Locy, R.D. (2013a) GABA shunt mediates thermotolerance in *Saccharomyces cerevisiae* by reducing reactive oxygen production. *Yeast* 30, 129–144.

Cao J., Barbosa, J.M., Singh, N. and Locy, R.D. (2013b) GABA transaminases from *Saccharomyces cerevisiae* and *Arabidopsis thaliana* complement function in cytosol and mitochondria. *Yeast* 30, 279–289.

Cao, J., Singh, N.K. and Locy, R.D. (2014) Characterization of the recombinant succinic semi-aldehyde dehydrogenase from *Saccharomyces cerevisiae*. *Yeast* 31, 411–420.

Castanie-Cornet, M.P., Penfound, T.A., Smith, D., Elliott, J.F. and Foster, J.W. (1999) Control of acid resistance in *Escherichia coli*. *Journal of Bacteriol*ogy 181, 3525–3535.

Chen, Y., Baum, G. and Fromm, H. (1994) The 58-kilodalton calmodulin-binding glutamate decarboxylase is a ubiquitous protein in petunia organs and its expression is developmentally regulated. *Plant Physiology* 106, 1381–1387.

Cherry, J.M., Hong, E.L., Amundsen, C., Balakrishnan, R., Binkley, G., Chan, E.T., Christie K.R., Costanzo, M.C., Dwight, S.S., Engel, S.R. *et al.* (2012) Saccharomyces Genome Database: the genomics resource of budding yeast. *Nucleic Acids Research* 40, D700–D705.

Christensen, R.L. and Schmit, J.C. (1980) Regulation and glutamic acid decarboxylase during *Neurospora crassa* conidial germination. *Journal of Bacteriol*ogy 144, 983–990.

Clark, S.M., Di Leo, R., Dhanoa, P.K., Van Cauwenberghe, O.R., Mullen, R.T. and Shelp, B.J. (2009a) Biochemical characterization, mitochondrial localization, expression, and potential functions for an *Arabidopsis* gamma-aminobutyrate transaminase that utilizes both pyruvate and glyoxylate. *Journal of Experimental Botany* 60, 1743–1757.

Clark, S.M., Di Leo, R., Van Cauwenberghe, O.R., Mullen, R.T. and Shelp, B.J. (2009b) Subcellular localization and expression of multiple tomato gamma-aminobutyrate transaminases that utilize both pyruvate and glyoxylate. *Journal of Experimental Botany* 60, 3255–3267.

Coleman, S.T., Fang, T.K., Rovinsky, S.A., Turano, F.J. and Moye-Rowley, W.S. (2001) Expression of a glutamate decarboxylase homologue is required for normal oxidative stress tolerance in *Saccharomyces cerevisiae*. *The Journal of Biological Chemistry* 276, 244–250.

Cooper, A.J.L. (1985) Glutamate-gamma-aminobutyrate transaminase. *Methods in Enzymology* 113, 80–82.

Davidson, J.F. and Schiest, R.H. (2001) Mitochondrial respiratory electron carriers are involved in oxidative stress during heat stress in *Saccharomyces cerevisiae*. *Molecular and Cellular Biology* 21, 8483–8489.

Davies, D.D. (1980) Anaerobic metabolism and the production of organic acids. In: Stumpf, P.K. and Conn, E.E. (eds) *The Biochemistry of Plants: A Comprehensive Treatise, Vol 2*. Academic Press, New York, pp 581–607.

De Biase, D. and Pennacchietti, E. (2012) Glutamate decarboxylase-dependent acid resistance in orally acquired bacteria: function, distribution and biomedical implications of the gadBC operon *Molecular Microbiology* 86, 770–786.

De Biase, D., Tramonti, A., John, R.A. and Bossa, F. (1996) Isolation, overexpression, and biochemical characterization of the two isoforms of glutamic acid decarboxylase from *Escherichia coli*. *Protein Expression and Purification* 8, 430–438.

Delaunay, A., Pflieger, D., Barrault, M.B., Vinh, J. and Toledano, M.B. (2002) A thiol peroxidase is an H_2O_2 receptor and redox-transducer in gene activation. *Cell* 111, 471–481.

Denner, L.A., Wei, S.C., Lin, H.S., Lin, C.T. and Wu, J.Y. (1987) Brain L-glutamate decarboxylase: purification and subunit structure. *Proceedings of the National Academy of Sciences of the United States of America* 84, 668–672.

Donnelly, M.I. and Cooper, R.A. (1981) Two succinic semialdehyde dehydrogenases are induced when *Escherichia coli* K-12 is grown on gamma-aminobutyrate. *Journal of Bacteriology* 145, 1425–1427.

Erdö, S.L., Kiss, B., Riesz, M. and Szporny, L. (1986) Stimulus-evoked efflux of GABA from preloaded slices of the rabbit oviduct. *European Journal of Pharmacology* 130, 295–303.

Erlander, M.G., Tillakaratne, N.J., Feldblum, S., Patel, N. and Tobin, A.J. (1991) Two genes encode distinct glutamate decarboxylases. *Neuron* 7, 91–100.

Feehily, C. and Karatzas, K.A. (2013) Role of glutamate metabolism in bacterial responses towards acid and other stresses. *Journal of Applied Microbiology* 114, 11–24.

Fiorina, P. (2013) GABAergic systems in β-cells: from autoimmunity target to regeneration tool. *Diabetes* 62, 3674–3676.

Gasch, A.P., Spellman, P.T., Kao, C.M., Carmel-Harel, O., Eisen, M.B., Storz, G., Botstein, D. and Brown, P.O. (2000) Genomic expression programs in the response of yeast cells to environmental changes. *Molecular Biology of the Cell* 11, 4241–4257.

Goffeau, A., Barrell, B.G., Bussey, H., Davis, R.W., Dujon, B., Feldmann, H., Galibert, F., Hoheisel, J.D., Jacq, C., Johnston, M. *et al.* (1996) Life with 6000 genes. *Science* 274, 563–567.

Grant, C.M. (2001) Role of the glutathione/glutaredoxin and thioredoxin systems in yeast growth and response to stress conditions. *Molecular Microbiology* 39, 533–541.

Grant, C.M. (2008) Metabolic reconfiguration is a regulated response to oxidative stress. *Journal of Biology* 7(1): 1.

Guengerich, F.P. and Broquist, H.P. (1976) Lysine catabolism in *Rhizoctonia leguminicola* and related fungi. *Journal of Bacteriology* 126, 338–347.

Gut, H., Dominici, P., Pilati, S., Astegno, A., Petoukhov, M.V., Svergun, D.I., Gruetter, M.G. and Capitani, G. (2009) A common structural basis for pH- and calmodulin-mediated regulation in plant glutamate decarboxylase *Journal of Molecular Biology* 392, 334–351.

Hao, R. and Schmit, J.C. (1991) Purification and characterization of glutamate decarboxylase from *Neurospora crassa* conidia. *The Journal of Biological Chemistry* 266, 5135–5139.

Howlett, N.G. and Avery, S.V. (1997) Induction of lipid peroxidation during heavy metal stress in *Saccharomyces cerevisiae* and influence of plasma membrane fatty acid unsaturation. *Applied and Environmental Microbiology* 63, 2971–2976.

Idicula, A.M., Blatch, G.L., Cooper, T.G. and Dorrington, R.A. (2002) Binding and activation by the zinc cluster transcription factors of *Saccharomyces cerevisiae*. Redefining the UASGABA and its interaction with Uga3p. *The Journal of Biological Chemistry* 277, 45977–45983.

Ilg, T., Berger, M., Noack, S., Rohwer, A. and Gassel, M. (2013) Glutamate decarboxylase of the parasitic arthropods *Ctenocephalides felis* and *Rhipicephalus microplus*: gene identification, cloning, expression, assay development, identification of inhibitors by high throughput screening and comparison with the orthologs. *Insect Biochemistry and Molecular Biology* 43, 162–177.

Inoue, Y., Matsuda, T., Sugiyama, K., Izawa, S. and Kimura, A. (1999) Genetic analysis of glutathione peroxidase in oxidative stress response of *Saccharomyces cerevisiae*. *The Journal of Biological Chemistry* 274, 27002–27009.

Izawa, S., Inoue, Y. and Kimura, A. (1996) Importance of catalase in the adaptive response to hydrogen peroxide: analysis of acatalasaemic *Saccharomyces cerevisiae*. *Biochemical Journal* 320, 61–67.

Izawa, S., Maeda, K., Miki, T., Mano, J., Inoue, Y. and Kimura, A. (1998) Importance of glucose-6-phosphate dehydrogenase in the adaptive response to hydrogen peroxide in *Saccharomyces cerevisiae*. *Biochemical Journal* 330, 811–817.

Izawa, S., Maeda, K., Sugiyama, K.-I., Mano, J., Inoue, Y. and Kimura, A. (1999) Thioredoxin deficiency causes the constitutive activation of Yap1, an AP-1-like transcription factor in *Saccharomyces cerevisiae*. *The Journal of Biological Chemistry* 274, 28459–28465.

Jang, E.H., Park, S.A., Chi, Y.M. and Lee, K.S. (2015) Structural insight into the substrate inhibition mechanism of NADP(+)-dependent succinic semialdehyde dehydrogenase from *Streptococcus pyogenes*. *Biochemical and Biophysical Research Communications* 461, 487–493.

Jarori GK and Maitra PK (1991) Nature of primary product(s) of D-glucose 6-phosphate dehydrogenase reaction. 13C and 31P NMR study. *FEBS Letters* 278, 247–251.

Jeong, H.K., Yong, B.P., Tae, L.H., Won, H.L., Myung, S.C. and Oh, S.K. (2005) High-level expression and characterization of the recombinant enzyme and tissue distribution of human succinic semialdehyde dehydrogenase. *Protein Expression and Purification* 44, 16–22.

Juhnke, H., Krems, B., Kotter, P. and Entian, K.D. (1996) Mutants that show increased sensitivity to hydrogen peroxide reveal an important role for the pentose phosphate pathway in protection of yeast against oxidative stress. *Molecular and General Genetics* 252, 456–464.

Karatzas, K.A., Suur, L. and O'Byrne, C.P. (2012) Characterization of the intracellular glutamate decarboxylase system: analysis of its function, transcription, and role in the acid resistance of various strains of *Listeria monocytogenes*. *Applied and Environmental Microbiology* 78, 3571–3579.

Kass, I., Hoke, D.E., Costa, M.G., Reboul, C.F., Porebski, B.T., Cowieson, N.P., Leh, H., Pennacchietti, E., McCoey, J., Kleifeld, O. *et al.* (2014) Cofactor-dependent conformational heterogeneity of GAD65 and its role in autoimmunity and neurotransmitter homeostasis. *Proceedings of the National Academy of Sciences of the United States of America* 111, E2524–E2529.

Kim, K.J., Pearl, P.L., Jensen, K., Snead, O.C., Malaspina, P., Jakobs, C. and Gibson, K.M. (2011) Succinic semialdehyde dehydrogenase: biochemical-molecular-clinical disease mechanisms, redox regulation, and functional significance. *Antioxidants and Redox Signaling* 15, 691–718.

Kojer, K., Bien, M., Gangel, H., Morgan, B., Dick, T.P. and Riemer, J. (2012) Glutathione redox potential in the mitochondrial intermembrane space is linked to the cytosol and impacts the Mia40 redox state. *The EMBO Journal* 31, 3169–3182.

Kumar, S., Kumar, S. and Punekar, N.S. (2015) Characterization of succinic semialdehyde dehydrogenase from *Aspergillus niger*. *Indian Journal of Experimental Biology* 53, 67–74.

Kurihara, S., Oda, S., Kato, K., Kim, H.G., Koyanagi, T., Kumagai, H. and Suzuki, H. (2005) A novel putrescine utilization pathway involves gamma-glutamylated intermediates of *Escherichia coli* K-12. *The Journal of Biological Chemistry* 280, 4602–4608.

Langendorf, C.G., Key, T.L., Fenalti, G., Kan, W.T., Buckle, A.M., Caradoc-Davies, T., Tuck, K.L., Law, R.H. and Whisstock, J.C. (2010) The X-ray crystal structure of *Escherichia coli* succinic semialdehyde dehydrogenase; structural insights into NADP+/enzyme interactions. *PLoS One* 5(2): e9280

Langendorf, C.G., Tuck, K.L., Key, T.L., Fenalti, G., Pike, R.N., Rosado, C.J., Wong, A.S., Buckle, A.M., Law, R.H. and Whisstock, J.C. (2013) Structural characterization of the mechanism through which human glutamic acid decarboxylase auto-activates. *Bioscience Reports* 33, 137–144.

Lee, J., Godon, C., Lagniel, G., Spector, D., Garin, J., Labarre, J. and Toledano, M.B. (1999) Yap1 and Skn7 control two specialized oxidative stress response regulons in yeast. *The Journal of Biological Chemistry* 274, 16040–16046.

Leone, O., Blasi, P., Palmerio, F., Kozlov, A.I., Malaspina, P. and Novelletto, A. (2007) A human derived SSADH coding variant is replacing the ancestral allele shared with primates. *Annals of Human Biology* 33, 593–603.

Llobell, A., Lopez-Ruiz, A., Peinado, J. and Lopez-Barea, J. (1988) Glutathione reductase directly mediates the stimulation of yeast glucose-6-phosphate dehydrogenase by GSSG. *Biochemical Journal* 249, 293–296.

Locy, R.D., Wu, S.J., Barger, T.W., Barbosa, J.M., McNabb, D., Bisnette, J., Singh, N.K. and Cherry, J.H. (2000) The accumulation of 4-aminobutyric acid (GABA) during heat stress in *Arabidopsis thaliana*. In: J.H. Cherry (ed.) *NATO Advanced Research Workshop Series in Cell Biology*, Kluwer, New York, pp 131–137.

Ludewig, F., Hüser, A., Fromm, H., Beauclair, L. and Bouché, N. (2008) Mutants of GABA transaminase (POP2) suppress the severe phenotype of succinic *semialdehyde dehydrogenase* (*ssadh*) mutants in *Arabidopsis*. *PLoS One* 3(10): e3383.

Lutke-Eversloh, T. and Steinbuchel, A. (1999) Biochemical and molecular characterization of a succinate semialdehyde dehydrogenase involved in the catabolism of 4-hydroxybutyric acid in *Ralstonia eutropha*. *FEMS Microbiology Letters* 181, 63–71.

Ma, L.H., Takanishi, C.L. and Wood, M.J. (2007) Molecular mechanism of oxidative stress perception by the orp1 protein. *The Journal of Biological Chemistry* 282, 31429–31436.

Mackay, I.R., Christ, D., Perahia, D., Buckle, M., Paiardini, A., De Biase, D. and Buckle, A.M. (2014) Cofactor-dependent conformational heterogeneity of GAD65 and its role in autoimmunity and neurotransmitter homeostasis. *Proceedings of the National Academy of Sciences of the United States of America* 111, E2524–E2429.

Maclaren, N. (1992) Immunology of diabetes mellitus. *Annals of Allergy* 68, 5–9.

Marchler-Bauer, A., Derbyshire, M.K., Gonzales, N.R., Lu, S., Chitsaz, F., Geer, L.Y., Geer, R.C., He, J., Gwadz, M., Hurwitz, D.I. *et al.* (2014) CDD: NCBI's conserved domain database. *Nucleic Acids Research* 43, D222–D226.

Mehta, P.K., Hale, T.I. and Christen, P. (1993) Aminotransferases: demonstration of homology and division into evolutionary subgroups. *European Journal of Biochemistry* 214, 549–561.

Metzer, E., Levitz, R. and Halpern, Y.S. (1979) Isolation and properties of *Escherichia coli* K-12 mutants impaired in the utilization of gamma-aminobutyrate. *Journal of Bacteriology* 137, 1111–1118.

Michán, C., Martínez, J.L., Alvarez, M.C., Turk, M., Sychrova, H. and Ramos, J. (2013) Salt and oxidative stress tolerance in *Debaryomyces hansenii* and *Debaryomyces fabryi*. *FEMS Yeast Research* 13, 180–188.

Miller, S.M. and Magasanik, B. (1990) Role of NAD-linked glutamate dehydrogenase in nitrogen metabolism in *Saccharomyces cerevisiae*. *Journal of Bacteriology* 172, 4927–4935.

Minard, K.I. and McAlister-Henn, L. (2005) Sources of NADPH in yeast vary with carbon source. *The Journal of Biological Chemistry* 280, 39890–39896.

Minard, K.I., Jennings, G.T., Loftus, T.M., Xuan, D. and McAlister-Henn, L. (1998) Sources of NADPH and expression of mammalian NADP+-specific isocitrate dehydrogenases in *Saccharomyces cerevisiae*. *The Journal of Biological Chemistry* 273, 31486–31493.

Morano, K.A., Grant, C.M. and Moye-Rowley, W.S. (2012) The response to heat shock and oxidative stress in *Saccharomyces cerevisiae*. *Genetics* 190, 1157–1195.

Murphy, M.P. (2009) How mitochondria produce reactive oxygen species. *Biochemical Journal* 417, 1–13.

Niegemann, E., Schulz, A. and Bartsch, K. (1993) Molecular organization of the *Escherichia coli gab* cluster: nucleotide sequence of the structural genes *gabD* and *gabP* and expression of the GABA permease gene. *Archives of Microbiology* 160, 454–460.

Okazaki, S., Naganuma, A. and Kuge, S. (2005) Peroxiredoxin-mediated redox regulation of the nuclear localization of Yap1, a transcription factor in budding yeast. *Antioxidants and Redox Signaling* 7, 327–334.

Outten C.E. and Culotta, V.C. (2003) A novel NADH kinase is the mitochondrial source of NADPH in *Saccharomyces cerevisiae*. *The EMBO Journal* 22, 2015–2024.

Outten, C.E. and Culotta, V.C. (2004) Alternative start sites in the *Saccharomyces cerevisiae GLR1* gene are responsible for mitochondrial and cytosolic isoforms of glutathione reductase. *The Journal of Biological Chemistry* 279, 7785–7791.

Park, J. and Rhee, S. (2013) Structural basis for a cofactor-dependent oxidation protection and catalysis of cyanobacterial succinic semialdehyde dehydrogenase. *The Journal of Biological Chemistry* 288, 15760–15770.

Popa, C.V., Dumitru, I., Ruta, L.L., Danet, A.F. and Farcasanu, I.C. (2010) Exogenous oxidative stress induces Ca2+ release in the yeast *Saccharomyces cerevisiae*. *The FEBS Journal* 277, 4027–4038.

Ralser, M., Walmelink, M.M., Kowald, A., Gerisch, B., Heeren, G., Struys, E.A., Klipp, E., Jakobs, C., Breitenbach, M., Lehrach, H. and Krobitsch, S. (2007) Dynamic rerouting of the carbohydrate flux is key to counteracting oxidative stress. *Journal of Biology* 6(4): 10.

Ralser, M., Mirjam, M.C., Wamelink, M.M., Latkolik, S., Jansen, E.E.W., Lehrach, H. and Jakobs, C. (2009) Metabolic reconfiguration precedes transcriptional regulation in the antioxidant response. *Nature Biotechnology* 27, 604–605.

Ramos, F., El Guezzar, M., Grenson, M. and Wiame, J.M. (1985) Mutations affecting the enzymes involved in the utilization of 4-aminobutyric acid as nitrogen source by the yeast *Saccharomyces cerevisiae*. *European Journal of Biochemistry* 149, 401–404.

Raven, J.A. (1985) pH regulation in plants. *Science Progress* 69, 495–509.

Reetz, A., Solimena, M., Matteoli, M., Folli, F., Takei, K. and De Camilli, P. (1991) GABA and pancreatic beta-cells: colocalization of glutamic acid decarboxylase (GAD) and GABA with synaptic-like microvesicles suggests their role in GABA storage and secretion. *The EMBO Journal* 10, 1275–1284.

Renault, H., Roussel, V., El Amrani, A., Arzel, M., Renault, D. and Bouchereau, A. (2010) The *Arabidopsis* pop 2-1 mutant reveals the involvement of GABA transaminase in salt stress tolerance. *BMC Plant Biology* 10:20.

Rinnerthaler, M., Jarolim, S., Heeren, G., Palle, E., Perju, S., Klinger, H, Bogengruber, E., Madeo, F., Braun, R.J., Breitenbach-Koller, L. *et al.* (2006) MMI1 (YKL056c, TMA19), the yeast orthologue of the translationally controlled tumor protein (TCTP) has apoptotic functions and interacts with both microtubules and mitochondria. *Biochimica et Biophysica Acta* 1757, 631–638.

Rivett, A.J. and Tipton, K.F. (1981) Kinetics studies with rat-brain succinic semialdehyde dehydrogenase. *European Journal of Biochemistry* 117, 187–193.

Rothacker, B. and Ilg, T. (2008) Functional characterization of a *Drosophila melanogaster* succinic semialdehyde dehydrogenase and a non-specific aldehyde dehydrogenase. *Insect Biochemistry and Molecular Biology* 38, 354–366.

Ryzlak, M.T. and Pietruszko, R. (1988) Human brain high K_m aldehyde dehydrogenase: purification, characterization, and identification as NAD+-dependent succinic semialdehyde dehydrogenase. *Archives of Biochemistry and Biophysics* 266, 386–396.

Sanchez, M., Fernández, J., Martin, M., Gibello, A. and Garrido-Pertierra, A. (1989) Purification and properties of two succinic semialdehyde dehydrogenases from *Klebsiella pneumonia*. *Biochimica et Biophysica Acta* 990, 225–231.

Saint-Prix, F., Bönquist, L. and Dequin, S. (2004) Functional analysis of the *ALD* gene family of *Saccharomyces cerevisiae* during anaerobic growth on glucose: the NADP+-dependent Ald6p and Ald5p isoforms play a major role in acetate formation. *Microbiology* 150, 2209–22020.

Schneider, B.L., Ruback, S., Kiupakis, A.K., Kasbarian, H., Pybus, C. and Reitzer, L. (2002) The *Escherichia coli gabDTPC* operon: specific γ-aminobutyrate catabolism and nonspecific induction. *Journal of Bacteriology* 184, 6976–6986.

Shenton, D. and Grant, C.M. (2003) Protein *S*-thiolation targets glycolysis and protein synthesis in response to oxidative stress in the yeast *Saccharomyces cerevisiae*. *Biochemical Journal* 374, 513–519.

Shimajiri, Y., Ozaki, K., Kainou, K. and Akama, K., J. (2013) Differential subcellular localization, enzymatic properties and expression patterns of gamma-aminobutyric acid transaminases (GABA-Ts) in rice (*Oryza sativa*). *Plant Physiology* 170, 196–201.

Sinha, A. and Maitra, P.K. (1992) Induction of specific enzymes of the oxidative pentose phosphate pathway by glucono-delta-lactone in *Saccharomyces cerevisiae*. *Journal of General Microbiology* 138, 1865–1873.

Slekar, K.H., Kosman, D.J. and Culotta, V.C. (1996) The yeast copper/zinc superoxide dismutase and the pentose phosphate pathway play overlapping roles in oxidative stress protection. *The Journal of Biological Chemistry* 271, 28831–28836.

Strand, M.K., Stuart, G.R., Longley, M.J., Graziewicz, M.A., Dominick, O.C. and Copeland, W.C. (2003) *POS5* gene of *Saccharomyces cerevisiae* encodes a mitochondrial NADH kinase required for stability of mitochondrial DNA. *Eukaryotic Cell* 2, 809–820.

Strigacova, J., Chovanec, P., Liptaj, T., Hudecova, D., Tursky, T., Simkovic, M. and Varecka, L. (2001) Glutamate decarboxylase activity in *Trichoderma viride* conidia and developing mycelia. *Archives of Microbiology* 175, 32–40.

Talibi, D., Grenson, M. and Andre, B. (1995) *Cis*- and *trans*-acting elements determining induction of the genes of the gamma-aminobutyrate (GABA) utilization pathway in *Saccharomyces cerevisiae*. *Nucleic Acids Research* 23, 550–557.

Tatusova, T., Ciufo, S., Fedorov, B., O'Neill, K. and Tolstoy, I. (2014) RefSeq microbial genomes database: new representation and annotation strategy. *Nucleic Acids Research* 42, D553–D559.

Thorpe, G.W., Reodica, M., Davies, M.J., Heeren, G., Jarolim, S., Pillay, B., Breitenbach, M., Higgins, V.J. and Dawes, I.W. (2013) Superoxide radicals have a protective role during H_2O_2 stress. *Molecular Biology of the Cell* 24, 2876–2884.

Tsuchiya, K., Nishimura, K. and Iwahara, M. (2003) Purification and characterization of glutamate decarboxylase from *Aspergillus oryzae*. *Food Science and Technology Research* 9, 283–287.

Van Cauwenberghe, O.R. and Shelp, B.J. (1999) Biochemical characterization of partially purified GABA: pyruvate transaminase from *Nicotiana tabacum*. *Phytochemistry* 52, 575–581.

van Loon, A.P., Pesold-Hurt, B. and Schatz, G. (1986) A yeast mutant lacking mitochondrial manganese-superoxide dismutase is hypersensitive to oxygen. *Proceedings of the National Academy of Sciences of the United States of America* 83, 3820–3824.

Vissers, S., Andre, B., Muyldermans, F. and Grenson, M. (1989) Positive and negative regulatory elements control the expression of the *UGA4* gene coding for the inducible 4-aminobutyric-acid-specific permease in *Saccharomyces cerevisiae*. *European Journal of Biochemistry* 181, 357–361.

Vissers, S., Andre, B., Muyldermans, F. and Grenson, M. (1990) Induction of the 4-aminobutyrate and urea-catabolic pathways in *Saccharomyces cerevisiae*. *European Journal of Biochemistry* 187, 611–616.

Wang, Q., Xin, Y., Zhang, F., Feng, Z., Fu, J., Luo, L. and Yin, Z. (2011) Enhanced gamma-aminobutyric acid-forming activity of recombinant glutamate decarboxylase (gadA) from *Escherichia coli*. *World Journal of Microbiology and Biotechnology* 27, 693–700.

White, W.H., Gunyuzlu, P.L. and Toyn, J.H. (2001) *Saccharomyces cerevisiae* is capable of *de novo* pantothenic acid biosynthesis involving a novel pathway of beta-alanine production from spermine. *The Journal of Biological Chemistry* 276, 10794–10800.

Winter, W.E. (1996) Type I insulin-dependent diabetes mellitus: a model for autoimmune polygenic disorders. *Advances in Dental Research* 10, 81–88.

Wu, J.Y., Matsuda, T. and Roberts, E. (1973) Purification and characterization of glutamate decarboxylase from mouse brain. *The Journal of Biological Chemistry* 248, 3029–3034.

Yang, R., Feng, L., Wang, S., Yu, N. and Gu, Z. (2015) Accumulation of γ-aminobutyric acid in soybean by hypoxia germination and freeze–thawing incubation. *Journal of the Science of Food and Agriculture* 96, 2090–2096.

Yu, K., Lin, L., Hu, S., Huang, J. and Mei, L. (2012) *C*-terminal truncation of glutamate decarboxylase from *Lactobacillus brevis* CGMCC 1306 extends its activity toward near-neutral pH. *Enzyme and Microbial Technology* 50, 263–269.

Yuan, Z., Yin, B., Wei, D. and Yuan, Y.R. (2013) Structural basis for cofactor and substrate selection by cyanobacterium succinic semialdehyde dehydrogenase. *Structural Biology* 182, 125–135.

Yun, H., Lim, S., Cho, B.K. and Kim, B.G. (2004) Omega-amino acid:pyruvate transaminase from *Alcaligenes denitrificans* Y2k-2: a new catalyst for kinetic resolution of beta-amino acids and amines. *Applied and Environmental Microbiology* 70, 2529–2534.

Zik, M., Arazi, T., Snedden, W.A. and Fromm, H. (1998) Two isoforms of glutamate decarboxylase in *Arabidopsis* are regulated by calcium/calmodulin and differ in organ distribution. *Plant Molecular Biology* 37, 967–975.

4 Lysine Biosynthesis in Microorganisms

A.O. Hudson,[1]* M.A. Savka,[1] F.G. Pearce[2] and R.C.J. Dobson[2,3]
[1]*Thomas H. Gosnell School of Life Sciences, Rochester Institute of Technology, New York, USA;* [2]*Biomolecular Interaction Centre, School of Biological Sciences, University of Canterbury, Christchurch, New Zealand;* [3]*Department of Biochemistry and Molecular Biology, Bio21 Institute, University of Melbourne, Parkville, Australia*

4.1 Abstract

Lysine is an essential amino acid that is synthesized in plants and microorganisms via two metabolic routes which evolved independently. Fungi and some archaeal species employ the intermediate α-aminoadipate to anabolize lysine. Bacteria and photosynthetic cohorts, in contrast, employ the intermediate diaminopimelate. The biosynthesis of lysine is of continued interest, because animals do not possess the enzymatic machinery to mediate its synthesis. As such, lysine is deemed essential for humans, as it is one of the 20 ubiquitous proteogenic amino acids that must be acquired through dietary means to facilitate protein synthesis for growth and development. From an antimicrobial drug development perspective, the pathway is of interest given its involvement in the synthesis of the bacterial peptidoglycan cell wall. The penultimate compound, *meso*-diaminopimelate, serves as the cross-linking amino acid in the cell wall of most Gram-negative bacteria, while lysine serves the same purpose in the cell wall of most Gram-positive bacteria. The enzymes are therefore attractive targets for biocides (antibacterials, algicides and herbicides). The overarching goal of this chapter is to present an updated review of the various pathways for the synthesis of lysine. The α-aminoadipate pathway will only be briefly discussed, whereas an in-depth description of the four variants of the diaminopimelate pathways is included. In addition, structure–activity relationships and kinetic analyses of key steps involved in the pathways will also be featured.

4.2 Introduction

4.2.1 The amino acid lysine

Lysine contains a lysyl side chain, $(CH_2)_4NH_3^+$, which denotes the positive charge of the molecule at physiological pH, in addition to the α-amino and α-carboxylic acid groups. It is classified as an aliphatic amino acid, meaning that the carbon/hydrogen side chain is joined together in a straight chain, rather than being branched or aromatic (see Fig. 4.1). In nature, lysine is synthesized by only some organisms, which utilize either one of two pathways. The α-aminoadipic acid (AAA) pathway is predominantly employed by fungi, but is present in a few species belonging to the domain Archaea (Nishida *et al.*, 1999; Velasco *et al.*, 2002). The AAA intermediate is synthesized from homocitrate,

*Corresponding author. E-mail address: aohsbi@rit.edu

 The Handbook of Microbial Metabolism of Amino Acids (ed. J.P.F. D'Mello)

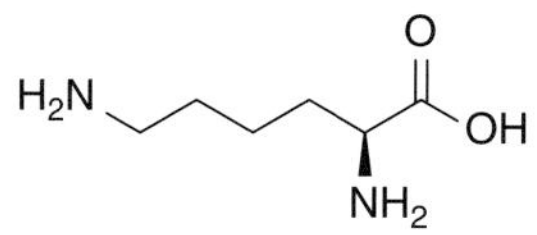

Fig. 4.1. The chemical structure of L-lysine.

which is a product from the citrate and glycolytic cycles. However, photosynthetic organisms and most bacteria employ the diaminopimelate (DAP) pathway. The diaminopimelate intermediate is derived from aspartate, an amino acid that is derived from α-ketoglutarate, which is also an intermediate from the citrate cycle (Fig. 4.2).

4.2.2 The aspartate-derived amino acids

In most organisms, the amino acids lysine, methionine, threonine and isoleucine are synthesized from aspartate. These four amino acids are commonly referred to as the aspartate-derived amino acids. The intrinsic regulation networks of the aspartate-derived amino acid pathways is of interest to the scientific community as two of the four amino acids that are anabolized from this pathway (lysine and threonine) are the most limiting essential amino acids in cereal grains (Galili, 1995). The regulation of the aspartate-derived amino acids proceeds through several negative feedback inhibition loops that control the expression of key enzymes involved in the anabolism of the these amino acids. This regulation is very interesting in photosynthetic organisms because it is intricately controlled through end product feedback regulation loops (Fig. 4.3). For example, aspartate kinase, the first enzyme of the pathway through which four of the 20 proteogenic amino acids are derived, is feedback inhibited by lysine and threonine (Galili, 1995). Lysine also inhibits the first enzyme that denotes the committed step in lysine biosynthesis, dihydrodipicolinate synthase, and threonine regulates the expression of homoserine dehydrogenase and aspartate kinase. Methionine regulates the enzyme cystathionine gamma synthase which catalyses the committed step in the synthesis of methionine (see Fig. 4.3) (Galili, 1995).

As a result of evolution, members of the animal kingdom have lost key enzymes for the synthesis of these amino acids. Thus, the amino acids that are derived from aspartate are deemed essential. Scientific interests in these pathways are twofold. First, by increasing the expression or changing the regulation of key enzymes in the pathway to manipulate the amount of available lysine, threonine and methionine, the nutritional value of cereal crops and other plants can be improved (Bright and Shewry, 1983). The development of antibiotics, algicides and herbicides is also of interest, because the enzymes involved in these pathways are putative and attractive targets for biocide development, given their absence in humans and all monogastric animals. For example, inhibition studies of dihydrodipicolinate synthase (Turner *et al.*, 2005; Boughton *et al.*, 2008a,b; Mitsakos *et al.*, 2008) and diaminopimelate transferase (McKinnie *et al.*, 2014; Triassi *et al.*, 2014) have been reported.

4.3 Lysine Biosynthesis in Microorganisms

Two divergent pathways have independently evolved for lysine biosynthesis: the DAP pathway, which is employed by plants and most bacteria, and the AAA acid pathway, which is employed by yeast, fungi and some Archaea (Nishida *et al.*, 1999; Velasco *et al.*, 2002). Four distinct variants of the DAP pathway have been identified to date: the succinylated pathway, the acetylated pathway, the dehydrogenase pathway and, very recently, the L,L-diaminopimelate

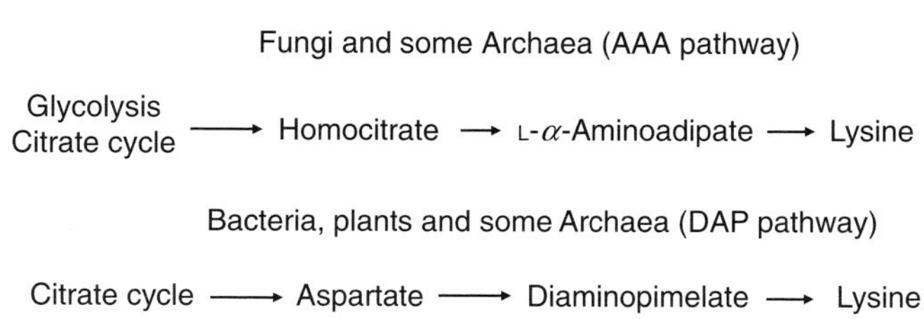

Fig. 4.2. The α-aminoadipate (AAA) and diaminopimelate (DAP) pathways for lysine biosynthesis.

aminotransferase pathway (Hudson *et al.*, 2006). The DAP pathway shows an evolutionary relationship with the arginine anabolic pathway, whereas the AAA pathway shares an evolutionarily origin with the leucine anabolic pathway (Velasco *et al.*, 2002).

4.3.1 The α-aminoadipate (AAA) pathway

The AAA pathway (found in yeasts, fungi and some Archaea) has eight enzymatic steps for lysine biosynthesis. The pathway proceeds through the citrate cycle metabolite α-ketoglutarate and the glycolytic metabolite acetyl-CoA (Fig. 4.4). The synthesis of lysine via the AAA intermediate is as follows (Fig. 4.4). The enzyme homocitrate synthase (EC 2.3.3.14) catalyses the synthesis of homocitrate using the substrates acetyl-CoA and α-ketoglutarate. Homocitrate is converted to *cis*-homocitrate by the hydrolyase homoaconitate hydratase (EC 4.2.1.36). The same enzyme catalyses the conversion *cis*-homocitrate to homoisocitrate. Homoisocitrate is then reduced by homoisocitrate dehydrogenase (EC 1.1.1.87) to synthesize α-ketoadipate. The amino acceptor α-ketoadipate is then used by the enzyme α-ketoadipate aminotransferase (EC 2.6.1.39) in a glutamate-dependent aminotransferase reaction

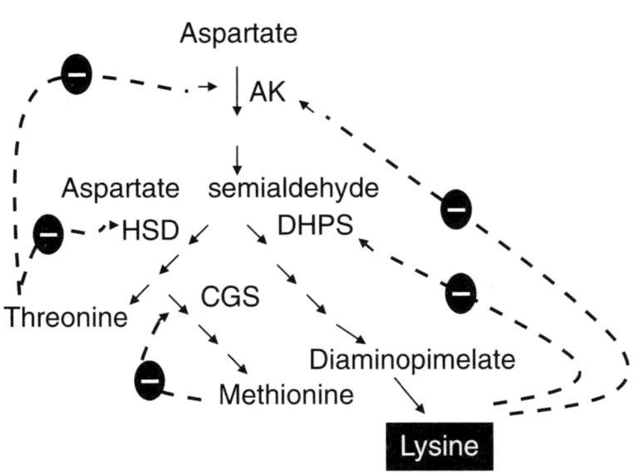

Fig. 4.3. End product feedback regulation of the aspartate-derived amino acids anabolic pathways. The abbreviations of the enzymes are as follows: aspartate kinase (AK), homoserine dehydrogenase (HSD), dihydrodipicolinate synthase (DHPS) and cystathionine gamma synthase (CGS). The black circles represent the negative feedback inhibition loops of the various pathways.

to synthesize α-aminoadipdate (Matsuda and Ogur, 1969). α-Aminoadipate is then reduced to synthesize α-aminoadipyl semialdehyde by the enzyme α-aminoadipate reductase (EC 1.2.1.31). The product of this reaction is further reduced by saccharopine reductase (EC 1.5.1.7) to synthesize saccharopine, the penultimate compound in the pathway, which is further reduced by the same enzyme to lysine.

4.4 Lysine Biosynthesis in Bacteria: the Diaminopimelate (DAP) Pathways

The DAP pathway, as already noted, is found in plants and most bacteria, and to date, four

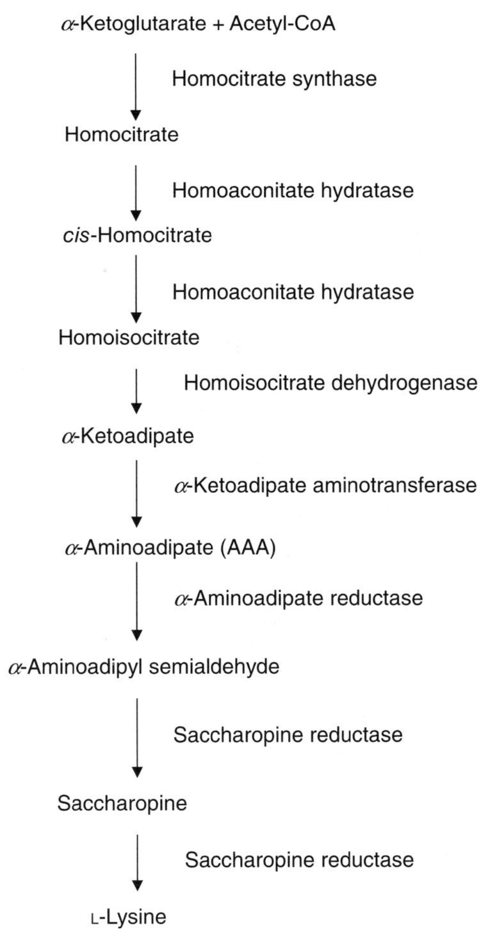

Fig. 4.4. Schematic representation of the α-aminoadipate (AAA) pathway.

variants of this lysine anabolic pathway have been elucidated and characterized. The acyl pathways, which are present in most bacterial species, utilize either succinylated (succinyl-CoA) or acetylated (acetyl-CoA) intermediates. The other pathways are the *meso*-diaminopimelate (*m*-DAP) dehydrogenase (Ddh) pathway, which was initially discovered in *Bacillus sphaericus*, *Corynebacterium glutamicum* and *Brevibacterium* sp. (Misono *et al.*, 1976, 1979, 1986; White, 1983), and the recently discovered L,L-diaminopimelate (L,L-DAP) aminotransferase (DapL) pathway (Hudson *et al.*, 2006, 2008; McCoy *et al.*, 2006).

The synthesis of lysine via the DAP pathways can be divided into three main stages. The first stage is characterized by the synthesis of tetrahydrodipicolinate (THDP) from aspartate (Asp), which is a general feature of all four variants (Fig. 4.5). This conversion is carried out in a series of reactions by the enzymes aspartate kinase (LysC; EC 2.7.2.4), aspartate semialdehyde dehydrogenase (Asd; EC 1.2.1.11), dihydrodipicolinate synthase (DHPS; EC 4.2.1.52) and dihydrodipicolinate reductase (DapB; EC 1.2.1.26). The second stage, which constitutes the conversion of tetrahydrodipicolinate acid to the penultimate intermediate *meso*-DAP, defines the uniqueness of the DAP variant pathways (Fig 4.6). The third and final stage, which is common to all DAP variant pathways, is the decarboxylation of *meso*-diaminopimelate to L-lysine, which is catalysed by *meso*-diaminopimelate decarboxylase (LysA; EC 4.1.1.20).

A general pattern for the regulation of the DAP biosynthetic pathway in bacteria is largely absent, although the repression of enzyme synthesis and enzyme activity has been observed in multiple systems (Rose, 1949; Yugari and Gilvarg, 1965; Neidhardt *et al.*, 1996; Grundy *et al.*, 2003). In *Escherichia coli*, the expression of genes encoding several lysine biosynthetic enzymes, including aspartate kinase, aspartate semialdehyde dehydrogenase (Neidhardt *et al.*, 1996) and dihydrodipicolinate reductase (Bouvier *et al.*, 1984), is regulated by lysine. This lysine-mediated gene regulation in bacteria has been proposed to operate via a unique RNA structural element, coined the 'LYS element' (Rodionov *et al.*, 2003) or 'L box' (Grundy *et al.*, 2003). In contrast, dihydrodipicolinate synthase is regulated at the enzyme activity level by lysine (Yugari and Gilvarg, 1965), which allosterically inhibits the enzyme. It has also been suggested that the gene encoding dihydrodipicolinate synthase is regulated by intracellular DAP levels, with expression from the *dap*A promoter increasing in response to DAP limitation (Acord and Masters, 2004).

4.4.1 The DAP acyl pathways

The DAP acyl pathways refer to diaminopimelate (DAP) intermediates that are either *N*-acetylated or *N*-succinylated, and use *N*-acetyl-CoA or *N*-succinyl-CoA intermediates. The Archaea and eubacteria employ *N*-succinylated intermediates, while the *N*-acetylated pathway is restricted to bacteria from the genus *Bacillus* (Gilvarg, 1959, 1961; Weinberger and Gilvarg, 1970; Bartlett and White, 1985; Velasco *et al.*, 2002).

From aspartate, the first step unique to lysine synthesis is the condensation of pyruvate and aspartate to form L-2,3-dihydrodipicolinate (DHDP) (Fig. 4.5) (Galili *et al.*, 1994), which is catalysed by the enzyme dihydrodipicolinate synthase. Dihydrodipicolinate synthase was initially purified to homogeneity and characterized from many bacterial species (Webster and Lechowich, 1970; Yamakura *et al.*, 1974;

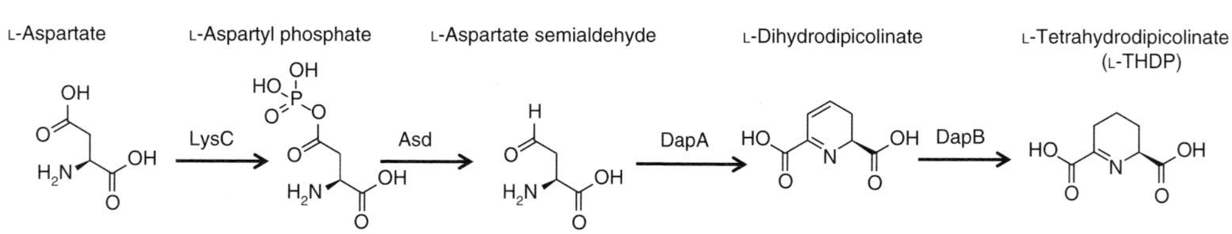

Fig. 4.5. Synthesis of tetrahydrodipicolinate (THDP) from aspartate. The symbols for the enzymes are as follows; aspartate kinase (LysC), aspartate semialdehyde dehydrogenase (Asd), dihydrodipicolinate synthase (DHPS, DapA), dihydrodipicolinate reductase (DapB).

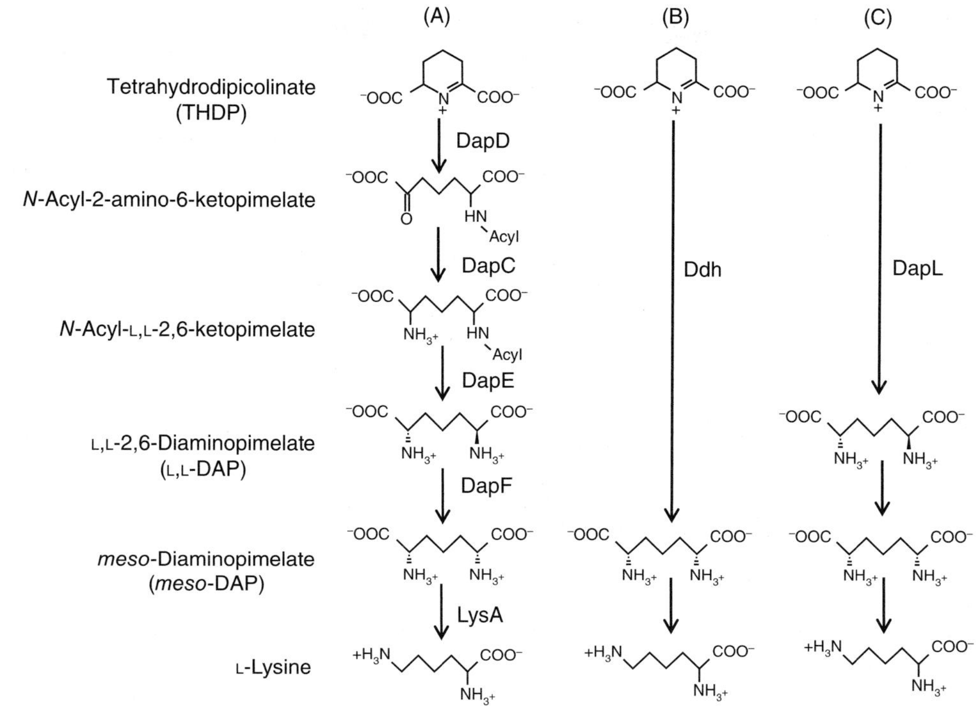

Fig. 4.6. Variants of the diaminopimelate (DAP) pathway for lysine biosynthesis from tetrahydrodipicolinate (THDP). (A) The acyl pathways; (B) the dehydrogenase pathway; and (C) the diaminopimelate aminotransferase pathway. Abbreviations: Dap C, acyl-diaminopimelate aminotransferase; DapE, acyl-diaminopimelate deacylase; DapD, 2,3,4,5-tetrahydropyridine-2,6-dicarboxylate *N*-acyl-transferase; DapF, acyl-diaminopimelate deacylase; DapL, L,L-diaminopimelate aminotransferase; Ddh, diaminopimelate dehydrogenase; LysA, diaminopimelate decarboxylase.

Halling *et al.*, 1976). In most cases, the enzyme is a homotetramer with a molecular weight between 28,000 and 35,000 Da. The apparent K_m for the *E. coli* enzyme is 0.57 mM for pyruvate and 0.55 mM for aspartate semialdehyde (ASA) (Bukhari and Taylor, 1971). This enzyme has been characterized in bacteria and it is the key regulatory enzyme in the pathway. It was first suggested that the product released by dihydrodipicolinate synthase was dihydrodipicolinate, but studies using ^{13}C-labelled pyruvate support the view that the product is the heterocyclic (4*S*)-4-hydroxy-2,3,4,5-tetrahydro-(2*S*)-dipicolinic acid (HTPA) (Blickling *et al.*, 1997a). This was confirmed more recently by the observation that the rapid decomposition of the ^{13}C-NMR signals of HTPA following its formation indicates that it is unstable; however, the formation of dihydrodipicolinate has not been detected (Devenish *et al.*, 2010), which suggests that the substrate of the next pathway enzyme, DapB, may not be dihydrodipicolinate.

In all cases examined, the dihydrodipicolinate synthase reaction proceeds via a ping-pong kinetic mechanism in which pyruvate binds the active site, followed by the release of a hydrogen ion. Aspartate β-semialdehyde then binds to the active site and the product tetrahydrodipicolinate is released (Blickling *et al.*, 1997a). For *E. coli* dihydrodipicolinate synthase, Schiff base formation is initiated by nucleophilic attack of the ε-amino group of Lys161 (Laber *et al.*, 1992) to the keto carbon of pyruvate, resulting in the formation of an imine. Formation of the Schiff base proceeds through a tetrahedral transition state. It is proposed that a catalytic triad of residues, Tyr133, Thr44 and Tyr107, act as a proton relay to transfer protons to and from the active site via a water-filled channel leading to bulk solvent (Dobson *et al.*, 2004a). The imine form of the Schiff base is converted to its enamine form for the addition of aspartate β-semialdehyde. The means of initiation of this tautomerization of the Schiff base is unclear. It has been demonstrated to proceed in the absence of aspartate β-semialdehyde, with the backbone oxygen of Ile203 believed to be important (Blickling *et al.*, 1997a; Dobson *et al.*, 2008). Aspartate β-semialdehyde adds with its aldehyde group to the enamine of the Schiff base, with the displacement of a water molecule. Isothermal titration microcalorimetry (ITC) studies show that aspartate β-semialdehyde does not interact with the enzyme in the absence of a Schiff base with pyruvate (Muscroft-Taylor *et al.*, 2010). Tetrahydrodipicolinate acid is formed by the nucleophilic attack of the amino group of aspartate β-semialdehyde to the Schiff base, leading to cyclization and the stepwise detachment of the product from the enzyme.

The structure of dihydrodipicolinate synthase from *E. coli* (Mirwaldt *et al.*, 1995; Dobson *et al.*, 2005) has been well characterized. It is a homotetramer of four identical $(\beta/\alpha)_8$-barrel monomers and can be described as a 'head-to-head' dimer of dimers (monomers A and B and monomers C and D; see Fig. 4.7) (Dobson *et al.*, 2005). The active site is located within the $(\beta/\alpha)_8$-barrel and allosteric site at the 'tight dimer interface' at the top and bottom of the tetramer (Dobson *et al.*, 2005) (Fig. 4.7). Each monomer comprises 292 amino acids and two domains (Mirwaldt *et al.*, 1995). The *N*-terminal domain is a $(\beta/\alpha)_8$-barrel (*E. coli* dihydrodipicolinate synthase: residues 1–224) with the active site within the centre of the barrel (Dobson *et al.*, 2005). The *C*-terminal domain (residues 225–292) consists of three α-helices that contain several of the key residues that form the 'weak dimer interface'. This interface has been shown to be important for both the activity and stability of the enzyme, with *E. coli* dihydrodipicolinate synthase truncation mutants showing significantly reduced solubility, stability and catalytic activity compared with the wild type enzyme (Guo *et al.*, 2009).

The dihydrodipicolinate synthase enzyme appears to be weakly allosterically regulated in Gram-negative bacteria, as it was shown that the bacterial enzyme is inhibited by high concentrations of lysine (Yugari and Gilvarg, 1965). In Gram-positive bacteria, however, the enzyme is only weakly inhibited by lysine. Interestingly, the opposite scenario has been demonstrated in photosynthetic cohorts as it has been shown that the enzyme is very sensitive to inhibition by low concentrations of lysine (Azevedo *et al.*, 1997).

The crystal structure of dihydrodipicolinate synthase from *E. coli* in the presence of lysine shows that two lysine molecules bind the lysine-binding site. ITC experiments (Phenix and Palmer, 2008; Muscroft-Taylor *et al.*, 2010) have shown that inhibition is cooperative and that the substrate has a substantial effect on the nature of enzyme–inhibitor association. The mechanism by which lysine exerts regulatory control over bacterial dihydrodipicolinate synthase is not well understood, although kinetic

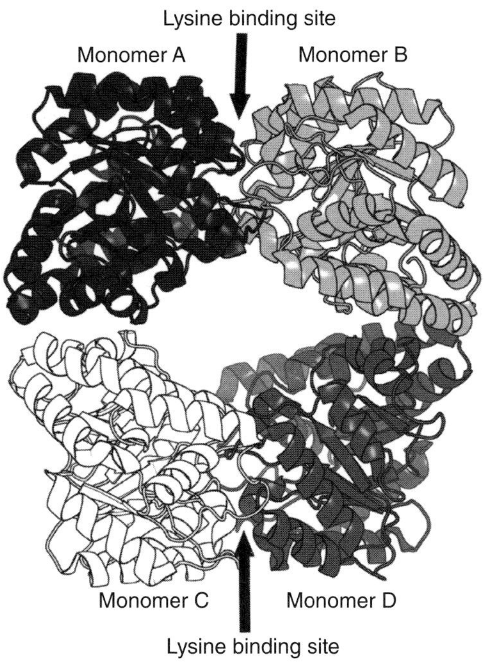

Fig. 4.7. The crystal structure of *Escherichia coli* Dihydrodipicolinate synthase (DHPS). Two monomers come together at the tight dimer interface (located at the allosteric lysine binding sites shown in the image) to form the dimeric unit, and then the two dimeric units dock at the weak dimer interface to form a tetramer. The active site is located within the monomeric TIM barrel conserved protein fold, whereas the allosteric binding sites are located at the tight dimer interface (PDB (Protein Data Bank) ID: 1YXC).

and structural studies suggest that it is an allosteric inhibitor, causing partial inhibition (90%) at saturating concentrations.

The mechanism of lysine inhibition of dihydrodipicolinate synthase from plants has been investigated to some extent in wheat (Kumpaisal *et al.*, 1987) and more recently in *Vitis vinifera* (grape) (Atkinson *et al.*, 2013). It was found that lysine is a non-competitive inhibitor relative to pyruvate and a competitive inhibitor relative to aspartate semialdehyde. Later kinetic studies also demonstrated that lysine is a partial inhibitor; the *E. coli* enzyme retains ~10% activity even at saturating lysine concentrations (Dobson *et al.*, 2004b). A structural study of dihydrodipicolinate synthase from the plant *Nicotiana sylvestris* (flowering tobacco) suggested a significant change in conformation upon lysine binding (Blickling *et al.*, 1997b), although structural studies of dihydrodipicolinate synthase from *Arabidopsis thaliana* (Griffin *et al.*, 2012) and *V. vinifera* (Atkinson *et al.*, 2013) did not indicate any large conformational changes. Several of the residues involved in contacts at the tight dimer interface are displaced when lysine is bound, dislocating the dimers in relation to each other. This translocation displaces the active site residues Arg138 and Tyr133, hindering coordination of the carboxyl group of aspartate semialdehyde during the cyclization of the substrate. Such an altered conformation was not observed in the *E. coli* dihydrodipicolinate synthase structure (Dobson *et al.*, 2005).

The next step in the lysine biosynthesis pathway is catalysed by the enzyme L-dihydrodipicolinate reductase (DapB; EC 1.31.26) (Fig. 4.5). This enzyme was initially thought to use dihydrodipicolinate as a substrate, but it has subsequently been suggested that dihydrodipicolinate reductase incorporates a dehydratase step, and that the true substrate is actually tetrahydrodipicolinate acid rather than dihydrodipicolinate (Devenish *et al.*, 2010). Dihydrodipicolinate reductase is somewhat unusual, in that while most pyridine-dependent dehydrogenases are specific for NADH or NADPH as a cofactor, some dihydrodipicolinate reductase enzymes utilize either nucleotide as an electron source, that is, they are dual specific. Dihydrodipicolinate reductase from both *E. coli* and *Mycobacterium tuberculosis* have similar specificity for both nucleotides, and mutations have been identified that alter the nucleotide specificity (Cirilli *et al.*, 2003; Reddy *et al.*, 1995, 1996). In contrast, dihydrodipicolinate reductase enzymes from *Thermotoga maritima*, *Staphylococcus aureus* and *A. thaliana* show a preference for NADPH, and often show inhibition by the tetrahydrodipicolinate acid substrate (Pearce *et al.*, 2008; Dommaraju *et al.*, 2011; Girish *et al.*, 2011; Griffin *et al.*, 2012).

Dihydrodipicolinate reductase was initially purified and characterized from *E. coli* (Farkas and Gilvarg, 1965) and from *Bacillus* spp. (Kimura and Goto, 1977). The *E. coli* enzyme is a homotetramer with a molecular weight of approximately 110,000 to 120,000 Da. Michaelis constants for the *E. coli* enzyme are 50 μM for L-2,3-dihydrodipicolinate and 8 μM for NADPH (Galili *et al.*, 1994). Another class of the dihydrodipicolinate reductase enzyme has been isolated from *B. subtilis* and is different from others in that it contains flavin mononucelotide (FMN). This enzyme reduces various compounds and may not be specific for lysine biosynthesis given its promiscuity in substrate specificity (Kimura and Goto, 1975).

Structures are now available for dihydrodipicolinate reductases from *E. coli* (Scapin *et al.*, 1995, 1997), *M. tuberculosis* (Cirilli *et al.*, 2003; Janowski *et al.*, 2010; Scapin *et al.*, 1998b), *T. maritima* (Pearce et *al.*, 2008) and *S. aureus* (Girish *et al.*, 2011). All of these enzymes have a homotetrameric structure comprising two domains, an *N*-terminal nucleotide-binding domain incorporating a Rossmann fold and a *C*-terminal tetrahydrodipicolinate acid binding domain (Fig. 4.8). During catalysis, it is thought that the *N*-terminal domain moves towards the *C*-terminal domain, allowing hydride transfer to occur (Scapin *et al.*, 1995, 1997; Wang *et al.*, 1997; Cirilli *et al.*, 2003; Janowski *et al.*, 2010; Girish *et al.*, 2011). The *C*-terminal domain also forms the interface between different subunits, with four *β*-sheets from each monomer coming together to form a 16 stranded *β*-barrel structure. While there are several structures available for bacterial dihydrodipicolinate reductases, little is known about the structure of the plant enzyme, which varies from the bacterial form in that while all bacterial enzymes characterized to date are tetrameric, the plant enzymes from maize (*Zea mays*) and *A. thaliana* have been shown to be dimeric (Tyagi *et al.*, 1983; Griffin *et al.*, 2012). As yet, the exact arrangement of

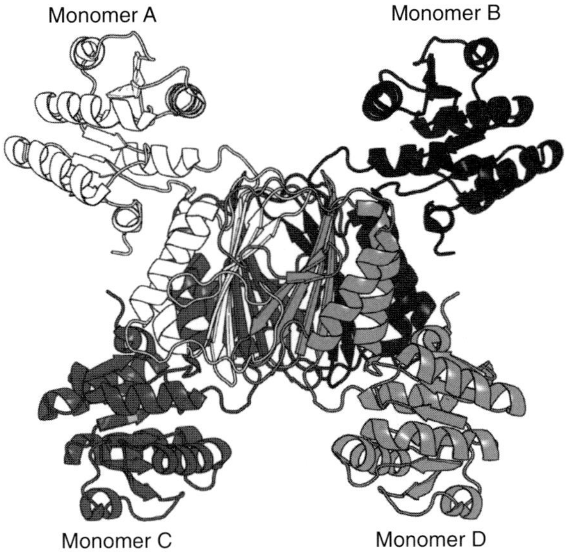

Fig. 4.8. The tetrameric structure of dihydrodipicolinate reductase (DapB) from *Mycobacterium tuberculosis* (PDB (Protein Data Bank) ID: 1YL6).

the plant dihydrodipicolinate reductase subunits is unknown.

The next step in the pathway is the addition of an acyl group to tetrahydrodipicolinate using succinyl-CoA or acetyl-CoA as substrates. The addition of the acyl group is catalysed by the enzyme L-tetrahydrodipicolinate *N*-acyl-transferase (DapD; EC 2.3.1.117) (Fig. 4.6A). This enzyme was initially purified and characterized from *E. coli* (Simms *et al.*, 1984), and structural analysis showed that it is a homodimer with a molecular weight of approximately 31,000 Da (Richaud *et al.*, 1984). The apparent K_m for tetrahydropicolinate is approximately 22 μM and for succinyl-CoA it is 15 μM (Simms *et al.*, 1984).

N-Acyl-diaminopimelate aminotransferase (DapC; EC 2.6.1.17), catalyses the next step in the acyl pathways. This enzyme transfers an amine group from L-glutamate to *N*-acyl-2-amino-6-ketopimelate (Fig. 4.5) and is involved in both lysine and arginine biosynthesis. The *N*-acyl-diaminopimelate aminotransferase enzyme was initially characterized from *E. coli* and its activity was subsequently demonstrated from a variety of Gram-positive and Gram-negative bacteria (Peterkofsky and Gilvarg, 1961). The enzyme requires pyridoxal 5′-phosphate (PLP) as a prosthetic group and L-glutamate as the amine donor, which is similar to the requirements of L,L-diaminopimelate aminotransferase (Hudson *et al.*, 2006). The apparent K_m for *N*-acyl-L,L-2,6-diaminopimelate is approximately 75 μM (Ledwidge and Blanchard, 1999) and L-glutamate the K_m is 1210 μM for (Cox, 1996). Like most aminotransferases, the *E. coli* enzyme is a homodimer and has a molecular weight of 39,900 Da/monomer (Peterkofsky and Gilvarg, 1961). The *dapC* gene has been cloned and sequenced as *argD*, which encodes acetyl-ornithine aminotransferase (ArgD), and is involved in arginine biosynthesis (Ledwidge and Blanchard, 1999). The acetyl-ornithine aminotransferase enzyme catalyses the fourth reaction in the biosynthesis of arginine. The enzyme transfers an amine group from L-glutamate to *N*-acetyl-ornithine to produce *N*-acetylglutamate semialdehyde (Ledwidge and Blanchard, 1999). The apparent K_m for *N*-acetyl-ornithine is approximately 150 μM (Ledwidge and Blanchard, 1999).

The product of the *N*-acyl-diaminopimelate aminotransferase catalysed reaction, *N*-acyl-L,L-2,6-diaminopimelate, is deacylated by the *N*-acyl-diaminopimelate deacylase enzyme (DapE; EC 3.5.1.18) (Fig. 4.6A). The *E. coli* enzyme was characterized and was found to have a molecular weight of 40,000 Da (Lin *et al.*, 1992). The K_m for *N*-acyl-diaminopimelate deacylase has been determined to be 0.4 mM (Bouvier *et al.*, 1992). The *E. coli* enzyme is a homodimer at low concentrations and forms a homotetramer at high concentrations (Bouvier *et al.*, 1992).

The penultimate step in the DAP pathway is the conversion of L,L-diaminopimelate to the isomer *meso*-diaminopimelate (*meso*-DAP). This reaction is catalysed by the enzyme *meso*-diaminopimelate epimerase (DapF; EC 5.1.1.7) (Fig. 4.6A), and activity of the enzyme has been detected in Gram-positive and Gram-negative bacteria (Antia *et al.*, 1957). Structural studies from various *meso*-diaminopimelate epimerase enzymes have been undertaken in the last decade (Cirilli *et al.*, 1998; Lloyd *et al.*, 2004; Pillai *et al.*, 2006, 2009; Usha *et al.*, 2009). More recently, structural studies of the enzyme from *E. coli*, using macromolecular crystallography and analytical ultracentrifugation, have revealed that the protein is a dimer in solution (Hor *et al.*, 2010, 2013). Each monomer of the dimeric *meso*-diaminopimelate epimerase adopts a symmetrical α/β domain structure (Fig. 4.9A), with each domain contributing one of the two key active site cysteine residues located in the monomeric interdomain cleft (Fig. 4.9B). This unique fold is now referred to as the DAP epimerase-like fold. The Michaelis constants for the *E. coli* enzyme have been determined to be 160 μM and 360 μM for L,L-diaminopimelate and *meso*-diaminopimelate, respectively (Higgins *et al.*, 1989). An *E. coli meso*-diaminopimelate epimerase mutant devoid of epimerase activity has been isolated (Richaud *et al.*, 1987); the mutant was shown to accumulate L,L-diaminopimelate, but it is not auxotrophic for *meso*-diaminopimelate. This may suggest that the racemization from L,L- to *meso*-DAP maybe a non-specific reaction and is possibly catalysed by numerous racemases in bacteria. In this mutant, the *meso*-diaminopimelate requirement for cell wall biosynthesis can be bypassed by the incorporation of L-lanthionine, a sulfur analogue of diaminopimelate. It has been shown that L-lanthionine is racemized to form *meso*-lanthionine, which is incorporated as the cross-linking amino acid in the biosynthesis of peptidoglycan (Richaud *et al.*, 1993). The relationship between lysine and bacterial peptidoglycan synthesis will be discussed further in Section 4.7.

The final step in the acyl pathways is the decarboxylation of *meso*-diaminopimelate to produce lysine. The decarboxylation reaction is catalysed by the enzyme *meso*-diaminopimelate decarboxylase (LysA) (Fig. 4.6A). In bacteria, the decarboxylase activity is specific for *meso*-diaminopimelate with an apparent K_m of ~1.7 μM for *meso*-diaminopimelate for most bacterial species (Stragier *et al.*, 1983). *meso*-Diaminopimelate decarboxylase enzymes from plants have not yet been well characterized (Oliver *et al.*, 2014). It should be noted that the decarboxylation of *meso*-diaminopimelate to form lysine is a common step in all of the variant pathways that utilize DAP as the intermediate.

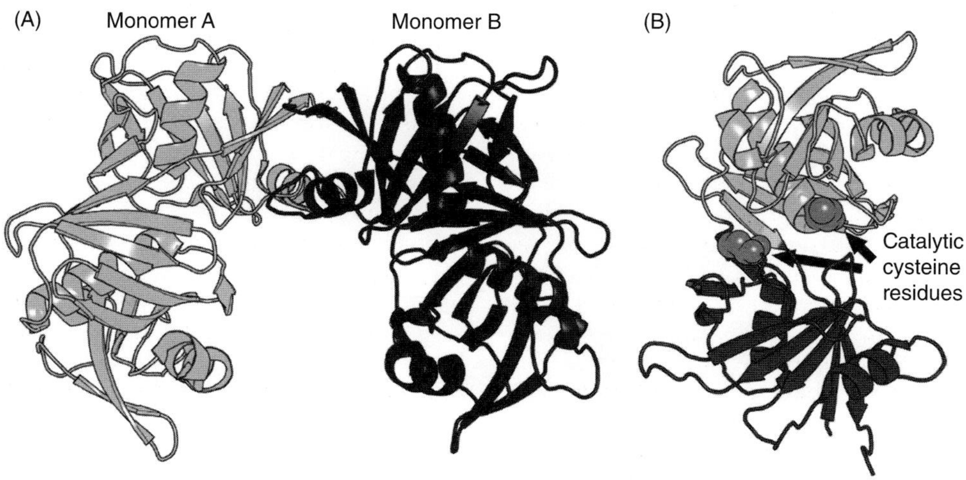

Fig. 4.9. The structure of (A) dimeric *Escherichia coli* diaminopimelate epimerase (DapF) and (B) the monomeric quaternary structure. The active site cysteine residues are highlighted by arrows (PDB (Protein Data Bank) ID: 4IJZ).

4.4.2 The *meso*-diaminopimelate (*meso*-DAP) dehydrogenase (Ddh) pathway

The synthesis of *meso*-diaminopimelate by acylated intermediates (acetylated or succinylated) that was discussed in Section 4.4.1 can be bypassed using a variant of the lysine biosynthesis pathway. In this pathway variant, L-tetrahydrodipicolinate is reduced to *meso*-DAP in a reaction catalysed by the NADPH dependent enzyme *meso*-diaminopimelate dehydrogenase (Ddh; EC 1.4.1.16) (Jouanneau *et al.*, 1985) (Fig 4.6B). This enzyme is present in *Bacillus sphaericus*, in some other Gram-positive bacteria and in some Gram-negative bacteria (Misono *et al.*, 1979). The existence of *meso*-diaminopimelate dehydrogenase in plants was reported by Wenko *et al.* (1985), but this was refuted in later reports (Chatterjee *et al.*, 1994). The enzyme has been purified from *C. glutamicum* (Misono *et al.*, 1986), *Brevibacterium* spp. (Misono *et al.*, 1986) and *B. sphaericus*, where it has been studied in depth, because the activity of the enzyme in this organism is relatively robust (Misono and Soda, 1980).

The bacterial *meso*-diaminopimelate dehydrogenases are homodimers with a molecular weight of approximately 39,000 Da and have two active sites per dimer. In the *B. sphaericus* enzyme, the apparent K_m for L-tetrahydrodipicolinate is 0.2 mM, for NADPH it is 0.2 mM and for NADP it is 0.83 μM (Cox, 1996). The *meso*-diaminopimelate dehydrogenase enzyme is capable of circumventing the L-tetrahydrodipicolinate *N*-acyl-transferase (DapD), *N*-acyl-diaminopimelate aminotransferase (DapC), *N*-acyl-diaminopimelate deacylase (DapE) and *meso*-diaminopimelate epimerase (DapF) enzymatic steps that are present in the acyl pathway variants (Fig. 4.6B).

As a way to access D-amino acids by engineering the *meso*-diaminopimelate dehydrogenase enzyme (Akita *et al.*, 2012), the three-dimensional structures for the dimeric *C. glutamicum* enzyme (Scapin *et al.*, 1996, 1998a; Cirilli *et al.*, 2000), the dimeric *Ureibacillus thermosphaericus* enzyme (Akita *et al.*, 2015), and the hexameric *Symbiobacterium thermophilum* enzyme (Liu *et al.*, 2014) have been solved (Fig. 4.10). The monomer is composed of three domains: a dinucleotide-binding domain, a dimerization domain and a third substrate-binding domain.

4.4.3 L,L-Diaminopimelate aminotransferase (DapL) Pathway: a novel variant of the DAP pathways

Plants and photosynthetic organisms were known to synthesize lysine *via* a pathway that

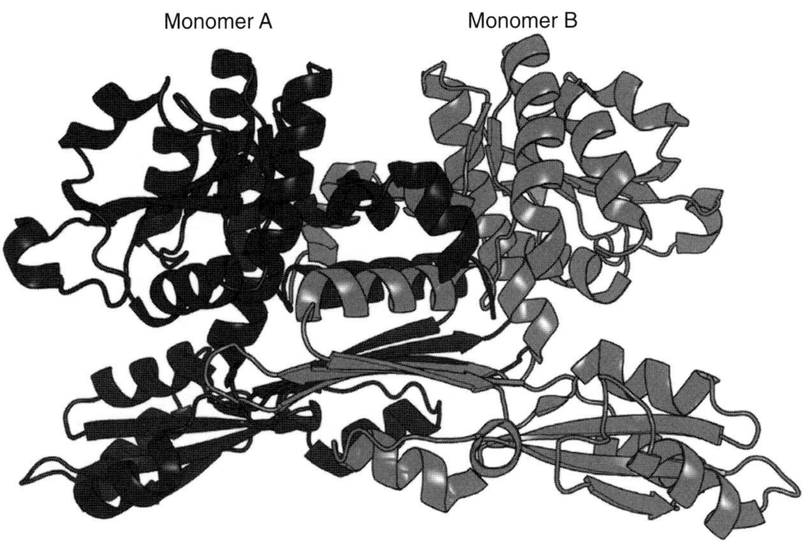

Fig. 4.10. The structure of dimeric *meso*-diaminopimelate dehydrogenase (Ddh) from *Ureibacillus thermosphaericus* (PDB (Protein Data Bank) ID: 3WYB).

utilizes diaminopimelate as an intermediate (Vogel, 1959; Bryan *et al.*, 1970; Glawischnig *et al.*, 2001), although it was unclear which DAP pathway they used. As stated earlier in Section 4.2.2, the aspartate-derived amino acids pathways are of interest to scientists, especially plant biologists, because some of the amino acids are deemed to be limited in certain cereal crops.

The lysine biosynthetic pathway in plants parallels the pathway in bacteria in that they utilize the acyl pathways and commence with the phosphorylation of aspartate, followed by a reduction to form aspartate semialdehyde (ASA) catalysed by the enzymes aspartate kinase (AK) and aspartate semialdehyde dehydrogenase (Asd), respectively. Like bacteria, the first reaction specific for diaminopimelate/lysine synthesis in plants is the condensation of aspartate β-semialdehyde and pyruvate catalysed by the enzyme dihydrodipicolinate synthase (DHPS). The enzyme has been isolated from a variety of plants (Azevedo *et al.*, 1997), including maize, wheat, spinach, tobacco (*N. sylvestris*) and pea (Dereppe *et al.*, 1992). Interestingly, while the dihydrodipicolinate synthase enzymes from *N. sylvestris*, *V. vinifera* and *A. thaliana* are all tetrameric, the arrangement of the subunits in the plant enzymes is different from that of the bacterial enzymes (Blickling *et al.*, 1997b; Griffin *et al.*, 2012; Atkinson *et al.*, 2013). Michaelis constants for the dihydrodipicolinate synthase enzyme have been determined for many species, and range from 0.5 to 11.8 mM for pyruvate (Eggeling, 1994) and 0.4 to 3.1 mM for L-aspartate semialdehyde (Eikmanns *et al.*, 1993). The dihydrodipicolinate synthase from *V. vinifera* was shown to be feedback regulated by exogenous lysine (Atkinson *et al.*, 2013).

The second reaction of lysine biosynthesis in plants is the formation of tetrahydrodipicolinate catalysed by dihydrodipicolinate reductase (DapB). Activity of this enzyme has been detected from crude extracts of certain plant species – in maize (Tyagi *et al.*, 1983) and in *Chlamydomonas*, maize, soybean (*Glycine max*) and *N. tabacum* (Chatterjee *et al.*, 1994).

However, activities of the enzymes tetrahydropyridine-2,6-dicarboxylate *N*-acyl-transferase (L-tetrahydrodipicolinate *N*-acyl-transferase; DapD), *N*-succinyl-α-amino-ε-ketopimelate-glutamate aminotransferase (*N*-acyl-diaminopimelate aminotransferase, DapC) and *N*-succinyldiaminopimelate deacylase (*N*-acyl-diaminopimelate deacylase, DapE) have never been reported in plants. A review describing the molecular evolution of lysine biosynthesis in plants (Velasco *et al.*, 2002) incorrectly reported that plants contain the enzyme *meso*-diaminopimelate dehydrogenase (Ddh). This statement was probably based on the report that a Ddh gene from soybean had been cloned and characterized (Wenko *et al.*, 1985). This was refuted in a later report and was shown to be an error by when additional experiments showed that *meso*-diaminopimelate dehydrogenase activity could not be detected in extracts from soybean. In addition, the gene for Ddh could not be identified in the genomes of plants and photosynthetic cohorts (Hudson *et al.*, 2005).

However, diaminopimelate epimerase (DapF) activity has been detected in plant extracts (Tyagi *et al.*, 1982; Chatterjee *et al.*, 1994). The gene responsible for DapF from *A. thaliana* was cloned and the recombinant enzyme was shown to possess epimerase activity (Hudson *et al.*, 2005). In addition, diaminopimelate decarboxylase (LysA) has been studied from *Lemna perpusilla* (Shimura and Vogel, 1966), *Vicia faba* and wheat (Mazelis and Creveling, 1978). Comparison with the bacterial LysA enzyme shows the plant enzyme to have similar properties. The K_m of the plant enzyme for *meso*-diaminopimelate is much lower that it is for bacterial enzyme; in plants, the K_m varies from 0.14 to 0.3 mM (Shimura and Vogel, 1966; Mazelis *et al.*, 1976; Sodek, 1978).

The sequencing and annotation of the *A. thaliana* genome provided the opportunity to identify the enzymes involved in lysine synthesis using the known bacterial or archaeal protein genes as queries to search and identify orthologous genes from *A. thaliana* (AGI, 2000). The search resulted in the identification of the functional enzymes diydrodipicolinate synthase (DapA, DHPS), dihydrodipicolinate reductase (DapB), diaminopimelate epimerase (DapF) and *meso*-diaminopimelate decarboxylase (LysA) (Hudson *et al.*, 2005). Suitable orthologues of enzymes that convert tetrahydrodipicolinate to diaminopimelate in *E. coli* (L-tetrahydrodipicolinate *N*-acyl-transferase, DapD; *N*-acyl-diaminopimelate aminotransferase, DapC; *N*-acyl-diaminopimelate deacylase, DapE) involved in the core of the pathway could not be detected in *A. thaliana* or in any other plant genomic databases (Hudson *et al.*, 2005). In addition, orthologues of the *meso*-diaminopimelate dehydrogenase (Ddh), the key enzyme in the dehydrogenase pathway, could not be identified. It should be noted that although putative orthologues of *N*-acyl-diaminopimelate aminotransferase (DapC), acetyl-ornithine aminotransferase (ArgD) and *N*-acyl-diaminopimelate deacylase (DapE) could be identified, none of these putative orthologues were considered to be involved in lysine biosynthesis, based on the fact the enzymes in question were predicted to be localized in the cytoplasm or other organelles (Hudson *et al.*, 2005). This is important because isolated chloroplasts from pea (*Pisum sativum*) were shown to be capable of synthesizing lysine from radiolabelled aspartate, indicating that all of the enzymes necessary for *de novo* synthesis of lysine synthesis are targeted and localized in the chloroplast (Mills and Wilson, 1978).

4.5 Insights on Bridging the Metabolic Gap Between Tetrahydrodipicolinate (THDP) and L,L-DAP in Plants and Bacteria That Do Not Contain Orthologues of the DAP Pathway Enzymes DapD, DapC and DapE

Because the four enzymes L-tetrahydrodipicolinate *N*-acyl-transferase (DapD), *N*-acyl-diaminopimelate aminotransferase (DapC), *N*-acyl-diaminopimelate deacylase (DapE) and *meso*-diaminopimelate dehydrogenase (Ddh) enzymes are necessary to bridge the metabolic gap between tetrahydropicolinate and L,L-diaminopimelate in the acyl and dehydrogenase pathways in bacteria, and the activities of these enzymes could not be detected in a variety of plant species (Chatterjee *et al.*, 1994; Hudson *et al.*, 2005), it became apparent that plants must synthesize lysine via a novel pathway. The simplest way to visualize the metabolic gap between tetrahydropicolinate and L,L-diaminopimelate without the activities of these four enzymes is the direct transamination of the acyclic form of tetrahydrodipicolinate, L-2-amino-6-oxopimelate, to synthesize L,L-diaminopimelate. Enzymes that catalyse the transfer of amine groups from one moiety to another are known as aminotransferases or transaminases (EC 2.6.1.x). As such, Hudson *et al.* (2005) hypothesized that plants and all photosynthetic organisms utilize a variant of the known DAP pathway to synthesize lysine.

4.6 The Discovery of the L,L-Diaminopimelate Aminotransferase (DapL) Variant Pathway

The availability of the genome of the model plant *A. thaliana* provided the opportunity to do comparative genomics. Hudson and colleagues hypothesised that a novel variant of DAP was employed by plants owing to the fact that orthologues of L-tetrahydrodipicolinate *N*-acyl-transferase (DapD), *N*-acyl-diaminopimelate aminotransferase (DapC), *N*-acyl-diaminopimelate deacylase (DapE) and *meso*-diaminopimelate dehydrogenase (Ddh) could not be identified in the genome of plants and, subsequently, in photosynthetic organisms by using known bacterial or archaeal protein sequences as queries against the plant database. In addition, the enzymatic activities of L-tetrahydrodipicolinate *N*-acyl-transferase (DapD) and *N*-acyl-diaminopimelate aminotransferase (DapC) could not be detected from crude protein extracts prepared from *G. max*, *Z. mays*, *N. tabacum* and *Chlamydomonas reinhardtii*, yet the extract from *Z. mays* was able to synthesize lysine when the precursor dihydrodipicolinate was provided as a substrate (Hudson *et al.*, 2005).

The analyses from these preliminary experiments suggested that plants and photosynthetic organisms bridge the metabolic gap between tetrahydropicolinate and L,L-diaminopimelate by employing an aminotransferase/transaminase enzyme. The aminotransferase responsible for this activity was subsequently identified and characterized from *A. thaliana* (locus tag At4g33680) and from the cyanobacterium *Syncechocystis* sp. PCC 6803 (locus tag sll0480) using *in vitro* and *in vivo* analyses (Hudson *et al.*, 2006). The discovery of L,L-diaminopimelate aminotransferase (DapL; EC 2.6.1.83) completed the metabolic steps from aspartate to lysine in plants and certain bacteria (see Fig. 4.6C). The *in vivo* analysis showed that the enzyme is able to functionally complement the *dapD* and *dapD/E*

E. coli mutants that are auxotrophic for diaminopimelate by circumventing the L-tetrahydrodipicolinate *N*-acyl-transferase (DapD), *N*-acyl-diaminopimelate aminotransferase (DapC) and *N*-acyl-diaminopimelate deacylase (DapE) reactions that are present in the *E. coli* acyl pathways. In addition, the enzyme was shown to be able to interconvert tetrahydropicolinate and L,L- diaminopimelate by using *in vitro* analyses. The identification of L,L-diaminopimelate aminotransferase (DapL) in a cyanobacterium meant that this pathway was not only a feature of photosynthetic cohorts but also a feature in lineages belonging to the bacterial domain (Hudson *et al.*, 2006). To that end, a number of orthologues of L,L-diaminopimelate aminotransferase have been identified and characterized from a variety of bacteria, including *Chlamydia trachomatis*, *Protochlamydia amoebophila*, *Leptospira interrogans*, *Desulfitobacterium hafniense* and *Morella thermoacetica*, and also from the archaeon *Methanothermobacter thermautotrophicus* (Triassi *et al.*, 2014). A recent study using the L,L-diaminopimelate aminotransferase protein signature IPR019942 as a query revealed that this enzyme was present in approximately 13% of the bacterial genomes that have been completely sequenced and annotated (Triassi *et al.*, 2014). It should be noted that L,L-diaminopimelate aminotransferase has been identified and characterized from the alga *Chlamydomonas reinhardtii*, thus demonstrating that this variant pathway is also a feature of all photosynthetic cohorts (plants, cyanobacteria and algae) (Dobson *et al.*, 2011).

The L,L-diaminopimelate aminotransferase (DapL) enzymes can be classified into two groups based on sequence similarity: Type I DapLs, which originate from plants and *Chlamydia*, and Type II DapLs, from bacteria. At the amino acid level, Types I and II have approximately 30% identity (Hudson *et al.*, 2008). Based on ligand-bound structures, the binding modes for the substrates have been detailed and such structural detail will be useful for inhibitor design. Indeed, inhibitors for the *A. thaliana* and *Chlamydomonas reinhardtii* enzymes have already been reported (Fan *et al.*, 2010; McKinnie *et al.*, 2014).

The structures of the L,L-diaminopimelate aminotransferase enzymes from various species have been solved recently: *Chlamydomonas reinhardtii* (Dobson *et al.*, 2011; Hudson *et al.*, 2011), *Chlamydia trachomatis* (Watanabe *et al.*, 2011) and *Arabidopsis thaliana* (Watanabe *et al.*, 2008). All of the enzymes that have been structurally determined to date are homodimers. Each monomer is an α/β protein in a V-shaped conformation (Fig. 4.11A) and is classified as a pyridoxal phosphate (PLP)-dependent transferase-like fold by SCOP. The overall fold of each monomer of L,L-diaminopimelate aminotransferase consists of two domains, a large domain and a small domain (Fig. 4.11B). The large domain belongs to an α/β class, which folds into an α-β-α sandwich. The small domain also belongs to the α/β class of protein fold and forms an α/β complex. In addition, the small domain also contains an 'arm' region at the N-terminus. The active site is situated in a very similar position to those

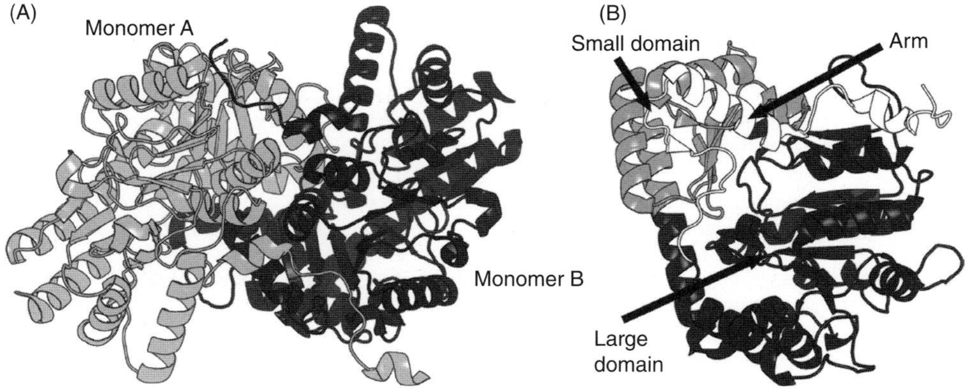

Fig. 4.11. The dimeric (A) and monomeric (B) structures of L,L-diaminopimelate aminotransferase (DapL) from *Chlamydomonas reinhardtii* (PDB (Protein Data Bank) ID: 3QGU).

of other aminotransferase enzymes and includes residues from both monomers within the tetramer.

The symbols, names, EC numbers and reactions catalysed by all of the enzymes involved in the synthesis of lysine via the diaminopimelate (DAP) pathway are summarized in Table 4.1.

4.7 The Link Between Lysine and Bacterial Peptidoglycan (PG) Biosynthesis

The diaminopimelate/lysine pathway plays an integral role in the synthesis of bacterial peptidoglycan (PG), commonly referred to as the bacterial cell wall. The peptidoglycan of bacteria has an important role in protecting the bacterial cell from internal and external forces and, as such, it is a vital and essential structure for bacterial growth and development. Peptidoglycan is primarily composed of a polymeric layer of the disaccharide *N*-acetylglucosamine (GlcNAc) and *N*-acetylmumaric acid (MurNAc) that are linked via a *β*-1,4-glyosidic bond. The disaccharide is cross-linked via a peptide stem that is connected to *N*-acetylmumaric acid with the general sequence of L-alanine-D-glutamate-X-D-alanine. The third position denoted by 'X' is either *meso*-diaminopimelate or lysine, and is important because it is this amino acid that is involved in the cross-linking of the polymer. *meso*-Diaminopimelate is normally the cross-linking amino acid in the peptidoglycan of most Gram-negative bacteria, while lysine is normally the cross-linking amino acid in the peptidoglycan of most Gram-positive bacteria (Hutton *et al.*, 2007) (Fig. 4.12). The addition of the *meso*-diaminopimelate or lysine to the third position of the peptide stem is facilitated by the enzyme

Table 4.1. The enzyme symbols, enzyme names, Enzyme Commission (EC) numbers and enzyme reaction definitions involved in the synthesis of lysine from the intermediate diaminopimelate (DAP).

Enzyme symbol	Enzyme name	EC No.	Reaction definition
Asd	Aspartate semialdehyde dehydrogenase	1.2.1.11	L-Aspartate 4-semialdehyde + Orthophosphate + NADP+ <=> 4-Phospho-L-aspartate + NADPH + H^+
DapA	Dihydrodipicolinate synthase	4.2.1.52	L-Aspartate 4-semialdehyde + Pyruvate <=> L-2,3-Dihydrodipicolinate + $2H_2O$
DapB	Dihydrodipicolinate reductase	1.3.1.26	2,3,4,5-Tetrahydrodipicolinate + NADP+ <=> L-2,3-Dihydrodipicolinate + NADPH + H^+
DapD	2,3,4,5-Tetrahydropyridine-2,6-dicarboxylate *N*-acyl-transferase	2.3.1.117	Acyl-CoA + 2,3,4,5-tetrahydrodipicolinate + H_2O <=> CoA + N-Acyl-2-L-amino-6-oxoheptanedioate
DapC	Acyl-diaminopimelate aminotransferase	2.6.1.17	*N*-Acyl-*L*,*L*-2,6-diaminoheptanedioate + 2-Oxoglutarate <=> *N*-Acyl-2-L-amino-6-oxoheptanedioate + L-glutamate
DapE	Acyl-diaminopimelate deacylase	3.5.1.18	*N*-Acyl-L,L-2,6-diaminoheptanedioate + H_2O <=> Acyl + L,L-2,6-Diaminoheptanedioate
DapF	Diaminopimelate epimerase	5.1.1.7	L,L-2,6-Diaminopimelate <=> *meso*-2,6-Diaminopimelate
DapL	L,L-Diaminopimelate aminotransferase	2.6.1.83	L,L-2,6-Diaminopimelate + 2-Oxoglutarate <=> 2,3,4,5-Tetrahydrodipicolinate + L-Glutamate + H_2O
Ddh	Diaminopimelate dehydrogenase	1.4.1.16	*meso*-2,6-Diaminoheptanedioate + NADP+ + H_2O <=> L-2-Amino-6-oxoheptanedioate + NH_3 + NADPH
LysA	Diaminopimelate decarboxylase	4.1.1.20	*meso*-2,6-Diaminopimelate <=> L-Lysine + CO_2
LysC	Aspartate kinase	2.7.2.4	ATP + L-Aspartate <=> ADP + 4-Phospho-L-aspartate

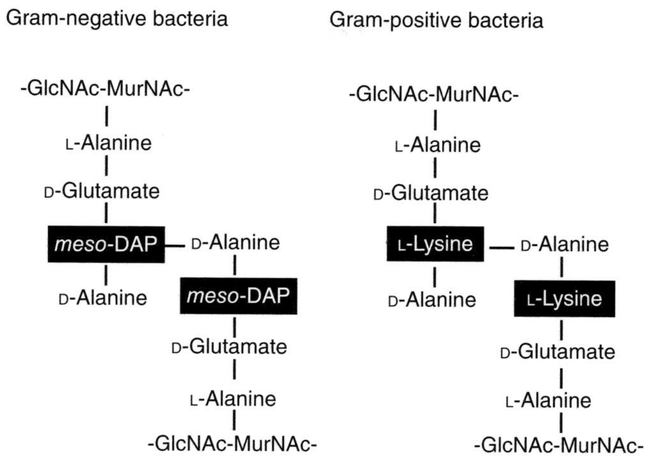

Fig. 4.12. Schematic representation of the peptidoglycan cell wall of Gram-negative and Gram-positive bacteria highlighting the role of *meso*-diaminopimelate (*meso*-DAP) and L-lysine in the cross-linking of the polymer.

UDP-*N*-acetylmuramoylalanyl-D-glutamate-2,6-diaminopimelate ligase (MurE; EC 6.3.2.13) in the cytoplasmic step of peptidoglycan synthesis.

Given the fact that peptidoglycan cross-linking is an important aspect for the integrity of the cell wall, the lack of or incorrect incorporation of either *meso*-diaminopimelate or lysine at the third position in the sequence L-alanine-D-glutamate-X-D-alanine would have a detrimental effect. The organism would not be able to withstand osmotic pressure, which would lead to lysis and eventually cell death (Triassi *et al.*, 2014). This is one of the reasons why the enzymes involved in the anabolism of diaminopimelate and lysine are attractive and putative targets for antibiotic development, along with the fact the lysine anabolic pathways are absent in all animals. As such, there may be opportunities to discover and or develop compounds that are specific for enzymes in the pathway that are not toxic to animals. From a bacterial perspective, the inhibition of enzymes in the diaminopimelate/lysine pathway will be detrimental for two reasons. First, the inhibition will prevent the synthesis of essential metabolites and subsequently lead to the improper construction of peptidoglycan as both *meso*- diaminopimelate and lysine are involved in peptidoglycan cross-linking. Secondly, enzymatic inhibition would also lead to a lack of protein synthesis because the substrate for lysine-tRNA synthetase (LysU; EC 6.1.1.6) would not be available to facilitate lysine incorporation into protein synthesis.

4.8 Conclusions

The anabolism of lysine in microorganisms is intriguing given that the amino acid is an essential metabolite for growth and development. The existence of variant pathways for lysine synthesis affords an opportunity for research and development on the identification and/or development of compounds that are deemed bactericidal because lysine is one of the 20 ubiquitous proteogenic amino acids and, along with the intermediate *meso*-diaminopimelate, is involved in peptidoglycan synthesis in bacteria. Protein synthesis and cell wall synthesis are two primary metabolic processes that are absolutely essential for bacterial growth and development, and there have been a number of research efforts aimed at their characterization – from both an enzymatic and structural viewpoint – to facilitate the development of antibiotics, especially because the genes for both the α-aminoadipate and diaminopimelate pathways are absent in the genomes of animals. One of the interesting features of the lysine biosynthetic pathway is the diversity of and employment of the pathways. For instance, organisms from the kingdom fungi employ the α-aminoadipate (AAA) pathway, while the majority of bacteria and all photosynthetic cohorts employ variant(s) of either the acyl, dehydrogenase or aminotransferase DAP (diaminopimelate) pathways. As such, the lysine anabolic pathway is a plausible target for the development of fungicides, algicides and/or antibiotics.

Acknowledgements

AOH and MAS acknowledge the support of the College of Science and the Thomas H. Gosnell School of Life Sciences at the Rochester Institute of Technology. AOH acknowledges a National Science Foundation (NSF) MCB-1120541 award. RCJD acknowledges the following for funding support: (i) the Ministry of Business, Innovation and Employment of New Zealand (contract UOCX1208); (ii) the New Zealand Royal Society Marsden Fund (contract UOC1013); and (iii) the US Army Research Laboratory and US Army Research Office under contract/grant number W911NF-11-1-0481. We especially thank Jackie Healy for her digitigradient technical support.

References

Acord, J. and Masters, M. (2004) Expression from the *Escherichia coli dapA* promoter is regulated by intracellular levels of diaminopimelic acid. *FEMS Microbiology Letters* 235, 131–137.

AGI (Arabidopsis Genome Initiative) (2000) Analysis of the genome sequence of the flowering plant *Arabidopsis thaliana*. *Nature* 408, 796–815.

Akita, H., Doi, K., Kawarabayasi, Y. and Ohshima, T. (2012) Creation of a thermostable NADP$^+$-dependent D-amino acid dehydrogenase from *Ureibacillus thermosphaericus* strain A1 meso-diaminopimelate dehydrogenase by site-directed mutagenesis. *Biotechnology Letters* 34, 1693–1699.

Akita, H., Ohshima, T. and Sakuraba, H. (2015) Thermostable NADP(+)-dependent D-amino acid dehydrogenase created by site-directed mutation: characteristics and application. *The Journal of Japanese Biochemical Society* 87, 582–590.

Antia, M., Hoare, D.S. and Work, E. (1957) The stereoisomers of $\alpha\varepsilon$-diaminopimelic acid. III. Properties and distribution of diaminopimelic acid racemase, an enzyme causing interconversion of the LL and *meso* isomers. *Biochemical Journal* 65,448–459.

Atkinson, S.C., Dogovski, C., Downton, M.T., Czabotar, P.E., Dobson, R.C., Gerrard, J.A., Wagner, J. and Perugini, M.A. (2013) Structural, kinetic and computational investigation of *Vitis vinifera* DHDPS reveals new insight into the mechanism of lysine-mediated allosteric inhibition. *Plant Molecular Biology* 81, 431–446.

Azevedo, R.A., Arruda, P., Turner, W.L. and Lea, P.J. (1997) The biosynthesis and metabolism of the aspartate derived amino acids in higher plants. *Phytochemistry* 46, 395–419.

Bartlett, A.T.M. and White, P.J. (1985) Species of *Bacillus* that make vegetative peptidoglycan containing lysine lack diaminopimelate epimerase but have diaminopimelate dehydrogenase. *Microbiology* 131, 2145–2152.

Blickling, S., Renner, C., Laber, B., Pohlenz, H.D., Holak, T.A. and Huber, R. (1997a) Reaction mechanism of *Escherichia coli* dihydrodipicolinate synthase investigated by X-ray crystallography and NMR spectroscopy. *Biochemistry* 36, 24–33.

Blickling, S., Beisel, H.G., Bozic, D., Knäblein, J., Laber, B. and Huber, R. (1997b) Structure of dihydrodipicolinate synthase of *Nicotiana sylvestris* reveals novel quaternary structure. *Journal of Molecular Biology* 274, 608–621.

Boughton, B.A., Griffin, M.D.W., O'Donnell, P.A., Dobson, R.C.J., Perugini, M.A., Gerrard, J.A. and Hutton, C.A. (2008a) Irreversible inhibition of dihydrodipicolinate synthase by 4-oxo-heptenedioic acid analogues. *Bioorganic and Medicinal Chemistry* 16, 9975–9983.

Boughton, B.A., Dobson, R.C.J., Gerrard, J.A. and Hutton, C.A. (2008b) Conformationally constrained diketopimelic acid analogues as inhibitors of dihydrodipicolinate synthase. *Bioorganic and Medicinal Chemistry Letters* 18, 460–463.

Bouvier, J., Richaud, C., Richaud, F., Patte, J.-C. and Stragier, P. (1984) Nucleotide sequence and expression of the *Escherichia coli dapB* gene. *The Journal of Biological Chemistry* 259, 14829–14834.

Bouvier, J., Richaud, C., Higgins, W., Bögler, O. and Stragier, P. (1992) Cloning, characterization, and expression of the *dapE* gene of *Escherichia coli*. *Journal of Bacteriology* 174, 5265–5271.

Bright, S.W.J. and Shewry, P.R. (1983) Improvement of protein quality in cereals. *Critical Reviews in Plant Sciences* 1, 49–93.

Bryan, P.A., Cawley, R.D., Brunner, C.E. and Bryan, J.K. (1970) Isolation and characterization of a lysine-sensitive aspartokinase from a multicellular plant. *Biochemical and Biophysical Research Communications* 41, 1211–1217.

Bukhari, A.I. and Taylor, A.L. (1971) Genetic analysis of diaminopimelic acid and lysine requiring mutants of *Escherichia coli*. *Journal of Bacteriology* 105, 844–854.
Chatterjee, S.P., Singh, B.K. and Gilvarg, C. (1994) Biosynthesis of lysine in plants: the putative role of *meso*-diaminopimelate dehydrogenase. *Plant Molecular Biology* 26, 285–290.
Cirilli, M., Zheng, R., Scapin, G. and Blanchard, J.S. (1998) Structural symmetry: the three-dimensional structure of *Haemophilus influenzae* diaminopimelate epimerase. *Biochemistry* 37, 16452–16458.
Cirilli, M., Scapin, G., Sutherland, A., Vederas, J.C., Blanchard, J.S., Cirilli, M. and Scapin, G. (2000) The three-dimensional structure of the ternary complex of *Corynebacterium glutamicum* diaminopimelate dehydrogenase-NADPH-L-2-amino-6-methylene-pimelate. *Protein Science* 9, 2034–2037.
Cirilli, M., Zheng, R.J., Scapin, G. and Blanchard, J.S. (2003) The three-dimensional structures of the *Mycobacterium tuberculosis* dihydrodipicolinate reductase-NADH-2,6-PDC and NADPH-2,6-PDC complexes. Structural and mutagenic analysis of relaxed nucleotide specificity. *Biochemistry* 42, 10644–10650.
Cox, R.J. (1996) The DAP pathway to lysine as a target for antimicrobial agents. *Natural Products Reports* 13, 29–43.
Dereppe, C., Bold, G., Ghisalba, O., Ebert, E. and Schär, H.P. (1992) Purification and characterization of dihydrodipicolinate synthase from pea. *Plant Physiology* 98, 813–821.
Devenish, S.R.A., Blunt, J.W. and Gerrard, J.A. (2010) NMR studies uncover alternate substrates for dihydrodipicolinate synthase and suggest that dihydrodipicolinate reductase is also a dehydratase. *Journal of Medicinal Chemistry* 53,4808–4812.
Dobson, R.C.J., Valegård, K. and Gerrard, J.A. (2004a) The crystal structure of three site-directed mutants of *Escherichia coli* dihydrodipicolinate synthase: further evidence for a catalytic triad. *Journal of Molecular Biology* 338, 329–339.
Dobson, R.C.J., Griffin, M.D.W., Roberts, S.J. and Gerrard, J.A. (2004b) Dihydrodipicolinate synthase (DHDPS) from *Escherichia coli* displays partial mixed inhibition with respect to its first substrate, pyruvate. *Biochimie* 86, 311–315.
Dobson, R.C.J., Griffin, M.D., Jameson, G.B. and Gerrard, J.A. (2005) The crystal structures of native and (*S*)-lysine-bound dihydrodipicolinate synthase from *Escherichia coli* with improved resolution show new features of biological significance. *Acta Crystallographica D: Biological Crystallography* 61, 1116–1124.
Dobson, R.C.J., Griffin, M.D., Devenish, S.R.A., Pearce, G.F., Hutton, C.A., Gerrard, J.A., Jameson, G.B. and Perugini, M.A. (2008) Conserved main-chain peptide distortions: a proposed role for Ile203 in catalysis by dihydrodipicolinate synthase. *Protein Science* 17, 2080–2090.
Dobson, R.C.J., Girón, I. and Hudson, A.O. (2011) L,L-Diaminopimelate aminotransferase from *Chlamydomonas reinhardtii*: a target for algaecide development. *PLoS ONE* 6(5): e20439.
Dommaraju, S.R., Dogovski, C., Czabotar, P.E., Hor, L., Smith, B.J. and Perugini, M.A. (2011) Catalytic mechanism and cofactor preference of dihydrodipicolinate reductase from methicillin-resistant *Staphylococcus aureus. Archives of Biochemistry and Biophysics* 512, 167–174.
Eggeling, L. (1994) Biology of L-lysine overproduction by *Corynebacterium glutamicum*. *Amino Acids* 6, 261–272.
Eikmanns, B.J., Eggeling, L. and Sahm, H. (1993) Molecular aspects of lysine, threonine, and isoleucine biosynthesis in *Corynebacterium glutamicum*. *Antonie Van Leeuwenhoek* 64, 45–63.
Fan, C., Clay, M.D., Deyholos, M.K. and Vederas, J.C. (2010) Exploration of inhibitors for diaminopimelate aminotransferase. *Bioorganic and Medicinal Chemistry* 18, 2141–2151.
Farkas, W. and Gilvarg, C. (1965) The reduction step in diaminopimelic acid biosynthesis. *The Journal of Biological Chemistry* 240, 4717–4722.
Galili, G. (1995) Regulation of lysine and threonine synthesis. *The Plant Cell* 7, 899–906.
Galili, G., Karchi, H., Shaul, O., Perl, A., Cahana, A., Ben-Tzvi Tzchori, I., Zhu, X.Z. and Galili, S. (1994) Production of transgenic plants containing elevated levels of lysine and threonine. *Biochemical Society Transactions* 22, 921–925.
Gilvarg, C. (1959) *N*-succinyl L-diaminopimelic acid. *The Journal of Biological Chemistry* 243, 2955–2959.
Gilvarg, C. (1961) *N*-succinyl-alpha-amino-6-ketopimelic acid. *The Journal of Biological Chemistry* 236, 1429–1431.
Girish, T.S., Navratna, V. and Gopal, B. (2011) Structure and nucleotide specificity of *Staphylococcus aureus* dihydrodipicolinate reductase (DapB). *FEBS Letters* 585, 2561–2567.
Glawischnig, E., Gierl, A., Tomas, A., Bacher, A. and Eisenreich, W. (2001) Retrobiosynthetic nuclear magnetic resonance analysis of amino acid biosynthesis and intermediary metabolism. Metabolic flux in developing maize kernels. *Plant Physiology* 125, 1178–1186.

Griffin, M.D.W., Billakanti, J.M., Wason, A., Keller, S., Mertens, H.D.T., Atkinson, S.C., Dobson, R.C.J., Perugini, M.A., Gerrard, J.A. and Pearce, F.G. (2012) Characterisation of the first enzymes committed to lysine biosynthesis in *Arabidopsis thaliana. PLoS ONE* 7(7): e40318.

Grundy, F.J., Lehman, S.C. and Henkin, T.M. (2003) The L box regulon: lysine sensing by leader RNAs of bacterial lysine biosynthesis genes. *Proceedings of the National Academy of Sciences of the United States of America* 100, 12057–120762.

Guo, B.B.B., Devenish, S.R.A., Dobson, R.C.J., Muscroft-Taylor, A.C. and Gerrard, J.A. (2009) The C-terminal domain of *Escherichia coli* dihydrodipicolinate synthase (DHDPS) is essential for maintenance of quaternary structure and efficient catalysis. *Biochemical and Biophysical. Research Communications* 380, 802–806.

Halling, S.M. and Stahly, D.P. (1976) Dihydrodipicolinic acid synthase of *Bacillus licheniformis.* Quaternary structure, kinetics, and stability in the presence of sodium chloride and substrates. *Biochimica et Biophysica Acta* 452, 580–596.

Higgins, W., Tardif, C., Richaud, C., Krivanek, M. and Cardin, A.(1989) Expression of recombinant diaminopimelate epimerase in *Escherichia coli*. Isolation and inhibition with an irreversible inhibitor. *The FEBS Journal* 186, 137–143.

Hor, L., Dobson, R.C.J., Dogovski, C., Hutton, C.A. and Perugini, M.A. (2010) Crystallization and preliminary X-ray diffraction analysis of diaminopimelate epimerase from *Escherichia coli. Acta Crystallographica Section F: Structural Biology and Crystallization Communications* 66, 37–40.

Hor, L, Dobson, R.C.J., Downton, M.T., Wagner, J., Hutton, C.A. and Perugini, M.A. (2013) Dimerization of bacterial diaminopimelate epimerase is essential for catalysis. *The Journal of Biological Chemistry* 13, 9238–9248.

Hudson, A.O., Bless, C., Macedo, P., Chatterjee, S.P., Singh, B.K., Gilvarg, C. and Leustek, T. (2005) Biosynthesis of lysine in plants: evidence for a variant of the known bacterial pathways. *Biochimica et Biophysica Acta* 1721, 27–36.

Hudson, A.O., Singh, B.K. Leustek, T. and Gilvarg, C. (2006) An L,L-diaminopimelate aminotransferase defines a novel variant of the lysine biosynthesis pathway in plants. *Plant Physiology* 140, 292–301.

Hudson, A.O., Gilvarg, C. and Leustek, T. (2008) Biochemical and phylogenetic characterization of a novel diaminopimelate biosynthesis pathway in prokaryotes identifies a diverged form of L,L-diaminopimelate aminotransferase. *Journal of Bacteriology* 190, 3256–3263.

Hudson, A.O., Girón, I. and Dobson, R.C.J. (2011) Crystallization and preliminary X-ray diffraction analysis of L,L-diaminopimelate aminotransferase (DapL) from *Chlamydomonas reinhardtii. Acta Crystallographica Section F: Structural Biology and Crystal Communications* 67, 140–143.

Hutton, C.A., Perugini, M.A. and Gerrard, J.A. (2007) Inhibition of lysine biosynthesis: an evolving antibiotic strategy. *Molecular BioSystems* 3, 458–465.

Janowski, R., Kefala, G. and Weiss, M.S. (2010) The structure of dihydrodipicolinate reductase (DapB) from *Mycobacterium tuberculosis* in three crystal forms. *Acta Crystallographica Section D: Biological Crystallography*. 66, 61–72.

Jouanneau, J., Stragier, P., Bouvier, J., Patte, J.-C. and Yaniv, M. (1985) Expression in mammalian cells of the diaminopimelic acid decarboxylase of *Escherichia coli* permits cell growth in lysine-free medium. *The FEBS Journal* 146, 173–178.

Kimura, K. and Goto, T. (1975) A new flavin enzyme catalyzing the reduction of dihydrodipicolinate in sporulating *Bacillus subtilis*. II. Kinetics and regulatory function. *The Journal of Biochemistry* 77, 415–420.

Kimura, K. and Goto, T. (1977) Dihydrodipicolinate reductases from *Bacillus cereus* and *Bacillus megaterium. The Journal of Biochemistry* 81, 1367–1373.

Kumpaisal, R., Hashimoto, T. and Yamada, Y. (1987) Purification and characterization of dihydrodipicolinate synthase from wheat suspension cultures. *Plant Physiology* 85, 145–151.

Laber, B., Gomis-Rüth, F.X., Romão, M.J. and Huber, R. (1992) *Escherichia coli* dihydrodipicolinate synthase. Identification of the active site and crystallization. *Biochemical Journal* 288, 691–695.

Ledwidge, R. and Blanchard, J.S. (1999) The dual biosynthetic capability of *N*-acetylornithine aminotransferase in arginine and lysine biosynthesis. *Biochemistry* 38, 3019–3024.

Lin, R., Ernsting, B., Hirshfield, I.N., Matthews, R.G., Neidhardt, F.C., Clark, R.L. and Newman, E.B. (1992) The *lrp* gene product regulates expression of *lysU* in *Escherichia coli* K-12. *Journal of Bacteriology* 174, 2779–2784.

Liu, W., Li, Z., Huang, C.H., Guo, R.T., Zhao, L., Zhang, D., Chen, X., Wu, Q. and Zhu, D. (2014) Structural and mutational studies on the unusual substrate specificity of *meso*-diaminopimelate dehydrogenase from *Symbiobacterium thermophilum*. *Chemobiochemistry* 15, 217–222.

Lloyd, A.J., Huyton, T., Turkenburg, J. and Roper, D.I. (2004) Refinement of *Haemophilus influenzae* diaminopimelic acid epimerase (DapF) at 1.75 Å resolution suggests a mechanism for stereocontrol during catalysis. *Acta Crystallographica Section D: Biological Crystallography.* 60, 397–400.

Matsuda, M. and Ogur, M. (1969) Enzymatic and physiological properties of the yeast glutamate-α-ketoadipate transaminase. *The Journal of Biological Chemistry* 244, 5153–5158.

Mazelis, M. and Creveling, R.K. (1978) The enzymology of lysine biosynthesis in higher plants. Diaminopimelate decarboxylase from wheat germ. *Journal of Food Biochemistry* 2, 29–37.

Mazelis, M., Miflin, B.J. and Pratt, H.M. (1976) A chloroplast-localized diaminopimelate decarboxylase in higher plants. *FEBS Letters* 64, 197–200.

McCoy, A., Adams, N., Hudson, A.O., Gilvarg, C., Leustek, T. and Maurelli, A.T. (2006) *Chlamydia trachomatis* CT390 is a transkingdom L.L-diaminopimelate aminotransferase variant of the *meso*-diaminopimelate/lysine biosynthetic pathway. *Proceedings of the National Academy of Sciences of the United States of America* 47, 17909–17914.

McKinnie, S.M.K., Rodriguez-Lopez, E.M., Vederas, J.C., Crowther, J.M., Suzuki, H., Dobson, R.C.J., Leustek, T., Triassi, A.J., Wheatley, M.S. and Hudson, A.O. (2014) Differential response of orthologous L,L-diaminopimelate aminotransferases (DapL) to enzyme inhibitory antibiotic lead compounds. *Bioorganic and Medicinal Chemistry* 22, 523–530.

Mills, W.R. and Wilson, K.G. (1978) Amino acid biosynthesis in isolated pea chloroplasts: metabolism of labeled aspartate and sulfate. *FEBS Letters* 92, 129–132.

Mirwaldt, C., Korndörfer, I. and Huber, R. (1995) The crystal structure of dihydrodipicolinate synthase from *Escherichia coli* at 2.5 A resolution. *Journal of Molecular Biology* 246, 227–239.

Misono, H. and Soda, K. (1980) Properties of *meso*-α,ε-diaminopimelate D-dehydrogenase from *Bacillus sphaericus*. *The Journal of Biological Chemistry* 55, 10599–10605.

Misono, H., Togawa, H., Yamamoto, T. and Soda, K. (1976) Occurrence of *meso*-α,ε-diaminopimelate dehydrogenase in *Bacillus sphaericus*. *Biochemical and Biophysical Research Communications* 72, 89–93.

Misono, H., Togawa, H., Yamamoto, T. and Soda, K. (1979) *Meso*-α,ε-Diaminopimelate D-dehydrogenase: distribution and the reaction product. *Journal of Bacteriology* 137, 22–27.

Misono, H., Ogasawara, M. and Nagasaki, S. (1986) Purification and properties of *meso*-diaminopimelate dehydrogenase from *Brevibacterium* sp. *Agricultural and Biological Chemistry* 50, 1329–1330.

Mitsakos, V., Dobson, R.C.J., Pearce, F.G., Devenish, S.R., Evans, G.L., Burgess, B.R., Perugini, M.A., Gerrard, J.A. and Hutton, C.A. (2008) Inhibiting dihydrodipicolinate synthase across species: towards specificity for pathogens. *Bioorganic and Medicinal Chemistry Letters* 18, 842–844.

Muscroft-Taylor, A.C., Soares da Costa, T.P. and Gerrard, J.A. (2010) New insights into the mechanism of dihydrodipicolinate synthase using isothermal titration calorimetry. *Biochimie* 92, 254–262.

Neidhardt, F.C., Curtis, R. III, Ingraham, J.L., Lin, E.C.C., Low, K.B., Magasanik, B., Reznikoff, W.S., Riley, M., Schaechter, M. and Umbarger, H.E. (eds) (1996) Escherichia coli *and* Salmonella: *Cellular and Molecular Biology*, 2nd edn. ASM Press, Washington, DC.

Nishida, H., Nishiyama, M., Kobashi, N., Kosuge, T., Hoshino, T. and Yamane, H. (1999) A prokaryotic gene cluster involved in synthesis of lysine through the amino adipate pathway: a key to the evolution of amino acid biosynthesis. *Genome Research* 9, 1175–1183.

Oliver, M.R., Crowther, J.M., Leeman, M.M., Kessans, S.A., North, R.A., Danovan, K.A., Griffin, M.D., Suzuki, H., Hudson, A.O., Kasanmascheff, M. and Dobson, R.C. (2014) The purification, crystallization and preliminary X-ray diffraction analysis of two isoforms of *meso*-diaminopimelate decarboxylase from *Arabidopsis thaliana*. *Acta Crystallographica Section F: Structural Biology and Crystal Communications* 70, 663–668.

Pearce, F.G., Sprissler, C. and Gerrard, J.A. (2008) Characterization of dihydrodipicolinate reductase from *Thermotoga maritima* reveals evolution of substrate binding kinetics. *The Journal of Biochemistry* 143, 617–623.

Peterkofsky, B. and Gilvarg, C. (1961) *N*-Succinyl-L-diaminopimelic-glutamic transaminase. *The Journal of Biological Chemistry* 236, 1432–1438.

Phenix, C.P. and Palmer, D.R. (2008) Isothermal titration microcalorimetry reveals the cooperative and noncompetitive nature of inhibition of *Sinorhizobium meliloti* L5-30 dihydrodipicolinate synthase by (*S*)-lysine. *Biochemistry* 47, 7770–7781.

Pillai, B., Cherney, M.M., Diaper, C.M., Sutherland, A., Blanchard, J.S., Vederas, J.C. and James, M.N.G. (2006) Structural insights into stereo-chemical inversion by diaminopimelate epimerase: an antibacterial drug target. *Proceedings of the National Academy of Sciences of the United States of America* 103, 8668–8673.

Pillai, B., Moorthie, V.A., van Belkum, M.J., Marcus, S.L., Cherney, M.M., Diaper, C.M., Vederas, J.C. and James, M.N. (2009) Crystal structure of diaminopimelate epimerase from *Arabidopsis thaliana*, an amino acid racemase critical for L-lysine biosynthesis. *Journal of Molecular Biology* 385, 580–594.

Reddy, S.G., Sacchettini, J.C. and Blanchard, J.S. (1995) Expression, purification, and characterization of *Escherichia-coli* dihydrodipicolinate reductase. *Biochemistry* 34, 3492–3501.

Reddy, S.G., Scapin, G. and Blanchard, J.S. (1996) Interaction of pyridine nucleotide substrates with *Escherichia coli* dihydrodipicolinate reductase: thermodynamic and structural analysis of binary complexes. *Biochemistry* 35, 13294–13302.

Richaud, C., Richaud, F., Martin, C., Haziza, C. and Patte, J.C. (1984) Regulation of expression and nucleotide sequence of the *Escherichia coli dapD* gene. *The Journal of Biological Chemistry* 259, 14824–14828.

Richaud, C., Higgins, W., Mengin-Lecreulx, D. and Stragier, P. (1987) Molecular cloning, characterization and chromosomal localization of dapF, the *Escherichia coli* gene for diaminopimelate epimerase. *Journal of Bacteriology* 169, 1454–1459.

Richaud, C., Mengin-Lecreulx, D., Pochet, S. and Marlière, P. (1993) Directed evolution of biosynthetic pathways. Recruitment of cysteine thioethers for constructing the cell wall of *Escherichia coli*. *The Journal of Biological Chemistry* 268, 26827–26835.

Rodionov, D.A., Vitreschak, A.G., Mironov, A.A. and Gelfand, M.S. (2003) Regulation of lysine biosynthesis and transport genes in bacteria: yet another RNA riboswitch. *Nucleic Acids Research* 31, 6748–6757.

Rose, W.C. (1949) Amino acid requirements of man. *Federation Proceedings* 8, 546–552.

Scapin, G., Blanchard, J.S. and Sacchettini, J.C. (1995) Three-dimensional structure of *Escherichia coli* dihydrodipicolinate reductase. *Biochemistry* 34, 3502–3512.

Scapin, G., Reddy, S.G. and Blanchard, J.S. (1996) Three-dimensional structure of *meso*-diaminopimelic acid dehydrogenase from *Corynebacterium glutamicum*. *Biochemistry* 35, 13540–13551.

Scapin, G., Reddy, S.G., Zheng, R. and Blanchard, J.S. (1997) Three-dimensional structure of *Escherichia coli* dihydrodipicolinate reductase in complex with NADH and the inhibitor 2,6-pyridinedicarboxylate. *Biochemistry* 36, 15081–15088.

Scapin, G., Cirilli, M., Reddy, S.G., Gao, Y., Vederas, J.C. and Blanchard, J.S. (1998a) Substrate and inhibitor binding sites in *Corynebacterium glutamicum* diaminopimelate dehydrogenase. *Biochemistry* 37, 3278–3285.

Scapin, G., Zhang, R. and Blanchard, J.S. (1998b) The crystal structure of *M. tuberculosis* DHPR: analysis of cofactor and substrate binding in the dihydrodipicolinate reductase family of enzymes. *FASEB Journal* 12, A1452–A1452.

Shimura, Y. and Vogel, H.J. (1966) Diaminopimelate decarboxylase of *Lemna perpusilla*: partial purification and some properties. *Biochimica et Biophysica Acta* 118, 396–404.

Simms, S.A., Voige, W.H. and Gilvarg, C. (1984) Purification and characterization of succinyl-CoA: tetrahydrodipicolinate *N*-succinyltransferase from *Escherichia coli*. *The Journal of Biological Chemistry* 259, 2734–2741.

Sodek, L. (1978) Partial purification and properties of diaminopimelate decarboxylase from maize endosperm. *Revista Brasileira de Botanica* 1, 65–69.

Stragier, P., Richaud, F., Borne, F. and Patte, J.C. (1983) Regulation of diaminopimelate decarboxylase synthesis in *Escherichia coli*. I. Identification of a *lysR* gene encoding an activator of the *lysA* gene. *Journal of Molecular Biology* 168, 307–320.

Triassi, A.J., Wheatley, M.S., Savka, M.A., Gan, H.M., Dobson, R.C. and Hudson, A.O. (2014) L,L-diaminopimelate aminotransferase (DapL): a putative target for the development of narrow-spectrum antibacterial compounds. *Frontiers in Microbiology* 5: 509.

Turner, J., Healy, J.P., Dobson, R.C., Gerrard, J.A. and Hutton, C.A. (2005) Two new irreversible inhibitors of dihydrodipicolinate synthase: diethyl (*E*,*E*)-4-oxo-2,5-heptadienedioate and diethyl (*E*)-4-oxo-2-heptenedioate. *Bioorganic and Medicinal Chemistry Letters* 15, 995–998.

Tyagi, V.V., Henke, R.R. and Farkas, W.R. (1982) Occurrence of diaminopimelic epimerase in maize. *Biochimica et Biophysica Acta* 719, 363–369.

Tyagi, V,V., Henke, R.R. and Farkas, W.R. (1983) Partial purification and characterization of dihydrodipicolinic acid reductase from maize. *Plant Physiology* 73, 687–691.

Usha, V., Dover, L., Roper, D.I., Fütterer, K. and Besra, G.S. (2009) Structure of the diaminopimelate epimerase DapF from *Mycobacterium tuberculosis*. *Acta Crystallographica Section D: Biological Crystallography* 65, 383–387.

Velasco, A.M., Leguina, J.I. and Lazcano, A (2002) Molecular evolution: lysine formation in higher plants. *Journal of Molecular Evolution* 55, 445–459.

Vogel, H.J. (1959) On biochemical evolution: lysine formation in higher plants. *Proceedings of the National Academy of Sciences of the United States of America* 45, 1717–1721.

Wang, F., Blanchard, J.S. and Tang, X.J. (1997) Hydrogen exchange electrospray ionization mass spectrometry studies of substrate and inhibitor binding and conformational changes of *Escherichia coli* dihydrodipicolinate reductase. *Biochemistry* 36, 3755–3759.

Watanabe, N., Clay, M.D., van Belkum, M.J., Cherney, M.M., Vederas, J.C. and James, M.N. (2008) Mechanism of substrate recognition and PLP-induced conformational changes in LL-diaminopimelate aminotransferase from *Arabidopsis thaliana*. *Journal of Molecular Biology* 384, 1314–1329.

Watanabe, N., Clay, M.D., van Belkum, M.J., Fan, C., Vederas, J.C. and James, M.N.G. (2011) The structure of LL-diaminopimelate aminotransferase from *Chlamydia trachomatis*: implications for its broad substrate specificity. *Journal of Molecular Biology* 411, 649–660.

Webster, F.H. and Lechowich, R.V. (1970) Partial purification and characterization of dihydrodipicolinic acid synthetase from sporulating *Bacillus megaterium*. *Journal of Bacteriology* 1, 118–126.

Weinberger, S. and Gilvarg, C. (1970) Bacterial distribution of the use of succinyl and acetyl blocking groups in diaminopimelic acid biosynthesis. *Journal of Bacteriology* 101, 323–324.

Wenko, L.K., Treick, W.R. and Wilson, K.G. (1985) Isolation and characterization of a gene encoding *m*-diaminopimelate dehydrogenase from *Glycine max*. *Plant Molecular Biology* 4, 197–204.

White, P.J. (1983) The essential role of diaminopimelate dehydrogenase in the synthesis of lysine by *Bacillus sphaericus*. *Journal of General Microbiology* 129, 739–749.

Yamakura, F., Ikeda, Y., Kimura, K. and Sasakawa, T. (1974) Partial purification and some properties of pyruvate-aspartic semialdehyde condensing enzyme from sporulating *Bacillus subtilis*. *The Journal of Biochemistry* 76, 611–621.

Yugari, Y. and Gilvarg, C. (1965) The condensation step in diaminopimelate synthesis. *The Journal of Biological Chemistry* 240, 4710–4716.

5 Arginine Deiminase in Microorganisms

F. Leroy[1]* and D. Charlier[2]

[1]Research Group of Industrial Microbiology and Food Biotechnology, Vrije Universiteit Brussel, Brussels, Belgium; [2]Research Group of Microbiology, Vrije Universiteit Brussel, Brussels, Belgium

5.1 Abstract

The arginine deiminase (ADI) pathway comprises three central enzymes, namely arginine deiminase, ornithine transcarbamoylase and carbamate kinase, whose importance is often neglected in reviews of amino acid metabolism in microorganisms. The relevance of the ADI pathway is underscored by the early observations that under anaerobic conditions and in the absence of terminal electron acceptors such as nitrate and nitrite, fermentative arginine breakdown via the ADI system can be used as the sole energy source for bacteria such as *Pseudomonas aeruginosa*, *Bacillus licheniformis* and *Enterococcus faecalis*, where the pathway and its enzymes were originally discovered. The wide distribution of the ADI system further emphasizes its role as a secondary energy-providing pathway in suboptimal or harsh growth conditions, such as low-oxygen, low-pH, high-salt and low-carbohydrate environments. Moreover, the generation of ammonia via the ADI system may reduce pH stress in specific microniches containing high concentrations of organic acids. A role for ADI in bacterial survival and pathogenicity has been identified in clinical cultures, as shown for *Burkholderia pseudomallei* isolates from patients with melioidosis. In food matrices, the ADI system may play a role in the establishment of specific microbial communities, as exemplified by the adaptation of *Lactobacillus sakei* to fermented meat matrices. *Staphylococcus* spp., which have significance for both clinical settings and food technology, also seem to benefit from the ADI system, albeit in a species- and even strain-dependent manner.

5.2 Introduction

Arginine deiminase (ADI) activity is a relatively conserved metabolic characteristic that is present within a broad group of microorganisms that comprise representatives of the three domains of life. It is the most widespread anaerobic route for arginine degradation and has not only been found in several bacteria, both Gram negative and Gram positive (e.g. Stalon and Mercenier, 1984; Fernandez and Zuñiga, 2006; Ryan *et al.*, 2009; Choi *et al.*, 2012; Sánchez Mainar *et al.*, 2014; Xiong *et al.*, 2014), but also in protists (e.g. Ringqvist *et al.*, 2008), moulds (e.g. El-Sayed *et al.*, 2015) and *Archaea* (e.g. Ruepp and Soppa, 1996). In the archaeal domain of life, the presence of a complete and active ADI pathway has only been

*Corresponding author. E-mail address: fleroy@vub.ac.be

demonstrated in some halophilic Euryarchaeota. Genes coding for enzymes of the pathway other than ADI (i.e. ornithine transcarbamoylase and carbamate kinase, see Section 5.3.2) have been identified in the genomes of other, non-halophilic Archaea, but these appear to be involved in arginine biosynthesis rather than degradation (e.g. *Pyrococcus furiosus* CK (Uriarte *et al.*, 1999).

The ADI system offers a biological fitness advantage and plays an important role in the adaptation of microorganisms to their respective ecosystems. In addition, it can have implications for human health and well-being as it is often relevant for the competitiveness, virulence and metabolism of microorganisms encountered in both clinical and food settings; this is not only valid for pathogenic microorganisms such as *Listeria monocytogenes* (Ryan *et al.*, 2009), leading to matters of food safety, but also for some beneficial bacteria that have industrial applications (e.g. lactic acid bacteria used as starter cultures for food fermentations). Other mechanisms for microbial arginine metabolism are of course possible, such as through the activities of arginase, arginine succinyltransferase, arginine decarboxylase and nitric oxide synthase, but those will not be discussed in this chapter. More information on these enzymes is given by Sánchez Mainar *et al.* (2014), and for further information on arginase the reader is referred to Chapter 6.

Although the focus of this chapter will be mostly on aspects related to microbial metabolism and ecology, it is also of interest to point out that the ADI enzyme has potential as a therapeutic anticancer agent, though some issues of antigenicity and stability may still need further attention (El-Sayed *et al.*, 2015). This anticancer effect is related to the arginine-depleting effect of ADI, so that the cell cycle is arrested at the G1 phase, thus causing apoptosis. Furthermore, ADI-catalysed arginine catabolism is absent in animals; in contrast, arginine appears to be the sole energy source in many mycoplasmas (Miles, 1992) and a major energy source for protists such as *Giardia intestinalis* (Schofield *et al.*, 1992) and *Trichomonas vaginalis* (Linstead and Cranshaw, 1983). Therefore, the development of specific inhibitors of ADI activity may have therapeutic and diagnostic potential of medical interest.

5.3 General Mechanisms

In this section, an overview is given of the architecture of the ADI pathway – its enzymes, genes and regulatory mechanisms, as well as of its general role in ecological behaviour and adjustment to different environmental conditions and dynamics.

5.3.1 Overview of the pathway and its enzymes

In general, the ADI system consists of three cytosolic reactions based on the successive actions of the enzymes arginine deiminase (ADI, also called arginine dehydrolase), ornithine transcarbamoylase (OTC) and carbamate kinase (CK) (see Fig. 5.1) (Cunin *et al.*, 1986; Fernandez and Zuñiga, 2006). The first reaction is based on actual ADI activity and converts the initial substrate arginine into citrulline and ammonia. In the next step, a catabolic OTC causes the conversion of citrulline into ornithine and carbamoyl phosphate. Finally, the latter is dephosphorylated by CK activity into carbamate, which further decomposes chemically into ammonia and carbon dioxide, with the concomitant equimolar production of ATP. Taken as a whole, the ADI pathway thereby leads, per mole of arginine metabolized, to one mole of ornithine, one mole of carbon dioxide, two moles of ammonia and one mole of ATP by substrate-level phosphorylation. Besides providing energy through ATP formation and acid-neutralizing action through ammonia release, the generation of the intermediate carbamoyl phosphate also is of interest for microbial functioning as it serves in *de novo* pyrimidine biosynthesis (Kilstrup *et al.*, 2005).

5.3.2 Genetic make-up

The ADI pathway is encoded into the microbial genome by the *arcA*, *arcB* and *arcC* genes, which generate the three pathway enzymes ADI, OTC and CK, respectively. These *arc* operon genes are generally transcribed as a polycistronic mRNA, except in the archaeon *Halobacterium salinarum*, where each gene appears to have its own promoter (Ruepp and Soppa, 1996). This basic

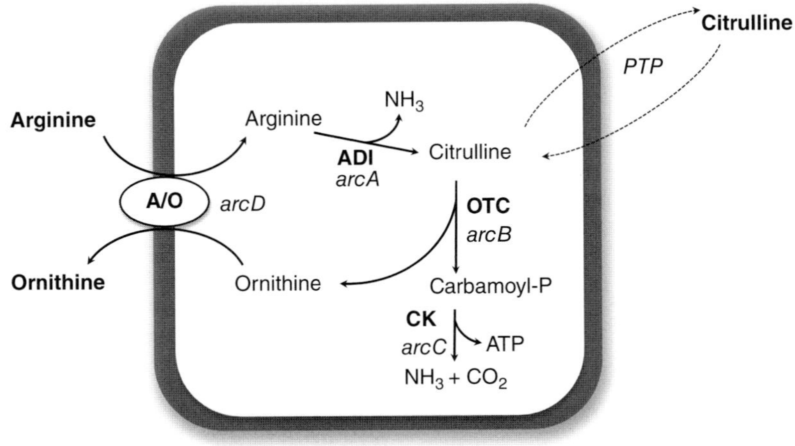

Fig. 5.1. Outline of the arginine deiminase pathway. Abbreviations: ADI, arginine deiminase; A/O, arginine/ornithine in/out; *arcA*, gene coding for ADI; *arcB*, gene coding for OTC; *arcC*, gene coding for CK; *arcD*, gene coding for an arginine/ornithine (A/O) antiporter; CK, carbamate kinase; OTC, ornithine transcarbamoylase; and *PTP*, gene coding for a putative citrulline/ornithine transporter (present in some bacteria).

tripartite gene cluster is frequently associated with an *arcD* gene that codes for a membrane-associated arginine/ornithine antiporter, an *arcR* gene that codes for a regulator of the Fnr/Anr/Crp family (see below) and an *arcT* gene that encodes a putative Xaa-His dipeptidase with a yet unclear function (Zuñiga *et al.*, 2002; Gupta *et al.*, 2013). In some bacteria, such as *E. faecalis* and *Streptococcus pneumoniae*, additional regulatory genes of the ArgR/AhrC family (see Section 5.3.3) involved in arginine-dependent regulation of the cluster are found in close proximity to the *arc* operon (Barcelona-Andrés *et al.*, 2002; Schulz *et al.*, 2014).

The genetic organization of the genes of the ADI pathway in an *arcDABC* operon was first characterized in *P. aeruginosa* (Baur *et al.*, 1989; Lüthi *et al.*, 1990). In other bacteria, the genes commonly exist in clustered configurations of variable composition and gene order. Thus, an *arcABDCR*, *arcABCRD* and *arcABCTD* operon are present in *B. licheniformis* (Maghnouj *et al.*, 1998, 2000), *E. faecalis* (Barcelona-Andrés, 2002) and *Clostridium perfringens* (Ohtani *et al.*, 1997), respectively, whereas the *arcABCTDR* configuration is found in strains of *L. sakei* where the operon is transcribed from a single *arcA* promoter (Rimaux *et al.*, 2013). In *L. sakei* CTC 494, an additional 1560 bp long gene (LSA0376) at 139 bp downstream of the *arc* operon has been identified, which probably yields a citrulline/ornithine transport protein (Rimaux *et al.*, 2013). Whether its presence and function are generally relevant to other microorganisms, or merely to a minority including *L. sakei*, is yet unknown and requires further investigation. Also, variations on the overall gene arrangements should be further explored within the heterogeneous group of ADI-positive microorganisms. In *Laribacter hongkongensis*, for instance, two functional and adjacent *arc* gene cassettes were found (Xiong *et al.*, 2014).

As another important example of genetic diversity and the spreading of ADI genes, staphylococci often contain a second copy of the *arc* operon on arginine catabolic mobile elements (ACMEs). The latter are pathogenicity islands that also carry, among others, genes for a spermine/spermidine *N*-acetyltransferase (*speG*) and several oligopeptide permeases (e.g. *opp-3*). Whole genomic analysis of the very persistent *Staphylococcus aureus* USA300 strain revealed that its ACME contains gene clusters related to *arc* and *opp-3* (Barbier *et al.*, 2011). Such ACMEs deserve further attention and may be of concern because they can be present in community-associated methicillin-resistant *S. aureus* strains (Diep *et al.*, 2008; Thurlow *et al.*, 2013), as well as in strains from other staphylococcal species that can act as opportunistic pathogens and may constitute a reservoir of ACMEs for transmission to more pathogenic strains (Pi *et al.*, 2009). Staphylococcal ACMEs probably originated from *S. epidermidis*, a major component of the human skin and mucosal microbiota, and a species in which about half of the strains seem to carry an ACME (Miragaia *et al.*, 2009; Barbier *et al.*, 2011). Experiments with *S. epidermidis* have suggested that native and ACME-derived ADIs can act in a compensatory manner (Lindgren *et al.*, 2014). Yet, the acquisition of an ACME may also result in extra enzymatic activity and, accordingly, confer a metabolic bonus on the microorganism (Thurlow *et al.*, 2013; Lindgren *et al.*, 2014). A preliminary study comparing two strains of *S. epidermidis* has led to the tentative conclusion that the presence of an ACME may indeed increase ADI pathway activity as compared with an ACME-negative strain cultivated under the same conditions (Janssens *et al.*, 2014). It is worth noticing that ACME-encoded *arc* operons are constitutively expressed, in contrast to the highly regulated *arc* operons of the core genome (Thurlow *et al.*, 2013).

5.3.3 Regulation of the pathway

As mentioned above, the expression of the ADI system depends at least in part on the regulatory protein encoded by the *arcR* gene, although a full comprehension of the regulation mechanism is still lacking. Nevertheless, some general lines have been sketched of the molecular underpinning of environmental effects on pathway functioning. At the molecular level, the best studied effects on activity and transcriptional regulation of the ADI pathway are oxygen tension and arginine availability, which are sensed and converted into transcriptional signals by regulatory proteins of the Fnr/Anr/Crp (fumarate nitrate reductase/anaerobic regulator of arginine catabolism and nitrate reduction/cAMP receptor protein) and ArgR/AhrC (arginine repressor/arginine hydroxamate resistance) families, respectively.

Anaerobiosis is known to stimulate the expression of the ADI system in different microorganisms (Gamper *et al.*, 1991; Champomier-Vergès *et al.*, 1999; Fernandez and Zuñiga, 2006). In some microorganisms, such as *P. aeruginosa* and *B. licheniformis*, the ADI pathway is used exclusively under anaerobic conditions and is at least partially dependent on a regulator of the Fnr/Anr type that binds to conserved ANR (anaerobic regulator) 'box' sequences in the control region of the *arc* operon (Gamper *et al.*, 1991; Maghnouj *et al.*, 1998, 2000). In *E. faecalis*, however, oxygen is a poor repressor, except in the presence of glucose (Simon *et al.*, 1982), while in electron-transport-deficient *Staphylococcus* small colony variants (SCV), the ADI pathway is stimulated aerobically in contrast to the parent strains, which only express it anaerobically (Proctor *et al.*, 2014). How this oxygen sensing occurs is not always clear. In *P. aeruginosa*, induction of the *arcDABC* operon under conditions of oxygen limitation or anaerobisosis relies on the Anr regulator that is not part of the *arc* operon, though in *B. licheniformis*, *L. sakei* and many other Gram-positive bacteria, the *arcR* regulatory gene is part of the operon. Generally, members of the Fnr/Crp family that sense oxygen or respond to oxidative stress use a thiol-disulfide switch or an iron-sulfur cluster (Bauer *et al.*, 1999). However, no iron-sulfur cluster is present in the ArcR regulator of *B. licheniformis* and, moreover, none of the two cysteine residues present in the ArcR protein are required for the correct functioning of the regulator (Wohlkönig *et al.*, 2004). Upregulation of the *arcRACB* cluster of the halophilic archaeon *H. salinarum* under fermentative conditions appears to be significantly different from the bacterial systems (Ruepp and Soppa, 1996). As indicated above, each *arc* gene appears to have its own promoter and there is no indication for the presence of an ANR box-like binding site in the *arc* gene cluster of *H. salinarum*. Also, the potential regulator ArcR protein of *H. salinarum* does not show significant sequence similarity with members of the Fnr/Anr family and how this protein exerts regulation is not known.

Arginine, as the initial substrate, is often another inducer of ADI pathway activity, but it has only a minor auxiliary effect on transcription of the *arc* operon in *P. aeruginosa* (Lu *et al.*, 1999), though it is a strong inducer of transcription in *B. licheniformis* (Maghnouj *et al.*, 1998). In the latter, arginine-bound ArgR binds to a single 18 bp ARG box sequence in the *arcA* control region, just upstream of the potential binding site for ArcR. This organization of regulatory binding sites might explain the observation that induction under oxygen-limiting conditions requires the simultaneous availability of arginine (Maghnouj *et al.*, 1998). In *E. faecalis*, two divergently transcribed *argR/ahrC*-like regulatory genes, *argR1* and *argR2*, are located immediately upstream of the *arcABCRD* operon (Barcelona-Andrés *et al.*, 2002). Arginine is a strong inducer of the ADI pathway in this microorganism, irrespective of the state of aeration. This effect appears to be ArgR2 mediated as this protein has a conserved arginine-binding pocket, and it is unlikely that ArgR1 would be able to bind arginine. Based on the distribution of potential binding sites for ArcR, ArgR1 and ArgR2, it has even been proposed that the expression of the ADI pathway by ArcR might occur via alterations in the expression of *argR2*. Furthermore, transcription of both *argR1* and *argR2* is affected by the availability of glucose and arginine. Such an organization of regulatory target sites provides a possibility for integrating all these signals via the production of ArgR/AhrC-like regulators Barcelona-Andrés *et al.*, 2002). In *S. pneumoniae*, different ArgR-type regulators (ArgR1, ArgR2 and AhrC) have been identified in the catabolic and anabolic pathways of arginine, of which ArgR2 positively regulates the ADI pathway, besides having other effects (Schulz *et al.*, 2014). Some of the genetic regulation mechanisms are still speculative and others are probably yet to be discovered. In *Burkholderia pseudomallei*, a hypothetical gene BPSS1356 has been shown to be involved in transcription regulation of the ADI pathway in addition to many other metabolic systems (Yam *et al.*, 2014). The *arcABC* operon of *S. aureus* was shown to be upregulated in biofilms (Resch *et al.*, 2005) and activated by Agr, the accessory gene regulator of *S. aureus* (Dunman *et al.*, 2001).

What is more, ADI pathway activity is usually under carbon catabolite repression (CCR), mediated by the catabolite control protein A (CcpA) and, as indicated above, induced by arginine (Dong *et al.*, 2004). Regulation thus depends on a balance between the level of repressive carbohydrates and substrate availability (Hitzmann *et al.*, 2013). Considering the link between the ADI pathway and ATP production, it is to be expected that substrate availability and energetic limitations are involved in the overall regulation of the ADI pathway (Pereira *et al.*, 2009). Additionally, the degree of CCR is dependent on the type of carbohydrate that is used as an energy source. Galactose is a stronger repressor of the pathway than glucose or lactose in *Lactobacillus leichmannii* (Manca de Nadra, 1986), but not so in *L. lactis* (Poolman *et al.*, 1987). Carbohydrate consumption by lactic acid bacteria commonly leads to the formation of organic acids affecting the pH, which, in its turn, will also play a role in the regulatory processes. For instance, at low pH values, the conversion of arginine by ADI-positive lactobacilli leads mainly to ornithine, whereas at moderate pH values a mixture of ornithine and the excreted intermediate citrulline is found, possibly to avoid too high intracellular concentrations of citrulline or to slow down ammonia accumulation (Vrancken *et al.*, 2009a; Rimaux *et al.*, 2011, 2012). How the pH exactly affects the mechanisms of ADI pathway activity is not clear-cut, as a decrease in intracellular pH may not only affect enzymatic functioning but also lead to a highly complex set of regulatory signals (Olsen, 1993).

On the whole, there is considerable interaction between the concentration changes in arginine and carbohydrates, the development of oxygen gradients and pH variations, and the dynamics of energy status and cell density. Further, a link with the growth phase seems to be at play, although it is not known how this is exactly operated. Within the species *L. sakei*, for instance, strain CTC 494 displays maximal *arc* gene expression at the end of its exponential growth phase, while for strain 23K this is already the case during the mid-exponential growth phase (Rimaux *et al.*, 2012). Such observations may indicate different and strain-specific adaptations to particular niches.

Besides the above-mentioned effects of oxygen, arginine, carbohydrates and pH, other factors that seem to affect ADI pathway activity include temperature and salt concentration (Vrancken *et al.*, 2009b), as well as variations in the contents of malic acid, citric acid, phenolic compounds and ethanol (Arena *et al.*, 2005, 2012; Araque *et al.*, 2013). Although it is not always straightforward to disentangle specific regulatory effects from growth-related kinetic effects, the general environmental situation can indeed interfere strongly with ADI pathway activity. This dependence may not be surprising, as it has been suggested that the regulation of *arc* genes can be affected by global regulators, for instance related to the interplay between the sigma factor σ^B and the virulence regulator PrfA in *L. monocytogenes* (Ryan *et al.*, 2009).

5.4 Ecological Situation and Examples

Below, a general overview is given of the ecological role of the ADI pathway in different microorganisms, followed by some specific examples from both non-food and food environments.

5.4.1 Role in ecological adaptation

The ADI pathway offers specific and meaningful advantages to microorganisms thriving in challenging environments. As outlined above, the pathway generates energy, yields a precursor for pyrimidine biosynthesis and serves in combating the acid stress that bacteria encounter in low pH environments. From experiments with *Streptococcus pyogenes*, the results indicated that the protection against low pH may occur in two ways: (i) by buffering of the extracellular environment through ammonia generation from arginine catabolism; or (ii) by ATP generation during citrulline conversion to activate F_1F_0-ATP synthase (Cusumano and Caparon, 2015). The combating of acid stress is particularly relevant upon prolonged carbohydrate depletion during fermentative processes. Whereas the enzymes of the glycolysis are eventually inhibited by acidity, the enzymes of the ADI pathway may still be active down to pH levels of 3.1 (De Angelis *et al.*, 2002; Lund *et al.*, 2014; Teixeira *et al.*, 2014). At still lower pH values and in some microorganisms, a complementary pH-balancing mechanism based on glutamate decarboxylation and glutamine deamidation may occur as the impact of ADI fades out (Teixeira *et al.*, 2014). Still, during fermentation, carbohydrate depletion will relieve CCR and contribute to ADI pathway activation. As a consequence, ecosystems that are characterized by simultaneous acidification and carbohydrate depletion may lead to ADI pathway activity and, thereby to extra ATP generation and pH balancing. Prolonged viability during the stressful conditions of the stationary growth phase may thus be obtained for ADI-positive microorganisms; this seems to be mainly due to the additional energy production rather than to the countering of acidic stress (Champomier-Vergès *et al.*, 1999, Fernandez and Zuñiga, 2006).

As life in high-salt environments is particularly energy demanding (Ma *et al.*, 2010), ATP production via the ADI pathway is an interesting means of generating extra energy for halophilic Euryarchaeota such as *H. salinarum*, especially under conditions of low aeration and low light intensity (where the light-driven proton pump is inactive). The capacity to grow by arginine fermentation in the absence of oxygen and light has even been developed into a procedure for the enrichment and isolation of *Halobacterium* strains from hypersaline environments (Oren and Litchfield, 1999).

The ADI-associated advantage for growth in biofilms (Resch *et al.*, 2005; Barbier *et al.*, 2011, Lindgren *et al.*, 2014; Proctor *et al.*, 2014), which are more tolerant to antibiotics than planktonic cells and frequently escape from the host immune responses, may also be of importance for the development of pathogens and opportunistic pathogens such as *P. aeruginosa*, *S. aureus* and *S. epidermidis* that have easy access to wounds and can cause chronic infections, such as cystic fibrosis and the growth of infections on catheters, shunts and other implanted devices (known as 'chronic-polymer-associated infections').

5.4.2 Examples from (clinical) non-food ecosystems

Arginine is present in the human body at concentrations of about 13 to 45 μM (human cerebrospinal fluid and human plasma, respectively) up to concentrations exceeding 1000 μM in muscle tissue (Schulz *et al.*, 2014). As a substrate, it is available to pathogenic and opportunistic pathogenic bacteria possessing the required ADI machinery, potentially reinforcing their virulence. This is not only the case in human hosts but also in animals.

Increased pathogenicity and virulence has, for instance, been shown for a range of streptococci, including strains of *S. pneumoniae* and *S. pyogenes* (Degnan *et al.*, 2000; Schulz *et al.*, 2014; Cusumano and Caparon, 2015). In dental health, it has been suggested that strain-specific variations in streptococcal ADI pathway activity play a role in the modulation and establishment of specific oral biofilms, which can have repercussions for the development of caries (Dong *et al.*, 2004; Huang *et al.*, 2015). Moreover, ADI pathway activity is of importance for the biological fitness of the zoonotic pathogen *Streptococcus suis* in pigs (Fulde *et al.*, 2014), as well as for strains of *S. canis* in both cats and dogs (Hitzmann *et al.*, 2013).

In pathogenic staphylococci, the ADI machinery may contribute to their persistence on the rather acidic mammal skin environment (pH 4.2–5.9), which may lead to skin infections and abscess formation (Lindgren *et al.*, 2014; Proctor *et al.*, 2014). In addition, the achievement of pH homeostasis through ADI pathway activity may contribute to an increase in staphylococcal biofilm maturation (Lindgren *et al.*, 2014). It is, however, important to realize that ADI pathway activity in staphylococci is not only a species-dependent but also often a strain-dependent property. It is, for example, very commonly present in *S. epidermidis* but much less so in species as *S. equorum* and *S. succinus* (Janssens *et al.*, 2014; Sánchez Mainar *et al.*, 2014).

In melioidosis-associated clinical isolates of the Gram-negative bacterium *B. pseudomallei*, colony morphology variants show variations in the expression of ADI enzymes at both the mRNA and protein levels, which indicate a role for the ADI pathway in survival under acidic conditions, but not for intracellular survival of the bacterium within the mouse macrophage cell (Chantratita *et al.*, 2012).

In addition to the above-mentioned protective effects, the role of the ADI system in pathogenesis can also be of a more active nature. In *Giardia*, it is believed to be part of the disease mechanism because arginine depletion through ADI pathway activity can cause apoptosis of human intestinal cells and disable the arginine-based innate host defence mechanism related to nitric oxide synthase activity (Ringqvist *et al.*, 2008).

In conclusion, ADI pathway activity may increase both the survival and virulence of several pathogens, which also encompass *L. hongkongensis* (Xiong *et al.*, 2014), *L. monocytogenes* (Ryan *et al.*, 2009) and *Salmonella enterica* serovar Typhimurium (Choi *et al.*, 2012).

5.4.3 Examples from food ecosystems

In food settings, the ADI system is particularly relevant during the production of fermented foods and beverages. Several industrially applied fermentative microorganisms display ADI pathway activity, which may be providing them with a competitive advantage over the background microbiota, thereby reinforcing process control. This is particularly the case for the group of lactic acid bacteria, as has for instance been shown for *L. sakei* in the context of fermented meat (Champomier-Vergès *et al.*, 1999; Rimaux *et al.*, 2011, 2012, 2013) and for *L. fermentum* and *L. reuteri* in sourdough environments (De Angelis *et al.*, 2002; Rollán *et al.*, 2003; Vrancken *et al.*, 2009a,b). In sourdough, the formation of ornithine has an additional technological benefit, as it acts as a precursor of 2-acetyl-1-pyrroline, contributing to the characteristic roast aroma of the crust of baked wheat bread (Gänzle *et al.*, 2007).

Other than lactic acid bacteria, certain ADI-positive strains of coagulase-negative staphylococci are also of technological bearing in food ecosystems, especially in fermented meats, where they are commonly used as starter cultures beside acidifying lactobacilli or pediococci (Ravyts *et al.*, 2012). As such, they contribute to the flavour, colour and oxidative stability of the end products. In a screening experiment using 86 staphylococcal strains from 17 different species, 45% of the strains possessed the *arcA* gene (Sánchez Mainar *et al.*, 2014). The staphylococcal ADI mechanism is especially active under the microaerobic conditions prevailing in a fermented meat matrix, whereas several strains switch to arginase activity when exposed to oxygen (Sánchez-Mainar *et al.*, 2014). Moreover, meat is naturally poor in carbohydrates but relatively rich in alternative energy sources such as arginine and nucleosides (Janssens *et al.*, 2014). In meat, arginine levels increase substantially post-mortem due to proteolysis of the muscle proteins. So, when the carbohydrates are depleted by the lactic acid bacteria, yielding lactic acid and a drop in pH, an important fallback is provided to the coagulase-negative staphylococci based on arginine conversion through the ADI system. Hypothetically, this may provide an explanation why *S. xylosus* is less commonly recovered from highly acidified fermented meats than when a starter culture of *S. carnosus* is used

(Ravyts *et al.*, 2012), because the former species generally possesses no or a rather low ADI pathway activity compared with the latter (Sánchez Mainar *et al.*, 2014).

Despite the above-mentioned advantages, some potential drawbacks of ADI pathway activity in food systems may come about. In acidified vegetable products, the generation of ammonia due to ADI pathway activity by *B. licheniformis* may cause an undesirable increase in pH, although this can be inhibited by the addition of appropriate concentrations of preservatives such as acetic acid (Yang *et al.*, 2015). More worrisome is the fact that ornithine, the end product of the ADI system, may act as a precursor of the biogenic amine putrescine (Araque *et al.*, 2013; Constantini *et al.*, 2013). Also, the release of citrulline by ADI-positive microorganisms in fermented foods and beverages may be alarming, because it can act as a precursor for the carcinogen ethyl carbamate. This has, for instance, been documented for lactic acid bacteria in wines (Arena and Manca de Nadra, 2005; Arena *et al.*, 2005, 2012; Araque *et al.*, 2013; Fang *et al.*, 2013) and soy sauce (Zhang *et al.*, 2014).

5.5 Conclusions

The fact that the ADI machinery is widespread among microorganisms is no coincidence: it confers a range of ecological and fitness-boosting advantages. Whether these advantages are based on competitive arginine depletion, energy generation or acid neutralization probably depends on the ecosystem. An improved knowledge of ADI mechanisms and their regulation can assist in exerting a larger control over the microbial world, to the benefit of man. This can be done by either an enhanced inhibition or stimulation of harmful and beneficial microorganisms, respectively. However, the interplay with other interfering metabolic features and the dynamic evolution of the environmental parameters of the different ecosystems make such a task convoluted.

Acknowledgements

FL acknowledges financial support of the Research Council of the Vrije Universiteit Brussel (OZR, SRP, IRP, HOA, and IOF projects), the Hercules Foundation and the Research Foundation-Flanders. DC acknowledges financial support of the Research Foundation Flanders, the Research Council of the Vrije Universiteit Brussel (OZR) and the Vlaamse Gemeenschapscommissie. Luc De Vuyst is acknowledged for his critical reading of the manuscript.

References

Araque, I., Bordons, A. and Reguant, C. (2013) Effect of ethanol and low pH on citrulline and ornithine excretion and *arc* gene expression by strains of *Lactobacillus brevis* and *Pediococcus pentosaceus*. *Food Microbiology* 33, 107–113.

Arena, M.E. and Manca de Nadra, M.C. (2005) Influence of ethanol and low pH on arginine and citrulline metabolism in lactic acid bacteria from wine. *Research in Microbiology* 156, 858–864.

Arena, M.E., Lisi, M.S., Manca de Nadra, M.C. and Alberto, M.R. (2012) Wine composition plays an important role in the control of carcinogenic precursor formation by *Lactobacillus hilgardii* X_1B. *Journal of the Science of Food and Agriculture* 93, 142–148.

Barbier, F., Lebeaux, D., Hernandez, D., Delannoy, A.S., Caro, V., François, P., Schrenzel, J., Ruppé, E., Gaillard, K., Wolff, M. *et al.* (2011) High prevalence of the arginine catabolic mobile element in carriage isolates of methicillin-resistant *Staphylococcus epidermidis*. *Journal of Antimicrobial Chemotherapy* 66, 29–36.

Barcelona-Andrés, B., Marina, A. and Rubio, V. (2002) Gene structure, organization, expression, and potential regulatory mechanisms of arginine catabolism in *Enterococcus faecalis*. *Journal of Bacteriology* 184, 6289–6300.

Bauer, C.E., Elsen, S. and Bird, T.H. (1999) Mechanisms for redox control of gene expression. *Annual Review of Microbiology* 53, 495–523.

Baur, H, Lüthi, E., Stalon, V., Mercenier, A. and Haas, D. (1989) Sequence analysis and expression of arginine-deiminase and carbamate-kinase genes of *Pseudomonas aeruginosa*. *European Journal of Biochemistry*, 179, 53–60.

Champomier-Vergès, M.C., Zuñiga, M., Morel-Deville, F., Perez-Martinez, G., Zagorec, M. and Ehrlich, S.D. (1999) Relationships between arginine degradation, pH and survival in *Lactobacillus sakei*. *FEMS Microbiology Letters* 180, 297–304.

Chantratita, N., Tandhavanani, S., Wikraiphat, C., Trunck, L.A., Rholl, D.A., Thanwisai, A., Saiprom, N., Limmathurotsakul, D., Korbsrisate, S., Day, N.P. *et al.* (2012) Proteomic analysis of colony morphology variants of *Burkholderia pseudomallei* defines a role for the arginine deiminase system in bacterial survival. *Journal of Proteomics* 75, 1031–1042.

Choi, Y., Choi, J., Groisman, E.A., Kang, D.H., Shin, D. and Ryu, S. (2012) Expression of STM4467-encoded arginine deiminase controlled by the STM4463 regulator contributes to *Salmonella enterica* serovar Typhimurium virulence. *Infection and Immunity* 80, 4291–4297.

Costantini, A., Pietroniro, R., Doria, F., Pessione, E. and Garcia-Moruno, E. (2013) Putrescine production from different amino acid precursors by lactic acid bacteria from wine and cider. *International Journal of Food Microbiology* 165, 11–17.

Cunin, R., Glansdorff, N., Piérard, A. and Stalon, V. (1986) Biosynthesis and metabolism of arginine in bacteria. *Microbiological Reviews* 50, 314–352.

Cusumano, Z.T. and Caparon, M.G. (2015) Citrulline protects *Streptococcus pyogenes* from acid stress using the arginine deiminase pathway and the F_1F_0-ATPase. *Journal of Bacteriology* 7, 1288–1296.

De Angelis, M., Mariotti, L., Rossi, J., Servili, M., Fox, P. F., Rollán, G. and Gobbetti, M. (2002) Arginine catabolism by sourdough lactic acid bacteria: purification and characterization of the arginine deiminase pathway enzymes from *Lactobacillus sanfranciscensis* CB1. *Applied and Environmental Microbiology* 68, 6193–6201.

Degnan, B.A., Fontaine, M.C., Doebereiner, A.H., Lee, J.J., Mastroeni, P., Dougan, G., Goodacre, J.A. and Kehoe, M.A. (2000) Characterization of an isogenic mutant of *Streptococcus pyogenes* Manfredo lacking the ability to make streptococcal acid glycoprotein. *Infection and Immunity* 68, 2441–2448.

Diep, B.A., Stone, G.G., Basuino, L., Graber, C.J., Miller, A., des Etages, S.A., Jones, A., Palazzolo-Balance, A.M., Perdreau-Remington, F., Sensabaugh, G.F. *et al.* (2008) The arginine catabolic mobile element and staphylococcal chromosomal cassette *mec* linkage: convergence of virulence and resistance in the USA300 clone of methicillin-resistant *Staphylococcus aureus*. *Journal of Infectious Diseases* 197, 1523–1530.

Dong, Y., Chen, Y.Y. and Burne, R.A. (2004) Control of expression of the arginine deiminase operon of *Streptococcus gordonii* by CcpA and Flp. *Journal of Bacteriology* 186, 2511–2514.

Dunman, P.M., Murphy, E., Haney, S., Palacios, D., Tucker-Kellogg, G., Wu, S., Brown, E.L., Zagursky, R.J., Shlaes, D. and Projan, S.J. (2001) Transcription profiling-based identification of *Staphylococcus aureus* gene regulated by the *agr* and/or *sarA* loci. *Journal of Bacteriology* 183, 7341–7353.

El-Sayed, A.S.A., Hassan, M.N. and Nada, H.M.S. (2015) Purification, immobilization, and biochemical characterization of L-arginine deiminase from thermophilic *Aspergillus fumigatus* KJ434941: anticancer activity in vitro. *Biotechnology Progress* 31, 396–405.

Fang, R.-S., Dong, Y.-C., Xu, T.-Y., He, G.-Q. and Chen, Q.H. (2013) Ethyl carbamate formation regulated by ornithine transcarbamylase and urea metabolism in the processing of Chinese yellow rice wine. *International Journal of Food Science and Technology* 48, 2551–2556.

Fernandez, M. and Zuñiga, M. (2006) Amino acid catabolic pathways of lactic acid bacteria. *Critical Reviews in Microbiology* 32, 155–183.

Fulde, M., Willenborg, J., Huber, C., Hitzmann, A., Willms, D., Seitz, M., Eisenreich, W., Valentin-Weigand, P. and Goethe, R. (2014) The arginine-ornithine antiporter ArcD contributes to biological fitness of *Streptococcus suis*. *Frontiers in Cellular and Infection Microbiology* 4: 107.

Gamper, M., Zimmermann, A. and Haas, D. (1991) Anaerobic regulation of transcription initiation in the *arcDABC* operon of *Pseudomonas aeruginosa*. *Journal of Bacteriology* 173, 4742–4750.

Gänzle, M.G., Vermeulen, N. and Vogel, R.F. (2007) Carbohydrate, peptide and lipid metabolism of lactic acid bacteria in sourdough. *Food Microbiology* 24, 128–138.

Gupta, R., Yang, J., Dong, Y., Swiatlo, E., Zhang, J.R., Metzger, D.W. and Bai, G. (2013) Deletion of *arcD* in *Streptococcus pneumoniae* D39 impairs its capsule and attenuates virulence. *Infection and Immunity* 81, 3903–3911.

Hitzmann, A., Bergmann, S., Rohde, M., Chhatwal, G.S. and Fulde, M. (2013) Identification and characterization of the arginine deiminase system of *Streptococcus canis*. *Veterinary Microbiology* 162, 270–277.

Huang, X.L., Schulte, R.M., Burne, R.A. and Nascimento, M.M. (2015) Characterization of the arginolytic microflora provides insights into pH homeostasis in human oral biofilms. *Caries Research* 49, 165–176.

Janssens, M., Van der Mijnsbrugge, A., Sánchez Mainar, M., Balzarini, T., De Vuyst, L. and Leroy, F. (2014) The use of nucleosides and arginine as alternative energy sources by coagulase-negative staphylococci in view of meat fermentation. *Food Microbiology* 39, 53–60.

Kilstrup, M., Hammer, K., Ruhdal Jensen, P. and Martinussen, J. (2005) Nucleotide metabolism and its control in lactic acid bacteria. *FEMS Microbiology Reviews* 29, 555–590.

Lindgren, J.K., Thomas, V.C., Olson, M.E., Chaudhari, S.S., Nuxoll, A.S., Schaeffer, C.R., Lindgren, K.E., Jones, J., Zimmerman, M.C., Dunman, P.M. *et al.* (2014) Arginine deiminase in *Staphylococcus epidermidis* functions to augment biofilm maturation through pH homeostasis. *Journal of Bacteriology* 196, 2277–2289.

Linstead, D. and Cranshaw, M.A. (1983) The pathway for arginine catabolism in the parasitic flagellate *Trichomonas vaginalis*. *Molecular and Biochemical Parasitology* 8, 241–252.

Lu, C.D., Winteler, H., Abdelal, A. and Haas, D. (1999) The ArgR regulatory protein, a helper to the anaerobic regulator ANR during transcriptional activation of the *arcD* promoter in *Pseudomonas aeruginosa*. *Journal of Bacteriology* 181, 2459–2464.

Lund, P., Tramonti, A. and De Biase, D. (2014) Coping with low pH: molecular strategies in neutralophilic bacteria. *FEMS Microbiology Reviews* 38, 1091–1125.

Lüthi, E., Baur, H., Gamper, M., Brunner, F., Villeval, D., Mercenier, A. and Haas, D. (1990) The *arc* operon for anaerobic arginine catabolism in *Pseudomonas aeruginosa* contains an additional gene, *arcD*, encoding a membrane protein. *Gene* 87, 37–43.

Ma, Y., Galinski, E.A., Grant, W.D., Oren, A. and Ventosa, A. (2010) Halophiles 2010: life in saline environments. *Applied and Environmental Microbiology* 76, 6971–6981.

Maghnouj, A., de Sousa Cabral, T.F., Stalon, V. and Vander Wauven, C. (1998) The *arcABDC* gene cluster, encoding the arginine deiminase pathway of *Bacillus licheniformis*, and its activation by the arginine repressor ArgR. *Journal of Bacteriology* 180, 6468–6475.

Maghnouj, A., Abu-Bakr, A.A.W., Baumberg, S., Stalon, V. and Vander Wauven, C. (2000) Regulation of anaerobic arginine catabolism in *Bacillus licheniformis* by a protein of the Crp/Fnr family. *FEMS Microbiology Letters* 191, 227–234.

Manca de Nadra, M. C., Nadra Chaud, C.A., Pesce de Ruiz Holgado, A. and Oliver, G. (1986) Arginine metabolism in *Lactobacillus leichmannii*. *Current Microbiology* 13, 155–158.

Miles, R.J. (1992) Catabolism in mollicutes. *Journal of General Microbiology* 138, 1773–1783.

Miragaia, M., de Lencastre, H., Perdreau-Remington, F., Chambers, H. F., Higashi, J., Sullam, P.M., Lin, J., Wong, K.I., King, K.A., Otto, M. *et al.* (2009) Genetic diversity of arginine catabolic mobile element in *Staphylococcus epidermidis*. *PLoS ONE* 4(11): 1–9.

Ohtani, K., Bando, M., Swe, T., Banu, S., Oe, M., Hayashi, H. and Shimizu, T. (1997) Collagenase gene (*colA*) is located in the 3′-flanking region of the perfringolysin O (*pfo*A) locus in *Clostridium perfringens*. *FEMS Microbiology Letters* 146, 155–159.

Olsen, E. (1993) Influence of pH on bacterial gene expression. *Molecular Microbiology* 8, 5–14.

Oren, A. and Litchfield, C.D. (1999) A procedure for the enrichment and isolation of *Halobacterium*. *FEMS Microbiology Letters* 173, 353–358.

Pereira, C.I., San Romao, M.V., Lolkema, J.S. and Crespo, M.T. (2009) *Weissella halotolerans* W22 combines arginine deiminase and ornithine decarboxylation pathways and converts arginine to putrescine. *Journal of Applied Microbiology* 107, 1894–1902.

Pi, B.R., Yu, M.H., Chen, Y.G., Yu, Y.S. and Li, L.J. (2009) Distribution of the ACME-*arcA* gene among meticillin-resistant *Staphylococcus haemolyticus* and identification of a novel *ccr* allotype in ACME-*arcA*-positive isolates. *Journal of Medical Microbiology* 58, 731–736.

Poolman, B., Driessen, A.J. and Konings, W.N. (1987) Regulation of arginine-ornithine exchange and the arginine deiminase pathway in *Streptococcus lactis*. *Journal of Bacteriology* 169, 5597–5604.

Proctor, R.A., Kriegeskorte, A., Kahl, B.C., Becker, K., Löffler, B. and Peters, G. (2014) *Staphylococcus aureus* small colony variants (SCVs): a road map for the metabolic pathways involved in persistent infections. *Frontiers in Cellular and Infection Microbiology* 4: 99.

Ravyts, F., De Vuyst, L. and Leroy, F. (2012) Bacterial diversity and functionalities in food fermentations. *Engineering in Life Sciences* 12, 1–12.

Resch, A., Rosenstein, R., Nerz, C. and Götz, F. (2005) Differential gene expression profiling of *Staphylococcus aureus* cultivated under biofilm and planktonic conditions. *Applied and Environmental Microbiology* 71, 2663–2676.

Rimaux, T., Vrancken, G., Pothakos, V., Maes, D., De Vuyst, L. and Leroy, F. (2011) The kinetics of the arginine deiminase pathway in the meat starter culture *Lactobacillus sakei* CTC 494 are pH-dependent. *Food Microbiology* 28, 597–604.

Rimaux, T., Rivière, A., Illeghems, K., Weckx, S., De Vuyst, L. and Leroy, F. (2012) Expression of the arginine deiminase pathway genes in *Lactobacillus sakei* is strain dependent and is affected by the environmental pH. *Applied and Environmental Microbiology 78*, 4874–4883.

Rimaux, T., Rivière, A., Hebert, E. M., Mozzi, F., Weckx, S., De Vuyst, L. and Leroy, F. (2013) A putative transport protein is involved in citrulline excretion and re-uptake during arginine deiminase pathway activity by *Lactobacillus sakei*. *Research in Microbiology* 164, 216–225.

Ringqvist, E., Palm, J.E.D., Skarin, H., Hehl, A.B., Weiland, M., Davids, B.J., Reiner, D.S., Griffiths, W.J., Eckmann, L., Gillin, F.D. and Svärd, S.G. (2008) Release of metabolic enzymes by *Giardia* in response to interaction with intestinal epithelial cells. *Molecular Biochemistry and Parasitology* 159, 85–91.

Rollán, G., Lorca, G.L. and Font de Valdez, G. (2003) Arginine catabolism and acid tolerance response in *Lactobacillus reuteri* isolated from sourdough. *Food Microbiology* 20, 313–319.

Ruepp, A. and Soppa, J. (1996) Fermentative arginine degradation in *Halobacterium salinarum* (formerly *Halobacterium halobium*): genes, gene products, and transcripts of the *arcRACB* gene cluster. *Journal of Bacteriology* 178, 4942–4947.

Ryan, S., Begley, M., Gahan, C.G. and Hill, C. (2009) Molecular characterization of the arginine deiminase system in *Listeria monocytogenes*: regulation and role in acid tolerance. *Environmental Microbiology* 11, 432–445.

Sánchez Mainar, M., Weckx, S. and Leroy, F. (2014) Coagulase-negative staphylococci favor conversion of arginine into ornithine despite a widespread genetic potential for nitric oxide synthase activity. *Applied and Environmental Microbiology* 80, 7741–7751.

Schofield, P.J., Edwards, M.R., Matthews, J. and Wilson, J.R. (1992) The pathway of arginine catabolism in *Giardia intestinalis*. *Molecular and Biochemical Parasitology* 51, 29–36.

Schulz, C., Gierok, P., Petruschka, L., Lalk, M., Mäder, U. and Hammerschmidt, S. (2014) Regulation of the arginine deiminase system by ArgR2 interferes with arginine metabolism and fitness of *Streptococcus pneumoniae*. *mBio* 5(6): e01858.

Simon, J.P., Wargnies, B. and Stalon, V. (1982) Control of enzyme synthesis in the arginine deiminase pathway of *Streptococcus faecalis*. *Journal of Bacteriology*, 150, 1085–1090.

Stalon, V. and Mercenier, A. (1984) L-arginine utilization by *Pseudomonas* species. *Journal of General Microbiology* 130, 69–76.

Teixeira, J.S., Seeras, A., Sanchez-Maldonado, A.F., Zhang, C., Su, M.S.-W. and Gänzle, M.G. (2014) Glutamine, glutamate, and arginine-based acid resistance in *Lactobacillus reuteri*. *Food Microbiology* 42, 172–180.

Thurlow, L.R., Joshi, G.S., Clark, J.R., Spontak, J.S., Neely, C.J., Maile, R. and Richardson, A.R. (2013) Functional modularity of the arginine catabolic mobile element contributes to the success of USA300 methicillin-resistant *Staphylococcus aureus*. *Cell Host and Microbe* 13, 100–107.

Uriarte, M., Marina, A., Ramón-Maiques, S., Fita, I. and Rubio, V. (1999) The carbamoyl-phosphate synthetase of *Pyrococcus furiosus* is enzymologically and structurally a carbamate kinase. *The Journal of Biological Chemistry* 274, 16295–16303.

Vrancken, G., Rimaux, T., Weckx, S., De Vuyst, L. and Leroy, F. (2009a) Environmental pH determines citrulline and ornithine release through the arginine deiminase pathway in *Lactobacillus fermentum* IMDO 130101. *International Journal of Food Microbiology* 135, 216–222.

Vrancken, G., Rimaux, T., Wouters, D., Leroy, F. and De Vuyst, L. (2009b) The arginine deiminase pathway of *Lactobacillus fermentum* IMDO 130101 responds to growth under stress conditions of both temperature and salt. *Food Microbiology* 26, 720–727.

Wohlkönig, A., Stalon, V. and Vander Wauven, C. (2004) Purification of ArcR, an oxidation-sensitive regulatory protein from *Bacillus licheniformis*. *Protein Expression and Purification* 37, 32–38.

Xiong, L., Teng, J.L.L., Watt, R.M., Kan, B., Lau, S.K.P. and Woo, P.C.Y. (2014) Arginine deiminase pathway is far more important than urease for acid resistance and intracellular survival in *Laribacter hongkongensis*: a possible result of *arc* gene cassette duplication. *BMC Microbiology* 14, 42.

Yam, H., Rahim, A.A., Mohamad, S., Mahadi, N.M., Manaf, U.A., Shu-Chien, A.C. and Najimudin, N. (2014) The multiple roles of hypothetical gene BPSS1356 in *Burkholderia pseudomallei*. *PLoS ONE* 9(7): e99218.

Yang, Z.Q., Meng, X., Breidt, F., Dean, L.L. and Arritt, F.M. (2015) Effects of acetic acid and arginine on pH elevation and growth of *Bacillus licheniformis* in an acidified cucumber juice medium. *Journal of Food Protection* 78, 728–737.

Zhang, J.R., Fang, F., Chen, J. and Du, G.C. (2014) The arginine deiminase pathway of koji bacteria is involved in ethyl carbamate precursor production in soy sauce. *FEMS Microbiology Letters* 358, 91–97.

Zuñiga, M., Perez, G. and Gonzalez-Candelas, F. (2002) Evolution of arginine deiminase (ADI) pathway genes. *Molecular Phylogenetics and Evolution* 25, 429–444.

6 Arginase and Microbial Pathogenesis in the Lungs

M.J. Romero Lucas,[1,2] R.W. Caldwell,[2] D. Fulton,[1,2] T. Chakraborty[3] and R. Lucas[1,2,4*]

[1]*Vascular Biology Center, Augusta University, Augusta, Georgia, USA;* [2]*Department of Pharmacology and Toxicology, Medical College of Georgia, Augusta University, Augusta, Georgia, USA;* [3]*Institute for Medical Microbiology, Justus-Liebig University Giessen, Germany;* [4]*Division of Pulmonary Medicine, Medical College of Georgia, Augusta University, Augusta, Georgia, USA*

6.1 Abstract

Strict regulation of pulmonary capillary barrier integrity is crucial for preventing capillary leak and ensuring optimal gaseous exchange at the alveolar–capillary interface. Bacterial infection of the lungs, as can be seen during severe pneumonia, can dramatically impair lung barrier function and as such contribute to the development of pulmonary permeability oedema. We will discuss here the role of the semi-essential amino acid L-arginine and its metabolites in pulmonary barrier integrity during bacterial infection, as well as in host–pathogen interactions. We will stress the importance of the competition between the enzymes arginase and the endothelial nitric oxide synthase for their common substrate L-arginine in the regulation of endothelial barrier integrity in the lungs. Bacterial toxins, such as pneumococcal pneumolysin, can be present in high local concentrations in the lungs of patients with acute lung injury and severe pneumonia, especially upon antibiotic treatment. These toxins have the capacity to shift the balance between nitric oxide synthase and arginase, resulting in dominant activation of the latter, with very different outcomes in the endothelial versus macrophage compartments of the lung. This chapter should thus reflect the growing body of evidence indicating the potential role of L-arginine metabolism in host–pathogen interactions and in lung injury.

6.2 Introduction

Over 4 million people develop pneumonia each year in the USA. Over a half a million of these people are admitted to a hospital for treatment. Community-acquired pneumonia (CAP) represents a major cause of morbidity and mortality in mainly elderly patients (Waterer *et al.*, 2011), whereas childhood pneumonia is the leading single cause of mortality worldwide in children aged less than 5 years old (Rudan *et al.*, 2008). The fatality rate associated with pneumonia caused by *Streptococcus pneumoniae*, which accounts for 45% of all CAP cases, still approximates 20%, despite the use of potent antibiotics and aggressive intensive-care support. Moreover, mortality after influenza is often due to secondary pneumococcal infection (Klugman *et al.*, 2009).

*Corresponding author E-mail address: rlucas@augusta.edu

Although traditionally considered to be a disease that is mainly confined to the lungs, pneumococcal pneumonia also has important effects on the cardiovascular system at all stages of infection, a fact which has been known for over 100 years, but which was only recently revived (Sprunt and Luetscher, 1912; Brown *et al.*, 2014). This is substantiated by the observation that about a quarter of adults admitted to hospital with pneumonia will develop major acute cardiac complications during their stay (Corrales-Medina *et al.*, 2013). Intriguingly, death in patients suffering from pneumococcal pneumonia can occur days after initiation of antibiotic therapy, when the lungs are essentially sterile and the pneumonia is clearing; it correlates with the presence of bacterial virulence factors, such as pneumococcal pneumolysin (Witzenrath *et al.*, 2006, Lucas *et al.*, 2012a). This cytoplasmic haemolytic protein is released from *S. pneumoniae* during autolysis, or upon antibiotic treatment, as occurs after treatment with beta lactams (Fukuda *et al.*, 2006; Anderson *et al.*, 2007). Pneumolysin-induced lung injury has been suggested to result from direct pneumotoxic effects on the alveolar–capillary barrier, rather than from resident or recruited phagocytic cells (Maus *et al.*, 2004). Pulmonary permeability oedema is characterized by dysfunctional alveolar liquid clearance, which is often linked to impaired sodium uptake capacity (Lucas *et al.*, 2012b, Czikora *et al.*, 2014), as well as by endothelial hyperpermeability (Lucas *et al.*, 2009), and represents a potentially lethal complication of severe pneumonia

As no standard therapy is currently available to treat pulmonary permeability oedema associated with severe pneumonia, identification of the mechanisms regulating capillary barrier function is very important. In this chapter, we will address the role of the non-essential amino acid L-arginine and its metabolites on bacterial fitness, host defence and barrier integrity during pneumococcal pneumonia.

6.3 Importance of Exotoxins From G$^+$ Bacteria in Permeability Oedema

Infections with *S. pneumoniae* have accounted historically for more morbidity and mortality than any other bacterium (Hendley *et al.*, 1975). Moreover, the increasing prevalence of antibiotic resistance necessitates the development of novel therapeutic strategies to combat pneumonia. Acute lung injury (ALI) and acute respiratory distress syndrome (ARDS) are sentinel features of the septic lung failure in patients with pneumogenic sepsis.

Bacterial virulence factors, such as pneumolysin secreted by *S. pneumoniae*, contribute to the transition from pneumonia to ALI to sepsis and, finally, to death. Pneumolysin is a 53 kDa cytoplasmic thiol-activated toxin released during pneumococcal lysis and is implicated in the pathogenesis of pneumococcal disease (Witzenrath *et al.*, 2006; Lucas *et al.*, 2012a). This protein toxin has cytolytic and complement-activating properties (Rubins *et al.*, 1996).

Apart from infections with *S. pneumoniae*, *Listeria monocytogenes* infections, which mostly start as an oral infection caused by contaminated food, can also cause severe lung complications in individuals with predisposing conditions, such as age, pregnancy, cancer, chronic diseases or organ transplantation (Lerolle *et al.*, 2002; Chaigne *et al.*, 2013). Moreover, airway infections with *L. monocytogenes* represent a serious problem in early onset neonatal listeriosis (Posfay-Barbe and Wald, 2004). In this case, listeriolysin O, a cholesterol-dependent cytolysin with homology to pneumolysin and streptolysin O, is the main virulence factor and can cause permeability oedema. Listeriolysin directly increases epithelial and endothelial permeability (Xiong *et al.*, 2010), which can, moreover, cause a dysfunction in the ion channels that are implicated in alveolar liquid clearance (Lucas *et al.*, 2012b).

In general, pore-forming toxins such as pneumolysin and listeriolysin damage the membrane of the infected cells, which could lead to the death of the cells and thus also of the invading bacteria. Conversely, it has recently been proposed that these toxins prevent death of the invaded host cells by means of inducing an efflux of K^+ into the infected cells, which induces the formation of the inflammasome and the activation of caspase 1. Caspase 1 then induces the activation of central regulators of membrane biogenesis, which facilitate membrane repair and hence promote cell survival (Gurcel *et al.*, 2006).

6.4 Importance of L-Arginine in Pneumococcal Physiological Fitness

S. pneumoniae is a commensal of the mucosal surfaces of the human nasopharynx and upper airways. Pneumococci can, however, spread from the upper respiratory tract to regions of the lower respiratory tract by their ability to evade the early components of the host immune response and through the actions of their virulence factors, thereby causing pneumonia (Kadioglu *et al.*, 2008). Pneumococci are highly versatile pathogens, adapted to the nutrients provided by the host niches. Colonization of the nasopharyngeal cavity and dissemination during invasive infections crucially depend upon the physiological fitness of the pathogen.

L-arginine (L-Arg) is a semi-essential amino acid, which in most mammals, is synthesized from glutamine, glutamate and proline via the intestinal–renal axis (Popolo *et al.*, 2014). Most strains of *S. pneumoniae* are auxotrophic for L-Arg, and the arginine pool of the host and its metabolites can actually be utilized as energy sources by these pathogens (Fig. 6.1) (Fulde *et al.*, 2014). The arginine-ornithine antiporter (ArcD) is part of the arginine deiminase system (ADS). This catabolic, energy-providing pathway can be found in a variety of bacterial species. ADS consists of arginine deiminase (AD), the catabolic ornithine carbamoyltransferase (cOCT), and carbamate kinase (CK).

Pneumococcal genomes contain three ArgR-type regulators (ArgR1, ArgR2 and AhrC), which are suggested to be involved in the regulation of arginine metabolism (Schulz *et al.*, 2014). ArgR2 binds to promoter sequences of the arc operon of TIGR4 as the regulator of the ADS and ArcD. ArgR2 deficiency abrogates expression of the ADS, including the arginine-ornithine antiporter (Schulz *et al.*, 2014). ArcD-associated uptake of arginine has been suggested to be required to maintain fitness of *S. pneumoniae* (Schulz *et al.*, 2014). Of note, a recent report using bacterial whole-genome sequencing of clinical isolates of *S. pneumoniae* identified pneumococcal arginine biosynthesis genes. Mouse models of pneumonia showed that pneumococcal strains without arginine biosynthesis genes were attenuated in growth or cleared, from the lungs, blood and cerebrospinal fluid (CSF) (Piet *et al.*, 2014).

Taken together, these data indicate that both increased arginine uptake by *S. pneumoniae* and/or augmented arginine synthesis by the pathogen promote its growth and virulence in invasive pneumococcal disease. It should also be noted that the arginine metabolite agmatine, which is generated in the arginine decarboxylase pathway, can serve as a precursor for the synthesis of polyamines, which in turn enhance biofilm development in several strains of respiratory pathogens (Allan *et al.*, 2014; Paulson *et al.*, 2014).

6.5 Role of L-Arginine in Alveolar Macrophage Polarization During Pneumococcal Pneumonia

Sharply opposed to the role of L-Arg and its metabolites as energy sources for *S. pneumoniae*, this amino acid also represents an integral part of the host defence during pneumococcal pneumonia. An important bactericidal metabolite of L-Arg is nitric oxide (NO), which is mainly generated by inducible nitric oxide synthase (iNOS) in alveolar macrophages during pneumonia (Satoh *et al.*, 2001; Simões *et al.*, 2013). Nitric oxide not only directly, but also indirectly, confers bactericidal properties, the latter upon complex formation with superoxide, resulting in the reactive nitrogen species peroxynitrite (Satoh *et al.*, 2001; Yadav *et al.*, 2014). The actions of nitric oxide during pneumococcal pneumonia are, however, a double-edged sword. Indeed, apart from its potent bactericidal activities, nitric oxide, when produced in excess, can also contribute to damage of host tissues (Bergeron *et al.*, 1999). It is important to keep in mind that both the host and the bacteria can produce oxygen and nitrogen radicals, and that the interplay of pneumococcal H_2O_2 and eukaryotic nitric oxide has been shown to be a major contributor to organ damage in pneumococcal meningitis (Hoffmann *et al.*, 2006).

Infection of the lung with *S. pneumoniae* results in the upregulation of iNOS in alveolar macrophages (Orman *et al.*, 1998; Braun *et al.*,

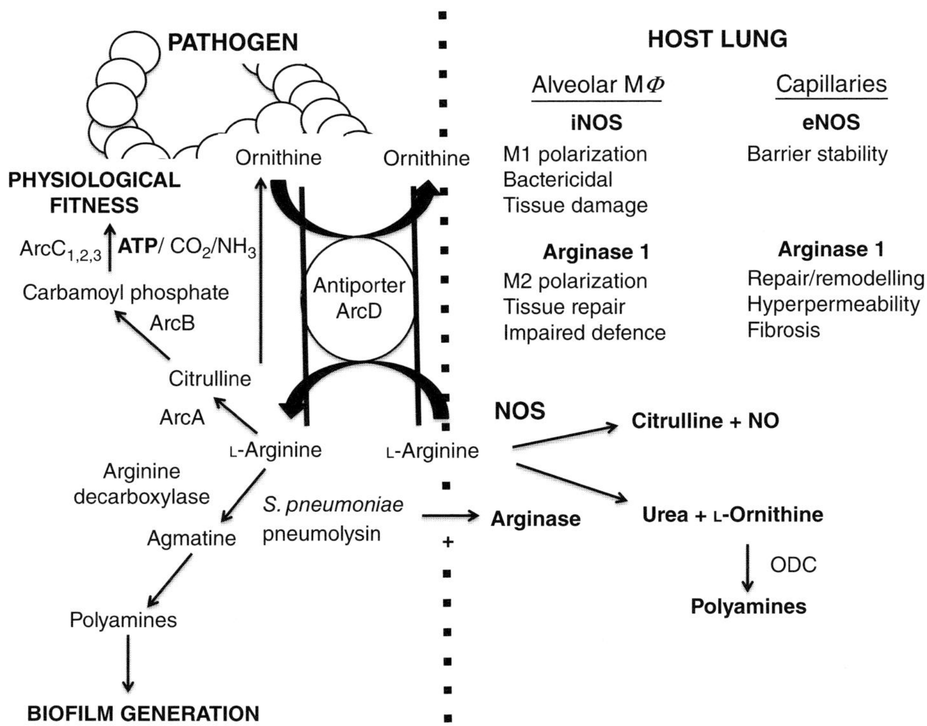

Fig. 6.1. Actions of L-arginine in the pathogen and in the lungs of the host during pneumococcal pneumonia. Left: L-Arginine (L-Arg) can be taken up by *Streptococcus pneumoniae* through the ornithine-arginine antiporter, ArcD. Once inside the pathogen, L-Arg can be converted to citrulline through the action of the enzyme ArcA (AD, arginine deiminase). Citrulline is then converted to carbamoyl phosphate in a reaction catalysed by ArcB (cOCT, ornithine carbamoyltransferase). Finally, carbamoyl phosphate is the substrate for the enzyme $ArcC_{1,2,3}$ (CK, carbamate kinase), which generates carbon dioxide and ammonia in a reaction that provides ATP as an energy source. As a result, this pathway improves energy generation and the physiological fitness of the pathogen. Another reaction is initiated by arginine decarboxylase, which generates agmatine, a precursor for polyamine synthesis. Polyamines improve the biofilm generation of pneumococci. Right: In the lungs of the host, L-Arg can affect alveolar macrophages as well as the capillary endothelium in different manners, depending on the activity of two competing enzymes: nitric oxide synthase (NOS) and arginase 1. NOS has different isoforms that are dominant in macrophages (inducible NOS; iNOS) and in endothelial cells (endothelial NOS, eNOS). Host defence against pneumococcal infection requires the activation of iNOS in the alveolar macrophages, which therefore have to polarize towards the M1 phenotype (which activates iNOS). When overproduced, NO can, however, also give rise to organ damage, either directly or indirectly. Pneumococcal pneumolysin activates the enzyme arginase 1, which polarizes macrophages towards the M2 phenotype. As a consequence, the bactericidal macrophage response is suppressed. In the capillaries, the activation of eNOS has barrier-strengthening functions, whereas arginase 1 activation increases capillary permeability. Arginase 1 activation generates urea and ornithine, and ornithine can increase polyamine formation through ornithine decarboxylase (ODC). This reaction has the capacity to improve tissue repair, but can also cause lung fibrosis in a more chronic setting.

1999) and it has been suggested the pneumococcal virulence factor pneumolysin plays a prominent role in this effect. Expression of iNOS is generally associated with M1 macrophage polarization. Intriguingly, *S. pneumoniae*, as well as its virulence factor pneumolysin, were also shown to induce activation of the enzyme arginase (Lucas *et al.*, 2012a; Knippenberg *et al.*, 2015a). This manganese-containing enzyme has two isoforms. The arginase 1 isoform is cytosolic and is mainly localized in the liver, where it eliminates nitrogen formed during amino acid and nucleotide metabolism via the urea cycle. Arginase 1 is also expressed in extra-hepatic tissues, including the vascular endothelium. Arginase 2 is a mitochondrial enzyme that is expressed in various cell types, including cells in the kidneys and the vasculature. In contrast to iNOS, which mediates bactericidal activities in M1 classically activated (polarized) alveolar macrophages, arginase 1 represents a hallmark of alternatively activated M2 macrophages and correlates with attenuated recruitment and function of pulmonary inflammatory cells (Wolf *et al.*, 2014).

The timing during which L-Arg is available to either iNOS or arginase 1 is crucial for the outcome in opportunistic pneumococcal infections in Influenza A. Secondary pneumococcal infection was shown to exacerbate disease in a mouse model of Influenza A Virus (IAV) infection and this correlated with the induction of M2 alveolar macrophages expressing arginase 1 (Chen *et al.*, 2012). These alternatively activated alveolar macrophages are induced by IAV and they likely contribute to hypersusceptibility to a secondary bacterial infection. In contrast, pneumococcal colonization prior to IAV rather protects from virus-induced morbidity and lung pathology by means of inducing arginase 1-expressing M2 alveolar macrophages in a pneumolysin-dependent manner. These alternatively activated alveolar macrophages, through their immunosuppressive nature, will then function as gatekeepers against virus-induced immunopathology (Wolf *et al.*, 2014).

In conclusion, the iNOS/arginase 1 balance in alveolar macrophages can determine the polarization of these cells, with important consequences for both bactericidal activity and lung injury. Apart from nitric oxide, another L-Arg metabolite – agmatine – was shown to activate the inflammatory response in the host, most likely through nuclear factor kappa B (NF-κB)-mediated upregulation of pro-inflammatory cytokines, such as tumour necrosis factor (TNF) (Paulson *et al.*, 2014). Another important L-Arg metabolite involved in host–pathogen interactions is ornithine, which has been demonstrated to be implicated in tissue repair and in anti-inflammatory actions, through the synthesis of polyamines (Gogoi *et al.*, 2016).

6.6 Opposing Actions of Arginase and Endothelial Nitric Oxide Synthase (eNOS) in Capillary Barrier Regulation

Endothelial barrier function in the lungs is determined by a balance between the adhesive forces at intercellular junctions and actin-generated centripetal forces (Bogatcheva *et al.*, 2002; Lucas *et al.*, 2009). When uncontrolled and lasting, the loss of adhesive forces can lead to impaired regulation of microvascular permeability which, in turn, causes oedema (Adyshev *et al.*, 2013). Oedema, in its turn, can lead to severe, and sometimes fatal, organ dysfunction. As such, increased paracellular flux of injury factors can aggravate alveolar liquid accumulation in the lungs, which then blunts oxygen exchange (Matthay, 2014). VE (vascular endothelial)-cadherin is the main component of adherens junctions (AJ) between cells and a crucial regulator of barrier integrity in the capillary endothelium (Usatyuk and Natarajan, 2005). A growing number of permeability inducing factors, including vascular endothelial growth factor (VEGF), reactive oxygen species (ROS), tumour necrosis factor (TNF), and the bacterial toxins lipopolysaccharide (LPS) and pneumolysin all signal through mechanisms that culminate in VE-cadherin destabilization (Dudek and Garcia, 2001; Lucas *et al.*, 2009).

The cytoplasmic haemolytic protein pneumolysin is released during the lysis of pneumococci, as occurs on treatment with beta lactams (Anderson *et al.*, 2007). Upon intratracheal instillation in mice, pneumolysin could be detected mainly in the endothelial cells of pulmonary arterial vessels, which concomitantly displayed strong vasoconstriction (Witzenrath *et al.*, 2006). Pneumolysin-induced lung injury was suggested to result from direct pneumotoxic effects on the alveolar–capillary barrier, rather than from resident or recruited phagocytic cells (Maus *et al.*, 2004). Upon binding of pneumolysin to cholesterol in cell membranes, oligomerization and pore formation occur, resulting in increased intracellular levels of Ca^{2+} (Lucas *et al.*, 2012a). A rise in cytosolic Ca^{2+} has been proposed to be the initial pivotal signal preceding endothelial cell contraction, including myosin light chain (MLC)-dependent mechanisms and microtubule rearrangement. Indeed, increased intracellular Ca^{2+} levels can activate key signalling pathways that mediate cytoskeletal reorganization (through MLC-dependent contraction) and disassembly of VE-cadherin at the adherens junctions (Yan *et al.*, 1996). Depolymerization of microtubules can cause disassembly of the adherens junction proteins with which they associate, such as VE-cadherin, thus increasing permeability (Petrache *et al.*, 2003). Tyr^{658} phosphorylation of VE-cadherin and the formation of stress fibres upon MLC phosphorylation provide a basis for rapid endocytosis and degradation of VE-cadherin and the loss of cell–cell contacts. The Ca^{2+}-dependent protein kinase C (PKC) α-isoform can be activated by pneumolysin and it has been suggested that it may play a critical role in initiating endothelial cell contraction and the disassembly of VE-cadherin-mediated cell–cell contacts (Lucas *et al.*, 2012a,b).

Rho (Ras homologous) GTP-binding proteins, which comprise multiple members of the Rho, Rac and Cdc42 subfamilies, are involved in the regulation of a variety of cellular processes (Sumi *et al.*, 1999). RhoA and Rho-associated kinase may directly catalyse MLC phosphorylation or act indirectly via the inactivation of MLC phosphatase (Kitazawa *et al.*, 2009) to induce cell contraction and endothelial barrier disruption, while endothelial barrier enhancement is associated with Rac 1-mediated formation of F-actin, increased association of focal adhesion proteins and the enlargement of intercellular adherens junctions (Vandenbroucke *et al.*, 2008). Therefore, a precise balance between RhoA- and Rac1-mediated signalling is essential for endothelial barrier regulation. Sublytic concentrations of pneumolysin were recently shown to induce cholesterol-dependent actin remodelling by interfering with the activity of RhoA (Iliev *et al.*, 2007). PKC α-activation can induce increased RhoA activity (Lucas *et al.*, 2012a).

Maintenance of the endothelial barrier requires a basal level of nitric oxide to be generated by endothelial NOS (eNOS) (Predescu *et al.*, 2005). Paradoxically, both the lack of nitric oxide (most likely eNOS derived) as well as excessively high levels of nitric oxide (most likely from iNOS) can destabilize interendothelial junctions (Cho *et al.*, 1999; Wong *et al.*, 2004). The availability of L-Arg is required for eNOS activity and nitric oxide production, and this is why L-Arg is essential for vascular integrity and function (Predescu *et al.*, 2005). Increased activity and/or overexpression of the enzyme arginase, which at least partially depends on p38 MAP (mitogen-activated protein) kinase (Shatanawi *et al.*, 2011) and RhoA activation (Chandra *et al.*, 2012), may impair the availability of L-Arg for eNOS and can thus increase the pathogenesis of vascular dysfunction (Romero *et al.*, 2008, 2012; Beleznai *et al.*, 2011).

In the vasculature, both isoforms 1 and 2 of arginase have been shown to be involved in the regulation of L-Arg homeostasis and in the production of L-ornithine for the subsequent synthesis of polyamines and proline, which are crucial components of physiological and pathological vascular remodelling (Cederbaum *et al.*, 2004; Lucas *et al.*, 2014; Caldwell *et al.*, 2015). Knockout of the arginase 1 gene in mice leads to the development of lethal hyperammonaemia within 2 weeks of birth. Arginase $1^{+/-}$/arginase $2^{-/-}$ transgenic mice were significantly protected from pneumolysin-induced lung capillary leak (Lucas *et al.*, 2012a, 2013a). Arginase competes with eNOS for their common substrate, L-Arg, thereby reducing the nitric oxide-generating capacity of the enzyme. eNOS catalyses the two-step conversion of substrate L-Arg into nitric oxide and utilizes electrons from NADPH to reduce molecular oxygen, followed by oxidation of the guanidino N group of arginine to form nitric oxide, L-citrulline and water (Caldwell *et al.*, 2015). It is this repeated activation of molecular oxygen that presents the opportunity for enzymatic missteps and superoxide production from eNOS, in the presence of reduced availability of L-Arg, a process that has been termed 'eNOS uncoupling'. All of these events will culminate in hyperpermeability in the capillaries (Lucas *et al.*, 2012a, 2013a). The activation of ornithine-mediated polyamine synthesis by arginase can promote lung tissue repair during severe pneumonia, which is a beneficial activity, but upon prolonged action it can also lead to lung fibrosis. Indeed, pneumococcal-derived pneumolysin was recently shown to promote pulmonary fibrosis (Knippenberg *et al.*, 2015b).

In summary, the inhibition of increased arginase activity, using newly developed specific arginase inhibitors, such as 2(*S*)-amino-6-hexanoic acid (ABH) (Steppan *et al.*, 2013), has the potential to restore dysfunctional eNOS, leading to reduced superoxide production and increased NO production by the enzyme within the endothelium, as well as blunted NADPH oxidase activation, so potentially improving endothelial barrier function. As no standard therapy is currently available to treat pulmonary permeability oedema, the need for novel substances that can improve pulmonary capillary barrier integrity in pneumonia patients is important. Based on our results and those of others, arginase 1 seems to be a promising therapeutic target for addressing capillary leakage in the context of pneumonia (Lucas *et al.*, 2012a; 2013a,b).

6.7 Conclusions

As summarized in Fig. 6.1, local concentrations of the semi-essential amino acid L-Arg and its metabolites define several important parameters in pneumococcal pneumonia. The effects on the pathogen, *S. pneumoniae*, include (i) increasing the physiological fitness of the invading pathogen, and (ii) promoting biofilm generation. In contrast, in the host lungs, L-Arg: (i) defines M1/M2 alveolar macrophage polarization, which is crucial in host defence, but also affects lung injury; (ii) improves tissue repair upon stimulating polyamine synthesis (although this can also promote lung fibrosis); and (iii) regulates capillary barrier function through switching of the eNOS/arginase balance.

Acknowledgement

RL and TC received funding from the German Research Foundation (DFG, TR84 'Innate Immunity of the Lung').

References

Adyshev, D.M., Dudek, S.M., Moldobaeva, N., Kim, K.M., Ma, S.F., Kasa, A., Garcia, J.G. and Verin, A.D. (2013) Ezrin/radixin/moesin proteins differentially regulate endothelial hyperpermeability after thrombin. *American Journal of Physiology Lung Cellular and Molecular Physiology* 305, L240–255.

Allan, R.N., Skipp, P., Jefferies, J., Clarke, S.C., Faust, S.N., Hall-Stoodley, L. and Webb, J. (2014) Pronounced metabolic changes in adaptation to biofilm growth by *Streptococcus pneumoniae*. *PLoS ONE* 9(9): e107015.

Anderson, R., Steel, H.C., Cockeran, R., von Gottberg, A., de Gouveia, L., Klugman, K.P., Mitchell, T.J. and Feldman C. (2007) Comparison of the effects of macrolides, amoxicillin, ceftriaxone, doxycycline, tobramycin and fluoroquinolones, on the production of pneumolysin by *Streptococcus pneumoniae in vitro*. *Journal of Antimicrobials and Chemotherapy* 60, 1155–1158.

Beleznai, T., Feher, A., Spielvogel, D., Lansman, S.L. and Bagi, Z. (2011) Arginase 1 contributes to diminished coronary arteriolar dilation in patients with diabetes. *American Journal of Physiology Heart and Circulatory Physiology* 300, H777–783.

Bergeron, Y., Ouellet, N., Simard, M., Olivier, M. and Bergeron, M. G. (1999) Immunomodulation of pneumococcal pulmonary infection with N^G-monomethyl-L-arginine. *Antimicrobial Agents and Chemotherapy*. 43, 2283–2290.

Bogatcheva, N.V., Garcia, J.G. and Verin, A.D. (2002) Role of tyrosine kinase signaling in endothelial cell barrier regulation. *Vascular Pharmacology* 39, 201–212.

Braun, J.S., Novak, R., Gao, G., Murray, P.J, and Shenep, J.L. (1999) Pneumolysin, a protein toxin of *Streptococcus pneumoniae*, induces nitric oxide production from macrophages. *Infection and Immunity* 67, 3750–3756.

Brown, A.O., Mann, B., Gao, G., Hankins, J.S., Humann, J., Giardina, J., Faverio, P., Restrepo, M.I., Halade, G.V., Mortensen, E.M. *et al.* (2014) *Streptococcus pneumoniae* translocates into the myocardium and forms unique microlesions that disrupt cardiac function. *PLoS Pathogens* 10(9): e1004383.

Caldwell, R.B., Toque, H.A., Narayanan, S.P. and Caldwell, R.W. (2015) Arginase: an old enzyme with new tricks. *Trends in Pharmacological Sciences* 36, 395–405.

Cederbaum, S.D., Yu, H., Grody, W.W., Kern, R.M., Yoo, P. and Iyer, R.K. (2004) Arginases I and II: do their functions overlap? *Molecular Genetics and Metabolism* 81(Suppl 1), S38–44.

Chaigne, B., Diot, B., Eudo, C., Maillot, F., Mondon, K., Hommet, C. and Constans, T. (2013) *Listeria monocytogenes* pleuritis: report of a geriatric case. *Geriatry and Gerontology International* 13, 505–506.

Chandra, S., Romero, M.J., Shatanawi, A., Alkilany, A.M., Caldwell, R.B. and Caldwell R.W. (2012) Oxidative species increase arginase activity in endothelial cells through the RhoA/Rho kinase pathway. *British Journal of Pharmacology* 165, 506–519.

Chen, W.H., Toapanta, F.R., Shirey, K.A., Zhang, L., Giannelou, A., Page, C., Frieman, M.B., Vogel, S.N. and Cross, A.S. (2012) Potential role for alternatively activated macrophages in the secondary bacterial infection during recovery from influenza. *Immunology Letters* 141, 227–234.

Cho, M.M., Ziats, N.P., Pal, D., Utian, W.H. and Gorodeski, G.I. (1999) Estrogen modulates paracellular permeability of human endothelial cells by eNOS- and iNOS-related mechanisms. *American Journal of Physiology* 276(2 Pt 1), C337–349.

Corrales-Medina, V.F., Musher, D.M., Shachkina, S. and Chirinos, J.A. (2013) Acute pneumonia and the cardiovascular system. *The Lancet* 381, 496–505.

Czikora, I., Alli, A., Bao, H.F., Kaftan, D., Apell, H.J., White, R., Gorshkov, B., Zimmermann, A., Pauly-Evers, M., Sridhar, S. *et al.* (2014) A novel TNF-mediated mechanism of direct epithelial sodium channel activation. *American Journal of Respiratory and Critical Care Medicine* 190, 522–532.

Dudek, S.M. and Garcia, J.G. (2001) Cytoskeletal regulation of pulmonary vascular permeability. *Journal of Applied Physiology* 91, 1487–1500.

Fukuda, Y., Yanagihara, K., Higashiyama, Y., Miyazaki, Y., Hirakata, Y., Mukae, H., Tomono, K., Mizuta, Y., Tsukamoto, K. and Kohno, S. (2006) Effects of macrolides on pneumolysin of macrolide-resistant *Streptococcus pneumoniae*. *European Respiratory Journal* 27, 1020–1025.

Fulde, M., Willenborg, J., Huber, C., Hitzmann, A., Willms, D., Seitz, M., Eisenreich, W., Valentin-Weigand, P. and Goethe, R. (2014) The arginine-ornithine antiporter ArcD contributes to biological fitness of *Streptococcus suis*. *Frontiers in Cellular and Infection Microbiology* 4:107.

Gogoi, M., Datey, A., Wilson, K.T. and Chakravortty, D. (2016) Dual role of arginine metabolism in establishing pathogenesis. *Current Opinions in Microbiology* 29, 43–48.

Gurcel, L., Abrami, L., Girardin, S., Tschopp, J. and van der Goot, F.G. (2006) Caspase-1 activation of lipid metabolic pathways in response to bacterial pore-forming toxins promotes cell survival. *Cell* 126, 1135–1145.

Hendley, J.O, Sande, M.A., Stewart, P.M. and Gwaltney, J.M. Jr (1975) Spread of *Streptococcus pneumoniae* in families. I. Carriage rates and distribution of types. *Journal of Infectious Diseases* 132, 55–61.

Hoffmann, O., Zweigner, J., Smith, S.H., Freyer, D., Mahrhofer, C., Dagand, E., Tuomanen, E.I. and Weber, J.R. (2006) Interplay of pneumococcal hydrogen peroxide and host-derived nitric oxide. *Infection and Immunity* 74, 5058–5066.

Iliev, A.I., Djannatian, J.R., Nau, R., Mitchell, T.J. and Wouters F.S. (2007) Cholesterol-dependent actin remodeling via RhoA and Rac1 activation by the *Streptococcus pneumoniae* toxin pneumolysin. *Proceedings of the National Academy of Sciences of the United States of America* 104, 2897–2902.

Kadioglu, A., Weiser, J.N., Paton, J.C. and Andrew, P.W. (2008) The role of *Streptococcus pneumoniae* virulence factors in host respiratory colonization and disease. *Nature Reviews in Microbiology* 6, 288–301.

Kitazawa, T., Semba, S., Huh, Y.H., Kitazawa, K. and Eto, M. (2009) Nitric oxide-induced biphasic mechanism of vascular relaxation via dephosphorylation of CPI-17 and MYPT1. *Journal of Physiology* 587, 3587–3603.

Klugman, K.P., Chien, Y.W. and Madhi, S.A. (2009) Pneumococcal pneumonia and influenza: a deadly combination. *Vaccine* 27(Suppl 3), C9–C14.

Knippenberg, S., Brumshagen, C., Aschenbrenner, F., Welte, T. and Maus, U.A. (2015a) Arginase 1 activity worsens lung-protective immunity against *Streptococcus pneumoniae* infection. *European Journal of Immunology* 45, 1716–1726.

Knippenberg, S., Ueberberg, B., Maus, R., Bohling, J., Ding, N., Tort Tarres, M., Hoymann, H.G., Jonigk, D., Izykowski, N., Paton, J.C. *et al.* (2015b) *Streptococcus pneumoniae* triggers progression of pulmonary fibrosis through pneumolysin. *Thorax* 70, 636–646.

Lerolle, N., Zahar, J.R., Duboc, V., Tissier, F. and Rabbat, A. (2002) Pneumonia involving *Legionella pneumophila* and *Listeria monocytogenes* in an immunocompromised patient: an unusual coinfection. *Respiration* 69, 359–361.

Lucas, R., Verin, A.D,. Black, S.M. and Catravas, J.D. (2009) Regulators of endothelial and epithelial barrier integrity and function in acute lung injury. *Biochemical Pharmacology* 77, 1763–1772.

Lucas, R., Yang, G., Gorshkov, B.A., Zemskov, E.A., Sridhar, S., Umapathy, N.S., Jezierska-Drutel, A., Alieva, I.B., Leustik, M., Hossain, H. *et al.* (2012a) Protein kinase C-alpha and arginase I mediate pneumolysin-induced pulmonary endothelial hyperpermeability. *American Journal of Respiratory Cell and Molecular Biology* 47, 445–453.

Lucas, R. Sridhar, S., Rick, F.G., Gorshkov, B., Umapathy, N.S., Yang, G., Oseghale, A., Verin, A.D., Chakraborty, T., Matthay, M.A. *et al.* (2012b) Agonist of growth hormone-releasing hormone reduces pneumolysin-induced pulmonary permeability edema. *Proceedings of the National Academy of Sciences of the United States of America* 109, 2084–2089.

Lucas, R., Czikora, I., Sridhar, S., Zemskov, E.A., Oseghale, A., Circo, S., Cederbaum, S.D., Chakraborty, T., Fulton, D.J., Caldwell, R.W. and Romero, M.J. (2013a) Arginase 1: an unexpected mediator of pulmonary capillary barrier dysfunction in models of acute lung injury. *Frontiers in Immunology* 4:228.

Lucas, R., Czikora, I., Sridhar, S., Zemskov, E., Gorshkov, B., Siddaramappa, U., Oseghale, A., Lawson, J., Verin, A., Rick F.G. *et al.* (2013b) Mini-review: novel therapeutic strategies to blunt actions of pneumolysin in the lungs. *Toxins* 5, 1244–1260.

Lucas, R., Fulton, D., Caldwell, R.W. and Romero M.J. (2014) Arginase in the vascular endothelium: friend or foe? *Frontiers in Immunology* 5:589.

Matthay, M.A. (2014) Resolution of pulmonary edema. Thirty years of progress. *American Journal of Respiratory and Critical Care Medicine* 189, 1301–1308.

Maus, U.A., Srivastava, M., Paton, J.C., Mack, M., Everhart, M.B., Blackwell, T.S., Christman, J.W., Schlöndorff, D., Seeger, W. and Lohmeyer, J. (2004) Pneumolysin-induced lung injury is independent of leukocyte trafficking into the alveolar space. *Journal of Immunology* 173, 1307–1312.

Orman, K.L., Shenep, J.L. and English, B.K. (1998) Pneumococci stimulate the production of the inducible nitric oxide synthase and nitric oxide by murine macrophages. *Journal of Infectious Diseases* 178, 1649–1657.

Paulson, N.B., Gilbertsen, A.J., Dalluge, J.J., Welchlin, C.W., Hughes, J., Han, W., Blackwell, T.S., Laguna, T.A. and Williams, B.J. (2014) The arginine decarboxylase pathways of host and pathogen interact to impact inflammatory pathways in the lung. *PLoS ONE* 9(10): e111441.

Petrache, I., Birukova, A., Ramirez, S.I., Garcia, J.G. and Verin, A.D. (2003) The role of the microtubules in tumor necrosis factor-alpha-induced endothelial cell permeability. *American Journal of Respiratory Cell and Molecular Biology* 28, 574–581.

Piet, J.R., Geldhoff, M., van Schaik, B.D., Brouwer, M.C., Valls Seron, M., Jakobs, M.E., Schipper, K., Pannekoek, Y., Zwinderman, A.H., van der Poll, T. *et al.* (2014) *Streptococcus pneumoniae* arginine synthesis genes promote growth and virulence in pneumococcal meningitis. *Journal of Infectious Diseases* 209, 1781–1791.

Popolo, A., Adesso, S., Pinto, A., Autore, G. and Marzocco, S. (2014) L-Arginine and its metabolites in kidney and cardiovascular disease. *Amino Acids* 46, 2271–2286.

Posfay-Barbe, K.M. and Wald, E.R. (2004) Listeriosis. *Pediatrics in Review* 25, 151–159.

Predescu, D., Predescu, S., Shimizu, J., Miyawaki-Shimizu, K. and Malik, A.B. (2005) Constitutive eNOS-derived nitric oxide is a determinant of endothelial junctional integrity. *American Journal of Physiology Lung Cellular and Molecular Physiology* 289, L371–L381.

Romero, M.J., Platt, D.H., Tawfik, H.E., Labazi, M., El-Remessy, A.B., Bartoli, M., Caldwell, R.B. and Caldwell, R.W. (2008) Diabetes-induced coronary vascular dysfunction involves increased arginase activity. *Circulation Research* 102, 95–102.

Romero, M.J., Iddings, J.A., Platt, D.H., Ali, M.I., Cederbaum, S.D., Stepp, D.W., Caldwell, R.B. and Caldwell, R.W. (2012) Diabetes-induced vascular dysfunction involves arginase I. *American Journal of Physiology Heart and Circulatory Physiology* 302, H159–H166.

Rubins, J.B., Charboneau, D., Fasching, C., Berry, A.M., Paton, J.C., Alexander, J.E., Andrew, P.W., Mitchell, T.J. and Janoff, E.N. (1996) Distinct roles for pneumolysin's cytotoxic and complement activities in the pathogenesis of pneumococcal pneumonia. *American Journal of Respiratory and Critical Care Medicine* 153, 1339–1346.

Rudan, I., Boschi-Pinto, C., Biloglav, Z., Mulholland, K. and Campbell, H. (2008) Epidemiology and etiology of childhood pneumonia. *Bulletin of the World Health Organization* 86, 408–416.

Satoh, S., Oishi, K., Iwagaki, A., Senba, M., Akaike, T., Akiyama, M., Mukaida, N., Atsushima, K.M. and Nagatake, T. (2001) Dexamethasone impairs pulmonary defence against *Pseudomonas aeruginosa* through suppressing iNOS gene expression and peroxynitrite production in mice. *Clinical and Experimental Immunology* 126, 266–273.

Schulz, C., Gierok, P., Petruschka, L., Lalk, M., Mäder, U. and Hammerschmidt, S. (2014) Regulation of the arginine deiminase system by ArgR2 interferes with arginine metabolism and fitness of *Streptococcus pneumoniae*. *MBio* 5(6): e01858-14.

Shatanawi, A., Romero, M.J., Iddings, J.A., Chandra, S., Umapathy, N.S., Verin, A.D., Caldwell, R.B. and Caldwell, R.W. (2011) Angiotensin II-induced vascular endothelial dysfunction through RhoA/Rho kinase/p38 mitogen-activated protein kinase/arginase pathway. *American Journal of Physiology Cell Physiology* 300, C1181–C1192.

Simões, R.L., Arruda, M.A., Canetti, C., Serezani, C.H., Fierro, I.M. and Barja-Fidalgo, C. (2013) Proinflammatory responses of heme in alveolar macrophages: repercussion in lung hemorrhagic episodes. *Mediators of Inflammation* 2013: 946878.

Sprunt, T.P. and Luetscher, J.A. (1912) Acute vascular lesions in mice following injections of pneumococci. *Journal of Experimental Medicine* 16, 443–451.

Steppan, J., Nyhan, D. and Berkowitz, D.E. (2013) Development of novel arginase inhibitors for therapy of endothelial dysfunction. *Frontiers in Immunology* 4:278.

Sumi, T., Matsumoto, K., Takai, Y. and Nakamura, T. (1999) Cofilin phosphorylation and actin cytoskeletal dynamics regulated by rho- and Cdc42-activated LIM-kinase 2. *Journal of Cell Biology* 147, 1519–1532.

Usatyuk, P.V. and Natarajan, V. (2005) Regulation of reactive oxygen species-induced endothelial cell–cell and cell–matrix contacts by focal adhesion kinase and adherens junction proteins. *American Journal of Physiology Lung Cellular and Molecular Physiology* 289, L999–L1010.

Vandenbroucke, E., Mehta, D., Minshall, R. and Malik, A.B. (2008) Regulation of endothelial junctional permeability. *Annals of the New York Academy of Sciences* 1123, 134–145.

Waterer, G.W., Rello, J. and Wunderink, R.G. (2011) Management of community-acquired pneumonia in adults. *American Journal of Respiratory and Critical Care Medicine* 183, 157–164.

Witzenrath, M., Gutbier, B., Hocke, A.C., Schmeck, B., Hippenstiel, S., Berger, K., Mitchell, T.J., de los Toyos, J.R., Rosseau, S., Suttorp, N. and Schütte, H. (2006) Role of pneumolysin for the development of acute lung injury in pneumococcal pneumonia. *Critical Care Medicine* 34, 1947–1954.

Wolf, A.I., Strauman, M.C., Mozdzanowska, K., Williams, K.L., Osborne, L.C., Shen, H., Liu, Q., Garlick, D., Artis, D., Hensley, S.E. *et al.* (2014) Pneumolysin expression by *Streptococcus pneumoniae* protects colonized mice from influenza virus-induced disease. *Virology* 462/463, 254–265.

Wong, D., Dorovini-Zis, K. and Vincent, S.R. (2004) Cytokines, nitric oxide, and cGMP modulate the permeability of an *in vitro* model of the human blood–brain barrier. *Experimental Neurology* 190, 446–455.

Xiong, C., Yang, G., Kumar, S., Aggarwal, S., Leustik, M., Snead, C., Hamacher, J., Fischer, B., Umapathy, N.S., Hossain, H. *et al.* (2010) The lectin-like domain of TNF protects from listeriolysin-induced hyperpermeability in human pulmonary microvascular endothelial cells – a crucial role for protein kinase C-alpha inhibition. *Vascular Pharmacology* 52, 207–213.

Yadav, R., Samuni, Y., Abramson, A., Zeltser, R., Casap, N., Kabiraj, T.K.L., Banach, M. and Samuni, U. (2014) Pro-oxidative synergic bactericidal effect of NO: kinetics and inhibition by nitroxides. *Free Radical Biology and Medicine* 67, 248–254.

Yan, W., Tiruppathi, C., Qiao, R., Lum, H. and Malik, A.B. (1996) Tumor necrosis factor decreases thrombin receptor expression in endothelial cells. *Journal of Cellular Physiology* 166, 561–567.

7 Arginine and Methionine as Precursors of Polyamines in Trypanosomatids

Y. Pérez-Pertejo,[1] J.M. Morán[1] and R. Balaña Fouce[1,2]*

[1]Departamento de Ciencias Biomédicas, Universidad de León, Campus de Vegazana, León, Spain; [2]Instituto de Biotecnología de León (INBIOTEC), León, Spain

7.1 Abstract

Trypanosomatids are protist parasites that are responsible for some serious diseases all over the world, mostly in developing countries, but also in countries of the northern shore of the Mediterranean Basin. *Trypanosoma* and *Leishmania* spp. infect millions of people every year and have become a major concern for international health organizations and the governments of more than 90 countries. Both genera are auxotrophic for some amino acids that they have to incorporate from the external medium to survive and proliferate. Besides the need for these amino acids for proteinogenesis, L-arginine and L-methionine are essential for polyamine synthesis, a series of aliphatic oligoamines that are required for cell growth. In trypanosomatids, polyamines play a key role in proliferation, growth and differentiation. Therefore, they have been described as potential therapeutic targets for treating infectious diseases such as leishmaniasis, Chagas disease or sleeping sickness. Spermidine can also participate in the synthesis of tripanothione, a specific thiol involved in free radical detoxification, which emerges as a relevant target for drug discovery. Finally, the fate of *T. cruzi* and *Leishmania* spp. parasites within macrophages is dependent on the host degradative pathways of L-arginine mediated by cytokines. This chapter highlights the importance of L-arginine and L-methionine in the synthesis of polyamines and the resources available up to now for developing new drugs, which could interact with these pathways in order to eliminate infection from patients suffering from these severe and neglected diseases.

7.2 Trypanosomatid-borne Diseases

Neglected tropical diseases (NTD) include 18 diseases affecting hundred millions of people. They are excluded from the commercial interests of the pharmaceutical industry owing to the poor conditions of the people that suffer them. Among the most neglected of these diseases – with the exception of malaria – are those produced by trypanosomatids. The Family Trypanosomatidae (order Kinetoplastida) includes unicellular flagellated protists that parasitize metazoans – including humans, and survive to produce severe life-threatening diseases. These diseases have high rates of morbidity and mortality, especially in low-income populations of the less developed and least economically developed countries (WHO, 2008).

*Corresponding author: E-mail address: rbalf@unileon.es

In Africa, species of the genus *Trypanosoma*, *T. brucei gambiense* and *T. brucei rhodesiense*, are responsible for sleeping sickness (human African trypanosomiasis), a lethal infection that is transmitted by tsetse flies (*Glossina*) in sub-Saharan countries. Infected people suffer an early haemolymphatic stage, which includes symptoms of headache, fever, joint pain and itching. The later meningoencephalic stage, which has a case fatality rate close to 100%, includes characteristic neurological symptoms as a consequence of the invasion of host central nervous system (CNS) by the parasite (Lejon *et al.*, 2013). Sleeping sickness is only prevalent in south Saharan countries where around 70 million people are at risk (Franco *et al.*, 2014). In 2009, the incidence of this disease as the number of cases reported dropped below 10,000 (to 9878) (WHO, 2016), and this probably represented about one third of the real infections that actually occurred. In addition, the estimated annual disability-adjusted life years (DALYs) of 1.7 million lost as a result of the disease has a profound impact in the economy of these countries (Simarro *et al.*, 2011).

Chagas disease is the name given in Central and South America to a specific zoonotic infection caused by *T. cruzi* and transmitted by the faeces of blood-sucking *Triatominae* 'kissing bugs'. Symptoms produced by Chagas disease may vary during the course of infection. Early local swelling during the acute stage can evolve to damage in the gastrointestinal tract with the presence of megacolon and megaoesophagus, so leading to an inevitable weight loss (Teixeira *et al.*, 2011). At the chronic stages, the occurrence of a severe cardiomyopathy may result in sudden death. American trypanosomiasis (i.e. Chagas disease) is prevalent in 18 Central and South American countries with 17 million people currently infected and 100 million people at risk. The incidence of 41,200 new cases every year, as well as 12,500 related deaths and estimated annual DALYs of 570,000 make this disease a formidable threat to social and economic development of the countries where it is endemic (Moncayo and Silveira, 2009).

Finally, leishmaniasis is a group of diseases transmitted by female sandflies of the genera *Phlebotomus* and *Lutzomyia*. Cutaneous leishmaniasis (CL) is the mildest form of the disease and is mainly produced by *Leishmania major* in the Old World (Asia, Africa and southern Europe) and by *L. panamensis* and *L. braziliensis* in the New World (Central and South America). CL is not fatal but can be extremely disfiguring when scars are located on a patient's face. A severe form of CL is American tegumentary leishmaniasis (mucocutaneous leishmaniasis), which arises as a consequence of an inadequately treated New World CL (WHO, 2007). This clinical form presents with severe facial disfiguration, including the nose and lips, which can evolve to cancer processes. Lastly, visceral leishmaniasis (VL) is the most serious form of the disease; this is produced by *L. donovani* in Asia and *L. infantum* on both shores of the Mediterranean Sea. VL produces organ swelling (specifically targeting the liver and spleen) and is fatal if left untreated (Murray *et al.*, 2005; Antinori *et al.*, 2012). The incidence of the different forms of leishmaniasis is near 1.6 new million cases every year, 20,000 to 40,000 of which become fatal. Over 50% of the 2 million worldwide appraised DALYs are concentrated in South-east Asia, with Bihar (India) as the epicentre of *Leishmania*-related deaths (Alvar *et al.*, 2012).

As no effective prophylaxis exists, chemotherapy is the most effective way to treat these diseases once symptoms appear. However, given the low interest that these diseases receive from Big Pharma, current treatments are outdated, have many side effects and often require parenteral administration, which encourages cessation of treatment and, thereby, the emergence of resistant strains (Leprohon *et al.*, 2011). Drugs of choice against African trypanosomiasis during the early blood stage are pentamidine (an aromatic diamidine) and suramin (Burri, 2010). By the time when the neurological stage has onset, only arsenic-derived drugs (such as melarsoprol), or the irreversible inhibitor of ornithine decarboxylase (ODC), eflornithine, alone or in combination with nifurtimox (NECT therapy), may improve symptomatology (Lütje *et al.*, 2013). For Chagas disease, the current pharmacopoeia includes two drugs, nifurtimox and benznidazole, which have to be administered for at least 60–90 days, and have significant side effects (Rassi *et al.*, 2010). The drugs of choice against leishmaniasis are the pentavalent antimony derivatives glucantime and pentostam, which must be parenterally administered and have numerous side effects (Croft and Olliaro, 2011).

In a second instance, macrolide antibiotics such as amphotericin B, especially in its liposomal formulation (AmBisome), are a good alternative to antimonials (Monge-Maillo and López-Vélez, 2013), although they are expensive and require intravenous administration (Mohamed-Ahmed *et al.*, 2012). Other promising compounds are miltefosine, a lipid ether synthesized primarily as an antitumour drug, and paromomycin, an aminoglycoside antibiotic.

Polyamine metabolism has been pointed as a promising target against proliferative processes since the late 1970s. After the description of the complete biosynthetic pathway of these compounds, many were synthesized as inhibitors of some of its key enzymes, with relevant antitumour and antiparasitic outcomes. Some of these compounds, including eflornithine, are current drugs in use against sleeping sickness in Africa. From a metabolic viewpoint, polyamine biosynthesis joins the degradative pathways of L-methionine and L-arginine. Trypanosomatids lack biosynthetic pathways for the *de novo* synthesis of these two amino acids, but they have powerful carriers to take them up from the surrounding media. In this chapter, we review the metabolism of L-arginine and L-methionine in trypanosomatids, not only as precursors of polyamines, but also within the context of the host–parasite relationship and as putative targets for drug discovery.

7.3 L-Methionine Metabolism

L-Methionine is one of the protein amino acids that contains a sulfur atom. *De novo* synthesis of this amino acid from L-aspartic acid is significantly costly to cells and trypanosomatids lack the enzymes involved in its biosynthesis (Vickers *et al.*, 2006). L-Methionine auxotrophy is characteristic of these parasites, and they have to take it up from the host's amino acid pool using suitable influx transporters (Reguera *et al.*, 2007). The transported L-methionine is rapidly trapped by trypanosomatids, which transform it into *S*-adenosyl-L-methionine (AdoMet) a 'crossroad' metabolite that involved in several key metabolic pathways: polyamine synthesis, the transmethylation of macromolecules and transsulfuration reactions. The key enzyme involved in L-methionine transformation to AdoMet is AdoMet synthetase (methionine adenosyltransferase, or MAT) a housekeeping enzyme that uses ATP and L-methionine as substrates. AdoMet is an extraordinarily versatile molecule that plays a role as a methyl donor in a wide variety of biosynthetic reactions, and as an aminopropyl donor that is used in polyamine biosynthesis. L-Methionine is also a source of L-cysteine via the transsulfuration pathway which arises from AdoMet-dependent methylations.

Figure 7.1 depicts comparative schemes for L-methionine metabolism in mammals and trypanosomatids, and provides a reference for sections 7.3.1–7.3.5.

7.3.1 L-Methionine and *S*-adenosyl-L-methionine (AdoMet) uptake

Trypanosomatids lack the ability to synthesize *de novo* L-arginine, L-histidine, L-isoleucine, L-leucine, L-phenylalanine, L-tryptophan, L-tyrosine, L-valine and homoserine, a common precursor of L-methionine and L-threonine. Therefore, they need to take them up from the environment. The amino acid transporters are the most important family of permeases in trypanosomatids. Examinations of the sequenced genomes of these parasites has shown that there are 29 predicted gene orthologues of this family in *L. major*, 38 in *T. brucei* and 42 in *T. cruzi*. Despite the interest in L-methionine influx in trypanosomes and leishmanias, there is no report describing the process at the molecular level. Biochemical characterization and the kinetics of L-methionine transport in *T. brucei* and *L. tropica* promastigotes has indicated that affinity constants for this amino acid are adapted to the naturally occurring concentrations of this amino acid in host plasma (Mukkada and Simon, 1977, Hasne and Barrett, 2000). In contrast, the transporters for purine nucleosides and AdoMet have been characterized in both trypanosomes and leishmanias. In African trypanosomes, a single transporter has been identified, which is not similar to the P1 or P2 adenosine or other purine nucleoside transporters (Goldberg *et al.*, 2000). In *Leishmania*, this has been cloned and the AdoMetT1 transporter characterized and shown to belong to the biopterin folate

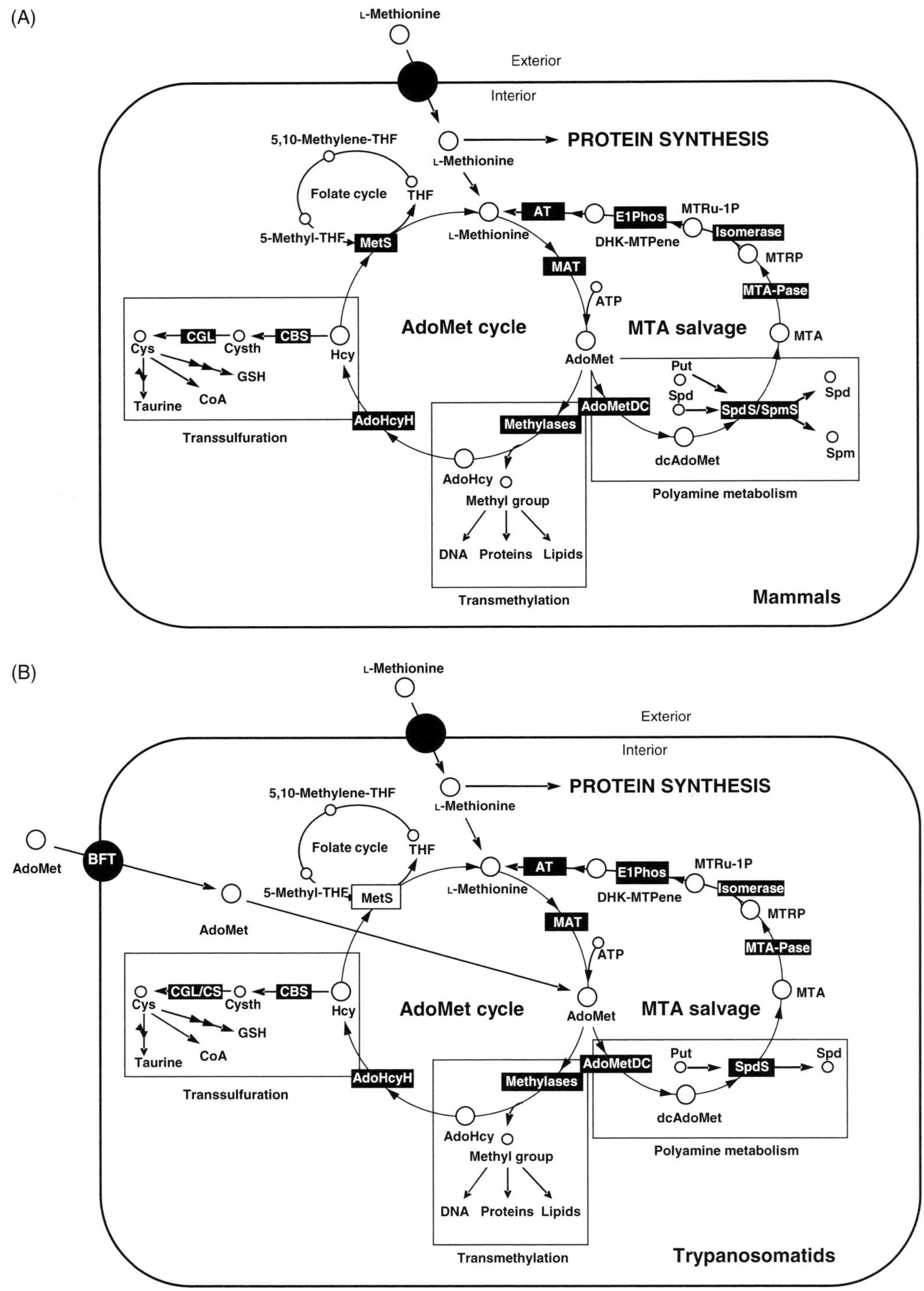

Fig. 7.1. Comparative schemes of L-methionine metabolism in mammalian cells (A) and trypanosomatid parasites (B). Enzymes are shown in dark boxes and metabolites in open circles; defective enzymes are marked in white boxes. Specific routes are shown with discontinuous arrows. Abbreviations of the

transporter (BFT) family. Sinefungin – an AdoMet analogue with strong antileishmanial activity – uses this carrier for internalization in the cell (Dridi *et al.*, 2010).

7.3.2 AdoMet synthesis

MAT is the enzyme responsible for AdoMet synthesis from L-methionine and ATP. AdoMet synthesis occurs after two sequential steps: first the sulfur atom of L-methionine attacks the 5′-carbon of ATP through an S_N2 reaction that yields AdoMet and tripolyphosphate. The tripoplyphosphate is then hydrolysed to pyro- and orthophosphoric acids. This last reaction is strongly induced by AdoMet. MAT shows sigmoidal kinetics for both reactions (Reguera *et al.*, 2007).

The MAT enzyme has been intensively studied in *Leishmania*. The primary structure of this protein is phylogenetically conserved in organisms ranging from protists to mammals. For this reason, MAT is being used as a marker for evolution studies (Sánchez-Pérez *et al.*, 2004). Unlike vertebrates, which contain two different genes able to synthesize three different active enzymes, the structure of MAT in trypanosomatids seems to be quite simple. According to our studies, leishmanial MAT is encoded by two genes organized in tandem that express a dimeric enzyme (Reguera *et al.*, 2002). For its part, *T. brucei* seems to have two distinct MAT isoforms (Yarlett *et al.*, 1993).

AdoMet is an essential metabolite for trypanosomatids and the operation of the MAT genes cannot be interrupted by current genetic engineering methods. All the efforts made by our group to produce MAT-defective *Leishmania* strains were unsuccessful (unpublished data). This fact points to fine regulation mechanisms that control the levels of AdoMet within the cell. These regulation mechanisms seem to differ between mammals and trypanosomatids. Several studies concluded that in mammals, MAT activity, and consequently AdoMet production, is post-translationally regulated by the allosteric kinetics and the polymerization of MAT subunits (Avila *et al.*, 2002). However, even though MAT has also sigmoidal kinetics in *L. donovani*, this is not enough to explain how intracellular AdoMet levels are finely tuned in strains that overproduce MAT. A plausible explanation is that excess AdoMet can be efficiently effluxed in the overloaded strain in comparison with the wild type (Pérez-Pertejo *et al.*, 2006). A second, and more striking, reason is that overproduced MAT protein can be efficiently demolished by the proteosome-like system that exists in trypanosomatids (Pérez-Pertejo *et al.*, 2011).

7.3.3 The transsulfuration pathway

S-Adenosylhomocysteine (AdoHcy) is the major by-product generated after the transmethylation of AdoMet. Under these conditions, AdoHcy becomes to be an important source of cysteine and its derivatives through the transsulfuration and folic acid synthesis pathways.

AdoHcy hydrolase (AdoHcyH) is the enzyme that cleaves AdoHcy to generate adenosine and homocysteine. This enzyme has been cloned and characterized in *T. cruzi* and *L. donovani*

Fig. 7.1. Continued.
enzymes and their corresponding EC numbers are given in alphabetic order: AdoMetDC, *S*-adenosylmethionine decarboxylase (EC 4.1.1.50); AdoHcyH, *S*-adenosylhomocysteine hydrolase (EC 3.3.1.1); AT, unspecific aminotransferase (EC 2.6.1.X); CBS, cystathionine β-synthase (EC 4.2.1.22); CGL, cystathionine γ-lyase (EC 4.4.1.1), which also has cysteine synthase (CS) activity in trypanosomatids; E1Phos, enolase-phosphatase E1 (EC 3.1.2.77); MAT, methionine adenosyltransferase (EC 2.5.1.6); MetS, methionine synthase (EC 2.1.1.13/14); MTA-Pase, methylthioadenosine phosphorylase (EC 2.4.2.28); SpdS, spermidine synthase (EC 2.5.1.16); SpmS, spermine synthase (EC 2.5.1.22). Abbreviations of the metabolites are also given in alphabetic order: AdoHcy, *S*-adenosylhomocysteine; AdoMet, *S*-adenosylmethionine; Cys, cysteine; Cysth, cystathionine; dcAdoMet, decarboxylated *S*-adenosylmethionine; DHK-MTPene, 1,2-dihydroxy-3-keto-5-methylthiopentene; GSH, reduced glutathione; Hcy, homocysteine; MTA, methylthioadenosine; MTRP, 5′-methylthioribose-1-phosphate; MTRu-1P, 5′-methylthioribulose-1-phosphate; Put, putrescine; Spd, spermidine; Spm, spermine; THF, tetrahydrofolate. Modified from Reguera *et al.* (2007).

(Yang and Borchardt, 2000; Parker *et al.*, 2003), and its crystallographic structure has been generated from *T. cruzi*. The primary structure of AdoHcyH is phylogenetically conserved, producing – after folding – a fully active homotetramer with a binding site for NAD^+ in each subunit. Kinetic studies have shown that NAD^+ interacts more weakly with the enzymes from *L. donovani* and *T. cruzi* than with the human enzyme, contrary to what happens with NADH, which binds more strongly and maintains the enzyme inactive for longer. This characteristic has been used to develop NAD^+ analogues as enzyme inhibitors of the parasite. In general, elevated AdoHcy levels induced by the inhibition of AdoHcyH result in a decrease of AdoMet, which will interrupt methylation reactions (Li *et al.*, 2009).

Homocysteine is a major source for the generation of cysteine and glutathione by transsulfuration reactions and to the recovery of L-methionine via the folate cycle (see Section 7.3.4). The reverse transsulfuration pathway transforms homocysteine into cysteine, through the participation of the enzymes cystathionine β-synthase (CBS) and cystathionine γ-lyase (CGL) (Williams *et al.*, 2009; Romero *et al.*, 2014). On the one hand, CBS catalyses the condensation of serine and homocysteine to yield cystathionine, and on the other, CGL hydrolyses cystathionine to yield α-ketobutyrate and cysteine. The functional conversion of homocysteine in cysteine has been characterized in trypanosomatids. Remarkably, unlike the mammalian enzyme, trypanosome CGL has cysteine synthase activity that points to a putative drug target (Giordana *et al.*, 2014).

7.3.4 L-Methionine regeneration: the folate cycle

L-Methionine can be salvaged by methionine synthase (MetS), an enzyme that uses homocysteine as substrate and requires 5-methyltetrahydrofolate as a cofactor. In *L. major*, two genes encoding for MetS have been found: one of them shows similarity to cobalamin-dependent MetS and the other to cobalamin-independent MetS, although no current data are available on the mode of action or location thereof. 5-Methyltetrahydrofolate can either be synthesized from methylene tetrahydrofolate by the action of methylene tetrahydrofolate reductase (MTHFR), or it can be taken up from the surrounding medium. *Leishmania* MTHFR uses NADH or NADPH and, unlike the enzyme from other species, is not an allosteric enzyme. Surprisingly, no genes encoding MetS or MTHFR have been found in the genome of African trypanosomes (Vickers *et al.*, 2006; Vickers and Beverley, 2011).

7.3.5 The 5′-methylthioadenosine salvage pathway

Some 3–5% of AdoMet serves as an aminopropyl donor to synthesize polyamines. 5′-Methylthioadenosine (MTA) is the major by-product of this route. The cell economy will recycle this useless by-product by transforming it into two important building blocks, adenine and L-methionine, via the MTA salvage pathway. When MTA is generated in the polyamine biosynthetic pathway, it is next hydrolysed to 5′-methylthioribose-1-phosphate (MTRP) and adenine. MTRP is further transformed into 5′-methylthioribulose-1-phosphate (MTRu-1-P), a molecule which after enzymatic dehydration becomes 2,3-diketo-5′-methylthiopentyl-1-phosphate (DK-MTP). In the next step, DK-MTP generates DK-1,2-dihidroxy-3-keto-5-methylthiopentene (DHK-MTPene) through the intermediate 1,2-dihydroxy-3-keto-5′-methylthiopentenyl-1-phosphate (HK-MTPenyl-1-P). DHK-MTPene is ultimately transformed into α-ketomethiobutyrate (KMTB), which participates in transamination reactions to regenerate L-methionine (Albers, 2009). This cycle has been partially characterized in trypanosomatids (Berger *et al.*, 2001; Sufrin *et al.*, 2008).

The enzymatic transformation of MTA to adenine and MTRP can be catalysed in just one step by MTA phosphorylase (MTA-Pase), or in two sequential steps catalysed by MTA nucleosidase (MTAN) and MTR kinase (MTRK). MTA-Pase activity has been identified in trypanosomatids and mammals, whereas other protozoans such as *Plasmodium falciparum* and *Giardia lamblia* use the alternative two-step route (Riscoe *et al.*, 1989). A 920 bp gene (LmjF.05.0830) encoding MTA-Pase has been described in chromosome 5 of the *L. major* genome and an orthologous 930 bp gene in chromosome 7 of *T. brucei* (Tbg972.7.8280). MTA-Pase activity has been assayed in cell extracts of *L. donovani* promastigotes (Koszalka and Krenitsky, 1986) and in purified extracts of *T. cruzi* and *T. brucei brucei* (Miller *et al.*, 1987; Ghoda *et al.*, 1988). Because trypanosomes lack *de novo* purine biosynthetic pathways, MTA-Pase has been pointed out as a potential drug target in trypanosomes. The purine nucleoside 5′-deoxy-5′-(hydroxyethylthio) adenosine (HETA) and some of its acetylated derivatives are rapidly internalized by the purine parasite transporter P2 (Sufrin *et al.*, 2008), and this permits high intracellular concentrations of the drug that are able to induce strong trypanocidal effects *in vitro* and curative effects in mouse models (Bacchi *et al.*, 1997). The data indicate that these compounds have higher affinity for trypanosomal MTA-Pase than for the mammalian enzyme, thereby generating toxic intermediates that prevent MTRP regeneration to L-methionine. In fact, the cytostatic *in vitro* effects of these compounds can be reversed by the addition of L-methionine or KMTB. Among the metabolic effects of HETA and derivatives are the inhibition of spermidine synthesis and increased levels of AdoMet, AdoHcy and MTA (Bacchi *et al.*, 1999).

Unlike their occurrence in mammals, neither the isomerase responsible for converting MTRP into MTRu-1-P nor the dehydratase that catalyses the transformation of MTRu-1-P to DK-MTP have yet been described in trypanosomatids.

The synthesis of DHK-MTPene requires two reactions that are often catalysed by the bifunctional enzyme enolase-phosphatase E1. Enolase activity transforms DK-MTP to HK-MTPenyl-1-P, this last compound serving as a substrate of the E1 phosphatase that generates DHK-MTPene (Heilbronn *et al.*, 1999). The gene encoding this enzyme has been identified in the genomes of *L. major* (LmjF36.5910) and *T. cruzi* but not in *T. brucei* (Shameer *et al.*, 2015).

Finally, the genes that encode the enzymes with aminotransferase (AT) activity involved in the MTA salvage cycle have been cloned and expressed in *Crithidia fasciculata*, *T. brucei brucei* and other major parasites. They all contain domains from the Ia aminotransferases subfamily and conserve high homology with cytosolic aspartate aminotransferases from eukaryotic cells. The major amino donors used as substrates for these enzymes are aromatic amino acids and L-glutamate (Berger *et al.*, 1996, 2001).

7.4 L-Arginine Metabolism

There is no specific metabolic pathway to synthesize L-arginine in trypanosomatids, so this amino acid must be taken up from the host. In addition to its role as a component of proteins and as a precursor for polyamine biosynthesis, L-arginine has different functions in the host and the parasite. In the host, L-arginine is involved in ammonia detoxification via the urea cycle and in the production of nitric oxide (NO), an important metabolite for killing parasites within macrophages (Morris, 2006; Wu and Morris, 1998). However, in trypanosomatids L-arginine serves as energy reservoir in form of phosphagens, and as a precursor of the free radical scavenger tripanothione (Colotti and Ilari, 2011; Balaña-Fouce *et al.*, 2012).

Figure 7.2 depicts comparative schemes for L-arginine metabolism in mammals and trypanosomatids, and provides a reference for sections 7.4.1–7.3.3.

7.4.1 L-Arginine uptake

L-Arginine is taken from the external medium by basic amino acid transporters. A wide range of L-arginine transporters has been described so far. The y+ transporter is an Na^+-independent transporter that belongs to the cationic amino acid transporters (CAT) family. Six types of proteins, CAT1, CAT2A, CAT2B, CAT3, CAT4 and CAT14 have been described and they all are coded by the *Slc7A* gene family. CAT2A and CAT2B are the principal isoforms expressed in macrophages, and these can be affected by a wide range of molecular mediators during the immune response (Closs *et al.*, 2006). The major amino acid transporter in trypanosomatids is the Amino Acid/Auxin Permease (AAAP) family. Members of this family were identified as transporting L-arginine in *T. cruzi* (TcAAP3) and *Leishmania* spp. (LAAP3). The transport of L-arginine is regulated at post-transcriptional level according to its availability in the external medium and its intracellular concentration (Castilho-Martins *et al.*, 2011; Miranda *et al.*, 2012).

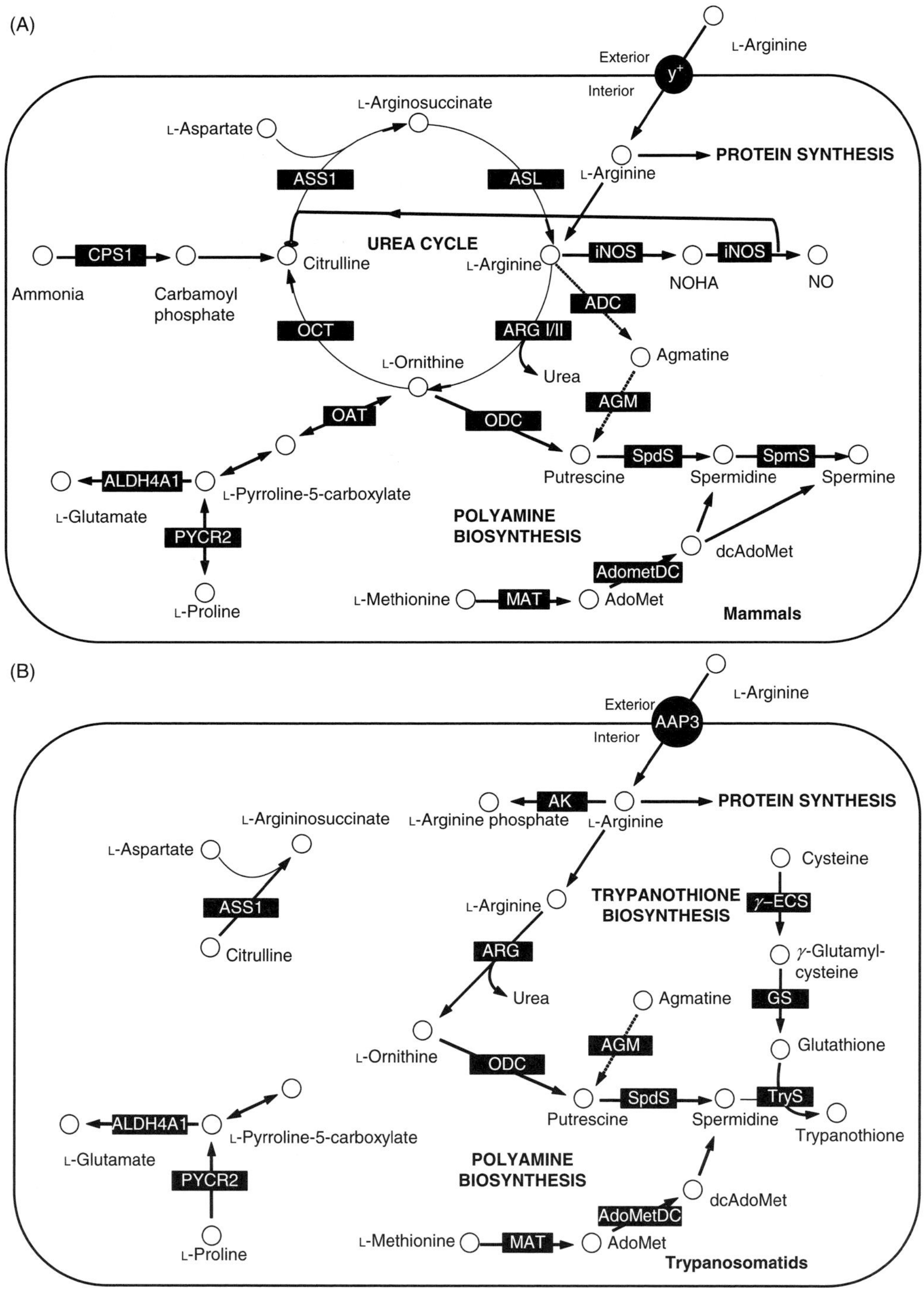

Fig. 7.2. Comparative schematic representation of L-arginine metabolism in mammalian cells (A) and trypanosomatid parasites (B). Enzymes are shown in dark boxes and metabolites in open circles. L-Arginine is taken up from the external medium by members of the cationic transporter families (CAT1) and then metabolized by different enzymes. (A) L-Arginine is hydrolysed by arginase in the liver, which

7.4.2 Arginase

Arginase is an Mn^{2+}-dependent metalloenzyme that catalyses the conversion of L-arginine to L-ornithine and urea. Most invertebrates, plants, bacteria and yeasts have only one form of arginase which is localized in the mitochondria, but ureotelic vertebrates metabolize excess nitrogen via a second cytosolic isoform. The cytosolic form is called arginase I, whereas the mitochondrial isoenzyme is called arginase II. Arginase activity has two major homeostatic purposes: first, to rid the body of ammonia via the urea cycle and second, to produce L-ornithine, the main precursor for polyamines and proline. The polyamines that are produced by the action of ODC are necessary for cell proliferation and the regulation of several ion channels. Proline produced via ornithine aminotransferase (OAT) is necessary for the production of collagen (Caldwell *et al.*, 2015).

In trypanosomatids, excess nitrogen is removed directly as ammonia, which makes unnecessary most of the enzymes of the urea cycle except for arginase, which is required to provide polyamines during parasite replication. Arginase is selectively expressed by some trypanosomatids (*Leishmania* sp., *Leptomonas* sp. or *Crithidia* sp.) but not by others, such as *T. brucei* and *T. cruzi*, despite the gene encoding being present in all the members of the family (Balaña-Fouce *et al.*, 2012). Nevertheless, it has been suggested that the trypanosomatids that do not express arginase instead produce putrescine as the result of agmatinase activity (da Silva and Floeter-Winter, 2014). In contrast to the arginase of mammals, the analysis of *Leishmania* arginase reveals the presence of specific signals that compartmentalize the protein within glycosomes – the organelles that are involved in energy production in trypanosomatids (Opperdoes and Szikora, 2006). Several studies in arginase-deficient *Leishmania*

Fig 7.2. Continued.
constitutes the key enzyme of the urea cycle. Mitochondrial arginase II, despite yielding the same products, is used to synthesize L-proline and creatine. L-Arginine is also cleaved by inducible NOS (iNOS) in a two-step mechanism, yielding nitric oxide (NO) and the intermediate N^{ω}-hydroxy-nor-L-arginine (NOHA), which has important regulatory functions in macrophages. L-Ornithine, the side product of arginase, is the precursor amino acid of polyamine biosynthesis. (B) L-Arginine metabolism in trypanosomatids is much simpler than that of mammals. First, the urea cycle is absent, but some protein orthologues remain in trypanosomatids placed at the same position as in the mammalian metabolic chart. Likewise, no gene encoding iNOS is currently listed in any trypanosomatid genomic database. Trypanosomatid arginase resembles mammalian arginase I, although the leishmanial enzyme is found in glycosomes, while that from *Trypanosoma cruzi* and *T. brucei* is not. The L-ornithine from L-arginine hydrolysis is the main source of polyamines in *T. brucei* and *Leishmania* spp. via ornithine decarboxylase (ODC); however, *T. cruzi* is a naturally occurring auxotroph in polyamines because it lacks this enzyme. Interestingly, spermidine is one of the precursors of trypanothione, a reactive oxygen species (ROS) scavenger compound that substitutes for glutathione in trypanosomatids (although they still contain glutathione). Arginine kinase, a specific phosphagen kinase that reversibly phosphorylates L-arginine from ATP, has been detected in the trypanosomatid flagellum and pointed out as potential target. Abbreviations of the enzymes/transporters and their corresponding EC numbers are given in alphabetic order: AAP3, Amino Acid/Auxin Permease (AAAP) family transporter; ADC, arginine decarboxylase (EC 4.1.1.19); AdoMetDC, *S*-adenosylmethionine decarboxylase (EC 4.1.1.50); AGM, agmatinase (EC 3.5.3.11); ALDH4A1, aldehyde dehydrogenase 4 (EC 1.5.1.12); AK, arginine kinase (EC 2.7.3.3); ARG, arginase (EC 3.5.3.1); ASL, argininosuccinate lyase (EC 4.3.2.1); ASS1, L-argininosuccinate synthase 1 (EC 6.3.4.5); CPS1, carbamoyl phosphate synthetase (EC 6.3.4.16); γ-ECS, γ-glutamylcysteine synthase (gamma-ECS) (EC 6.3.2.2) GS, glutathione synthetase (EC 6.3.2.3); iNOS, inducible nitric oxide synthase (EC 1.14.13.39); MAT, methionine adenosyltransferase (EC 2.5.1.6); OAT, ornithine aminotransferase (EC 2.6.1.13); OCT, ornithine carbamoyltransferase (EC 2.1.3.3); ODC, ornithine decarboxylase (EC 4.1.1.17); PYCR2, pyrroline-5-carboxylate reductase (EC 1.5.1.2); SpdS, spermidine synthase (EC 2.5.1.16); SpmS, spermine synthase (EC 2.5.1.22); TryS, trypanothione synthase (EC 6.3.1.9); y+, CAT1, cationic amino acid transporter 1. Abbreviations of the metabolites are also given in alphabetical order: AdoMet, *S*-adenosylmethionine; dcAdoMet, decarboxylated *S*-adenosylmethionine; NO, nitric oxide; NOHA, N^{ω}-hydroxy-nor-L-arginine. Modified from Balaña-Fouce *et al.*, 2012.

strains (*Δ-arg*) show that this enzyme is essential for cell proliferation (D'Antonio *et al.*, 2013). Leishmanial arginase is represented by a single gene in chromosome 35. Double gene replacement of both copies by antibiotic-resistant markers produced an auxotrophic strain that needs L-ornithine, putrescine or spermidine to survive, although this mutant did not lose the ability to infect mice (Muleme *et al.*, 2009). In light of these data, it was suggested that the polyamine transporters were upregulated in order to circumvent the auxotrophy, thereby providing polyamine precursors to the mutant parasites.

Last but not least, a *Leishmania* strain has been developed that overexpresses arginase. The arginase activity of this transgenic strain was more than 30-fold higher than that found in the wild type. As a consequence, some specific inhibitors of arginase and of ODC, which has an important role in polyamine metabolism, have lost their inhibitory effect. For example, eflornithine and *N*-hydroxy-L-arginine were unable to inhibit ODC and arginase, respectively (Reguera *et al.*, 2009).

7.4.3 Phosphagen production

Phosphagens play an important role as energy reserves because they are involved in ATP regeneration when it is needed (Ellington, 2001). Unlike vertebrates, in which the main phosphagen is phosphocreatine, invertebrates mainly use L-arginine, a guanidine moiety-bearing amino acid, as an energy reserve. Arginine kinase (AK) is a group of enzymes responsible for the reversible phosphorylation of arginine from ATP to yield phosphoryl arginine (Pereira, 2014), and genes encoding AK isozymes have been identified in the genome of all subspecies of *Trypanosoma*, but not in *Leishmania*. The number of gene copies present in the sequenced genomes differs among *Trypanosoma* spp. with *T. brucei* having three *ak* genes arranged in tandem on chromosome 9 (*ak*1, *ak*2, *ak*3). These enzymes have different cell compartmentalization: AK1 is cytosolic, AK2 is glycosomal and AK3 is flagellar.

Flagellum mobility in trypanosomatids is essential for parasite migration from the posterior midgut to the higher anterior midgut of insects during different stages of the promastigote form. *T. brucei* shows higher AK3 levels in procyclic (gut-invading) trypanosomes than in bloodstream trypanosomes. Inducible RNAi knockdowns on the *T. brucei ak*3 gene show that this enzyme is not essential for parasite growth but originates a disabled phenotype that makes attachment to the fly midgut difficult and prevents transmission to the host (Ooi *et al.*, 2015).

7.5 Polyamine Metabolism

Naturally occurring polyamines – putrescine, spermidine and spermine – are short-chain aliphatic polycationic compounds that are essential for cellular growth and proliferation in all living organisms (see reviews by: Müller *et al.*, 2001; Bacchi and Yarlett, 2002; Heby *et al.*, 2003; Wallace and Fraser, 2003). The role played by these compounds in cellular physiology can be attributed to the protonated nature of their amino/imino groups, which allows spontaneous interaction with anionic macromolecules, such as nucleic acids, proteins and lipids, thus affecting their function. The multifunctional role of polyamines in cellular physiology and their involvement in fast proliferation processes were detected early on and, therefore, their precursors and metabolites were pointed out as potential chemotherapeutic targets in tumoural or infectious processes. All vertebrates and lower eukaryotes, including fungi and protozoa, synthesize polyamines *de novo* from two amino acidic precursors: L-arginine and L-methionine.

Figure 7.3 depicts comparative schemes for polyamine metabolism in mammals and trypanosomatids, and provides a reference for sections 7.5.1 and 7.5.2.

7.5.1 Polyamine biosynthesis in mammals: the canonical pathway

Polyamine metabolism has been intensively studied by a group led by Anthony E. Pegg (PennState University College of Medicine, Hershey, Pennsylvania, USA), which provides a reference for many scientific groups worldwide. In order to make polyamine precursors available, it is mandatory that enzymes with arginase activity cleave L-arginine to produce L-ornithine and urea as by-product. Two arginases, a liver cytosolic enzyme or arginase I, and a cosmopolitan mitochondrial enzyme or arginase II are operative in mammals. Hepatic arginase I is mainly involved in ammonia detoxification, whereas arginase II is involved in the synthesis of proline, glutamate and creatine.

L-Ornithine is transformed to putrescine (1,4 diaminobutane) by the dimeric, pyridoxal phosphate-dependent ODC, an extremely regulated enzyme, whose activity varies in response to many stimuli, which enables it to adapt polyamine levels to cell needs (Pegg *et al.*, 1995; Pegg, 2006). Mammals control ODC activity by transcriptional and post-translational mechanisms. At the transcription level, the promoter of the *odc* gene is targeted by transcription factors in response to hormones, growth factors and tumour promoters (Shantz and Pegg, 1999). ODC turns over very rapidly because it binds to a specific ODC antizyme at the post-translational level that drives the protein to protease degradation by the 26S proteasome system (Coffino, 2001). Once putrescine is synthesized, L-methionine enters the scene.

To play a role in polyamine metabolism, L-methionine must be condensed with ATP to synthesize AdoMet and pyrophosphate as by-product in a reaction catalysed by MAT, which is an essential housekeeping cytosolic enzyme (Reguera *et al.*, 2002). AdoMet is the key methyl donor of myriads of transmethylation reactions, although a limited (but necessary) 2–5% of this intermediate is used to synthesize spermidine and spermine. AdoMet is metabolically incorporated into the polyamine pathway via decarboxylation by AdoMet decarboxylase (AdoMetDC), a heterodimeric, self-cleaving, short-lived cytosolic enzyme that is activated by putrescine (Stanley and Shantz, 1994; Pegg, 2009). AdoMetDC has a covalently bound pyruvate prosthetic group (Stanley *et al.*, 1989), which is essential for the decarboxylation reaction to yield decarboxylated AdoMet (dcAdoMet), an unstable intermediate that can transfer its aminopropyl moiety to putrescine or spermidine. These transfer reactions are sequentially catalysed by two highly specific aminopropyltransferases: spermidine synthase (SpdS), which uses putrescine as cosubstrate to produce spermidine; and spermine synthase (SpmS), which uses spermidine to synthesize spermine. Both of these reactions produce a still valuable methyl thioadenosine (MTA) by-product (Ikeguchi *et al.*, 2006; Pegg and Michael, 2010). In mammals, an interconversion pathway that transforms, in turn, spermine into spermidine and spermidine into putrescine, is carried out by enzymatic acetylation of spermidine/spermine-N^1-acetyltransferase (SSAT) and further oxidation by acetylpolyamine oxidase (Parry *et al.*, 1995). This interconversion pathway is aimed to regulating cellular polyamine content according to cell needs (Pegg, 2008).

7.5.2 Polyamine metabolism in trypanosomatids

L-Arginine is an essential amino acid for trypanosomatids because there is no specific metabolic pathway for its biosynthesis. Therefore, in trypanosomatids, L-arginine must be taken up from the external medium by the LdAAP3 transporter, which specifically binds L-arginine and no other basic amino acids (Kandpal *et al.*, 1995; Shaked-Mishan *et al.*, 2006). As in the host, cytoplasmic arginine can be cleaved by parasite arginase to yield L-ornithine for the production of the parasite's own polyamines. Competition for L-arginine between host and parasite, and the use of this amino acid, are crucial in the host–pathogen struggle, as discussed in Section 7.6.

L-Ornithine – the main product of arginase – is next used by ODC to synthesize putrescine, but this point is where differences among trypanosomatids are observed (Persson, 2007). *Leishmania* spp. and *T. brucei* have an active ODC enzyme, unlike *T. cruzi*, which has lost the *odc* gene and so is a putrescine auxotroph (Carrillo *et al.*, 1999; Algranati, 2010; Willert and Phillips, 2012). Trypanosomatid ODC is a long-life (more than 6 h) homodimeric protein that contains a pyridoxal-5′-phosphate (PLP) prosthetic moiety in the active centre (Phillips *et al.*, 1987). ODC is an enzyme target of crucial importance in *T. brucei* as it is irreversibly inhibited by eflornithine – an enzyme-activated inhibitor – that is currently used during the late neurological stages of sleeping sickness in Africa. The interest of trypanosomal ODC for drug targeting is based on turnover differences with the host (Persson *et al.*, 2003). Long-lasting inhibition of pathogen ODC by

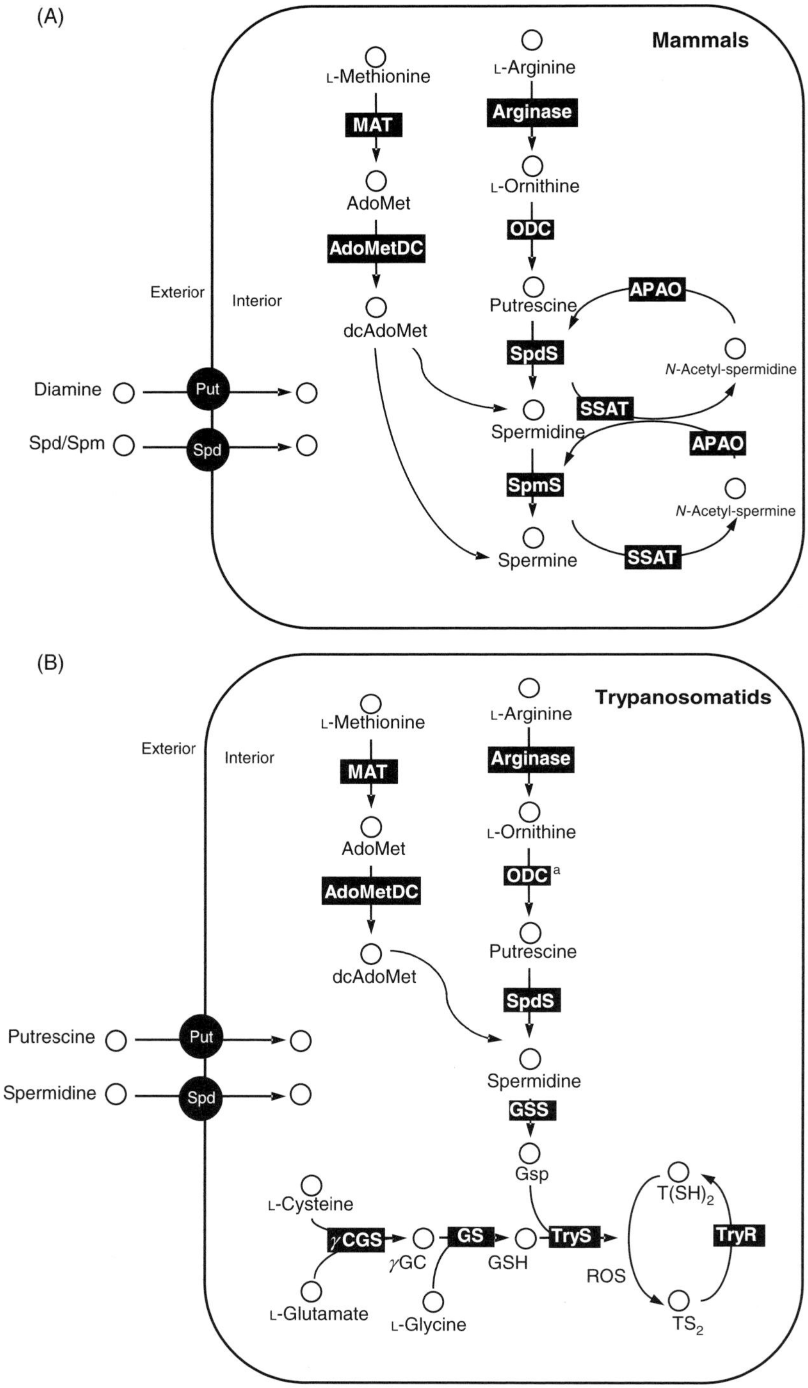

[a]ODC is not present in *T. cruzi*

Fig. 7.3. Differences between polyamine metabolism in mammalian cells (A) and trypanosomatid parasites (B). The enzymes involved in polyamine biosynthesis are shown in black boxes, polyamine transporters in filled circles and polyamine metabolites in empty circles). (A) The full polyamine

eflornithine may drive to permanent downstream effects, such as the blocking of spermidine and trypanothione synthesis, poisonous free radical build-up (Fairlamb, 2003) and the hypermethylation of macromolecules due to AdoMet accumulation (Byers *et al.*, 1991). However, because the host cells are mostly non-proliferative and the host ODC is rapidly demolished by the ODC-antizyme dependent proteasome, the effects of eflornithine can be readily circumvented. Leishmanial ODC has an *N*-terminal domain that differs from that of mammals and *T. brucei*, and a C-terminus end, which contains the active site, with a closer homology to other trypanosomes. Similarly to *T. brucei*, leishmanial ODC lacks the PEST sequence (a recognition signal for proteasome degradation) in its primary structure and no antizyme gene has been found in its genome (Hua *et al.*, 1995).

Ullman and coworkers have knocked down all genes and carriers involved in polyamine metabolism in *Leishmania* parasites. According to their results, the *Δ-odc L. donovani* strain is auxotrophic for putrescine and spermidine (Jiang *et al.*, 1999) and has a similar response to eflornithine to that *T. cruzi in vitro*. Exogenous spermine cannot rescue this auxotrophy and yet can be toxic (Merkel *et al.*, 2007), which explains the lack of the SpmS-encoding gene in the genome of trypanosomatids (Berriman *et al.*, 2005; Ivens *et al.*, 2005). The question raised at this point is why eflornithine is effective against sleeping sickness (a trypanosomiasis) but not against leishmaniasis if ODC is targeted with similar inhibitory constants in both pathogens and their half-lives are not different. A plausible answer to this question is that different polyamine concentrations are found in the blood – the living environment of African trypanosomes, and in the parasitophorous vacuoli of macrophages – where leishmanial amastigotes grow and develop. Because eflornithine transforms both African trypanosomes and leishmanias in polyamine auxotrophs, external polyamine concentration and the activity of polyamine pumps are of paramount importance for pathogen survival during ODC targeting by drugs (Hasne and Ullman, 2005; Reguera *et al.*, 2005).

AdoMetDC is the second key enzyme of polyamine synthesis, and in this case trypanosomatids produce an active self-cleaving prozyme, which after dimerization, is involved in spermidine synthesis. As in the rest of eukaryotes, AdoMetDC is a putrescine-activated enzyme that contains a pyruvoyl residue – created during proenzyme autoprocessing – in the active site, which is responsible for the electrophilic attack to the α carbon of AdoMet to produce dcAdoMet as aminopropyl donor (Willert and Phillips, 2012). Like ODC, trypanosomatid-borne AdoMetDC is a long-lasting enzyme that lacks PEST regions for proteasomal degradation. Double-targeted replacement of both AdoMetDC-encoding gene alleles served to

Fig. 7.3. Continued.
biosynthesis pathway of mammals involves the stepwise transformation of ornithine to putrescine, spermidine and spermine with the help of the propylamino donor intermediate dcAdoMet. The back conversion of spermine to spermidine and putrescine is carried out by deacetylases and oxidases that are not present in trypanosomatids. Note that there are no links between polyamine and glutathione metabolism in mammals. (B) Polyamine metabolism in trypanosomatids is simpler (spermine is not synthesized) and it is linked to sulfur redox reactions and the detoxification of reactive oxygen species (ROS). The lack of an ODC encoding gene characterizes the *Trypanosoma cruzi* polyamine biosynthesis pathway, which make this organism an auxotroph of putrescine and/or spermidine. Abbreviations of the enzymes/transporters (where used) and their corresponding EC numbers are given in alphabetic order: AdoMetDC, *S*-adenosylmethionine decarboxylase (EC 4.1.1.50); APAO, N^1-acetyl polyamine oxidase (EC 1.5.3.13); Arginase (EC 3.5.3.1); γGCS, γ-glutamylcysteine synthetase (EC 6.3.2.2); GS, glutathione synthetase (EC 6.3.2.3); GSS, glutathionylspermidine synthetase (EC 6.3.1.8); MAT, methionine adenosyltransferase (EC 2.5.1.6); ODC, ornithine decarboxylase (EC 4.1.1.17); Put, putrescine transporter; Spd, spermidine transporter; SpdS, spermidine synthase (EC 2.5.1.16); SpmS, spermine synthase (EC 2.5.1.22); SSAT, spermidine/spermine N^1-acetyltransferase (EC 2.3.1.57); TryR, trypanothione reductase (EC 1.8.1.12); TryS, trypanothione synthase (EC 6.3.1.9). Abbreviations of the metabolites are also given in alphabetic order: AdoMet, *S*-adenosylmethionine; dcAdoMet, decarboxylated *S*-adenosylmethionine; γGC, γ-glutamylcysteine; GSH, reduced glutathione; Gsp, glutathionylspermidine; ROS, reactive oxygen species; Spm, spermine; $T(SH)_2$, reduced trypanothione; TS_2, oxidized trypanothione.

create a *Δ-adometdc L. donovani* strain, which contains high intracellular putrescine levels and has to be rescued by exogenous spermidine but not by putrescine or spermine, thus ruling out any hypothetical auxotrophy for spermidine in these parasites (Roberts *et al.*, 2002).

The enzyme AdoMetDC constitutes the key point of polyamine biosynthesis regulation in trypanosomatids. Early studies with recombinant trypanosomatid AdoMetDC showed that the resultant homodimer was active and required putrescine for maximal enzyme activity, but not for generating the pyruvoyl cofactor (Beswick *et al.*, 2006). However, the activity of this homodimer was several orders of magnitude lower than that found in other eukaryotic sources. To reach the catalytic activity that is expected within parasite cell extracts, the AdoMetDC monomer must be dimerized with a specific prozyme that has a primary structure close to that of AdoMetDC, but lacking the amino acids involved in its autoprocessing and not enzymatically active on its own. This protein plays the role of allosteric activator when it binds to 'true' AdoMetDC to form an active heterodimer, whose activity is three orders of magnitude higher than that of the homodimer (Willert *et al.*, 2007; Willert and Phillips, 2009). Silencing of the AdoMetDC prozyme-encoding gene produces a strain with putrescine overload and decreased spermidine and trypanothione, resulting in parasite cell growth arrest and cell death. As for *Δ-adometdc*, the silencing effect of the prozyme can be rescued by exogenous spermidine (Willert and Phillips, 2008). Similarly to ODC, AdoMetDC was pointed out early on as a putative target for antitrypanosomal drugs (see Section 7.7).

Finally, the aminopropyl group of dcAdoMet is enzymatically bound to putrescine to yield spermidine by the catalytic action of SpdS, generating 5′-deoxy-5′-methylthioadenosine (MTA) as by-product. SpdS is present in all trypanosomatids and its genetic knockdown originates auxotrophic strains for spermidine that have to be rescued by adding this polyamine, but not putrescine or spermine, to the culture medium (Roberts *et al.*, 2001). As mentioned before, no SpmS encoding gene is present in any trypanosomatid genome database, thereby confirming the well-documented absence of this polyamine in lower eukaryotes, where spermidine plays the role that spermine plays in vertebrates.

Due to the absence of a polyamine interconversion pathway to adapt intracellular polyamine pools to optimal growth rates, trypanosomatids have developed a fine-tuning mechanism to take polyamines from the external medium (Reguera *et al.*, 2005). This process is of paramount importance in *T. cruzi*, because the species is a putrescine auxotroph and therefore depends on an external source of putrescine to survive (Carrillo *et al.*, 2006). Early reports showed that *Leishmania* parasites could drive extracellular polyamines from the host's blood or parasitophorous vacuoli by an active, inducible and pH-sensitive system (Balaña-Fouce *et al.*, 1989). The Buddy Ullman group (at Oregon Health and Science University, USA) initiated the molecular characterization of polyamine transporters in trypanosomatids, reporting the existence of a gene encoding the putative carrier LmPOT1 in *L. major* chromosome 14. When *LmPOT1* was functionally expressed in *Xenopus laevis* oocytes, the protein showed a high affinity for putrescine and spermidine, and was inhibited by pentamidine (an antiparasitic drug), so confirming previous reports carried out in *Leishmania* promastigotes (Hasne and Ullman, 2005).

7.6 Role of L-Arginine in Host–Parasite Interactions

Apart from being essential for the above-mentioned pathways, L-arginine is also an indispensable factor for the innate immune host response (see Fig. 7.4). Under permissive host response, *Leishmania* and *T. cruzi* can proliferate within parasitophorous vacuoles, thereby evading the cytolytic mechanisms displayed by macrophages. Macrophages are attracted to amastigote forms, which expose phosphatidylserine on the outer face of the plasma membrane (a typical feature of early apoptotic cells) interacting with a specific macrophage receptor, which leads to the endocytic pathway for apoptotic cell removal (Gironès *et al.*, 2005). L-Arginine is vital for the efficient activation of macrophages, which will decide the fate of intracellular parasites such as American trypanosomes (*T. cruzi*) and leishmanias (Wanasen and Soong, 2008).

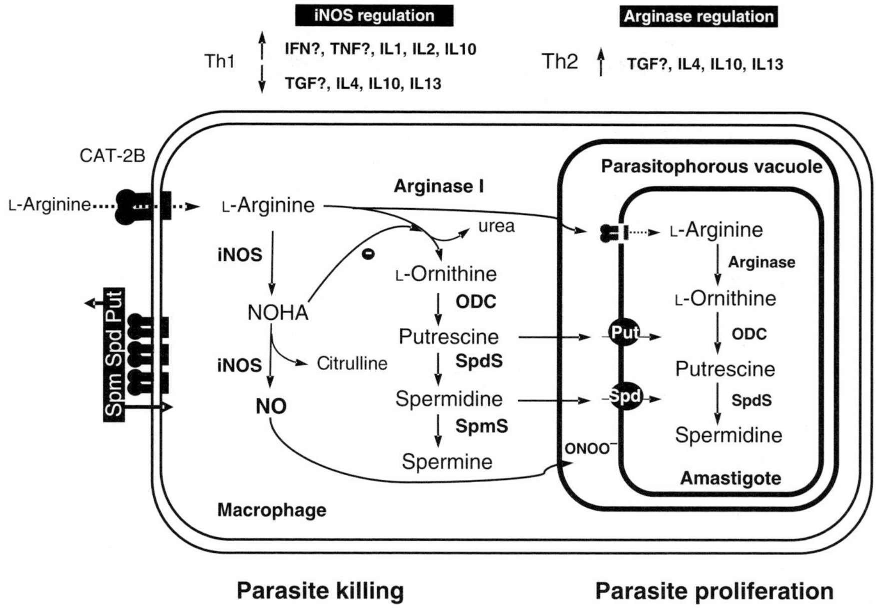

Fig. 7.4. Host and parasite L-arginine metabolism play a key role in macrophage activation and parasite survival during infection. The classic activation mechanism of M1 (pro-inflammatory) macrophages responds to Th1 (T1 helper cell) cytokines such as tumour necrosis factor α (TNF-α) and interferon γ(IFN-γ) and the interleukins IL1, IL2 and IL10. The alternative activation of M2 (anti-inflammatory) macrophages is by Th2 (T2 helper cell) cytokines such as TGF-β and interleukins IL4, IL10 and IL13. IL10 can be produced by various subpopulations of T helper cells (Th2, Th1, Th3 and Treg) and by infected or activated macrophages. Similarly, the cytokine TNF-α can be produced by T cells and activated macrophages. M1 activation induces inducible nitric oxide synthase (iNOS) activity in macrophages, and this is responsible for nitric oxide (NO) production and citrulline formation from L-arginine. NO synthesis will induce the synthesis of other reactive nitrogen and oxygen (RNOS) species, such as $ONOO^-$ (peroxynitrite). In addition to the transcriptional regulation of iNOS/arginase by cytokines and interleukins, there is metabolic regulation at the biochemical level that depends on the kinetic properties of both enzymes (affinity for L-arginine and enzyme capacity) as well as on the inhibitory effect of N^{ω}-hydroxy-L-arginine (NOHA) – an intermediary of L-arginine hydrolysis by iNOS – on arginase. Th1 activation ultimately results in parasite killing, the collapse of the infection and self-healing. In contrast, macrophage Th2 activation causes an increase in macrophage arginase I activity in the host, resulting in the formation of L-ornithine – one of the precursors of polyamines in the parasite. Both host and parasite-synthesized polyamines are used for the proliferation of amastigotes within the parasitophorous vacuole and as a result of the progress of infection and death of the host. *Trypanosoma cruzi* is innately auxotrophic for L-ornithine and putrescine, and therefore has an absolute dependence on these two intermediates of the host for survival. These responses should be controlled by the mammalian host to limit the pathology and are likely to allow an effective immune memory response. Abbreviations not yet mentioned above: CAT-2B, cationic amino acid transporter 2B; ODC, ornithine decarboxylase; Put, putrescine transporter; Spd, spermidine transporter; SpdS, spermidine synthase; Spm, spermine transporter; SpmS, spermine synthase. Modified from Balaña-Fouce *et al.* (2012).

M1 macrophages are pro-inflammatory, produce NO from L-arginine, and also superoxide radicals, which react together to produce the free radical peroxynitrite, leading to a Th1 (T1 helper cell) adaptive immune response. In contrast, M2 macrophages are anti-inflammatory in nature and produce L-ornithine via arginase, which drives L-arginine to polyamine synthesis and gives rise to a Th2 (T2 helper cell) response. Therefore, a correct switch of the use of this amino acid is responsible for an adequate immune response against parasitic infection (Colotti and Ilari, 2011).

Cytokines can control the transcriptional activation of one specific pathway to the detriment of others. Inducible nitric oxide synthase (iNOS or NOS2) is induced in M1 macrophages by the type I pro-inflammatory cytokines tumour necrosis factor α (TNF-α) and interferon γ (IFN-γ) (Popovic *et al.*, 2007), ligands of Toll-like receptors such as lipopolysaccharide (LPS) (Kropf *et al.*, 2004) and the IL12 and IL10 interleukins (Green *et al.*, 1990; Liew *et al.*, 1997). In this way, experimental infections of macrophages obtained from C57BL/6J mice with *L. major* are unsuccessful owing to the large NO production (in the millimolar range) and consistently Th1 priming (Miralles *et al.*, 1994; Taylor-Robinson, 2001; Lee *et al.*, 2003). In contrast, macrophages obtained from BALB/c mice are alternatively activated with transforming growth factor β (TGF-β) (Iniesta *et al.*, 2005), or type II anti-inflammatory interleukins (IL4, IL10 and IL13), producing high arginase I activity that facilitates the course of *Leishmania* infections (Iniesta *et al.*, 2002). The products of arginase I are in turn incorporated to the parasite polyamine milieu.

Because arginase I and iNOS compete each other for L-arginine, stimulation of one of them involves inhibition of the other and vice versa. L-Arginine is a preferred substrate for iNOS due its low affinity constant, but arginase has a V_{max} value 1000-fold higher than that of iNOS (Wu and Morris, 1998). As a result, the induction of arginase by Th2 cytokines depletes the L-arginine pool to be used by iNOS, thus downregulating the production of NO at safe levels for the parasite and producing L-ornithine, which fuels polyamine biosynthesis in amastigotes. iNOS induction by Th1-type cytokines produces an enzymatic intermediate, N^{ω}-hydroxy-L-arginine (designated both LOHA and NOHA by different authors), which may inhibit arginase I competitively at low micromolar concentrations (Iniesta *et al.*, 2001).

In order to study the role played by *Leishmania* arginase in the establishment and progression of experimental infections, arginase null mutants (Δarg) were generated of *L. mexicana* (Gaur *et al.*, 2007) and *L. major* (Reguera *et al.*, 2009). Beside the need of exogenous putrescine or spermidine, there was a significant reduction in parasite proliferation, but not in the ability to infect macrophages. Neither did experimental infections of mice with *L. major* Δarg and wild type (WT) strains result in changes in the cytokine profile and levels (Muleme *et al.*, 2009). Furthermore, no differences in NO production were found in mice infected with Δarg compared with those infected with WT strains. It appears then that the collapse of the *L. major* Δarg mutant in establishing the infection is mostly attributable to the drop in polyamines (due to a drop in arginase activity) rather than to a compensatory induction of iNOS.

7.7 Polyamines as Potential Targets for Drug Development

Polyamines are essential metabolites for the survival of trypanosomatids, hence the depletion of intracellular levels of these compounds by preventing their synthesis or uptake from the surrounding environment has long been proposed as a promising target for drug development. Since the early 1980s, hundreds of compounds have been tested against the enzymes involved in polyamine biosynthesis, including the free radical scavenger trypanothione (Olin-Sandoval *et al.*, 2010).

ODC represents the first step in polyamine synthesis. Except for *T. cruzi*, which is a putrescine auxotroph, different species of *Leishmania* and *T. brucei* have an ODC orthologue that is much more stable than the mammalian protein (Sánchez *et al.*, 1989). Eflornithine is an irreversible inhibitor of ODC, which early proved its effectiveness to cure acute infections of experimental mice infections (Bacchi *et al.*, 1980). Soon after, eflornithine was successfully tested in field cases of human sleeping sickness caused by *T. brucei gambiense* (Taelman *et al.*, 1987), though it was ineffective against the *rhodesiense* variant (Iten *et al.*, 1995). ODC inhibition in *T. brucei* produces a depression of more than 90% of the intracellular content of putrescine and spermidine (Burri and Brun, 2003). Intracellular polyamine depletion induces increased oxidative stress, due to the inhibition of trypanothione synthesis (Fairlamb *et al.*, 1987), and over a 100-fold increase in cytoplasmic AdoMet levels. The latter results in an increased rate of protein, DNA and lipid methylation, and it is these that are the actual killers of the parasite (Bacchi and Yarlett, 1993).

In 1990, the US Food and Drug Administration approved the administration of eflornithine by intravenous infusion as a treatment for the meningoencephalic stage of human African trypanosomiasis. This treatment involves daily administration of 400 mgkg^{-1} body weight for 14 days. Intravenous administration of eflornithine is usually troublesome to health services from underdeveloped countries where the disease is endemic. This problem has been solved in part by using combination therapy of eflornithine with the anti-Chagas drug nifurtimox (NECT therapy). This treatment has been shown to be safe and permits shortening by half of the 2 weeks infusion period required for eflornithine alone (Yun *et al.*, 2010). Oral administration of eflornithine has no bioavailability problems, but due to its short half-life it is necessary to administer high drug doses to achieve a therapeutic effect, which increases the cost of treatment. Due to the relevance of this molecule, the need for the development of chemical analogues or suitable formulations that can overcome these drawbacks is urgent (Yang *et al.*, 2012).

Despite eflornithine inhibiting the growth of *Leishmania* parasites (Kaur *et al.*, 1986; Balaña-Fouce *et al.*, 1989), it has no effect on experimental infections (McCann *et al.*, 1981). This may be due to the intracellular nature of the amastigote form within host macrophages. It is plausible to speculate that the host can provide polyamines to the parasite via arginase, thereby circumventing the effect of the drug. In addition, the emergence of eflornithine-resistant promastigotes is common. Resistance mechanisms appear to be associated with increased ODC activity (Coons *et al.*, 1990), but an increase in the expression of proteins involved in free radicals detoxification, in polyamines and in trypanothione metabolism has also been observed (Singh *et al.*, 2014). Arginase, the enzyme that catalyses the flow of L-ornithine to polyamines in trypanosomatids, also plays an essential role in parasite survival (Das *et al.*, 2005). The arginase inhibitors N$^{\omega}$-hydroxy-L-arginine (NOHA) and nor-NOHA reduce the growth of *L. major* and *L. infantum* amastigotes within macrophages in both *in vitro* and *in vivo* experimental infections in BALB/c mice. Other reports have shown similar effects with other trypanosomatids, such as *T. cruzi* amastigotes in J774 mouse cells treated with NOHA (Stempin *et al.*, 2004) and *T. brucei* infecting macrophages treated with nor-NOHA (Duleu *et al.*, 2004). These promising results point to arginase as an interesting therapeutic target for drug design (Iniesta *et al.*, 2005).

AdoMetDC controls spermidine synthesis in trypanosomatids, and is a long half-life enzyme that from the beginning was addressed as a potential drug target. Compounds bearing guanidine and amidine substituents with proven antiparasitic properties, such as pentamidine or berenil, have been shown to be strong reversible AdoMetDC inhibitors, producing spermidine depletion and an increase of intracellular putrescine (Balaña-Fouce *et al.*, 1986; Mukhopadhyay and Madhubala, 1995). What is more, diamidines had an interesting synergic effect in preventing the uptake of external polyamines (Balaña-Fouce *et al.*, 1989). CGP 40215A (bis {[3-aminoiminomethyl) phenyl] methylene} carbonimidic dihydrazide trihydrochloride) was the best inhibitor of this series of compounds (Brun *et al.*, 1996). This drug showed good results *in vitro* against *T. b. rhodesiense*, multidrug-resistant *T. b. brucei*, and clinical isolates of *T. b. rhodesiense* and *T. b. gambiense*, but because the compound is unable to cross the blood–brain barrier during the CNS stage of the infection, the further development of the molecule was discontinued (Brun *et al.*, 2001). CGP 40215A was active against *L. donovani* promastigotes, but not against amastigotes, and showed no effect against any stage of *T. cruzi* (Mukhopadhyay *et al.*, 1996). Chemical analogues of dcAdoMet were also developed as irreversible inhibitors of AdoMetDC. MDL 73811 (AbeAdo) (5′-{[(Z)-4-amino-2-butenyl] methylamino}-5′-deoxyadenosine) and its 8-methyl derivative produced healing of mice infected with *T. brucei brucei* (Barker et al., 2009), although the antiparasitic effects of this compound appeared not to be due to its inhibition of spermidine depletion, but instead to a hypermethylation status produced by the AdoMet accumulated in the parasite (Byers *et al.*, 1991). These compounds have shown inhibitory effects on free-living forms of *T. cruzi* and *Leishmania* spp. *in vitro*, but have failed to cure experimental infections.

SpdS inhibition is another possible target to prevent spermidine synthesis (Taylor *et al.*, 2008). *T. brucei* RNAi-silenced in SpdS shows decreased spermidine levels and, as a consequence, a cessation of parasite growth. Remarkably, this situation cannot be reversed by the addition of spermidine to the culture medium, because unlike *L. major* and *T. cruzi*, *T. brucei* lacks a highly efficient transporter (Carrillo *et al.*, 2006; Hasne *et al.*, 2010). *S*-adenosyl-1,8-diamino-3-thiooctane (AdoDATO), a multi-substrate adduct inhibitor of SpdS, and cyclohexylamine and dicyclohexylamine (Bitonti *et al.*, 1984), the latter being two competing substrates with putrescine, did not show healing effects on *T. brucei* infections in rodents.

The enzymes involved in trypanothione synthesis, particularly trypanothione synthetase (TryS), and reduction (trypanothione reductase, TryR) have been pointed out as robust therapeutic targets against trypanosomatids (Müller *et al.*, 2003; Flohé, 2012), and except for the case of *T. brucei* TryR-KO mutants, which cease to be viable when the enzymatic activity is reduced to less than 5% (Krieger *et al.*, 2000), the incomplete suppression of TryS implies parasite death (Comini *et al.*, 2004). Using high-throughput-target-based drug screening, several TryS inhibitors have been identified with activity against trypanosomatids, but none of them have passed to preclinical studies (Spinks *et al.*, 2012).

Attempts to achieve polyamine deprivation by inhibiting the biosynthetic enzymes for these compounds have failed because the compensatory mechanisms available in the parasites can circumvent these pharmacological effects. The inducible polyamine transporter is extremely sensitive to intracellular polyamine levels and can be upregulated as a primary response to polyamine antimetabolites (Balaña-Fouce *et al.*, 1991). Several polyamine analogues have been developed to target polyamine transporters and biosynthetic enzymes, which deplete cells of their polyamine requirements and deprive them of an external supply. The first generation of these compounds consisted of symmetrically or not terminally alkylated polyamine analogues (Zou *et al.*, 2001). More recently, substituted polyamine derivatives have been synthesized, including long-chain and cyclic oligoamine analogues. Numerous *in vitro* studies have been done to assess the activity of such compounds against *Trypanosoma* and *Leishmania*, and it has been observed that these molecules are multisite inhibitors and can act specifically on different stages of the parasite (Labadie *et al.*, 2004; Tavares *et al.*, 2005, Yamanaka *et al.*, 2013).

7.8 Conclusions

Because there are no suitable synthetic pathways for L-methionine and L-arginine in trypanosomatids, these organisms are dependent on their host metabolism. Specific amino acid transporters provide sufficient quantities of these molecules for the parasite requirements, including protein synthesis and polyamine metabolism, both processes being highly active during the rapid proliferation of these pathogens within the host. With the exception of the agent responsible for Chagas disease, *T. cruzi*, L-methionine and L-arginine are the precursors of putrescine and spermidine. From the beginning, ODC was identified as a singular target to prevent the proliferation of African trypanosomes. A long-lived protein and a poor content of circulating polyamines in the blood were the relevant conditions that gave rise to the early success of eflornithine against these parasites. However, these conditions were not present in the surrounding environment of leishmanias, which can take up polyamines from the parasitophorous vacuole. After its approval, eflornithine was one of the second-choice drugs against *T. brucei gambiense* infections in Africa – a singular success for the Pharma industries because it represented the only molecule developed against these diseases in modern era. Despite eflornithine being efficient against late stages of the disease, the galenic formulation of the drug is far from optimum. Large volumes of daily repeated infusions over weeks imply painful and time-consuming treatments, which are hardly successfully completed and administered by deficient health services often far away from the villages concerned. The combination therapy with oral nifurtimox (NECT therapy) has significantly reduced the timespan of eflornithine infusions to 3 weeks with similar healing outcomes to the longer single-drug treatment.

Other attempts to target polyamine enzymes have only worked *in vitro* or at early preclinical stages. Cationic diamidines, such as berenil, pentamidine or CGP40215A, behaved as polyamine antimetabolites by inhibiting AdoMetDC and polyamine uptake, but they have other mechanisms of action. Further remodelling of the molecules does not improve their antiparasitic efficacy and they are not able to cross the blood–brain barrier. The relevance of spermidine in preventing oxidative stress within the tripanothione molecule opened new perspectives to pharmaceutical chemists. Hundreds of new small molecules aimed against TryS and TryR have shown that polyamines are still in the pipeline of drug discovery. Last, but not least, L-arginine plays a pivotal role in the host response to intracellular infections. Macrophage response to T lymphocyte activation may result in parasite proliferation (linked to host arginase induction and polyamine build-up after Th2 response) or death (linked to NOS induction and polyamine deprivation after the Th1 response). Compounds that can change this response are now under study and could show the revival of polyamines as strong targets against these diseases.

Acknowledgements

Our research group is being supported by Ministerio de Economía y Competitividad (MINECO; AGL2010-16078/GAN and CYTED 214RT0482), Instituto de Salud Carlos III (Feder PI12/00104) and Junta de Castilla y León (grants Gr238 and LE182U13).

References

Albers, E. (2009) Metabolic characteristics and importance of the universal methionine salvage pathway recycling methionine from 5´-methylthioadenosine. *IUBMB Life* 61, 1132–1142.

Algranati, I.D. (2010) Polyamine metabolism in *Trypanosoma cruzi*: studies on the expression and regulation of heterologous genes involved in polyamine biosynthesis *Amino Acids* 38, 645–651.

Alvar, J., Vélez, I.D., Bern, C., Herrero, M., Desjeux, P., Cano, J., Jannin, J., den Boer, M. and WHO Leishmaniasis Control Team (2012) Leishmaniasis worldwide and global estimates of its incidence. *PLoS ONE* 7(5): e35671.

Antinori, S., Schifanella, L. and Corbellino, M. (2012) Leishmaniasis: new insights from an old and neglected disease. *European Journal of Clinical Microbiology and Infectious Diseases* 31, 109–111.

Avila, M.A., García-Trevijano, E.R., Martínez-Chantar, M.L., Latasa, M.U., Pérez-Mato, I., Martínez-Cruz, L.A., del Pino, M.M., Corrales, F.J. and Mato, J.M. (2002) *S*-Adenosylmethionine revisited: its essential role in the regulation of liver function. *Alcohol* 27, 163–167.

Bacchi, C.J. and Yarlett, N. (1993) Effects of antagonists of polyamine metabolism on African trypanosomes. *Acta Tropica* 54, 225–236.

Bacchi, C.J. and Yarlett, N. (2002) Polyamine metabolism as chemotherapeutic target in protozoan parasites. *Mini-Reviews in Medical Chemistry* 2, 553–563.

Bacchi, C.J., Nathan, H.C., Hutner, S.H., McCann, P.P. and Sjoerdsma, A. (1980) Polyamine metabolism: a potential therapeutic target in trypanosomes. *Science* 210, 332–334.

Bacchi, C.J., Sanabria, K., Spiess, A.J., Vargas, M., Marasco, C.J. Jr, Jimenez, L.M., Goldberg, B. and Sufrin, J.R. (1997) *In vivo* efficacies of 5´-methylthioadenosine analogs as trypanocides. *Antimicrobial Agents and Chemotherapy* 41, 2108–2112.

Bacchi, C.J., Goldberg, B., Rattendi, D., Gorrell, T.E., Spiess, A.J. and Sufrin, J.R. (1999) Metabolic effects of a methylthioadenosine phosphorylase substrate analog on African trypanosomes. *Biochemical Pharmacology* 57, 89–96.

Balaña-Fouce, R., Garzón Pulido, T., Ordóñez-Escudero, D. and Garrido-Pertierra, A. (1986) Inhibition of diamine oxidase and *S*-adenosylmethionine decarboxylase by diminacene aceturate (berenil). *Biochemical Pharmacology* 35, 1597–1600.

Balaña-Fouce, R., Ordóñez, D. and Alunda, J.M. (1989) Putrescine transport system in *Leishmania infantum* promastigotes. *Molecular and Biochemical Parasitology* 35, 43–50.

Balaña Fouce, R., Escribano, M.I. and Alunda, J.M. (1991) Putrescine uptake regulation in response to alpha-difluoromethylornithine treatment in *Leishmania infantum* promastigotes. *Molecular and Cellular Biochemistry* 107, 127–133.

Balaña-Fouce, R., Calvo-Álvarez, E., Álvarez-Velilla, R., Prada, C.F., Pérez-Pertejo, Y. and Reguera, R.M. (2012) Role of trypanosomatid's arginase in polyamine biosynthesis and pathogenesis. *Molecular and Biochemical Parasitology* 181, 85–93.

Barker, R.H. Jr, Liu, H., Hirth, B., Celatka, C.A., Fitzpatrick, R., Xiang, Y., Willert, E.K., Phillips, M.A., Kaiser, M., Bacchi, C.J. *et al.* (2009) Novel *S*-adenosylmethionine decarboxylase inhibitors for the treatment of human African trypanosomiasis. *Antimicrobial Agents and Chemotherapy* 53, 2052–2058.

Berger, B.J., Dai, W.W., Wang, H., Stark, R.E. and Cerami, A. (1996) Aromatic amino acid transamination and methionine recycling in trypanosomatids. *Proceedings of the National Academy of Sciences of the United States of America* 93, 4126–4130.

Berger, L.C., Wilson, J., Wood, P. and Berger, B.J. (2001) Methionine regeneration and aspartate aminotransferase in parasitic protozoa. *Journal of Bacteriology* 183, 4421–4434.

Berriman, M., Ghedin, E., Hertz-Fowler, C., Blandin, G., Renauld, H., Bartholomeu, D.C., Lennard, N.J., Caler, E., Hamlin, N.E., Haas, B. *et al.* (2005) The genome of the African trypanosome *Trypanosoma brucei. Science* 309, 416–422.

Beswick, T.C., Willert, E.K. and Phillips, M.A. (2006) Mechanisms of allosteric regulation of *Trypanosoma cruzi S*-adenosylmethionine decarboxylase. *Biochemistry* 45, 7797–807.

Bitonti, A.J., Kelly, S.E. and McCann, P.P. (1984) Characterization of spermidine synthase from *Trypanosoma brucei brucei. Molecular and Biochemical Parasitology* 13, 21–28.

Brun, R., Bühler, Y., Sandmeier, U., Kaminsky, R., Bacchi, C.J., Rattendi, D., Lane, S., Croft, S.L., Snowdon, D., Yardley, V. *et al.* (1996) *In vitro* trypanocidal activities of new *S*-adenosylmethionine decarboxylase inhibitors. *Antimicrobial Agents and Chemotherapy* 40, 1442–1447.

Brun, R., Burri, C. and Gichuki, C.W. (2001) The story of CGP 40215: studies on its efficacy and pharmacokinetics in African green monkey infected with *Trypanosoma brucei rhodesiense. Tropical Medicine and International Health* 6, 362–368.

Burri, C. (2010) Chemotherapy against human African trypanosomiasis: is there a road to success? *Parasitology* 137, 1987–1994.

Burri, C. and Brun, R. (2003) Eflornithine for the treatment of human African trypanosomiasis. *Parasitology Research* 90(Supp 1), S49–S52.

Byers, T.L., Bush, T.L., McCann, P.P. and Bitonti, A.J. (1991) Antitrypanosomal effects of polyamine biosynthesis inhibitors correlate with increases in *Trypanosoma brucei brucei S*-adenosyl-L-methionine. *The Biochemical Journal* 274, 527–533.

Caldwell, R.B., Toque, H.A., Narayanan, S.P. and Caldwell, R.W. (2015) Arginase: an old enzyme with new tricks. *Trends in Pharmacological Sciences* 36, 395–405.

Carrillo, C., Cejas, S., González, N.S. and Algranati, I.D. (1999) *Trypanosoma cruzi* epimastigotes lack ornithine decarboxylase but can express a foreign gene encoding this enzyme. *FEBS Letters* 454, 192–196.

Carrillo, C., Canepa, G.E., Algranati, I.D. and Pereira, C.A. (2006) Molecular and functional characterization of a spermidine transporter (TcPAT12) from *Trypanosoma cruzi. Biochemical and Biophysical Research Communications* 344, 936–940.

Castilho-Martins, E.A., Laranjeira da Silva, M.F., dos Santos, M.G., Muxel, S.M. and Floeter-Winter, L.M. (2011) Axenic *Leishmania amazonensis* promastigotes sense both the external and internal arginine pool distinctly regulating the two transporter-coding genes. *PLoS ONE* 6(11): e27818.

Closs, E.I., Boissel, J.P., Habermeier, A. and Rotmann, A. (2006) Structure and function of cationic amino acid transporters (CATs). *Journal of Membrane Biology* 213, 67–77.

Coffino, P. (2001) Regulation of cellular polyamines by antizyme. *Nature Reviews Molecular Cell Biology* 2, 188–194.

Colotti, G. and Ilari, A. (2011) Polyamine metabolism in *Leishmania*: from arginine to trypanothione. *Amino Acids* 40, 269–285.

Comini, M.A., Guerrero, S.A., Haile, S., Menge, U., Lünsdorf, H. and Flohé, L. (2004) Validation of *Trypanosoma brucei* trypanothione synthetase as drug target. *Free Radical Biology and Medicine* 36, 1289–1302.

Coons, T., Hanson, S., Bitonti, A.J., McCann, P.P. and Ullman, B. (1990) Alpha-difluoromethylornithine resistance in *Leishmania donovani* is associated with increased ornithine decarboxylase activity. *Molecular and Biochemical Parasitology* 39, 77–89.

Croft, S.L. and Olliaro, P. (2011) Leishmaniasis chemotherapy – challenges and opportunities. *Clinical Microbiology and Infection* 17, 1478–1483.

D'Antonio, E.L., Ullman, B., Roberts, S.C., Dixit, U.G., Wilson, M.E., Hai, Y. and Christianson, D.W. (2013) Crystal structure of arginase from *Leishmania mexicana* and implications for the inhibition of polyamine biosynthesis in parasitic infections. *Archives of Biochemistry and Biophysics* 535, 163–176.

da Silva, M.F. and Floeter-Winter, L.M. (2014) Arginase in *Leishmania*. *Subcellular Biochemistry* 74, 103–117.

Das, P., Lahiri, A., Lahiri, A. and Chakravorty, D. (2005) Arginase and polyamine synthesis are key factors in the regulation of experimental leishmaniasis *in vivo*. *The FASEB Journal* 19, 1000–1002.

Dridi, L., Ahmed Ouameur, A. and Ouellette, M. (2010) High affinity *S*-adenosylmethionine plasma membrane transporter of *Leishmania* is a member of the folate biopterin transporter (FBT) family. *The Journal of Biological Chemistry* 285, 19767–19775.

Duleu, S., Vincendeau, P., Courtois, P., Semballa, S., Lagroye, I., Daulouède, S., Boucher, J.L., Wilson, K.T., Veyret, B. and Gobert, A.P. (2004) Mouse strain susceptibility to trypanosome infection: an arginase-dependent effect. *Journal of Immunology* 172, 6298–6303.

Ellington, W.R. (2001) Evolution and physiological roles of phosphagen systems. *Annual Review of Physiology* 63, 289–325.

Fairlamb, A.H. (2003) Chemotherapy of human African trypanosomiasis: current and future prospects. *Trends in Parasitology* 19, 488–494.

Fairlamb, A.H., Henderson, G.B., Bacchi, C.J. and Cerami, A. (1987) *In vivo* effects of difluoromethylornithine on trypanothione and polyamine levels in bloodstream forms of *Trypanosoma brucei*. *Molecular and Biochemical Parasitology* 24, 185–191.

Flohé, L. (2012) The trypanothione system and its implications in the therapy of trypanosomatid diseases. *International Journal of Medical Microbiology* 302, 216–220.

Franco, J.R., Simarro, P.P., Diarra, A. and Jannin, J.G. (2014) Epidemiology of human African trypanosomiasis. *Clinical Epidemiology* 6, 257–275.

Gaur, U., Roberts, S.C., Dalvi, R.P., Corraliza, I., Ullman, B. and Wilson, M.E. (2007) An effect of parasite-encoded arginase on the outcome of murine cutaneous leishmaniasis. *Journal of Immunology* 179, 8446–8453.

Ghoda, L.Y., Savarese, T.M., Northup, C.H., Parks, R.E. Jr, Garofalo, J., Katz, L., Ellenbogen, B.B. and Bacchi, C.J. (1988) Substrate specificities of 5′-deoxy-5′-methylthioadenosine phosphorylase from *Trypanosoma brucei brucei* and mammalian cells. *Molecular and Biochemical Parasitology* 27, 109–118.

Giordana, L., Mantilla, B.S., Santana, M., Silber, A.M. and Nowicki, C. (2014) Cystathionine γ-lyase, an enzyme related to the reverse transsulfuration pathway, is functional in *Leishmania spp*. *Journal of Eukaryotic Microbiology* 61, 204–213.

Gironès, N., Cuervo, H. and Fresno, M. (2005) *Trypanosoma cruzi*-induced molecular mimicry and Chagas' disease. *Current Topics in Microbiology and Immunology* 296, 89–123.

Goldberg, B., Rattendi, D., Lloyd, D., Yarlett, N. and Bacchi, C.J. (2000) Kinetics of methionine transport and metabolism by *Trypanosoma brucei brucei* and *Trypanosoma brucei rhodesiense*. *Archives of Biochemistry and Biophysics* 377, 49–57.

Green, S.J., Crawford, R.M., Hockmeyer, J.T., Meltzer, M.S. and Nacy, C.A. (1990) *Leishmania major* amastigotes initiate the L-arginine-dependent killing mechanism in IFN-gamma-stimulated macrophages by induction of tumor necrosis factor-alpha. *Journal of Immunology* 145, 4290–4297.

Hasne, M.P. and Barrett, M.P. (2000) Transport of methionine in *Trypanosoma brucei brucei*. *Molecular and Biochemical Parasitology* 111, 299–307.

Hasne, M.P. and Ullman, B. (2005) Identification and characterization of a polyamine permease from the protozoan parasite *Leishmania major*. *The Journal of Biological Chemistry* 280, 15188–15194.

Hasne, M.P., Coppens, I., Soysa, R. and Ullman, B. (2010) A high-affinity putrescine-cadaverine transporter from *Trypanosoma cruzi*. *Molecular Microbiology* 76, 78–91.

Heby, O., Roberts, S.C. and Ullman, B. (2003) Polyamine biosynthetic enzymes as drug targets in parasitic protozoa. *Biochemical Society Transactions* 31, 415–419.

Heilbronn, J., Wilson, J. and Berger, B.J. (1999) Tyrosine aminotransferase catalyzes the final step of methionine recycling in *Klebsiella pneumoniae*. *Journal of Bacteriology* 181, 1739–1747.

Hua, S.B., Li, X., Coffino, P. and Wang, C.C. (1995) Rat antizyme inhibits the activity but does not promote the degradation of mouse ornithine decarboxylase in *Trypanosoma brucei*. *The Journal of Biological Chemistry* 270, 10264–10271.

Ikeguchi, Y., Bewley, M.C. and Pegg, A.E. (2006) Aminopropyltransferases: function, structure and genetics. *Journal of Biochemistry* 139, 1–9.

Iniesta V., Gómez-Nieto, L.C. and Corraliza, I. (2001) The inhibition of arginase by N^{ω}-hydroxy-L-arginine controls the growth of *Leishmania* inside macrophages. *Journal of Experimental Medicine* 193, 777–784.

Iniesta, V., Gómez-Nieto, L.C., Molano, I., Mohedano, A., Carcelén, J., Mirón, C., Alonso, C. and Corraliza, I. (2002) Arginase I induction in macrophages, triggered by Th2-type cytokines, supports the growth of intracellular *Leishmania* parasites. *Parasite Immunology* 24, 113–118.

Iniesta, V., Carcelén, J., Molano, I., Peixoto, P.M., Redondo, E., Parra, P., Mangas, M., Monroy, I., Campo, M.L., Nieto, C.G. and Corraliza, I. (2005) Arginase I induction during *Leishmania major* infection mediates the development of disease. *Infection and Immunity* 73, 6085–6090.

Iten, M., Matovu, E., Brun, R. and Kaminsky, R. (1995) Innate lack of susceptibility of Ugandan *Trypanosoma brucei rhodesiense* to DL-alpha-difluoromethylornithine (DFMO). *Tropical Medicine and Parasitology* 46, 190–194.

Ivens, A.C., Peacock, C.S., Worthey, E.A., Murphy, L., Aggarwal, G., Berriman, M., Sisk, E., Rajandream, M.A., Adlem, E., Aert, R. *et al.* (2005) The genome of the kinetoplastid parasite, *Leishmania major. Science* 309, 436–442.

Jiang, Y., Roberts, S.C., Jardim, A., Carter, N.S., Shih, S., Ariyanayagam, M., Fairlamb, A.H. and Ullman, B. (1999) Ornithine decarboxylase gene deletion mutants of *Leishmania donovani. The Journal of Biological Chemistry* 274, 3781–3788.

Kandpal, M., Balaña-Fouce, R., Pal, A., Guru, P.Y. and Tekwani, B.L. (1995) Kinetics and molecular characteristics of arginine transport by *Leishmania donovani* promastigotes. *Molecular and Biochemical Parasitology* 71, 193–201.

Kaur, K., Emmett, K., McCann, P.P., Sjoerdsma, A. and Ullman, B. (1986) Effects of DL-alpha-difluoromethylornithine on *Leishmania donovani* promastigotes. *Journal of Protozoology* 33, 518–521.

Koszalka, G.W. and Krenitsky, T.A. (1986) 5´-Methylthioadenosine (MTA) phosphorylase from promastigotes of *Leishmania donovani. Advances in Experimental Medicine and Biology* 195, 559–563.

Krieger, S., Schwarz, W., Ariyanayagam, M.R., Fairlamb, A.H., Krauth-Siegel, R.L. and Clayton, C. (2000) Trypanosomes lacking trypanothione reductase are avirulent and show increased sensitivity to oxidative stress. *Molecular Microbiology* 35, 542–552.

Kropf, P., Freudenberg, M.A., Modolell, M., Price, H.P., Herath, S., Antoniazi, S., Galanos, C., Smith, D.F. and Müller, I. (2004) Toll-like receptor 4 contributes to efficient control of infection with the protozoan parasite *Leishmania major. Infection and Immunity* 72, 1920–1928.

Labadie, G.R., Choi, S.R. and Avery, M.A. (2004) Diamine derivatives with antiparasitic activities. *Bioorganic and Medicinal Chemistry Letters* 14, 615–619.

Lee, J., Ryu, H., Ferrante, R.J., Morris Jr., S.M. and Ratan, R.R. (2003) Translational control of inducible nitric oxide synthase expression by arginine can explain the arginine paradox. *Proceedings of the National Academy of Sciences of the United States of America* 100, 4843–4848.

Lejon, V., Bentivoglio, M. and Franco, J.R. (2013) Human African trypanosomiasis. *Handbook of Clinical Neurology* 114, 169–181.

Leprohon, P., Légaré, D. and Ouellette, M. (2011) ABC transporters involved in drug resistance in human parasites. *Essays in Biochemistry* 50, 121–144.

Li, Q.S., Cai, S., Fang, J., Borchardt, R.T., Kuczera, K., Middaugh, C.R. and Schowen, R.L. (2009) Evaluation of NAD(H) analogues as selective inhibitors for *Trypanosoma cruzi S*-adenosylhomocysteine hydrolase. *Nucleosides, Nucleotides Nucleic Acids* 28, 473–484.

Liew, F.Y., Wei, X.Q. and Proudfoot, L. (1997) Cytokines and nitric oxide as effector molecules against parasitic infections. *Philosophical Transactions of the Royal Society London B: Biological Sciences* 352, 1311–1315.

Lütje, V., Seixas, J. and Kennedy, A. (2013) Chemotherapy for second-stage human African trypanosomiasis. *Cochrane Database of Systematic Reviews* 6: CD006201.

McCann, P.P. Bacchi, C.J., Hanson, W.L. Cain, G.D., Nathan, H.C., Hunter S.H. and Sjoerdsma, A. (1981) Effect on parasitic protozoa of alpha-difluoromethylornithine, an inhibitor of ornithine decarboxylase. In: Calderera, C.M., Zuppia, V. and Bachrach, U. (eds) *Advances in Polyamine Research, Vol. 3*. Raven Press, New York, pp. 97–100.

Merkel, P., Beck, A., Muhammad, K., Ali, S.A., Schönfeld, C., Voelter, W. and Duszenko, M. (2007) Spermine isolated and identified as the major trypanocidal compound from the snake venom of *Eristocophis macmahoni* causes autophagy in *Trypanosoma brucei. Toxicon* 50, 457–469.

Miller, R.L., Sabourin, C.L. and Krenitsky, T.A. (1987) *Trypanosoma cruzi* adenine nucleoside phosphorylase. Purification and substrate specificity. *Biochemical Pharmacology* 36, 553–560.

Miralles, G.D., Stoeckle, M.Y., McDermott, D.F., Finkelman, F.D. and Murray, H.W. (1994) Th1 and Th2 cell-associated cytokines in experimental visceral leishmaniasis. *Infection and Immunity* 62, 1058–1063.

Miranda, M.R., Sayé, M., Bouvier, L.A., Cámara, M. de los M., Montserrat, J. and Pereira, C.A. (2012) Cationic amino acid uptake constitutes a metabolic regulation mechanism and occurs in the flagellar pocket of *Trypanosoma cruzi. PLoS ONE* 7(2): e32760.

Mohamed-Ahmed, A.H., Brocchini, S. and Croft, S.L. (2012) Recent advances in development of amphotericin B formulations for the treatment of visceral leishmaniasis. *Current Opinion in Infectious Diseases* 25, 695–702.

Moncayo, A. and Silveira, A.C. (2009) Current epidemiological trends for Chagas disease in Latin America and future challenges in epidemiology, surveillance and health policy. *Memorias do Instituto Oswaldo Cruz* 104(Suppl 1), 17–30.

Monge-Maillo, B. and López-Vélez, R. (2013) Therapeutic options for Old World cutaneous leishmaniasis and New World cutaneous and mucocutaneous leishmaniasis. *Drugs* 73, 1889–1920.

Morris, S.M. Jr (2006) Arginine: beyond protein. *American Journal of Clinical Nutrition* 83, 508S–512S.

Mukhopadhyay, R. and Madhubala, R. (1995) Antileishmanial activity of berenil and methylglyoxal bis(guanylhydrazone) and its correlation with *S*-adenosylmethionine decarboxylase and polyamines. *International Journal of Biochemistry and Cell Biology* 27, 55–59.

Mukhopadhyay, R., Kapoor, P. and Madhubala, R. (1996) Antileishmanial effect of a potent *S*-adenosylmethionine decarboxylase inhibitor: CGP 40215A. *Pharmacological Research* 33, 67–70.

Mukkada, A.J. and Simon, M.W. (1977) *Leishmania tropica*: uptake of methionine by promastigotes. *Experimental Parasitology* 42, 87–96.

Muleme, H.M., Reguera R.M., Berard A., Azinwi, R., Jia, P., Okwor, I.B., Beverley, S. and Uzonna, J.E. (2009) Infection with arginase-deficient *Leishmania major* reveals a parasite number-dependent and cytokine-independent regulation of host cellular arginase activity and disease pathogenesis. *Journal of Immunology* 183, 8068–8076.

Müller, S., Coombs, G.H. and Walter, R.D. (2001) Targeting polyamines of parasitic protozoa in chemotherapy. *Trends in Parasitology* 17, 242–249.

Müller, S., Liebau, E., Walter, R.D. and Krauth-Siegel, R.L. (2003) Thiol-based redox metabolism of protozoan parasites. *Trends in Parasitology* 19, 320–328.

Murray, H.W., Berman, J.D., Davies, C.R. and Saravia, N.G. (2005) Advances in leishmaniasis. *The Lancet* 366, 1561–1577.

Olin-Sandoval, V., Moreno-Sánchez, R. and Saavedra, E. (2010) Targeting trypanothione metabolism in trypanosomatid human parasites. *Current Drug Targets* 11, 1614–1630.

Ooi, C.P., Rotureau, B., Gribaldo, S., Georgikou, C., Julkowska, D., Blisnick, T., Perrot, S., Subota, I. and Bastin, P. (2015) The flagellar arginine kinase in *Trypanosoma brucei* is important for infection in tsetse flies. *PLoS ONE* 10(7): e0133676.

Opperdoes, F.R. and Szikora, J.P. (2006) *In silico* prediction of the glycosomal enzymes of *Leishmania major* and trypanosomes. *Molecular and Biochemical Parasitology* 147, 193–206.

Parker, N.B., Yang, X., Hanke, J., Mason, K.A., Schowen, R.L., Borchardt, R.T. and Yin, D.H. (2003) *Trypanosoma cruzi*: molecular cloning and characterization of the *S*-adenosylhomocysteine hydrolase. *Experimental Parasitology* 105, 149–158.

Parry, L., Balaña-Fouce, R. and Pegg, A.E. (1995) Post-transcriptional regulation of the content of spermidine/spermine N^1-acetyltransferase by N^1N^{12}-bis(ethyl)spermine. *The Biochemical Journal* 305, 451–458.

Pegg, A.E. (2006) Regulation of ornithine decarboxylase. *The Journal of Biological Chemistry* 281, 14529–14532.

Pegg, A.E (2008) Spermidine/spermine-N^1-acetyltransferase: a key metabolic regulator. *American Journal of Physiology Endocrinology and Metabolism* 294, E995–E1010.

Pegg, A.E. (2009) *S*-Adenosylmethionine decarboxylase. *Essays in Biochemistry* 46, 25–45.

Pegg, A.E. and Michael, A.J. (2010) Spermine synthase. *Cellular and Molecular Life Sciences* 67, 113–121.

Pegg, A.E., Shantz, L.M. and Coleman, C.S. (1995) Ornithine decarboxylase as a target for chemoprevention. *Journal of Cellular Biochemistry* 22, 132–138.

Pereira, C.A. (2014) Arginine kinase: a potential pharmacological target in trypanosomiasis. *Infectious Disorders – Drug Targets* 14, 30–36.

Pérez-Pertejo, Y., Reguera, R.M., Ordóñez, D. and Balaña-Fouce, R. (2006) Characterization of a methionine adenosyltransferase over-expressing strain in the trypanosomatid *Leishmania donovani. Biochimica et Biophysica Acta* 1760, 10–19.

Pérez-Pertejo, Y., Alvarez-Velilla, R., Estrada, C.G., Balaña-Fouce, R. and Reguera, R.M. (2011) *Leishmania donovani*: proteasome-mediated down-regulation of methionine adenosyltransferase. *Parasitology* 138, 1082–1092.

Persson, L., Jeppsson, A. and Nasizadeh, S. (2003) Turnover of trypanosomal ornithine decarboxylases. *Biochemical Society Transactions* 31, 411–414.

Persson, L. (2007) Ornithine decarboxylase and *S*-adenosylmethionine decarboxylase in trypanosomatids. *Biochemical Society Transactions* 35, 314–317.

Phillips, M.A., Coffino, P. and Wang, C.C. (1987) Cloning and sequencing of the ornithine decarboxylase gene from *Trypanosoma brucei*. Implications for enzyme turnover and selective difluoromethylornithine inhibition. *The Journal of Biological Chemistry* 262, 8721–8727.

Popovic, P.J., Zeh, H.J. and Ochoa, J.B. (2007) Arginine and immunity. *Journal of Nutrition* 137, 1681S–1686S.

Rassi, A. Jr, Rassi, A. and Marin-Neto, J.A. (2010) Chagas disease. *The Lancet* 375, 1388–1402.

Reguera, R.M., Balaña-Fouce, R., Pérez-Pertejo, Y., Fernández, F.J., García-Estrada, C., Cubría, J.C., Ordóñez, C. and Ordóñez, D. (2002) Cloning expression and characterization of methionine adenosyltransferase in *Leishmania infantum* promastigotes. *The Journal of Biological Chemistry* 277, 3158–3167.

Reguera, R.M., Tekwani, B.L. and Balaña-Fouce, R. (2005) Polyamine transport in parasites: a potential target for new antiparasitic drug development. *Comparative Biochemistry and Physiology C: Toxicology and Pharmacology* 140, 151–164.

Reguera, R.M., Redondo, C.M., Pérez-Pertejo, Y. and Balaña-Fouce, R. (2007) *S*-Adenosylmethionine in protozoan parasites: functions, synthesis and regulation. *Molecular and Biochemical Parasitology* 152, 1–10.

Reguera, R.M., Balaña-Fouce, R., Showalter, M., Hickerson, S. and Beverley, S.M. (2009) *Leishmania major* lacking arginase are auxotrophic for polyamines but retain infectivity to susceptible BALB/c mice. *Molecular and Biochemical Parasitology* 165, 48–56.

Riscoe, M.K., Ferro, A.J. and Fitchen, J.H. (1989) Methionine recycling as a target for antiprotozoal drug development. *Parasitology Today* 5, 330–333.

Roberts, S.C., Jiang, Y., Jardim, A., Carter, N.S., Heby, O. and Ullman B. (2001) Genetic analysis of spermidine synthase from *Leishmania donovani*. *Molecular and Biochemical Parasitology* 115, 217–226.

Roberts, S.C., Scott, J., Gasteier, J.E., Jiang, Y., Brooks, B., Jardim, A., Carter, N.S., Heby, O. and Ullman, B. (2002) *S*-Adenosylmethionine decarboxylase from *Leishmania donovani*. Molecular, genetic, and biochemical characterization of null mutants and overproducers. *The Journal of Biological Chemistry* 277, 5902–5909.

Romero, I., Téllez, J., Yamanaka, L.E., Steindel, M., Romanha, A.J. and Grisard, E.C. (2014) Transsulfuration is an active pathway for cysteine biosynthesis in *Trypanosoma rangeli*. *Parasites and Vectors* 7, 197.

Sánchez, C.P., González, N.S. and Algranati, I.D. (1989) Stable ornithine decarboxylase in promastigotes of *Leishmania mexicana mexicana*. *Biochemical and Biophysical Research Communications* 161, 754–761.

Sánchez-Pérez, G.F., Bautista, J.M. and Pajares, M.A. (2004) Methionine adenosyltransferase as a useful molecular systematics tool revealed by phylogenetic and structural analyses. *Journal of Molecular Biology* 335, 693–706.

Shaked-Mishan, P., Suter-Grotemeyer, M., Yoel-Almagor, T., Holland, N., Zilberstein, D. and Rentsch, D. (2006) A novel high-affinity arginine transporter from the human parasitic protozoan *Leishmania donovani*. *Molecular Microbiology* 60, 30–38.

Shameer, S., Logan-Klumpler, F.J., Vinson, F., Cottret, L., Merlet, B., Achcar, F., Boshart, M., Berriman, M., Breitling, R., Bringaud, F. *et al.* (2015) TrypanoCyc: a community-led biochemical pathways database for *Trypanosoma brucei*. *Nucleic Acids Research* 43, 637–644.

Shantz, L.M. and Pegg, A.E. (1999) Translational regulation of ornithine decarboxylase and other enzymes of the polyamine pathway. *International Journal of Biochemistry and Cell Biology* 31, 107–122.

Simarro, P.P., Diarra, A., Ruiz Postigo, J.A., Franco, J.R. and Jannin, J.G. (2011) The Human African Trypanosomiasis Control and Surveillance Programme of the World Health Organization 2000–2009: the way forward. *PLoS Neglected Tropical Diseases* 5(2): e1007.

Singh, A.K., Roberts, S., Ullman, B. and Madhubala, R. (2014) A quantitative proteomic screen to identify potential drug resistance mechanism in α-difluoromethylornithine (DFMO) resistant *Leishmania donovani*. *Journal of Proteomics* 102, 44–59.

Spinks, D., Torrie, L.S., Thompson, S., Harrison, J.R., Frearson, J.A., Read, K.D., Fairlamb, A.H., Wyatt, P.G. and Gilbert, I.H. (2012) Design, synthesis and biological evaluation of *Trypanosoma brucei* trypanothione synthetase inhibitors. *ChemMedChem* 7, 95–106.

Stanley, B.A. and Shantz, L.M. (1994) *S*-Adenosylmethionine decarboxylase structure–function relationships. *Biochemical Society Transactions* 22, 863–869.

Stanley, B.A., Pegg, A.E. and Holm, I. (1989) Site of pyruvate formation and processing of mammalian *S*-adenosylmethionine decarboxylase proenzyme. *The Journal of Biological Chemistry* 264, 21073–21079.

Stempin, C.C., Tanos, T.B., Coso, O.A. and Cerbán, F.M. (2004) Arginase induction promotes *Trypanosoma cruzi* intracellular replication in Cruzipain-treated J774 cells through the activation of multiple signaling pathways. *European Journal of Immunology* 34, 200–209.

Sufrin, J.R., Spiess, A.J., Marasco, C.J. Jr, Rattendi, D. and Bacchi, C.J. (2008) Novel trypanocidal analogs of 5′-(methylthio)-adenosine. *Antimicrobial Agents and Chemotherapy* 52, 211–219.

Taelman, H., Schechter, P.J., Marcelis, L., Sonnet, J., Kazyumba, G., Dasnoy, J., Haegele, K.D., Sjoerdsma, A. and Wery M. (1987) Difluoromethylornithine, an effective new treatment of Gambian trypanosomiasis. Results in five patients. *American Journal of Medicine* 82, 607–614.

Tavares, J., Ouaissi, A., Lin, P.K., Tomás, A. and Cordeiro-da-Silva, A. (2005) Differential effects of polyamine derivative compounds against *Leishmania infantum* promastigotes and axenic amastigotes. *International Journal of Parasitology* 35, 637–646.

Taylor, M.C., Kaur, H., Blessington, B., Kelly, J.M. and Wilkinson, S.R. (2008) Validation of spermidine synthase as a drug target in African trypanosomes. *The Biochemical Journal* 409, 563–569.

Taylor-Robinson, A. (2001) Th1/Th2-regulated arginase availability modulates *Leishmania* infection. *Trends in Parasitology* 17, 262.

Teixeira, A.R., Hecht, M.M., Guimaro, M.C., Sousa, A.O. and Nitz, N. (2011) Pathogenesis of Chagas' disease: parasite persistence and autoimmunity. *Clinical Microbiology Reviews* 24, 592–630.

Vickers, T.J. and Beverley, S.M. (2011) Folate metabolic pathways in *Leishmania*. *Essays in Biochemistry* 51, 63–80.

Vickers, T.J., Orsomando, G., de la Garza, R.D., Scott, D.A., Kang, S.O., Hanson, A.D. and Beverley, S.M. (2006) Biochemical and genetic analysis of methylenetetrahydrofolate reductase in *Leishmania* metabolism and virulence. *The Journal of Biological Chemistry* 281, 38150–38158.

Wallace, H.M. and Fraser, A.V. (2003) Polyamine analogues as anticancer drugs. *Biochemical Society Transactions* 31, 393–396.

Wanasen, N. and Soong, L. (2008) L-Arginine metabolism and its impact on host immunity against *Leishmania infection. Immunologic Research* 41, 15–25.

WHO (2007) *Global Plan to Combat Neglected Tropical Diseases 2008–2015*. Document WHO/CDS/NTD/2007.3, World Health Organization, Geneva, Switzerland. Available at: http://apps.who.int/iris/bitstream/10665/69708/1/WHO_CDS_NTD_2007.3_eng.pdf (accessed 12 August 2016).

WHO (2008) *The Global Burden of Disease: 2004 Update*. World Health Organization, Geneva, Switzerland. Available at: http://www.who.int/healthinfo/global_burden_disease/GBD_report_2004update_full.pdf?ua=1 (accessed 12 August 2016).

WHO (2016) Trypanosomiasis, human African (sleeping sickness): February 2016 update. World Health Organization, Geneva, Switzerland. Available at: http://www.who.int/mediacentre/factsheets/fs259/en/ (accessed 11 August 2016).

Willert, E.K. and Phillips, M.A. (2008) Regulated expression of an essential allosteric activator of polyamine biosynthesis in African trypanosomes. *PLoS Pathogens* 4(10): e1000183.

Willert, E.K. and Phillips, M.A. (2009) Cross-species activation of trypanosome *S*-adenosylmethionine decarboxylase by the regulatory subunit prozyme. *Molecular and Biochemical Parasitology* 168, 1–6.

Willert, E. and Phillips, M.A. (2012) Regulation and function of polyamines in African trypanosomes. *Trends in Parasitology* 28, 66–72.

Willert, E.K., Fitzpatrick, R. and Phillips, M.A. (2007) Allosteric regulation of an essential trypanosome polyamine biosynthetic enzyme by a catalytically dead homolog. *Proceedings of the National Academy of Sciences of the United States of America* 104, 8275–8280.

Williams, R.A., Westrop, G.D. and Coombs, G.H. (2009) Two pathways for cysteine biosynthesis in *Leishmania major*. *The Biochemical Journal* 420, 451–462.

Wu, G. and Morris, S.M. Jr (1998) Arginine metabolism: nitric oxide and beyond. *The Biochemical Journal* 336, 1–17.

Yamanaka, C.N., Giordani, R.B., Rezende, C.O. Jr, Eger, I., Kessler, R.L., Tonini, M.L., de Moraes, M.H., Araújo, D.P., Zuanazzi, J.A., de Almeida, M.V. and Steindel, M. (2013) Assessment of leishmanicidal and trypanocidal activities of aliphatic diamine derivatives. *Chemical Biology and Drug Design* 82, 697–704.

Yang, X. and Borchardt, R.T. (2000) Overexpression, purification, and characterization of *S*-adenosylhomocysteine hydrolase from *Leishmania donovani. Archives of Biochemistry and Biophysics* 383, 272–280.

Yang, Y., Voak, A., Wilkinson, S.R. and Hu, L. (2012) Design, synthesis, and evaluation of potential prodrugs of DFMO for reductive activation. *Bioorganic and Medicinal Chemistry Letters* 22, 6583–6586.

Yarlett, N., Garofalo, J., Goldberg, B., Ciminelli, M.A., Ruggiero, V., Sufrin, J.R. and Bacchi, C.J. (1993) *S*-Adenosylmethionine synthetase in bloodstream *Trypanosoma brucei. Biochimica et Biophysica Acta* 1181, 68–76.

Yun, O., Priotto, G., Tong, J., Flevaud, L. and Chappuis, F. (2010) NECT is next: implementing the new drug combination therapy for *Trypanosoma brucei gambiense* sleeping sickness. *PLoS Neglected Tropical Diseases* 4(5): e720.

Zou, Y., Wu, Z., Sirisoma, N., Woster, P.M., Casero, R.A. Jr, Weiss, L.M., Rattendi, D., Lane, S. and Bacchi, C.J. (2001) Novel alkylpolyamine analogues that possess both antitrypanosomal and antimicrosporidial activity. *Bioorganic and Medicinal Chemistry Letters* 11, 1613–1617.

8 Ornithine and Lysine Decarboxylation in Bacteria

P.M. Lucas*

Institut des Sciences de la Vigne et du Vin (ISVV), University of Bordeaux, Villenave d'Ornon, France

8.1 Abstract

The ornithine and lysine decarboxylation systems are involved in the production of putrescine and cadaverine in a number of enterobacteria, lactic acid bacteria and other Gram-positive and Gram-negative bacteria. Typically, they are composed of a specific amino acid decarboxylase and a membrane exchanger catalysing the entry of the amino acid substrate and the extrusion of the biogenic amine product. Their combined activities contribute to cytosolic pH homeostasis when cells are exposed to acidic environments. While these systems were initially investigated in bacteria of the gastrointestinal tract to elucidate the mechanism of bacterial putrefaction and the concomitant production of putrescine and cadaverine, they proved to be important for enteric bacteria to survive acid stress during their passage through the stomach and for food bacteria during the production or spoilage of food products. This chapter is a review of the genetic and biochemical organization of bacterial ornithine and lysine decarboxylation systems, their occurrence and distribution in bacteria, their physiological roles and the implications that their activities have for the quality of food.

8.2 Introduction

The search for bacteria and enzymes that can decarboxylate amino acids started at the beginning of the 20th century with the aim of understanding the phenomenon of putrefaction. In 1910, Ackermann (1910) showed that bacteria from pancreatic extracts produce putrescine, cadaverine, histamine, γ-aminobutyric acid and β-alanine from the corresponding amino acids. In the following years, the use of pure bacteria isolated from the intestines or stomach has revealed that the capacity to decarboxylate amino acids is strain dependent. A '*Bacillus aminophilus intestinalis*' that produces histamine, tryptamine and tyramine was isolated by Berthelot and Bertrand (1912) and Hirai (1936) reported a '*Bacterium coli*' that produced putrescine from arginine. In the early 1940s, the development of analytical methods and the use of a variety of *Escherichia coli* strains allowed Gale to describe the bacteria that perform the decarboxylation, the enzymes involved and the environmental conditions required for their activity (Gale, 1940a–d). While these first studies focused on Gram-negative bacteria, later studies by Rodwell (1953) investigated the potential of Gram-positive cells. Rodwell isolated several lactic

*E-mail address: patrick.lucas@u-bordeaux.fr

acid bacteria (LAB) from the gastrointestinal tract and showed that they carry amino acid decarboxylases producing putrescine, cadaverine and other biogenic amines. Since these first studies, putrescine- and cadaverine-producing bacteria have been investigated with the aim of understanding their pathogenic properties and avoiding their negative impacts on food products, and the genes and enzymes involved in ornithine and lysine decarboxylation have been identified in strains of numerous species. It is now clear that these decarboxylation systems contribute primarily to the survival of bacteria to acid stress, but that they are also involved in other cellular functions. Knowledge of these systems is crucial to controlling the bacteria used in food production, as well as enteric pathogens.

8.3 Definitions

8.3.1 L-Ornithine

L-Ornithine (2,5-diaminopentanoate) is a non-protein amino acid produced from L-arginine by many eukaryotic cells and by some bacteria carrying an L-arginine deiminase pathway (ADI). The ADI pathway comprises three main enzymes, an arginine deiminase (ADI), an ornithine carbamoyltransferase (OTC) and a carbamate kinase (CK), whose combined activities result in the production of 1 mole of ornithine and 1 mole of ATP from the consumption of 1 mole of arginine (Fernández and Zúñiga, 2006). Ornithine can also become a substrate for bacteria. Some of them carry an anabolic OTC which works in the opposite direction to that of the ADI pathway and uses ornithine and carbamoyl phosphate to synthesize citrulline, a key step in the arginine biosynthetic pathway and the first step of the urea cycle (Zúñiga *et al.*, 2002). An L-ornithine decarboxylase (ODC; EC 4.1.1.17) can also be present in bacteria, which converts ornithine into putrescine and carbon dioxide. As in the case of OTC, two types of ODC are differentiated depending whether they function in a biodegradative or a biosynthetic mode. Biosynthetic ODCs form low amounts of putrescine in the cytosol, while the biodegradative enzymes produce large amounts of putrescine that is rejected from the cell.

8.3.2 L-Lysine

L-Lysine (2,6-diaminohexanoate) is an essential amino acid that is catabolized by diverse bacteria. Clostridia can ferment lysine to acetate, butyrate and ammonia. They contain a lysine 2,3-aminomutase, which is a member of the *S*-adenosylmethionine superfamily and converts L-lysine into β-lysine in the first step of the reaction (Chirpich *et al.*, 1970). This metabolic pathway is rarely encountered is other bacteria. In contrast, numerous species possess an L-lysine decarboxylase (LDC; EC 4.1.1.18) which decarboxylates lysine into cadaverine. As with the biodegradative ODC, LDC activity can result in the production of large amounts of extracellular cadaverine.

8.3.3 Putrescine

Putrescine (1,4-diaminobutane) is a low molecular weight aliphatic diamine characterized by a rotting flesh odour. It belongs to the two groups of biogenic amines and polyamines (molecules carrying two or more amino groups), together with cadaverine, spermine and spermidine. Intracellular putrescine may act as a growth cell factor or play other cellular roles. It also has physiological functions in the human body, although the intake of food containing excessive amounts of putrescine can lead to serious toxicological consequences. The accumulation of putrescine in food is mainly due to bacteria, and at least three biosynthetic pathways exist in Gram-negative bacteria and two in Gram-positive bacteria (Wunderlichova *et al.*, 2014). Bacteria contain either a biodegradative ODC that produces putrescine directly from ornithine or diverse enzymes such as an arginine decarboxylase and/or an agmatine deiminase that form the amine indirectly from arginine.

8.3.4 Cadaverine

Cadaverine (1,5-diaminopentane) has a similar unpleasant odour to putrescine and belongs to the same two groups of biogenic amines and polyamines. Like putrescine, its accumulation in food is due to bacteria carrying a specific amino acid decarboxylase (LDC). Cadaverine is mainly expelled from the cell, but in some bacteria it contributes to cell envelope integrity and other cellular processes.

8.4 The Ornithine Decarboxylase (ODC) and Lysine Decarboxylase (LDC) Systems

8.4.1 Putrescine and cadaverine production systems

In bacteria, only the LDC pathway is used for cadaverine production. In contrast, there are two major pathways for putrescine production, the ODC and the agmatine deiminase (AGDI) pathways (Fig. 8.1) (Morris and Pardee, 1966; Morris and Koffron, 1969). The most direct pathway is the decarboxylation of ornithine by ODC. Although ornithine may result from the degradation of arginine by the ADI pathway, very few bacteria contain both ADI and ODC. In fact, this has been reported in only one strain of *Weissella halotolerans*, which indicates that the two enzymes function independently and are not evolutionarily related (Pereira *et al.*, 2009). Bacteria can also produce large amounts of putrescine by converting agmatine using the AGDI pathway (Fig. 8.1). Agmatine is either available in the bacterial environment or is produced by bacteria that have arginine decarboxylase (ADC). As for the ODC and ADI pathways, the ADC and AGDI pathways are independent and are rarely detected together in the same bacteria. Similarly, the presence of both ODC and AGDI pathways is exceptional, and has been reported only in two strains of *Lactobacillus brevis* that produce putrescine from both ornithine and agmatine (Romano *et al.*, 2014).

8.4.2 The ODC system

8.4.2.1 Genetic organization of the odc *operon*

The ODC pathway is the same in whichever bacterial family it is found. It was described first in *E. coli* (Kashiwagi *et al.*, 1991) and later in other *Enterobacteriaceae* (de las Rivas *et al.*, 2007), LAB (Marcobal *et al.*, 2006; Romano *et al.*, 2012) and staphylococci (Coton *et al.*, 2010a; Tsoi and Tse, 2011). It comprises only two genes, named *speF* or *odc*, and *potE*, which encode ODC and an ornithine/putrescine exchanger, respectively. No specific transcriptional regulator is associated with this operon. In *Oenococcus oeni*, the flanking DNA regions contain mobile genetic elements such as phage genes and insertion sequence elements, suggesting that the operon was acquired by horizontal gene transfer (Marcobal *et al.*, 2006). *L. saerimneri* 30a shows an atypical ODC locus (Romano *et al.*, 2013b). Its *odc* gene has no adjacent *potE*, but another *potE* is present in the *L. saerimneri* 30a genome and associated with an LDC gene. As indicated below in Section 8.4.3.4, this second *potE* encodes a transporter that performs both lysine/cadaverine and ornithine/putrescine exchanges and allows the ODC system of this bacterium to function. It is noteworthy that genes encoding putative ODCs (i.e. enzymes sharing sequence similarities) are often found in bacterial genomes. Some of these enzymes could be biosynthetic ODCs that produce low

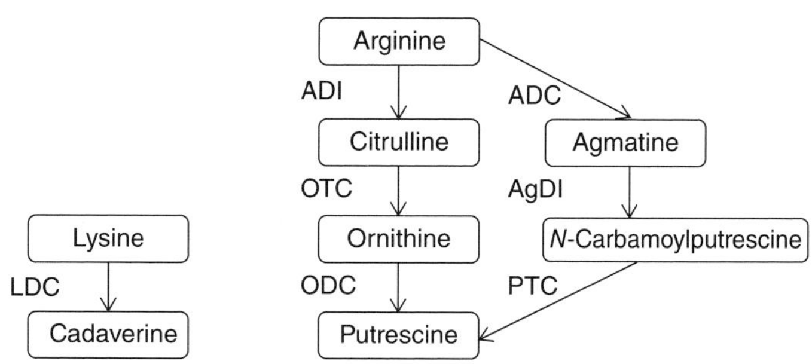

Fig. 8.1. Cadaverine and putrescine production pathways in bacteria. Abbreviations: ADC, arginine decarboxylase; ADI, arginine deiminase; AgDI, agmatine deiminase; LDC, lysine decarboxylase; ODC, ornithine decarboxylase; OTC, ornithine transcarbamylase; PTC, putrescine transcarbamylase.

levels of intracellular putrescine to perform various physiological functions (Tabor and Tabor, 1985). Others are not able to decarboxylate ornithine and must be involved in different reactions (Romano *et al.*, 2012).

8.4.2.2 L-Ornithine decarboxylase

ODCs were initially characterized in *E. coli* and *L. saerimneri* 30a (Applebaum *et al.*, 1975; Guirard and Snell, 1980). The two enzymes share a high protein sequence identity. They share also sequence similarities with LDC. They require pyridoxal-5′-phosphate as a cofactor for activity and thus belong to the large family of pyridoxal-5′-phosphate (PLP)-dependent amino acid decarboxylases along with LDC (Hackert *et al.*, 1994). The *L. saerimneri* 30a enzyme is a dodecamer composed of six symmetric dimers associated in a ring-shaped molecule (Momany *et al.*, 1995). It adopts its functional dodecameric structure at its optimal pH (5.8), but is found in monomeric or dimeric forms at different pHs. ODC is highly specific for ornithine and has only a weak affinity for lysine. Although it can convert lysine into cadaverine, its catalytic efficiency for this amino acid is 250 times less than with ornithine (Romano *et al.*, 2013b). As mentioned in Section 8.4.2.1, there can be confusion between these typical biodegradative ODCs and putative biosynthetic ODCs. While they share protein sequence similarities, the latter form a specific phylogenetic group that is clearly separated from the group of the *E. coli*/*L. saerimneri* ODCs (Romano *et al.*, 2012).

8.4.2.3 The ornithine/putrescine exchanger

The amino acid transporter encoded by *potE* in the ODC system of *E. coli* is a specific ornithine/putrescine exchanger (Kashiwagi *et al.*, 1992). It contains 12 transmembrane segments and belongs to the same basic APA family in the amino acid/polyamine/organocation superfamily of transporters (Saier *et al.*, 2006) as the lysine/cadaverine exchanger. A similar enzyme is present in LAB carrying an ODC system, except that in *L. saerimneri* 30a there is a transporter capable of exchanging both ornithine/putrescine and lysine/cadaverine (see Sections 8.4.2.1 and 8.4.3.2). Some LAB carrying a putative biosynthetic ODC also have a transporter gene associated with their *odc* gene. However, not only are these transporters phylogenetically distant from the ornithine/putrescine exchangers, they also have different transport capacities. For example, the enzyme of *L. gasseri* ATCC 33323 is not able to catalyse the exchange of ornithine against putrescine, but instead performs the unidirectional transport of ornithine inside the cell (Romano *et al.*, 2012). This supports the idea that there are two types of ODC systems in LAB: a typical biodegradative system that excretes large amounts of putrescine and a putative biosynthetic system that produces small amounts of putrescine or other compounds that remain inside the cell (Romano *et al.*, 2012).

8.4.3 The LDC system

8.4.3.1 The *cad* operon of enterobacteria

The nucleotide sequences of the genes involved in the LDC system was reported for the time in *E. coli* (Meng and Bennett, 1992a). This so-called '*cad* operon' comprises two genes, *cadA* and *cadB*, that encode the lysine decarboxylase and a lysine-cadaverine membrane exchanger, respectively. A third gene, *cadC*, is located upstream of *cadAB* on the same DNA strand and is transcribed independently. This encodes a transcriptional regulator, CadC, which belongs to the ToxR family and acts as a positive activator in response to low pH and the presence of lysine (Meng and Bennett, 1992b). Low pH and lysine have no impact on the transcriptional regulation of *cadC* itself, but they stimulate the activity of CadC (Dell *et al.*, 1994). Two other proteins contribute to the regulation of expression of the *cad* operon, the nucleoid protein H-NS and the lysine transporter LysP. H-NS represses the transcription of *cadAB* and other amino acid decarboxylation systems in *E. coli* under non-inducing pH conditions. H-NS is able to bind on to the promoter of *cadAB* when cells are exposed to a basic pH, thus avoiding the binding of CadC and the activation of *cadAB* transcription (Shi *et al.*, 1993; Kuper and Jung, 2005). LysP acts as a repressor of *cadAB* expression in the absence of exogenous lysine (Neely *et al.*, 1994).

8.4.3.2 The LDC locus of the LAB strain *Lactobacillus saerimneri* 30a

Although the well characterized, *cad* operon has been detected in a number of *Enterobacteriaceae*, little is known about the genetics of lysine decarboxylation in other bacteria. Several studies have suggested that LAB contribute to the accumulation of cadaverine in food, though this phenotype has been rarely detected in LAB (Guerrini *et al.*, 2002; Coton *et al.*, 2010b). *L. saerimneri* 30a is the single strain carrying an LDC described so far (Rodwell, 1953; Miller and Rodwell, 1971), and its LDC locus has recently been identified (Romano *et al.*, 2013a). The locus comprises only two genes that probably form an operon (Romano *et al.*, 2013b), and code for the LDC and a lysine/cadaverine membrane exchanger, which also has the capacity to exchange ornithine and putrescine (see Section 8.4.2.3). Despite its apparent similarity to the *cad* locus, the LDC locus of *L. saerimneri* differs significantly. First, the two genes are not organized similarly: the LDC gene is transcribed first in *L. saerimneri*, whereas this is the membrane exchanger gene in *E. coli*. Second, there is no gene encoding a transcriptional regulator in *L. saerimneri*. These differences, as well as functional divergences in the encoded enzymes (see Section 8.4.3.3), suggest that the two systems are not evolutionarily related.

8.4.3.3 L-Lysine decarboxylase

The inducible LDC encoded by the *cad* operon of *E. coli* was first characterized by Sabo *et al.* (Sabo *et al.*, 1974; Sabo and Fischer, 1974). Like many other amino acid decarboxylases, it uses PLP as a cofactor to decarboxylate the α-carbonyl group of lysine in a reaction that consumes a cytosolic proton and releases cadaverine and carbon dioxide. The enzyme is able to decarboxylate hydroxylysine and *S*-aminoethylcysteine but not arginine, ornithine or a number of lysine analogues. It is a double-ring shaped decamer formed by the association of five symmetric dimers. Structural analyses have revealed that it can bind the alarmome ppGpp (5′-diphosphate-3′-diphosphate guanosine) in each of its ten active sites (Kanjee *et al.*, 2011). The alarmome is normally expressed in bacteria under nutrient-limiting conditions and is involved in physiological changes such as the repression of rRNA synthesis, gene activation/repression or enzyme activation/inactivation. Binding of the alarmome strongly inhibits LDC activity, which is important for modulating lysine consumption under nutrient-limiting conditions. The more recently characterized LDC of *L. saerimneri* 30a belongs to the same family of PLP-decarboxylases as the *E. coli* enzyme. However, it has more sequence similarity with the ODCs of LAB than with the LDC of *E. coli* (Romano *et al.*, 2013b). In addition, both enzymes have a similar pH optimum (5.2–5.7), but unlike the LDC of *E. coli*, the LDC of *L. saerimneri* can decarboxylate ornithine.

8.4.3.4 The lysine/cadaverine exchanger

The membrane exchanger encoded by the *E. coli cadB* gene belongs to the same basic APA as the ornithine/putrescine exchanger. It has significant sequence identity with the ornithine/putrescine exchanger PotE, but biochemical characterization has demonstrated that it is a specific lysine/cadaverine membrane exchanger, which has a high specificity for lysine and cadaverine and not for putrescine (Soksawatmaekhin *et al.*, 2004). As for the LDCs, the lysine/cadaverine exchanger of *L. saerimneri* has more sequence identity with PotE exchangers than with CadB (Romano *et al.*, 2013b). In addition, it has the unprecedented capacity to catalyse both lysine/cadaverine and ornithine/putrescine exchanges, which makes it able to work with both LDC and ODC.

8.5 ODC- and LDC-Positive Bacteria

8.5.1 Origin and diversity of bacteria

The first studies on putrescine- and cadaverine-producing bacteria were in digestive tracts, with the aim of understanding their contribution to the phenomenon of putrefaction, but today there is still a strong interest in enteric bacteria because of their implied role(s) in human diseases. The ODC and LDC systems are particularly important in these bacteria as they allow the bacteria to better withstand acid stress during passage through the stomach. It is also of great interest to search for putrescine- and cadaverine-producing bacteria in foods in order to control the accumulation of these (biogenic) amines and limit their negative impacts. It is likely that the ODC and LDC pathways are present in a large number of Gram-positive and Gram-negative bacteria, but the bacteria that have these pathways are difficult to detect because the relevant genes are generally present only in some strains of a species, with their distribution resulting from horizontal gene transfer events.

8.5.2 Detection methods

Diverse methods have been employed to detect bacteria that produce biogenic amines. The bacteria are grown in a suitable medium supplemented with precursor amino acids and the production of biogenic amines analysed by a method based on high-pressure liquid chromatography, capillary electrophoresis or thin layer chromatography (Latorre-Moratalla *et al.*, 2009). This method has been widely used and is still required to confirm the bacterial phenotype (Coton *et al.*, 2010b). To avoid the step of biogenic amine determination, screening procedures based on colorimetric plate media have been proposed. The media consist of a basal composition supplemented with the amino acid precursor and a pH indicator, such as bromocresol purple. After plating the bacteria and the growth of colonies, those that can decarboxylate the amino acid cause a pH shift, which is detected by a colour change in the medium around the colonies (Bover-Cid and Holzapfel, 1999). This method has the advantage of being easier than the previous one, but it often leads to false positive or false negative results. More recently, the identification of ODC and LDC genes from diverse species has made it possible to develop specific PCR tests. Different sets of PCR primers have been developed to detect these genes in Gram-negative bacteria and in LAB (de las Rivas *et al.*, 2006; Wunderlichova *et al.*, 2013). Multiplex PCR tests combining different primer sets have also been proposed (de Las Rivas *et al.*, 2005; Coton *et al.*, 2010b). These allow the rapid and reliable detection of bacteria, although a phenotypic validation of positive strains is still required. In addition, no primers targeting the LDC gene of LAB have been developed to date. Methods of quantitative PCR have also been proposed to determine the population levels of ODC-containing LAB in foods (Nannelli *et al.*, 2008; Ladero *et al.*, 2011, 2012). Finally, next-generation sequencing methods have made it possible to access the genome sequence of most bacteria. This is certainly the best option for identifying the ODC and LDC pathway genes in new species (Tsoi and Tse, 2011; Romano *et al.*, 2013a).

8.5.3 Lactic acid bacteria

Ever since they have contributed to the production of fermented foods, numerous studies have been conducted to analyse the potential of LAB to form biogenic amines. Diverse species can form putrescine and/or cadaverine, but this is always a characteristic of strains, not of species. The detection of LAB carrying an LDC system is performed by phenotypic analysis, because there are no specific PCR primers for the lysine decarboxylase gene of LAB. Inventories have revealed a number of cadaverine-producing strains of *L. brevis* and *L. curvatus* isolated from sausage, *Oenococcus oeni* from wine, *L. curvatus* from mayonnaise, and *L. acidophilus*, *L. brevis*, *L. casei*, *L. curvatus* and *L. hilgardii* from diverse origins (Leuschner and Hammes, 1999; Bover-Cid *et al.*, 2001; Guerrini *et al.*, 2002). Despite the variety of LAB species identified, the LDC system of strain *L. saerimneri* 30a is the only one that has been described to date (Romano *et al.*, 2013b). In contrast, ODC pathway genes have been described in a several strains of *L. brevis* and *O. oeni*, as well as in *L. saerimneri* 30a (Hackert *et al.*, 1994; Marcobal *et al.*, 2004; Romano *et al.*, 2012, 2014), which has made it possible to develop PCR primers for their detection in other LAB (Nannelli *et al.*, 2008; Coton *et al.*, 2010b; Costantini *et al.*, 2013). Inventories based on phenotypic and/or molecular methods have suggested the presence of an ODC system in some strains of *L. hilgardii*, *L. plantarum*, *L. brevis* and *O. oeni* in sources such as orange, wine, cider and sugarcane molasses (Arena and Manca de Nadra, 2001; Coton *et al.*, 2010b; Ladero *et al.*, 2011). This system seems relatively rare among LAB, but it plays a predominant role in the production of putrescine in certain products, such as wine (Nannelli *et al.*, 2008). In other products, such as cider and cheese, the AGDI system is the most important (Ladero *et al.*, 2011).

8.5.4 Enterobacteria

The LDC and ODC systems are widely distributed among enterobacteria. However, as for LAB, they are present in a limited number of strains of each species. The ODC system has been described or detected in numerous species, such as *E. coli*, *Salmonella enterica*, *Morganella morganii*, *Enterobacter cloacae*, *E. aerogenes*, *Citrobacter rodentium*, *C. koseri* and *Yersinia enterocolitica* (Kashiwagi *et al.*, 1991; de las Rivas *et al.*, 2007; Wunderlichova *et al.*, 2013). The LDC system has been studied in *E. coli*, *S. enterica*, *S. typhimurium*, *Vibrio parahaemolyticus*, *V. cholerae* and *V. vulnificus* (Park *et al.*, 1996; Rhee *et al.*, 2002; de las Rivas *et al.*, 2007; Tanaka *et al.*, 2008). Several of these bacteria are opportunistic pathogens that contaminate foods and cause food-borne diseases. Enterobacteria are strong producers of biogenic amines that can cause the accumulation of high levels of putrescine and cadaverine in contaminated foods. They are often detected in products such as cheese, fermented sausages, ground beef, salmon and vegetables (Bover-Cid *et al.*, 2001; Coton *et al.*, 2012; Maifreni *et al.*, 2013; Wunderlichova *et al.*, 2014).

8.5.5 Other bacteria

Besides LAB and enterobacteria, ODC or LDC systems are also present in other food-borne bacteria or pathogens such as *Aeromonas sp.*, *Micrococcus sp.* and staphylococci (Straub *et al.*, 1995; Wunderlichova *et al.*, 2014). The ODC systems of *Staphylococcus epidermidis* and *S. lugdunensis* have been identified recently (Coton *et al.*, 2010a; Tsoi and Tse, 2011). Interestingly, while the ODC and LDC pathway genes are usually present in a limited number of strains in enterobacteria and LAB species, the ODC system has been detected in all strains the *S. lugdunensis*, suggesting that it represents a rare case of a species-specific feature.

8.6 Physiological Role of ODC and LDC

8.6.1 Acid resistance and energy production

The production of putrescine and cadaverine by the ODC and LDC systems can be a source of energy or a form of acid stress resistance for bacteria. The lysine and ornithine decarboxylases consume an intracellular proton in the decarboxylation reaction, thereby generating a proton gradient between the inside and outside cell compartments. Furthermore, because putrescine and cadaverine contain one more positive charge than ornithine and lysine, the membrane transporters catalyse an electrogenic subtrate/product exchange, which results in the generation of a membrane potential. The combined activities of the decarboxylase and membrane exchanger thus result in alkalinization of the cytosol, which can be exploited for acid stress resistance, and/or production of a proton motive force that can be converted into metabolic energy in the form of ATP (Konings *et al.*, 1997).

It is clear that LAB and enterobacteria use these systems primarily to cope with the acidic environments in which they develop. Studies performed on ODC-positive strains of *L. brevis*, *O. oeni* and *L. saerimneri* have shown that putrescine production is concomitant with cytosolic pH alkalinization and acid shock survival (Romano *et al.*, 2014). In contrast, diverse situations are observed in enterobacteria depending on the species and their decarboxylation systems. In *S. enterica* serovar Typhimurium, the LDC system confers both survival at pH 2.3 and growth improvement at pH 4.5, while the ODC system only improves growth at a moderate acidic pH (Park *et al.*, 1996). The LDC system also plays an important role in *V. cholerae*; the expression of its *cad* operon is induced at an acidic pH (Rhee *et al.*, 2002). In *E coli*, the *cad* operon is expressed only at a moderately acidic pH and seems to be less efficient in protecting cells against acidity (Prosseda *et al.*, 2007). However, a protective action of the *cad* operon has been measured on starved *E. coli* cells incubated under conditions mimicking their passage through the gut, and this suggests that the LDC system is crucial for their survival (Moreau, 2007).

8.6.2 *cad* Genes and the virulence of pathogenic strains of *Escherichia coli*

The LDC system is of particular interest in *Shigella*, a group of pathogenic strains that belong to the *E. coli* species. It has been demonstrated that the virulence of *Shigella* is associated with the emergence of a *cad*-negative phenotype (Day *et al.*, 2001; Torres, 2009). Restoration of cadaverine production causes attenuation in the virulence of and inhibition of enterotoxin production by these strains. It is cadaverine which is the inhibitor, and it acts by inhibiting the virulence of *Shigella* enterotoxins and by retarding the lysis of the vacuole containing the *Shigella* species, which prevents the pathogen from adequately interacting with the cellular cytoskeleton. Inactivation of *cadA* also increases the adherence properties of the bacteria, which are then better suited to colonizing the intestine, suggesting that cadaverine may play a role in adhesion as well. Analysis of the genomic sequence of the *cad* locus revealed large genomic deletions as well as insertion sequences and prophages that disrupted the locus. Interestingly, the inactivation has occurred independently in diverse *Shigella* spp., which indicates that this is a case of convergent evolution in which gene disruption has increased the pathogenicity of the bacteria. The concept of 'black holes' has been proposed to suggest that the deletion of genes that are detrimental for pathogenicity confers an evolutionary advantage (Maurelli *et al.*, 1998).

8.7 Food Safety Implications

8.7.1 Origins of biogenic amines in foods

Biogenic amines can be expected in virtually all fermented and non-fermented foods containing proteins or amino acids, such as fish products, meat products, vegetables, dairy products, cheese, wine or beer (ten Brink *et al.*, 1990). Small quantities of the amines occur naturally in raw materials, but their presence in higher amounts is always associated with the development of decarboxylating microorganisms. In fermented foods, the microorganisms may belong to the complex microbiota required for the production of such foods. In contrast, in unfermented foods, the accumulation of biogenic amines indicates the development of contaminating microorganisms (Silla Santos, 1996). This has led to the proposal of several biogenic amine indices based on the amounts of putrescine, cadaverine and (eventually) other biogenic amines to evaluate the level of contamination and freshness of fish and fish products (Al Bulushi *et al.*, 2009).

8.7.2 Toxicological effects of food-borne biogenic amines

Biogenic amines are normal constituents of the human body. They play important roles in brain activity, regulation of body temperature and stomach pH, gastric acid secretion, immune response, etc. (Ladero *et al.*, 2010). Given their physiological importance, their concentration is tightly regulated by mechanisms of biosynthesis and degradation. The intake of food rich in biogenic amines can alter this equilibrium and induce adverse reactions such as headache, nausea, facial flushing, blood pressure disorders, itchy rashes, vomiting or respiratory distress. The severity of clinical symptoms is variable depending on individual susceptibility and the amount and type of biogenic amine ingested. Detoxification is mainly performed by mono- and di-amine oxidases present in the systemic circulation, although the activity of these enzymes is inhibited in individuals with genetic predisposal or in the presence of inhibitory compounds such as alcohol and a variety of drugs used for the treatment of stress, depression, and Alzheimer's or Parkinson's diseases (Ladero *et al.*, 2010).

The most notorious food-borne intoxications are due to histamine. This has been implicated in outbreaks caused by the ingestion of contaminated fish and detected in large amounts (100 to 8000 mg kg^{-1}) in numerous fish products for which consumer complaints have been notified (BIOHAZ, 2011). Histamine is the only biogenic amine for which maximum limits have been set. The European legislation limits histamine levels to 200 and 400 mg kg^{-1} in fresh fish and cured fishery products, respectively. Some countries have established a maximum level in wine of 2 mg l^{-1}, but for imported products only. Tyramine has a less important impact than histamine on human health, though it is often associated with cheese intoxications and might be responsible for headaches and migraines.

The literature gives very few examples of toxicological disorders directly caused by putrescine or cadaverine, probably because their own toxicological effects are low. However, they can potentiate the adverse effects of other biogenic amines by acting as competitive inhibitors of mono- and di-amine oxidases. It is also suggested that they alter the physical barrier function of the small intestine and allow greater quantities of histamine to enter into the circulation (BIOHAZ, 2011). Another serious aspect of food-borne putrescine and cadaverine is their capacity to form nitrosamines by reaction with nitrites, which are used as preservatives in some fermented meat products. Nitrosamines are known to have carcinogenic, mutagenic and teratogenic activities (Ladero *et al.*, 2010). In addition, putrescine, which is naturally involved in cell growth and proliferation, can cause the development of neoplasms (Wunderlichova *et al.*, 2014). Tumour cells can benefit from the intake of putrescine in contaminated food.

8.7.3 Control of biogenic amine formation in food

A number of analytical methods have been developed for determining the content of biogenic amines in foods. They vary depending on the target products, but are generally based on high-pressure liquid chromatography, capillary electrophoresis or thin layer chromatography, and are used to quantify at least histamine, tyramine, putrescine and cadaverine (Wunderlichova *et al.*, 2014). Although they are necessary for monitoring the quality of foods, these methods only give information about the amount of biogenic amines that has already been produced in them. As indicated above, molecular methods based on PCR or quantitative PCR provide another kind of information. They inform on the presence of biogenic amine-producing bacteria and the potential formation of biogenic amines. PCR methods are also useful for the characterization and selection of the biogenic amine-deficient starter cultures that are used for the production of fermented foods.

8.8 Conclusions

From the numerous works reported to date, the ODC and LDC systems have been well described in a number of bacterial species. A general description of their organization, functioning and role can be proposed. Indeed, these are two-component systems, including a specific decarboxylase and a substrate/product membrane exchanger, which improve the survival of bacteria in acidic environments. However, this general pattern is far from explaining all of the specific features observed in the diverse bacteria that are concerned. The genetic organization and regulation of the pathways may differ depending on the species, the membrane exchanger may have a broad specificity and be shared between different decarboxylases, the improvement of survival at an acidic pH is minor for some bacteria, etc. These subtle variations show that it is crucial to identify and characterize similar systems in new species. This is particularly important for the LDC system of LAB which, until now, has been described only in a single strain. The identification of new LDC systems in other LAB will make it possible to understand its evolutionary origin and phylogenetic dissemination. In addition, it will be possible to develop PCR tests for detecting LDC-positive LAB in food and for selecting starter cultures used to produce safer foods. Moreover, as putrescine and cadaverine are involved in various cellular and physiological processes, the role(s) of the LDC and ODC systems may have been underestimated in many species and be more than just a strategy for surviving the acid stress, as has been demonstrated in pathogenic *E. coli Shigella* strains with inactivation of the LDC system.

References

Ackermann, D. (1910) Über ein neues, auf bakteriellem Wege gewinnbares, Aporrhegma. *Hoppe-Seyler's Zeitschrift für Physiologische Chemie* 69, 273–281.

Al Bulushi, I., Poole, S., Deeth, H.C. and Dykes, G.A. (2009) Biogenic amines in fish: roles in intoxication, spoilage, and nitrosamine formation – a review. *Critical Reviews in Food Science and Nutrition* 49, 369–377.

Applebaum, D., Sabo, D.L., Fischer, E.H. and Morris, D.R. (1975) Biodegradative ornithine decarboxylase of *Escherichia coli*. Purification, properties, and pyridoxal 5′-phosphate binding site. *Biochemistry* 14, 3675–3681.

Arena, M.E. and Manca de Nadra, M.C. (2001) Biogenic amine production by *Lactobacillus*. *Journal of Applied Microbiology* 90, 158–162.

Berthelot, A. and Bertrand, D.M. (1912) Sur quelques propriétés biochimiques du *Bacillus aminophilus intestinalis*. *Comptes Rendus de l'Académie des Sciences Paris* 154, 1826–1828.

BIOHAZ (EFSA Panel on Biological Hazards) (2011) Scientific opinion on risk based control of biogenic amine formation in fermented foods. *EFSA Journal* 9(10):2393.

Bover-Cid, S. and Holzapfel, W.H. (1999) Improved screening procedure for biogenic amine production by lactic acid bacteria. *International Journal of Food Microbiology* 53, 33–41.

Bover-Cid, S., Hugas, M., Izquierdo-Pulido, M. and Vidal-Carou, M.C. (2001) Amino acid-decarboxylase activity of bacteria isolated from fermented pork sausages. *International Journal of Food Microbiology* 66, 185–189.

Chirpich, T.P., Zappia, V., Costilow, R.N. and Barker, H.A. (1970) Lysine 2,3-aminomutase. Purification and properties of a pyridoxal phosphate and *S*-adenosylmethionine-activated enzyme. *The Journal of Biological Chemistry* 245, 1778–1789.

Costantini, A., Pietroniro, R., Doria, F., Pessione, E. and Garcia-Moruno, E. (2013) Putrescine production from different amino acid precursors by lactic acid bacteria from wine and cider. *International Journal of Food Microbiology* 165, 11–17.

Coton, E., Mulder, N., Coton, M., Pochet, S., Trip, H. and Lolkema, J.S. (2010a) Origin of the putrescine-producing ability of the coagulase-negative bacterium *Staphylococcus epidermidis* 2015B. *Applied and Environmental Microbiology* 76, 5570–5576.

Coton, M., Romano, A., Spano, G., Ziegler, K., Vetrana, C., Desmarais, C., Lonvaud-Funel, A., Lucas, P. and Coton, E. (2010b) Occurrence of biogenic amine-forming lactic acid bacteria in wine and cider. *Food Microbiology* 27, 1078–1085.

Coton, M., Delbes-Paus, C., Irlinger, F., Desmasures, N., Le Fleche, A., Stahl, V., Montel, M.C. and Coton, E. (2012) Diversity and assessment of potential risk factors of Gram-negative isolates associated with French cheeses. *Food Microbiology* 29, 88–98.

Day, W.A. Jr, Fernández, R.E. and Maurelli, A.T. (2001) Pathoadaptive mutations that enhance virulence: genetic organization of the *cadA* regions of *Shigella* spp. *Infection and Immunity* 69, 7471–7480.

de las Rivas, B., Marcobal, A. and Muñoz, R. (2005) Improved multiplex-PCR method for the simultaneous detection of food bacteria producing biogenic amines. *FEMS Microbiology Letters* 244, 367–372.

de las Rivas, B., Marcobal, A., Carrascosa, A.V. and Muñoz, R (2006) PCR detection of foodborne bacteria producing the biogenic amines histamine, tyramine, putrescine, and cadaverine. *Journal of Food Protection* 69, 2509–2514.

de las Rivas, B., Marcobal, A. and Muñoz, R. (2007) Gene organization of the ornithine decarboxylase-encoding region in *Morganella morganii*. *Journal of Applied Microbiology* 102, 1551–1560.

Dell, C.L., Neely, M.N. and Olson, E.R. (1994) Altered pH and lysine signalling mutants of *cadC*, a gene encoding a membrane-bound transcriptional activator of the *Escherichia coli cadBA* operon. *Molecular Microbiology* 14, 7–16.

Fernández, M. and Zúñiga, M. (2006) Amino acid catabolic pathways of lactic acid bacteria. *Critical Reviews in Microbiology* 32, 155–183.

Gale, E.F. (1940a) Enzymes concerned in the primary utilization of amino acids by bacteria. *Bacteriology Reviews* 4, 135–176.

Gale, E.F. (1940b) The production of amines by bacteria: the decarboxylation of amino-acids by strains of *Bacterium coli*. *The Biochemical Journal* 34, 392–413.

Gale, E.F. (1940c) The production of amines by bacteria: the production of putrescine from L(+)-arginine by *Bacterium coli* in symbiosis with *Streptococcus faecalis*. *The Biochemical Journal* 34, 853–857.

Gale, E.F. (1940d) The production of amines by bacteria: the production of tyramine by *Streptococcus faecalis*. *The Biochemical Journal* 34, 846–852.

Guerrini, S., Mangani, S., Granchi, L. and Vincenzini, M. (2002) Biogenic amine production by *Oenococcus oeni*. *Current Microbiology* 44, 374–378.

Guirard, B.M. and Snell, E.E. (1980) Purification and properties of ornithine decarboxylase from *Lactobacillus* sp. 30a. *The Journal of Biological Chemistry* 255, 5960–5964.

Hackert, M.L., Carroll, D.W., Davidson, L., Kim, S.O., Momany, C., Vaaler, G.L. and Zhang, L. (1994) Sequence of ornithine decarboxylase from *Lactobacillus* sp. strain 30a. *Journal of Bacteriology* 176, 7391–7394.

Hirai, K. (1936) Über die Bildung von Putrescin aus a-Arginin durch Bakterieneinwirkung. *Biochemische Zeitschrift* 283, 390–392.

Kanjee, U., Gutsche, I., Alexopoulos, E., Zhao, B., El Bakkouri, M., Thibault, G., Liu, K., Ramachandran, S., Snider, J., Pai, E.F. and Houry, W.A. (2011) Linkage between the bacterial acid stress and stringent responses: the structure of the inducible lysine decarboxylase. *The EMBO Journal* 30, 931–944.

Kashiwagi, K., Suzuki, T., Suzuki, F., Furuchi, T., Kobayashi, H. and Igarashi, K. (1991) Coexistence of the genes for putrescine transport protein and ornithine decarboxylase at 16 min on *Escherichia coli* chromosome. *The Journal of Biological Chemistry* 266, 20922–20927.

Kashiwagi, K., Miyamoto, S., Suzuki, F., Kobayashi, H. and Igarashi, K. (1992) Excretion of putrescine by the putrescine-ornithine antiporter encoded by the *potE* gene of *Escherichia coli*. *Proceedings of the National Academy of Sciences of the United States of America* 89, 4529–4533.

Konings, W.N., Lolkema, J.S., Bolhuis, H., van Veen, H.W., Poolman, B. and Driessen, A.J. (1997) The role of transport processes in survival of lactic acid bacteria. Energy transduction and multidrug resistance. *Antonie Van Leeuwenhoek* 71, 117–128.

Kuper, C. and Jung, K. (2005) CadC-mediated activation of the *cadBA* promoter in *Escherichia coli*. *Journal of Molecular Microbiology and Biotechnology* 10, 26–39.

Ladero, V., Calles-Enríquez, M. and Fernández, M. and Alvarez M.A. (2010) Toxocological effects of dietary biogenic amines. *Current Nutrition and Food Science* 6, 145–156.

Ladero, V., Coton, M., Fernández, M., Buron, N., Martin, M.C., Guichard, H., Coton, E. and Alvarez, M.A. (2011) Biogenic amines content in Spanish and French natural ciders: application of qPCR for quantitative detection of biogenic amine-producers. *Food Microbiology* 28, 554–561.

Ladero, V., Cañedo, E., Pérez, M., Cruz Martin, M., Fernández, M. and Alvarez M.A. (2012) Multiplex qPCR for the detection and quantification of putrescine-producing lactic acid bacteria in dairy products. *Food Control* 27, 307–313.

Latorre-Moratalla, M.L., Bover-Cid, S., Veciana-Nogues, T. and Vidal-Carou, M.C. (2009) Thin-layer chromatography for the identification and semi-quantification of biogenic amines produced by bacteria. *Journal of Chromatography A* 1216, 4128–4132.

Leuschner, R.G. and Hammes, W.P. (1999) Formation of biogenic amine in mayonnaise, herring and tuna fish salad by lactobacilli. *International Journal of Food Sciences and Nutrition* 50, 159–164.

Maifreni, M., Frigo, F., Bartolomeoli, I., Innocente, N., Biasutti, M. and Marino, M. (2013) Identification of the *Enterobacteriaceae* in Montasio cheese and assessment of their amino acid decarboxylase activity. *Journal of Dairy Research* 80, 122–127.

Marcobal, A., de las Rivas, B., Moreno-Arribas, M.V. and Muñoz, R. (2004) Identification of the ornithine decarboxylase gene in the putrescine-producer *Oenococcus oeni* BIFI-83. *FEMS Microbiology Letters* 239, 213–220.

Marcobal, A., de las Rivas, B., Moreno-Arribas, M.V. and Muñoz, R. (2006) Evidence for horizontal gene transfer as origin of putrescine production in *Oenococcus oeni* RM83. *Applied and Environmental Microbiology* 72, 7954–7958.

Maurelli, A.T., Fernández, R.E., Bloch, C.A., Rode, C.K. and Fasano, A. (1998) "Black holes" and bacterial pathogenicity: a large genomic deletion that enhances the virulence of *Shigella* spp. and enteroinvasive *Escherichia coli*. *Proceedings of the National Academy of Sciences of the United States of America* 95, 3943–3948.

Meng, S.Y. and Bennett, G.N. (1992a) Nucleotide sequence of the *Escherichia coli cad* operon: a system for neutralization of low extracellular pH. *Journal of Bacteriology* 174, 2659–2669.

Meng, S.Y. and Bennett, G.N. (1992b) Regulation of the *Escherichia coli cad* operon: location of a site required for acid induction. *Journal of Bacteriology* 174, 2670–2678.

Miller, D.L. and Rodwell, V.W. (1971) Inactivation of L-lysine decarboxylase by L-lysine bromomethyl ketone. *Biochemical and Biophysical Research Communications* 44, 1227–1233.

Momany, C., Ernst, S., Ghosh, R., Chang, N.L. and Hackert, M.L. (1995) Crystallographic structure of a PLP-dependent ornithine decarboxylase from *Lactobacillus* 30a to 3.0 A resolution. *Journal of Molecular Biology* 252, 643–655.

Moreau, P.L. (2007) The lysine decarboxylase CadA protects *Escherichia coli* starved of phosphate against fermentation acids. *Journal of Bacteriology* 189, 2249–2261.

Morris, D.R. and Koffron, K.L. (1969) Putrescine biosynthesis in *Escherichia coli*. Regulation through pathway selection. *The Journal of Biological Chemistry* 244, 6094–6099.

Morris, D.R. and Pardee, A.B. (1966) Multiple pathways of putrescine biosynthesis in *Escherichia coli*. *The Journal of Biological Chemistry* 241, 3129–3135.

Nannelli, F., Claisse, O., Gindreau, E., de Revel, G., Lonvaud-Funel, A. and Lucas, P.M. (2008) Determination of lactic acid bacteria producing biogenic amines in wine by quantitative PCR methods. *Letters in Applied Microbiology* 47, 594–599.

Neely, M.N., Dell, C.L. and Olson, E.R. (1994) Roles of LysP and CadC in mediating the lysine requirement for acid induction of the *Escherichia coli* cad operon. *Journal of Bacteriology* 176, 3278–3285.

Park, Y.K., Bearson, B., Bang, S.H., Bang, I.S. and Foster, J.W. (1996) Internal pH crisis, lysine decarboxylase and the acid tolerance response of *Salmonella typhimurium*. *Molecular Microbiology* 20, 605–611.

Pereira, C.I., San Romao, M.V., Lolkema, J.S. and Barreto Crespo, M.T. (2009) *Weissella halotolerans* W22 combines arginine deiminase and ornithine decarboxylation pathways and converts arginine to putrescine. *Journal of Applied Microbiology* 107, 1889–1902.

Prosseda, G., Carmela Latella, M., Barbagallo, M., Nicoletti, M., Al Kassas, R., Casalino, M. and Colonna, B. (2007) The two-faced role of cad genes in the virulence of pathogenic *Escherichia coli*. *Research in Microbiology* 158, 487–493.

Rhee, J.E., Rhee, J.H., Ryu, P.Y. and Choi, S.H. (2002) Identification of the *cadBA* operon from *Vibrio vulnificus* and its influence on survival to acid stress. *FEMS Microbiology Letters* 208, 245–251.

Rodwell, A.W. (1953) The occurrence and distribution of amino-acid decarboxylases within the genus *Lactobacillus*. *Journal of General Microbiology* 8, 224–232.

Romano, A., Trip, H., Lonvaud-Funel, A., Lolkema, J.S. and Lucas, P.M. (2012) Evidence of two functionally distinct ornithine decarboxylation systems in lactic acid bacteria. *Applied and Environmental Microbiology* 78, 1953–1961.

Romano, A., Trip, H., Campbell-Sills, H., Bouchez, O., Sherman, D., Lolkema, J.S. and Lucas, P.M. (2013a) Genome sequence of *Lactobacillus saerimneri* 30a (formerly *Lactobacillus* sp. strain 30a), a reference lactic acid bacterium strain producing biogenic amines. *Genome Announcement* 1(1): e00097-12.

Romano, A., Trip, H., Lolkema, J.S. and Lucas, P.M. (2013b) A three-component lysine/ornithine decarboxylation system in *Lactobacillus saerimneri* 30a. *Journal of Bacteriology* 195, 1249–1254.

Romano, A., Ladero, V., Alvarez, M.A. and Lucas, P.M. (2014) Putrescine production via the ornithine decarboxylation pathway improves the acid stress survival of *Lactobacillus brevis* and is part of a horizontally transferred acid resistance locus. *International Journal of Food Microbiology* 175, 14–19.

Sabo, D.L. and Fischer, E.H. (1974) Chemical properties of *Escherichia coli* lysine decarboxylase including a segment of its pyridoxal 5′-phosphate binding site. *Biochemistry* 13, 670–676.

Sabo, D.L., Boeker, E.A., Byers, B., Waron, H. and Fischer, E.H. (1974) Purification and physical properties of inducible *Escherichia coli* lysine decarboxylase. *Biochemistry* 13, 662–670.

Saier, M.H. Jr, Tran, C.V. and Barabote, R.D. (2006) TCDB: the Transporter Classification Database for membrane transport protein analyses and information. *Nucleic Acids Research* 34, D181–D186.

Shi, X., Waasdorp, B.C. and Bennett, G.N. (1993) Modulation of acid-induced amino acid decarboxylase gene expression by *hns* in *Escherichia coli*. *Journal of Bacteriology* 175, 1182–1186.

Silla Santos, M.H. (1996) Biogenic amines: their importance in foods. *International Journal of Food Microbiology* 29, 213–231.

Soksawatmaekhin, W., Kuraishi, A., Sakata, K., Kashiwagi, K. and Igarashi, K. (2004) Excretion and uptake of cadaverine by CadB and its physiological functions in *Escherichia coli*. *Molecular Microbiology* 51, 1401–1412.

Straub, B.W., Kicherer, M., Schilcher, S.M. and Hammes, W.P. (1995) The formation of biogenic amines by fermentation organisms. *Zeitschrift für Lebensmittel-Untersuchung und -Forschung* 201, 79–82.

Tabor, C.W. and Tabor, H. (1985) Polyamines in microorganisms. *Microbiological Reviews* 49, 81–99.

Tanaka, Y., Kimura, B., Takahashi, H., Watanabe, T., Obata, H., Kai, A., Morozumi, S. and Fujii, T. (2008) Lysine decarboxylase of *Vibrio parahaemolyticus*: kinetics of transcription and role in acid resistance. *Journal of Applied Microbiology* 104, 1283–1293.

ten Brink, B., Damink, C., Joosten, H.M. and Huis in 't Veld, J.H. (1990) Occurrence and formation of biologically active amines in foods. *International Journal of Food Microbiology* 11, 73–84.

Torres, A.G. (2009) The *cad* locus of *Enterobacteriaceae*: more than just lysine decarboxylation. *Anaerobe* 15, 1–6.

Tsoi, H.W. and Tse, H. (2011) *Staphylococcus lugdunensis* is the likely origin of the ornithine decarboxylase operon in *Staphylococcus epidermidis* 2015B. *Applied and Environmental Microbiology* 77, 392–393.

Wunderlichova, L., Bunkova, L., Koutny, M., Valenta, T. and Bunka, F. (2013) Novel touchdown-PCR method for the detection of putrescine producing Gram-negative bacteria in food products. *Food Microbiology* 34, 268–276.

Wunderlichova, L., Bunkova, L., Koutny, M., Valenta, T. and Bunka, F. (2014) Formation, degradation, and detoxification of putrescine by foodborne bacteria: a review. *Comprehensive Reviews in Food Science and Food Safety* 13, 1012–1030.

Zúñiga, M., Pérez, G. and González-Candelas, F. (2002) Evolution of arginine deiminase (ADI) pathway genes. *Molecular Phylogenetics and Evolution* 25, 429–444.

9 The Role of Nitric Oxide Signalling in Yeast Stress Response and Cell Death

P. Ludovico,* B. Sampaio-Marques, N. Osório and F. Rodrigues
Life and Health Sciences Research Institute (ICVS), University of Minho, Braga, Portugal; ICVS/3B's – PT Government Associate Laboratory, Guimarães, Braga, Portugal

9.1 Abstract

Nitric oxide (NO) is a small gaseous molecule and free radical with important and distinct roles in diverse physiological functions in biological systems. The relevance of NO as a signalling molecule is well known, particularly in neurons as a neurotransmitter, in blood vessels as a regulator of aggregation and in cells of the immune system, where it participates in the host response to invading pathogenic microbes. NO is synthesized by the conversion of L-arginine to L-citrulline, a reaction that is catalysed by nitric oxide synthase (NOS). Once generated, NO may rapidly react with superoxide anions forming peroxynitrite ($OONO^-$); both molecules are reactive nitrogen species (RNS). NO can also react in a slower rate with O_2 and with metal ions such as iron and copper. Reactive oxygen species (ROS) play a major role in NO signalling and can affect NO availability from its production to post-production scavenging. Moreover, ROS/RNS are able to promote specific post-translational modifications, including *S*-nitrosation and tyrosine nitration, which both have an impact on protein function.

The synthesis of NO and orthologues of mammalian NOS enzymes have been described in eukaryotes, including birds, fish and amphibians, and in prokaryotes. However, plants and the higher fungal species – representatives of the ascomycetes and zygomycetes – do not have a protein with sequence similarity to known mammalian type NOS enzymes, although they do display a NOS-like activity that indicates the presence of an enzyme(s) that is structurally unrelated to the mammalian counterparts. While the origin of NO in yeast cells is still debatable, several reports have implicated NO signalling in stress response and programmed cell death mechanisms in yeast. In this chapter, the state of the art of the biosynthesis and functions of NO in the yeast *Saccharomyces cerevisiae* is reviewed, with particular emphasis on stress responses and cell death.

9.2 Introduction

Nitric oxide (NO) is an 'ancient' molecule that readily diffuses through biological membranes. With its derivatives, it acted as an oxidizing substrate in the archaeal world, driving the evolution of energetic pathways related to modern dissimilatory denitrification (Tillmann *et al.*, 2011). NO was first recognized as a messenger molecule in the mammalian central nervous system (CNS) in the late 1980s (Moncada *et al.*, 1991; Griffiths and Garthwaite, 2001). At the time, the finding of a messenger molecule that

*Corresponding author. E-mail address: pludovico@ecsaude.uminho.pt

was not constrained by membranes, or transporters drastically changed the classical view of cell signalling. The discovery that NO had a signalling role also raised awareness for the first time that free radicals could be more than toxic by-products of oxidative metabolism; they could also function as endogenous physiological signals. Subsequent research has exposed the complexity and relevance of NO signalling. NO can have beneficial or detrimental effects within cells depending on its concentration. At low concentrations, it can act as a second messenger regulating an impressive variety of physiological processes, including respiration, programmed cell death, mitochondrial biogenesis, cell motility, blood flow, immune function, neurodevelopment and neuronal function. At high concentrations, NO is cytotoxic and is exploited as a weapon in host–pathogen defences (Nathan and Shiloh, 2000).

NO synthases (NOS) are complex and highly regulated enzymes that catalyse the oxidation of L-arginine to NO and L-citrulline; they are the primary producers of NO in mammalian cells. The three isoforms of NOS (endothelial NOS, eNOS; neuronal NOS, nNOS; and inducible NOS, iNOS) are expressed in different tissues and under different conditions. Nevertheless, NO production does not seem to be exclusively mediated by NOS, as there are arginine-, nitrite- and nitrate-dependent mechanisms that are not catalysed by these enzymes. It has been shown that both nitrate and nitrite can be converted to NO in the blood and tissues in the course of physiological processes. The generation of NO by these pathways is greatly enhanced during hypoxia and acidosis, thereby possible ensuring NO production in situations when the oxygen-dependent NOS enzyme activities are compromised.

In mammals and in the yeast *Saccharomyces cerevisiae*, cytochrome *c* oxidase can utilize nitrite instead of oxygen as an electron acceptor in hypoxia, resulting in NO generation. In addition, compelling evidence suggests that *S. cerevisiae* is able to produce NO by an L-arginine-dependent mechanism similar to the one involved in classical NOS activity (Kanadia *et al.*, 1998; Domitrovic *et al.*, 2003; Almeida *et al.*, 2007; Osório *et al.*, 2007). However, contrarily to other eukaryotic species (Ghosh *et al.*, 2006), and even some bacteria (Adak *et al.*, 2002), no orthologues of mammalian NOS have been found in the *S. cerevisiae* genome. None the less, cross-reactive proteins with mammalian NOS antibodies have been detected in cellular extracts of *S. cerevisiae* (Kanadia *et al.*, 1998). It is conceivable that the L-arginine-dependent NO synthesis in *S. cerevisiae* involves an unknown enzyme with NOS-like activity that is unrelated to mammalian NOS, which is similar to what occurs in plants (Guo and Crawford, 2005; Foresi *et al.*, 2010). It has been suggested that this enzyme in yeast might be an essential flavoprotein that has been denoted Tah18 (Nishimura *et al.*, 2013).

NO has distinct roles in diverse physiological functions in biological systems, among them the regulation of stress responses and the control of the cell death signalling cascade. In *S. cerevisiae*, low levels of NO are protective and seem to be involved in various stress response pathways, while high NO levels promote cell death. In this chapter, we will review the state of the art of NO biosynthesis and functions in *S. cerevisiae*. Where pertinent, other fungal species and eukaryotic organisms will be mentioned, but the main goal is to provide a comprehensive overview of NO metabolism in this unicellular eukaryotic model organism.

9.3 Nitric Oxide Biosynthesis in Yeast

NO is a simple small gaseous molecule and radical that is formed by the bonding between one oxygen and one nitrogen atom in a +2 oxidation state. It has an unpaired electron, which confers upon it the properties of a free radical. The Gibbs energy necessary for NO formation from molecular oxygen and nitrogen is very high ($\Delta G = 86.3$ kJ mol^{-1}) and its production in biological systems is normally catalysed by NOS enzymes. In mammals, these complex and highly regulated enzymes have at least three isoforms. These are distinguished by their encoding gene, expression profile, regulation catalytic properties and sensitivity to inhibitors (Alderton *et al.*, 2001). NO is known to induce a negative feedback control mechanism in NOS, most probably by its interaction with the haem group of the enzyme (Griscavage *et al.*, 1995). The different NOS isoforms found in mammals are expressed in different tissues. The constitutively expressed NOS isoforms are found primarily in endothelial cells (the eNOS isoform) and in neuronal cells (the nNOS isoform). The third isoform, inducible NOS (iNOS), is transcriptionally controlled and produces cytotoxic concentrations of NO at sites of infection or inflammation (Campbell *et al.*, 2014). The regulation of enzyme activity at the expression level is not as relevant in the isoforms eNOS and nNOS (which are considered to be constitutively expressed) as it is in the iNOS isoform. The activity of the constitutive NOS isoforms is mainly dependent on the complex of Ca^{2+} and calmodulin, which occurs when the intracellular Ca^{2+} concentration is high, whereas iNOS is independent of the intracellular Ca^{2+} concentration.

NOS are only active in the homodimeric form and harbour the following cofactors: protoporphyrin IX a haem group (McMillan *et al.*, 1992), one FAD and one FMN (Bredt *et al.*, 1992), and one (6R)-5,6,7,8-tetrahydrobiopterin (BH_4) (Hevel and Marletta, 1992). Four structural characteristics are shared by all NOS isoforms in mammals: (i) the amino terminal end forms the oxygenase catalytic domains and binds the cofactors haem, BH_4 and the substrate L-arginine; (ii) the carboxyl terminal end forms the reductase domain that binds the remaining cofactors (FMN, FAD) and the co-substrate NADPH; (iii) between these two domains there is a binding site for calmodulin (Ghosh and Stuehr, 1995); and (iv) there is a binding site for zinc, which is most likely involved in the stabilization of the homodimeric form of the enzyme (Hemmens *et al.*, 2000).

The NOS enzymes catalyse the reaction of oxidation of L-arginine to NO and L-citrulline. This reaction requires O_2 and NADPH as co-substrates and occurs in two steps: first, L-arginine is hydroxylated to *N*-hydroxy-L-arginine. In the second step of the reaction, this intermediate is oxidized to NO and L-citrulline. The reductase domain functions as an acceptor of the electrons from NADPH, transferring the electrons to the haem group, in a manner similar to that performed by P450 reductases. The transfer of electrons between FAD and FMN requires the presence of Ca^{2+}-calmodulin (Matsuda and Iyanagi, 1999). The electrons from the reductase domain of one subunit are transferred to the haem group of the oxygenase domain of the other subunit (Siddhanta *et al.*, 1996).

Orthologues to mammalian NOS have been found in other eukaryotes, including birds, fish and amphibians, and in prokaryotes such as *Bacillus subtilis* and *Deinococcus radiodurans* (Adak *et al.*, 2002; Ghosh and Salerno, 2003). Albeit, plants and higher fungal species – representatives of ascomycetes and zygomycetes – do not have a protein with sequence similarity to known mammalian-type NOS, but they do display a NOS-like activity that indicates the presence of an enzyme that is structurally unrelated to those of their mammalian counterparts.

NO production has been investigated in yeast, in particular in the budding yeast, *S. cerevisiae*. However, the origin of NO in yeast cells is still a matter of debate, essentially because of the lack of mammalian NOS orthologues in the yeast genome. Previous studies have shown the existence of proteins with cross-reactivity with the mammalian NOS antibodies in budding yeast (Kanadia *et al.*, 1998; Domitrovic *et al.*, 2003), although the immunoreactive bands visualized were of different molecular sizes and the identity of these NOS-immunoreactive molecules is uncertain. NO production observed in budding yeast was also associated with a NOS-like activity by the use of a classical NOS kit assays and was inhibited by the use of classical inhibitors of NO synthesis, such as N^{ω}-nitro-L-arginine methyl ester (L-NAME) (Almeida *et al.*, 2007; Osório *et al.*, 2007). In addition, NO production was shown to be dependent on intracellular L-arginine levels, which once again suggested the existence of an enzyme with NOS-like activity (Almeida *et al.*, 2007; Osório *et al.*, 2007) (Fig. 9.1).

Another possible source of NO in yeast cells under hypoxic or anoxic conditions is the mitochondria (Castello *et al.*, 2006). Mitochondria can produce NO in a nitrite (NO_2^-) dependent pathway that requires an electron donor, and is carried out by cytochrome *c* oxidase (COX, complex IV) in a pH-dependent fashion (Castello *et al.*, 2006) (Fig. 9.1). Mitochondrial NO production seems to be influenced by the Yhb flavohaemoglobin NO oxidoreductase (Castello *et al.*, 2006), though this NO production pathway, being a part of hypoxic signalling, can proceed only under hypoxic and anoxic conditions and therefore seems to be not relevant to normoxic conditions. A recent study using *S. cerevisiae* mutants with increased NO levels demonstrated a correlation between increased NO, mitochondrial respiration and calorific restriction (CR)-mediated longevity (Li *et al.*, 2011). Interestingly, CR-increased

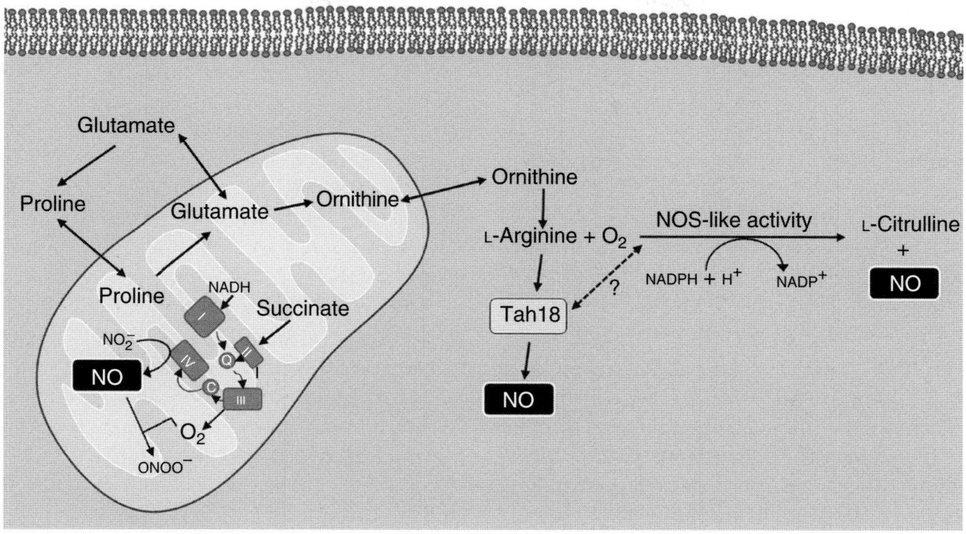

Fig. 9.1. Different pathways of nitric oxide (NO) biosynthesis in *Saccharomyces cerevisiae*. The nitric oxide synthase (NOS)-like activity generates NO in the cytosol. Tah18 also mediates NO generation in the cytosol that is dependent on L-arginine and so it is suggested that it is an enzyme with classical NOS-like activity. In the mitochondria (on the left), NO is generated under hypoxic conditions via cytochrome *c* oxidase (complex IV) in a reaction that is dependent on nitrite (NO_2^-) levels. Once generated, NO may rapidly react with superoxide anions (O_2^-), forming peroxynitrite ($OONO^-$).

NO is largely dependent on mitochondrial electron transport and COX activity (Li *et al.*, 2011). These results suggest that CR may derepress some hypoxic genes, thereby promoting mitochondrial functioning and NO production, which extends yeast lifespan. Thus, the relevance of the mechanisms of NO production under hypoxic conditions to normoxic conditions remains to be clarified but cannot be disregarded.

Recently, it was reported that the yeast essential protein Tah18, which is a putative oxidoreductase, is involved in NO production under high-temperature stress conditions in an L-arginine-dependent manner, and it has been suggested that Tah18 is the enzyme responsible for the NOS-like activity observed in yeast cells (Nishimura *et al.*, 2013) (Fig. 9.1). Further studies are needed to clarify whether Tah18 really is an enzyme with functional homology to mammalian NOS enzymes.

Another NO production pathway operating under oxidative stress conditions has been described that involves NO production from proline via proline oxidase (Put1) and *N*-acetyltransferase (Mpr1) in *S. cerevisiae* Σ1278b strain (Nishimura *et al.*, 2010) (Fig. 9.1). These NO production mechanisms will be further explored in Section 9.4.

In spite of the accumulated evidence on NO production in yeast, the synthetic mechanism and the physiological role of NO in *S. cerevisiae* remain uncertain, mainly because of the lack of mammalian and bacterial NOS gene orthologues in the genome.

9.4 Nitric Oxide Signalling

NO is a highly diffusible free radical with dichotomous regulatory roles in numerous physiological and pathological events, and is recognized as an intra- and intercellular signalling molecule in both animals and plants. In mammals, NO is involved in stress tolerance through the activation of the soluble form of the haem-containing enzyme guanylate cyclase and diverse post-translational modifications that affect protein function (Jobgen *et al.*, 2006; Sun *et al.*, 2006).

Diverse cellular functions can be directly or indirectly affected by NO through the post-translational modification of proteins. The most widespread and functionally relevant modification is *S*-nitrosation, which is defined as the covalent attachment of NO to the thiol side chain of a cysteine residue (Hess *et al.*, 2005). NO and NO metabolites can oxidize, nitrate or nitrosate proteins, and they are responsible for tyrosine nitration, which is a covalent protein modification resulting from the addition of a nitro (NO_2) group on the aromatic ring of tyrosine residues (Ischiropoulos, 2003).

NO might also act as a signalling molecule in unicellular eukaryotes like yeast (Almeida *et al.*, 2007; Osório *et al.*, 2007; Kig and Temizkan, 2009). However, the cellular responses to NO are complex, with considerable crosstalk between responses to several stimuli. Brief summaries are given of the potential roles of NO during various stresses in *S. cerevisiae* and the signalling pathways that are activated in the two following sections.

9.4.1 The nitrosative stress response in yeast

In contrast to mammalian cells, yeast cells do not express any obvious NO sensor and thus, it is not clear how yeast cells detect NO and which signalling pathways mediate NO responses. Curiously, in *Candida albicans*, a pathway of NO sensing has already been defined. Chiranand *et al.* (2008) found that Cta4, a Zn(II)2-Cys6 transcription factor, is necessary for initiating the NO response in *C. albicans*. The regulator Cta4 is able to associate with the NO-responsive element (NORE) present in the *YHB1* gene (which codifies a flavohaemoglobin), leading to the activation of its expression (Chiranand *et al.*, 2008). The closest homologue to the *C. albicans CTA4* gene in *S. cerevisiae* is *OAF1*, an oleate receptor that so far has been shown to function independently of NO sensing and response.

The transcriptional response of *S. cerevisiae* to exogenously supplied NO, which results in both a general stress response and a nitrosative stress-specific response, revealed that the flavohaemoglobin gene *YHB1*, *SSU1* (which encodes a CTA4-dependent plasma membrane sulfite pump involved in sulfite metabolism) and three other uncharacterized open reading frames are upregulated (Sarver and DeRisi, 2005) (Fig. 9.2). The data generated showed that the induction of the *YHB1* and *SSU1* genes is dependent on the zinc finger transcription factor Fzf1 (Sarver and DeRisi, 2005). These authors also showed that both Yhb1 and Ssu1 contribute to nitrosative stress resistance, depending on growth conditions, while the abrogation of *FZF1* results in hypersensitivity to nitrosative stress (Sarver and DeRisi, 2005). In the same work, it was also shown that the deletion of *FZF1* eliminated the nitrosative stress-specific transcriptional response, whereas *FZF1* overexpression restored nitrosative stress response even in the absence of NO exposure, implicating Fzf1 in NO sensing in *S. cerevisiae* (Sarver and DeRisi, 2005).

More recently, the response of *S. cerevisiae* cells to nitrosative stress was shown to involve the nuclear translocation of Yap1, a leucine zipper (bZIP) transcription factor required for oxidative stress tolerance, and consequent upregulation of the antioxidant enzymes superoxide dismutase and catalase (Lushchak *et al.*, 2010). Inactivation of the *YAP1* gene abrogates the upregulation of superoxide dismutase and catalase induced by nitrosative stress (Lushchak *et al.*, 2010) (Fig. 9.2). These results suggest that Yap1 could be one of the effector mechanisms underlying NO-regulated transcription.

The role of the yeast flavohaemoglobin (Yhb1) on the inducible response to nitrosative stress cannot be disregarded either. Lewinska and Bartos (2006) showed that the growth of *YHB1* mutant cells is impaired by nitrosoglutathione and peroxynitrite, and that increased levels of NO are detected after incubation with NO donors. Another study used a Yhb1-GFP fusion protein to determine alterations on Yhb1 levels after cell treatment with oxidants and NO donors (Lewinska *et al.*, 2008). The authors found a specific dose-dependent increase in *YHB1*-GFP expression in response to NO donors, while H_2O_2, menadione or diamide caused a moderate diminution of Yhb1-GFP fluorescence, suggesting that Yhb1 does not contribute to cross-protection against oxidative and nitrosative stress (Lewinska *et al.*, 2008). The *S. cerevisiae* Yhb1 protein was also shown to be translocated to the mitochondria under hypoxic conditions, so promoting the detoxification of NO (Cassanova *et al.*, 2005).

A genome-wide transcriptional response analysis performed in *YHB1*-deleted *S. cerevisiae* cells also indicated that major ROS scavenging enzymes have a negligible role in the NO response (Horan *et al.*, 2006). Nevertheless, the

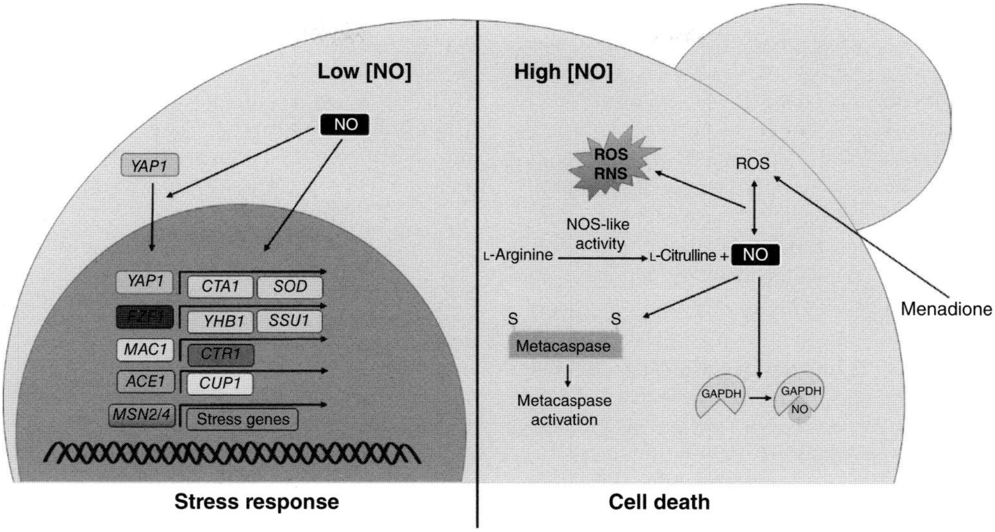

Fig. 9.2. The signalling functions of nitric oxide (NO) in *Saccharomyces cerevisiae*. NO promotes transcriptional and post-transcriptional responses which, depending on the NO concentration, are either protective or pro-death. The transcriptional response to NO is suggested to be protective and to confer stress resistance, while the vast majority of the post-translational alterations described so far are related to the promotion of cell death. Abbreviations: *ACE1*, gene for the copper-responsive *CUP1* transcriptional activator, Ace1; *CTA1*, gene for a peroxisomal catalase involved in copper regulation; *CTR1*, gene encoding a high-affinity copper transporter whose induction is promoted by Mac1 activation by NO; *CUP1*, gene for a yeast metallothionein playing an important role in copper detoxification; *FZF1*, zinc finger transcription factor gene; GAPDH, glyceraldehyde-3-phosphate dehydrogenase; *MAC1*, gene for the transcription factor controlling copper metabolism, Mac1, which is activated by NO; NOS, nitric oxide synthase; RNS, reactive nitrogen species; ROS, reactive oxygen species; *SOD*, gene for superoxide dismutase; *SSU1*, encodes a CTA4-dependent plasma membrane sulfite pump involved in sulfite metabolism (CTA4 is the gene for a transcription factor necessary for initiating the NO response); *YAP1*, gene for a leucine zipper (bZIP) transcription factor; *YHB1*, gene for a flavohaemoglobin that includes an NO-responsive element (NORE). See text for further details.

key regulators of the general and oxidative stress responses, Msn2/4 and Yap1, are apparently also implicated in mediating the NO response (Horan *et al.*, 2006). The genes coding respiratory chain components and regulated by the haem-responsive transcription factor Hap1 were found repressed on treatment with RNS, suggesting that nitrosative stress could also interfere with Hap1 activity (Horan *et al.*, 2006). Curiously, *S. cerevisiae HAP1* is a close homologue of *C. albicans CTA4*, though whether they share functional similarity remains to be clarified. Interestingly, the genes involved in iron homeostasis and transport, as well as in the glutathione and DNA repair systems, were consistently found to be upregulated, implicating these processes in NO responses in *S. cerevisiae* (Horan *et al.*, 2006).

Further, the *S. cerevisiae* Fre1 protein, an important component of the metal metabolism system, loses its activity on exposure to NO under aerobic conditions (Shinyashiki *et al.*, 2004). Apparently, NO inhibits Fre1 and, consequently, the reduction of extracellular iron and copper, their solubility and the high-affinity uptake of these metals (Dancis *et al.*, 1992). In agreement with its role in metal metabolism, the observed inhibition of Fre1 by NO also inhibited yeast growth in a low-iron medium (Shinyashiki *et al.*, 2004). In addition, these data show that some of the NO targets of regulation are conserved, particularly those involving haem group regulation.

A recent study showed that exogenous NO treatment induced the expression of the genes responsible for copper metabolism under the control of the transcription factor Mac1, including the *CTR1* gene encoding high-affinity copper transporter (Nasuno *et al.*, 2014). Apparently, NO does not affect the expression level of *MAC1*, thus Mac1 is likely to be activated through post-translational modifications (Nasuno *et al.*, 2014). Consistently with the hypothesis that NO functions as a disrupter of yeast copper metabolism, it has been previously shown that NO inhibits the copper-dependent induction of the yeast *CUP1* gene (which codes for a yeast metallothionein that plays an important role in copper detoxification) through its action on the copper-responsive *CUP1* transcriptional activator Ace1 (Shinyashiki *et al.*, 2000) (Fig. 9.2). The yeast metal-responsive transcription factors Ace1 and Mac1 are responsible for the expression of a variety of metal regulatory proteins (Thiele, 1988; Shinyashiki *et al.*, 2000). Considering the ability that NO and NO metabolites have to modify thiol functions, it is to be expected that NO may be capable of regulating yeast metal metabolism by reacting with the metal-binding thiol ligands in proteins (Shinyashiki *et al.*, 2000).

Comparisons of the NO-transcriptional response in *S. cerevisiae* and *C. albicans* are intriguing and show that they are dissimilar, most probably reflecting their evolutionary adaptation to environmental requirements (Hromatka *et al.*, 2005; Horan *et al.*, 2006). However, the genes encoding the proteins involved in iron homeostasis and transport seem to be consistently upregulated in both species. Altogether, the findings suggest that NO regulation of metal metabolism and metal-dependent transcription is an important role of NO that is worthy of exploration in other contexts.

Summing up, the accumulated findings implicate different players in *S. cerevisiae* specific response to nitrosative stress, including: (i) the flavohaemoglobin (Yhb1), which converts nitric oxide to nitrate, and the plasma membrane sulfite pump Ssu1; (ii) Yap1, Fzf1 and Msn2/4 as involved in the mechanisms underlying NO-regulated transcription; and (iii) regulation of the yeast metal-responsive transcription factors Ace1 and Mac1, most probably by modifications of the metal-binding thiol ligands in these proteins (Shinyashiki *et al.*, 2000).

Apparently, when facing nitrosative stress, *S. cerevisiae* cells also upregulate the antioxidant systems involved in general oxidative stress response, such as superoxide dismutase, catalases and glutathione (Costa and Moradas-Ferreira, 2001; Horan *et al.*, 2006; Lushchak *et al.*, 2010), but their role in the NO-specific response and on the ROS/RNS crosstalk is still poorly understood in yeast (Missall *et al.*, 2004).

9.4.2 Nitric oxide signalling in the yeast stress response

In *S. cerevisiae*, low levels of NO are indicated to be involved in various stress response pathways, including responses to oxidative stress, high osmotic pressure, dehydration, high hydrostatic pressure, high temperature and copper metabolism. A study that focused on the responses of an *S. cerevisiae* yeast strain to air drying and freeze–thaw stress (as commonly found during the bread-making process) revealed that NO produced under oxidative stress conditions from proline via proline oxidase Put1 and *N*-acetyltransferase Mpr1 confers stress tolerance. This is a novel antioxidant mechanism that enhances fermentation ability under multiple baking-associated stress conditions of the baker's yeast (Sasano *et al.*, 2012) (Fig. 9.3).

NO produced under high-temperature stress conditions also constitutes an antioxidant mechanism with different protective effects for *S. cerevisiae* cells. It leads to an increase of copper levels, the activity of superoxide dismutases, particularly Cu,Zn-superoxide dismutase Sod1, and cell viability in a manner that is dependent on Mac1 (Nasuno *et al.*, 2014). These authors propose that the NO-mediated antioxidant mechanism is based on the induction of the *CTR1* gene promoted by Mac1 activation by NO, leading to an increase in the cellular copper level, and then the activation of Sod1 (Nasuno *et al.*, 2014) (Fig. 9.3).

A study by Domitrovic *et al.* (2003) showed that NO promotes the cytoprotection of *S. cerevisiae* cells subjected to high hydrostatic pressure. In agreement with others (Chiang *et al.*, 2000), at low levels, NO interacts with the transcription factor Ace1, enhancing its activity and promoting stress tolerance (Shinyashiki *et al.*, 2000) (Fig. 9.3).

The accumulated evidence indicates that NO signalling is an important component of the

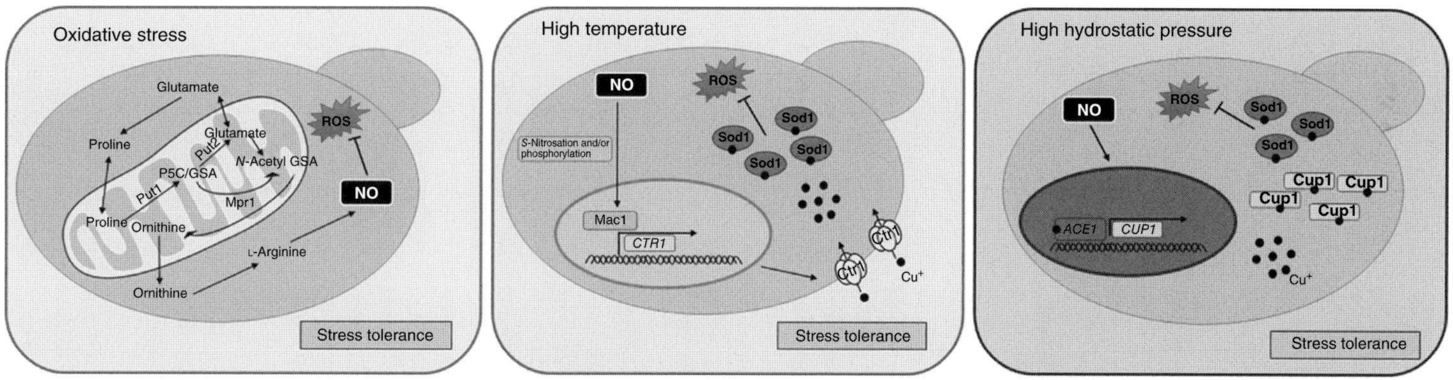

Fig. 9.3. Nitric oxide (NO)-elicited responses to different stress stimuli in *Saccharomyces cerevisiae*. These responses are cytoprotective and result in stress tolerance that is suggested to be mainly mediated by decreasing the accumulation of reactive oxygen species (ROS). Abbreviations: *ACE1*, gene for the copper-responsive *CUP1* transcriptional activator, Ace1; *CTR1*, gene encoding a high-affinity copper transporter, Ctr1; *CUP1*, gene for a yeast metallothionein, Cup1, that plays an important role in copper detoxification; GSA, glutamate semialdehyde; Mac1, a transcription factor controlling copper metabolism that is activated by NO; Mpr1, *N*-acetyltransferase; P5C, pyrroline-5-carboxylate; Put1, Put2, proline oxidase 1, 2; Sod1, Cu,Zn-superoxide dismutase 1.

general stress response in *S. cerevisiae*. Apparently, low levels of NO promote stress resistance by different reported mechanisms that often involve the upregulation of antioxidant systems such as superoxide dismutases and catalases. This phenomenon resembles a hormetic-like (beneficial) effect, which may be based on the intricate crosstalk between ROS and RNS. This crosstalk between ROS and RNS is dependent on the concentration of NO and NO metabolites and on their effects on metal homeostasis, but requires further investigation in yeast cells.

9.5 Nitric Oxide Signalling in Yeast Cell Death and Ageing

NO also controls the apoptotic and necrotic signalling cascade by regulating the expression of several genes, mitochondrial dysfunction and caspase activity/activation (Brune *et al.*, 1995). However, the role of NO in apoptosis and/or necrosis is still controversial, with reports showing that it has both pro-survival and pro-death functions. Furthermore, even though NO has also been assigned a pro-death role, as described in Section 9.5, low NO levels seem to confer stress resistance.

Our team has previously shown that H_2O_2 apoptosing cells synthesize NO that is associated with a NOS-like activity as demonstrated by the use of a classical NOS kit assays (Almeida *et al.*, 2007). The results of this study showed that NO induces glyceraldehyde-3-phosphate dehydrogenase (GAPDH) *S*-nitrosation during yeast apoptosis and that this is prevented by the blocking of NO synthesis with L-NAME (Almeida *et al.*, 2007) (Fig. 9.2). The endogenous NO synthesis during H_2O_2-induced apoptosis in yeast cells presents similarities with previous reports in mammalian cells, in which H_2O_2 activates e-NOS (Thomas *et al.*, 2002). More recent work has shown that GAPDH interacts with glutathione peroxidase (Gpx3), which reduces GAPDH *S*-nitrosation, thus increasing cell viability under NO stress (Lee *et al.*, 2011).

NO homeostasis was also implicated in the context of the yeast model of Batten's disease (the *BTN1* deleted strain). Batten's disease is a juvenile form of neuronal ceroid lipofuscinoses resulting from mutations in the *CLN3* gene, a homologue of yeast *BTN1*, which seems to be involved in pH homeostasis. We have shown that the decreased levels of L-arginine in *BTN1*-deleted cells limit the synthesis of NO (Osório *et al.*, 2007). This defect in NO synthesis seems to suppress the signalling required for yeast menadione-induced apoptosis by decreasing ROS accumulation that is partially dependent on endogenous NO production (Osório *et al.*, 2007) (Fig. 9.2).

The relevance of NO to yeast cell death has also been highlighted by its role in the regulation of metacaspase activity. Metacaspases are suggested to be the ancestors of metazoan caspases and of the executing caspase-like proteases in yeast (Madeo *et al.*, 2002) and plants (Suarez *et al.*, 2004). *Arabidopsis thaliana* metacaspase 9 can be kept inactive by the *S*-nitrosation of one cysteine residue (Belenghi *et al.*, 2007). However, the mature metacaspase processed form is not susceptible to NO inhibition due to the presence of a second *S*-nitrosation-insensitive cysteine residue that can replace the *S*-nitrosated cysteine residue within the catalytic centre of the processed metacaspase (Belenghi *et al.*, 2007). This second cysteine residue seems to be absent in caspases and paracaspases, but is conserved in all reported metacaspases. In yeast, our team found that menadione or H_2O_2 treatment, conditions associated with NO generation, induce metacaspase activity that is increased in the presence of L-arginine and decreased in the presence of L-NAME (Almeida *et al.*, 2007; Osório *et al.*, 2007). Overall, the data suggest that NO is involved in metacaspase activation and that while NO could have pro-survival functions at low concentrations, it results in increased metacaspase activity at high concentrations and, consequently, in yeast cell death (Fig. 9.2).

Given that a common feature of apoptotic cell death induced by H_2O_2 or menadione, or even during age-associated cell death, is the generation of ROS, the relevance of NO in yeast cell death programmes is reinforced, pointing to the conservation of links between ROS and RNS (Espey *et al.*, 2000). In addition, these findings suggesting the involvement of NO in the programmed cell death of yeast induced by pro-oxidative agents such as H_2O_2 and menadione are consistent with studies on the yeast flavohaemoglobin (Yhb1) that imply the production of NO in response to oxidative stress (Cassanova *et al.*, 2005).

NO has also a role in mediating cell death during the chronological lifespan (CLS; a measure of the time a non-dividing cell population can remain viable in liquid media) (Almeida *et al.*, 2007). Due to experimental limitations on the use of an NO-selective electrode, NO levels were indirectly measured by the Griess reaction, and it was shown that increased NO is produced during CLS (Almeida *et al.*, 2007). The presence of oxyhaemoglobin (OxyHb), a compound that scavenges NO and is considered a major route of its catabolism, resulted in a decreased accumulation of superoxide anions during CLS, suggesting that NO could mediate superoxide anion production, most probably by the action of intracellular RNS (Almeida *et al.*, 2007).

Recent work has assessed the CLS of yeast cells lacking: (i) *YHB1*, the gene for the flavohaemoglobin involved in NO detoxification that plays a role in oxidative and nitrosative stress responses; (ii) *FZF1*, the gene for a transcription factor involved in sulfite metabolism; and (iii) *SFA1*, the gene for a bifunctional alcohol dehydrogenase with formaldehyde dehydrogenase activity that is glutathione dependent (whose functions are annotated in the *Saccharomyces* Genome Database (SGD, available at http://www.yeastgenome.org). Based on observations made showing a strain-independent decrease rather than increase in NO production associated with increased apoptosis during CLS, Lewinska *et al.* (2011) concluded that NO does not mediate apoptosis during CLS of yeast cells. These authors also claimed that during CLS, yeast cells are not facing nitrosative stress, given the slightly decreased pattern of protein tyrosine nitration and the decreased expression profile of the flavohaemoglobin. As mentioned in the previous section, the NO signalling response is not limited to the protein nitration profile and flavohaemoglobin expression, but has other broader implications in cell homeostasis and in transcriptional regulation. Although the 'double edged sword' nature of NO deserves further investigation and is an emerging topic, one cannot forget that DAF-FM (4-amino-5-methylamino-2′,7′-difluorofluorescein), the dye used to produce the above mentioned observations (Lewinska *et al.*, 2011), is a moderately specific NO probe that provides an unreliable measurement of NO in the presence of ROS/RNS (Balcerczyk *et al.*, 2005), a condition that is known to occur in aged yeast cells.

In spite of the many techniques developed to provide an adequate methodology for measuring physiological levels of NO, the electrochemical methods using NO selective electrodes are still the most reliable techniques for measuring NO (Davies and Zhang, 2008). The different experimental protocols used to assess CLS and the limitations in measuring NO under these conditions indicate that further investigation is required of the role of NO signalling during yeast ageing. A very elegant approach to this issue was made by Li *et al.* (2011) who showed that CR induces NO generation in yeast cells under normoxic conditions. Castello *et al.* (2006) proposed that NO production during CR conditions is most probably dependent on mitochondria and occurs by derepressing the hypoxic subunit isoforms of COX and cytochrome *c*. CR most likely confers its beneficial effects via a mitochondrial-NO-mediated adaptive metabolic shift, which optimizes metabolism and at the same time improves cellular defence systems against the oxidative stress that accumulates with age (Castello *et al.*, 2006; Li *et al.*, 2011). Nevertheless, they do not exclude the possibility that CR conditions may activate NO production and regulate mitochondrial respiration and longevity via different mechanisms. The fact that NO production is associated with a longevity-promoting condition like CR suggests a hormesis-like effect for this free radical as has been described for other ROS (Mesquita *et al.*, 2010).

It is proposed that the differential role of Yhb1 and Sfa1, which are both enzymes involved in NO detoxification, on replicative lifespan (RLS) and CLS under CR conditions might be related to the presence of different levels of NO metabolites such as peroxynitrites (Li *et al.*, 2011). This NO metabolite is generated by the reaction of NO with the superoxide anion, which is found at lower levels in CLS under CR as a result of the upregulation of superoxide dismutases (Li *et al.*, 2011).

Intriguingly, studies on NO metabolism in yeast, and particularly those studies on CLS (Lewinska *et al.*, 2011; Li *et al.*, 2011), commonly abrogate genes encoding the proteins involved in nitrosative stress responses. However, a detailed search on the literature concerning Yhb1, Fzf1 and Sfa1 showed that these proteins are also involved in DNA replication stress and genome integrity (Cassanova *et al.*, 2005; Tkach et al., 2012), processes that are implicated in yeast ageing and also in the longevity-promoting effects of CR (Weinberger et al., 2013). Thus, further studies should address the complex scenario of ageing in yeast and the possible dual role of NO in this physiological process in order to obtain new insights on the role of NO in more complex mammalian systems.

9.6 Conclusions

The accumulated evidence indicates that yeast cells synthesize NO by different mechanisms but certainly including a NOS-like activity that requires further investigation. Similar to mammalian cells, NO also has important signalling functions in yeast cells. NO definitely seems to be protective and to confer stress resistance in the main stressful conditions that have been explored, but it also has an important role in cell death. This dichotomous role of NO is, most probably, dependent on its intracellular concentration, but other mechanisms need to be considered as well, such as the crosstalk between NO, NO metabolites and ROS. An important feature of NO signalling is its ability to regulate cellular metal homeostasis, indicating that yeast could also be a good model for clarifying this complex role of NO in mammalian systems. Although NO is an 'ancient' molecule, the investigation and understanding of its metabolism in yeast is still far from giving us a detailed picture, but shows further exploration would be worthwhile.

References

Adak, S., Bilwes, A.M., Panda, K., Hosfield, D., Aulak, K.S., McDonald, J.F., Tainer, J.A., Getzoff, E.D., Crane, B.R. and Stuehr, D.J. (2002) Cloning, expression, and characterization of a nitric oxide synthase protein from *Deinococcus radiodurans*. *Proceedings of the National Academy of Sciences of the United States of America* 99, 107–112.

Alderton, W.K., Cooper, C.E. and Knowles, R.G. (2001) Nitric oxide synthases: structure, function and inhibition. *The Biochemical Journal* 357, 593–615.

Almeida, B., Buttner, S., Ohlmeier, S., Silva, A., Mesquita, A., Sampaio-Marques, B., Osório, N.S., Kollau, A., Mayer, B., Leão, C., Laranjinha, J., Rodrigues, F., Madeo, F. and Ludovico, P. (2007) NO-mediated apoptosis in yeast. *Journal of Cell Science* 120, 3279–3288.

Balcerczyk, A., Soszynski, M. and Bartosz, G. (2005) On the specificity of 4-amino-5-methylamino-2′,7′-difluorofluorescein as a probe for nitric oxide. *Free Radical Biology and Medicine* 39, 327–335.

Belenghi, B., Romero-Puertas, M.C., Vercammen, D., Brackenier, A., Inze, D., Delledonne, M. and Van Breusegem, F. (2007) Metacaspase activity of *Arabidopsis thaliana* is regulated by *S*-nitrosylation of a critical cysteine residue. *The Journal of Biological Chemistry* 282, 1352–1358.

Bredt, D.S., Ferris, C.D. and Snyder, S.H. (1992) Nitric oxide synthase regulatory sites. Phosphorylation by cyclic AMP-dependent protein kinase, protein kinase C, and calcium/calmodulin protein kinase; identification of flavin and calmodulin binding sites. *The Journal of Biological Chemistry* 267, 10976–10981.

Brune, B., Messmer, U.K. and Sandau, K. (1995) The role of nitric oxide in cell injury. *Toxicology Letters* 82/83, 233–237.

Campbell, M.G., Smith, B.C., Potter, C.S., Carragher, B. and Marletta, M.A. (2014) Molecular architecture of mammalian nitric oxide synthases, *Proceedings of the National Academy of Sciences of the United States of America* 111, E3614–E3623.

Cassanova, N., O'Brien, K.M., Stahl, B.T., McClure, T. and Poyton, R.O. (2005) Yeast flavohemoglobin, a nitric oxide oxidoreductase, is located in both the cytosol and the mitochondrial matrix: effects of respiration, anoxia, and the mitochondrial genome on its intracellular level and distribution. *The Journal of Biological Chemistry* 280, 7645–7653.

Castello, P.R., David, P.S., McClure, T., Crook, Z. and Poyton, R.O. (2006) Mitochondrial cytochrome oxidase produces nitric oxide under hypoxic conditions: implications for oxygen sensing and hypoxic signaling in eukaryotes. *Cell Metabolism* 3, 277–287.

Chiang, K.T., Shinyashiki, M., Switzer, C.H., Valentine, J.S., Gralla, E.B., Thiele, D.J. and Fukuto, J.M. (2000) Effects of nitric oxide on the copper-responsive transcription factor Ace1 in *Saccharomyces cerevisiae*: cytotoxic and cytoprotective actions of nitric oxide, *Archives of Biochemistry and Biophysics* 377, 296–303.

Chiranand, W., McLeod, I., Zhou, H., Lynn, J.J., Vega, L.A., Myers, H., Yates, J.R. 3rd, Lorenz, M.C. and Gustin, M.C. (2008) *CTA4* transcription factor mediates induction of nitrosative stress response in *Candida albicans*. *Eukaryotic Cell* 7, 268–278.

Costa, V. and Moradas-Ferreira, P. (2001) Oxidative stress and signal transduction in *Saccharomyces cerevisiae*: insights into ageing, apoptosis and diseases. *Molecular Aspects of Medicine* 22, 217–246.

Dancis, A., Roman, D.G., Anderson, G.J., Hinnebusch, A.G. and Klausner, R.D. (1992) Ferric reductase of *Saccharomyces cerevisiae*: molecular characterization, role in iron uptake, and transcriptional control by iron. *Proceedings of the National Academy of Sciences of the United States of America* 89, 3869–3873.

Davies, I.R. and Zhang, X. (2008) Nitric oxide selective electrodes. *Methods in Enzymology* 436, 63–95.

Domitrovic, T., Palhano, F.L., Barja-Fidalgo, C., DeFreitas, M., Orlando, M.T. and Fernandes, P.M. (2003) Role of nitric oxide in the response of *Saccharomyces cerevisiae* cells to heat shock and high hydrostatic pressure. *FEMS Yeast Research* 3, 341–346.

Espey, M.G., Miranda, K.M., Feelisch, M., Fukuto, J., Grisham, M.B., Vitek, M.P. and Wink, D.A. (2000) Mechanisms of cell death governed by the balance between nitrosative and oxidative stress. *Annals of the New York Academy of Sciences* 899, 209–221.

Foresi, N., Correa-Aragunde, N., Parisi, G., Calo, G., Salerno, G. and Lamattina, L. (2010) Characterization of a nitric oxide synthase from the plant kingdom: NO generation from the green alga *Ostreococcus tauri* is light irradiance and growth phase dependent. *The Plant Cell* 22, 3816–3830.

Ghosh, D.K. and Salerno, J.C. (2003) Nitric oxide synthases: domain structure and alignment in enzyme function and control. *Frontiers in Bioscience* 8, d193–d209.

Ghosh, D.K. and Stuehr, D.J. (1995) Macrophage NO synthase: characterization of isolated oxygenase and reductase domains reveals a head-to-head subunit interaction. *Biochemistry* 34, 801–807.

Ghosh, D.K., Holliday, M.A., Thomas, C., Weinberg, J.B., Smith, S.M. and Salerno, J.C. (2006) Nitric-oxide synthase output state. Design and properties of nitric-oxide synthase oxygenase/FMN domain constructs. *The Journal of Biological Chemistry* 281, 14173–14183.

Griffiths, C. and Garthwaite, J. (2001) The shaping of nitric oxide signals by a cellular sink. *Journal of Physiology* 536, 855–862.

Griscavage, J.M., Hobbs, A.J. and Ignarro, L.J. (1995) Negative modulation of nitric oxide synthase by nitric oxide and nitroso compounds. *Advances in Pharmacology* 34, 215–234.

Guo, F.Q. and Crawford, N.M. (2005) *Arabidopsis* nitric oxide synthase1 is targeted to mitochondria and protects against oxidative damage and dark-induced senescence. *The Plant Cell* 17, 3436–3450.

Hemmens, B., Goessler, W., Schmidt, K. and Mayer, B. (2000) Role of bound zinc in dimer stabilization but not enzyme activity of neuronal nitric-oxide synthase. *The Journal of Biological Chemistry* 275, 35786–35791.

Hess, D.T., Matsumoto, A., Kim, S.O., Marshall, H.E. and Stamler, J.S. (2005) Protein *S*-nitrosylation: purview and parameters. *Nature Reviews Molecular Cell Biology* 6, 150–166.

Hevel, J.M. and Marletta, M.A. (1992) Macrophage nitric oxide synthase: relationship between enzyme-bound tetrahydrobiopterin and synthase activity. *Biochemistry* 31, 7160–7165.

Horan, S., Bourges, I. and Meunier, B. (2006) Transcriptional response to nitrosative stress in *Saccharomyces cerevisiae*. *Yeast* 23, 519–535.

Hromatka, B.S., Noble, S.M. and Johnson, A.D. (2005) Transcriptional response of *Candida albicans* to nitric oxide and the role of the *YHB1* gene in nitrosative stress and virulence. *Molecular Biology of the Cell* 16, 4814–4826.

Ischiropoulos, H. (2003) Biological selectivity and functional aspects of protein tyrosine nitration. *Biochemical and Biophysical Research Communications* 305, 776–783.

Jobgen, W.S., Fried, S.K., Fu, W.J., Meininger, C.J. and Wu, G. (2006) Regulatory role for the arginine-nitric oxide pathway in metabolism of energy substrates. *The Journal of Nutritional Biochemistry* 17, 571–588.

Kanadia, R.N., Kuo, W.N., McNabb, M. and Botchway, A. (1998) Constitutive nitric oxide synthase in *Saccharomyces cerevisiae*. *Biochemistry and Molecular Biology International* 45, 1081–1087.

Kig, C. and Temizkan, G. (2009) Nitric oxide as a signaling molecule in the fission yeast *Schizosaccharomyces pombe*. *Protoplasma* 238, 59–66.

Lee, P.Y., Bae, K.H., Jeong, D.G., Chi, S.W., Moon, J.H., Kang, S., Cho, S., Lee, S.C., Park, B.C. and Park, S.G. (2011) The *S*-nitrosylation of glyceraldehyde-3-phosphate dehydrogenase 2 is reduced by interaction with glutathione peroxidase 3 in *Saccharomyces cerevisiae*. *Molecules and Cells* 31, 255–259.

Lewinska, A. and Bartosz, G. (2006) Yeast flavohemoglobin protects against nitrosative stress and controls ferric reductase activity. *Redox Report* 11, 231–239.

Lewinska, A., Grzelak, A. and Bartosz, G. (2008) Application of a *YHB1*-GFP reporter to detect nitrosative stress in yeast. *Redox Report* 13, 161–171.

Lewinska, A., Macierzynska, E., Grzelak, A. and Bartosz, G. (2011) A genetic analysis of nitric oxide-mediated signaling during chronological aging in the yeast. *Biogerontology* 12, 309–320.

Li, B., Skinner, C., Castello, P.R., Kato, M., Easlon, E., Xie, L., Li, T., Lu, S.P., Wang, C., Tsang, F. *et al.* (2011) Identification of potential calorie restriction-mimicking yeast mutants with increased mitochondrial respiratory chain and nitric oxide levels. *Journal of Aging Research* 2011: 673185.

Lushchak, O.V., Inoue, Y. and Lushchak, V.I. (2010) Regulatory protein Yap1 is involved in response of yeast *Saccharomyces cerevisiae* to nitrosative stress. *Biochemistry (Moscow)* 75, 629–664.

Madeo, F., Herker, E., Maldener, C., Wissing, S., Lächelt, S., Herlan, M., Fehr, M., Lauber, K., Sigrist, S.J., Wesselborg, S. and Fröhlich, K.U. (2002) A caspase-related protease regulates apoptosis in yeast. *Molecular Cell* 9, 911–917.

Matsuda, H. and Iyanagi, T. (1999) Calmodulin activates intramolecular electron transfer between the two flavins of neuronal nitric oxide synthase flavin domain. *Biochimica et Biophysica Acta* 1473, 345–355.

McMillan, K., Bredt, D.S., Hirsch, D.J., Snyder, S.H., Clark, J.E. and Masters, B.S. (1992) Cloned, expressed rat cerebellar nitric oxide synthase contains stoichiometric amounts of heme, which binds carbon monoxide. *Proceedings of the National Academy of Sciences of the United States of America* 89, 11141–11145.

Mesquita, A., Weinberger, M., Silva, A., Sampaio-Marques, B., Almeida, B., Leão, C., Costa, V., Rodrigues, F., Burhans, W.C. and Ludovico, P. (2010) Caloric restriction or catalase inactivation extends yeast chronological lifespan by inducing H_2O_2 and superoxide dismutase activity *Proceedings of the National Academy of Sciences of the United States of America* 107, 15123–15128.

Missall, T.A., Lodge, J.K. and McEwen, J.E. (2004) Mechanisms of resistance to oxidative and nitrosative stress: implications for fungal survival in mammalian hosts. *Eukaryotic Cell* 3, 835–846.

Moncada, S., Palmer, R.M. and Higgs, E.A. (1991) Nitric oxide: physiology, pathophysiology, and pharmacology. *Pharmacological Reviews* 43, 109–142.

Nasuno, R., Aitoku, M., Manago, Y., Nishimura, A., Sasano, Y. and Takagi, H. (2014) Nitric oxide-mediated antioxidative mechanism in yeast through the activation of the transcription factor Mac1. *PloS ONE* 9(11): e113788.

Nathan, C. and Shiloh, M.U. (2000) Reactive oxygen and nitrogen intermediates in the relationship between mammalian hosts and microbial pathogens. *Proceedings of the National Academy of Sciences of the United States of America* 97, 8841–8848.

Nishimura, A., Kotani, T., Sasano, Y. and Takagi, H. (2010) An antioxidative mechanism mediated by the yeast *N*-acetyltransferase Mpr1: oxidative stress-induced arginine synthesis and its physiological role. *FEMS Yeast Research* 10, 687–698.

Nishimura, A., Kawahara, N. and Takagi, H. (2013) The flavoprotein Tah18-dependent NO synthesis confers high-temperature stress tolerance on yeast cells., *Biochemical and Biophysical Research Communications* 430, 137–143.

Osório, N.S., Carvalho, A., Almeida, A.J., Padilla-Lopez, S., Leão, C., Laranjinha, J., Ludovico, P., Pearce, D.A. and Rodrigues, F. (2007) Nitric oxide signaling is disrupted in the yeast model for Batten disease. *Molecular Biology of the Cell* 18, 2755–2767.

Sarver, A. and DeRisi, J. (2005) Fzf1p regulates an inducible response to nitrosative stress in *Saccharomyces cerevisiae*. *Molecular Biology of the Cell* 16, 4781–4791.

Sasano, Y., Haitani, Y., Hashida, K., Ohtsu, I., Shima, J. and Takagi, H. (2012) Enhancement of the proline and nitric oxide synthetic pathway improves fermentation ability under multiple baking-associated stress conditions in industrial baker's yeast. *Microb Cell Factories* 11: 40.

Shinyashiki, M., Chiang, K.T., Switzer, C.H., Gralla, E.B., Valentine, J.S., Thiele, D.J. and Fukuto, J.M. (2000) The interaction of nitric oxide (NO) with the yeast transcription factor Ace1: a model system for NO-protein thiol interactions with implications to metal metabolism. *Proceedings of the National Academy of Sciences of the United States of America* 97, 2491–2496.

Shinyashiki, M., Pan, C.J., Lopez, B.E. and Fukuto, J.M. (2004) Inhibition of the yeast metal reductase heme protein Fre1 by nitric oxide (NO): a model for inhibition of NADPH oxidase by NO. *Free Radical Biology and Medicine* 37, 713–723.

Siddhanta, U., Wu, C., Abu-Soud, H.M., Zhang, J., Ghosh, D.K. and Stuehr, D.J. (1996) Heme iron reduction and catalysis by a nitric oxide synthase heterodimer containing one reductase and two oxygenase domains. *The Journal of Biological Chemistry* 271, 7309–7312.

Suarez, M.F., Filonova, L.H., Smertenko, A., Savenkov, E.I., Clapham, D.H., von Arnold, S., Zhivotovsky, B. and Bozhkov, P.V. (2004) Metacaspase-dependent programmed cell death is essential for plant embryogenesis. *Current Biology* 14, R339–R340.

Sun, J., Steenbergen, C. and Murphy, E. (2006) *S*-nitrosylation: NO-related redox signaling to protect against oxidative stress. *Antioxidants and Redox Signaling* 8, 1693–1705.

Thiele, D.J. (1988) *ACE1* regulates expression of the *Saccharomyces cerevisiae* metallothionein gene. *Molecular and Cellular Biology* 8, 2745–2752.

Thomas, S.R., Chen, K. and Keaney, J.F. Jr (2002) Hydrogen peroxide activates endothelial nitric-oxide synthase through coordinated phosphorylation and dephosphorylation via a phosphoinositide 3-kinase-dependent signaling pathway. *The Journal of Biological Chemistry* 277, 6017–6024.

Tillmann, A., Gow, N.A. and Brown, A.J. (2011) Nitric oxide and nitrosative stress tolerance in yeast. *Biochemical Society Transactions* 39, 219–223.

Tkach, J.M., Yimit, A., Lee, A.Y., Riffle, M., Costanzo, M., Jaschob, D., Hendry, J.A., Ou, J., Moffat, J., Boone, C. *et al.* (2012) Dissecting DNA damage response pathways by analysing protein localization and abundance changes during DNA replication stress. *Nature Cell Biology* 14, 966–976.

Weinberger, M., Sampaio-Marques, B., Ludovico, P. and Burhans, W.C. (2013) DNA replication stress-induced loss of reproductive capacity in S. cerevisiae and its inhibition by caloric restriction, *Cell Cycle*, 12 (8), 1189–200.

10 Hydroxyproline Metabolism in Microorganisms

S. Watanabe*
Department of Bioscience, Graduate School of Agriculture, Ehime University, Matsuyama, Japan

10.1 Abstract

L-Hydroxyproline (L-Hyp) has been detected in certain proteins, particularly collagen, as well as in some peptide antibiotics. In mammalian systems, the L-proline residue in procollagen is post-translationally hydroxylated to *trans*-4-hydroxy-L-proline (T4LHyp) or *trans*-3-hydroxy-L-proline (T3LHyp). Free T4LHyp and T3LHyp, produced by the degradation of collagen, are then converted into pyruvate and glyoxylate, and L-proline, respectively, through different pathways. In microorganisms, the metabolic pathways of T4LHyp, T3LHyp, *cis*-3-hydroxy-L-proline (C3LHyp) and T4LHyp betaine (T4LHyp-B) have recently been examined at the molecular level. In the T4LHyp pathway from bacteria, T4LHyp is converted into α-ketoglutarate through the stepwise actions of four enzymes: hydroxyproline 2-epimerase, D-hydroxyproline dehydrogenase, Δ^1-pyrroline-4-hydroxy-2-carboxylate deaminase and α-ketoglutaric semialdehyde dehydrogenase. There are two types of D-hydroxyproline dehydrogenase, one with a homomeric structure and the other with a heteromeric structure. Thus, mammals and bacteria catabolize L-Hyp through different pathways.

In contrast, the metabolism of T3LHyp may be common between mammals, bacteria and archaea. This involves T3LHyp dehydratase and Δ^1-pyrroline-2-carboxylate (Pyr2C) reductase, through which T3LHyp is converted into L-proline. T3LHyp dehydratase belongs to the same protein family as hydroxyproline 2-epimerase, and their different catalytic reactions depend on a single amino acid substitution in the active sites. Pyr2C reductase is a novel member of the ornithine cyclodeaminase (OCD) family: the archetypical OCD synthesizes L-proline in a single step by deaminating L-ornithine. The C3LHyp pathway from bacteria contains two enzymes, C3LHyp dehydratase and Pyr2C reductase, through which C3LHyp is metabolized to L-proline via Pyr2C. However, these enzymes are not evolutionarily related to the T3LHyp dehydratase or Pyr2C reductase that are involved the metabolism of T3LHyp. T4LHyp-B ultimately enters the bacterial T4LHyp pathway as T4LHyp or C4DHyp, and T4LHyp-B 2-epimerase is homologous to C3LHyp dehydratase.

An enzymatic method for the detection of L-Hyp (collagen)-containing biological samples is currently being developed as an alternative to an amino acid analyser based on the high substrate specificities of several enzyme(s) and is also described in this chapter.

*E-mail address: irab@agr.ehime-u.ac.jp

10.2 Introduction

Hydroxyproline (L-Hyp), one of the major compounds analogous to proline, is a non-essential amino acid. Among its eight stereoisomers (see Fig. 10.1), several L-hydroxyproline (L-Hyp) stereoisomers have been detected in certain proteins, particularly collagen. Approximately 25% of the amino acid residues in mammalian collagens are L-proline, with nearly 40% of these being post-translationally converted into *trans*-4-hydroxy-L-proline (T4LHyp) and *trans*-3-hydroxy-L-proline (T3LHyp). T4LHyp is synthesized by the action of prolyl 4-hydroxylase (EC 1.14.11.2), which is both α-ketoglutarate and Fe(II)-dependent. T3LHyp is synthesized via prolyl 3-hydroxylase (EC 1.14.11.7). As free L-proline is not accepted as a substrate of these enzymes, free L-Hyp in the serum and urine is provided by the degradation of collagen through collagenase, a matrix metalloproteinase. The direct conversion of free L-proline to T4LHyp (Shibasaki *et al.*, 1999), T3LHyp (Petersen, *et al.*, 2003), *cis*-4-hydroxy-L-proline (C4LHyp) (Hara and Kino, 2009) and *cis*-3-hydroxy-L-proline (C3LHyp) (Mori *et al.*, 1997) has been identified in a few bacteria and fungi. One of the physiological roles of the hydroxylases concerned may be in the biosynthesis of peptide antibiotics containing L-Hyp, such as microcolin A (C4LHyp), telomycin (T3LHyp and C3LHyp) and pneumocandins (T4LHyp and T3LHyp). The catabolism of L-Hyp (by (micro)organisms), and also that of T4LHyp, T3LHyp and C3LHyp, has recently been investigated at the molecular level.

10.3 Metabolic Pathway of *trans*-4-Hydroxy-L-Proline (T4LHyp) in Mammals

In the mammalian pathway (Cooper *et al.*, 2008; Riedel *et al.*, 2011), free T4LHyp is metabolized to pyruvate and glyoxylate via the intermediates Δ^1-pyrroline-3-hydroxy-5-carboxylate, 4-hydroxyglutamate and 4-hydroxy-2-oxoglutarate, through the stepwise actions of four mitochondrial enzymes: L-Hyp oxidase, Δ^1-pyrroline-5-carboxylate dehydrogenase (EC 1.5.1.12), aspartate aminotransferase (EC 2.6.1.1) and 4-hydroxy-2-oxoglutarate aldolase (EC 4.1.3.16). This pathway is partially linked to the general L-proline pathway: L-Hyp oxidase is the same as L-proline dehydrogenase (L-ProDH), and Δ^1-pyrroline-5-carboxylate reductase has the ability to reduce either Δ^1-pyrroline-5-carboxylate or Δ^1-pyrroline-3-hydroxy-5-carboxylate.

10.4 Metabolic Pathway of T4LHyp in Bacteria

In contrast to mammals, bacteria metabolize free T4LHyp to α-ketoglutarate through four enzymatic steps via *cis*-4-hydroxy-D-proline (C4DHyp), Δ^1-pyrroline-4-hydroxy-2-carboxylate (Pyr4H2C) and α-ketoglutaric semialdehyde (αKGSA) intermediates. Therefore, bacteria with this pathway have the ability to grow using not only T4LHyp, but also C4DHyp as their sole carbon source (Fig. 10.1). The genes encoding these four enzymes are often clustered together with those encoding the putative L-Hyp transporter

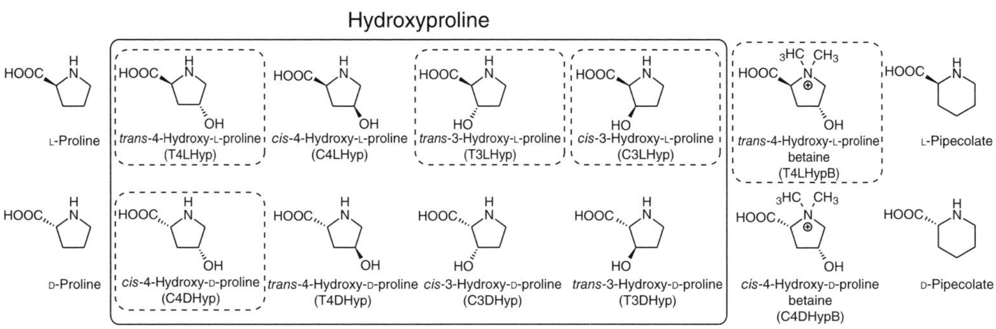

Fig. 10.1. Chemotypes of hydroxyproline analogues and related compounds. The metabolic pathways in microorganisms of the compounds shown in the dashed boxes are described in the chapter.

on bacterial genomes (referred to as the T4LHyp gene cluster) (Fig. 10.2).

10.4.1 Hydroxyproline 2-epimerase

Hydroxyproline 2-epimerase (HypE; EC 5.1.1.8), which catalyses the isomerization of T4LHyp to C4DHyp, is a member of the pyridoxal 5′-phosphate (PLP)-independent proline racemase superfamily (pfam05544), along with the archetypical proline racemase (ProR; EC 5.1.1.4), and differs from other amino acid racemases such as alanine racemase. ProR is involved in the Stickland reaction (fermentation) in certain clostridia, as well as in the mechanisms underlying the escape of parasites in eukaryotic *Trypanosoma* from host immunity. The stereoinversion of the chiral Cα centre in hydroxyproline epimerization and proline racemization involves the abstraction of a proton from the carbon atom and reprotonation of the resulting 'carbanion intermediate' in the opposite stereochemical sense using two general acidic/basic cysteine residues (C-C type) (Buschiazzo *et al.*, 2006). The substrate specificities of HypE and ProR are generally very strict: T4LHyp and C4LHyp for HypE, and proline for ProR. Zhao *et al.* (2014) characterized 51 HypE-like proteins from bacteria, and the k_{cat}/K_m value for the T4LHyp of HypE from *Pseudomonas putida* KT2440, for example, was 3430- and 192-fold higher than the values for L-proline and T3LHyp, respectively. This strict substrate specificity of HypE has the potential to detect T4LHyp enzymatically (see Section 10.10).

10.4.2 D-Hydroxyproline dehydrogenase

D-Hydroxyproline dehydrogenase (D-HypDH; EC 1.4.99.–), a so-called 'dye-linked flavin-containing

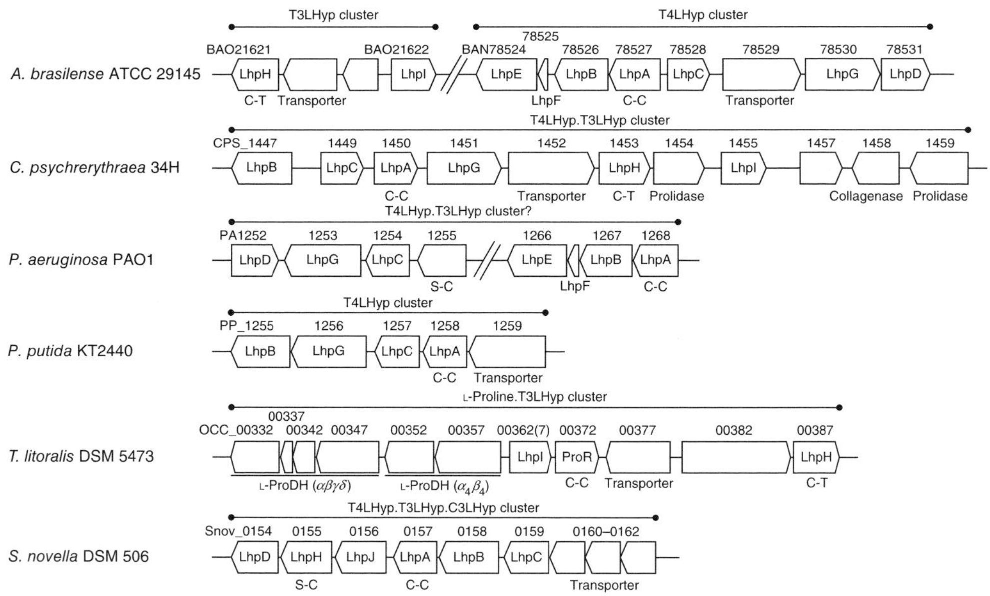

Fig. 10.2. Schematic gene clusters related to the L-hydroxyproline (L-Hyp) metabolism of microorganisms: *Azospirillum brasilense* ATCC 29145; *Colwellia psychrerythraea* 34H; *Pseudomonas aeruginosa* PAO1; *P. putida* KT2440; *Thermococcus litoralis* DSM 5473; and *Starkeya novella* DSM 506. LhpA, LhpB (or LhpB, LhpE and LhpF), LhpC, LhpD, LhpG, LhpH and LhpI encode HypE (hydroxyproline 2-epimerase), D-HypDH (D-hydroxyproline dehydrogenase), Pyr4H2C (Δ^1-pyrroline-4-hydroxy-2-carboxylate) deaminase, Pyr2CR (Δ^1-pyrroline-2-carboxylate reductase, DpkA-type), αKGSADH (α-ketoglutaric semialdehyde dehydrogenase), T3LHypD (*trans*-3-hydroxy-L-proline dehydratase) and Pyr2CR (Δ^1-pyrroline-2-carboxylate reductase, OCD-type), respectively. C-C (two cysteine residues), C-T (cysteine and threonine residues) and S-C (serine and cysteine) indicate a pair of catalytic amino acid residues of the ProR (proline racemase) superfamily enzymes. See text for further details.

dehydrogenase', has the ability to utilize artificial electron acceptors that include 2,6-dichloroindophenol (Bater *et al.*, 1977). Such enzymes are frequently associated with cell and/or organelle membranes and are unstable in solution, which has made their purification, preservation and characterization very difficult. In contrast, the gene(s) encoding the enzyme that is annotated as 'D-amino acid oxidase' has been located within the flanking regions of the HypE and αKGSADH (αKGSA dehydrogenase) genes (see Section 10.4.4) on bacterial chromosome DNA, and the subsequent characterization of the recombinant protein(s) has established its function as that of D-HypDH (Watanabe *et al.*, 2007, 2012).

There are two different types of D-HypDH, one with a heteromeric structure and the other with a homomeric structure, and both of these are completely different from the bacterial L-ProDH (EC 1.5.5.2), which is known as PutA (proline utilization A). *P. putida* KT2440 and *Sinorhizobium meliloti* 1021 possess the homomeric type of enzyme, in which only FAD is contained as a prosthetic group. The heteromeric enzyme is found in *Pseudomonas aeruginosa* PAO1 and *Azospirillum brasilense* and this has a novel heterododecameric structure consisting of three different subunits ($\alpha_4\beta_4\gamma_4$), two FAD, FMN and a (2Fe-2S)-iron sulfur cluster contained in the $\alpha\beta\gamma$ of the heterotrimeric unit. The β-subunit (containing only FAD) exhibits full dehydrogenase activity by itself, but assembly with the other subunit(s) improves the kinetics of C4DHyp (and its thermostability) (Watanabe *et al.*, 2015b). In spite of these differences, the two types of D-HypDH show similar high specificities for C4DHyp: other substrates include only *trans*-4-hydroxy-D-proline and D-proline (see Fig. 10.1), and their k_{cat}/K_m values are more than 100-fold lower than that of C4DHyp. *p*-Iodonitrotetrazolium violet (INT) and nitroblue tetrazolium (NBT), together with phenazine methosulfate (PMS) (an electron-transfer intermediate), are available artificial electron acceptors, but 2,6-dichloroindophenol (Cl2Ind), ferricyanide, cytochrome *c* and $NAD(P)^+$ are not.

The β-subunit of heteromeric D-HypDH (D-HypDHβ) belongs to the D-amino acid oxidase superfamily (pfam01266). This superfamily contains several D-amino acid oxidases (EC 1.4.3.3), including homomeric D-HypDH and sarcosine oxidase (EC 1.5.3.1), but a poor phylogenetic relationship exists between them, suggesting that the T4LHyp pathway evolved in a convergent manner in different bacterial species. A relatively high homology with D-HypDHβ has been reported in hydrogen cyanide synthase (Hcn; EC 1.4.99.5) and opine dehydrogenase (OpnDH) from bacteria, and heteromeric L-ProDH from archaea. The L-ProDH of *Thermococcus profundus* has the heterotetrameric structure of $\alpha\beta\gamma\delta$ containing 2 FAD (in the α- and β-subunits), the (2Fe-2S)-iron sulfur cluster (α) and an (8Fe-8S)-iron sulfur cluster (γ), whereas the enzyme of *Pyrococcus horikoshii* (isozyme 1) has the heterooctameric structure of $\alpha_4\beta_4$ containing 1 FAD (β), 1 FMN (between the α- and β-subunits) and 1 ATP (α) (Kawakami *et al.*, 2012). Although the subunit assembly and cofactor components of OpnDH (and probably also Hcn) are the same as those of D-HypDH (Watanabe *et al.*, 2015c), their sequential homology is not typically high, suggesting the acquisition of substrate specificity after segmentation. The γ-subunit of these enzymes corresponds to the *N*-terminal ~120 amino acid residues of the α-subunit of archaeal L-ProDH.

10.4.3 Δ1-Pyrroline-4-hydroxy-2-carboxylate (Pyr4H2C) deaminase

A potential reaction product of D-HypDH is Δ^1-pyrroline-4-hydroxy-2-carboxylate (Pyr4H2C), but the spontaneously hydrolysed 2-oxo-4-hydroxy-5-aminovalerate may be a substrate for Pyr4H2C deaminase (EC 3.5.4.22). Even though the enzyme from *Pseudomonas striata* was partially characterized more than 50 years ago (Singh and Adams, 1965), the corresponding gene has not yet been isolated. The dihydrodipicolinate synthase/*N*-acetylneuraminate lyase (DHDPS/NAL) protein superfamily (COG0329) includes the archetypical DHDPS (EC 4.2.1.52), NAL (EC 4.1.3.3), D-2-keto-3-deoxygluconate aldolase (EC 4.1.2.51), D-5-keto-4-deoxyglucarate (D-KDG) dehydratase (decarboxylating) (EC 4.2.1.41), *trans-o*-hydroxybenzylidenepyruvate hydrolase-aldolase (EC 4.1.2.45), *trans*-2′-carboxybenzalpyruvate hydratase-aldolase (EC 4.1.2.34), 4-hydroxy-2-oxoglutarate aldolase (EC 4.1.3.16) and L-2-keto-3-deoxyarabinonate (L-KDA) dehydratase (EC 4.2.1.43). 4-Hydroxy-2-oxoglutarate aldolase is involved in the metabolism of T4LHyp by mammals (see Section 10.3).

It is noteworthy that while most of these enzymes produce pyruvate and several aldehydes in the aldol-cleavage reaction, the reaction involving L-KDA dehydratase consists of only dehydration, without the cleavage of C–C or C=C bonds, in order to produce αKGSA (Fig. 10.3). Based on this insight, a putative DHDPS/NAL-like gene, located within the flanking regions of the HypE, αKGSADH and D-HypDH genes, was first identified as Pyr4H2C deaminase (Watanabe *et al.*, 2012). Members of the DHDPS/NAL family generally possess a common $(\beta/\alpha)_8$ barrel fold and share a common reaction step in their reactions, namely, the formation of a Schiff base intermediate between a strictly conserved lysine residue and the C2 carbon of the common α-keto acid moiety of the substrate. The phenolic hydroxyl group of the tightly conserved tyrosine residue forms double-hydrogen bonds with the carboxyl and hydroxyl groups at position 4 of the substrate. In spite of the lack of aldolase activity, these two residues are completely conserved in

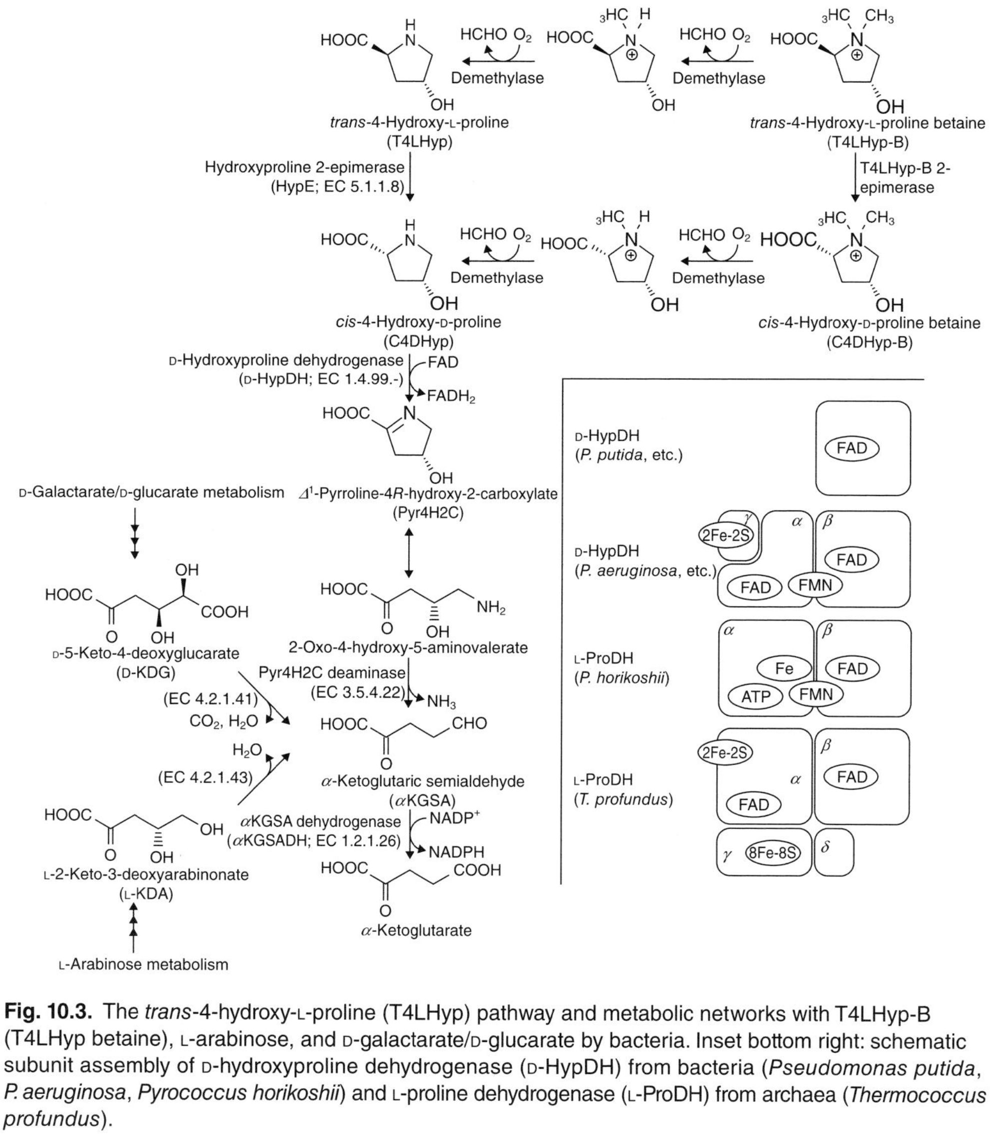

Fig. 10.3. The *trans*-4-hydroxy-L-proline (T4LHyp) pathway and metabolic networks with T4LHyp-B (T4LHyp betaine), L-arabinose, and D-galactarate/D-glucarate by bacteria. Inset bottom right: schematic subunit assembly of D-hydroxyproline dehydrogenase (D-HypDH) from bacteria (*Pseudomonas putida*, *P. aeruginosa*, *Pyrococcus horikoshii*) and L-proline dehydrogenase (L-ProDH) from archaea (*Thermococcus profundus*).

Pyr4H2C deaminase. A treatment with sodium borohydride ($NaBH_4$) in the presence of pyruvate, a competitive inhibitor, leads to irreversible inactivation and covalent binding between this lysine residue and pyruvate, suggesting that at least a Schiff base intermediate is involved in the catalytic mechanism.

10.4.4 α-Ketoglutaric semialdehyde dehydrogenase

The last step in the metabolic pathway of T4lHyp in bacteria involves the NAD(P)$^+$-dependent dehydrogenation of αKGSA to α-ketoglutarate by αKGSA dehydrogenase (αKGSADH; EC 1.2.1.26). αKGSADH is involved in two other metabolic pathways in which αKGSA is provided from different precursors (Watanabe *et al.*, 2006) (Fig. 10.3). In the L-arabinose and D-galactarate/D-glucarate pathways in bacteria, αKGSA is provided by the enzymatic dehydration of L-KDA and the decarboxylating dehydration of D-5-keto-4-deoxyglucarate (D-KDG) by L-KDA dehydratase and D-KDG dehydratase, respectively. Pyr4H2C deaminase, L-KDA dehydratase and D-KDG dehydratase belong to the same protein family (see Section 10.4.3). *A. brasilense* is a unique bacterium that has three different αKGSADH isoenzymes induced by T4LHyp, D-galactarate/D-glucarate or L-arabinose (Watanabe *et al.*, 2007). Although these commonly belong to the aldehyde dehydrogenase superfamily (pfam00171), the two isoenzymes induced by T4LHyp or D-galactarate/D-glucarate are closely related to each other (they have 64% identity) and show similar high substrate specificities for αKGSA: their activities with a number of aliphatic aldehydes from C_3 to C_8 are only 30% less than those with αKGSA.

10.5 Metabolic Pathway of *trans*-3-Hydroxy-L-Proline (T3LHyp)

Though its degradation by organism(s) has been poorly understood for a long period of time, a human C14orf149 protein has been reported to catalyse the dehydration of T3LHyp to Δ^1-pyrroline-2-carboxylate (Pyr2C) (Visser *et al.*, 2012), and *A. brasilense* has been shown to possess the ability to grow using not only T4LHyp but also T3LHyp as the sole carbon source (Watanabe *et al.*, 2014a). In contrast to T4LHyp, organisms may commonly metabolize T3LHyp to L-proline through two enzymatic steps via a Pyr2C intermediate (Fig. 10.4), and metabolic genes (enzymes) are often clustered together with gene(s) encoding putative amino acid transporters and/or L-ProDH on the genomes of bacteria and archaea (referred to as the T3LHyp gene cluster) (Fig. 10.2). In the case of *A. brasilense*, the T4LHyp and T3LHyp gene clusters are separately located on the genome, and neither L-Hyp has any significant impact on the transcriptional induction of the other. Conversely, *P. aeruginosa* PAO1 and *Colwellia psychrerythraea* 34H may possess a single gene cluster related to the metabolism of both T4LHyp and T3LHyp.

10.5.1 T3LHyp dehydratase

T3LHyp dehydratase (T3LHypD; EC 4.2.1.77) catalyses the dehydration of T3LHyp to Pyr2C. In spite of differences in their reactions, the enzyme belongs to the same ProR superfamily as HypE and ProR, and possesses the catalytic residues of cysteine and threonine (C-T type), which differ from those of ProR and HypE (C-C type) (see Section 10.4.1) (Visser *et al.*, 2012). The conversion of T3LHyp to Pyr2C has been suggested to occur via the abstraction of a hydrogen from the Cα of the substrate, followed by the spontaneous hydrolysis and tautomerization of the Δ^2-pyrroline-2-carboxylate intermediate product. The substitution of threonine by cysteine in the human C14orf149 protein leads to the acquisition of similar epimerization activity to that of T3LHyp, suggesting that the activities of the ProR superfamily of enzymes are dictated by a single residue. In the phylogenetic tree of the ProR superfamily, the T3LHypD enzymes from all domains of life form a distinct subclass from ProR and HypE, indicating that dehydratase activity with T3LHyp in this superfamily was acquired at an early evolutionary stage (possibly by a single mutation in the active site, as described above).

10.5.2 Δ1-Pyrroline-2-carboxylate (Pyr2C) reductase

In *Pseudomonas* strains including *P. putida* (not capable of growing on T3LHyp), D-lysine is

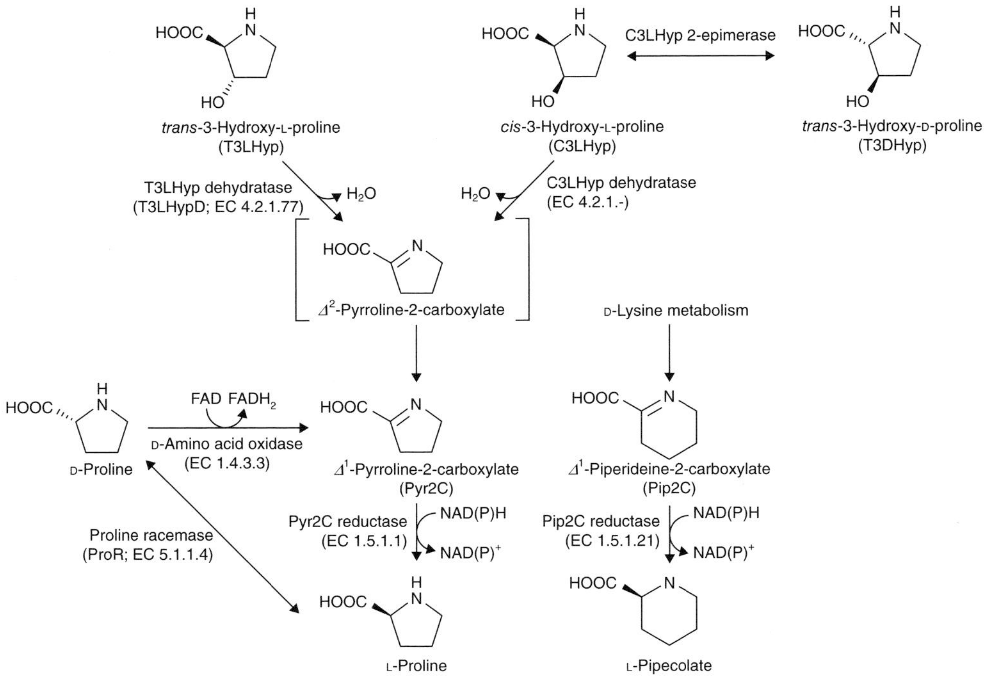

Fig. 10.4. The *trans*-3-hydroxy-L-proline (T3LHyp) pathway and metabolic networks with *cis*-3-hydroxy-L-proline (C3LHyp), D-proline and D-lysine in microorganisms.

metabolized to α-ketoadipate through the so-called 'L-pipecolate pathway', in which the NADPH-dependent DpkA protein catalyses the second step (the conversion of Δ^1-piperidine-2-carboxylate (Pip2C) to L-pipecolate) as a Pip2C reductase (EC 1.5.1.21) (Muramatsu *et al.*, 2005) (Fig. 10.4). DpkA also functions as a Pyr2C reductase (EC 1.5.1.1; Pyr2CR) involving (a hypothetical) D-proline, in which Pyr2C may be produced from D-proline by (an unknown) D-amino acid oxidase.

Although *A. brasilense* possesses an orthologue of the DpkA protein with strict NADPH-dependent Pyr2CR activity, the corresponding gene, located within the T4LHyp gene cluster, is not upregulated by T3LHyp. In the genomes of bacteria (*A. brasilense* and *C. psychrerythraea*) and archaea (*Thermococcus litoralis*), the T3LHypD gene is frequently clustered together with a function-unknown protein (gene) annotated as an ornithine cyclodeaminase (OCD; EC 4.3.1.12) (Fig. 10.2): OCD catalyses an unusual cyclization to convert L-ornithine to L-proline via a Pyr2C intermediate, with the release of ammonia. These recombinant proteins exhibit NAD(P)H-dependent reduction, but no OCD activities towards Pyr2C (Watanabe *et al.*, 2014a,b).

The three Pyr2CR enzymes from *A. brasilense*, *C. psychrerythraea* and *T. litoralis* (referred to as AbPyr2CR, CpPyr2CR and TlPyr2CR, respectively) display several different properties. There is no preference for co-enzyme utilization between NADPH and NADH in AbPyr2CR, whereas CpPyr2CR and TlPyr2CR are NADP(H) preferring and strictly NAD(H)-dependent enzymes, respectively. Furthermore, AbPyr2CR and TlPyr2CR utilize Pip2C (and L-pipecolate) in a similar manner to Pyr2C (L-proline). Disruption of the *AbPyr2CR* gene eliminates the ability to grow on not only on T3LHyp, but also on D-proline and D-lysine, suggesting a physiological function(s) as a bifunctional Pyr2C/Pip2C reductase. In addition, only TlPyr2CR exhibits the two additional enzymatic activities of L-alanine dehydrogenase (EC 1.4.1.1) in the reaction L-alanine + NAD^+ + H_2O $\leftrightarrow$ pyruvate + NH_3 + NADH + H^+, and of *N*-methyl-L-alanine dehydrogenase (EC 1.4.1.17) in the reaction *N*-methyl-L-alanine + NAD^+ + H_2O $\leftrightarrow$ pyruvate + methylamine + NADH + H^+.

Pyr2CR belongs to the ornithine cyclodeaminase/μ-crystallin (OCD/CRYM) superfamily (pfam02423), which consists of several subclasses, including the archetype OCD, CRYM/ketimine reductase and L-alanine dehydrogenase. Of these, μ-crystallin (CRYM) from mammals, known as an NADPH-dependent T3 thyroid hormone (triiodothyronine)-binding protein without enzymatic catalysis, was recently shown to function as an NAD(P)H-dependent ketimine reductase (EC 1.5.1.25) (Hallen *et al.*, 2011). Pyr2C is one of the substrates of this enzyme, suggesting the potential involvement of the metabolism of T3LHyp together with the C14orf149 protein (see Section 10.5.1). L-Alanine dehydrogenase is only found in archaea (Schröder *et al.*, 2004) and is derived from well-known bacterial enzymes, exhibiting the same activity as found in TlPyr2CR.

10.6 Metabolic Pathway of *cis*-3-Hydroxy-L-Proline (C3LHyp)

C3LHyp is found in some peptide antibiotics, such as telomycin, but as its amount in nature is markedly less than those of T4LHyp and T3LHyp, the possibility of catabolism by (micro) organisms had not previously been considered. However, Zhang *et al.* (2015) recently reported that *Labrenzia aggregata* IAM 12614, a marine *Agrobacterium* species, had the ability to grow on C3LHyp as the sole carbon source, and that C3LHyp was converted into L-proline via Pyr2C through the activities of C3LHyp dehydratase and Pyr2CR (see Fig. 10.4).

10.6.1 C3LHyp dehydratase

The Snov_0156 protein from *L. aggregata* IAM 12614 belongs to the *cis,cis*-muconate cycloisomerase (MLE) subclass of the enolase superfamily, which also contains the 2-epimerase for T4LHyp-B (see Section 10.7). The corresponding gene is clustered together with (putative) HypE, D-HypDH, Pyr4H2C deaminase and T3LHypD genes on the bacterial genome (Fig. 10.2). This protein uses only C3LHyp as a substrate, and catalyses (i) reversible 2-epimerization to T3DHyp and (ii) irreversible dehydration to Pyr2C, though the physiological role of the former reaction currently remains unclear (Zhang *et al.*, 2015) (Fig. 10.4). Both reactions are initiated by the abstraction of the α-proton of the substrate, similar to the action of T3LHypD, and the resulting intermediate is an enolate anion stabilized by Mg^{2+}, but not by a carbanion. *L. aggregata* IAM 12614 has the ability to grow on C3LHyp as the sole carbon source, and transcription of the gene cluster containing the Snov_0156 gene is significantly upregulated not only by C3LHyp, but also by T4LHyp and T3LHyp, suggesting physiological role(s) in hydroxyproline metabolism.

10.6.2 Pyr2C reductase

Although the metabolic fate of the Pyr2C intermediate may be its reduction to L-proline, as well as T3LHyp, an orthologue of the *dpkA* gene (but not the OCD-type gene) is contained in the C3LHyp gene cluster of *L. aggregata* IAM 12614. Thus, this gene (enzyme) functions as a Pyr2CR involved not only in the metabolism of C3LHyp, but also in that of T3LHyp.

10.7 T4LHyp Betaine Metabolism

trans-4-Hydroxy-L-proline betaine (T4LHyp-B) is one of the quaternary amine derivatives of amino acids that often serve as osmoprotectants, and the bacterial metabolic pathway is (partially) linked to that of T4LHyp in bacteria (see Section 10.4) (Zhao *et al.*, 2013). T4LHyp-B is commonly converted into C4DHyp through two parallel routes, and its metabolic fate is the same as that of T4LHyp from bacteria (see Fig. 10.3). In one route, in *Pelagibaca bermudensis* HTCC2601, a marine bacterium, T4LHyp-B undergoes two *N*-demethylation reactions to T4LHyp, which is subsequently epimerized to C4DHyp by HypE. Another route, in *Paracoccus denitrificans* PD1222, is initiated by the epimerization of T4LHyp-B to C4DHyp betaine, followed by two *N*-demethylation reactions to C4DHyp. The genes encoding these enzymes related to the metabolism of T4LHyp-B and T4LHyp are clustered on the genomes of *Pelagibaca* species. Proline betaine (Pro-B) is also an active substrate for

T4LHyp-B 2-epimerase (Kumar *et al.*, 2014). Among the several structural analogues of proline, T4LHyp-B (Pro-B) strongly resembles T4LHyp (proline) (see Fig. 10.1), whereas the 2-epimerase is homologous to C3LHyp epimerase/dehydratase (see Section 10.6.1), but not to HypE (and ProR).

10.8 Hydroxyproline Metabolism in Archaea

The metabolism of T3LHyp in archaea is similar to that in bacteria (and mammals) (see Section 10.5). The substrate specificities of HypE and ProR were previously considered to be very strict, and no racemase/epimerase has been detected in archaea. However, the putative function of another ProR-like gene (OCC_00372) contained in the T3LHyp gene cluster of *T. litoralis* is not the dehydration of T3LHyp because of the two cysteine residues at the active sites (Watanabe *et al.*, 2014b). The OCC_00372 protein has the ability to catalyse the stereo conversion not only of proline, but also of 4-hydroxyproline and 3-hydroxyproline (Watanabe *et al.*, 2015a). In spite of significant HypE activity, the IC_{50} value for pyrrole-2-carboxylate, a transition state analogue of proline, is ~28-fold higher than that of native HypE (from *A. brasilense*), so confirming dual activity towards proline and hydroxyproline(s).

Among several (putative) ligand-binding sites, the OCC_00372 protein possesses a tryptophan residue at position 241, which is substituted for phenylalanine in all known ProR. The W241F mutant shows a significantly stronger preference for proline than for hydroxyproline, suggesting that this (hydrophobic and bulky) tryptophan residue plays an important role in the recognition of hydroxyproline (which is more hydrophilic and bulky than proline). A phylogenetic analysis indicated that such unique broad substrate specificity was derived from an ancestral enzyme of this superfamily. While the physiological role of the OCC_00372 protein has not yet been established, significantly, *T. litoralis* accumulates T4LHyp as an intercellular organic solute (Lamosa *et al.*, 1998), and archaeal D-proline dehydrogenase is capable of converting C4DHyp to Pyr4H2C (Satomura *et al.*, 2002): so the homologous T4LHyp pathway in bacteria may also be operative in *T. litoralis* (see Section 10.4).

10.9 Hydroxyproline Metabolism in Fungi

Although the fungal metabolism of hydroxyproline has not yet been examined, many species of *Aspergillus*, and *Penicillium chrysogenum* and *Nectria haematococca*, possess orthologues of T3LHypD and the OCD-type Pyr2CR gene.

10.10 Enzymatic Detection of Hydroxyproline

L-Hyp is an important marker for directly measuring the content of collagen in several biological samples, including foods and tissue samples from tissue fibrosis. Furthermore, serum and urinary L-Hyp have attracted attention as important biomarkers of bone resorption and human diseases (Simsek *et al.*, 2004). So a simple, fast, inexpensive and sensitive method needs to be established to determine the content of L-Hyp. Here, enzymatic detection represents a useful alternative to employing an amino acid analyser (high-performance liquid chromatography, HPLC).

HypE and T3LHypD only utilize T4LHyp and T3LHyp (and not any other natural L-amino acid, such as L-proline), respectively, and difficulties have been associated with directly assaying their activities. In contrast, the activities of D-HypDH and Pyr2CR may be readily monitored using a spectrophotometer. Therefore, coupling the systems of HypE/D-HypDH and of T3LHypD/Pyr2CR enable the contents of T4LHyp and T3LHyp, respectively, to be estimated enzymatically. By using a spectrophotometric assay with different wavelengths, the contents of T4LHyp and T3LHyp were successfully estimated within the ranges of 0.004~1 mM and 0.05~1 mM, respectively, and were consistent with those determined by HPLC (Watanabe *et al.*, 2015b). Even when T4LHyp and T3LHyp coexisted in the same sample, neither had any significant impact on estimating the contents of the other. Furthermore, other L-amino acids in the acid-hydrolysate of collagen did not have any effect on the estimation of T4LHyp contents. This rapid and convenient method may have further advantages for clinical diagnoses using a microplate reader, as described by Watanabe and Tanimoto (2014).

10.11 Conclusions

The metabolic pathways of several L-hydroxyproline (L-Hyp) compounds in microorganisms are currently being examined at the molecular level. A direct relationship has not yet been detected between 4-hydroxyproline and L-proline, whereas 3-hydroxyproline is ultimately converted to L-proline. It is important to note that microorganisms possess several enzymes with the same functions: for D-hydroxyproline dehydrogenase (D-HypDH), homomeric and heteromeric types; for Δ^1-pyrroline-2-carboxylate reductase (Pyr2CR), the DpkA (an NADPH-dependent protein) and OCD (ornithine cyclodeaminase) types; and for T3LHypD (*trans*-3-hydroxy-L-proline dehydratase), the C-T (catalytic residues of cysteine and threonine) and S-C (catalytic residues of serine and cysteine) types. These findings suggest strongly that the same pathway(s) evolved convergently. Characterization of the (putative) pathways by archaea (for 4-hydroxyproline) and fungi (for 4- and 3-hydroxyprolines) will be useful for understanding not only metabolic diversity among microorganisms, but also the molecular evolution of enzyme functions.

References

Bater, A.J., Venables, W.A. and Thomas, S. (1977) Allohydroxy-D-proline dehydrogenase: an inducible membrane-bound enzyme in *Pseudomonas aeruginosa* PA01. *Archives of Microbiology* 112, 287–289.

Buschiazzo, A., Goytia, M., Schaeffer, F., Degrave, W., Shepard, W., Grégoire, C., Chamond, N., Cosson, A., Berneman, A., Coatnoan, N. *et al.* (2006) Crystal structure, catalytic mechanism, and mitogenic properties of *Trypanosoma cruzi* proline racemase. *Proceedings of the National Academy of Sciences of the United States of America* 103, 1705–1710.

Cooper, S.K., Pandhare, J., Donald, S.P. and Phang, J.M. (2008) A novel function for hydroxyproline oxidase in apoptosis through generation of reactive oxygen species. *The Journal of Biological Chemistry* 283, 10485–10492.

Hallen, A., Cooper, A.J., Jamie, J.F., Haynes, P.A. and Willows, R.D. (2011) Mammalian forebrain ketimine reductase identified as μ-crystallin; potential regulation by thyroid hormones. *Journal of Neurochemistry* 118, 379–387.

Hara, R. and Kino, K. (2009) Characterization of novel 2-oxoglutarate dependent dioxygenases converting L-proline to *cis*-4-hydroxy-L-proline. *Biochemical and Biophysical Research Communications* 379, 882–886.

Kawakami, R., Satomura, T., Sakuraba, H. and Ohshima, T. (2012) L-Proline dehydrogenases in hyperthermophilic archaea: distribution, function, structure, and application. *Applied Microbiology and Biotechnology* 93, 83–93.

Kumar, R., Zhao, S., Vetting, M.W., Wood, B.M., Sakai, A., Cho, K., Solbiati, J., Almo, S.C., Sweedler, J.V., Jacobson, M.P. *et al.* (2014) Prediction and biochemical demonstration of a catabolic pathway for the osmoprotectant proline betaine. *MBio* 5(1): e00933-13.

Lamosa, P., Martins, L.O., Da Costa, M.S. and Santos, H. (1998) Effects of temperature, salinity, and medium composition on compatible solute accumulation by *Thermococcus* spp. *Applied and Environmental Microbiology* 64, 3591–3598.

Mori, H., Shibasaki, T., Yano, K. and Ozaki, A. (1997) Purification and cloning of a proline 3-hydroxylase, a novel enzyme which hydroxylates free L-proline to *cis*-3-hydroxy-L-proline. *Journal of Bacteriology* 179, 5677–5683.

Muramatsu, H., Mihara, H., Kakutani, R., Yasuda, M., Ueda, M., Kurihara, T. and Esaki, N. (2005) The putative malate/lactate dehydrogenase from *Pseudomonas putida* is an NADPH-dependent Δ^1-piperideine-2-carboxylate/Δ^1-pyrroline-2-carboxylate reductase involved in the catabolism of D-lysine and D-proline. *The Journal of Biological Chemistry* 280, 5329–5335.

Petersen, L., Olewinski, R., Salmon, P. and Connors, N. (2003) Novel proline hydroxylase activities in the pneumocandin-producing fungus *Glarea lozoyensis* responsible for the formation of *trans* 3- and *trans* 4-hydroxyproline. *Applied Microbiology and Biotechnology* 62, 263–267.

Riedel, T.J., Johnson, L.C., Knight, J., Hantgan, R.R., Holmes, R.P. and Lowther, W.T. (2011) Structural and biochemical studies of human 4-hydroxy-2-oxoglutarate aldolase: implications for hydroxyproline metabolism in primary hyperoxaluria. *PLoS ONE* 6(10): e26021.

Satomura, T., Kawakami, R., Sakuraba, H. and Ohshima, T. (2002) Dye-linked D-proline dehydrogenase from hyperthermophilic archaeon *Pyrobaculum islandicum* is a novel FAD-dependent amino acid dehydrogenase. *The Journal of Biological Chemistry* 277, 12861–12867.

Schröder, I., Vadas, A., Johnson, E., Lim, S. and Monbouquette, H.G. (2004) A novel archaeal alanine dehydrogenase homologous to ornithine cyclodeaminase and μ-crystallin. *Journal of Bacteriology* 186, 7680–7689.

Shibasaki, T., Mori, H., Chiba, S. and Ozaki, A. (1999) Microbial proline 4-hydroxylase screening and gene cloning. *Applied and Environmental Microbiology* 65, 4028–4031.

Simsek, B., Karacaer, O. and Karaca, I. (2004) Urine products of bone breakdown as markers of bone resorption and clinical usefulness of urinary hydroxyproline: an overview. *Chinese Medical Journal* 117, 291–295.

Singh, R.M. and Adams, E. (1965) Enzymatic deamination of Δ^1-pyrroline-4-hydroxy-2-carboxylate to 2,5-dioxovalerate (α-ketoglutaric semialdehyde). *The Journal of Biological Chemistry* 240, 4344–4451.

Visser, W.F., Verhoeven-Duif, N.M. and de Koning, T.J. (2012) Identification of a human *trans*-3-hydroxy-L-proline dehydratase, the first characterized member of a novel family of proline racemase-like enzymes. *The Journal of Biological Chemistry* 287, 21654–21662.

Watanabe, S. and Tanimoto, Y. (2014) *Method for Analysis for 3-Hydroxyproline, Method for Examining Collagen, and Novel δ1-Pyrroline-2-Carboxylic Acid Reductase for Use Therein: WO 2015033636 A1*. International Patent Application No. PCT/JP2014/065019. Available at: https://www.google.com/patents/WO2015033636A1?cl=en (accessed 16 August 2016).

Watanabe, S., Shimada, N., Tajima, K., Kodaki, T. and Makino, K. (2006) Identification and characterization of L-arabonate dehydratase, L-2-keto-3-deoxyarabonate dehydratase and L-arabinolactonase involved in an alternative pathway of L-arabinose metabolism: novel evolutionary insight into sugar metabolism. *The Journal of Biological Chemistry* 281, 33521–33536.

Watanabe, S., Yamada, M., Ohtsu, I. and Makino, K. (2007) α-Ketoglutaric semialdehyde dehydrogenase isozymes involved in metabolic pathways of D-glucarate, D-galactarate and hydroxy-L-proline: molecular and metabolic convergent evolution. *The Journal of Biological Chemistry* 282, 6685–6695.

Watanabe, S., Morimoto, D., Fukumori, F., Shinomiya, H., Nishiwaki, H., Kawano-Kawada, M., Sasai, Y., Tozawa, Y. and Watanabe, Y. (2012) Identification and characterization of D-hydroxyproline dehydrogenase and Δ^1-pyrroline-4-hydroxy-2-carboxylate deaminase involved in novel L-hydroxyproline metabolism of bacteria: metabolic convergent evolution. *The Journal of Biological Chemistry* 287, 32674–32688.

Watanabe, S., Tanimoto, Y., Yamauchi, S., Tozawa, Y., Sawayama, S. and Watanabe, Y. (2014a) Identification and characterization of *trans*-3-hydroxy-L-proline dehydratase and Δ^1-pyrroline-2-carboxylate reductase involved in *trans*-3-hydroxy-L-proline metabolism of bacteria. *FEBS Open Bio* 4, 240–250.

Watanabe, S., Tozawa, Y. and Watanabe, Y. (2014b) Ornithine cyclodeaminase/μ-crystallin homolog from hyperthermophilic archaeon *Thermococcus litoralis* functions as a novel Δ^1-pyrroline-2-carboxylate reductase involved in putative *trans*-3-hydroxy-L-proline metabolism. *FEBS Open Bio* 4, 617–626.

Watanabe, S., Tanimoto, Y., Nishiwaki, H. and Watanabe, Y. (2015a) Identification and characterization of bifunctional proline racemase/hydroxyproline epimerase from archaea: discrimination of substrates and molecular evolution. *PLoS ONE* 10(3): e0120349.

Watanabe, S., Hiraoka, Y., Endo, S., Tanimoto, Y., Tozawa, Y. and Watanabe, Y. (2015b) An enzymatic method to estimate the content of L-hydroxyproline. *Journal of Biotechnology* 199, 9–16.

Watanabe, S., Sueda, R., Fukumori, F. and Watanabe, Y. (2015c) Characterization of flavin-containing opine dehydrogenase from bacteria. *PLoS ONE* 10(9): e0138434.

Zhang, X., Kumar, R., Vetting, M.W., Zhao, S., Jacobson, M.P., Almo, S.C. and Gerlt, J.A. (2015) A unique *cis*-3-hydroxy-L-proline dehydratase in the enolase superfamily. *Journal of the American Chemical Society* 137, 1388–1391.

Zhao, S., Kumar, R., Sakai, A., Vetting, M.W., Wood, B.M., Brown, S., Bonanno, J.B., Hillerich, B.S., Seidel, R.D., Babbitt, P.C. *et al.* (2013) Discovery of new enzymes and metabolic pathways by using structure and genome context. *Nature* 502, 698–702.

Zhao, S., Sakai, A., Zhang, X., Vetting, M.W., Kumar, R., Hillerich, B., San Francisco, B., Solbiati, J., Steves, A., Brown, S. *et al.* (2014) Prediction and characterization of enzymatic activities guided by sequence similarity and genome neighborhood networks. *eLife* 3: e03275.

11 Cellular Responses to Serine in Yeast

I.W. Dawes* and G.D. Kornfeld
School of Biotechnology and Biomolecular Sciences, The University of New South Wales, Sydney, Australia

11.1 Abstract

The serine–glycine interconversion has been the subject of considerable research over several decades and it involves series of reactions of relevance to bacteria and fungi as well as higher organisms. Much has been published on the role of cofactors and one-carbon units in this reversible relationship. Current studies are aimed at understanding the regulation and the overall systems biology of the synthesis of serine and glycine and their roles in one-carbon metabolism (and hence methylation reactions), in the synthesis of other important metabolites critical for DNA replication, in membrane lipid synthesis and amino acid synthesis, in oxidative and heat stress responses, and also as sources of nitrogen, as exemplified by evidence obtained with yeast (*Saccharomyces cerevisiae*). The transcriptional and metabolic responses of yeast to added serine and glycine are discussed. It should be noted that serine can be synthesized from a variety of sources, including glycolysis via 3-phosphoglycerate and threonine via threonine aldolase, which converts threonine to acetaldehyde and glycine, the latter being the precursor of serine through the action of the glycine decarboxylase complex.

11.2 Introduction

11.2.1 Metabolic roles of L-serine

In addition to its role in protein synthesis, L-serine is involved in a wide range of processes that are critical to the survival of the cell, including the synthesis of membrane lipids and the purines needed for the synthesis of ATP, DNA replication and repair, and RNA synthesis (Fig. 11.1). In the yeast *Saccharomyces cerevisiae*, it is the major source of one-carbon units for many metabolic reactions, including the synthesis of L-methionine and hence all methylation reactions in the cell (Schirch, 1984). Folate-dependent one-carbon metabolism is also crucial to the synthesis of purines, thymidylate, phospholipids (Kanfer and Kennedy, 1964; Bae-Lee and Carman, 1984; Summers *et al.*, 1988), sphingolipids (Montefusco *et al.*, 2012), and pantothenic acid and *N*-formyl methionine (Blakley, 1969). It is also critical for regulatory processes including the methylation of proteins (Winter *et al.*, 2015; Yagoub *et al.*, 2015) and, in higher organisms, the methylation of cytosines in DNA, which plays an important role in the regulation of gene expression (Smith and Meissner, 2013). In *S. cerevisiae*, L-serine provides a substrate for the synthesis of other

*Corresponding author. E-mail address: i.dawes@unsw.edu.au

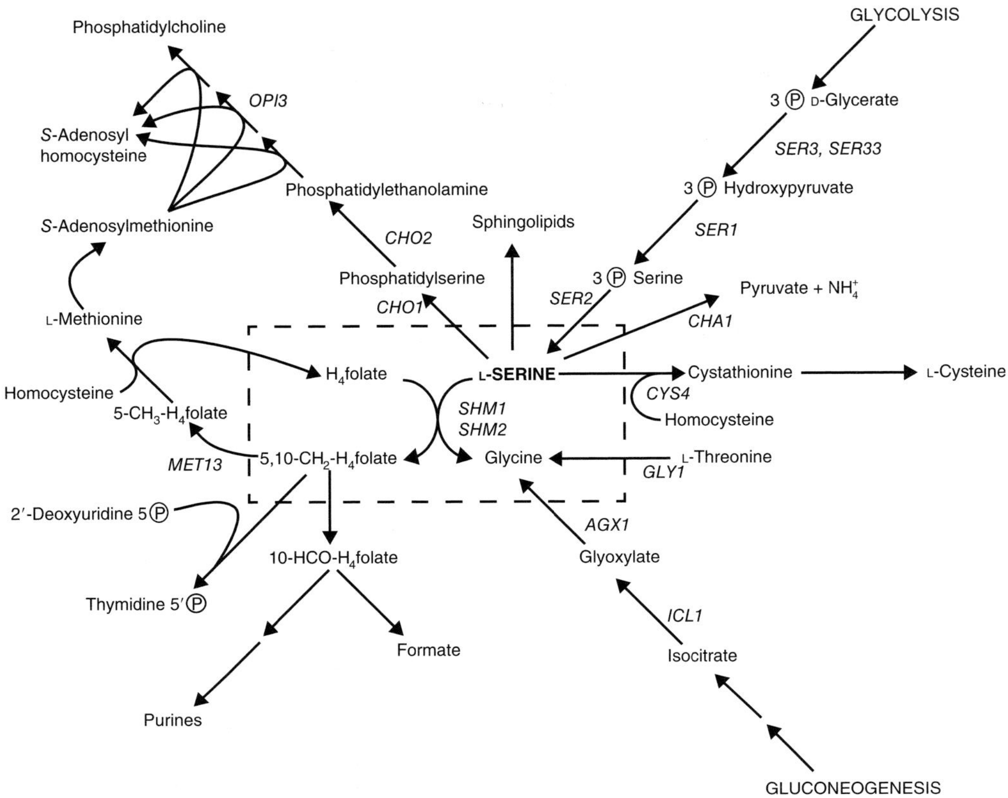

Fig. 11.1. The roles of L-serine in yeast (*Saccharomyces cerevisiae*) metabolism. The relevant genes are indicated in italics. The dotted lines indicate the one-carbon metabolic functions. Ⓟ indicates a phospho group. The genes designations are as follows: *AGX1*, alanine:glyoxylate aminotransferase; *CHA1*, L-serine (threonine) deaminase; *CHO1*, phosphatidylserine synthase; *CHO2*, phosphatidylethanolamine *N*-methyltransferase; *CYS4*, cystathionine *β*-synthase; *GLY1*, threonine aldolase; *ICL1*, isocitrate lyase; *OP13*, phospholipid methyltransferase; *MET13*, 5,10-CH_2-H_4folate reductase; *SER1*, phosphoserine transaminase; *SER2*, phosphoserine phosphatase; *SER3* and *SER33*, phosphoglycerate dehydrogenase isoenzymes; *SHM1* and *SHM2*, serine hydroxymethyltransferases.

amino acids, including glycine, L-methionine and cysteine (Cherest and Surdin-Kerjan, 1992; Cherest *et al.*, 1993). It is also, therefore, needed for the synthesis of the redox buffer glutathione (*γ*-glutamylcysteinylglycine) and hence for the maintenance of resistance to oxidative stress and many aspects of redox homeostasis (Grant *et al.*, 1996), especially those relating to FeS metabolism (Kumar *et al.*, 2011) and the maintenance of mitochondrial integrity (Ayer *et al.*, 2010). The role of L-serine in sphingolipid metabolism also implicates it in cellular responses to heat shock (Cowart and Hannun, 2007; Montefusco *et al.*, 2012). If present in excess, it can be a source of nitrogen and pyruvate in the mitochondrion, which can also contribute to L-aspartate synthesis (Ramos and Wiame, 1982; Lee *et al.*, 2013).

In humans, L-serine is not an essential amino acid, but it does play an additional role to those outlined above as it is a precursor of the neurotransmitters glycine and D-serine. There is a rare serine deficiency syndrome caused by mutations affecting L-serine synthesis, which leads to microcephaly, severe psychomotor retardation and seizures; the disorder is treatable by the addition of L-serine to the diet (Jaeken *et al.*, 1996; de Koning and Klomp, 2004). There are a number of genetic disorders that arise from mutations or deficiency in one-carbon metabolism. Folate deficiency leads to megaloblastic anaemia as a result of impaired cell division (Chanarin, 1969) and supplementation with low levels of folate during pregnancy reduces the incidence of neural tube closure defects (Czeizel and Dudas, 1992; Pitkin, 2007). The dependence of methylation reactions on one-carbon metabolism has far-reaching outcomes when the methylation balance is affected, especially because of the role of methylation in the regulation of gene expression; this may account for why individuals with folate deficiency or some polymorphisms in folate metabolic genes have an increased risk of developmental defects or a range of cancers (Houlston and Tomlinson, 2001; Larsson *et al.*, 2006; Marchal *et al.*, 2008), epilepsy (Dean *et al.*, 2008) or psychiatric disorders (Muskiet and Kemperman, 2006; López-León *et al.*, 2008).

11.2.2 Role of L-serine in the cell walls of yeast and other fungi

In *S. cerevisiae*, there is a major requirement for L-serine for the synthesis of cell wall mannoproteins, which are highly enriched in serine. This is especially the case under anaerobic conditions because there are a large number of mannoproteins (including Tir1p-4p) that are essential under anaerobiosis, with some containing up to 30% of the amino acyl residues as serine. Mutants lacking *SER1* (the gene for phosphoserine transaminase) cannot synthesize L-serine via its major synthetic pathway (glycolysis) and are unable to grow anaerobically even on a rich medium (Reiner *et al.*, 2006). Under anaerobic conditions the cell not only needs to produce L-serine from glycolysis by the pathway requiring *SER1*, but also by increased synthesis from L-aspartate via L-threonine and glycine, which requires the *GLY1* gene (which encodes threonine aldolase). Any shortfall of L-serine synthesis during the process of adaptation from aerobic growth to anaerobic growth leads to an extended lag phase while the cells adjust their cell wall mannoproteins to those needed for anaerobic growth (Tsoi *et al.*, 2009). While *S. cerevisiae* has ~20 cell wall genes, some fungal pathogens have many more serine-rich cell wall proteins with homology to the *S. cerevisiae* Tir mannoproteins (which are induced by anaerobiosis or other shocks). For example, *Candida albicans* has many genes encoding serine-rich proteins and many of these are annotated as cell wall proteins induced by stress or antibiotics, or are involved in cell adhesion or pathogenicity. Tsoi *et al.* (2009) speculated that the adaptation of these pathogens to tolerate antibiotics may also depend on an increased supply of L-serine, and that an inhibitor of seryl-tRNA synthetase may have antifungal activity; in combination, these could increase the efficacy of available antifungal antibiotics.

11.2.3 Role of L-serine in the heat-shock response

The sudden exposure of yeast cells leads to a sublethal heat stress causing transient cell cycle arrest (Rowley *et al.*, 1993), the accumulation of unfolded proteins (Vidair *et al.*, 1996), altered gene expression, degradation of nutrient permeases, trehalose accumulation, induction of sphingolipid synthesis (Cowart and Hannun, 2007), impaired coupling of oxidative phosphorylation (Patriarca and Maresca, 1990), defective secretory pathway trafficking and damage to cellular structures, including membranes and the cytoskeleton (Richter *et al.*, 2010). In response, cells have an evolutionarily conserved response involving the increased transcription of a subset of 'heat-shock' response genes involved in repair and survival, while that of other genes encoding more highly expressed proteins is reduced (McAlister *et al.*, 1979; Miller *et al.*, 1979). This heat-shock response affects the transcription of a substantial number of genes (Gasch *et al.*, 2000). Moreover, in *S. cerevisiae*, more than a 1000 genes are important for the normal survival of heat shock (in the exponential and stationary phase cells) and these are largely independent of those whose transcription is altered following heat shock (Gibney *et al.*, 2013; Jarolim *et al.*, 2013).

The role of L-serine in the heat-shock response is illustrated by the importance for heat-shock survival of sphingolipids and phospholipids derived from L-serine or one-carbon metabolism (Dickson, 1998; Skrzypek *et al.*, 1999; Cowart and Hannun, 2007; Jarolim *et al.*, 2013). Uptake assays have indicated that heat treatment increased L-serine uptake from the medium by 100% during the early response to heat stress. It is particularly interesting that sphingolipid synthesis, which is critical for heat-shock survival, is strongly affected by the ability of cells to take up L-serine from the medium, and that there is an interdependence on L-serine and sphingolipids for the control of the levels of both L-serine and sphingolipids in cells (Montefusco *et al.*, 2012).

11.3 Uptake and Synthesis of L-Serine and Glycine and the Central Role of One-Carbon Metabolism

11.3.1 Uptake of L-serine and glycine

L-Serine can be transported into the cell via the general amino acid permease (Gap1p), which transports all of the natural amino acids (Grenson *et al.*, 1970; Jauniaux and Grenson, 1990; Regenberg *et al.*, 1999) or via other more specific permeases, including Dip5p, Gnp1p and Agp1p. The transport is active, driven by the proton gradient across the plasma membrane (Ballarin-Denti *et al.*, 1984). Of the permeases that are involved, Agp1p is less specific, and transports most of the uncharged amino acids; it has high affinity for L-asparagine and L-glutamine, but can also transport L-serine if it is present in the medium at high levels (Schreve *et al.*, 1998). Gnp1p, a paralogue of Agp1p, has a slightly greater degree of specificity for Leu, Ser, Thr, Cys, Met, Gln and Asn, while Dip5p transports Leu, Ser, Thr, Cys, Met, Gln and Asn. Gnp1p and Dip5p have moderate activity in the uptake of L-serine (Regenberg *et al.*, 1999). Glycine transport is also mediated via the Gap1p permease, as well as by the transporters Put4p, Agp1p and Dip5p (Regenberg *et al.*, 1999).

11.3.2 Synthesis of L-serine and glycine

S. cerevisiae has two main biosynthetic pathways and a potential third minor one for L-serine (Fig. 11.1). The glycolytic pathway from 3-phospho-D-glycerate via 3-phosphoserine is the main one that operates when glucose is the carbon source. The first step of this pathway is catalysed by two phosphoglycerate dehydrogenase isozymes encoded by the *SER3* and *SER33* genes; of these, the *SER3* gene is the more active (Albers *et al.*, 2003). Subsequently, the 3-phosphohydroxypyruvate formed is converted to 3-phosphoserine by the action of phosphoserine transaminase, encoded by *SER1* (*ADE9*), and then to L-serine by phosphoserine phosphatase (encoded by *SER2*) (Ulane and Ogur, 1972; Belhumeur *et al.*, 1994). In the presence of glucose, the other pathways are repressed, and the deletion of *SER1*, *SER2* or both *SER3* and *SER33* leads to auxotrophy for L-serine under these conditions (Melcher and Entian, 1992).

In cells growing on non-fermentable substrates, such as glycerol, ethanol or acetate, L-serine and glycine are synthesized via the gluconeogenic pathway, which involves the action of the glyoxylate cycle and one-carbon metabolism. Isocitrate derived initially from the tricarboxylic acid cycle (TCA) is converted in the peroxisome to glyoxylate and succinate by isocitrate lyase, the product of the *ICL1* gene (Scholer and Schuller, 1993). Glyoxylate is converted to glycine via the activity of alanine:glyoxylate aminotransferase (encoded by *AGX1*) (Schlösser *et al.*, 2004), and the glycine can be converted to L-serine by serine hydroxymethyltransferase via operation of the one-carbon metabolic pathway discussed below. L-threonine is also a potential source of L-serine in yeast as glycine can be formed in yeast from threonine via the action of threonine aldolase encoded by *GLY1* (McNeil *et al.*, 1994; Monschau *et al.*, 1997). In yeast, L-threonine is derived from pyruvate via oxaloacetate, L-aspartate and L-homoserine. Under glycolytic conditions, the *GLY1* route is the major one for the synthesis of glycine rather than that from L-serine via the serine hydroxymethyltransferases encoded by *SHM1* and *SHM2* (see Section 11.3.3 on the interconversion of glycine and L-serine).

It has been speculated that a further minor pathway for the synthesis of L-serine from cysteine may occur in yeast via the action of a vanadate-activated serine sulfhydrase catalysing the conversion of L-cysteine to L-serine and H_2S (Meisch and Kappesser, 1987). The existence of this pathway was considered to be supported by the demonstration that a *ser1*, *ser2*, *cat1* triple mutant can grow, albeit slowly, on non-fermentable substrates in the absence of added L-serine (Melcher and Entian, 1992). Under these conditions, the *cat1* mutation abolishes the glucose-repressible gluconeogenic pathway, and the glycolytic pathway is deleted by the *ser1 ser2* mutations. The authors used this triple mutant to isolate a mutation in a gene which they designated *SER10*, which they suggested defined a third pathway, possibly involving the serine sulfhydrase activity reported by Meisch and Kappesser (1987). No gene sequence for *SER10* or a serine sulfhydrase has been identified, and another possible explanation for this possible third pathway is that the activity in yeast that generates L-serine and H_2S is cystathionine *β*-synthase encoded by the *CYS4* gene. This enzyme, which is normally involved in the assimilation of sulfur from sulfate to cysteine, can (among other reactions) convert L-cysteine to L-serine (Singh *et al.*, 2009). Given that L-cysteine is relatively toxic, is normally present only in low concentrations in yeast cells (Ono *et al.*, 1999) and that the usual operation of Cys4p is to generate rather than degrade L-cysteine, this pathway is unlikely to be a significant producer of L-serine under most conditions. Another possibility for the third pathway is from L-threonine via the activity of threonine aldolase (the Gly1p protein), as indicated above, because the synthetic complete medium used by Melcher and Entian probably contained L-threonine.

11.3.3 Central role of one-carbon metabolism in synthesis and interconversion of glycine and L-serine

The metabolism of L-serine and glycine is intricately involved with the cellular process of folate-dependent one-carbon metabolism. Both amino acids, as well as formate, can act as donors of the one-carbon moieties at different oxidation states that can be transferred to synthetic reactions. L-Serine and glycine can be interconverted via this pathway, depending on the nutritional conditions. The pathways involved in one-carbon metabolism, which are to a great extent duplicated between the cytoplasm and the mitochondrion, are indicated in Fig. 11.2 with the genes and enzymes involved. The one-carbon units at a range of oxidation states from methyl to formyl are attached to tetrahydrofolate (H_4folate).

When glucose is the carbon source, L-serine is the main source for the synthesis of glycine and one-carbon units, through cleavage catalysed by either the cytoplasmic or mitochondrial serine hydroxymethyltransferase (Shm2p or Shm1p) to form glycine and the one-carbon donor 5,10-methylene-H_4folate (5,10-CH_2-H_4folate) (Schirch, 1984). The cytoplasmic Shm2p is the principle catalyst for serine catabolism in *S. cerevisiae*, although if required, the mitochondrial Shm1p can provide the 1-C units required for growth (McNeil *et al.*, 1994; Kastanos *et al.*, 1997). However, when cells are growing on non-fermentable carbon sources or in the presence of excess glycine, glycine can be a source of 5,10-CH_2-H_4folate and nitrogen as NH_4^+ via the oxidative decarboxylation of glycine catalysed by the glycine cleavage complex (GCV) encoded by the *GCV1*, *GCV2*, *GCV3* and *LPD1* genes (Ogur *et al.*, 1977; Sinclair and Dawes, 1995). While glycine could potentially be formed by reversal of the GCV reaction (Pasternack *et al.*, 1992), growth experiments have shown that in yeast the GCV does not play a significant role in glycine biosynthesis (Schlösser *et al.*, 2004), rather, its function is to provide the cell with the ability to use glycine as sole one-carbon or nitrogen source (Piper *et al.*, 2002).

Because the glycine to L-serine conversion catalysed by either Shm1p or Shm2p is reversible, two molecules of glycine can lead to the synthesis of L-serine through the oxidation of one glycine via the GCV in the mitochondrion and the reaction of the 5,10-CH_2-H_4folate formed with a second glycine to produce L-serine via either Shm1p or Shm2p. This is the pathway by which L-serine is synthesized under gluconeogenic conditions when cells are growing by respiration on non-fermentable carbon sources, or possibly when glycine is formed from L-threonine under conditions in which L-serine synthesis from other sources is restricted (e.g. in a *ser1* mutant growing on a fermentable carbon source).

5,10-CH_2-H_4folate formed in the mitochondrion cannot cross the mitochondrial inner membrane, but L-serine, glycine and formate can (Pasternack *et al.*, 1992, 1994). Pasternack *et al.* (1992), using 13C-NMR, showed that in yeast the major flow of one-carbon units formed from GCV-catalysed glycine cleavage to the cytoplasm occurs through conversion of the 5,10-CH_2-H_4folate to formate in the mitochondrion via the oxidative trifunctional C_1-tetrahydrofolate synthase (Mis1p). Formate can then cross the mitochondrial inner membrane, and one-carbon units at various oxidation states are made available in the cytoplasm via the reverse reactions

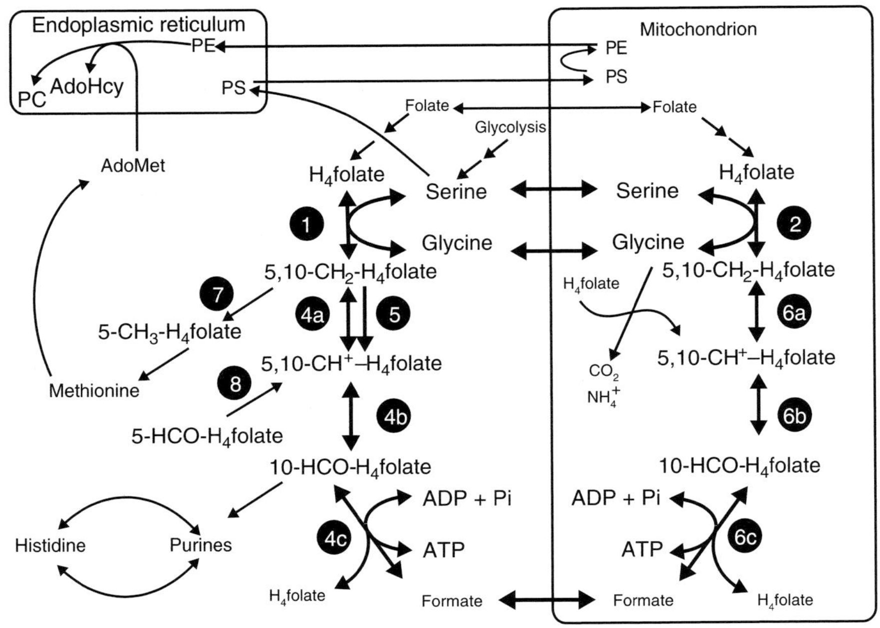

Fig. 11.2. The connection between L-serine and glycine metabolism in *Saccharomyces cerevisiae*. The bold numbers that follow (shown in the figure in black circles) indicate enzymes; the genes that encode them are in parentheses: **1**, cytoplasmic serine hydroxymethyltransferase (*SHM2*); **2**, mitochondrial serine hydroxymethyltransferase (*SHM1*); **3**, glycine decarboxylase complex (*GCV1*, *GCV2*, *GCV3*, *LPD1*); **4**, cytoplasmic trifunctional C1-H_4folate synthase (*ADE3*); **5**, NAD^+-dependent 5,10-CH_2-H_4folate dehydrogenase (*MTD1*); **6**, mitochondrial trifunctional C1-H_4folate synthase (*MIS1*); **7**, 5,10-CH_2-H_4folate reductase (*MET13*); **8**, 5,10-CH^+-H_4folate synthetase, MTHFS (*FAU1*). Abbreviations: AdoHcy, *S*-adenosylhomocysteine; AdoMet, *S*-adenosylmethionine; PC, phosphatidylcholine; PE, phosphatidylethanolamine; Pi, phosphate; PS, phosphatidylserine. Figure was adapted from West *et al.* (1993).

catalysed by the cytoplasmic trifunctional C_1-tetrahydrofolate synthase (Ade3p). In the cytoplasm, there is also a second monofunctional NAD^+-dependent 5,10-CH_2-H_4folate dehydrogenase that can convert 5,10-CH_2-H_4folate to 5,10-methenyl-H_4folate.

L-Methionine is produced in the cytoplasm from 5,10-CH_2-H_4folate, which is converted to 5-CH_3-H_4folate by 5,10-CH_2-H_4folate reductase (encoded by *MET13*). The methyl group of 5-CH_3-H_4folate is then transferred to L-homocysteine by the Met13p transferase to form L-methionine. 5,10-CH_2-H_4folate is also directly involved in the synthesis and repair of DNA through the production of thymidine-5′-phosphate from 2′-deoxyuridine-5′-phosphate.

Purine biosynthesis depends on one-carbon metabolism for the incorporation of glycine as well as two other steps requiring 10-formyl-H_4folate (5-HCO-H_4folate) as a formyl donor. The function of the cytoplasmic route for the synthesis of 5-HCO-H_4folate from L-serine is highlighted by the adenine dependence of mutants lacking the *ADE3* gene product. None the less, a quarter of the one-carbon units for purine biosynthesis are derived from glycine via the mitochondrial pathway to formate, even when L-serine is abundant in the medium (Pasternack *et al.*, 1992); however, this pathway is not sufficient for the growth of the cell on glycine when the pathway to 5-HCO-H_4folate from L-serine is blocked (McNeil *et al.*, 1996). From a combined genetic and biochemical approach, Piper *et al.* (2000) proposed that under normal conditions, cytoplasmic 5,10-CH_2-H_4folate is mainly directed to generating methyl groups via L-methionine, whereas one-carbon units generated from glycine in the mitochondria are more directed towards purine biosynthesis. When glycine is in excess, 5,10-CH_2-H_4folate is decreased, and the regulation loop shifts the balance of generation of one-carbon units into the mitochondria.

11.4 Cellular Responses to Excess L-Serine and Glycine

Excess L-serine in the cytoplasm is either transferred to the vacuole (Kitamoto *et al.*, 1988), or when it is present in the medium at relatively high levels, it is converted to pyruvate and NH_4^+ in the mitochondrion via the action of the L-serine (threonine) deaminase encoded by *CHA1* (Ramos and Wiame, 1982; Petersen *et al.*, 1988). Amino acid analyses of the response to the addition of excess L-serine to the medium have shown that there is a transient increase in the intracellular levels of cysteine, L-asparagine, L-glutamate and, to a much greater extent, of L-aspartate (Lee *et al.*, 2013). One-carbon metabolism is so tightly controlled that the level of glycine does not rise more than twofold in response to excess intracellular L-serine.

L-Aspartate appears to be an important metabolite for the short-term storage of NH_4^+ from the degradation of L-serine because the excess NH_4^+ was metabolized faster than was the excess L-glutamate produced. The pathway for the incorporation of the NH_4^+ into L-aspartate was shown to occur via Cha1p (L-serine (threonine) deaminase)-mediated mitochondrial metabolism of L-serine to pyruvate, but subsequently in the cytoplasm via pyruvate carboxylase (mainly Pyc1p but also Pyc2p) and cytoplasmic aspartate aminotransferase (Aat2p) (Lee *et al.*, 2013). This raises the intriguing possibility that no carbon atoms from L-serine are incorporated into the L-aspartate, because the mitochondrial pyruvate produced from L-serine is metabolized via the TCA cycle, thus sparing cytoplasmic pyruvate formed from glycolysis so that it can be converted to L-aspartate as an interim measure to store excess NH_4^+.

The addition of excess L-serine (100 mM) to the medium leads to a substantial transcriptional response at the genomic level, with more than 700 genes showing significant changes in their expression. One specific response to L-serine is induction of the *CHA1* gene due to activation of the Cha4p transcription factor. Other functions that are induced include a general stimulation of amino acid synthesis via the action of the Gcn4p transcription factor, iron-sulfur metabolism and some stress responses; there is also a downregulation of the genes involved in protein synthesis and hexose transport (Lee *et al.*, 2013). Interestingly more than 90% of the responses (both induced and repressed) could be accounted for by a network of ten transcription factors: Sfp1p involved in ribosome biogenesis; Rpn4p in proteasome synthesis; Gcn4p regulating nitrogen metabolism, Met4p for sulfur metabolism; Aft1p for iron metabolism; Yap1p and Msn2p for stress responses; Ste12p and Sok2p involved in some cell differentiation; and the general transcription factor Rap1p. These form a set of highly co-regulated transcription factors as indicated in Fig. 11.3, and the same ten could account for the changes seen when cells were treated with other amino acids (glycine, tryptophan, threonine) or with the superoxide-generating compound menadione, or following heat shock (Gasch *et al.*, 2000; Alic *et al.*, 2004; Gelling *et al.*, 2004; Lee *et al.*, 2013). This sheds a new light on the concept of an environmental stress response (Gasch *et al.*, 2000), which appears to involve a system of highly interacting transcription factors acting to provide a generalized response to a wide range of cell stresses or insults, including the addition of amino acids to the medium. The addition of L-serine appears to activate the Gcn4p system first, and this then leads to the responses orchestrated by the other factors (Lee *et al.*, 2013). For other types of stress (e.g. oxidative stress), another transcription factor (e.g. Yap1p) may be activated first, but the overall system would then be activated with a similar outcome for most of the response.

Sudden addition of glycine (10 mM) to *S. cerevisiae* cells also leads to a major transcriptional response. Initially it was shown that glycine specifically induces three of the four genes (*GCV1*, *GCV2* and *GCV3*) encoding the glycine decarboxylase complex (Sinclair and Dawes, 1995; Sinclair *et al.*, 1996; McNeil *et al.*, 1997; Nagarajan and Storms, 1997) and that this was in some part a response to a decrease in the cytoplasmic levels of 5,10-CH_2-H_4folate caused by the inhibition of cytoplasmic serine hydroxymethyltransferase (Piper *et al.*, 2000). Promoter analysis of the *GCV2* gene identified a 17 bp region upstream from the start site for transcription that included a binding motif for the Gcn4p transcription factor and/or the Bas1p transcription factor (which regulates the expression of adenine biosynthetic genes), and a CTTCTT

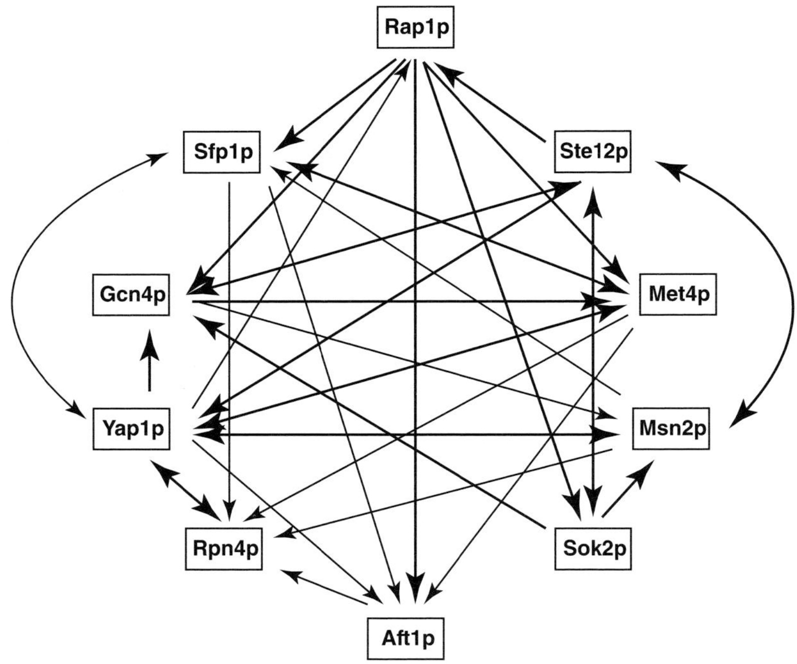

Fig. 11.3. Interactions between *Saccharomyces cerevisiae* transcription factors regulating more than 90% of the genes whose expression is significantly increased by the addition of excess L-serine. Data are based on the YEASTRACT database (Abdulrehman *et al.*, 2011). Single (small)-headed arrows indicate that the transcription factor regulates the target gene for the other factor. Double (large)-headed arrows indicate that each factor regulates the gene for the other. Heavy arrows represent direct evidence based on chromatin immunoprecipitation data; lighter arrows represent indirect evidence based on microarray data using mutants lacking the transcription factor concerned. See text for details of the functions of each of the ten transcription factors. Reproduced with permission from Lee *et al.* (2013).

motif; this region was necessary for much of the glycine response. A protein was identified that bound to the region in a manner that was increased by the presence of tetrahydrofolate. The involvement of a folate derivative in this control was confirmed by the response of a *fol1* deletion mutant (which is inhibited in H_4folate biosynthesis) to glycine (Hong *et al.*, 1999).

Subsequent genome-wide transcript analysis indicated that the addition of glycine stimulates a more general one-carbon transcriptional control system, which includes the *GCV* genes (Gelling *et al.*, 2004). As with the addition of L-serine to cells, sudden glycine addition led to the altered expression of about 950 genes, including the transient induction of those involved in respiration and amino acid synthesis, and the repression of those involved in ribosome biogenesis and protein synthesis. Most of these changes were due to the general stress response described above, but in this case it was possible to clearly delineate the specific changes that represented the one-carbon regulon because the genes whose expression was first to change were all involved in one-carbon metabolism and purine biosynthesis, and these were co-regulated over time following addition of glycine. Moreover, the availability of two mutants that affect 5,10-CH_2-H_4folate levels in the cytoplasm (and the transcriptional response to glycine) confirmed the identification of the one-carbon regulon responsive to 5,10-CH_2-H_4folate. The *shm2* mutation leads to decreased levels of 5,10-CH_2-H_4folate for the reasons discussed below, while the *met13* mutation increases the cytoplasmic levels of 5,10-CH_2-H_4folate by blocking its reduction to CH_3-H_4folate. As indicated in Fig. 11.4, these activities include the central one-carbon metabolic genes including *GCV1–3*, and those involved in adenine, L-methionine, *S*-adenosylmethionine

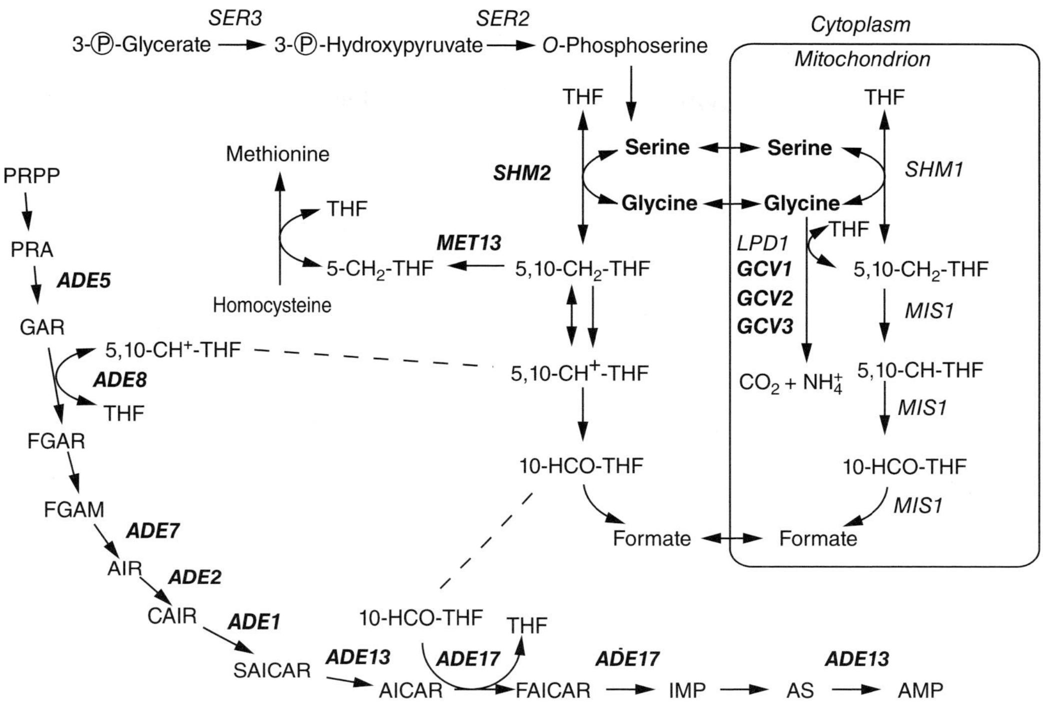

Fig. 11.4. The main genes of the one-carbon regulon. The pathways represented include the AMP biosynthetic pathway and those involved in central one-carbon metabolism. Genes that are highly regulated are in boldface italic type. Those that are less highly induced by glycine are in regular (not boldface) italic type and those that are not induced significantly are omitted. The gene designations for the one-carbon pathway are: *GCV1*, *GCV2*, *GCV3*, *LPD1*, glycine decarboxylase complex; *MET13*, 5,10-CH_2-H_4folate reductase; *MIS1*, mitochondrial trifunctional C1-H_4folate synthase; *SER1*, phosphoserine transaminase; *SER2*, phosphoserine phosphatase; *SHM1*, mitochondrial serine hydroxymethyltransferase; *SHM2*, cytoplasmic serine hydroxymethyltransferase. The abbreviations used for compounds of the adenine biosynthetic pathway are: AICAR, 5-amino-1-(5′-phosphoribosyl)-imidazole-4-carboxamide; AIR, 1-(5′-phosphoribosyl)-5-aminoimidazole; AMP, adenosine 5′-monophosphate; AS, adenylosuccinate; CAIR, 1-(5′-phosphoribosyl)-5-aminoimidazole-4-carboxylate; FAICAR, 5-formamido-1-(5′-phosphoribosyl)-imidazole 4-carboxamide; FGAM, 5′-phosphoribosyl-*N*′-formylglycinamidine; FGAR, 5′-phosphoribosyl-*N*′-formylglycinamide; GAR, 5′-phosphoribosylglycinamide; IMP, inosine 5′-monophosphate; PRA, 5-phosphoribosylamine; PRPP, 5-phosphoribosyl diphosphate; SAICAR, 1-(5′-phosphoribosyl)-4-(*N*-succinocarboxamido)-5-aminoimidazole; THF, tetrahydrofolate. The gene designations for the adenine biosynthetic pathway are: *ADE1*, phosphoribosyl amino imidazolesuccinocarbozamide synthetase; *ADE2*, phosphoribosylamino-imidazole-carboxylase; *ADE5*, *ADE7*, glycinamide ribotide synthetase and aminoimidazole ribotide synthetase; *ADE8*, glycinamide ribotide transformylase; *ADE13*, adenylosuccinate lyase; *ADE17*, 5-aminoimidazole-4-carboxamide ribonucleotide (AICAR) transformylase/IMP cyclohydrolase. Redrawn from figure originally published in Gelling *et al.* (2004) Identification of a novel one-carbon metabolism regulon in *Saccharomyces cerevisiae*. *The Journal of Biological Chemistry* 279, 7072–7081. © The American Society for Biochemistry and Molecular Biology.

and dTMP (deoxythymidine monophosphate) synthesis.

Hong *et al.* (1999) concluded that the Bas1p transcription factor was not needed for the glycine response, although Subramanian *et al.* (2005) showed that loss of Bas1p significantly reduced the repression of these genes in response to glycine removal, thereby indicating that Bas1p is involved in the regulation of the one-carbon metabolism regulon. The glycine response is, however, independent of adenine repression because glycine still provokes a response in *ade* mutant strains. The regulatory mechanisms for the specific responses to glycine (especially of the *GCV* genes) are complex as genetic analysis has shown that there are two components that overlap, one being the 'one-carbon metabolon response' that is most readily identified using the *shm2* mutant (Gelling *et al.*, 2004; Gelling, 2008). This response is dependent on Bas1p, and for the *GCV2* gene it does not require the 17 bp promoter region that contains the Bas1p-binding motif, and the mutation of this motif does not affect the one-carbon response of this gene. The other component of the 'glycine response' of the *GCV2* gene (and presumably other *GCV* genes) does depend on the CTTCTT motif and not on the Bas1p binding site of the 17 bp region of the *GCV2* promoter (Gelling, 2008). Mobility shift analysis of the promoter region has identified four protein–DNA complexes, one of which is abolished by mutation of the CTTCTT motif. Clearly, more work needs to be done to resolve the complexity of this control system.

11.5 Regulation of L-Serine and Glycine Metabolism

While some aspects of transcriptional regulation are covered above, the overall regulation of this area of metabolism is complex, involving the control over uptake, synthesis and degradation of both L-serine and glycine – further complicated by the demands on one-carbon metabolism and other biosynthetic processes that are dependent on these amino acids.

11.5.1 Control of L-serine and glycine uptake

Given its central role in metabolism, it is not surprising that the concentration of L-serine in the cell is fairly tightly controlled in comparison with that of other amino acids such as glycine. The steady-state intracellular level of L-serine in *S. cerevisiae* cells grown on a variety of different media (including rich and defined minimal media) is maintained in the range of 10–16 nmol/10^8 cells (Kitamoto *et al.*, 1988). Moreover, under conditions in which there is an increase in internal L-serine levels (e.g. when cells are grown with glycine in the medium), most of the excess is transported to the vacuole, maintaining the cytoplasmic level within a narrow range (Kitamoto *et al.*, 1988). The peak intracellular response to the sudden addition of L-serine is linear in the range of 10–100 mM external L-serine, and addition to cells of the high level of 100 mM L-serine leads to only a transient 12-fold increase in the intracellular level (Lee *et al.*, 2013). In *S. cerevisiae*, the sudden addition of 10 mM L-serine to the medium leads to only a temporary threefold increase in the intracellular concentration, with little effect on genome-wide transcriptional responses. In contrast, the addition of 10 mM glycine to the medium leads to saturation, with a substantial 20-fold increase in the intracellular level of glycine and a major transcriptional response affecting at least 949 genes (Sinclair *et al.*, 1996; Hong *et al.*, 1999; Piper *et al.*, 2000; Gelling *et al.*, 2004). The cell yield increases for a serine auxotroph as more serine is added to the medium, and it is difficult to determine an extracellular concentration at which the system becomes saturated, unlike the situation with other amino acids (unpublished observations), possibly because excess L-serine can act as a supplementary carbon source through conversion to pyruvate (Ramos and Wiame, 1982).

Serine uptake is positively regulated by the cellular amino acid sensing system. External amino acids in the medium are sensed by the SPS nutrient sensor system, which comprises the Ssy1p-Ptr3p-Ssy5p complex in the plasma membrane. When activated, this transmits the signal leading to the induction of a number of amino acid and peptide transporter genes, including all of those involved in L-serine uptake (Forsberg and Ljungdahl, 2001). The presence of amino acids activates the proteolytic activity of Ssy5p, which cleaves the *N*-terminal inhibitory domains of the two redundant transcription factors Stp1p and Stp2p. These shortened forms of Stp1p and Stp2p can then enter the nucleus and activate expression of genes encoding a subset of amino acid permeases, including those transporting L-serine (Ljungdahl, 2009; Omnus and Ljungdahl, 2013). The inducibility of the permeases transporting serine differs with the nitrogen source in the medium. On a non-repressive nitrogen source, *AGP1* is induced, while *GNP1* and *DIP5* are not (Regenberg *et al.*, 1999).

What regulates entry when serine or glycine is present in excess, as the unlimited uptake of amino acids is toxic to cells? The Gap1p general amino acid transporter activity at the plasma membrane is regulated at both transcriptional and post-translational levels. The transcription of *GAP1* (and also *AGP1*, *DIP5* and *PUT4*) is regulated by the much studied nitrogen catabolite repression mechanism mediated by the combined action of the four transcription factors, Gln3p, Gat1p, Dal80p and Deh1p; growth on glutamine or asparagine as nitrogen source strongly represses these genes (Cooper and Sumrada, 1983; Hofman-Bang, 1999; Regenberg *et al.*, 1999). At the post-translational level, Gap1p is downregulated when amino acids are abundant by two distinct mechanisms: sorting from the *trans*-Golgi through the multivesicular endosome to the vacuole rather than to the plasma membrane; and inactivation of the permease located at the plasma membrane. When external amino acid levels are low, Gap1p is sorted back to the cell surface. This post-translational regulation requires the catalytic activity of Gap1p (Cain and Kaiser, 2011).

11.5.2 Regulation of L-serine and glycine metabolism

Flux through the glycolytic pathway for L-serine synthesis is controlled by L-serine levels, and by the carbon and nitrogen sources in the medium. In cells growing on glucose, transcription of the first two genes (*SER3* and *SER1*) in the pathway is repressed specifically by the addition of L-serine to the medium as well as by the general control of amino acid synthesis mediated by the Gcn4p transcription factor (Ulane and Ogur, 1972; Melcher and Entian, 1992; Melcher *et al.*, 1995; Natarajan *et al.*, 2001). The serine-specific response of the *SER3* gene is controlled by an interesting mechanism, which depends on the non-coding RNA *SRG1* (Martens *et al.*, 2004; Martens *et al.*, 2005) transcribed from the intergenic region upstream of *SER3*. At high L-serine concentrations, the transcription of *SRG1* is activated by the Cha4p transcription factor (Holmberg and Schjerling, 1996) and proceeds across the *SER3* promoter, blocking the access of transcription factors needed to activate *SER3* transcription.

In cells growing on non-fermentable substrates such as ethanol or acetate, both the expression and the activity of all proteins of the phosphoglycerate pathway are low, and the genes of the glyoxylate pathway encoding the glyoxylate cycle enzyme isocitrate lyase (*ICL1*) and alanine:glyoxylate aminotransferase (*AGX1*) are derepressed by the system involving the Snf1 kinase, the transcriptional repressor Mig1 and the zinc finger transcription factors Cat8p, Sip4p and Rds2p (for a review see Turcotte *et al.*, 2010). This pathway is sufficient to sustain the growth of yeast cells independently of added glycine in the absence of the other two pathways, e.g. in a strain with disruption of the genes for cytosolic and mitochondrial serine hydroxymethyltransferase (SHMT), as well as the *GLY1* gene encoding threonine aldolase (Schlösser *et al.*, 2004).

The Cha4p transcription factor has a major function in regulating intracellular L-serine levels. In the presence of L-serine (or L-threonine) it activates the Cha1p L-serine (threonine) deaminase that catabolizes excess L-serine to pyruvate and NH_4^+, thereby helping to maintain the cytoplasmic levels of L-serine within a narrow range (Petersen *et al.*, 1988; Bornaes *et al.*, 1993). The precise mechanism whereby L-serine signals to Cha4p is not fully understood, although there is recent evidence that sphingolipids, through the involvement of the protein kinases Pkh1p and Pkh2p, may mediate the induction of *CHA1* in response to serine availability. The inhibition of sphingolipid synthesis inhibits the induction of *CHA1* in response to increased serine levels, and phytosphingosine and dihydrosphingosine have been identified as the potential mediators of *CHA1* upregulation (Montefusco *et al.*, 2012). This process is part of a feedforward/feedback control system in which sphingoid bases act as sensors of L-serine availability and upregulate *CHA1* expression in response to L-serine concentration, which in turn regulates sphingolipid levels by limiting L-serine accumulation.

The one-carbon metabolic system is highly regulated at both enzyme activity and transcriptional levels to ensure that an adequate supply of the H_4folate derivatives is available for important reactions, as well as to balance the interconversion of L-serine and glycine, depending on the relative availability of these amino acids (Stover and Schirch, 1991; Piper *et al.*, 2000, 2002; Gelling *et al.*, 2004). The complexity of regulation varies between organisms and those aspects of the control relevant to the metabolic products of one-carbon metabolism other than glycine, L-serine and L-methionine have been discussed in more detail by Piper (2000). During growth on glucose, the predominant pathway for glycine synthesis and charged folates is via L-serine and the action of the cytoplasmic Shm2p. The activity of Shm2p is tightly regulated in a number of organisms – glycine and 5-CH_3-H_4folate can bind tightly to the enzyme, leading to a dead-end complex that is catalytically inactive, thus preventing the breakdown of L-serine, and essentially reflecting the oversupply of both L-methionine (and methylation capacity) and glycine (Schirch *et al.*, 1977; Matthews *et al.*, 1982). Glycine and 5-HCO-H_4folate also bind together to Shm2p to form a dead-end complex. Initially, the role and origin of 5-HCO-H_4folate was not understood, but it is now clear that it is formed as a side reaction of Shm2p in the presence of excess glycine and regulating the L-serine to glycine conversion (Stover and Schirch, 1991). In yeast, L-methionine inhibits not only Shm2p, but also the methylene tetrahydrofolate reductase converting 5,10-CH_2-H_4folate to 5-CH_3-H_4folate (Botsford and Parks, 1969; Lor and Cossins, 1972). Similarly, a combination of adenine, histidine, methionine and pantothenate represses the activity of the C1-H_4folate synthase to the same extent as occurs in cells grown on rich medium (Appling and Rabinowitz, 1985). All of these metabolites are needed for this regulation, which may reflect the accumulation of a folate derivative, as antifolates lead to derepression of the activity.

Excess glycine also controls its own breakdown by the inducing the glycine decarboxylase complex in the mitochondrion, thereby sparing some of the requirement for synthesis of 5,10-CH_2-H_4folate in the cytoplasm (Sinclair and Dawes, 1995; Sinclair *et al.*, 1996). Pasternack *et al.* (1992) used ^{13}C-NMR to show that in *S. cerevisiae* a quarter of the one-carbon units used for purine biosynthesis were derived from glycine via the mitochondrial production of formate, even when cells had excess L-serine. Despite this, glycine is unable to supply all of the purine requirement, because strains that cannot catabolize serine for one-carbon metabolism have a requirement for adenine or formate, even when glycine is present (McNeil *et al.*, 1996), which indicates that the mitochondrial pathway of one-carbon metabolism may play a regulatory role in connecting the availability of glycine to the availability of charged folates, rather than being able to provide all of the cytoplasmic one-carbon requirements (Piper, 2000).

The transcriptional control system interconverting L-serine and glycine via the one-carbon metabolic pathway has been described above. It consists of two components – a one-carbon metabolic regulon and an additional glycine response. The one-carbon regulon is responsive to levels of 5,10-CH_2-H_4folate, reflecting the availability of charged folates, while the interconnected glycine response is regulated in response to the demand for one-carbon units as well as any requirement for a nitrogen source. Based on transcriptional data and enzyme inhibition analyses from many authors, Piper *et al.* (2000) proposed a model for how cells balance the need to supplement their one-carbon pools when charged folates are limiting or when glycine is in excess. They suggested that under growth on glucose, cytoplasmic 5,10-CH_2-H_4folate is mainly directed towards generating methyl groups via L-methionine, whereas one-carbon units generated from glycine in mitochondria are more directed towards purine biosynthesis. In the presence of excess glycine, the level of 5,10-CH_2-H_4folate in the cytoplasm is decreased, and the transcriptional regulation loop shifts the balance of generation of one-carbon units into the mitochondrion.

As glycine can also be a nitrogen source, it is not surprising that the genes encoding the glycine decarboxylase complex (as determined using *GCV2* reporter constructs) are also subject to regulation by the Gln3p and Nil1p transcription factors of the nitrogen catabolite repression system. The *GCV2* gene is also regulated by a novel rich medium repression mechanism, which depends on a palindromic sequence adjacent to the GATA (DNA) sequence bound by the Gln3p/Nil1p factors. This mechanism almost completely shuts down *GCV2* transcription in cells grown on a rich medium containing yeast extract and peptone (Piper *et al.*, 2002).

11.5.3 Differences in L-serine metabolic networks and regulation between aerobic and anaerobic growth

As indicated above, *S. cerevisiae* and other fungi contain cell wall mannoproteins that are highly enriched in seryl residues, and these mannoproteins can make up to 40% of the protein mass in the cells. When L-serine levels are low, translation of the relatively highly expressed cell wall genes is favoured over that for other genes, including the ribosomal genes that are important for cell growth – probably due to the strong codon bias towards abundant seryl-tRNA species found in the sequences encoding cell wall proteins (Tsoi *et al.*, 2009). The limitation of L-serine synthesis during adaptation to anaerobic growth leads to a substantially extended lag phase before growth can recommence, and the deletion of both serine hydroxymethyl transferases (Shm1p and Shm2p) prevents anaerobic growth completely. In the absence of the Gcn4p transcription factor mediating the general control of amino acid metabolism, the transition to anaerobic growth is extensively delayed.

Thus under anaerobic conditions, the direction of one-carbon metabolic flux differs from that in aerobic *S. cerevisiae* cells, which use L-serine formed from glycolysis as the main one-carbon donor. Under anaerobic conditions, there is a need for glycine to be converted to L-serine. The role of glycine is highlighted by the observation that in minimal medium, the *gly1* mutant, which is unable to synthesize glycine from L-threonine, is unable to adapt to anaerobiosis in minimal medium unless this is supplemented with L-serine. Both the Gcn4p and Bas1p transcription factors are important for the transition to anaerobiosis: Bas1p through upregulating the one-carbon metabolic pathway enzymes, including the genes encoding Shm1p and Shm2p, and Gcn4p and Bas1p together for the induction of *GLY1* (Tsoi *et al.*, 2009). These authors concluded that: 'from a systems biology perspective ... there are substantial differences between aerobic and anaerobic growth in gene regulation and metabolic networks'.

11.6 Conclusions

While there is now a mature understanding of the metabolic pathways that connect L-serine to one-carbon metabolism and other cellular reactions, there remain a number of aspects that provide considerable scope for further analysis. These include:

- Systems biology analysis and modelling of the effects of L-serine changes, given the important roles of this amino acid in cell growth, DNA replication and gene regulation.
- The differences in regulation of metabolism between anaerobic and aerobic growth conditions, as there is a substantial extra demand for L-serine in the very serine-rich mannoproteins in the cell wall of yeast and other fungi under anaerobic conditions. This has relevance because many of the mutations in the mannoprotein genes of pathogenic fungi confer a phenotype of altered antibiotic sensitivity.
- Aspects of the transcriptional regulation systems remain to be explored more fully, especially those involving the differences and overlap of the glycine and one-carbon metabolism responses, and that activating Cha1p and its relationship to sphingolipid metabolism.
- Translation of the control mechanisms and metabolic systems in *S. cerevisiae* through to other organisms.

Acknowledgements

The authors are very grateful for the invaluable discussion and research impetus from a succession of PhD students, including David Sinclair, Seung-Pyo (Raymond) Hong, Matthew Piper and Cristy Gelling.

References

Abdulrehman, D., Monteiro, P.T., Teixeira, M.C., Mira, N.P., Lourenço, A.B., dos Santos, S.S., Cabrito, T.R., Francisco, A.P., Madeira, S.C., Aires, R.S. *et al.* (2011) YEASTRACT: providing a programmatic access to curated transcriptional regulatory associations in *Saccharomyces cerevisiae* through a web services interface. *Nucleic Acids Research* 39, D136–D140.

Albers, E., Laize, V., Blomberg, A., Hohmann, S. and Gustafsson, L. (2003) Ser3p (Yer081wp) and Ser33p (Yil074cp) are phosphoglycerate dehydrogenases in *Saccharomyces cerevisiae*. *The Journal of Biological Chemistry* 278, 10264–10272.

Alic, N., Felder, T., Temple, M.D., Gloeckner, C., Higgins, V.J., Briza, P. and Dawes, I.W. (2004) Genome-wide transcriptional responses to a lipid hydroperoxide: adaptation occurs without induction of oxidant defenses. *Free Radical Biology and Medicine* 37, 23–35.

Appling, D.A. and Rabinowitz, J.C. (1985) Regulation of expression of the *ADE3* gene for yeast tetrahydrofolate synthase, a trifunctional enzyme involved in one-carbon metabolism. *The Journal of Biological Chemistry* 260, 1248–1256.

Ayer, A., Tan, S.X., Grant, C.M., Meyer, A.J., Dawes, I.W. and Perrone, G.G. (2010) The critical role of glutathione in maintenance of the mitochondrial genome. *Free Radical Biology and Medicine* 49, 1956–1968.

Bae-Lee, M.S. and Carman, G.M. (1984) Phosphatidylserine synthesis in *Saccharomyces cerevisiae.* Purification and characterization of membrane-associated phosphatidylserine synthase. *The Journal of Biological Chemistry* 259, 10857–10862.

Ballarin-Denti, A., Den Hollander, J.A., Sanders, D., Slayman, C.W. and Slayman, C.L. (1984) Kinetics and pH-dependence of glycine-proton symport in *Saccharomyces cerevisiae*. *Biochimica et Biophysica Acta* 778, 1–16.

Belhumeur, P., Fortin, N. and Clark, M.W. (1994) A gene from *Saccharomyces cerevisiae* which codes for a protein with significant homology to the bacterial 3-phosphoserine aminotransferase. *Yeast* 10, 385–389.

Blakley, R.L. (1969) *The Biochemistry of Folic Acid and Related Pteridines*. North Holland, Amsterdam.

Bornaes, C., Ignjatovic, M.W., Schjerling, P., Kielland-Brandt, M.C. and Holmberg, S. (1993) A regulatory element in the *CHA1* promoter which confers inducibility by serine and threonine on *Saccharomyces cerevisiae* genes. *Molecular Cell Biology* 13, 7604–7611.

Botsford, J.L. Jr and Parks, L.W. (1969) Serine transhydroxymethylase in methionine biosynthesis in *Saccharomyces cerevisiae*. *Journal of Bacteriology* 97, 1176–1183.

Cain, N.E. and Kaiser, C.A. (2011) Transport activity-dependent intracellular sorting of the yeast general amino acid permease. *Molecular Biology of the Cell* 22, 1919–1929.

Chanarin, I. (1969) *The Megaloblastic Anaemias*. Blackwell Scientific, Oxford, UK.

Cherest, H. and Surdin-Kerjan, Y. (1992) Genetic analysis of a new mutation conferring cysteine auxotrophy in *Saccharomyces cerevisiae*: updating of the sulfur metabolism pathway. *Genetics* 130, 51–58.

Cherest, H., Thomas, D. and Surdin-Kerjan, Y. (1993) Cysteine biosynthesis in *Saccharomyces cerevisiae* occurs through the transsulfuration pathway which has been built up by enzyme recruitment. *Journal of Bacteriology* 175, 5366–5374.

Cooper, T.G. and Sumrada, R.A. (1983) What is the function of nitrogen catabolite repression in *Saccharomyces cerevisiae*? *Journal of Bacteriology* 155, 623–627.

Cowart, L.A. and Hannun, Y.A. (2007) Selective substrate supply in the regulation of yeast *de novo* sphingolipid synthesis. *The Journal of Biological Chemistry* 282, 12330–12340.

Czeizel, A.E. and Dudas, I. (1992) Prevention of the first occurrence of neural-tube defects by periconceptional vitamin supplementation. *The New England Journal of Medicine* 327, 1832–1835.

de Koning, T.J. and Klomp, L.W. (2004) Serine-deficiency syndromes. *Current Opinion in Neurology* 17, 197–204.

Dean, J.C., Robertson, Z., Reid, V., Wang, Q., Hailey, H., Moore, S., Rasalam, A.D., Turnpenny, P., Lloyd, D., Shaw, D. and Little, J. (2008) A high frequency of the MTHFR 677C>T polymorphism in Scottish women with epilepsy: possible role in pathogenesis. *Seizure*, 17, 269–275.

Dickson, R.C. (1998) Sphingolipid functions in *Saccharomyces cerevisiae:* comparison to mammals. *Annual Review of Biochemistry* 67, 27–48.

Forsberg, H. and Ljungdahl, P.O. (2001) Genetic and biochemical analysis of the yeast plasma membrane Ssy1p-Ptr3p-Ssy5p sensor of extracellular amino acids. *Molecular Cell Biology* 21, 814–826.

Gasch, A.P., Spellman, P.T., Kao, C.M., Carmen-Harel, O., Eisen, M.B., Storz, G., Botstein, D. and Brown, P.O. (2000) Genomic expression programs in the response of yeast cells to environmenal changes. *Molecular Biology of the Cell* 11, 4241–4257.

Gelling, C.L. (2008) Tetrahydrofolate and iron-sulfur metabolism in *Saccharomyces cerevisiae.* PhD thesis, University of New South Wales, Sydney, Australia.

Gelling, C.L., Piper, M.D.W., Hong, S.-P., Kornfeld, G.D. and Dawes, I.W. (2004) Identification of a novel one-carbon metabolism regulon in *Saccharomyces cerevisiae*. *The Journal of Biological Chemistry* 279, 7072–7081.

Gibney, P.A., Lu, C., Caudy, A.A., Hess, D.C. and Botstein, D. (2013) Yeast metabolic and signaling genes are required for heat-shock survival and have little overlap with the heat-induced genes. *Proceedings of the National Academy of Sciences od the United States of America* 110, E4393–E4402.

Grant, C.M., MacIver, F.H. and Dawes, I.W. (1996) Glutathione is an essential metabolite required for resistance to oxidative stress in the yeast *Saccharomyces cerevisiae*. *Current Genetics* 29, 511–515.

Grenson, M., Hou, C. and Crabeel, M. (1970) Multiplicity of the amino acid permeases in *Saccharomyces cerevisiae*. IV. Evidence for a general amino acid permease. *Journal of Bacteriology* 103, 770–777.

Hofman-Bang, J. (1999) Nitrogen catabolite repression in *Saccharomyces cerevisiae*. *Molecular Biotechnology* 12, 35–73.

Holmberg, S. and Schjerling, P. (1996) Cha4p of *Saccharomyces cerevisiae* activates transcription via serine/threonine response elements. *Genetics* 144, 467–478.

Hong, S.-P., Piper, M.D., Sinclair, D.A. and Dawes, I.W. (1999) Control of expression of one-carbon metabolism genes of *Saccharomyces cerevisiae* is mediated by a tetrahydrofolate-responsive protein binding to a glycine regulatory region including a core 5′-CTTCTT-3′ motif. *The Journal of Biological Chemistry* 274, 10523–10532.

Houlston, R.S. and Tomlinson, I.P. (2001) Polymorphisms and colorectal tumor risk. *Gastroenterology* 121, 282–301.

Jaeken, J., Detheux, M., Van Maldergem, L., Foulon, M., Carchon, H. and Van Schaftingen, E. (1996) 3-Phosphoglycerate dehydrogenase deficiency: an inborn error of serine biosynthesis. *Archives of Disease in Childhood* 74, 542–545.

Jarolim, S., Ayer, A., Pillay, B., Gee, A.C., Phrakaysone, A., Perrone, G.G., Breitenbach, M. and Dawes, I.W. (2013) *Saccharomyces cerevisiae* genes involved in survival of heat shock. *Genes, Genomes and Genetics* 2321–2333.

Jauniaux, J.C. and Grenson, M. (1990) *GAP1*, the general amino acid permease gene of *Saccharomyces cerevisiae*. Nucleotide sequence, protein similarity with the other bakers yeast amino acid permeases, and nitrogen catabolite repression. *European Journal of Biochemistry* 190, 39–44.

Kanfer, J. and Kennedy, E.P. (1964) Metabolism and function of bacterial lipids. II. Biosynthesis of phospholipids in *Escherichia coli*. *The Journal of Biological Chemistry* 239, 1720–1726.

Kastanos, E.K., Woldman, Y.Y. and Appling, D.R. (1997) Role of mitochondrial and cytoplasmic serine hydroxymethyltransferase isozymes in *de novo* purine synthesis in *Saccharomyces cerevisiae*. *Biochemistry* 36, 14956–14964.

Kitamoto, K., Yoshizawa, K., Ohsumi, Y. and Anraku, Y. (1988) Dynamic aspects of vacuolar and cytosolic amino acid pools of *Saccharomyces cerevisiae*. *Journal of Bacteriology* 170, 2683–2686.

Kumar, C., Igbaria, A., D'Autreaux, B., Planson, A.G., Junot, C., Godat, E., Bachhawat, A.K., Delaunay-Moisan, A. and Toledano, M.B. (2011) Glutathione revisited: a vital function in iron metabolism and ancillary role in thiol-redox control. *The EMBO Journal* 30, 2044–2056.

Larsson, S.C., Giovannucci, E. and Wolk, A. (2006) Folate intake, MTHFR polymorphisms, and risk of esophageal, gastric, and pancreatic cancer: a meta-analysis. *Gastroenterology* 131, 1271–1283.

Lee, J.C., Tsoi, A., Kornfeld, G.D. and Dawes, I.W. (2013) Cellular responses to L-serine in *Saccharomyces cerevisiae*: roles of general amino acid control, compartmentalization, and aspartate synthesis. *FEMS Yeast Research* 13, 618–634.

Ljungdahl, P.O. (2009) Amino-acid-induced signalling via the SPS-sensing pathway in yeast. *Biochemical Society Transactions* 37, 242–247.

López-León, S., Janssens, A.C., González-Zuloeta Ladd, A.M., Del-Favero, J., Claes, S.J., Oostra, B.A. and van Duijn, C.M. (2008) Meta-analyses of genetic studies on major depressive disorder. *Molecular Psychiatry* 13, 772–785.

Lor, K.L. and Cossins, E.A. (1972) Regulation of C metabolism by L-methionine in *Saccharomyces cerevisiae*. *The Biochemical Journal* 130, 773–783.

Marchal, C., Redondo, M., Reyes-Engel, A., Perea-Milla, E., Gaitan, M.J., Machuca, J., Diaz, F., Caballero, J. and Carnero, J. (2008) Association between polymorphisms of folate-metabolizing enzymes and risk of prostate cancer. *European Journal of Surgical Oncology* 34, 805–810.

Martens, J.A., Laprade, L. and Winston, F. (2004) Intergenic transcription is required to repress the *Saccharomyces cerevisiae SER3* gene. *Nature*, 429, 571–74.

Martens, J.A., Wu, P.Y. and Winston, F. (2005) Regulation of an intergenic transcript controls adjacent gene transcription in *Saccharomyces cerevisiae. Genes and Development* 19, 2695–2704.

Matthews, R.G., Ross, J., Baugh, C.M., Cook, J.D. and Davis, L. (1982) Interactions of pig liver serine hydroxymethyltransferase with methyltetrahydropteroylpolyglutamate inhibitors and with tetrahydropteroylpolyglutamate substrates. *Biochemistry* 21, 1230–1238.

McAlister, L., Strausberg, S., Kulaga, A. and Finkelstein, D.B. (1979) Altered patterns of protein synthesis induced by heat shock of yeast. *Current Genetics*, 1, 63–74.

McNeil, J.B., McIntosh, E.M., Taylor, B.V., Zhang, F.R. and Bognar, A.L. (1994) Cloning and molecular characterization of three genes, including two genes encoding serine hydroxymethyltransferases, whose inactivation is required to render yeast auxotrophic for glycine. *The Journal of Biological Chemistry* 1994, 9155–9165.

McNeil, J.B., Bognar, A.L. and Pearlman, R.E. (1996) *In vivo* analysis of folate coenzymes and their compartmentation in *Saccharomyces cerevisae*. *Genetics* 142, 371–381.

McNeil, J.B., Zhang, F., Taylor, B.V., Sinclair, D.A., Pearlman, R.E. and Bognar, A.L. (1997) Cloning, and molecular characterization of the *GCV1* gene encoding the glycine cleavage T-protein from *Saccharomyces cerevisiae*. *Gene* 186, 13–20.

Meisch, H.U. and Kappesser, S. (1987) Activation of serine sulphydrase from baker's yeast (*Saccharomyces cerevisiae*) by vanadate. *Biochimica et Biophysica Acta* 925, 234–237.

Melcher, K. and Entian, K.D. (1992) Genetic analysis of serine biosynthesis and glucose repression in yeast. *Current Genetics* 21, 295–300.

Melcher, K., Rose, M., Kunzler, M., Braus, G.H. and Entian, K.D. (1995) Molecular analysis of the yeast *SER1* gene encoding 3-phosphoserine aminotransferase: regulation by general control and serine repression. *Current Genetics* 27, 501–508.

Miller, M.J., Xuong, N.H. and Geiduschek, E.P. (1979) A response of protein synthesis to temperature shift in the yeast *Saccharomyces cerevisiae*. *Proceedings of the National Academy of Sciences of the United States of America* 76, 5222–5225.

Monschau, N., Stahmann, K.P., Sahm, H., McNeil, J.B. and Bognar, A.L. (1997) Identification of *Saccharomyces cerevisiae* GLY1 as a threonine aldolase: a key enzyme in glycine biosynthesis. *FEMS Microbiology Letters* 150, 55–60.

Montefusco, D.J., Newcomb, B., Gandy, J.L., Brice, S.E., Matmati, N., Cowart, L.A. and Hannun, Y.A. (2012) Sphingoid bases and the serine catabolic enzyme CHA1 define a novel feedforward/feedback mechanism in the response to serine availability. *The Journal of Biological Chemistry* 287, 9280–9289.

Muskiet, F.A. and Kemperman, R.F. (2006) Folate and long-chain polyunsaturated fatty acids in psychiatric disease. *The Journal of Nutritional Biochemistry* 17, 717–727.

Nagarajan, L. and Storms, R.K. (1997) Molecular characterization of *GCV3*, the *Saccharomyces cerevisiae* gene coding for the glycine cleavage system hydrogen carrier protein. *The Journal of Biological Chemistry* 272, 4444–4450.

Natarajan, K., Meyer, M.R., Jackson, B.M., Slade, D., Roberts, C., Hinnebusch, A.G. and Marton, M.J. (2001) Transcriptional profiling shows that Gcn4p is a master regulator of gene expression during amino acid starvation in yeast. *Molecular Cell Biology* 21, 4347–4368.

Ogur, M., Liu, T.N., Cheung, I., Paulavicus, I., Wales, W., Mehnert, D. and Blaise, D. (1977) "Active" one-carbon generation in *Saccharomyces cerevisiae*. *Journal of Bacteriology* 129, 926–933.

Omnus, D.J. and Ljungdahl, P.O. (2013) Rts1-protein phosphatase 2A antagonizes Ptr3-mediated activation of the signaling protease Ssy5 by casein kinase I. *Molecular Biology of the Cell* 24, 1480–1492.

Ono, B.I., Hazu, T., Yoshida, S., Kawato, T., Shinoda, S., Brzvwczy, J. and Paszewski, A. (1999) Cysteine biosynthesis in *Saccharomyces cerevisiae*: a new outlook on pathway and regulation. *Yeast* 15, 1365–1375.

Pasternack, L.B., Laude, D.A.J. and Appling, D.R. (1992) ^{13}C NMR detection of folate-mediated serine and glycine synthesis *in vivo* in *Saccharomyces cerevisiae*. *Biochemistry* 31, 8713–8719.

Pasternack, L.B., Laude, D.A.J. and Appling, D.R. (1994) ^{13}C NMR analysis of intercompartmental flow of one-carbon units into choline and purines in *Saccharomyces cerevisiae*. *Biochemistry* 33, 74–82.

Patriarca, E.J. and Maresca, B. (1990) Acquired thermotolerance following heat shock protein synthesis prevents impairment of mitochondrial ATPase activity at elevated temperatures in *Saccharomyces cerevisiae*. *Experimental Cell Research* 190, 57–64.

Petersen, J.G., Kielland-Brandt, M.C., Nilsson-Tillgren, T., Bornaes, C. and Holmberg, S. (1988) Molecular genetics of serine and threonine catabolism in *Saccharomyces cerevisiae*. *Genetics* 119, 527–534.

Piper, M.D.W. (2000) Regulation of one-carbon metabolism in *Saccharomyces cerevisiae.* PhD thesis, University of New South Wales, Sydney, Australia.

Piper, M.D.W., Hong, S.-P., Ball, G.E. and Dawes, I.W. (2000) Regulation of the balance of one-carbon metabolism in *Saccharomyces cerevisiae*. *The Journal of Biological Chemistry* 275, 30987–30995.

Piper, M.D., Hong, S.-P., Eissing, T., Sealey, P. and Dawes, I.W. (2002) Regulation of the yeast glycine cleavage genes is responsive to the availability of multiple nutrients. *FEMS Yeast Research* 2, 59–71.

Pitkin, R.M. (2007) Folate and neural tube defects. *American Journal of Clinical Nutrition* 85, 285S–288S.

Ramos, F. and Wiame, J.M. (1982) Occurrence of a catabolic L-serine (L-threonine) deaminase in *Saccharomyces cerevisiae. European Journal of Biochemistry* 123, 571–576.

Regenberg, B., During-Olsen, L., Kielland-Brandt, M.C. and Holmberg, S. (1999) Substrate specificity and gene expression of the amino-acid permeases in *Saccharomyces cerevisiae*. *Current Genetics* 36, 317–328.

Reiner, S., Micolod, D., Zellnig, G. and Schneiter, R. (2006) A genomewide screen reveals a role of mitochondria in anaerobic uptake of sterols in yeast. *Molecular Biology of the Cell* 17, 90–103.

Richter, K., Haslbeck, M. and Buchner, J. (2010) The heat shock response: life on the verge of death. *Molecular Cell* 40, 253–266.

Rowley, A., Johnston, G.C., Butler, B., Werner-Washburne, M. and Singer, R.A. (1993) Heat shock-mediated cell cycle blockage and G1 cyclin expression in the yeast *Saccharomyces cerevisiae*. *Molecular Cell Biology* 13, 1034–1041.

Schirch, L.V. (1984) Folates in serine and glycine metabolism. In: Blakley, R.L. and Benkovic, S.J. (eds) *Folates and Pterins*. Wiley, New York, pp. 399–431.

Schirch, L.V., Tatum, C.M. Jr and Benkovic, S.J. (1977) Serine transhydroxymethylase: evidence for a sequential random mechanism. *Biochemistry* 16, 410–419.

Schlösser, T., Gatgens, C., Weber, U. and Stahmann, K.P. (2004) Alanine: glyoxylate aminotransferase of *Saccharomyces cerevisiae* – encoding gene *AGX1* and metabolic significance. *Yeast* 21, 63–73.

Scholer, A. and Schuller, H.J. (1993) Structure and regulation of the isocitrate lyase gene *ICL1* from the yeast *Saccharomyces cerevisiae*. *Current Genetics* 23, 375–381.

Schreve, J.L., Sin, J.K. and Garrett, J.M. (1998) The *Saccharomyces cerevisiae YCC5* (*YCL025c*) gene encodes an amino acid permease, Agp1, which transports asparagine and glutamine. *Journal of Bacteriology* 180, 2556–2559.

Sinclair, D.A. and Dawes, I.W. (1995) Genetics of the synthesis of serine from glycine and the utilization of glycine as sole nitrogen source by *Saccharomyces cerevisiae*. *Genetics* 140, 1213–1222.

Sinclair, D.A., Hong, S.-P. and Dawes, I.W. (1996) Specific induction by glycine of the gene for the P-subunit of glycine decarboxylase from *Saccharomyces cerevisiae*. *Molecular Microbiology* 19, 611–623.

Singh, S., Padovani, D., Leslie, R.A., Chiku, T. and Banerjee, R. (2009) Relative contributions of cystathionine β-synthase and γ-cystathionase to H_2S biogenesis via alternative trans-sulfuration reactions. *The Journal of Biological Chemistry* 284, 22457–22466.

Skrzypek, M.S., Nagies, M.M., Lester, R.L. and Dickson, R.C. (1999) Analysis of phosphorylated sphingolipid long-chain bases reveals potential roles in heat stress and growth control in *Saccharomyces*. *Journal of Bacteriology* 181, 1134–1140.

Smith, Z.D. and Meissner, A. (2013) DNA methylation: roles in mammalian development. *Nature Reviews Genetics* 14, 204–220.

Stover, P. and Schirch, V. (1991) 5-Formyltetrahydrofolate polyglutamates are slow tight binding inhibitors of serine hydroxymethyltransferase. *The Journal of Biological Chemistry* 266, 1543–1550.

Subramanian, M., Qiao, W.B., Khanam, N., Wilkins, O., Der, S.D., Lalich, J.D. and Bognar, A.L. (2005) Transcriptional regulation of the one-carbon metabolism regulon in *Saccharomyces cerevisiae* by Bas1p. *Molecular Microbiology* 57, 53–69.

Summers, E.F., Letts, V.A., McGraw, P. and Henry, S.A. (1988) *Saccharomyces cerevisiae cho2* mutants are deficient in phospholipid methylation and cross-pathway regulation of inositol synthesis. *Genetics* 120, 909–922.

Tsoi, B.M., Beckhouse, A.G., Gelling, C.L., Raftery, M.J., Chiu, J., Tsoi, A.M., Lauterbach, L., Rogers, P.J., Higgins, V.J. and Dawes, I.W. (2009) Essential role of one-carbon metabolism and Gcn4p and Bas1p transcriptional regulators during adaptation to anaerobic growth of *Saccharomyces cerevisiae*. *The Journal of Biological Chemistry* 284, 11205–11215.

Turcotte, B., Liang, X.B., Robert, F. and Soontorngun, N. (2010) Transcriptional regulation of nonfermentable carbon utilization in budding yeast. *FEMS Yeast Research* 10, 2–13.

Ulane, R. and Ogur, M. (1972) Genetic and physiological control of serine and glycine biosynthesis in *Saccharomyces*. *Journal of Bacteriology* 109, 34–43.

Vidair, C.A., Huang, R.N. and Doxsey, S.J. (1996) Heat shock causes protein aggregation and reduced protein solubility at the centrosome and other cytoplasmic locations. *International Journal of Hyperthermia* 12, 681–695.

West, M.G., Barlowe, C.K. and Appling, D.R. (1993) Cloning and characterization of the *Saccharomyces cerevisiae* gene encoding NAD-dependent 5,10-methylene-tetrahydrofolate dehydrogenase. *The Journal of Biological Chemistry* 268, 153–160.

Winter, D.L., Abeygunawardena, D., Hart-Smith, G., Erce, M.A. and Wilkins, M.R. (2015) Lysine methylation modulates the protein–protein interactions of yeast cytochrome *c* Cyc1p. *Proteomics* 15, 2166–2176.

Yagoub, D., Hart-Smith, G., Moecking, J., Erce, M.A. and Wilkins, M.R. (2015) Yeast proteins Gar1p, Nop1p, Npl3p, Nsr1p and Rps2p are natively methylated and are substrates of the arginine methyltransferase Hmt1p. *Proteomics* 15, 3209–3218.

12 Threonine Degradation in Hyperthermophilic Organisms

Q. Bashir,[1] N. Rashid[1]* and M. Akhtar[1,2]
[1]School of Biological Sciences, University of the Punjab, Quaid-e-Azam Campus, Lahore, Pakistan; [2]School of Biological Sciences, University of Southampton, Southampton, UK

12.1 Abstract

Threonine degradation involves three key enzymes that lead to three different pathways. These are threonine dehydrogenase, threonine dehydratase and threonine aldolase. A search for homologues of these enzymes has revealed the presence of threonine dehydratase and threonine aldolase in the genomes of hyperthermophilic bacteria. Among the hyperthermophilic archaea, threonine dehydrogenase is present in members of *Euryarchaeota*, whereas threonine dehydratase is found in members of *Crenarchaeota*. In fact, threonine dehydrogenase has been characterized from several members of the *Euryarchaeota*. Based on the enzymes studied as well as on the presence of gene homologues, it can be postulated that all three pathways exist in hyperthermophiles. In this chapter, these three metabolic pathways, their presence or absence in a particular hyperthermophile and an account of the thermostable enzymes involved in the pathways are described.

12.2 Introduction

In recent years, a large number of microorganisms have been isolated from environments that were previously considered too harsh to sustain microbial life. These include hot springs at temperatures up to or even exceeding the boiling point of water, cold habitats such as the Arctic and Antarctic regions, highly acidic niches such as acid mine drainage springs or sulfuric pools, alkaline hot springs or soda lakes, and salt-saturated waters or salt mines. The microorganisms isolated from these environments are termed (hyper)thermophiles, psychrophiles, acidophiles, alkalophiles and halophiles on the basis of their habitats. These organisms thrive in their harsh environments and exhibit unusual biological properties that had not been found in previously known forms of life. Therefore, they are attractive research targets.

Hyperthermophiles are microorganisms with optimal growth temperatures above 80°C (Adams and Kelly, 1998; Stetter, 1999). They form a cluster around the root of the tree of life and are likely to be the most primitive organisms in existence. It is probable that the last common ancestor of all present-day life was a hyperthermophile. All hyperthermophiles known to date except for *Thermotoga* and *Aquifex* belong to the archaea. The complete genome sequences of more than 235 archaean species have been determined and this has contributed enormously to the identification of the presence or absence of a particular metabolic pathway in these organisms.

*Corresponding author. E-mail address: naeem.ff.sbs@pu.edu.pk; naeemrashid37@hotmail.com

Threonine is a polar amino acid with a side chain containing a hydroxyl group. It is synthesized by plants and microorganisms. In humans, it is an indispensable amino acid and must be derived from a dietary source. Three pathways for the degradation of threonine are reported to exist. However, there are conflicting reports on the predominance of particular pathways. In the present chapter, we will elucidate the threonine degradation pathways in hyperthermophiles based on the enzymes that have been studied from them, as well as the presence or absence of homologous genes in their genomes.

12.3 Threonine Degradation Pathways

There are three pathways for threonine degradation in various organisms. The first pathway involves the conversion of threonine to 2-amino-3-ketobutyrate and is catalysed by an NAD⁺-dependent threonine dehydrogenase. The second is catalysed by threonine aldolase, which converts threonine into glycine and acetaldehyde. The third pathway employs threonine dehydratase, which catalyses the conversion of threonine into 2-ketobutyrate, from which propionate or isoleucine are eventually formed (Fig. 12.1).

The first pathway, involving the conversion of threonine to 2-amino-3-ketobutyrate, is two-step conversion of L-threonine into glycine and acetyl-CoA (Bird and Nunn, 1983). The initial step of the pathway is the NAD⁺-dependent conversion of L-threonine into 2-amino-3-ketobutyrate catalysed by the enzyme L-threonine dehydrogenase. Another enzyme, 2-amino-3-ketobutyrate CoA ligase, converts 2-amino-3-ketobutyrate and CoA into glycine and acetyl-CoA. The intermediate in the conversion from threonine, 2-amino-3-ketobutyrate, is unstable and can spontaneously decarboxylate, liberating CO_2 plus aminoacetone (Fig. 12.2). The acetyl-CoA produced as a result of the above two reactions is used in a variety of metabolic reactions, while the glycine that is produced can be used for serine biosynthesis (Karp *et al.*, 2000). Aminoacetone formed is thought to be incorporated into vitamin B_{12} after being reduced to D-1-amino-2-propanol (Campbell and Dekker, 1973).

The reactions catalysed by L-threonine dehydrogenase and 2-amino-3-ketobutyrate CoA ligase are coupled, both *in vivo* and *in vitro* (Marcus and Dekker, 1993; Bashir *et al.*, 2006). The two enzymes may form a complex, most likely containing one tetramer of L-threonine dehydrogenase and two dimers of 2-amino-3-ketobutyrate CoA ligase (Tressel *et al.*, 1986), though this feature has not been demonstrated unambiguously. The coordinated activity of these two

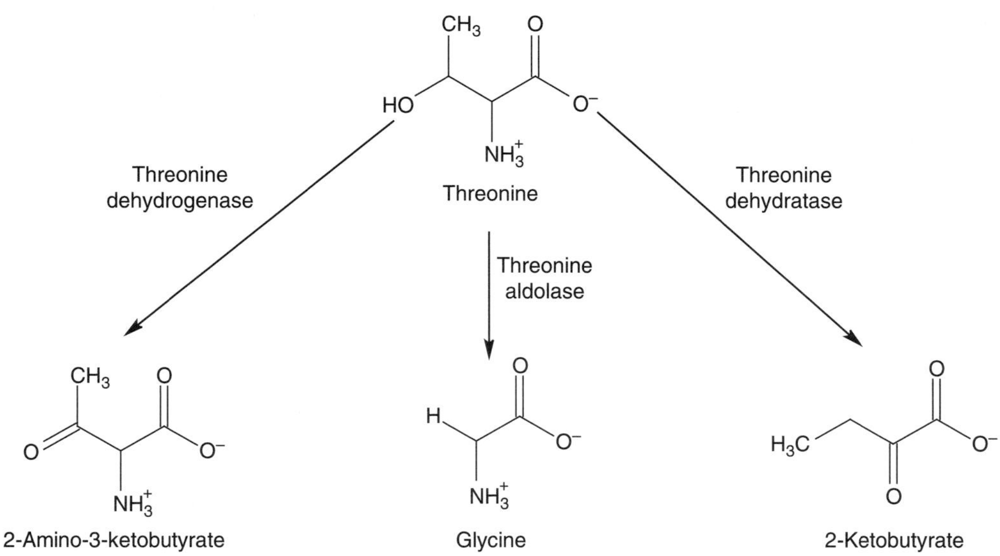

Fig. 12.1. The three threonine degradation pathways leading to the synthesis of 2-amino-3-ketobutyrate, glycine and 2-ketobutyrate.

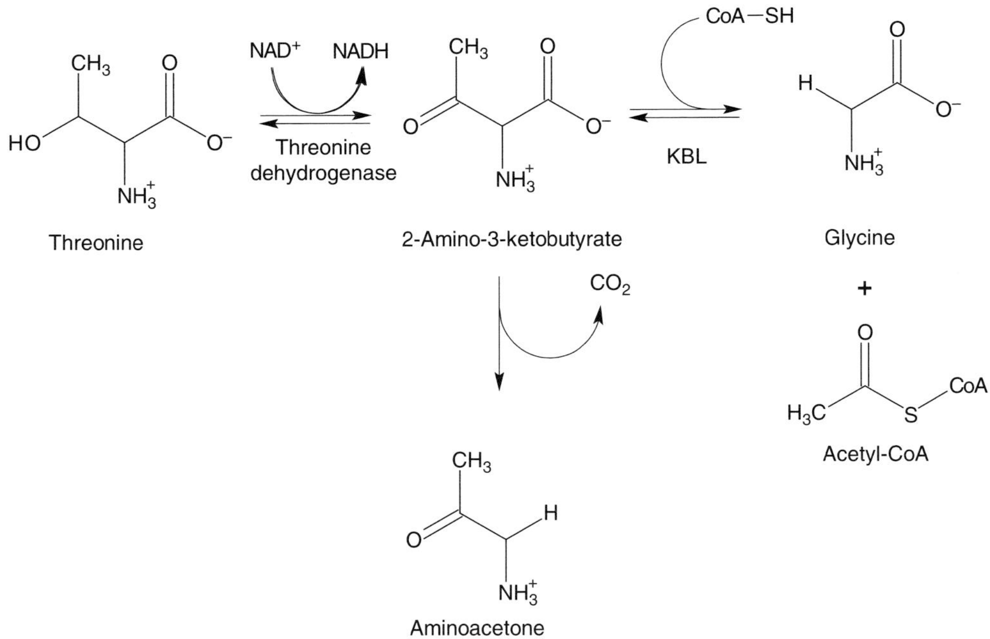

Fig. 12.2. The threonine degradation pathway involving L-threonine dehydrogenase and 2-amino-3-ketobutyrate CoA ligase (KBL) for the synthesis of glycine and acetyl-CoA.

enzymes ensures that the highly reactive 2-amino-3-ketobutyrate intermediate is minimally decarboxylated to yield aminoacetone and CO_2. The equilibrium among threonine, 2-amino-3-ketobutyrate, glycine and aminoacetone is determined by the intracellular concentrations of threonine, NAD^+, glycine and CoA.

The pathway initiated by L-threonine dehydrogenase is recognized as a major route for threonine utilization in several prokaryotic (Komatsubara *et al.*, 1978; Boylan and Dekker, 1981) and eukaryotic (Dale, 1978) organisms. The coupled reactions catalysed by L-threonine dehydrogenase and 2-amino-3-ketobutyrate CoA ligase make it possible for some strains of *Escherichia coli* and *Pseudomonas aeruginosa* to utilize threonine as their sole carbon and energy source (Chan and Newman, 1981; Boylan and Dekker, 1983; Mancini and Castric, 1989); they also provide an efficient alternate route for glycine and serine biosynthesis (Chan and Newman, 1981; Ravnikar and Somerville, 1987). This pathway accounts for the majority of the threonine catabolized in rat liver (Bird and Nunn, 1983), and L-threonine dehydrogenase activity is the only threonine catabolic reaction that is detected in chicken liver extracts (Aoyama and Motokawa, 1981).

Expression of the L-threonine dehydrogenase gene in *E. coli* has been shown to be under the control of leucine responsive protein (Willins *et al.*, 1991), which induces the enzymes in this pathway seven- to nine-fold (Newman *et al.*, 1976; Potter *et al.*, 1977). Although most of the studies on the L-threonine dehydrogenase and 2-amino-3-ketobutyrate CoA ligase-initiated pathway have focused on its role in providing acetyl-CoA and glycine to the cell, i.e. threonine breakdown, some researchers (Marcus and Dekker 1993; Bashir *et al.*, 2006) have shown that the reaction can also occur in the reverse direction, i.e. threonine can be formed by the action of these two enzymes. The dehydrogenation catalysed by L-threonine dehydrogenase occurs at the β-position of L-threonine, whereas other amino acid dehydrogenases catalyse α-deamination reactions (Asano *et al.*, 1987; DeLuna *et al.*, 2001). In this respect, L-threonine dehydrogenase can be regarded as a kind of alcohol dehydrogenase.

The second pathway for threonine degradation is catalysed by threonine aldolase. In this pathway, the threonine is converted into glycine and acetaldehyde (Fig. 12.3). Threonine aldolase is a pyridoxal phosphate-dependent enzyme. It is a member of lyase family and is also

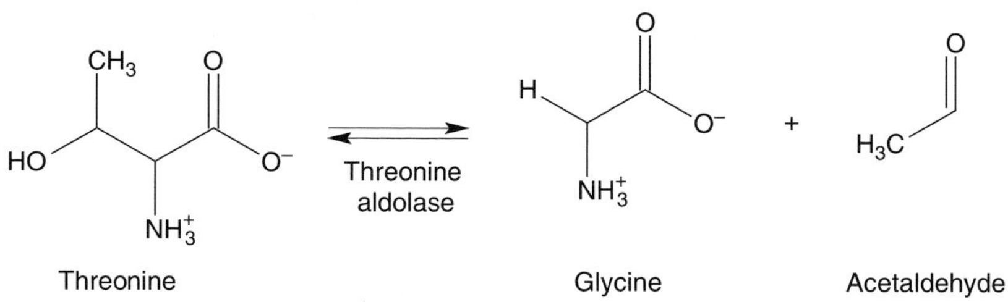

Fig. 12.3. The threonine aldolase pathway leading to the synthesis of glycine and acetaldehyde.

called as L-threonine acetaldehyde lyase. Threonine aldolase-encoding genes have been reported from several organisms (Liu *et al.*, 2000; Jander *et al.*, 2004; Edgar, 2005; Liu *et al.*, 2015). The X-ray structure of threonine aldolase from *Thermotoga maritima*, in its apo and substrate-bound form, has revealed that the enzyme exists as a homotetramer with each subunit containing pyridoxal phosphate bound to a lysine residue through Schiff base formation. The structure has also shown the presence of bound calcium and chloride ions, which have a possible role in catalysis and oligomerization (Kielkopf and Burley, 2002).

The third pathway for threonine degradation is catalysed by threonine dehydratase, also known as threonine ammonia-lyase. This involves degradation of threonine into 2-ketobutyrate and ammonia (Fig. 12.4). This is the first reaction of isoleucine biosynthesis pathway. Isoleucine is formed from 2-ketobytrate via methylation by hydroxyethyl thiamine pyrophosphate (TPP). Thus, threonine is a key enzyme in branched chain amino acid synthesis. Threonine dehydratase was the first enzyme that was shown to exhibit allosteric feedback inhibition by the end product (Umbarger and Brown, 1957). It is a homotetrameric enzyme with the monomeric subunits arranged as a dimer of dimers. Each monomer consists of two domains, one with the catalytic active site and the other containing allosteric modulation sites (Eisenstein, 1991; Gallagher *et al.*, 1998). The enzyme is a pyridoxal phosphate dependent, with the pyridoxal phosphate binding to a lysine residue through the formation of a Schiff base.

While the importance of the L-threonine dehydrogenase pathway in threonine catabolism is being increasingly recognized, the threonine aldolase-catalysed reaction seems to be much less significant than was once thought. Although threonine aldolase has been detected in a large number of microorganisms, its significance for threonine degradation is uncertain (Karasek and Greenberg, 1957; Newman *et al.*, 1976; Bird and Nunn 1983; Yeung, 1986).

The relative importance of threonine dehydratase pathway to the overall degradation of threonine is also debatable, because the biosynthetic form of this enzyme is inhibited by small amounts of isoleucine (Umbarger, 1973, 1978) and biodegradative threonine dehydratase is detectable only when cells are grown anaerobically or in the absence of glucose (Potter *et al.*, 1977; Goss and Datta, 1985). This pathway is not common in microorganisms growing on threonine as sole carbon source.

12.4 Threonine Degradation in Hyperthermophiles

Experimental evidence for the presence of L-threonine aldolase or dehydratase in hyperthermophiles is scanty, whereas L-threonine dehydrogenases have been characterized from several hyperthermophiles, including *Thermococcus kodakarensis* (formerly *T. kodakaraensis*; Bashir *et al.*, 2009; Bowyer *et al.*, 2009), *Pyrococcus horikoshii* (Higashi *et al.*, 2005, 2008; Shimizu *et al.*, 2005), *P. furiosus* (Machielsen and van der Oost, 2006) and *Thermoplasma volcanium* (Yoneda *et al.*, 2012). Furthermore, their crystal structures have been solved and this has provided insights into their structural and functional mechanisms (Bowyer *et al.*, 2009; Yoneda *et al.*, 2010, 2012; Nakano *et al.*, 2014). The hyperthermophilic enzymes characterized so far have subunit molecular masses of 35–38 kDa and exist in tetrameric form.

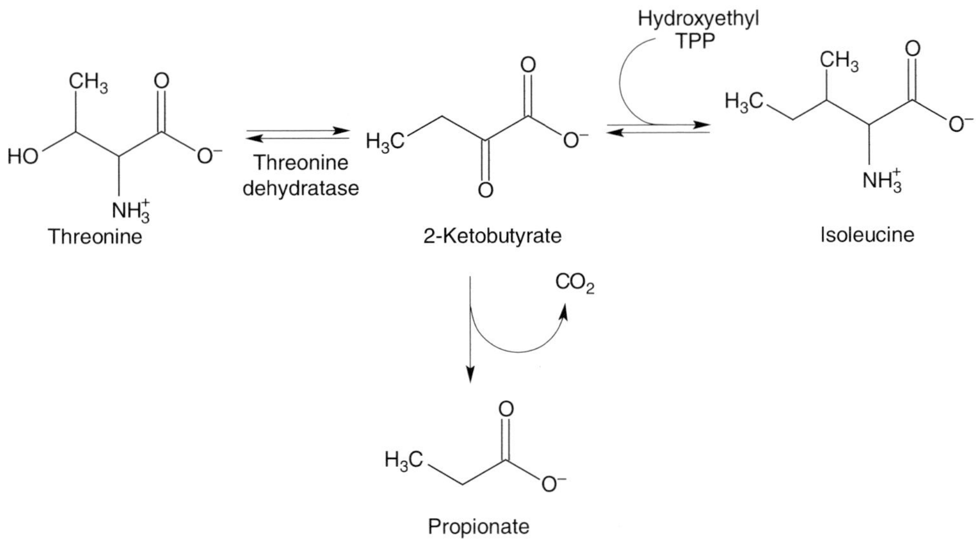

Fig. 12.4. The threonine dehydratase pathway leading to the synthesis of isoleucine via 2-ketobutyrate, which is methylated by hydroxyethyl thiamine pyrophosphate (TPP).

The enzyme from *P. horikoshii* was the first L-threonine dehydrogenase whose crystal structure was solved, in 2007. The homotetrameric enzyme shared high structural homology with medium-chain NAD-dependent alcohol dehydrogenase despite the fact that the two have very low sequence identity and have different substrate specificities. Each monomer consisted of an NAD-binding domain and a catalytic domain. An α/β Rossmann fold motif, characteristic of NAD-binding proteins, was present in the NAD-binding domain. Alcohol dehydrogenases usually contain two Zn ions, one playing a structural role and the other a catalytic role, but only a single Zn ion was found per molecule of L-threonine dehydrogenase from *P. horikoshii*. The single Zn ion is coordinated to four cysteine ligands and is thought to play a structural role, but there is controversy about whether it has a structural or catalytic significance. Each monomer contains a large interdomain cleft that is able to accommodate the threonine substrate and NAD(H).

The crystal structure of L-threonine dehydrogenase from *T. kodakarensis* demonstrates high homology with the L-threonine dehydrogenase from *P. horikoshii* and with alcohol dehydrogenase, with the same overall fold of monomer and tetramer. Each monomer of the tetrameric enzyme contains a catalytic domain and an NAD-binding domain with a Rossmann fold motif. Interestingly, each monomer contained a bound NAD cofactor, although NAD was not present in the purification or crystallization procedures. This implies that the cofactor has strong binding mediated by a large number of contacts with the enzyme and water molecules. It was also found that each pair of monomers associated to form an extended 12-stranded β-sheet. The researchers performed docking studies to investigate the interaction of L-threonine dehydrogenase with 2-amino-3-ketobutyrate CoA ligase, the second enzyme in the pathway of conversion of threonine into glycine. The docking studies suggested that L-threonine dehydrogenase forms a functional complex with 2-amino-3-ketobutyrate CoA ligase, thereby shielding the unstable intermediate formed in this pathway.

L-Threonine dehydrogenase from the thermophilic archaeon *Thermoplasma volcanium* is quite distinct. Its amino acid sequence remarkably resembles that of UDP (uridine diphosphate)-galactose 4-epimerase. By site-directed mutagenesis, it has been determined that Tyr137 is essential for catalysis. The crystal structures of wild type and mutant (Tyr137Phe) were determined in the presence of the cofactor (NAD) and cofactor plus substrate or analogue of the substrate. The monomer contained a Rossmann fold domain and a C-terminal catalytic domain. The catalytic domain was remarkably similar to UDP-galactose 4-epimerase like L-threonine dehydrogenase from the pychrophile *Flavobacterium frigiimaris*, which has no sequence similarity with the conventional L-threonine dehydrogenases. Their substrate binding models suggest that the enzymatic reaction proceeds via removal of β-hydroxyl proton of L-threonine by proton transfer driven by Tyr137. Close comparison of thermophilic and psychrophilic L-threonine dehydrogenases reveals the presence of extensive ion pair networks within and among monomers, suggesting their possible role in the thermostability of *T. volcanium* L-threonine dehydrogenase.

As the reports on threonine degradation in hyperthermophiles are scanty, we have analysed the annotation of the sequences of the hyperthermophilic organisms whose complete genomes have been determined. Based on the annotation, even though all three threonine degradation pathways have been detected in various hyperthermophilic organisms, they have not been found simultaneously in any hyperthermophile. Genomes analyses of various phyla of hyperthermophiles have revealed the presence of one or more threonine degradation pathways that are specific for each phylum (see Table 12.1).

Table 12.1. Homologues of threonine-degrading enzymes found in the genomes of hyperthermophilic organisms.

		Threonine degradation enzyme		
Domain	Organism	Dehydrogenase	Dehydratase	Aldolase
Archaea: *Euryarchaeota*	*Pyrococcus abyssi*	WP_010868502	Not found	Not found
	P. furiosus	WP_011012128	Not found	Not found
	P. horikoshii	WP_010884751	Not found	Not found
	P. yayanosii	WP_013905768	Not found	Not found
	Thermococcus barophilus	WP_013466679	Not found	Not found
	T. eurythermalis	AIU70147	Not found	Not found
	T. gammatolerans	WP_015858763	Not found	Not found
	T. kodakarensis	WP_011249867	Not found	Not found
	T. litoralis	WP_004069171	Not found	Not found
	T. nautili	WP_042689212	Not found	Not found
	T. onnurineus	WP_012571354	Not found	Not found
	T. sibiricus	WP_048160234	Not found	Not found
Archaea: *Crenarchaeota*	*Pyrobaculum aerophilum*	Not found	NP_559924	Not found
	P. arsenaticum	Not found	YP_001153739	Not found
	P. calidifontis	Not found	WP_011850336	Not found
	P. islandicum	Not found	YP_930546	Not found
	Sulfolobus acidocaldarius	Not found	AAY80092	Not found
	S. islandicus	Not found	WP_012711865	Not found
	S. solfataricus	Not found	WP_009990508	Not found
	S. tokodaii	Not found	BAB65265	Not found
	Thermoproteus neutrophilus	Not found	YP_001794838	Not found
	T. tenax	Not found	YP_004892655	Not found
	T. uzoniensis	Not found	YP_004337344	Not found
Bacteria	*Thermotoga maritima*	Not found	AKE28154	AKE29499
	T. naphthophila	Not found	WP_011943186	WP_011943587
	T. neapolitana	Not found	ACM22460	WP_038067327
	T. petrophila	Not found	ABQ46585	ABQ47057

Organisms of the phylum Euryarchaeota contain L-threonine dehydrogenase as the sole threonine-degrading enzyme, while members of the phylum Crenarchaeota have threonine dehydratase as their only threonine-degrading enzyme. Hyperthermophiles of the archaeal domain show the existence of either of the two above-mentioned pathways, whereas hyperthermophilic bacteria contain two threonine degradation pathways, i.e. those via threonine dehydratase and threonine aldolase. Threonine dehydrogenase has not been found in the genomes of hyperthermophilic bacteria.

12.5 Conclusions

The evidence reviewed above supports an emerging consensus that, among hyperthermophilic archaea, the members of the Euryarchaeota harbour the genes of the threonine dehydrogenase pathway in contrast to the Crenarchaeota, where threonine dehydratase pathway is found. However, the hyperthermophilic bacteria possess the genes for both threonine dehydratase and threonine aldolase. Further studies are needed to establish the presence of a particular metabolic pathway in a particular type of organism.

References

Adams, M.W.W. and Kelly, R.M. (1998) Finding and using hyperthermophilic enzymes. *Trends in Biotechnology* 16, 329–332.

Aoyama, Y. and Motokawa, Y. (1981) L-threonine dehydrogenase of chicken liver. Purification, characterization and physiological significance. *The Journal of Biological Chemistry* 256, 12367–12373.

Asano, Y., Nakazawa, A. and Endo, K. (1987) Novel phenylalanine dehydrogenases from *Sporosarcina ureae* and *Bacillus sphaericus*. Purification and characterization. *The Journal of Biological Chemistry* 262, 10346–10354.

Bashir, Q., Rashid, N. and Akhtar, M. (2006) Mechanism and substrate stereochemistry of 2-amino-3-oxobutyrate CoA ligase: implications for 5-aminolevulinate synthase and related enzymes. *Chemical Communications* 48, 5065–5067.

Bashir, Q., Rashid, N., Jamil, F., Imanaka, T. and Akhtar, M. (2009) Highly thermostable L-threonine dehydrogenase from the hyperthermophilic archaeon *Thermococcus kodakaraensis*. *The Journal of Biochemistry* 146, 95–102.

Bird, M.I. and Nunn, P.B. (1983) Metabolic homoeostasis of L-threonine in the normally-fed rat. Importance of liver threonine dehydrogenase activity. *Biochemical Journal* 214, 687–694.

Bowyer, A., Mikolajek, H., Stuart, J.W., Wood, S.P., Jamil, F., Rashid, N., Akhtar, M. and Cooper, J.B. (2009) Structure and function of the L-threonine dehydrogenase (TkTDH) from the hyperthermophilic archaeon *Thermococcus kodakaraensis*. *Journal of Structural Biology* 168, 294–304.

Boylan, S.A. and Dekker, E.E. (1981) L-threonine dehydrogenase. Purification and properties of the homogeneous enzyme from *Escherichia coli* K-12. *The Journal of Biological Chemistry* 256, 1809–1815.

Boylan, S.A. and Dekker, E.E. (1983) Growth, enzyme levels, and some metabolic properties of an *Escherichia coli* mutant grown on L-threonine as the sole carbon source. *Journal of Bacteriology* 156, 273–280.

Campbell, R.L. and Dekker, E.E. (1973) Formation of D-1-amino-2-propanol from L-threonine by enzymes from *Escherichia coli* K-12. *Biochemical and Biophysical Research Communications* 53, 432–438.

Chan, T.T. and Newman, E.B. (1981) Threonine as a carbon source for *Escherichia coli*. *Journal of Bacteriology* 145, 1150–1153.

Dale, R.A. (1978) Catabolism of threonine in mammals by coupling of L-threonine 3-dehydrogenase with 2-amino-3-oxobutyrate-CoA ligase. *Biochimica et Biophysica Acta* 544, 496–503.

DeLuna, A., Avendaño, A., Riego, L. and González, A. (2001) NADP-glutamate dehydrogenase isoenzymes of *Saccharomyces cerevisiae*. Purification, kinetic properties and physiological roles. *The Journal of Biological Chemistry* 276, 43775–43783.

Edgar, A.J. (2005) Mice have a transcribed L-threonine aldolase/*GLY1* gene, but the human *GLY1* gene is a non-processed pseudogene. *BMC Genomics* 6:32.

Eisenstein, E. (1991) Cloning, expression, purification, and characterization of biosynthetic threonine deaminase from *Escherichia coli*. *The Journal of Biological Chemistry* 266, 5801–5807.

Gallagher, D.T., Gilliland, G.L., Xiao, G., Zondlo, J., Fisher, K.E., Chinchilla, D. and Eisenstein, E. (1998) Structure and control of pyridoxal phosphate dependent allosteric threonine deaminase. *Structure* 6, 465–475.

Goss, T.J. and Datta, P. (1985) Molecular cloning and expression of the biodegradative threonine dehydratase gene (*tdc*) of *Escherichia coli* K-12. *Molecular Genetics and Genomics* 201, 308–314.

Higashi, N., Fukada, H. and Ishikawa, K. (2005) Kinetic study of thermostable L-threonine dehydrogenase from an archaeon *Pyrococcus horikoshii*. *The Journal of Biochemistry* 99, 175–180.

Higashi, N., Tanimoto, K., Nishioka, M., Ishikawa, K. and Taya, M. (2008) Investigating a catalytic mechanism of hyperthermophilic L-threonine dehydrogenase from *Pyrococcus horikoshii*. *The Journal of Biochemistry* 144, 77–85.

Jander, G., Norris, S.R., Joshi, V., Fraga, M., Rugg, A., Yu, S., Li, L. and Last, R.L. (2004) Application of a high-throughput HPLC-MS/MS assay to *Arabidopsis* mutant screening; evidence that threonine aldolase plays a role in seed nutritional quality. *The Plant Journal* 39, 465–475.

Karasek, M.A. and Greenberg, D.M. (1957) Studies on the properties of threonine aldolases. *The Journal of Biological Chemistry* 227, 191–205.

Karp, P.D., Riley, M., Saier, M., Paulsen, I.T., Paley, S.M. and Pellegrini-Toole, A. (2000) The EcoCyc and MetaCyc databases. *Nucleic Acids Research* 28, 56–59.

Kielkopf, C.L. and Burley, S.K. (2002) X-ray structures of threonine aldolase complexes: structural basis of substrate recognition. *Biochemistry* 41, 11711–11720.

Komatsubara, S., Murata, K., Kisumi, M. and Chibata, I. (1978) Threonine degradation by *Serratia marcescens*. *Journal of Bacteriology* 135, 318–323.

Liu, G., Zhang, M., Chen, X., Zhang, W., Ding, W. and Zhang, Q. (2015) Evolution of threonine aldolases, a diverse family involved in the second pathway of glycine biosynthesis. *Journal of Molecular Evolution* 80, 102–107.

Liu, J.Q., Dairi, T., Itoh, N., Kataoka, M., Shimizu, S. and Yamada, H. (2000) Diversity of microbial threonine aldolases and their application. *Journal of Molecular Catalysis B: Enzymatic* 10, 107–115.

Machielsen, R. and van der Oost, J. (2006) Production and characterization of a thermostable L-threonine dehydrogenase from the hyperthermophilic archaeon *Pyrococcus furiosus*. *The FEBS Journal* 273, 2722–2729.

Mancini, M.A. and Castric, P.A. (1989) Threonine metabolism by *Pseudomonas aeruginosa*. *Current Microbiology* 18, 105–108.

Marcus, J.P. and Dekker, E.E. (1993) Threonine formation via the coupled activity of 2-amino-3-ketobutyrate coenzyme A lyase and threonine dehydrogenase. *Journal of Bacteriology* 175, 6505–6511.

Nakano, S., Okazaki, S., Tokiwa, H. and Asano, Y. (2014) Binding of NAD^+ and L-threonine induces stepwise structural and flexibility changes in *Cupriavidus nector* L-threonine dehydrogenase. *The Journal of Biological Chemistry* 289, 10445–10454.

Newman, E.B., Kapoor, V. and Potter, R. (1976) Role of L-threonine dehydrogenase in the catabolism of threonine and synthesis of glycine by *Escherichia coli*. *Journal of Bacteriology* 126, 1245–1249.

Potter, R., Kapoor, V. and Newman, E.B. (1977) Role of threonine dehydrogenase in *Escherichia coli* threonine degradation. *Journal of Bacteriology* 132, 385–391.

Ravnikar, P.D. and Somerville, R.L. (1987) Genetic characterization of a highly efficient alternate pathway of serine biosynthesis in *Escherichia coli*. *Journal of Bacteriology* 169, 2611–2617.

Shimizu, Y., Sakuraba, H., Kawakami, R., Goda, S., Kawarabayasi, Y. and Ohshima, T. (2005) L-Threonine dehydrogenase from the hyperthermophilic archaeon *Pyrococcus horikoshii* OT3: gene cloning and enzymatic characterization. *Extremophiles* 9, 317–324.

Stetter, K.O. (1999) Extremophiles and their adaptation to hot environments. *FEBS Letters* 452, 22–25.

Tressel, T., Thompson, R., Zieske, L.R., Menendez, M.I. and Davis, L. (1986) Interaction between L-threonine dehydrogenase and aminoacetone synthetase and mechanism of aminoacetone production. *The Journal of Biological Chemistry* 261, 16428–16437.

Umbarger, H.E. (1973) Threonine deaminases. *Advances in Enzymology and Related Areas of Molecular Biology* 37, 349–395.

Umbarger, H.E. (1978) Amino acid biosynthesis and its regulation. *Annual Review of Biochemistry* 47, 533–606.

Umbarger, H.E. and Brown, B. (1957) Threonine deamination in *Escherichia coli*. II. Evidence for two L-threonine deaminases. *Journal of Bacteriology* 73, 105–112.

Willins, D.A., Ryan, C.W., Platko, J.V. and Calvo, J.M. (1991) Characterization of Lrp, an *Escherichia coli* regulatory protein that mediates a global response to leucine. *The Journal of Biological Chemistry* 266, 10768–10774.

Yeung, Y.G. (1986) L-Threonine aldolase is not a genuine enzyme in rat liver. *Biochemical Journal* 237, 187–190.

Yoneda, K., Sakuraba, H., Muraoka, I., Oikawa, T. and Ohshima, T. (2010) Crystal structure of UDP-galactose 4-epimerase-like L-threonine dehydrogenase belonging to the intermediate short chain dehydrogenase reductase superfamily. *The FEBS Journal* 277, 5124–5132.

Yoneda, K., Sakuraba, H., Araki, T. and Ohshima, T. (2012) Crystal structure of binary and ternary complexes of archaeal UDP-galactose 4-epimerase-like L-threonine dehydrogenase from *Thermoplasma volcanium*. *The Journal of Biological Chemistry* 287, 12966–12974.

13 Methionine Synthesis in Microbes

F. Wencker and W. Ziebuhr*
Institut für Molekulare Infektionsbiologie, Universität Würzburg, Würzburg, Germany

13.1 Abstract

The *de novo* biosynthesis of methionine is widespread in many microorganisms, as consistent with the vital role of methionine in the initiation of protein synthesis and in diverse metabolic pathways. In this chapter, the general biochemistry of bacterial methionine biosynthesis and the generic types of its regulation are outlined. Common features and differences of bacterial methionine metabolism are exemplified by illustrating the pathway in the model organisms *Escherichia coli*, *Corynebacterium glutamicum*, *Bacillus subtilis* and *Staphylococcus aureus*. Finally, the exploitation of methionine biosynthesis as an anti-infective drug target is discussed.

13.2 Introduction

Methionine and its chemical derivatives fulfil important functions in any living cell. Thus, methionine is the universal *N*-terminal amino acid of nearly all newly synthesized proteins. In prokaryotic proteins, the *N*-terminal methionine is *N*-formylated and plays an eminent role in the initiation of protein biosynthesis. The methionine derivative *S*-adenosylmethionine (SAM) is the precursor molecule of polyamine synthesis and acts as a methyl group donor in a variety of cellular processes. Methionine is an unbranched, non-polar amino acid containing a sulfur atom in a characteristic *S*-methyl thioether (C-S-C) side chain, which renders the molecule hydrophobic (Fig. 13.1). In contrast to humans and animals, in which the amino acid is essential and has to be taken up by the diet, the vast majority of bacteria are capable of synthesizing methionine *de novo*. The main routes of methionine consumption in bacteria are protein biosynthesis, SAM production and polyamine synthesis (via SAM), while recycling of the amino acid occurs by the methionine salvage pathway (Fig. 13.2). Overall, methionine metabolism is a costly process that undergoes individual regulation by a variety of mechanisms. Especially in bacteria lacking pathways to recycle excess methionine, the regulation seems to be specifically tight and elaborate.

Currently, there is an increasing interest in better understanding bacterial methionine metabolism. This is because methionine has economic importance as a food additive in plant-derived animal nutrition and many other applications. For industrial-scale purposes, methionine is still produced by cost-intensive chemical synthesis, as progress in engineering

*Corresponding author. E-mail address: w.ziebuhr@mail.uni-wuerzburg.de

 The Handbook of Microbial Metabolism of Amino Acids (ed. J.P.F. D'Mello)

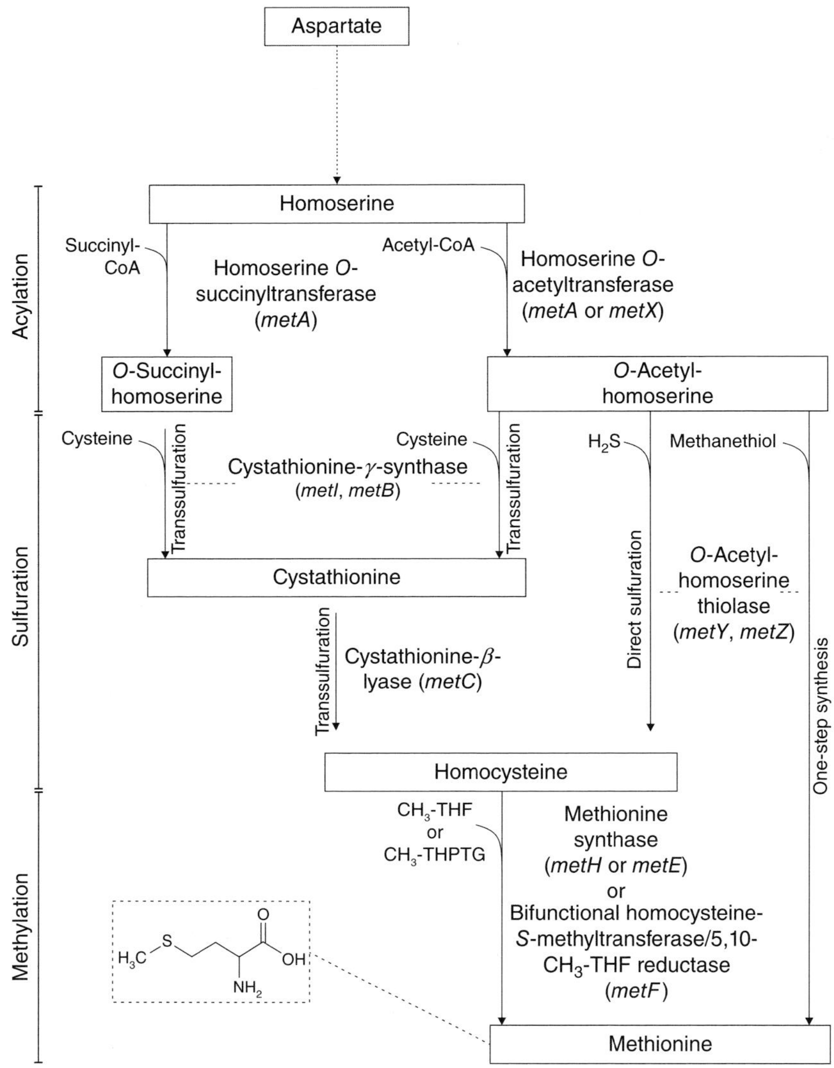

Fig. 13.1. Methionine biosynthesis in bacteria. Key: CH_3-THF, *N*[5]-methyltetrahydrofolate; CH_3-THPTG, *N*[5]-methyltetrahydropteroyl-(tri)glutamate. The genes encoding each enzyme are given in italics in parentheses after the name of the enzyme.

suitable microorganisms for methionine production by fermentation has been achieved only very recently (Bolten *et al.*, 2010; Qin *et al.*, 2015). In human and veterinary medicine, the ongoing development of antibiotic resistance in bacteria is a matter of major concern, and the existing links between bacterial metabolism and virulence have sparked interest in exploiting metabolic pathways as targets for future anti-infectives (Somerville and Proctor, 2009); methionine metabolism might represent a promising candidate in this respect. This chapter outlines the major routes of methionine biosynthesis in bacteria. As biotechnological applications of bacterial methionine biosynthesis have been excellently covered elsewhere (Figge, 2006; Kromer *et al.*, 2006; Ferla and Patrick, 2014), special emphasis will be put on advances in understanding the regulation of methionine biosynthesis and the potential of the pathway as a future anti-infective drug target.

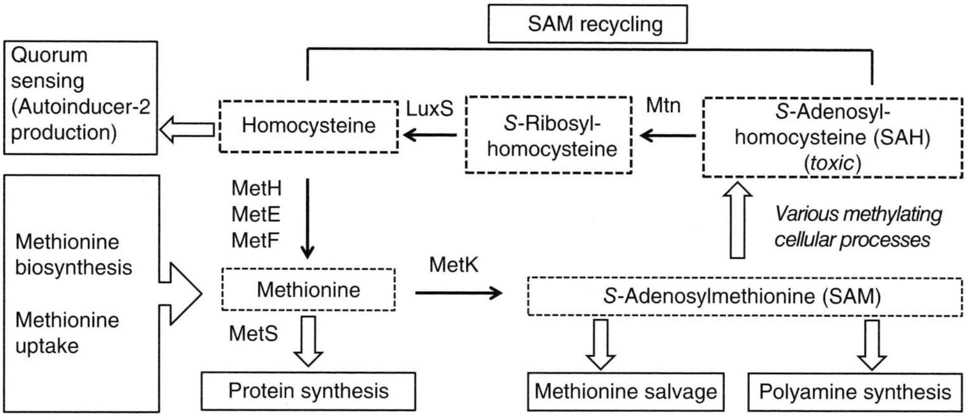

Fig. 13.2. The main routes of methionine conversion and integration into other metabolic pathways. Abbreviations: LuxS, *S*-ribosylhomocysteinase; MetE and MetH, methionine synthase; MetF, methylene-tetrahydrofolate reductase; MetK, *S*-adenosylmethionine (SAM) synthase; MetS, methionyl-tRNA synthase; Mtn, *S*-adenosylhomocysteine (SAH)-nucleosidase.

13.3 Bacterial Methionine Biosynthesis

Methionine biosynthesis in bacteria usually employs homoserine (derived from aspartate) as the main building block (Fig. 13.1). Other required factors are sulfur- and methyl-group donors derived from various sources, as well as a number of vitamins and cofactors. Although methionine biosynthesis pathways may vary in their details among bacteria, the main biochemical steps are widely identical and comprise: (i) the activation of homoserine by acylation; (ii) the sulfuration of the activated homoserine yielding homocysteine; and finally, (iii) the methylation of homocysteine resulting in methionine. Some bacteria, however, are able to add an entire S-CH_3 group directly on to activated homoserine, thereby generating methionine in one step by a single transfer reaction.

13.3.1 Acylation: activation of homoserine

Homoserine is activated either by adding an *O*-succinyl group derived from succinyl-CoA or an *O*-acetyl group derived from acetyl-CoA (Fig. 13.1). This reaction is catalysed by homoserine *O*-succinyltransferases (HSTs) and homoserine *O*-acetyltransferases (HATs), respectively. HSTs are encoded by *metA*, whereas HATs can be encoded by both the *metA* and *metX* genes. The distribution of these genes among bacteria is highly diverse, and many species harbour the *metA* and *metX* genes simultaneously (Ferla and Patrick, 2014). In *Bacillus cereus*, the substrate specificity of the homoserine *O*-succinyltransferase MetA was shown to depend on the amino acid residue present at position 111 in the active site of the enzyme. Thus, a single amino acid exchange from glycine to glutamate (G111E) in MetA was sufficient to switch the substrate specificity from succinyl-CoA to acetyl-CoA, turning the enzyme into a functional HAT (Zubieta *et al.*, 2008). Depending on the amino acid present in the active site, MetA enzymes may therefore represent either HAT or HST enzymes. Based on the amino acid that determines the specificity, the majority of bacterial MetA proteins are considered to be HAT enzymes, while only a minor proportion represent *bona fide* HSTs (Ferla and Patrick, 2014). MetX is an alternative HAT that shares no sequence or structural homology with MetA. As all *metX*-encoded HAT enzymes analysed so far exclusively use acetyl-CoA as substrate, *O*-acetylation is supposed to be the preferred reaction to activate homoserine in bacteria.

13.3.2 Sulfuration: from homoserine to homocysteine

During sulfuration, the acylated hydroxyl group of homoserine is replaced by a thiol group to produce homocysteine. Apart from direct sulfuration (employing hydrogen sulfide to provide sulfur), bacteria may use a methionine one-step synthesis (with organosulfur compounds as sulfur source), or the transsulfuration pathway (using cysteine as sulfur donor) to generate homocysteine. All bacterial species capable of *de novo* methionine biosynthesis harbour enzymes for at least one or the other pathway. Some species are able to use both the direct and the transsulfuration route, and a few species, such as *Corynebacterium glutamicum*, additionally have the one-step synthesis pathway at their disposal. Which one of these pathways is preferred depends on the substrate availability and the environmental conditions that the bacteria live in, reflecting a certain flexibility in executing this step of methionine biosynthesis.

13.3.2.1 Direct sulfuration

Most bacteria employ direct sulfuration to generate homocysteine. The reaction involves the exchange of the *O*-acetyl group of activated homoserine by a thiol group using hydrogen sulfide as the sulfur donor (Fig. 13.1). The enzymes catalysing this reaction are *O*-acetylhomoserine thiolases (also referred to as *O*-acetyl thiol-lyases or *O*-acetyl aminocarboxypropyl transferases), and these are encoded by *metY* genes which are detectable in nearly all bacterial classes (Ferla and Patrick, 2014). The *metZ* gene is homologous to *metY* and has probably evolved by a gene duplication event (Gophna *et al.*, 2005). *metZ* is rare in bacterial genomes and it is supposed to encode *O*-succinylhomoserine thiolases, which catalyse the generation of homocysteine from *O*-succinyl-activated homoserine (Gophna *et al.*, 2005). However, as the majority of bacteria build methionine from *O*-acetyl homoserine rather than *O*-succinyl homoserine (see above), it is suggested that *metZ*-encoded enzymes represent *O*-acetyl homoserine thiolases (Ferla and Patrick, 2014).

13.3.2.2 One-step synthesis

A shortcut in synthesizing methionine is the addition of the terminal S-CH_3 group to *O*-acetyl homoserine in a single transfer reaction that omits homocysteine as an intermediate and yields methionine directly in one step (see Fig. 13.1). The reaction is catalysed by the *metY*- and *metZ*-encoded O-acetylhomoserine thiolases discussed above, although the enzymes utilize organosulfur compounds such as methanethiol or dimethyldisulfide instead of hydrogen sulfide as sulfur donors (Bolten *et al.*, 2010). One-step methionine synthesis is mainly used by bacteria that produce methanethiol or other sulfur compounds as by-products of catabolism. *C. glutamicum* has been shown to employ the pathway when fed with exogenous methanethiol, making this an interesting approach for the design of bacterial strains that produce a high yield of methionine for biotechnological applications (Kromer *et al.*, 2006).

13.3.2.3 Transsulfuration

Transsulfuration during methionine biosynthesis is understood as the conversion of activated homoserine into homocysteine, generating the thioether cystathionine as intermediate (Fig. 13.1). The reaction occurs in two steps and employs cysteine as the sulfur source. First, cystathionine is built by transferring cysteine to activated homoserine, thereby removing the *O*-succinyl or *O*-acetyl group from homoserine. The reaction is catalysed by cystathionine-γ-synthases (CGSs), which are encoded by *metB* and *metI* genes. CGSs display a certain flexibility in substrate specificity and may accept both *O*-succinyl homoserine and *O*-acetyl homoserine, or only one of these compounds, as reaction partners (Hacham *et al.*, 2003). Thus, MetB from *E. coli* uses *O*-succinyl homoserine as a substrate for cystathionine synthesis, while the bifunctional MetI enzymes present in, for example, *B. subtilis*, synthesize cystathionine from *O*-acetyl homoserine and cysteine or catalyse a direct sulfuration reaction (with hydrogen sulfide as sulfur donor) to yield homocysteine (Auger *et al.*, 2002). This versatility is supposed to reflect the origin of the CGS and CBL (cystathionine-β-lyase) enzymes (see below) from a common evolutionary ancestor (Gophna *et al.*, 2005). Involving cysteine as a precursor in the transsulfuration pathway is metabolically costly, at least in comparison with the low-molecular-weight sulfur compounds employed in one-step synthesis and direct sulfuration. Nevertheless, cysteine is likely to be better available than hydrogen sulfide or methanethiol, which are both rather volatile compounds. In the second step of transsulfuration, cystathionine is cleaved by *metC*-encoded CBLs yielding homocysteine and releasing pyruvate and ammonia as by-products. Transsulfuration is not very common in the bacterial world, though it is the preferred pathway among the *Alphaproteobacteria* and *Gammaproteobacteria* (Ferla and Patrick, 2014). Also, a number of Gram-positive bacteria such as *B. subtilis* and staphylococci use transsulfuration to synthesize methionine.

13.3.3 Methylation: from homocysteine to methionine

The final step in methionine biosynthesis comprises the methylation of the thiol group of homocysteine (Fig. 13.1). The methionine synthases mediating this reaction are highly conserved. They are present even in organisms that do not synthesize methionine *de novo*, because methionine synthases are also capable of regenerating the vital methyl-group donor SAM, which is ubiquitous in any living cell (Fig. 13.2). The methylation of homocysteine in bacteria is mediated by two non-homologous methionine synthases, MetE and MetH, which use tetrahydrofolate (THF) compounds as methyl-group donors. MetH is a cobalamin-dependent enzyme that obtains the methyl group required for homocysteine *S*-methylation from the monoglutamate present in N^5-methyl-tetrahydrofolate (CH_3-THF). In contrast, MetE, which is cobalamin independent, employs N^5-methyl-tetrahydropteroyl-(tri)glutamate (CH_3-THPTG) with three or more glutamate molecules as methyl group donors (Figge, 2006). The synthesis of the folate-derived methyl group donors, in turn, is ensured by the MetF enzyme, which represents a methylene-tetrahydrofolate reductase catalysing the conversion of N^5,N^{10}-methylene-tetrahydrofolate or N^5,N^{10}-methylene-tetrahydropteroyl-(tri)glutamate into CH_3-THF and CH_3-THPTG, respectively. MetH, which uses methylcobalamin (MeCbl) as intermediate methyl carrier, has been shown to catalyse the methylation reaction of homocysteine 100-fold faster than MetE (Rodionov *et al.*, 2004; Figge, 2006). MetH accepts both CH_3-THF and CH_3-THPTG as substrates, whereas MetE is restricted to CH_3-THPTG.

With a few exceptions, the majority of bacteria harbour both MetE and MetH enzymes, ensuring a certain flexibility with respect to substrate availability and reflecting the general essentiality of the reaction (Gophna *et al.*, 2005; Ferla and Patrick, 2014). Some Gram-positive bacteria, such as *B. subtilis* and staphylococci, harbour, in addition to MetE, a bifunctional homocysteine *S*-methyltransferase/5,10-methylene-tetrahydrofolate reductase enzyme with both methionine synthase and methylene-THF reductase activities (Grundy and Henkin, 1998). Based on similarities and shared domain architectures in both enzyme classes, the protein is alternatively (and confusingly) referred to as MetH or MetF. Due to their structural and functional features, the bifunctional enzymes apparently represent a special subclass of methionine synthases, which would warrant their own nomenclature.

13.4 Regulation of Methionine Biosynthesis

The regulation of methionine biosynthesis in bacteria is as complex and diverse as the pathways ensuring the conversion of homoserine into methionine. Methionine metabolism is interconnected with other crucial metabolic pathways, such as SAM synthesis and recycling, as well as polyamine synthesis (Fig. 13.2). These processes are often (co)regulated by shared factors and mechanisms.

13.4.1 Mechanisms of control of methionine biosynthesis

13.4.1.1 Protein-mediated transcription control (transcription factors)

Transcription factors (TFs) are DNA-binding proteins that recognize specific DNA target sequences in the promoter regions of genes. Depending on the nature of the TF, this interaction may repress or activate the transcription of downstream genes. The affinity of TFs for their target DNA sequences is often modulated by the binding of cofactors, which are usually metabolites or end products of the TF-controlled pathway. TF-mediated transcription control allows for the simultaneous regulation of numerous genes, preceded by a specific target DNA sequence scattered throughout the genome and belonging to the same cellular response programme. Bacterial methionine biosynthesis genes undergo regulation by a variety of TFs, which are summarized in Table 13.1. The most common TFs involved in methionine metabolism regulation are introduced separately below.

METJ. MetJ is a transcriptional repressor that was originally identified in *E. coli* and represents the prototype of its own TF family (Weissbach and Brot, 1991). MetJ governs the biosynthesis and transport of methionine in *Gammaproteobacteria*. The protein binds as a homodimer to a specific 8 bp DNA sequence dubbed the 'Met-box'(Somers and Phillips, 1992). MetJ binds SAM as a cofactor and this interaction enhances the affinity of MetJ for its Met-box targets (Phillips *et al.*, 1989).

Table 13.1. The genes involved in bacterial methionine biosynthesis and their regulatory mechanisms.

Gene (synonym/s)	Enzyme name (short name; EC. No.)[a]	Methionine metabolism-specific transcription factors[b]	Global transcription factors[b]	Riboswitch type[b]
Acylation				
metA (*metB*)	Homoserine-*O*-succinyltransferase (HST; 2.3.1.46) Homoserine-*O*-acetyltransferase (HAT; 2.3.1.31)	MetJ, MetR, MtaR	CmbR, CodY	SAM-I, SAM-II, MET-T-box
metX (*metA*)	Homoserine-*O*-acetyltransferase (HAT; 2.3.1.31)	McbR, MetJ	CodY, GlxR	SAM-I, SAM-II, SAM-IV
Sulfuration				
metY/metZ (*cysD*, *oahS*)	*O*-acetylhomoserine thiolase (2.5.1.49) *O*-succinylhomoserine thiolase	McbR, MetJ, MetR	CymR	SAM-I, SAM-II, SAM-IV, MET-T-box
metB (*metB1*)	Cystathionine-γ-synthase (CGS; 2.5.1.48)	McbR, MetJ, MetR, SamR	CmbR, HomR	SAM-II, MET-T-box
metI (*yjcI*)		MetJ, MetR	CodY	SAM-I, MET-T-box
metC (*yjcJ*, *aecD*, *metB2*, *cbl*)	Cystathionine-β-lyase (CBL; 4.4.1.8)	MetJ, MetR	CmbR, CodY, HomR	SAM-I, MET-T-box
Methylation				
metH	Methionine synthase I (2.1.1.13)	McbR, MetJ, MetR, MtaR, SamR/SahR	CmbR, HomR	SAM-I, SAM-IV
metE (*metH*, *yxyH*)	Methionine synthase II (2.1.1.14)	McbR, MetJ, MetR, MtaR, SamR/SahR	CmbR, CodY, GlxR, HomR, NrdR	SAM-I, MET-T-box, Coenzyme B_{12}-riboswitch
metF	5,10-Methylene-tetrahydrofolate reductase (1.5.1.20)	McbR, MetJ, MetR, MtaR, SamR/SahR	CmbR, CodY, HomR	SAM-I, SAM-IV, MET-T-box, Coenzyme B_{12}-riboswitch
metF (bifunctional) (*metH*)	5,10-Methylene-tetrahydrofolate reductase (1.5.1.20)/ homocysteine *S*-methyltransferase (2.1.1.10)			

[a]Based on information deposited in the KEGG database (http://www.genome.jp/kegg/pathway.html) (Kanehisa, 1997)
[b]Based on information available in the RegPrecise database (http://regprecise.lbl.gov/RegPrecise/index.jsp) (Novichkov *et al.*, 2010).

METR. MetR is a widespread regulator of methionine biosynthesis in *Proteobacteria* (*Alphaproteobacteria*, *Betaproteobacteria* and *Gammaproteobacteria*). MetR belongs to the LysR family of TFs and it binds homocysteine as a cofactor (Weissbach and Brot, 1991). It has been shown to activate genes of the methionine biosynthesis pathway (e.g. *metH*, *metE*), while repressing its own transcription (Cai *et al.*, 1989; Maxon *et al.*, 1989).

MTAR. MtaR is also a member of the LysR type TF family, and represents a repressor of methionine biosynthesis and transport genes in the *Lactobacillaceae* and *Streptococcaceae*. It requires homocysteine as a cofactor and binds to specific Met-box sequences that differ from those recognized by MetJ in *E. coli* (Shelver *et al.*, 2003; Rodionov *et al.*, 2004).

MCBR. McbR is a transcriptional repressor belonging to the TetR family of TFs. This protein regulates methionine and cysteine biosynthesis genes in the *Corynebacteriaceae* and employs SAM and *S*-adenosylhomocysteine (SAH) as cofactors (Rey *et al.*, 2003, 2005).

CODY. CodY is a global transcription factor that represses many amino acid biosynthesis pathways and a number of other processes in Gram-positive bacteria. The binding of GTP and branched-chain amino acids (BCAAs) enhances CodY target sequence affinity. The protein has been comprehensively studied in *B. subtilis* and *S. aureus* (Ratnayake-Lecamwasam *et al.*, 2001; Shivers and Sonenshein, 2004; Majerczyk *et al.*, 2010). In *S. aureus*, it was recently shown to be significantly involved in the control of methionine biosynthesis gene transcription (Schoenfelder *et al.*, 2013).

13.4.1.2 RNA-mediated transcription control (riboswitches)

A protein-independent way of the control of gene expression is the employment of RNA elements located in the 5′-untranslated regions (5′-UTRs) of nascent mRNA molecules. These *cis*-encoded and *cis*-acting leader RNAs are called riboswitches. They are characterized by the specific binding of ligands, which triggers a conformational change of the RNA molecule. In its repressing form, the riboswitch/ligand interaction results in the premature termination of transcription or inhibition of the initiation of translation of the downstream mRNA. The ligands are often metabolites of the respective riboswitch-controlled pathways or, as in case of 'T-box' riboswitches, tRNAs. Riboswitches are considered to be phylogenetically ancient and robust mechanisms of gene expression control that enable bacteria to sense physiological signals directly, without making the intricate detour of TF synthesis (Vitreschak *et al.*, 2004; Smith *et al.*, 2010). The riboswitches involved in bacterial methionine synthesis control are SAM-binding riboswitches, SAH-binding riboswitches, Coenzyme B_{12}-binding riboswitches and T-box riboswitches (Nahvi *et al.*, 2004; Rodionov *et al.*, 2004; Leyn *et al.*, 2014).

SAM-BINDING RIBOSWITCHES. SAM (*S*-adenosylmethionine) is a key metabolite of methionine metabolism and recycling (Fig. 13.2). It is the major methyl group donor in various cellular processes, such as DNA methylation. SAM demethylation generates SAH, a toxic compound that is detoxified in the SAM recycling pathway, thereby generating homocysteine, which eventually fuels the methylation step of methionine biosynthesis. SAM is also the precursor for the synthesis of polyamines and of organic amino compounds that play an important role in nucleic acid and protein functions. Finally, SAM is an intermediate in the regeneration of methionine by the methionine salvage pathway (Fig. 13.2). In line with this central position in methionine metabolism, it is no surprise that SAM acts as a common ligand of riboswitches residing in the 5′-UTRs of genes for cysteine and methionine metabolism (Table 13.2). Currently, a number of different SAM-binding riboswitches are known (Batey, 2011). Below, we will discuss the S-box, SAM-alpha and SMK-box riboswitches, which are the most widespread in the control of methionine metabolism.

S-box riboswitches (SAM-I). S-box riboswitches were among the first riboswitches to be identified. In a seminal study published in 1998, Grundy and Henkin analysed a conserved RNA element (called the S-box) upstream of a variety of cysteine and methionine metabolic genes. Depending on methionine supply, the leader RNA formed either a transcription terminator or anti-terminator for downstream gene expression (Grundy and

Table 13.2. Riboswitches involved in control of the metabolism of methionine (Met, MET), *S*-adenosyl methionine (SAM) and *S*-adenosylhomocysteine (SAH).

Riboswitch type	Ligand	Mechanism	Bacteria	Genes and pathways controlled	Reference
S-box (SAM-I)	SAM	Transcription attenuation	Gram-positives and others	Met biosynthesis, Met transport, Met salvage, SAM synthesis	Grundy and Henkin, 1998
SAM-alpha (SAM-II)		Translation attenuation	*Alphaproteobacteria*	Met biosynthesis	Corbino *et al.*, 2005
SMK-box (SAM-III)		Translation attenuation	*Lactobacillales*	SAM synthesis (*metK*)	Fuchs *et al.*, 2006
SAH-riboswitch (SAM-IV)	SAH	Translation activation	*Actinobacteria* *Proteobacteria*	SAM recycling (SAH degradation), Met biosynthesis (methylation step)	Wang *et al.*, 2008
MET-T-box	$tRNA_i^{fMet}$ [a]	Transcription attenuation	*Firmicutes*	Methionyl-tRNA synthase (*metS*) Met biosynthesis, Met transport	Schoenfelder *et al.*, 2013
Coenzyme-B_{12}-riboswitch	Adenosyl-cobalamin	Transcription/ translation attenuation	*Actinobacteria*, *Proteobacteria*, *Firmicutes* and others	SAM recycling, Met biosynthesis	Nahvi *et al.*, 2004

[a]Initiator formylmethionyl-tRNA

Henkin, 1998). As no protein factor was found to be involved in the process, it was postulated that regulation occurs at the level of RNA by employing an unknown factor to trigger the conformational change required. Later, it was shown that SAM was indeed the ligand that binds to S-box riboswitches and determines terminator or anti-terminator formation (Epshtein *et al.*, 2003; McDaniel *et al.*, 2003; Winkler *et al.*, 2003). S-box riboswitches are most common in Gram-positive bacteria. They act as transcription attenuators upon SAM binding. In addition to methionine and SAM metabolism, S-box riboswitches are also involved in the control of methionine salvage and transport genes (Table 13.2).

SAM-alpha riboswitches (SAM-II). The SAM-alpha riboswitch regulates methionine biosynthesis genes in *Alphaproteobacteria* that lack S-box riboswitches (Corbino *et al.*, 2005). The occurrence of both SAM-I and SAM-II riboswitches in the same genome is considered to be mutually exclusive (Barrick and Breaker, 2007). SAM-alpha differs in its sequence, length and structure from the classical S-box element. Upon SAM binding, SAM-alpha adopts a structure that sequesters the ribosomal binding site residing in the 5′-UTR of controlled genes, resulting in translational attenuation and the repression of downstream gene expression.

SMK-box riboswitch (SAM-III). The SAM-III riboswitch is an element that was identified in lactic acid bacteria such as lactococci, streptococci and enterococci, in which it controls SAM synthesis mediated by MetK (*S*-adenosylmethionine synthetase) (Fuchs *et al.*, 2006). The consensus sequence varies from that of the two other SAM-binding riboswitches, indicating that it recognizes SAM through different structural features (Batey, 2011). Similar to SAM-II, however, the SMK-box riboswitch acts as translational attenuator by sequestering the ribosomal binding site, thereby preventing access of the 30S ribosomal subunit to the mRNA (Fuchs *et al.*, 2007; Lu *et al.*, 2008).

SAH-BINDING RIBOSWITCHES (SAM-IV). SAH differs from SAM solely by the lack of a methyl group at the sulfur atom. In the light of this structural similarity, it was an intriguing finding that

SAM-binding riboswitches are highly discriminatory of these two compounds and clearly prefer SAM over SAH (Fuchs *et al.*, 2006). SAH riboswitches reverse this preference in favour of SAH (Wang and Breaker, 2008). SAH constantly accumulates during the demethylation of SAM, and as it is toxic for the cell, its detoxification is a vital requirement. Accordingly, SAH-binding riboswitches are found upstream of the genes involved in SAM recycling and SAH detoxification in *Proteobacteria* and some *Actinobacteria* (Wang *et al.*, 2008; Edwards *et al.*, 2010). In contrast to most other riboswitches, which usually downregulate gene expression in response to ligand binding, SAH riboswitches induce downstream gene expression when SAH is bound. This is most likely accomplished by stabilizing a secondary structure that exposes the Shine–Dalgarno sequence (a ribosomal binding site in prokaryotic messenger RNA) and promotes ribosome binding and subsequent translation. In this manner, the riboswitch ensures the upregulation of SAH hydrolysing enzymes when the concentration of the toxic metabolite is high.

COENZYME B_{12}-BINDING RIBOSWITCHES. As described in Section 13.3.3, methylcobalamin (MeCbl) is the cofactor of the methionine synthase MetH. The closely related compound coenzyme B_{12} (adenosylcobalamin; AdoCbl) serves as the ligand of riboswitches located in the 5′-UTR of genes involved in the synthesis and transport of vitamin B_{12} (cobalamin; Cbl), ribonucleotide reductase and *metE* genes (cobalamin-independent synthases) (Nahvi *et al.*, 2004).

The B_{12}-box motif which had already been identified by Richter-Dahlfors and Andersson (1992) comprises only a short fragment of the highly conserved B_{12} element at both a sequence and a structural level (Vitreschak *et al.*, 2003). These coenzyme B_{12} riboswitches regulate gene expression either by inhibition of the initiation of translation or by the termination of transcription, with the first predominantly occurring in Gram-negative *Proteobacteria*, the *Cytophaga-Flavobacterium-Bacteroides* (CFB) group as well as in *Cyanobacteria*, and the second being prevalent in Gram-positive bacteria. Ligand binding induces a conformational change that causes the formation of an RBS (ribosome-binding site) sequestor hairpin or an anti-terminator structure, depending on the regulation type.

A coenzyme B_{12} riboswitch regulates *metE* gene expression in various taxonomic groups, including the *Bacillus*/*Clostridium* group, *Actinobacteria* and *Alphaproteobacteria* (Vitreschak *et al.*, 2003). Control by an AdoCbl-responsive riboswitch permits expression of the less efficient MetE synthase only in the absence of cobalamin, when MetH is catalytically inactive due to the lack of its cofactor MeCbl.

T-BOX RIBOSWITCHES. T-box riboswitches represent a specific class of regulatory RNA elements that recognize tRNAs as ligands (Henkin, 2014). They occur mainly in Gram-positive bacteria and a few other bacterial groups (Vitreschak *et al.*, 2008). T-box riboswitches were initially identified in *B. subtilis* as elements controlling tRNA synthase genes and other genes involved in amino acid metabolism and transport (Henkin *et al.*, 1992). They are transcription control units that exert their function by specific interaction and differential binding to either charged or uncharged cognate tRNAs, in this way indirectly sensing the amino acid levels of the bacterial cell (Fig. 13.3). Despite a certain flexibility in length and primary nucleotide sequence, T-box riboswitches are generally characterized by the presence of conserved structural features. Thus, towards the 3′-end of the T-box riboswitch is located the T-box sequence motif that gives the riboswitch its name; this is part of a Rho-independent transcription terminator and contains a nucleotide stretch that can base pair with the free 3′-CCA end of an uncharged tRNA. When the supply of the amino acid under control is high and the cognate tRNA is therefore charged, no T-box/tRNA interaction occurs and a transcription terminator is formed that prevents downstream gene transcription. In contrast, when the level of this amino acid declines and the tRNA becomes uncharged, interaction of the free 3′-CCA tRNA end with the T-box motif takes place and stabilizes an anti-terminator, resulting in transcriptional read-through into downstream genes (see Fig. 13.3).

T-box riboswitches are highly discriminatory for interaction with a specific cognate tRNA species, though as all uncharged tRNAs share an identical 3′-CCA end to interact with the T-box motif, additional structural features are required to confer specificity to the system. This is accomplished by the so-called specifier

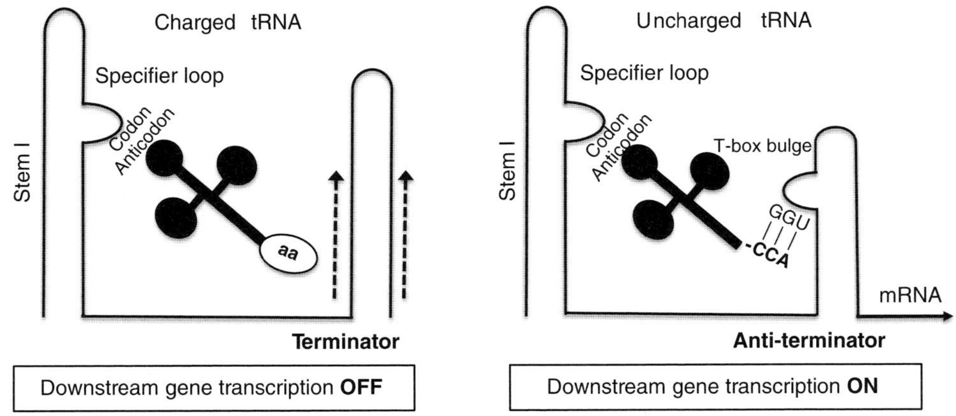

Fig. 13.3. The general function of T-box riboswitch transcription control systems. The scheme illustrates the respective interactions of a charged or uncharged cognate tRNA molecule with the T-box riboswitch. The presence of an amino acid (aa) at high levels results in a charged tRNA and terminator formation so preventing downstream gene transcription, while the free 3′-CCA end of an uncharged tRNA (at a low amino acid level) base pairs with a conserved GGU within the T-box bulge, so favouring anti-terminator formation and transcriptional read-through.

loop, which is regularly found in a stem structure (i.e. stem I) and resides in the 5′ portion of the riboswitch. The specifier loop contains an unpaired base triplet that recognizes the anticodon of the cognate tRNA, in this way defining the amino acid specificity of the system (Fig. 13.3). T-box riboswitches are predicted to exist for nearly all individual amino acids and tRNAs (Novichkov *et al.*, 2010). However, the frequency and distribution of distinct T-box riboswitch species varies significantly among bacterial groups.

Methionyl-tRNA-specific riboswitches (MET-T-box riboswitches). MET-T-box riboswitches seem to be rare in the bacterial world. According to bioinformatic predictions, existence of the element is restricted to a few bacterial families of the orders *Bacillales* and *Lactobacillales* (see Table 13.3.) (Novichkov *et al.*, 2010). Most commonly, MET-T-box riboswitches are involved in the control of the methionyl-tRNA synthase gene *metS* and of methionine transporters in these bacteria, but only a few species use the mechanism to regulate methionine biosynthesis (staphylococci, *Leuconostoc mesenteroides*, *Lactobacillus* spp.) (Novichkov *et al.*, 2010).

In contrast to other T-box riboswitches, which have been intensively studied in the past, MET-T-box riboswitches have only recently been experimentally addressed. Thus, the MET-T-box riboswitch residing in the 5′-UTR (*met* leader RNA) of the *metICFE-mdh* biosynthesis operon in *S. aureus* has been shown to interact specifically with uncharged initiator formylmethionyl-tRNA (tRNA$_i^{fMet}$), while the binding of elongator methionyl-tRNA proved to be weak (Schoenfelder *et al.*, 2013). In prokaryotes, each newly synthesized protein initiates with *N*-formyl-methionine, which is carried to the ribosome by the initiator tRNA. Initiator tRNAs differ by a number of characteristic sequence features from the elongator methionyl-tRNAs. Currently, the molecular background and biological function of the selectivity of the staphylococcal MET-T-box riboswitch for the initiator tRNA is not known. Both the initiator and elongator tRNAs share, in their uncharged form, the free 3′-CCA end for T-box interaction. They also harbour the same anticodon that would mediate binding to the specifier loop. Indeed, the specifier loop in the staphylococcal MET-T-box riboswitch was demonstrated to be dispensable for initiator tRNA recognition, suggesting that other, so far undefined, structures of the riboswitch are involved in the interaction with the initiator tRNA (Schoenfelder *et al.*, 2013). The structural features of the initiator tRNA are likely to contribute to discrimination as well, as shown previously for the recognition of the molecule by the methionyl-tRNA-formyltransferase (MTFT) enzyme (Mayer and RajBhandary, 2002). Both initiator and elongator methionyl-tRNAs are charged with

Table 13.3. The existence of MET-T-box riboswitch regulators in bacteria.[a]

Phylum (Class)	Order	Family	Genus	Enzyme (gene)/Pathway
Firmicutes (*Bacilli*)	*Bacillales*	*Bacillaceae*	*Bacillus*	Methionyl-tRNA synthase (*metS*) Met transport
			Paenibacillus	Methionyl-tRNA synthase (*metS*)
		Staphylococcaceae	*Staphylococcus*	Met biosynthesis
			Macrococcus	*metE2*
	Lactobacillales	*Clostridiaceae*	*Clostridium*	Methionyl-tRNA synthase (*metS*) Met transport
		Lactobacillaceae	*Lactobacillus*	Met biosynthesis Met transport
			Oenococcus	Met transport (*metE2*)
			Leuconostoc	Met biosynthesis Met transport
			Pediococcus	Met transport

[a]Based on information available in the RegPrecise database (http://regprecise.lbl.gov/RegPrecise/index.jsp) (Novichkov *et al.*, 2010).

methionine by the same *metS*-encoded tRNA synthase, but only the charged initiator tRNA is subsequently employed as substrate for the MTFT enzyme, which mediates *N*-formylation of methionine to yield fMet. The MTFT/initiator tRNA interaction was found to be facilitated by unique structural features in the acceptor stem of the initiator tRNA which are specifically recognized by the MTFT protein (Mayer and RajBhandary, 2002). It is reasonable to speculate that these structural features of the initiator tRNA also contribute to the specific interaction with the staphylococcal MET-T-box riboswitch. However, it is currently not known exactly how the RNA-based riboswitch may accomplish this selectivity. It is also unknown whether the initiator tRNA preference is a general feature of all MET-T-box riboswitches or represents an attribute that is unique to the element controlling methionine biosynthesis in staphylococci.

13.4.2 Methionine metabolism and its control in selected bacterial species

13.4.2.1 Escherichia coli

E. coli is a model organism in which methionine biosynthesis and its regulation have been intensively studied. The bacterium activates homoserine by MetA-mediated *O*-succinylation, followed by MetB- and MetC-mediated transsulfuration, which yield cystathionine and homocysteine as intermediates, respectively. Methionine synthesis is then finalized by the methylation of homocysteine through MetE or MetH, supported by MetF, which facilitates the synthesis of tetrahydrofolate-derived methyl group donors (Ferla and Patrick, 2014). The *E. coli* methionine biosynthesis genes are scattered throughout the chromosome. To ensure a coordinated expression of these genes, they share control by the same transcription factors; regulation occurs almost exclusively via the repressor protein MetJ and the transcription activator protein MetR (Novichkov *et al.*, 2010). MetJ binds to Met-box consensus sequences present in the promoter regions of its target genes *metA*, *metB*, *metC* and *metE*, as well as *metF*, and prevents their transcription. In this respect, SAM acts as an efficient co-repressor of MetJ, as binding of SAM to the transcription factor increases the affinity of MetJ for its target DNAs by at least 100-fold (Figge, 2006). MetR is the functional antagonist of MetJ by activating the transcription of *metA*, *metE*, *metH* and *metF* upon binding of homocysteine as a co-activator (Weissbach and Brot, 1991).

Apart from the control of methionine biosynthesis, MetJ and MetR are additionally involved in the regulation of *metNIQ*-mediated methionine transport (MetNIQ is an ABC family Met transporter) as well as in *metK*-mediated SAM synthesis and autoinducer-2 production by LuxS (*S*-ribosylhomocysteinase) (Table 13.4). The autoregulation of MetJ and MetR completes the regulatory network. Thus, MetJ represses its own transcription and that of the *metR* gene, which is additionally downregulated by its gene product MetR (Cai *et al.*, 1989; Maxon *et al.*, 1989). As effector binding of SAM and homocysteine modulates the affinity of the transcription factors for their targets, an efficient feedback control is implemented that allows *E. coli* to manage with a few transcription factors to adjust expression of a plethora of methionine metabolic genes to its current requirements.

13.4.2.2 Corynebacterium glutamicum

C. glutamicum exhibits high metabolic flexibility with respect to methionine biosynthesis. Depending on substrate availability, the species is able to build methionine by direct sulfuration, by transsulfuration or by one-step synthesis (Hwang *et al.*, 2002; Bolten *et al.*, 2010). Upon MetX-mediated *O*-acetylation of homoserine, direct sulfuration by MetY is employed when hydrogen sulfide is available as a sulfur donor to produce homocysteine. In the absence of hydrogen sulfide, *C. glutamicum* generates homocysteine by the transsulfuration route employing the MetB and AecD (cystathionine-β-lyase) enzymes and cysteine as a sulfur source. Finally, when methanethiol or dimethyldisulfide is fed exogenously, *C. glutamicum* is able to generate methionine in one step from *O*-acetyl-homoserine by again employing MetY (Kromer *et al.*, 2006). Methylation of homocysteine is then catalysed by the MetH, MetE and MetF enzymes.

As in *E. coli*, the genes for methionine biosynthesis in *C. glutamicum* are scattered throughout the chromosome. Also similar to *E. coli*, regulation exclusively depends on transcription factors, although while *E. coli* has both an activator and a repressor protein at its disposal, *C. glutamicum* relies on a single protein, McbR, as a master regulator (Rey *et al.*, 2003). McbR is a transcriptional repressor that binds to a consensus sequence located in the 5′-UTR of nearly all *C. glutamicum* methionine biosynthesis genes. Exceptions are *aecD*, which encodes a CBL that is non-homologous to other *metC*-encoded CBLs, and *metF*, whose control mechanism is currently unknown. The affinity of the McbR repressor to its targets decreases upon SAH cofactor binding (Rey *et al.*, 2005). SAH accumulation signals methionine consumption via SAM synthesis and SAM demethylation, a process that requires the expression of methionine synthases (MetH, MetF) to convert SAH-derived homocysteine into methionine (see Fig. 13.2). Monitoring the SAH concentration by McbR is therefore an efficient means to adapt methionine biosynthesis gene expression. The McbR transcription factor additionally controls methionine transport genes and *metK*-mediated SAM synthesis, as well as a number of other genes involved in sulfur metabolism and stress response in *C. glutamicum* (Figge, 2006).

13.4.2.3 Bacillus subtilis

B. subtilis is a Gram-positive model bacterium whose genome structure and metabolic capacities have been comprehensively analysed. Methionine biosynthesis in *B. subtilis* starts with homoserine *O*-acetylation through the homoserine *O*-succinyltransferase MetA which uses acetyl-CoA as substrate. Depending on the available sulfur source, *B. subtilis* can employ direct sulfuration or transsulfuration to add a thiol group to *O*-acetylhomoserine. Both reactions are mediated by MetI, which is a cystathionine-γ-synthase that also possesses thiolase activity (Auger *et al.*, 2002). Transsulfuration is completed by MetC, which catalyses the conversion of cystathionine to homocysteine (see Fig. 13.1). Finally, methylation of homocysteine is performed by MetE and/or MetF, the latter representing a bifunctional enzyme with both methionine synthase and methylene-THF reductase activities (Grundy and Henkin, 1998).

Except for *metI* and *metC*, which are organized in an operon structure, the methionine biosynthesis genes are scattered throughout the *B. subtilis* chromosome. In contrast to *E. coli* and *C. glutamicum*, however, regulation is not mediated by transcription factors. Instead, *B. subtilis* employs RNA-based S-box (SAM-I) riboswitches residing in the 5′-UTRs of *metIC*, *metF* and *metE* to control the transcription of these genes (see

Table 13.4. Overview of methionine (Met) metabolism and its regulation in selected bacterial species.[a]

Organism (Order, Class)	Met biosynthesis genes and regulators[b]	Type of homoserine activation	Route of sulfuration	Met transporter genes and regulators[b]	Genes involved in SAM-recycling, tRNA charging and their regulators[a]
Escherichia coli (*Enterobacteriales*, *Gammaproteobacteria*)	▼Δ *metA*, ▼ *metB-metL*, ▼*metC*, ▼Δ *metE*, Δ *metH*, ▼Δ *metF*	Succinyl-CoA	Transsulfuration	▼*metN-metI-metQ*	▼◊ *metK*, Δ *luxS*, *mtn(pfs)*, *metS* (*metG*)
Corynebacterium glutamicum (*Actinomycetales*, *Actinobacteria*)	● *metX*, ● *metB*, *metC* (*aecD*)-*brnQ*, *metF*, ● *metE*, ● *metH*, ● *metY*	Acetyl-CoA	Transsulfuration Direct sulfuration One-step synthesis	● *metQ-metN-metI* *metS-metP*	● *metK-priA*, *sahH*, *metS*
Bacillus subtilis (*Bacillales*, *Bacilli*)	*metA*, ■-**S**-*metI-metC*(*yjcI-yjcJ*), **S**-*metF*[c] (*yitJ*), **S**-*metE*	Acetyl-CoA[d]	Transsulfuration, Direct sulfuration	**S**-*metN-metP-metQ*	**S**-*metK*, ♦ *yrrT-mtn-mccAB-yrhC*
Staphylococcus aureus (*Bacillales*, *Bacilli*)	■-**S**-*metX*, ■-**T**-*metI-metC-metF*[c]-*metE-mdh*	Acetyl-CoA	Transsulfuration	**S**-*metT* *metN1-metP1-metQ1* ■-**S**-*metN2-metP2-metQ2*	**S**-*metK*, *mtn(pfs)*, ♦ *luxS*, *metS*

[a]Further details of some of the gene encoding are given in the text.
[b]Symbols: **S**, *S*-adenosylmethionine (SAM) riboswitch; **T**, T-box riboswitch; ■, CodY, global transcription factor; ●, McbR, transcriptional repressor; ▼, MetJ, transcriptional repressor; Δ, MetR, widespread Met synthesis regulator; ♦, CymR, global transcription factor; ◊, Crp. global transcription factor.
[c]Bifunctional homocysteine *S*-methyl-transferase/5,10-methylene-tetrahydrofolate reductase with methionine synthase and methylene- tetrahydrofolate reductase domains.
[d]Homoserine acylation catalysed by an *O*-succinyl-homoserinetransferase.

Table 13.4.). S-box riboswitches undergo a conformational change upon binding SAM, resulting in premature transcription termination of downstream gene expression (Grundy and Henkin, 1998). Other S-box controlled methionine metabolic genes comprise the methionine transporter operon *metNPQ* and the SAM synthase gene *metK*, completing a regulatory network in *B. subtilis* that is primarily responsive to the central methionine metabolite SAM. Thus, high SAM concentrations are a signal of excess methionine, resulting in downregulation of methionine biosynthesis, methionine uptake and SAM production (Fig. 13.2). Interestingly, the S-box regulated *metIC* operon, encoding the sulfuration enzymes, is additionally under control of the CodY global repressor protein whose consensus binding sequence is present upstream of the *metIC* S-box riboswitch (Novichkov *et al.*, 2010). CodY is embedded in the stringent response circuit (see Section 13.4.2.4) and controls the biosynthesis of various amino acids and many other genes in low GC Gram-positive bacteria (Geiger and Wolz, 2014). The involvement of CodY in the regulation of methionine biosynthesis suggests that the methionine pathway has been integrated into the general energy and carbon metabolism of *B. subtilis*.

13.4.2.4 Staphylococcus aureus

S. aureus and *B. subtilis* are related species belonging to the order *Bacillales*, so the methionine biosynthesis pathway of *S. aureus* resembles that of *B. subtilis*. Homoserine activation is achieved by MetX-mediated *O*-acetylation, followed by the transsulfuration pathway that employs MetI and MetC to build homocysteine. Methylation of homocysteine is then catalysed by MetE and/or MetF, the latter representing, as in *B. subtilis*, a bifunctional methionine synthase/methylene THF-reductase enzyme.

At this point, the analogy with *B. subtilis* ends, as the two species differ significantly with respect to gene order and regulation. Except for *metX*, all *S. aureus* methionine biosynthesis genes are organized in a single polycistronic *metICFE-mdh* operon, with *mdh* encoding a metal-dependent hydrolase of unknown function. While *metX* is under S-box control, the *metICFE-mdh* operon is preceded by an extraordinarily long 5′-UTR, dubbed *met* leader RNA, which harbours a MET-T-box riboswitch (Rodionov *et al.*, 2004; Novichkov *et al.*, 2010; Schoenfelder *et al.*, 2013). As explained in Section 13.4.1.2, under T-box riboswitches, this MET-T-box riboswitch is sensitive to uncharged initiator tRNA binding, which triggers antiterminator formation and transcriptional read-through into the *metICFE-mdh* genes when methionine becomes scarce (Schoenfelder *et al.*, 2013). Interestingly, the MET-T-box riboswitch was found to be integrated into a complex hierarchical regulatory network that links methionine biosynthesis with the general metabolic and energy status of the cell. This circuit involves the so called stringent response, in concert with downstream action of the CodY repressor. Thus, CodY (which employs GTP and BCAAs as cofactors) binds upstream of the *met* leader RNA and controls its transcription in a stringent response-dependent manner. The stringent response is understood as the RSH enzyme (named after RelA/SpoT homologue, where RelA/SpoT are *E. coli* proteins) mediated synthesis of the alarmone (p)ppGpp from GTP upon starvation signals, resulting in a drop of intracellular GTP levels and release of CodY from its DNA targets (Geiger and Wolz, 2014). Methionine depletion was demonstrated to activate RSH-dependent (p)ppGpp synthesis, releasing CodY from its binding site and thereby activating the *met* leader RNA promoter. Other amino acid starvation signals were also shown to activate the system, resulting in *met* leader RNA transcription. However, only methionine depletion was able to trigger MET-T-box/initiator tRNA interaction and subsequent expression of the *metICFHE-mdh* mRNA, which was found to be rapidly degraded by RNases (Schoenfelder *et al.*, 2013). This rare combination of stringent response-controlled CodY activity, MET-T-box riboswitch and mRNA decay mechanisms seems to be unique and specific to staphylococci in their control of methionine biosynthesis. Notably, the involvement of a MET-T-box riboswitch is unusual in this context, as other *Bacillales* exclusively use S-box riboswitches to control this pathway.

The reasons for this complex control are currently poorly understood, although staphylococci do lack important recycling pathways such as methionine salvage and polyamine synthesis

(Kanehisa, 1997; Joshi *et al.*, 2011). These metabolic constraints might impose the need for a strict control of methionine biosynthesis, which is apparently met by the hierarchical network employed by staphylococci.

13.5 Methionine Biosynthesis as a Target for Novel Antibacterial Strategies

Increasing antimicrobial drug resistance in numerous bacterial pathogens is a significant threat to public health. The search for new anti-infective compounds, along with the identification of novel prokaryotic target structures for future anti-infectives, is therefore an urgent need. The perfect anti-infective drug would be highly selective and not interfere with host structures, and the ideal targets would be essential and unique to the microorganism concerned.

Methionine metabolism, for the most part, fulfils these criteria. The enzymes and regulators involved are largely specific to the pathogens. Whether or not the pathway is essential often depends on the environment the bacteria live in. For example, prior to causing infection, *S. aureus* colonizes the human nasal mucosa. In this habitat, methionine is scarce and *de novo* methionine biosynthesis is required to enable the bacterium to establish in the niche (Krismer *et al.*, 2014). In contrast, when *S. aureus* invades the host by degrading tissue structures, methionine will become abundantly available. For this situation, *S. aureus* has evolved the sophisticated regulatory network described above to meticulously shut down the methionine biosynthesis pathway (Schoenfelder *et al.*, 2013). As methionine biosynthesis is metabolically costly and *S. aureus* has limited capacities to reuse excess methionine, it was speculated that the pathogen employs this efficient control mechanism to avoid metabolic exhaustion (Schoenfelder *et al.*, 2013). Thus, both inhibition of the pathway and deregulated, permanent induction of methionine biosynthesis might represent useful approaches to tackling bacterial growth. Future drugs that interfere with methionine metabolism may target the methionine biosynthesis and regulatory proteins as well as the riboswitches involved in control of the pathway.

13.5.1 Methionine biosynthesis enzyme and regulator protein inhibition

The screening of natural compounds and small molecule libraries has identified various inhibitors of methionine biosynthesis proteins. For example, β-lactone natural products were shown to inactivate the homoserine transacetylase (HTA) from *Haemophilus influenzae* (De Pascale *et al.*, 2011). Similarly, the *Salmonella enterica* cystathionine-β-lyase (CBL), which is important for virulence, was successfully inhibited by a number of lead structures identified in a small molecule library screen (Ejim *et al.*, 2004, 2007). Also, the MtaR transcription factor, which controls methionine transport and synthesis in group-B-streptococci, was identified as an essential target that mediates *in vivo* survival of this pathogen (Shelver *et al.*, 2003). Another promising structure, indirectly involved in methionine metabolism, is the methionyl-tRNA synthase MetS, which charges the initiator and elongator tRNAs with methionine. MetS inactivation by the compound REP8839 resulted in excellent antibacterial efficacy in *S. aureus* and *Streptococcus pyogenes* and prompted further efforts to exploit MetS as a future antibacterial target (Ochsner *et al.*, 2005; Liu *et al.*, 2013).

13.5.2 Riboswitches as drug targets

Employing riboswitches as drug targets is enjoying a remarkable boost, based largely on the recent progress in understanding the RNA biology of prokaryotes (Blount and Breaker, 2006; Matzner and Mayer, 2015). Interestingly, some antibiotics with a so far unknown mode of action were found to interact with riboswitches, and a number of novel compounds have recently been identified to target riboswitches (Sudarsan *et al.*, 2005; Ster *et al.*, 2013; Howe *et al.*, 2015). Riboswitches occur in bacteria, plants and fungi, but they are not present in humans, making them selective anti-infective targets (Matzner and Mayer, 2015). Moreover, riboswitches recognize metabolites or tRNAs with an amazing specificity. This selectivity is certainly an advantage for the toxicity of future drugs, but at the same time it requires detailed knowledge of the structure and dynamics of the metabolite/riboswitch interactions to efficiently screen for suitable metabolite analogues (Matzner and Mayer, 2015).

So far, no compound(s) interfering exclusively with methionine metabolism-controlling riboswitches are under investigation. However, a number of promising studies have been done that may create the basis for such an approach. The high-resolution crystal structures of the S-box and SMK-box riboswitches have been solved (Montange and Batey, 2006; Lu *et al.*, 2008), and a protocol for structure-guided design of SAM analogues was recently developed to screen for S-box interference (Hickey and Hammond, 2014). T-box riboswitches have also been demonstrated to be structurally accessible for rational drug design (Jentzsch and Hines, 2012). In this context, certain oxazolidinone derivatives have been found that have the potential for specific structural recognition of T-box antiterminator RNA (Anupam *et al.*, 2008; Orac *et al.*, 2011). These 4,5-substituted oxazolidinones differ structurally from other oxazolidinone antibiotics such as linezolid, which bind to ribosomal RNA and prevent mRNA translation. Two different oxazolidinone-derivatives have been designed that were able to either support or inhibit the T-box anti-terminator function of the *B. subtilis glyQS* T-box riboswitch (Anupam *et al.*, 2008). Lately, the polyamine compound spermidine has also been demonstrated to bind to T-box anti-termiantor RNA. Surprisingly, this interaction occurs in a highly sequence-specific manner at the conserved T-box bulge, although polyamines were previously thought to bind to nucleic acids in an unspecific manner by electrostatic interactions (Liu *et al.*, 2016). Taken together, these are encouraging data that highlight the potential of riboswitches as future drug targets and an evolving field of study.

13.6 Conclusions

Methionine metabolism in bacteria is a promising area of research. It bears huge potential, not only for better understanding of the details of microbial lifestyles, but also in terms of biotechnological and pharmaceutical applications. In light of the ongoing antibiotic resistance crisis, the latter aspect is of particular interest. Here, methionine metabolism offers a wide range of opportunities to interfere with bacterial growth and survival in distinct ecological niches. Notably, the RNA-based regulatory mechanisms represent a so far unexploited field to identify novel anti-infective drugs.

References

Anupam, R., Nayek, A., Green, N.J., Grundy, F.J., Henkin, T.M., Means, J.A., Bergmeier, S.C. and Hines, J.V. (2008) 4,5-Disubstituted oxazolidinones: high affinity molecular effectors of RNA function. *Bioorganic and Medicinal Chemistry Letters* 18, 3541–3544.

Auger, S., Yuen, W.H., Danchin, A. and Martin-Verstraete, I. (2002) The *metIC* operon involved in methionine biosynthesis in *Bacillus subtilis* is controlled by transcription antitermination. *Microbiology* 148, 507–518.

Barrick, J.E. and Breaker, R.R. (2007) The distributions, mechanisms, and structures of metabolite-binding riboswitches. *Genome Biology* 8:R239.

Batey, R.T. (2011) Recognition of *S*-adenosylmethionine by riboswitches. *Wiley Interdisciplinary Reviews: RNA* 2, 299–311.

Blount, K.F. and Breaker, R.R. (2006) Riboswitches as antibacterial drug targets. *Nature Biotechnology* 24, 1558–1564.

Bolten, C.J., Schröder, H., Dickschat, J. and Wittmann, C. (2010) Towards methionine overproduction in *Corynebacterium glutamicum* – methanethiol and dimethyldisulfide as reduced sulfur sources. *Journal of Microbiology and Biotechnology* 20, 1196–1203.

Cai, X.Y., Maxon, M.E., Redfield, B., Glass, R., Brot, N. and Weissbach, H. (1989) Methionine synthesis in *Escherichia coli*: effect of the MetR protein on *metE* and *metH* expression. *Proceedings of the National Academy of Sciences of the United States of America* 86, 4407–4411.

Corbino, K.A., Barrick, J.E., Lim, J., Welz, R., Tucker, B.J., Puskarz, I., Mandal, M., Rudnick, N.D. and Breaker, R.R. (2005) Evidence for a second class of *S*-adenosylmethionine riboswitches and other regulatory RNA motifs in alpha-proteobacteria. *Genome Biology* 6:R70.

De Pascale, G., Nazi, I., Harrison, P.H. and Wright, G.D. (2011) β-Lactone natural products and derivatives inactivate homoserine transacetylase, a target for antimicrobial agents. *The Journal of Antibiotics* 64, 483–487.

Edwards, A.L., Reyes, F.E., Heroux, A. and Batey, R.T. (2010) Structural basis for recognition of *S*-adenosylhomocysteine by riboswitches. *RNA* 16, 2144–2155.

Ejim, L.J., D'Costa, V.M., Elowe, N.H., Loredo-Osti, J.C., Malo, D. and Wright, G.D. (2004) Cystathionine β-lyase is important for virulence of *Salmonella enterica* serovar Typhimurium. *Infection and Immunity* 72, 3310–3314.

Ejim, L.J., Blanchard, J.E., Koteva, K.P., Sumerfield, R., Elowe, N.H., Chechetto, J.D., Brown, E.D., Junop, M.S. and Wright, G.D. (2007) Inhibitors of bacterial cystathionine β-lyase: leads for new antimicrobial agents and probes of enzyme structure and function. *Journal of Medicinal Chemistry* 50, 755–764.

Epshtein, V., Mironov, A.S. and Nudler, E. (2003) The riboswitch-mediated control of sulfur metabolism in bacteria. *Proceedings of the National Academy of Sciences of the United States of America* 100, 5052–5056.

Ferla, M.P. and Patrick, W.M. (2014) Bacterial methionine biosynthesis. *Microbiology* 160, 1571–1584.

Figge, R.M. (2006) Methionine biosynthesis in *Escherichia coli* and *Corynebacterium glutamicum*. In: Wendisch, V.F. (ed.) *Amino Acid Biosynthesis*. Springer, Berlin, Heidelberg, Germany, pp. 163–193.

Fuchs, R.T., Grundy, F.J. and Henkin, T.M. (2006) The S(MK) box is a new SAM-binding RNA for translational regulation of SAM synthetase. *Nature Structural and Molecular Biology* 13, 226–233.

Fuchs, R.T., Grundy, F.J. and Henkin, T.M. (2007) *S*-adenosylmethionine directly inhibits binding of 30S ribosomal subunits to the SMK box translational riboswitch RNA. *Proceedings of the National Academy of Sciences of the United States of America* 104, 4876–4880.

Geiger, T. and Wolz, C. (2014) Intersection of the stringent response and the CodY regulon in low GC Gram-positive bacteria. *International Journal of Medical Microbiology* 304, 150–155.

Gophna, U., Bapteste, E., Doolittle, W.F., Biran, D. and Ron, E.Z. (2005) Evolutionary plasticity of methionine biosynthesis. *Gene* 355, 48–57.

Grundy, F.J. and Henkin, T.M. (1998) The S box regulon: a new global transcription termination control system for methionine and cysteine biosynthesis genes in Gram-positive bacteria. *Molecular Microbiology* 30, 737–749.

Hacham, Y., Gophna, U. and Amir, R. (2003) *In vivo* analysis of various substrates utilized by cystathionine γ-synthase and *O*-acetylhomoserine sulfhydrylase in methionine biosynthesis. *Molecular Biology and Evolution* 20, 1513–1520.

Henkin, T.M. (2014) The T box riboswitch: a novel regulatory RNA that utilizes tRNA as its ligand. *Biochimica et Biophysica Acta* 1839, 959–963.

Henkin, T.M., Glass, B.L. and Grundy, F.J. (1992) Analysis of the *Bacillus subtilis tyrS* gene: conservation of a regulatory sequence in multiple tRNA synthetase genes. *Journal of Bacteriology* 174, 1299–1306.

Hickey, S.F. and Hammond, M.C. (2014) Structure-guided design of fluorescent *S*-adenosylmethionine analogs for a high-throughput screen to target SAM-I riboswitch RNAs. *Chemistry and Biology* 21, 345–356.

Howe, J.A., Wang, H., Fischmann, T.O., Balibar, C.J., Xiao, L., Galgoci, A.M., Malinverni, J.C., Mayhood, T., Villafania, A., Nahvi, A. *et al.* (2015) Selective small-molecule inhibition of an RNA structural element. *Nature* 526, 672–677.

Hwang, B.-J., Yeom, H.-J., Kim, Y. and Lee, H.-S. (2002) *Corynebacterium glutamicum* utilizes both transsulfuration and direct sulfhydrylation pathways for methionine biosynthesis. *Journal of Bacteriology* 184, 1277–1286.

Jentzsch, F. and Hines, J.V. (2012) Interfacing medicinal chemistry with structural bioinformatics: implications for T box riboswitch RNA drug discovery. *BMC Bioinformatics* 13(Suppl 2): S5.

Joshi, G.S., Spontak, J.S., Klapper, D.G. and Richardson, A.R. (2011) Arginine catabolic mobile element encoded *speG* abrogates the unique hypersensitivity of *Staphylococcus aureus* to exogenous polyamines. *Molecular Microbiology* 82, 9–20.

Kanehisa, M. (1997) A database for post-genome analysis. *Trends in Genetics* 13, 375–376.

Krismer, B., Liebeke, M., Janek, D., Nega, M., Rautenberg, M., Hornig, G., Unger, C., Weidenmaier, C., Lalk, M. and Peschel, A. (2014) Nutrient limitation governs *Staphylococcus aureus* metabolism and niche adaptation in the human nose. *PLoS Pathogens* 10(1): e1003862.

Kromer, J.O., Wittmann, C., Schroder, H. and Heinzle, E. (2006) Metabolic pathway analysis for rational design of L-methionine production by *Escherichia coli* and *Corynebacterium glutamicum*. *Metabolic Engineering* 8, 353–369.

Leyn, S.A., Suvorova, I.A., Kholina, T.D., Sherstneva, S.S., Novichkov, P.S., Gelfand, M.S. and Rodionov, D.A. (2014) Comparative genomics of transcriptional regulation of methionine metabolism in *Proteobacteria*. *PLoS One* 9(11): e113714.

Liu, C., He, G., Jiang, Q., Han, B. and Peng, C. (2013) Novel hybrid virtual screening protocol based on molecular docking and structure-based pharmacophore for discovery of methionyl-tRNA synthetase inhibitors as antibacterial agents. *International Journal of Molecular Sciences* 14, 14225–14239.

Liu, J., Zeng, C., Hogan, V., Zhou, S., Monwar, M.M. and Hines, J.V. (2016) Identification of spermidine binding site in T-box riboswitch antiterminator RNA. *Chemical Biology and Drug Design* 87, 182–189.

Lu, C., Smith, A.M., Fuchs, R.T., Ding, F., Rajashankar, K., Henkin, T.M. and Ke, A. (2008) Crystal structures of the SAM-III/S(MK) riboswitch reveal the SAM-dependent translation inhibition mechanism. *Nature Structural and Molecular Biology* 15, 1076–1083.

Majerczyk, C.D., Dunman, P.M., Luong, T.T., Lee, C.Y., Sadykov, M.R., Somerville, G.A., Bodi, K. and Sonenshein, A.L. (2010) Direct targets of CodY in *Staphylococcus aureus*. *Journal of Bacteriology* 192, 2861–2877.

Matzner, D. and Mayer, G. (2015) (Dis)similar analogues of riboswitch metabolites as antibacterial lead compounds. *Journal of Medicinal Chemistry* 58, 3275–3286.

Maxon, M.E., Redfield, B., Cai, X.Y., Shoeman, R., Fujita, K., Fisher, W., Stauffer, G., Weissbach, H. and Brot, N. (1989) Regulation of methionine synthesis in *Escherichia coli*: effect of the MetR protein on the expression of the *metE* and *metR* genes. *Proceedings of the National Academy of Sciences of the United States of America* 86, 85–89.

Mayer, C. and RajBhandary, U.L. (2002) Conformational change of *Escherichia coli* initiator methionyl-tRNA(fMet) upon binding to methionyl-tRNA formyl transferase. *Nucleic Acids Research* 30, 2844–2850.

McDaniel, B.A., Grundy, F.J., Artsimovitch, I. and Henkin, T.M. (2003) Transcription termination control of the S box system: direct measurement of *S*-adenosylmethionine by the leader RNA. *Proceedings of the National Academy of Sciences of the United States of America* 100, 3083–3088.

Montange, R.K. and Batey, R.T. (2006) Structure of the *S*-adenosylmethionine riboswitch regulatory mRNA element. *Nature* 441, 1172–1175.

Nahvi, A., Barrick, J.E. and Breaker, R.R. (2004) Coenzyme B_{12} riboswitches are widespread genetic control elements in prokaryotes. *Nucleic Acids Research* 32, 143–150.

Novichkov, P.S., Laikova, O.N., Novichkova, E.S., Gelfand, M.S., Arkin, A.P., Dubchak, I. and Rodionov, D.A. (2010) RegPrecise: a database of curated genomic inferences of transcriptional regulatory interactions in prokaryotes. *Nucleic Acids Research* 38, D111–D118.

Ochsner, U.A., Young, C.L., Stone, K.C., Dean, F.B., Janjic, N. and Critchley, I.A. (2005) Mode of action and biochemical characterization of REP8839, a novel inhibitor of methionyl-tRNA synthetase. *Antimicrobial Agents and Chemotherapy* 49, 4253–4262.

Orac, C.M., Zhou, S., Means, J.A., Boehm, D., Bergmeier, S.C. and Hines, J.V. (2011) Synthesis and stereospecificity of 4,5-disubstituted oxazolidinone ligands binding to T-box riboswitch RNA. *Journal of Medicinal Chemistry* 54, 6786–6795.

Phillips, S.E., Manfield, I., Parsons, I., Davidson, B.E., Rafferty, J.B., Somers, W.S., Margarita, D., Cohen, G.N., Saint-Girons, I. and Stockley, P.G. (1989) Cooperative tandem binding of *met* repressor of *Escherichia coli*. *Nature* 341, 711–715.

Qin, T., Hu, X., Hu, J. and Wang, X. (2015) Metabolic engineering of *Corynebacterium glutamicum* strain ATCC13032 to produce L-methionine. *Biotechnology and Applied Biochemistry* 62, 563–573.

Ratnayake-Lecamwasam, M., Serror, P., Wong, K.W. and Sonenshein, A.L. (2001) *Bacillus subtilis* CodY represses early-stationary-phase genes by sensing GTP levels. *Genes and Development* 15, 1093–1103.

Rey, D.A., Puhler, A. and Kalinowski, J. (2003) The putative transcriptional repressor McbR, member of the TetR-family, is involved in the regulation of the metabolic network directing the synthesis of sulfur containing amino acids in *Corynebacterium glutamicum*. *Journal of Biotechnology* 103, 51–65.

Rey, D.A., Nentwich, S.S., Koch, D.J., Rückert, C., Pühler, A., Tauch, A. and Kalinowski, J. (2005) The McbR repressor modulated by the effector substance *S*-adenosylhomocysteine controls directly the transcription of a regulon involved in sulphur metabolism of *Corynebacterium glutamicum* ATCC 13032. *Molecular Microbiology* 56, 871–887.

Richter-Dahlfors, A.A. and Andersson, D.I. (1992) Cobalamin (vitamin B_{12}) repression of the Cob operon in *Salmonella typhimurium* requires sequences within the leader and the first translated open reading frame. *Molecular Microbiology* 6, 743–749.

Rodionov, D.A., Vitreschak, A.G., Mironov, A.A. and Gelfand, M.S. (2004) Comparative genomics of the methionine metabolism in Gram-positive bacteria: a variety of regulatory systems. *Nucleic Acids Research* 32, 3340–3353.

Schoenfelder, S.M., Marincola, G., Geiger, T., Goerke, C., Wolz, C. and Ziebuhr, W. (2013) Methionine biosynthesis in *Staphylococcus aureus* is tightly controlled by a hierarchical network involving an initiator tRNA-specific T-box riboswitch. *PLoS Pathogens* 9(9): e1003606.

Shelver, D., Rajagopal, L., Harris, T.O. and Rubens, C.E. (2003) MtaR, a Regulator of methionine transport, is critical for survival of group B *Streptococcus in vivo*. *Journal of Bacteriology* 185, 6592–6599.

Shivers, R.P. and Sonenshein, A.L. (2004) Activation of the *Bacillus subtilis* global regulator CodY by direct interaction with branched-chain amino acids. *Molecular Microbiology* 53, 599–611.

Smith, A.M., Fuchs, R.T., Grundy, F.J. and Henkin, T.M. (2010) Riboswitch RNAs: regulation of gene expression by direct monitoring of a physiological signal. *RNA Biology* 7, 104–110.

Somers, W.S. and Phillips, S.E. (1992) Crystal structure of the *met* repressor-operator complex at 2.8 A resolution reveals DNA recognition by beta-strands. *Nature* 359, 387–393.

Somerville, G.A. and Proctor, R.A. (2009) At the crossroads of bacterial metabolism and virulence factor synthesis in staphylococci. *Microbiology and Molecular Biology Reviews* 73, 233–248.

Ster, C., Allard, M., Boulanger, S., Lamontagne Boulet, M., Mulhbacher, J., Lafontaine, D.A., Marsault, E., Lacasse, P. and Malouin, F. (2013) Experimental treatment of *Staphylococcus aureus* bovine intramammary infection using a guanine riboswitch ligand analog. *Journal of Dairy Sciences* 96, 1000–1008.

Sudarsan, N., Cohen-Chalamish, S., Nakamura, S., Emilsson, G.M. and Breaker, R.R. (2005) Thiamine pyrophosphate riboswitches are targets for the antimicrobial compound pyrithiamine. *Chemistry and Biology* 12, 1325–1335.

Vitreschak, A.G., Rodionov, D.A., Mironov, A.A. and Gelfand, M.S. (2003) Regulation of the vitamin B_{12} metabolism and transport in bacteria by a conserved RNA structural element. *RNA* 9, 1084–1097.

Vitreschak, A.G., Rodionov, D.A., Mironov, A.A. and Gelfand, M.S. (2004) Riboswitches: the oldest mechanism for the regulation of gene expression? *Trends in Genetics* 20, 44–50.

Vitreschak, A.G., Mironov, A.A., Lyubetsky, V.A. and Gelfand, M.S. (2008) Comparative genomic analysis of T-box regulatory systems in bacteria. *RNA* 14, 717–735.

Wang, J.X. and Breaker, R.R. (2008) Riboswitches that sense *S*-adenosylmethionine and *S*-adenosylhomocysteine. *Biochemistry and Cell Biology* 86, 157–168.

Wang, J.X., Lee, E.R., Morales, D.R., Lim, J. and Breaker, R.R. (2008) Riboswitches that sense *S*-adenosylhomocysteine and activate genes involved in coenzyme recycling. *Molecular Cell* 29, 691–702.

Weissbach, H. and Brot, N. (1991) Regulation of methionine synthesis in *Escherichia coli*. *Molecular Microbiology* 5, 1593–1597.

Winkler, W.C., Nahvi, A., Sudarsan, N., Barrick, J.E. and Breaker, R.R. (2003) An mRNA structure that controls gene expression by binding *S*-adenosylmethionine. *Nature Structural Biology* 10, 701–707.

Zubieta, C., Arkus, K.A.J., Cahoon, R.E. and Jez, J.M. (2008) A single amino acid change is responsible for evolution of acyltransferase specificity in bacterial methionine biosynthesis. *The Journal of Biological Chemistry* 283, 7561–7567.

14 Regulation of Sulfur Amino Acid Metabolism in Fungi

J.V. Paietta*

Department of Biochemistry and Molecular Biology, Wright State University, Dayton, Ohio, USA

14.1 Abstract

The fungal biosynthesis of cysteine and methionine requires the ability to accumulate sulfur atoms from the environment and to convert the sulfur into a usable form (e.g. sulfide). In particular, the processes involved in sulfur amino acid biosynthesis, including sulfur acquisition, the generation of sulfide and cysteine, and homocysteine and methionine metabolism are considered. A particular focus is on how fungi use a complex regulatory network to monitor their cellular sulfur status and respond with appropriate acquisition and/or metabolic remodelling strategies. A common theme is the versatility and resourcefulness of fungi. For example, a spectrum of transporters allow fungi to acquire a variety of different sulfur compounds. Another common feature is the tight control and coordinated transcriptional upregulation that is observed in the sulfur multigene network. A mechanistic understanding of these sulfur regulatory networks is likely to have widespread biological applications.

14.2 Introduction

Sulfur is an important low-abundance element that is found in a number of compounds and macromolecules that are needed for the growth and function of cells (e.g. proteins, lipids, intermediary metabolites and electron carriers). The lack of availability of an essential macroelement, such as sulfur, can be an important limiting factor in the growth of organisms. Fungi are quite versatile in accumulating sulfur atoms from a variety of compounds in their environment. As a specific example, they can respond to sulfur limitation by upregulating (or derepressing) the production of transporters (e.g. sulfate permeases) and enzymes (e.g. sulfatases) that liberate sulfur from extracellular or intracellular compounds (Marzluf, 1997; Paietta, 2010). The sulfur that is acquired can then be converted into a reduced state (i.e. sulfide) and used for the eventual production of the sulfur-containing amino acids cysteine and methionine.

Fungi have provided a useful model for unravelling the mechanisms of how organisms adjust to a constantly changing environment in order to sustain levels of key nutrients. In particular, they use complex regulatory networks that allow the monitoring of cellular sulfur levels and then the up- or downregulation of sulfur acquisition and other processes to achieve stable levels of the element. The regulatory system of *Neurospora crassa* sulfur will be used as the main example, with consideration of other fungal species at key points. A number

*E-mail address: john.paietta@wright.edu

of reviews have examined aspects of the major fungal model systems for sulfur metabolism in detail, e.g. Paszewski *et al.* (2000) for *Aspergillus nidulans*, Thomas and Surdin-Kerjan (1997) for the yeast *Saccharomyces cerevisiae* and Paietta (2004, 2010) for *N. crassa*.

The insights obtained from studies of sulfur regulatory systems such as those found in *N. crassa* and other model species appear to be applicable to a diverse group of fungi and relevant to various biological situations. Examples include: (i) the essential role of methionine synthesis in infection by the rice blast fungus *Magnaporthe oryzae* (Saint-Macary *et al.*, 2015); (ii) the essential role of the regulation of sulfur assimilation in the virulence of the human pathogen *Aspergillus fumigatus* (Amich *et al.*, 2013); (iii) the sulfur-regulated transporter genes of the penicillin producer *Penicillium chrysogenum* (van de Kamp *et al.*, 2000); (iv) the extracellular protease released in sulfur-limited growth conditions by the dermatophyte *Trichophyton rubrum* (Apodaca and McKerrow, 1989); (v) the methionine-repressible arylsulfatase in infection of mallow by the plant pathogen *Colleotrichum gloeosporiodes* (Goodwin *et al.*, 2000); and (vi) the role of SconCp, a negative regulator of inorganic sulfur assimilation, as a virulence determinant of the causative agent of paracoccidioidomycosis, *Paracoccidioides brasiliensis* (Menino *et al.*, 2013).

14.3 Useful Sources of Sulfur

Fungi are capable of accumulating sulfur atoms from a wide variety of sulfur-containing compounds in the environment. Subsequently, as will be discussed, a primary use of the acquired sulfur from these compounds is to convert it into a reduced state (a sulfide), which is then utilized for the synthesis of cysteine and methionine (Fig. 14.1). Transsulfuration reactions then allow the conversion of cysteine to homocysteine through the intermediate cystathionine. Homocysteine can subsequently be methylated to generate methionine (Fig. 14.2). Reverse transsulfuration allows homocysteine (via methionine and *S*-adenosylmethionine), to be used for the production of cysteine.

Soil-living fungi potentially have available a selection of sulfur compounds, such as sulfate esters, sulfonates and inorganic sulfate, along with much lower concentrations of the sulfur-containing amino acids, to transport and assimilate. Soil analyses typically show that these compounds are present, but at quite variable levels, in different samples (Autry and Fitzgerald, 1990). Fungi will generally respond to sulfur limitation by coordinatively producing a battery of the proteins that are necessary to transport and metabolize these alternative sulfur compounds.

In the case of *N. crassa* and many other fungi, inorganic sulfate is a preferred sulfur source. This is obtained from the environment via specific membrane transporters and then enters the sulfur assimilation pathway (Fig. 14.1). Sulfur that is available as aliphatic and aromatic sulfate esters (e.g. choline-*O*-sulfate and others) can be acted upon by sulfatases such as choline-*O*-sulfatase and arylsulfatase (Marzluf, 1997; Paietta, 2010) (Fig. 14.1) and provides an additional source of sulfate. If present, thiosulfate, sulfide and sulfite can also be directly routed into the assimilation pathway, thus generating cysteine for many fungi (Thomas and Surdin-Kerjan 1997; Paszewski *et al.*, 2000). In addition, sulfite can be directly generated by alkanesulfonate monooxygenases from sulfonate compounds such as ethanedisulfonic acid and isoethionic acid, and from cysteic acid and taurine by the action of cysteic acid decarboxylase and taurine dioxygenase, and routed into the sulfur assimilation pathway that way (Fig. 14.1).

A systematic study of 13 ascomycetous yeast species revealed substantial variability in their capability to use alternative sulfur sources (Linder, 2012). The results of tests of the growth of these species on sulfate esters, sulfamates, sulfones, sulfonates and sulfoxides showed that *Schizosaccharomyces pombe* was capable of using only aromatic sulfate esters, while species such *Komagataella pastoris* (syn. *Pichia pastoris*) and *Yarrowia lipolytica* could use a wide variety of sulfur sources. However, even though some fungal species may lack the particular enzymes needed to handle a variety of sulfur compounds, the basic sulfur regulatory mechanisms (as described below) that control the sulfur-related enzymes in a particular species do appear to be evolutionarily conserved (Hebert *et al.*, 2011; Linder, 2012).

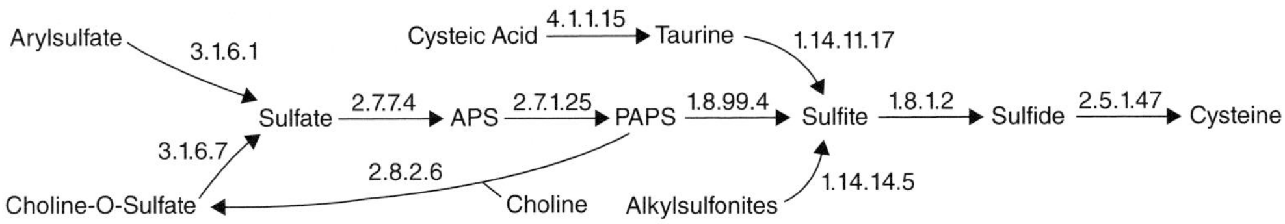

Fig. 14.1. Outline of the diverse sulfur sources acquired by *Neurospora crassa* and the pathways leading to their assimilation and the synthesis of cysteine. Sulfur is potentially available both from the environment (e.g. as inorganic sulfate, choline-*O*-sulfate, arylsulfate) and from internal stores (e.g. choline-*O*-sulfate). Enzyme Commission (E.C.) designations are given for the enzymes that act at each step in the processes involved in the assimilation of these (and other) sulfur compounds: EC 3.1.6.1, arylsulfatase; EC 3.1.6.7, choline sulfatase; EC 2.7.7.4, ATP sulfurylase; EC 2.1.1.25, adenylyl sulfate kinase; EC 1.8.99.4, PAPS reductase; EC 2.8.2.6, choline sulfotransferase; EC 4.1.1.15 cysteic acid decarboxylase; EC 1.14.14.5, alkanesulfonate monooxygenase; EC 1.14.11.17, taurine dioxygenase; EC 1.8.1.2 sulfite reductase; EC 2.5.1.47, cysteine synthase. Abbreviations: APS, adenosine-5′-phosphosulfate; PAPS, 3′-phosphoadenosine-5′-phosphosulfate.

14.4 Response to Sulfur Limitation Related to Acquisition

Fungi have versatile, highly regulated systems for the acquisition of sulfur and its subsequent use for the synthesis of cysteine and methionine as required. *N. crassa* is used here as a focus for outlining the typical cell response to sulfur limitation. The most striking response is the coordinated expression (often termed derepression) of a battery of genes involved in acquiring sulfur and its subsequent metabolism. The genes include an extensive variety of proteins that hydrolyse or transport sulfur compounds. Examples of these genes are those that encode aromatic sulfate permease, arylsulfatase, choline-*O*-sulfatase, choline-*O*-sulfate permease, extracellular protease, methionine permease and sulfate permeases I and II (reviewed by Paietta, 2004, 2010 and Marzluf, 1997). Because *N. crassa* can use a variety of sulfur sources (see Section 14.3), the simultaneous coordinated expression of these genes provides a rapid means of obtaining sulfur in whatever usable form is present in the organism's environment. In contrast, under conditions of high sulfur availability, the sulfur-related genes are tightly regulated and there is little to no detectable expression observed of the network of sulfur-related genes. The effects of sulfur limitation have been studied experimentally in *N. crassa*, by growing it on low levels of methionine (0.25 mM), while its response to sulfur-sufficient (or high sulfur) conditions was investigated by growing it on 5.0 mM methionine (Paietta, 1990).

An example of the coordinated response to sulfur limitation is the acquisition of sulfur as sulfate derived from arylsulfate compounds, which has been extensively studied in *N. crassa* (Paietta, 1989). After growth on a low sulfur medium, the fungus derepresses the expression of arylsulfatase, and the enzyme that is then produced has extracellular and intracellular activity in both mycelia and conidia (Scott and Metzenberg, 1970). In particular, the extracellular arylsulfatase can act on naturally occurring arylsulfate compounds in the immediate environment of the fungal cell to release free sulfate. Note that the production of arylsulfatase is tightly regulated and is not detectable under sulfur-sufficient (or repressive) conditions. An important aspect of the action of the regulatory system is the simultaneous derepression of the production of two sulfate transporters (permeases) (Paietta, 2010). These two sulfate permeases, which are encoded by the genes *cys-13*$^+$ (expressed in conidia; '+' designates a positive effect) and *cys-14*$^+$ (expressed in mycelia), can then transport arylsulfatase-liberated sulfate into the cell. An arylsulfate transporter is produced as well (Marzluf, 1997), and so any arylsulfate compounds present can also be transported and hydrolysed by the intracellular arylsulfatase present. The accumulated sulfate, which is efficiently scavenged from the environment by the coordinated derepression response, can then enter the sulfur assimilation pathway (Fig. 14.1) and be reduced to sulfide and subsequently used to generate cysteine.

Fungi may also store reserves of sulfur in compounds such as choline-*O*-sulfate, which can be accessed under sulfur-limiting growth conditions. Figure 14.1 shows that a portion of the sulfur being routed through the sulfur assimilation pathway can be used to produce choline-*O*-sulfate, which is synthesized from 3′-phosphoadenosine-5′-phosphosulfate (PAPS) and choline through the action of PAPS-choline sulfotransferase. The internal stores of choline-*O*-sulfate so produced are accessed when the expression of choline sulfatase is derepressed under sulfur-limiting conditions; the enzyme then acts on the internal stores of choline-*O*-sulfate to release sulfate for use in the assimilation pathway leading to cysteine (Fig. 14.1). A further aspect of choline-*O*-sulfate metabolism is that *N. crassa* produces a choline-*O*-sulfate transporter; the choline-*O*-sulfate that is acquired via the transporter can either be stored or acted upon immediately by choline-*O*-sulfatase. Some examples of the mobilization of stored sulfur in fungi include the depression of choline sulfatase when *Paracoccidiodes brasiliensis* undergoes a transition from the mycelial to the yeast form (Ferreira *et al.*, 2006), and the germination of conidia in *N. crassa* (McGuire and Marzluf, 1974).

When fungal cells are sulfur limited, besides the coordinated derepression of the genes directly involved in sulfur acquisition, there appears to be additional metabolic remodelling that takes place. For example, cystathionine occupies a central point in the transsulfuration pathway between cysteine and homocysteine, as outlined in Fig. 14.2. In *N. crassa*, the enzymes converting cystathionine to homocysteine (cystathionine β-lyase) and cystathionine to cysteine (cystathionine γ-lyase) are depressed under sulfur-limiting growth conditions (Reveal and Paietta, 2012, 2013). Under such conditions, the intracellular pool of cystathionine is accessed and routed towards cysteine and methionine. Further, the gene encoding methionine synthase, which converts homocysteine into methionine, is also derepressed under sulfur limitation (Jensen and Paietta, unpublished data). Clearly, the cellular mechanisms acting to supply the essential sulfur involve not only the coordinated production of proteins acting to acquire environmental sulfur but the efficient co-regulation of internal metabolic pathways.

14.4.1 Direct acquisition of cysteine and methionine

Fungi can directly acquire cysteine and methionine from the extracellular environment. In particular, the release of cysteine and methionine from exogenous protein contributes to the maintenance of sulfur homeostasis in the cell. This process has been a well characterized case in *N. crassa*, which secretes an extracellular protease under conditions of sulfur limitation (Hanson and Marzluf, 1975). The uptake of the resulting cysteine and methionine (or of peptides with cysteine or methionine residues) can satisfy the sulfur requirements of the fungus. Similarly, dermatophytes such as *Trichophyton rubrum* will synthesize and release proteases that are responsive to sulfur limitation (Apodaca and McKerrow, 1989) and which allow digested host proteins to provide the needed sulfur. Related to this process in dermatophytes and other fungi such as *Aspergillus fumigatus* is the secretion of sulfite as a reducing agent to aid in the digestion of keratinous tissues. The excreted sulfite cleaves the disulfide bonds of keratin and therefore makes the protein accessible to proteolytic degradation (Monod, 2008). The presence or absence of sulfur compounds in the growth medium is also well known to affect the mycelial-to-yeast-phase transition of certain dimorphic fungi. *P. brasiliensis* (Ferreira *et al.*, 2006) is an example in which the presence of cysteine or cystine is required for the transition from mycelial to yeast phase, but inorganic sulfate as a sole sulfur source is insufficient for this process.

Related to the direct acquisition of cysteine and methionine is the production of specific transporters that allow uptake of these amino acids resulting from the hydrolysed exogenous protein. For example, in *S. cerevisiae*, there is a high affinity cysteine-specific transporter (Yct1p; Kaur and Bachawat, 2007) which serves a major role under sulfur-limiting growth conditions. Although not yet thoroughly characterized, the homology of Yct1p to similar proteins from other fungi indicates the likely widespread role of cysteine-specific transport. In contrast, methionine transport has already been well characterized. For example, *N. crassa* coordinatively derepresses a methionine permease (Pall and Robertson, 1988) along with producing an extracellular protease to acquire exogenous methionine, whereas in yeast, the MUP1, MUP2, and

MUP3 (methionine uptake) series of transporters plays this role (Isnard *et al.*, 1996).

14.5 Adjustments to Cellular Protein Sulfur Composition

Based on studies in yeast, fungal cells demonstrate a number of potential adaptations to sulfur limitation. On one level, the process of tRNA thiolation is controlled by the level of intracellular cysteine and methionine (Laxman *et al.*, 2013). When sulfur amino acids are in short supply, tRNA thiolation (which is of importance in nutritionally difficult conditions) is downregulated and this allows the organism to modulate translation in order to decrease sulfur consumption and control cell growth. Further, in response to sulfur demand, there can be a specific reduction in the level of sulfur-rich proteins produced, with some of these being replaced by sulfur-depleted proteins (e.g. the isoenzymes of certain glycolytic enzymes), leading to a 30% saving of sulfur-containing amino acids (Fauchon *et al.*, 2002). The sulfur that is spared can then redirected to cellular needs such as the enhanced production of glutathione for cadmium detoxification. This adjustment is coordinated by the yeast regulator Met4p (methionine-requiring protein 4, a transcriptional activator of MET4, a leucine zipper protein), which is a homologue of the *N. crassa* CYS3 (cystathionine γ-lyase) sulfur regulator (Fauchon *et al.*, 2002). When sulfur sparing in yeast is considered at the proteome level in terms of amino acid composition, there are reduced levels of sulfur amino acids present in the sulfur assimilatory enzymes (Baudouin-Cornu *et al.*, 2001). For organisms like *S. cerevisiae*, for which environmental sulfur sources may have generally been limiting, evolutionary adaptation appears to have resulted in the replacement of dispensable sulfur amino acids in proteins, thus leading to an overall reduction in the sulfur (as cysteine or methionine) that is present in cellular proteins (Scheibel *et al.*, 1997).

14.6 Sulfur Assimilation and the Synthesis of Cysteine and Methionine

The routing of potential sulfur sources and the sulfur assimilation pathway for *N. crassa* is shown in Fig. 14.1. Sulfate is activated in a two-step process to 3′-phosphoadenosine-5′-phosphosulfate (PAPS). To carry this out, ATP sulfurylase catalyses the production of adenosine-5′-phosphosulfate (APS) from sulfate and then adenylyl sulfate kinase acts to produce PAPS. The sulfur moiety is then reduced to sulfite by the action of PAPS reductase, and to sulfide by the action of sulfite reductase. In *N. crassa*, cysteine is then synthesized from *O*-acetylserine and sulfide by the action of cysteine synthase (Marzluf, 1997).

The transsulfuration reactions connecting cysteine to homocysteine and thereby to methionine metabolism in *N. crassa* are shown in Fig. 14.2. The key intermediate, cystathionine, is formed from *O*-acetyl homoserine (itself derived from aspartate-β-semialdehyde by the action of homoserine dehydrogenase and homoserine *O*-acetyltransferase) in a condensation reaction with cysteine which is catalysed by cystathionine γ-synthase. Cystathionine is then cleaved to generate homocysteine by the action of cystathionine β-lyase. Homocysteine is then methylated using the folate derivative 5-methyltetrahydrofolate by the action of methionine synthase, to yield methionine (Borkovich, *et al.*, 2004).

Fungi can convert methionine into cysteine using a different set of enzymes. The first reaction is the formation of *S*-adenosyl methionine (SAM) from methionine. SAM is then used in a transmethylation reaction and cycles back via *S*-adenosylhomocysteine (SAH) to homocysteine (Fig. 14.2). The homocysteine so generated can either lead back to methionine or be routed to cysteine through reverse transsulfuration. Although, common intermediates are involved in this part of the cycle, a separate set of enzymes is involved in the conversion of homocysteine to cysteine. Homocysteine is first converted to cystathionine by the action of cystathionine β-synthase, and cystathionine is then converted to cysteine by the action of cystathionine γ-lyase. The intracellular pool of cystathionine, which sits at the central point in the transsulfuration pathway, is available for either cysteine or methionine synthesis. As noted previously, studies show that in *N. crassa* the pool of cystathionine is accessed under sulfur-limiting conditions by depressing (or inducing) the production of both cystathionine

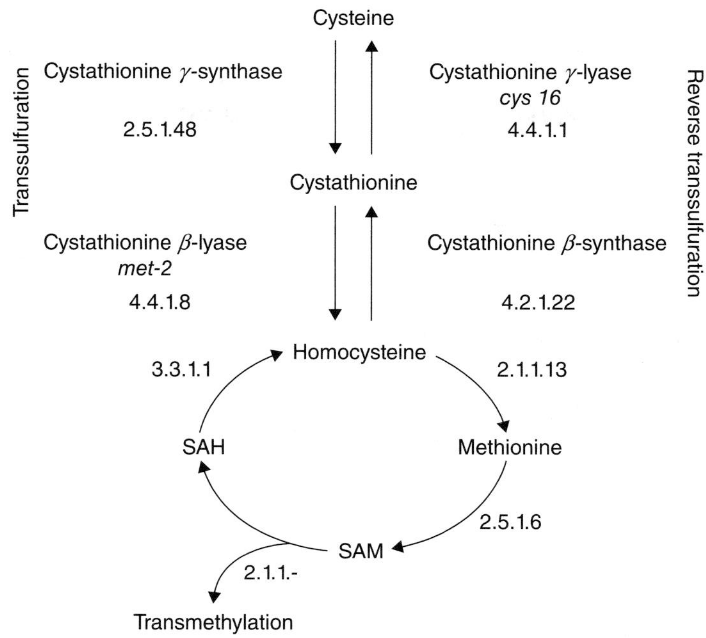

Fig. 14.2. The interconnection of cysteine and methionine metabolism via transsulfuration in *Neurospora crassa*. Expression of the genes for cystathionine γ-lyase (*cys-16*) and cystathionine β-lyase (*met-2*) is controlled by the sulfur regulatory system by derepression under conditions of sulfur limitation and repression under sulfur sufficient conditions (Reveal and Paietta, 2012, 2013). The pathway for the generation and usage of *S*-adenosylmethionine (SAM, aka AdoMet) is also shown. Enzyme Commission (E.C.) designations and the corresponding enzyme names are given for the steps of the SAM cycle and for cysteine metabolism: EC 2.1.1.13, methionine synthase; EC 2.5.1.6, SAM synthetase; EC 2.1.1.- methyltransferase (various); EC 3.3.1.1, adenosylhomocysteinase. Abbreviation: SAH, *S*-adenosylhomocysteine.

β-lyase (producing homocysteine) and cystathionine γ-lyase (producing cysteine) (Reveal and Paietta, 2012, 2013). Methionine synthase catalyses the conversion of homocysteine into methionine and is upregulated in a coordinated fashion (along with the cystathionine β-lyase and γ-lyase) (Jensen and Paietta, unpublished data). These findings provide a preliminary characterization of the metabolic remodelling in transsulfuration that occurs under sulfur-eficient growth conditions.

14.7 Components of the Sulfur Regulatory System

The complex regulatory network involved in the control of sulfur metabolism monitors the cellular sulfur status and coordinatively upregulates the production of the necessary proteins as required for sulfur acquisition and the maintenance of adequate internal levels of sulfur. The *N. crassa* regulatory system consists of the following major components: (i) the bZIP transcriptional activator CYS3; (ii) the F-box protein sulfur controller-2 (SCON2) and the SCF^{SCON2} machinery; and (iii) the putative sulfur sensor (the nature of which is speculative at present). A schematic diagram of the regulatory system is shown in Fig. 14.3.

14.7.1 The CYS3 regulator

The most proximal component of the *N. crassa* sulfur regulatory system is the *cys-3*$^{+}$ regulatory gene (Paietta *et al.*, 1987; Paietta, 1992). Δ*cys-3*

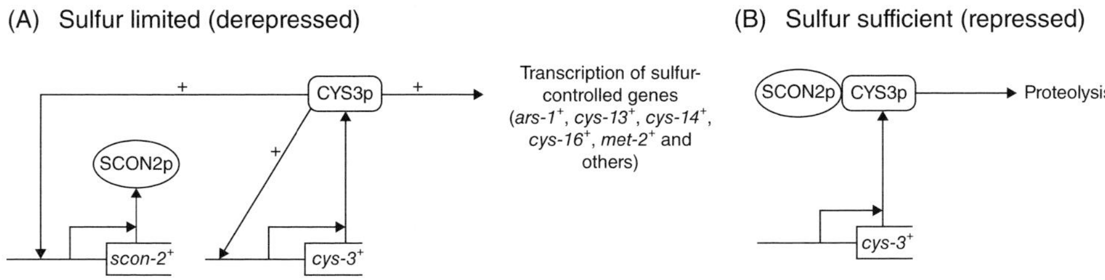

Fig. 14.3. Schematic diagram of the principal components and interactions involved in the *Neurospora crassa* sulfur regulatory system. (A) Under sulfur-limiting or derepressing growth conditions, the CYS3 transcriptional activator (designated CYS3p) binds to the promoters of the genes *cys-3*$^+$ (a positive sulfur regulatory gene that is switched on during conditions of sulfur limitation), *scon-2*$^+$ (a negative regulatory gene sulfur controller gene which encodes an F-box protein designated SCON2p) and a large set of genes encoding sulfur-related proteins (e.g. sulfatases, transporters, proteases). The plus symbol (+) designates a positive effect. (B) In contrast, under sulfur-sufficient or repressing growth conditions, targeted proteolysis quickly reduces the level of the CYS3 transcriptional activator (CYS3p). The targeted proteolysis is carried out through the binding of the F-box protein SCON2p to CYS3p as part of the action of the SCF^{SCON2} complex (the Skp/Cullin/F-box) (Paietta, 2010).

mutants are sulfur auxotrophs and are unable to express the entire set of sulfur-regulated genes (e.g. arylsulfatase) under high or low sulfur growth conditions. Expression of the *cys-3*$^+$ gene itself is switched on during conditions of sulfur limitation (Paietta *et al.*, 1987). Both *cys-3*$^+$ transcripts and the CYS3 protein turn over rapidly under conditions of sulfur sufficiency (i.e. repressing conditions) (Tao and Marzluf, 1998). The CYS3 regulator is a basic region-leucine zipper (bZIP) DNA binding protein whose level is autoregulated. An important factor in how the sulfur regulatory system ultimately controls sulfur gene expression is based on how the action and level of CYS3 is controlled.

The coordinated upregulation of the set of sulfur-related genes under sulfur-limited conditions is due to the binding of the CYS3 activator to a promoter sequence termed the Sulfur Response Element or SuRE (Paietta, 2000). For example, the cystathionine γ-lyase gene (*cys-16*$^+$) shows a typical expression profile with high transcript levels present only under sulfur-limited conditions and no expression in a *Δcys-3* mutant strain (Reveal and Paietta, 2012). Paietta (2008) defined a consensus sequence of ATGGCGCCAT for CYS3 binding based on binding-site selection from random-sequence oligonucleotides with a central core characterized by alternating purines and pyrimidines (RYRY), as in ATGRYRYCAT (Shuler, 1993; Li and Marzluf, 1996). A sulfur-related gene under the control of CYS3, such as *cys-16*$^+$, demonstrates the presence of CYS3 binding sites within its promoter region as detected by gel-mobility shift analysis (Reveal and Paietta, 2012). Met4p is the protein in *S. cerevisiae* with homology to CYS3 in *Neurospora*, but the promoter binding site for Met4p, TCACGTG, is not similar (Thomas and Surdin-Kerjan, 1997).

14.7.2 Sulfur controller regulators

A second group of regulatory components in *N. crassa* were identified using mutants that showed constitutive derepression of all the sulfur-regulated genes (e.g. arylsulfatase, sulfate permease). These were designated negative regulators, and the group was given the name of 'sulfur controller', hence the name *scon*. The group includes *scon-1* (Burton and Mezenberg 1972), *scon-2* (Paietta, 1990; Kumar and Paietta, 1998) and *scon-3* (Sizemore and Paietta, 2002). In *A. nidulans*, the name *scon* is also used to designate a similar group of genes, *sconA*, *sconB*, *sconC* and *sconD* (Natorff *et al.*, 1993, Paszewski *et al.*, 2000), and *scon* genes are also found in other fungal species (Menino, *et al.*, 2013; Uthman *et al.*, 2005). The *scon-2* gene was cloned (Paietta, 1990) by sib selection screening of a cosmid-based gene library based on the use of chromate (a toxic sulfate analogue; Holland and Avery, 2011), which is transported into scon-2 cells grown on high sulfur but not into cells that have regained normal (wild type) sulfur regulation in which sulfate uptake is repressed under sulfur-sufficient conditions. The encoded SCON2 and SCON3 proteins have key regulatory roles and are discussed further below in terms of the F-box domain. *scon-1*$^{+}$ may encode the putative sulfur sensor in the system.

Upon characterization of the SCON2 protein of *N. crassa*, a conserved domain designated the '*N*-terminal domain' was discovered (Kumar and Paietta, 1995). In cell cycle studies with yeast, this was later designated the homologous 'F-box' domain (Bai *et al.*, 1996). A model was proposed that involved ubiquitin-mediated proteolysis, with the F-box proteins serving as an adaptor between the target protein to be regulated and the proteolytic machinery. The F-box motif has been identified in a number of regulatory proteins (Kipreos and Pagano, 2000; Jonkers and Rep, 2009). The SCF complex (Skp/Cullin/F-box) acts as E3 ubiquitin ligases to target certain proteins for proteolysis; it directs the ligation of ubiquitin to the target protein, with the F-box component providing target specificity, and leads to its destruction within the proteosome. In yeast, a number of major metabolic pathways are regulated by SCF complexes, including glucose induction (SCF^{Grr1}) and sulfur metabolism (SCF^{Met30}) (Patton *et al.*, 1998). The SCON2 F-box domain is required for its role as a regulator within the *N. crassa* sulfur control system as demonstrated by Kumar and Paietta (1998). Further, the capability of SCON2 to form a functional SCF complex is reliant on protein–protein interactions between the F-box motif of SCON2 and the *N. crassa* homologue of yeast Skp1p, which has been designated SCON3 (Sizemore and Paietta, 2002). An SCF^{SCON2} complex in the current model would exert its effect by the controlling the level of the CYS3 transcriptional activator.

14.7.3 The sulfur signal and sensor

Presumably, a 'sensor' protein (currently unidentified) is used to detect the sulfur status of the cell and signals through a transduction chain (perhaps by a phosphorylation event). In turn, the signal would have its most proximal effect in controlling the action of the CYS3 regulator. Fungi, through this mechanism, would monitor their sulfur levels and adjust their sulfur-related gene expression profile accordingly to ensure that the required amounts of cysteine and methionine are synthesized.

Experiments with *N. crassa* have suggested that the level of cysteine or a related metabolite is monitored by the sulfur regulatory system. Jacobson and Metzenberg (1977) showed that the effector metabolite in the system appears to be cysteine rather than methionine (high levels of which are typically used to repress the system). In a *ser-l* mutant, they observed that high levels of methionine (5 mM) did not repress arylsulfatase expression as would be seen in the wild type. Due to the biochemical defect in *ser-l*, this mutant strain is starved of cysteine even if methionine is present. When the *ser-1* mutant is provided with supplemental serine, it is no longer starved of cysteine and repression occurs (Jacobson and Metzenberg 1977). Experimental addition of exogenous cysteine to more directly define its role is problematic owing to the growth inhibition that it causes (Adiga *et al.*, 1962; Jacobson and Metzenberg, 1977). Growing *N. crassa* on a cysteine pro-drug (such as N-acetyl-L-cysteine) gets around the toxicity problem and does result in repression of sulfur gene expression (Paietta, unpublished data).

For other fungi, there is also support for cysteine as an effector molecule. In *A. nidulans*, *cys* mutants blocked in cysteine synthesis, as well as starvation for cysteine in a *cysAmecB* strain, showed derepression of the sulfur-related enzymes (Paszewski *et al.*, 2000). For *S. cerevisiae*, the level of SAM has been hypothesized to be the repression signal. Mutants that are defective in the synthesis of SAM, when grown with supplemental methionine, did not show repression effects (Thomas and Surdin-Kerjan, 1997). Other work, however, by Hansen and Johannesen (2000) indicates a regulatory role for cysteine in *S. cerevisiae* in that the formation of cysteine from homocysteine was necessary for methionine repression. Experiments with the yeast *Hansenula polymorpha* also demonstrate a primary role for cysteine based on repression of sulfur-related genes (Sohn *et al.*, 2014).

In any case, a clear resolution of the identity of the effector molecule, which is likely to be cysteine or a derivative, will be needed for studies of interactions with the putative sulfur sensor. Currently, for fungi, the putative *sulfur controller-1* (*scon-1*) gene product of *N. crassa* may be the best candidate for a sulfur sensor. *scon-1* mutants demonstrate constitutive expression of the sulfur-related genes, such as arylsulfatase (Burton and Metzenberg, 1972). *scon-1* is also epistatic to other known elements in the sulfur regulatory system (Paietta, 1990) and shows a novel 'nuclear limited' effect in heterokaryotic fusions (Burton and Metzenberg, 1972). Essentially, the nuclear limitation effect was detected in a heterokaryon containing wild type nuclei (i.e. *scon-1*$^+$) and *scon*-1 nuclei in which the wild type nuclei showed normal sulfur regulation while the *scon*-1 nuclei showed constitutive sulfur gene expression (Burton and Metzenberg, 1972). The cloning of the *scon-1* gene, which is in progress (Paietta, unpublished data), would appear to be essential to determining the mechanism of how the cellular sulfur status is monitored.

14.8 Operation of the Sulfur Regulatory System

In *N. crassa*, the sulfur regulatory system represents a complex control network that assesses the cellular sulfur level and specifically upregulates the expression of a set of sulfur-related genes under sulfur limitation in order to achieve optimal internal sulfur levels for use in the synthesis of cysteine, methionine and a variety of other compounds. When operating under sulfur-sufficient conditions, the sulfur-related genes are subject to repression. This regulatory system allows fungi to efficiently adjust to a constantly changing nutrient environment.

The established portions of the sulfur regulatory system are shown in Fig. 14.3. Not shown in the figure is the hypothetical *scon-1*$^+$ gene product, which may act as the sulfur sensor. A possible end result of the transduction pathway involving the sensor would be to change the phosphorylation state of the CYS3 protein. A potential model, based on the regulation seen in other SCF-controlled systems, would have the CYS3 protein existing in two states. For example, under sulfur-sufficient (i.e. repressed) conditions (Fig. 14.3B, the CYS3 protein might be phosphorylated and this modification would render binding by the SCON2 F-box protein possible. According to this model, CYS3 is a dual target of both the signal transduction chain and, subsequently, of the F-box receptor protein. When SCON2 binds CYS3, there would be an immediate sequestration of CYS3, so that transcriptional activation is blocked. Next, SCON2 (with bound CYS3) interacts with SCON3 to form the SCF^{SCON2} complex, which targets CYS3 for degradation. The double tier of SCON2 regulation by sequestration and the targeted degradation of CYS3 would effectively block transcriptional activation of the sulfur-related genes. This model is supported by SCON2 mutagenesis experiments (Kumar and Paietta, 1998). Further, the CYS3 autoregulatory loop that is in operation under sulfur-limiting conditions (Paietta, 1992) to upregulate *cys-3*$^+$ expression would be blocked.

For the contrasting situation (shown in Fig. 14.3A), under sulfur-limiting conditions CYS3 would be in an alternative state (perhaps dephosphorylated) that would not favour binding to SCON2. SCON2 would not block transcriptional activation by CYS3, and nor would there be targeting of CYS3 to the SCF^{SCON2} complex for degradation. The CYS3 autoregulatory loop would upregulate *cys-3*$^+$ expression and provide sufficient levels of CYS3 to coordinatively induce the expression of the entire set of sulfur-related genes. As the cells transition to sulfur sufficiency, CYS3 would be modified and subjected to the control mechanism outlined above. A final note on the model is that SCON2 is present at its highest level when the system is fully derepressed (Paietta, 1990), due to a feedback loop in which CYS3 activates expression of the *scon-2*$^+$ gene (Kumar and Paietta, 1995). As a consequence, the system is poised for immediate repression by SCON2 when a state of cellular sufficiency is signalled.

14.9 Regulatory Comparison with Other Fungi

The *A. nidulans* sulfur regulatory system resembles that of *N. crassa* in some respects, but with certain substantial differences. In *A. nidulans*, METR represents the transcriptional activator (Natorff *et al.*, 2003; Brzywczy *et al.*, 2011), and this corresponds to CYS3 in *N. crassa* and Met4p in *S. cerevisiae*. METR and CYS3 share substantial homology only in the DNA binding and dimerization domain. Given that a chimeric *metR* gene carrying a *cys-3* bZIP domain is functional in *A. nidulans*, this suggests that the promoter binding sites are similar (Natorff *et al.*, 2003). However, in contrast to the tightly regulated expression of *cys-3*$^+$, in which extremely low transcript levels are detectable under high sulfur levels (Paietta *et al.*, 1987), the *metR* transcript is apparently not regulated by the sulfur source and not subject to autoregulation (Natorff *et al.*, 2003). A further difference is that *A. nidulans* encodes a second regulator, MetZ, which *is* subject to sulfur regulation (as with *cys-3* in *N. crassa*), has two splice variants and whose expression is MetR-dependent (Pilsyk *et al.*, 2015).

Both the *sconB* and *sconC* genes of *A. nidulans* have been characterized. *sconB* encodes a homologue of *N. crassa scon-2*$^+$ that is functionally interchangeable (Natorff *et al.*, 1998). The *scon-2*$^+$ gene is highly regulated and its transcript is present at low levels under repressing conditions (Paietta, 1990). In contrast, the *A. nidulans sconB* transcript increases about twofold under sulfur limitation and is only slightly reduced under sulfur repression (Natorff *et al.*, 1998). *sconC* encodes a Skp1 family protein that shares a high degree of homology with SCON3. For both *sconC* and *scon-3*$^+$, expression can be observed under sulfur-limiting or sulfur-sufficient conditions (Piotrowska *et al.*, 2000; Sizemore and Paietta, 2002).

The sulfur regulatory system of *S. cerevisiae* demonstrates similarities to that of *N. crassa* as well as having some significant differences. Met4p, which contains the transcriptional activation and bZIP domains, is the equivalent of CYS3 in *N. crassa*, but the homology is weak, with the greatest similarity in the leucine zipper dimerization domain. A second bZIP protein has been identified in yeast and designated Met28p, although this lacks a transcriptional activation domain (Kuras *et al.*, 1996). The yeast system appears to represent a case of combinatorial control by the activator Met4p, along with Met31p, Met32p and Cbf1p (DNA binding factors), and with a cofactor Met28p and an F-box protein Met30p (Carillo *et al.*, 2012; McIsaac *et al.*, 2012; Sadhu *et al.*, 2014). Neither *A. nidulans* nor *N. crassa* appears to have this level of complexity in the proximal steps of the control system.

SCON2 in *N. crassa* is homologous to Met30p in yeast and both have F-box and WD-40 domains (Thomas, *et al.*, 1995). A functional difference between these F-box proteins is that deletion of MET30 revealed that it was an essential gene (Thomas *et al.*, 1995), while in *N. crassa* the deletion of *scon-2*$^+$ resulted in derepressed sulfur gene expression (Kumar and Paietta, 1998). A substantial amount of data supports the involvement of Met30p and SCON2 in an SCF complex. SCON3 appears to play a similar role to that of Skp1p in the SCF control mechanism (Sizemore and Paietta, 2002). A final note is that there are no candidate genes that appear in either *A. nidulans* or *S. cerevisiae* with similar properties to the *scon-1*$^+$ of *N. crassa*.

14.10 Conclusions

The fungi have efficient and complex regulatory systems in place to achieve relatively stable levels of sulfur for which a primary use is for the synthesis of cysteine and methionine. Versatility in both the uptake of a variety of sulfur-containing compounds and the capability to extract or convert the sulfur into a usable state appears to be a common feature of many species. Based on studies of the *N. crassa* sulfur regulatory system, along with a comparison with other fungal model systems, we can define the necessary regulatory components as follows: (i) the detection of the effector metabolite by the sulfur sensor and transduction of the sulfur status of the cell (which is hypothetical at this point); (ii) coordinated transcriptional activation of a set of sulfur-related genes by bZIP activator(s); and (iii) control of the activator through target proteolysis by the action of the SCF complex and associated F-box protein. As noted in Section 14.2, a thorough mechanistic understanding of these widespread sulfur regulatory networks is likely to have biological applicability to a number of plant and animal pathogens, as well as biotechnological significance. Another example is that the development of antifungal drugs based on the inhibition of amino acid biosynthesis (Jastrzebowska and Gabriel, 2015) is likely to be aided by such regulatory studies.

References

Adiga, P.R., Sastry, K.S. and Sarma P.S. (1962) Amino acid relationships in cysteine toxicity in *Neurospora crassa*. *Journal of General Microbiology* 29, 149–155.

Amich, J., Schafferer, L., Haas, H. and Krappmann, S. (2013) Regulation of sulphur assimilation is essential for virulence and affects iron homeostasis of the human-pathogenic mould *Aspergillus fumigatus*. *PLOS Pathogens* 9, e1003573.

Apodaca, G. and McKerrow, J.H. (1989) Regulation of *Trichophyton rubrum* proteolytic activity. *Infection and Immunology* 57, 3081–3090.

Autry, A.R. and Fitzgerald, J.W. (1990) Sulfonate S: a major form of forest soil organic sulfur. *Biology and Fertility of Soils* 10, 50–56.

Bai, C., Sen, P., Hofmann, K., Ma, L., Goebl, M., Harper, J.W. and Elledge, S.J. (1996) SKP 1 connects cell cycle regulators to the ubiquitin proteolysis machinery through a novel motif, the F-box. *Cell* 86, 263–274.

Baudouin-Cornu, P., Surdin-Kerjan, Y., Marliere, P. and Thomas, D. (2001) Molecular evolution of protein atomic composition. *Science* 293, 297–300.

Borkovich, K.A., Alex, L.A., Yarden, O., Freitag, M., Turner, G.E., Read, N.D., Seiler, S., Bell-Pederson, D., Paietta, J.V., Plesofsky, N. *et al.* (2004) Lessons from the genome sequence of *Neurospora crassa*: tracing the path from genomic blueprint to multicellular organism. *Microbiology and Molecular Biology Reviews* 68, 1–108.

Brzywczy, J., Kacprzak, M.M. and Paszewski, A. (2011) Novel mutations reveal two important regions in *Aspergillus nidulans* transcriptional activator MetR. *Fungal Genetics and Biology* 48, 104–112.

Burton, E.G. and Metzenberg, R.L. (1972) Novel mutation causing derepression of several enzymes of sulfur metabolism of several enzymes of sulfur metabolism in *Neurospora crassa*. *Journal of Bacteriology* 109, 140–151.

Carrillo, E., Ben-Ari, G., Wildenhain, J., Tyers, M., Grammentz, D. and Lee, T.A. (2012) Characterizing the roles of Met31 and Met32 in coordinating Met4-activated transcription in the absence of Met30. *Molecular Biology of the Cell* 23, 1928–1942.

Fauchon, M., Lagniel, G., Aude, J.-C., Lombardis, L., Soularue, P., Petat, C., Marguerie, G., Senenac, A., Werner, M. and Labarre, J. (2002) Sulfur sparing in the yeast proteome in response to sulfur demand. *Molecular Cell* 9, 713–723.

Ferreira, M.E. da S., Marques, E. dos R., Malavazi, I., Torres, I., Restrepo, A., Numes, L.R., de Oliveira, R.C., Goldman, M.H.S. and Goldman, G.H. (2006) Transcriptome analysis and molecular studies on sulfur metabolism in the human pathogenic fungus *Paracoccidioides brasiliensis*. *Molecular Genetics and Genomics* 276, 450–463.

Goodwin, P.H., Li, J. and Jin, S.M. (2000) Evidence for sulfate derepression of an arylsulfatase gene of *Colletotrichum gloeosporioides* f. sp. *malvae* during infection of round-leaved mallow, *Malva pusilla*. *Physiological and Molecular Plant Pathology* 57, 169–176.

Hansen, J. and Johannesen, P.F. (2000) Cysteine is essential for transcriptional regulation of the sulfur assimilation genes in *Saccharomyces cerevisiae*. *Molecular General Genetics* 263, 535–542.

Hanson, M.A. and Marzluf, G.A. (1975) Control of the synthesis of a single enzyme by multiple regulatory circuits in *Neurospora crassa*. *Proceedings of the National Academy of Sciences of the United States of America* 72, 1240–1244.

Hebert, A., Casaregola, S. and Beckerich, J.-M. (2011) Biodiversity in sulfur metabolism in hemiascomycetous yeasts. *FEMS Yeast Research* 11, 366–378.

Holland, S.L. and Avery, S.V. (2011) Chromate toxicity and the role of sulfur. *Metallomics* 3, 1119–1123.

Isnard, A.D., Thomas, D., and Surdin-Kerjan, Y. (1996) The study of methionine uptake in *Saccharomyces cerevisiae* reveals a new family of amino acid permeases. *Journal of Molecular Biology* 262, 473–484.

Jacobson, E.S. and Metzenberg, R.L. (1977) Control of arylsulfatase in a serine auxotroph of *Neurospora*. *Journal of Bacteriology* 130, 1397–1398.

Jastrzebowska, K. and Gabriel, I. (2015) Inhibitors of amino acid biosynthesis as antifungal agents. *Amino Acids* 47, 227–249.

Jonkers, W. and Rep, M. (2009) Lessons from fungal F-box proteins. *Eukaryotic Cell* 8, 677–695.

Kaur, J. and Bachawati, A.K. (2007) Yct1p, a novel, high-affinity, cysteine-specific transporter from the yeast *Saccharomyces cerevisiae*. *Genetics* 176, 877–890.

Kipreos, E.T. and Pagano, M. (2000) The F-box protein family. *Genome Biology* 1, 3002.1–3002.7.

Kumar, A. and Paietta, J.V. (1995) The sulfur controller-2 negative regulatory gene of *Neurospora crassa* encodes a protein with β-transducin repeats. *Proceedings of the National Academy of Sciences of the United States of America* 92, 3343–3347.

Kumar, A. and Paietta, J.V. (1998) An additional role for the F-box motif: gene regulation within the *Neurospora crassa* sulfur control network. *Proceedings of the National Academy of Sciences of the United States of America* 95, 2417–2422.

Kuras, L., Cherest, H., Surdin-Kerjan, Y. and Thomas, D. (1996) A heteromeric complex containing the centromere binding factor 1 and two basic leucine zipper factors, met4 and me28, mediates the transcription activation of yeast sulfur metabolism. *The EMBO Journal* 15, 2519–2529.

Laxman, S., Sutter, B.M., Wu, X.., Kumar, S., Guo, X., Trudgian, D.C., Mirzaei, H. and Tu, B.P. (2013) Sulfur amino acids regulated translational capacity and metabolic homeostasis through modulation of tRNA thiolation. *Cell* 154, 416–429.

Li, Q. and Marzluf, G.A. (1996) Determination of the *Neurospora crassa* CYS3 sulfur regulatory protein consensus DNA-binding site: amino acid substitutions in the CYS3 bZIP domain that alter DNA-binding specificity. *Current Genetics* 30, 298–304.

Linder, T. (2012) Genomics of alternative sulfur utilization in ascomycetous yeasts. *Microbiology* 158, 2585–2597.

Marzluf, G.A. (1997) Molecular genetics of sulfur assimilation in filamentous fungi and yeast. *Annual Review of Microbiology* 51, 73–96.

McGuire, W.G. and Marzluf, G.A. (1974) Developmental regulation of choline sulfatase and arylsulfatase in *Neurospora crassa*. *Archives of Biochemistry and Biophysics* 161, 360–368.

McIsaac, R.S., Petti, A.A., Bussemaker, H.J. and Botstein, D. (2012) Perturbation-based analysis and modeling of combinatorial regulation in the yeast sulfur assimilation pathway. *Molecular Biology of the Cell* 23, 2993–3007.

Menino, J.F., Saraiva, M., Gomes-Rezende, J., Sturme, M., Pedrosa, J., Castro, A.G., Ludovico, P., Goldman, G.H. and Rodrigues, F. (2013) *P. brasiliensis* virulence is affected by SconC, the negative regulator of inorganic sulfur assimilation. *PLoS ONE* 8(9): e74725.

Monod, M. (2008) Secreted proteases from dermatophytes. *Mycopathologica* 166, 285–294.

Natorff, R., Balinska, M. and Paszewski, A. (1993) At least four regulatory genes control sulphur metabolite repression in *Aspergillus nidulans*. *Molecular and General Genetics* 238, 185–192.

Natorff, R., Piotrowska, M. and Paszewski, A. (1998) The *Aspergillus nidulans* sulphur regulatory gene sconB encodes a protein with WD40 repeats and an F-box. *Molecular and General Genetics* 257, 255–263.

Natorff, R., Sienko, M., Brzywczy, J. and Paszewski, A. (2003) The *Aspergillus nidulans metR* gene encodes a bZIP protein which activates transcription of sulphur metabolism genes. *Molecular Microbiology* 49, 1081–1094.

Paietta, J.V. (1989) Molecular cloning and regulatory analysis of the arylsulfatase structural gene of *Neurospora crassa*. *Molecular and Cellular Biology* 9, 3630–3637.

Paietta, J.V. (1990) Molecular cloning and analysis of the *scon-2* negative regulatory gene of *Neurospora crassa*. *Molecular and Cellular Biology* 10, 5207–5214.

Paietta, J.V. (1992) Production of the CYS3 regulator, a bZIP DNA-binding protein is sufficient to induce sulfur gene expression in *Neurospora crassa*. *Molecular and Cellular Biology* 12, 1568–1577.

Paietta, J.V. (2000) Regulation of sulfur metabolism in *Neurospora crassa*. *Fungal Genetics Newsletter* 47S, 8.

Paietta, J.V. (2004) Regulation of sulfur metabolism in mycelial fungi. In: Brambl, R. and Marzluf G.A. (eds) *The Mycota. III. Biochemistry and Molecular Biology*, 2nd edn. Springer, Berlin, pp.369–383.

Paietta, J.V. (2008) DNA-binding specificity of the CYS3 transcription factor of *Neurospora crassa* defined by binding-site selection. *Fungal Genetics and Biology* 45, 1166–1171.

Paietta, J.V. (2010) Sulfur, phosphorus and iron metabolism. In: Borkovich, K.A. and Ebbole, D.J. (eds) *Cellular and Molecular Biology of Filamentous Fungi*. ASM Press, Washington, DC, pp 359–375.

Paietta, J.V., Akins, R.A., Lambowitz, A.M. and Marzluf, G.A. (1987) Molecular cloning and characterization of the *cys-3* regulatory gene of *Neurospora crassa*. *Molecular and Cellular Biology* 7, 2506–2511.

Pall, M.L. and Robertson, C.K. (1988) Growth-regulation by GTP: regulation of nucleotide pools in *Neurospora* by nitrogen and sulfur control systems. *The Journal of Biological Chemistry* 263, 11168–11174.

Paszewski, A., Natorff, R., Piotrowska, M., Brywczy, J., Sienko, M., Grynberg, M., Pizzinini, E. and Turner, G. (2000) Regulation of sulfur amino acid biosynthesis in *Aspergillus nidulans*: physiological and genetic aspects. In: Brunhold, C., Rennenberg, H., De Kok, L.J., Stulen, I. and Davidian, J.-C. (eds) *Sulfur Nutrition and Sulfur Assimilation in Higher Plants.* Haupt, Bern, pp. 93–105.

Patton, E.E., Willems, A.R. and Tyers, M. (1998) Combinatorial control in ubiquitin-dependent proteolysis: don't Skp the F-box hypothesis. *Trends in Genetics* 14, 236–243.

Pilsyk, S., Natorff, R., Sienko, M., Skoneczny, M. and Paszewski, A. (2015) The *Aspergillus nidulans metZ* gene encodes a transcription factor involved in the regulation of sulfur metabolism in this fungus and other *Eurotiales*. *Current Genetics* 61, 115–125.

Piotrowska, M., Natorff, R. and Paszewski, A. (2000) *sconC*, a gene involved in the regulation of sulphur metabolism in *Aspergillus nidulans*, belongs to the *SKP1* gene family. *Molecular and General Genetics* 264, 276–282.

Reveal, B.S. and Paietta, J.V. (2012) Analysis of the sulfur-regulated control of the cystathionine γ-lyase gene of Neurospora crassa. *BMC Research Notes* 5:339.

Reveal, B.S. and Paietta, J.V. (2013) Sulfur-regulated control of the *met-2*$^+$ gene of *Neurospora crassa* encoding cystathionine β-lyase. *BMC Research Notes* 6:259.

Sadhu, M.J., Moresco, J.J., Zimmer, A.D., Yates, J.R. and Rine, J. (2014) Multiple inputs control sulfur-containing amino acid synthesis in *Saccharomyces cerevisiae*. *Molecular Biology of the Cell* 25, 1653–1665.

Saint-Macary, M.E., Barbisan, C., Gagey, M.J., Frelin, O., Beffa, R., Lebrun, M.H. and Droux, M. (2015) Methionine biosynthesis is essential for infection in the rice blast fungus *Magnaporthe oryzae*. *PLoS ONE* 10(4): e0111108.

Scheibel, T., Bell, S. and Walke, S. (1997) *S. cerevisiae* and sulfur: a unique way to deal with the environment. *The FASEB Journal* 11, 917–921.

Scott, W.A. and Metzenberg, R.L. (1970) Location of arylsulfatase in conidia and young mycelia of *Neurospora crassa*. *Journal of Bacteriology* l04, 1254–1265.

Shuler, J.L. (1993) Analysis of the DNA binding of the *Neurospora crassa* CYS3 sulfur regulatory protein to the *ars-1*$^+$ and *cys-3*$^+$ promoters and determination of the consensus binding site sequence. PhD thesis, Wright State University, Dayton, Ohio.

Sizemore, S.T. and Paietta, J.V. (2002) Cloning and characterization of *scon-3*$^+$: a new member of the *Neurospora crassa* sulfur regulatory system. *Eukaryotic Cell* 1, 875–883.

Sohn, M.J., Yoo, S.J., Oh, D.B., Kwon, O., Lee, S.Y., Sibirny, A.A. and Kang, H.A. (2014) Novel cysteine-centered sulfur metabolic pathway in the thermotolerant methyltrophic yeast *Hansenula polymorpha*. *PLoS ONE* 9(6), e100725.

Tao, Y. and Marzluf, G.A. (1998) Synthesis and differential turnover of the CYS3 regulatory protein of *Neurospora crassa* are subject to sulfur control. *Journal of Bacteriology* 180, 478–482.

Thomas, D. and Surdin-Kerjan, Y. (1997) Metabolism of sulfur amino acids in *Saccharomyces cerevisiae*. *Microbiology and Molecular Biology Reviews* 61, 503–532.

Thomas, D., Kuras, L., Barbey, R., Cheres, H., Blaiseau, P.-L. and Surdin-Kerjan, Y. (1995) Met30p, a yeast transcriptional inhibitor that responds to *S*-adenosylmethionine, is an essential protein with WD40 repeats. *Molecular and Cellular Biology* 15, 6526–6534.

Uthman, A., Dockal, M., Söltz-Szöts, J. and Tschachler, E. (2005) Fluconazole upregulates *sconC* expression and inhibits sulphur metabolism in *Microsporum canis*. *Fungal Genetics and Biology* 42, 719–725.

van de Kamp, M., Schuurs, T.A., Vos, A., van der Lende, T.R., Konings, W.N. and Driessen, A.J.M. (2000) Sulfur regulation of the sulfate transporter genes *sutA* and *sutB* in *Penicillium chrysogenum*. *Applied and Environmental Microbiology* 66, 4536–4538.

15 Insights on *O*-Acetylserine Sulfhydrylase Structure, Function and Biopharmaceutical Applications

B. Campanini* and A. Mozzarelli
Department of Pharmacy, University of Parma, Parma, Italy

15.1 Abstract

Bacteria assimilate sulfur via a reductive pathway that leads to L-cysteine synthesis. The last step is catalysed by the pyridoxal 5′-phosphate dependent enzyme *O*-acetylserine sulfhydrylase (OASS). Two isoenzymes, OASS-A and OASS-B, have been identified that share similar binding sites, although their specific functions are still under investigation. The activity of OASS-A is regulated by its interaction with the preceding enzyme in the pathway, serine acetyltransferase (SAT). The two enzymes interact within the cysteine synthase complex, where the active site of OASS is occupied by the *C*-terminus of SAT. A similar interaction involving OASS has been observed to occur with different binding partners, leading to additional moonlighting functions of the enzyme. OASS, which is involved in the adaptation to the oxidative stress, antibiotic resistance and biofilm formation, is absent in mammals. Therefore, the enzyme is a potential target for enhancers of the antibiotic therapy.

15.2 Introduction

Sulfur is an essential element of living systems because many different biological molecules contain sulfur atoms. These include the amino acids cysteine/cystine and methionine, the coenzymes biotin and thiamine pyrophosphate, the iron–sulfur clusters, and the reducing agents glutathione and mycothiols. Eukaryotic cells are unable to assimilate sulfur from inorganic molecules, whereas bacteria and plants possess systems for sulfur uptake. Bacteria assimilate sulfur via a reductive pathway that begins with either sulfate or thiosulfate transport within the cell. This step is followed by a series of reactions culminating in cysteine synthesis. Under cysteine-rich growth conditions, the reductive sulfate assimilation pathway is repressed and cysteine is imported as cystine by two import (transporter) systems, namely fliY-YecSC and the secondary YdjN (Kredich, 1996; Deutch *et al.*, 2014; Chonoles Imlay *et al.*, 2015). Sulfur assimilation in plants is similar to its bacterial counterpart, but with differences in the import of sulfated molecules and in the regulatory mechanisms that operate (Hell and Wirtz, 2011; Koprivova and Kopriva, 2014).

Cysteine has several functions: (i) it serves as a building block of protein primary structures and, by forming disulfide bridges, increases protein thermal stability; (ii) it is the sulfur source for all the sulfur-containing molecules; and (iii) it plays a role in maintaining the cellular redox state in pathogens such as *Trichomonas*

*Corresponding author. E-mail address: barbara.campanini@unipr.it

vaginalis, *Mycobacterium tuberculosis* and *Salmonella enterica* serovar Typhimurium. These microorganisms spend part of their life cycle in highly oxidizing environments, such as human macrophages, where cysteine and its derivatives counteract the host's oxidative defences during the latent phase.

Cysteine biosynthesis is located at the branch point of nitrogen, carbon and sulfur metabolism (Fig. 15.1), and two linked pathways lead to it. In the first, the imported sulfate is activated and reduced to bisulfide (HS^-) via several steps requiring the consumption of one mole of GTP, two moles of ATP and three moles of NADPH for each mole of synthesized cysteine. The coupling of sulfate reduction with ATP/GTP hydrolysis is required by the low reactivity of sulfate, which needs to be activated prior to reduction (Sekowska *et al.*, 2000; Schelle and Bertozzi, 2006). The intermediate 3′-phosphoadenosine 5′-phosphosulfate (PAPS) is the product of this activation process and represents the universal sulfuryl group donor.

In the second pathway, L-serine, originating from glycolysis, reacts with acetyl-CoA to form activated *O*-acetylserine (OAS), a reaction catalysed by serine acetyltransferase (SAT; EC 2.3.1.30). The final step of cysteine biosynthesis is catalysed by *O*-acetylserine sulfhydrylase (OASS; 2.5.1.47; accepted name: cysteine synthase) which, via a β-replacement reaction, substitutes the acetoxy moiety of OAS with bisulfide. Interestingly, at neutral pH, OAS spontaneously converts to *N*-acetylserine (NAS), which is the inducer of the cysteine operon acting on the transcriptional activator CysB, a member of the LysR-type regulators (Kredich, 1996). Thus, OAS availability, coupled with sulfide limitation, leads to

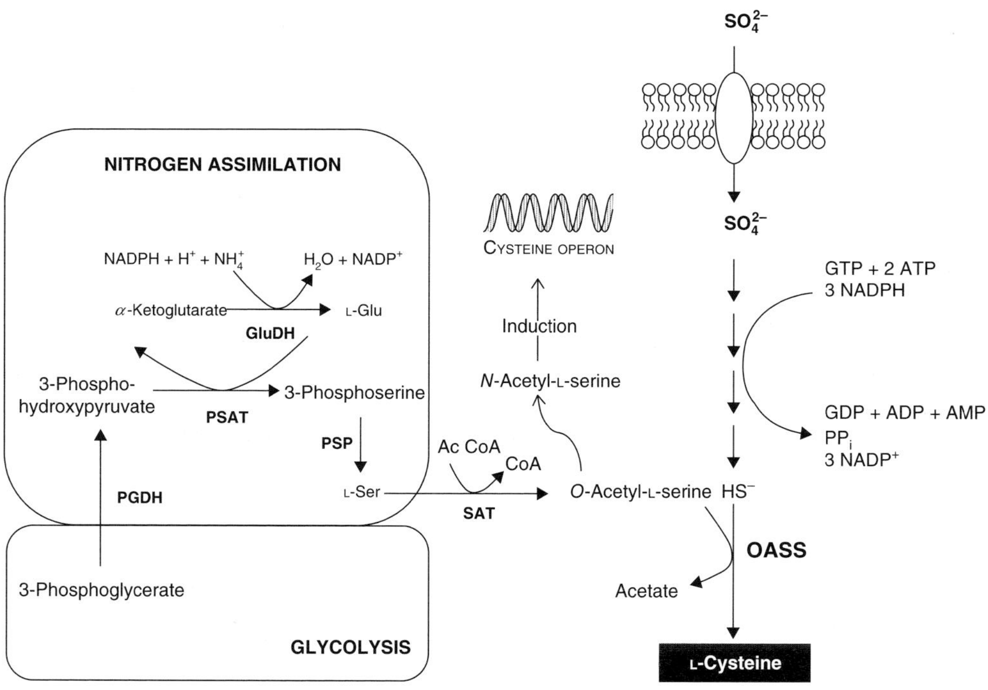

Fig. 15.1. Diagrammatic representation of the location of the cysteine biosynthesis at the branch point of nitrogen, carbon and sulfur metabolism, showing the two pathways that lead to cysteine biosynthesis in *Escherichia coli* and *Salmonella* Typhimurium: (i) the reductive sulfate assimilation pathway to bisulfide requiring the consumption of GTP, ATP and NADPH; and (ii) the synthesis of *O*-acetylserine from the glycolysis intermediate 3-phosphoglycerate. The final step of sulfur assimilation from bisulfide and *O*-acetylserine is catalysed by *O*-acetylserine sulfhydrase (**OASS**). Abbreviations: **GluDH**, glutamate dehydrogenase; **PGDH**, 3-phosphoglycerate dehydrogenase; **PSAT**, 3-phosphoserine aminotransferase; **PSP**, 3-phosphoserine phosphatase; **SAT**, serine acetyltransferase.

an increase in the expression of the cysteine operon, i.e. of the sulfate transporters and of the enzymes involved in sulfate reduction and the incorporation of sulfate into cysteine. OASS is a dimeric pyridoxal 5′-phosphate (PLP)-dependent enzyme that belongs to the fold type II family of the PLP-dependent enzymes (Burkhard *et al.*, 1998).

Two OASS isoforms have generally been detected in bacteria, OASS-A and OASS-B, the products of *cysK* and *cysM* genes, respectively. Redundancy in enzymatic function points to the central role played by cysteine biosynthesis in bacteria, especially for those pathogens, like *S.* Typhimurium and *M. tuberculosis*, that have to face highly oxidizing environments during the persistent phase of their life cycle (Schelle and Bertozzi, 2006; Agren *et al.*, 2009). OASS-B is present in lower amounts and seems to be upregulated under anaerobic conditions (Filutowicz *et al.*, 1982). The two isozymes share 43% sequence identity and show a very similar three-dimensional structure (Burkhard *et al.*, 1998, 2000; Heine *et al.*, 2004; Claus *et al.*, 2005; Oda *et al.*, 2005; Chattopadhyay *et al.*, 2007; Schnell *et al.*, 2007; Zocher *et al.*, 2007; Chinthalapudi *et al.*, 2008). However, OASS-B has a larger active site than OASS-A, which probably accounts for the more pronounced substrate promiscuity of this isoform (Maier, 2003; Claus *et al.*, 2005). For example, OASS-B is capable of using the more oxidized thiosulfate in place of bisulfide. In *M. tuberculosis*, an OASS isozyme has been identified that does not incorporate bisulfide directly into OAS but rather interacts with CysO, which carries a thiocarboxylated *C*-terminus (Burns *et al.*, 2005; O'Leary *et al.*, 2008). This isoform also uses *O*-phosphoserine in place of OAS. Unlike OASS-B, OASS-A interacts with the preceding enzyme of the sulfur assimilation pathway, SAT, to form the so-called cysteine synthase complex (Kredich *et al.*, 1969; Huang *et al.*, 2005). When OASS is within the cysteine synthase complex, its activity is competitively inhibited by the *C*-terminal sequence of SAT that binds to the OASS active site (Mino *et al.*, 2000a,b; Huang *et al.*, 2005). In plants, cysteine synthase formation has been proposed as a sensor of sulfur levels inside the cell (Hell and Hillebrand, 2001). Whether the same role also applies to bacterial cysteine synthase needs further investigation.

This chapter specifically focuses on the structure, function and regulation of OASS, as well as on the recently discovered moonlighting functions that the enzyme exhibits in bacteria (Tanous *et al.*, 2008; Diner *et al.*, 2012; Campanini *et al.*, 2015a). The information accumulated over the years on OASS has triggered a series of studies aimed at identifying inhibitors of this enzyme, which, by decreasing the concentration of cysteine in bacteria, should reduce bacterial fitness at infection loci.

15.3 Structure

Both OASS-A and OASS-B are homodimers, with a buried surface area at the subunit interface of about 1300 Å^2 (Protein Data Bank (PDB) code 1OAS, Hara *et al.*, 1990; 2BHT, Claus *et al.*, 2005; and PDB code 2JC3, Chattopadhyay *et al.*, 2007). Each subunit is formed by a large and a small domain. The large domain contains a β-sheet formed by five parallel strands and one antiparallel strand, surrounded on both sides by three and two α-helices. The smaller domain contains an internal β-sheet contoured by four α-helices (Chattopadhyay *et al.*, 2007). The active site is located at the interface between the two domains with PLP bound via a Schiff base to Lys41. In OASS-A from *S.* Typhimurium (Tai *et al.*, 1995), PLP forms eight hydrogen bonds, six with active site residues and two with water molecules. Moreover, the *N*1 of the coenzyme pyridine ring is within hydrogen bonding distance of the hydroxyl oxygen of Ser272 and is probably deprotonated, and the 3′-oxygen interacts with the side-chain amide nitrogen of Asn71.

The structure of OASS-B from *S.* Typhimurium (PDB code 2JC3, Chattopadhyay *et al.*, 2007) exhibits a very similar fold to the A isoform, with a large and a small domain containing PLP at the domain interface. The main structural difference is represented by the volume of the active site, which is larger than that in OASS-A. Furthermore, the B isozyme seems to be more flexible, as suggested by the higher crystallographic B-factors observed for OASS-B than for OASS-A (Chattopadhyay *et al.*, 2007).

The comparison of the structures of OASS-A and OASS-B solved in the absence and presence of ligands indicates that the active site can adopt either an open or closed conformation. The closed conformation is associated with a relative rotation of the large and small domains. This event: (i) decreases the size of the entrance tunnel to the active site; (ii) changes the polarity of the active site, excluding water molecules, in agreement with coenzyme fluorescence enhancement upon OAS and ligand binding (see Section 15.4); (iii) affects the conformation of the substrate binding loop formed by residues 67–71, which is highly conserved in bacteria and plants (Tian *et al.*, 2010); (iv) repositions the catalytic residues; and (v) causes a tilt of the coenzyme of 13°. A semi-closed, inhibited conformation is stabilized by chloride binding at an allosteric site at the dimer interface (Burkhard *et al.*, 2000).

No structure is available for the cysteine synthase complex. However, the interaction between SAT and OASS-A has been investigated by a variety of biochemical and biophysical techniques (Bogdanova and Hell, 1997; Droux *et al.*, 1998; Zhu *et al.*, 1998; Mino *et al.*, 1999, 2000a,b, 2001; Wirtz *et al.*, 2001, 2010; Bonner *et al.*, 2005; Campanini *et al.*, 2005; Huang *et al.*, 2005; Francois *et al.*, 2006; Wirtz and Hell, 2006; Zhao *et al.*, 2006; Kumaran and Jez, 2007; Kumaran *et al.*, 2009; Salsi *et al.*, 2010a,b; Kumar *et al.*, 2011; Spyrakis *et al.*, 2013a; Yi *et al.*, 2013). A key feature that has emerged from these studies is that the *C*-terminal four amino acid residues of SAT are critical for driving complex formation (Mino *et al.*, 2000b). Furthermore, the last residue of the *C*-terminal sequence of SAT, an isoleucine, plays an essential role in binding to OASS-A (Mino *et al.*, 2000b; Huang *et al.*, 2005; Salsi *et al.*, 2010a), and it has been demonstrated that removal of the *C*-terminal Ile from SAT impairs complex formation. The solution of the three-dimensional structure of the complex between OASS-A from *Haemophilus influenzae* and the isolated *C*-terminal decapeptide showed that only the last four residues of the peptide were detectable in the electron density map (Huang *et al.*, 2005). The carboxylic moiety of Ile binds to the carboxylate subsite of the OASS-A active site and its side chain fits into the hydrophobic pocket normally occupied by the acetyl group of OAS.

While these structural studies shed light on the interaction of SAT with the OASS-A active site, insufficient information is available on the interactions outside the active site that contribute significantly to the stabilization of cysteine synthase. In fact, the dissociation constant of SAT for OASS increases to 580 nM for the decapeptide (Campanini *et al.*, 2005) and to 50 μM for the pentapeptide (Salsi *et al.*, 2010a), indicating that other residues at the protein–protein interface are relevant for the binding energetics.

15.4 Function and Regulation

OASS-A catalyses the β-replacement of the acetoxy moiety of OASS with sulfide to form L-cysteine. The reaction takes place in two half reactions that comprise several steps (see Fig. 15.2). In the first step of the first half reaction (Fig. 15.2), a transaldimination reaction between OAS and the internal aldimine of OASS leads to the formation of an external aldimine. This reaction is accompanied by a change in the absorption maximum of PLP from 412 to 422 nm. Unlike the internal aldimine, the external aldimine is a highly fluorescent species, with a maximum emission at 500 nm upon excitation at the peak wavelength of PLP. This observation suggests that the formation of the external aldimine triggers the closing of the active site. The second step of the first half reaction is the removal of an α-carbon proton by a basic residue, probably Lys41, with the transient formation of a very unstable quinonoid species that has never been detected in the OASS-catalysed reaction. The subsequent release of the β-acetoxy moiety concludes the first half-reaction leading to the α-aminoacrylate Schiff base (Fig. 15.2). This species absorbs at 470 nm and is poorly fluorescent. In the absence of a nucleophilic agent, such as sulfide, the α-aminoacrylate accumulates and is stable for several hours, although as a parasitic reaction, α-aminoacrylate can undergo hydrolysis to pyruvate and ammonia. The second half reaction (Fig. 15.2) starts with the fast sulfide attack on the α-aminoacrylate β-carbon, leading to the external aldimine of L-cysteine, and then to L-cysteine itself. This catalytic mechanism is supported by

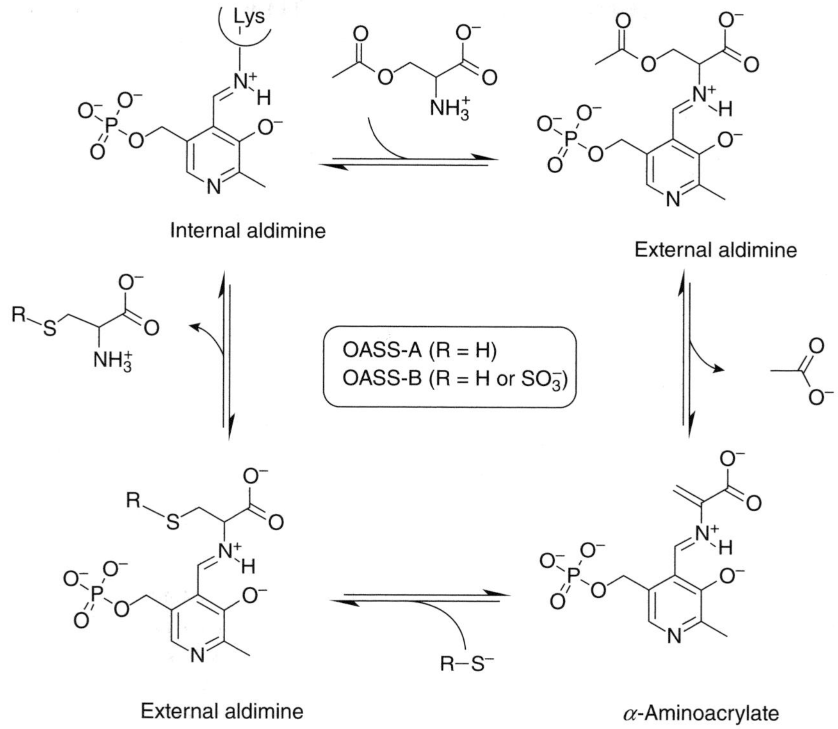

Fig. 15.2. The reaction mechanism of *O*-acetylserine sulfhydrase (OASS). The rate-limiting step in the reaction catalysed by the OASS-A isozyme is the formation of α-aminaoacrylate. This intermediate undergoes a nuclophilic attack by bisulfide or, in the case of OASS-B, by thiosulfate. When thiosulfate is the nucleophile, the product *S*-sulfocysteine (SSC) is reduced by reductases to cysteine (Nakatani *et al.*, 2012).

stopped-flow kinetic studies and isotope effects. Kinetically, the data are consistent with a ping–pong mechanism with competitive substrate inhibition by OAS and sulfide (Cook and Wedding, 1976; Tai *et al.*, 1993), the latter showing a strong competitive substrate inhibition ($K_{I,SH}$ = 50 μM; $K_{M,SH}$ = 6 μM) (Tai *et al.*, 1993). The inhibition was originally attributed to the binding of sulfide to the free enzyme, but the three-dimensional structure of OASS-A in the presence of chloride suggested that sulfide might act as an allosteric effector (Burkhard *et al.*, 2000; Tai *et al.*, 2001). Furthermore, the product cysteine binds to both the free enzyme and the α-aminoacrylate intermediate, as demonstrated by its non-competitive inhibition behaviour vs OAS and sulfide (Mozzarelli *et al.*, 2011).

The turnover number for OASS-A is 130 s^{-1} with formation of α-aminoacrylate being the rate-limiting step of the overall reaction. The maximum rate of formation of the α-aminoacrylate, k_{max}, is 300 s^{-1}, similar to the turnover number obtained from steady-state studies. Rapidly mixing the α-aminoacrylate with sulfide results in formation of the free enzyme and L-cysteine within the dead time of the instrument, even at low (μM) concentrations of sulfide. This finding suggests that the rates of all processes within the second half reaction are characterized by rate constants equal to or higher than 1000 s^{-1} (Woehl *et al.*, 1996).

Analysis of the rate dependence on pH using substrate analogues, such as β-chloroalanine, indicates that the reaction is controlled by two pK_a values, 6.7 and 7.4. The higher pK_a is attributed to the α-amine of the substrate, and the lower pK_a to an enzyme residue. The pH dependence of V/K using thionitrobenzoic acid, an analogue of sulfide endowed with lower reactivity, reflects the properties of the groups involved in the second half reaction. It was found that this step is controlled by pK_a values of 6.5 and 8.2. The pK_a of 8.2 was associated with the ε-amine of Lys41, whereas the group with a value of pK_a 6.5 is probably the same as that observed in the pH-rate profile of the first half-reaction.

The regulation of OASS activity is achieved via either product inhibition or interaction with SAT. Cysteine accumulation leads to the inhibition of both OASS and SAT, so cysteine acts as a feedback inhibitor of the serine acetylation reaction. However, as is common for bacterial biosynthetic pathways, the main mechanism for the control of enzymatic activity is at the transcription level. Genes belonging to the cysteine regulon, with the exception of *cysE*, are under the positive control of the LysR-type regulator CysB (Kredich, 1996; Tyrrell *et al.*, 1997; Paritala and Carroll, 2013), which also represses its own synthesis (Kredich, 1996). LysR-type regulators require both inducers and anti-inducers to fulfil their function. In the case of the cysteine regulon, NAS, which is generated from the unstable OAS by a non-enzymatic *O*- to *N*-acetyl migration at basic pH (Flavin and Slaughter, 1965; Kredich, 1971), acts as an inducer of the four transcriptional units. The accumulation of NAS inside the cell signals a deficiency in the sulfur supply, ultimately of sulfide. Sulfate and thiosulfate are believed to act as anti-inducers of the cysteine regulon (Sekowska *et al.*, 2000).

15.5 Moonlighting Activities

As is the case with a number of metabolic enzymes, moonlighting activities were recently also discovered for OASS, thus widening our understanding of the multiple roles played by OASS in diverse processes such as transcription regulation and biofilm formation. The mechanisms by which moonlighting activities can emerge have been summarized by Jeffery (1999) and comprise the ability of proteins to form complexes with different binding partners, as observed for bacterial thioredoxin. OASS is no exception and it is able to perform different functions, depending on its binding partner (Campanini *et al.*, 2015a). It is useful to recall that the natural binding partner of OASS is SAT and that the formation of the cysteine synthase complex is driven by the insertion of the SAT *C*-terminal peptide into the OASS active site (Kredich, 1992), with two main outcomes: complex stabilization and OASS inhibition.

In the last 5 years, at least two different binding partners have been identified for OASS in bacteria: a toxin involved in contact-dependent growth inhibition (CDI) (Diner *et al.*, 2012); and a transcriptional regulator involved in the modulation of cysteine biosynthesis as a function of sulfur availability (Tanous *et al.*, 2008). In both cases, it has been proved – or strongly suggested – that the *C*-terminus of the binding partner is actually responsible for complex formation, which is similar to what has been observed for the cysteine synthase complex. Notably, moonlighting functions have been described only for the OASS-A isoform and, similarly to what observed for cysteine synthase, the OASS-B isoform is unable to bind to any of the newly identified effectors.

OASS-A identified was as the permissive factor responsible for toxin activation in the uropathogenic *Escherichia coli* strain 536 (UPEC536) (Diner *et al.*, 2012). UPEC536 is a strain that is able to inhibit the growth of non-isogenic bacteria in a contact-dependent manner (Aoki *et al.*, 2005, 2009, 2010) through the delivery, stimulated by direct cell-to-cell contact, of the *C*-terminal (CT) domain of the CdiA toxin, namely CdiA-CT. Isolated CdiA-CT has no intrinsic activity and needs to be activated by OASS to express its anticodon nuclease activity. Many results suggest that the binding of CdiA-CT to OASS is directed to the active site: (i) the CdiA-CT *C*-terminal sequence (KGYGI) is a degenerated SAT *C*-terminal sequence (YGDGI); (ii) mutation or deletion of the *C*-terminal Ile hampers complex formation; and (iii) OAS competes with the toxin for binding to OASS and is able to dissociate the preformed complex. The activation of CdiA-CT by OASS is truly moonlighting because the inactivation of sulfhydrylase activity does not have any effect on toxin activation (Diner *et al.*, 2012).

In *Bacillus subtilis*, OASS, which is unable to form a functional cysteine synthase complex, is a trigger enzyme for controlling the transcription of the cysteine regulon in response to fluctuations in sulfur availability (Tanous *et al.*,

2008; Hullo *et al.*, 2010; Shepard *et al.*, 2011). OASS is in fact able to bind to the transcription regulator, CymR, and to increase its affinity for the operator sequence. The *C*-terminal sequence of CymR, YMFYI, suggests that this protein also binds to OASS by direct interaction of the *C*-terminal residues with the enzyme active site (Campanini *et al.*, 2015a). Indeed, under conditions of decreased sulfur availability, OAS accumulates and effectively competes with CymR for OASS binding. This relieves the repression on the cysteine regulon and leads to transcription of the enzymes involved in sulfate activation and reduction (Tanous *et al.*, 2008). This last moonlighting function has led to the inclusion of OASS among the moonlighting proteins in the MoonProt database (see http://www.moonlightingproteins.org) (Mani *et al.*, 2015).

15.6 Biopharmaceutical Applications

Many enzymes of the reductive sulfate assimilation pathway have been considered for therapeutic intervention (Campanini *et al.*, 2015b). The exploitation of OASS itself for biotechnological and pharmaceutical purposes was proposed more than 10 years ago given its reasonably wide substrate specificity and its potential use for the enantioselective synthesis of unnatural amino acids (Maier, 2003). This application is particularly well suited to the OASS-B isoform of the enzyme, which is especially promiscuous, most likely because of its wider and more polar active site.

More recently, the possible exploitation of OASS for therapeutic intervention in antimicrobial therapy has been pointed out (Schelle and Bertozzi, 2006; Schnell *et al.*, 2007; Salsi *et al.*, 2010a; Campanini *et al.*, 2015b), based on two main observations: (i) OASS is absent in mammals that perform the reverse transsulfuration pathway for cysteine biosynthesis; and (ii) a decrease in cysteine biosynthetic potential invariably leads to changes in the fitness of bacteria that range from decreased infectivity to defects in biofilm formation (Lestrate *et al.*, 2000; Senaratne *et al.*, 2006; Schnell *et al.*, 2015; Singh *et al.*, 2015). Inhibitors of OASS have been developed, with dissociation constants ranging from millimolar to nanomolar, and targeting the enzymes from *M. tuberculosis* (Schnell *et al.*, 2007; Nagpal *et al.*, 2012; Kumar *et al.*, 2013; Poyraz *et al.*, 2013), *Entamoeba histolytica* (Nagpal *et al.*, 2012), *H. influenzae* (Spyrakis *et al.*, 2013a) and *S.* Typhimurium (Amori *et al.*, 2012; Spyrakis *et al.*, 2013a,b).

Inhibitors have been identified using different approaches, from rational design to the screening of natural compound libraries or virtual screening. Most of the inhibitors identified to date share some common structural features, the most relevant being a carboxylate or carboxylate bioisostere. It is tempting to speculate that the carboxylic moieties bind to the carboxylate sub-site in the active site, as is the case of peptides (Huang *et al.*, 2005; Salsi *et al.*, 2010a) and thiazolidine inhibitors (Poyraz *et al.*, 2013). Notably, the complex between OASS from *M. tuberculosis* (MtOASS) and a potent thiazolidine inhibitor is the only example of three-dimensional structures of OASS in complex with a small molecule inhibitor.

The inhibitors that have now been identified are mainly based on the structure of SAT *C*-terminal peptides of *H. influenzae* (Huang *et al.*, 2005; Salsi *et al.*, 2010a; Spyrakis *et al.*, 2013a) and *M. tuberculosis* (Schnell *et al.*, 2007), which have inspired both rational design (Amori *et al.*, 2012) and virtual screening campaigns (Kumar *et al.*, 2013; Poyraz *et al.*, 2013; Spyrakis *et al.*, 2013b). Structure–activity relationship studies with pentapeptides of the general sequence MNXXI on the OASS from *H. influenzae* (HiOASS) have confirmed the role played by the *C*-terminal Ile in driving complex formation and identified two additional hydrophobic pockets that could be exploited for drug design (Salsi *et al.*, 2010a). One of these pockets is occupied by the hydrophobic moiety of an aromatic residue at position P3, while the other is unoccupied. How binding of a hydrophobic moiety to the first pocket can positively affect affinity is demonstrated by a structure–activity relationship study carried out with MNXXI peptides on OASS from *S.* Typhimurium (StOASS-A), in which the MNYDI peptide was identified as the best binder, with a dissociation constant, K_d, of about 200 nM (Spyrakis *et al.*, 2013a) (Fig. 15.3). Non-peptidic inhibitors for HiOASS-A (Amori *et al.*, 2012) and StOASS-A/StOASS-B (Spyrakis *et al.*, 2013b) have been identified by rational design and virtual screening, respectively (Fig. 15.3).

MNLNI	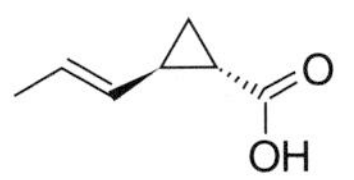	MNYDI	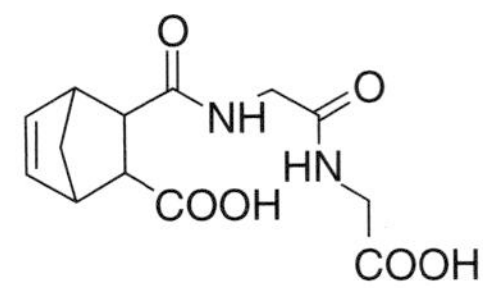
K_d HiOASS-A = 44 μM	K_d HiOASS-A = 1.46 μM	K_d StOASS-A = 0.22 μM K_d StOASS-B = 96 μM	K_d StOASS-A = 3.7 μM K_d StOASS-B = 50 μM

Fig. 15.3. The best peptidic and non-peptidic inhibitors identified for *O*-acetylserine sulfhydrase (OASS) from *Haemophilus influenzae* (HiOASS) and *Salmonella enterica* serovar Typhimurium (StOASS). Peptidic inhibitors of HiOASS and StOASS were identified by virtual screening of a peptide library containing the permutations of the general sequence MNXXI (Salsi *et al.*, 2010a; Spyrakis *et al.*, 2013a). Non-peptidic inhibitors were identified either by rational design based on the structure of the MNLNI peptide bound to HiOASS-A (Amori *et al.*, 2012) or virtual screening using the three-dimensional structures of StOASS-A and StOASS-B as templates (Spyrakis *et al.*, 2013b).

Inhibitors identified by rational design are *trans*-2-substituted-cyclopropane-1-carboxylic acids and the substituent at position 2 is postulated to occupy the hydrophobic cleft delimited by Phe144 (Amori *et al.*, 2012) that accommodates the Ile side chain in the peptide–HiOASS complex structure (Huang *et al.*, 2005; Salsi *et al.*, 2010a). The most active compound identified by structure-based virtual screening and ligand-based virtual screening on StOASS-A and StOASS-B is a norbornene derivative that displays K_d values for StOASS-A and StOASS-B of about 3 and 50 μM, respectively. Unfortunately, most of the identified inhibitors of OASS target only one OASS isoform, e.g. MtCysK1 and StOASS-A.

In the case of StOASS-B the identification of inhibitors is a difficult task, with the most potent molecules identified to date displaying a K_d of about 33 μM (Spyrakis *et al.*, 2013b). In addition, it was demonstrated for peptidic inhibitors that their affinity for the StOASS-A isoform correlates negatively with the affinity for the StOASS-B isoform. This suggests that there is some sort of structural and/or dynamic control over peptide binding to OASS-B, as is also suggested by the observation that SAT does not bind to the B isoform of OASS. Therefore, much effort should be put into the development of inhibitors targeting multiple isoforms and the evaluation of the differential role of isoforms during infection and long-term survival.

15.7 Conclusions

O-acetylserine sulfhydrylase (OASS) was identified in *S.* Typhimurium more than 40 years ago (Kredich and Tomkins, 1966; Becker *et al.*, 1969). Over the years, different isoforms of the enzyme have been identified and their structure, reaction mechanism, stability and substrate specificity have been investigated in great detail. The acquisition of knowledge of OASS function, and of its interaction with serine acetyltransferase (SAT) has allowed the development of inhibitors that might be suitable as antibiotic enhancers. Moreover, the recently discovered moonlighting functions of OASS suggest a more intimate relationship between cysteine biosynthesis and clinically relevant aspects of bacterial biology, like biofilm formation and antibiotic resistance, that warrant in-depth investigations.

References

Agren, D., Schnell, R. and Schneider, G. (2009) The *C*-terminal of CysM from *Mycobacterium tuberculosis* protects the aminoacrylate intermediate and is involved in sulfur donor selectivity. *FEBS Letters* 583, 330–336.

Amori, L., Katkevica, S., Bruno, A., Campanini, B., Felici, P., Mozzarelli, A. and Costantino, G. (2012) Design and synthesis of *trans*-2-substituted-cyclopropane-1-carboxylic acids as the first non-natural small molecule inhibitors of *O*-acetylserine sulfhydrylase. *MedChemComm* 3, 1111–1116.

Aoki, S.K., Pamma, R., Hernday, A.D., Bickham, J.E., Braaten, B.A. and Low, D.A. (2005) Contact-dependent inhibition of growth in *Escherichia coli*. *Science* 309, 1245–1248.

Aoki, S.K., Webb, J.S., Braaten, B.A. and Low, D.A. (2009) Contact-dependent growth inhibition causes reversible metabolic downregulation in *Escherichia coli*. *Journal of Bacteriology* 191, 1777–1786.

Aoki, S.K., Diner, E.J., de Roodenbeke, C.T., Burgess, B.R., Poole, S.J., Braaten, B.A., Jones, A.M., Webb, J.S., Hayes, C.S., Cotter, P.A. and Low, D.A. (2010) A widespread family of polymorphic contact-dependent toxin delivery systems in bacteria. *Nature* 468, 439–442.

Becker, M.A., Kredich, N.M. and Tomkins, G.M. (1969) The purification and characterization of *O*-acetylserine sulfhydrylase-A from Salmonella typhimurium. *The Journal of Biological Chemistry* 244, 2418–2427.

Bogdanova, N. and Hell, R. (1997) Cysteine synthesis in plants: protein–protein interactions of serine acetyltransferase from *Arabidopsis thaliana*. *The Plant Journal* 11, 251–262.

Bonner, E.R., Cahoon, R.E., Knapke, S.M. and Jez, J.M. (2005) Molecular basis of cysteine biosynthesis in plants: structural and functional analysis of *O*-acetylserine sulfhydrylase from *Arabidopsis thaliana*. *The Journal of Biological Chemistry* 280, 38803–38813.

Burkhard, P., Rao, G.S., Hohenester, E., Schnackerz, K.D., Cook, P.F. and Jansonius, J.N. (1998) Three-dimensional structure of *O*-acetylserine sulfhydrylase from *Salmonella typhimurium*. *Journal of Molecular Biology* 283, 121–133.

Burkhard, P., Tai, C.H., Jansonius, J.N. and Cook, P.F. (2000) Identification of an allosteric anion-binding site on *O*-acetylserine sulfhydrylase: structure of the enzyme with chloride bound. *Journal of Molecular Biology* 303, 279–286.

Burns, K.E., Baumgart, S., Dorrestein, P.C., Zhai, H., McLafferty, F.W. and Begley, T.P. (2005) Reconstitution of a new cysteine biosynthetic pathway in *Mycobacterium tuberculosis*. *Journal of the American Chemical Society* 127, 11602–11603.

Campanini, B., Speroni, F., Salsi, E., Cook, P.F., Roderick, S.L., Huang, B., Bettati, S. and Mozzarelli, A. (2005) Interaction of serine acetyltransferase with *O*-acetylserine sulfhydrylase active site: evidence from fluorescence spectroscopy. *Protein Science* 14, 2115–2124.

Campanini, B., Benoni, R., Bettati, S., Beck, C.M., Hayes, C.S. and Mozzarelli, A. (2015a) Moonlighting *O*-acetylserine sulfhydrylase: new functions for an old protein. *Biochimica et Biophysica Acta (BBA) – Proteins and Proteomics* 1854, 1184–1189.

Campanini, B., Pieroni, M., Raboni, S., Bettati, S., Benoni, R., Pecchini, C., Costantino, G. and Mozzarelli, A. (2015b) Inhibitors of the sulfur assimilation pathway in bacterial pathogens as enhancers of antibiotic therapy. *Current Medicinal Chemistry* 22, 187–213.

Chattopadhyay, A., Meier, M., Ivaninskii, S., Burkhard, P., Speroni, F., Campanini, B., Bettati, S., Mozzarelli, A., Rabeh, W.M., Li, L. and Cook, P.F. (2007) Structure, mechanism, and conformational dynamics of *O*-acetylserine sulfhydrylase from *Salmonella typhimurium*: comparison of A and B isozymes. *Biochemistry* 46, 8315–8330.

Chinthalapudi, K., Kumar, M., Kumar, S., Jain, S., Alam, N. and Gourinath, S. (2008) Crystal structure of native *O*-acetyl-serine sulfhydrylase from *Entamoeba histolytica* and its complex with cysteine: structural evidence for cysteine binding and lack of interactions with serine acetyl transferase. *Proteins* 72, 1222–1232.

Chonoles Imlay, K.R., Korshunov, S. and Imlay, J.A. (2015) Physiological roles and adverse effects of the two cystine importers of *Escherichia coli*. *Journal of Bacteriology* 197, 3629–3644.

Claus, M.T., Zocher, G.E., Maier, T.H. and Schulz, G.E. (2005) Structure of the *O*-acetylserine sulfhydrylase isoenzyme CysM from *Escherichia coli*. *Biochemistry* 44, 8620–8626.

Cook, P.F. and Wedding, R.T. (1976) A reaction mechanism from steady state kinetic studies for *O*-acetylserine sulfhydrylase from *Salmonella typhimurium* LT-2. *The Journal of Biological Chemistry* 251, 2023–2029.

Deutch, C.E., Spahija, I. and Wagner, C.E. (2014) Susceptibility of *Escherichia coli* to the toxic L-proline analogue L-selenaproline is dependent on two L-cystine transport systems. *Journal of Applied Microbiology* 117, 1487–1499.

Diner, E.J., Beck, C.M., Webb, J.S., Low, D.A. and Hayes, C.S. (2012) Identification of a target cell permissive factor required for contact-dependent growth inhibition (CDI). *Genes and Development* 26, 515–525.

Droux, M., Ruffet, M.L., Douce, R. and Job, D. (1998) Interactions between serine acetyltransferase and *O*-acetylserine (thiol) lyase in higher plants – structural and kinetic properties of the free and bound enzymes. *European Journal of Biochemistry* 255, 235–245.

Filutowicz, M., Wiater, A. and Hulanicka, D. (1982) Delayed inducibility of sulphite reductase in cysM mutants of *Salmonella typhimurium* under anaerobic conditions. *Journal of General Microbiology* 128, 1791–1794.

Flavin, M. and Slaughter, C. (1965) Synthesis of the succinic ester of homoserine, a new intermediate in the bacterial biosynthesis of methionine. *Biochemistry* 4, 1370–1375.

FranÇois, J.A., Kumaran, S. and Jez, J.M. (2006) Structural basis for interaction of *O*-acetylserine sulfhydrylase and serine acetyltransferase in the *Arabidopsis* cysteine synthase complex. *The Plant Cell* 18, 3647–3655.
Hara, S., Payne, M.A., Schnackerz, K.D. and Cook, P.F. (1990) A rapid purification procedure and computer-assisted sulfide ion selective electrode assay for *O*-acetylserine sulfhydrylase from *Salmonella typhimurium*. *Protein Expression and Purification* 1, 70–76.
Heine, A., Canaves, J.M., von Delft, F., Brinen, L.S., Dai, X., Deacon, A.M., Elsliger, M.A., Eshaghi, S., Floyd, R., Godzik, A. *et al.* (2004) Crystal structure of *O*-acetylserine sulfhydrylase (TM0665) from *Thermotoga maritima* at 1.8 Å resolution. *Proteins* 56, 387–391.
Hell, R. and Hillebrand, H. (2001) Plant concepts for mineral acquisition and allocation. *Current Opinion in Biotechnology* 12, 161–168.
Hell, R. and Wirtz, M. (2011) Molecular biology, biochemistry and cellular physiology of cysteine metabolism in *Arabidopsis thaliana*. *The Arabidopsis Book* 9: e0154.
Huang, B., Vetting, M.W. and Roderick, S.L. (2005) The active site of *O*-acetylserine sulfhydrylase is the anchor point for bienzyme complex formation with serine acetyltransferase. *Journal of Bacteriology* 187, 3201–3205.
Hullo, M.F., Martin-Verstraete, I. and Soutourina, O. (2010) Complex phenotypes of a mutant inactivated for CymR, the global regulator of cysteine metabolism in *Bacillus subtilis*. *FEMS Microbiology Letters* 309, 201–207.
Jeffery, C.J. (1999) Moonlighting proteins. *Trends in Biochemical Sciences* 24, 8–11.
Koprivova, A. and Kopriva, S. (2014) Molecular mechanisms of regulation of sulfate assimilation: first steps on a long road. *Frontiers in Plant Science* 5: 589.
Kredich, N.M. (1971) Regulation of L-cysteine biosynthesis in *Salmonella typhimurium*. I. Effects of growth of varying sulfur sources and *O*-acetyl-L-serine on gene expression. *The Journal of Biological Chemistry* 246, 3474–3484.
Kredich, N.M. (1992) The molecular basis for positive regulation of *cys* promoters in *Salmonella typhimurium* and *Escherichia coli*. *Molecular Microbiology* 6, 2747–2753.
Kredich, N.M. (1996) Biosynthesis of cysteine. In: Neidhardt, F.C., Curtiss, R. III, Ingraham, J.L., Lin, E.C.C., Low, K.B., Magasanik, B., Renznikoff, W.S., Riley, M., Schaechter, M. and Umbarger, H.E. (eds) Escherichia coli *and* Salmonella*: Cellular and Molecular Biology*, 2nd edn. ASM Press, Washington, DC, pp. 514–527.
Kredich, N.M. and Tomkins, G.M. (1966) The enzymic synthesis of L-cysteine in *Escherichia coli* and *Salmonella typhimurium*. *The Journal of Biological Chemistry* 241, 4955–4965.
Kredich, N.M., Becker, M.A. and Tomkins, G.M. (1969) Purification and characterization of cysteine synthetase, a bifunctional protein complex, from *Salmonella typhimurium*. *The Journal of Biological Chemistry* 244, 2428–2439.
Kumar, J.V.U., Poyraz, O., Saxena, S., Schnell, R., Yogeeswari, P., Schneider, G. and Sriram, D. (2013) Discovery of novel inhibitors targeting the *Mycobacterium tuberculosis* *O*-acetylserine sulfhydrylase (CysK1) using virtual high-throughput screening. *Bioorganic and Medicinal Chemistry Letters* 23, 1182–1186.
Kumar, S., Raj, I., Nagpal, I., Subbarao, N. and Gourinath, S. (2011) Structural and biochemical studies of serine acetyltransferase reveal why the parasite *Entamoeba histolytica* cannot form a cysteine synthase complex. *The Journal of Biological Chemistry* 286, 12533–12541.
Kumaran, S. and Jez, J.M. (2007) Thermodynamics of the interaction between *O*-acetylserine sulfhydrylase and the *C*-terminus of serine acetyltransferase. *Biochemistry* 46, 5586–5594.
Kumaran, S., Yi, H., Krishnan, H.B. and Jez, J.M. (2009) Assembly of the cysteine synthase complex and the regulatory role of protein–protein interactions. *The Journal of Biological Chemistry* 284, 10268–10275.
Lestrate, P., Delrue, R.M., Danese, I., Didembourg, C., Taminiau, B., Mertens, P., De Bolle, X., Tibor, A., Tang, C.M. and Letesson, J.J. (2000) Identification and characterization of *in vivo* attenuated mutants of *Brucella melitensis*. *Molecular Microbiology* 38, 543–551.
Maier, T.H. (2003) Semisynthetic production of unnatural L-alpha-amino acids by metabolic engineering of the cysteine-biosynthetic pathway. *Nature Biotechnology* 21, 422–427.
Mani, M., Chen, C., Amblee, V., Liu, H.P., Mathur, T., Zwicke, G., Zabad, S., Patel, B., Thakkar, J. and Jeffery, C.J. (2015) MoonProt: a database for proteins that are known to moonlight. *Nucleic Acids Research* 43, D277–D282.
Mino, K., Yamanoue, T., Sakiyama, T., Eisaki, N., Matsuyama, A. and Nakanishi, K. (1999) Purification and characterization of serine acetyltransferase from *Escherichia coli* partially truncated at the *C*-terminal region. *Bioscience, Biotechnology, and Biochemistry* 63, 168–179.

Mino, K., Yamanoue, T., Sakiyama, T., Eisaki, N., Matsuyama, A. and Nakanishi, K. (2000a) Effects of bienzyme complex formation of cysteine synthetase from *Escherichia coli* on some properties and kinetics. *Bioscience, Biotechnology, and Biochemistry* 64, 1628–1640.

Mino, K., Hiraoka, K., Imamura, K., Sakiyama, T., Eisaki, N., Matsuyama, A. and Nakanishi, K. (2000b) Characteristics of serine acetyltransferase from *Escherichia coli* deleting different lengths of amino acid residues from the *C*-terminus. *Bioscience, Biotechnology, and Biochemistry* 64, 1874–1880.

Mino, K., Imamura, K., Sakiyama, T., Eisaki, N., Matsuyama, A. and Nakanishi, K. (2001) Increase in the stability of serine acetyltransferase from *Escherichia coli* against cold inactivation and proteolysis by forming a bienzyme complex. *Bioscience, Biotechnology, and Biochemistry* 65, 865–874.

Mozzarelli, A., Bettati, S., Campanini, B., Salsi, E., Raboni, S., Singh, R., Spyrakis, F., Kumar, V.P. and Cook, P.F. (2011) The multifaceted pyridoxal 5′-phosphate-dependent *O*-acetylserine sulfhydrylase. *Biochimica et Biophysica Acta* 1814, 1497–1510.

Nagpal, I., Raj, I., Subbarao, N. and Gourinath, S. (2012) Virtual screening, identification and *in vitro* testing of novel inhibitors of *O*-acetyl-L-serine sulfhydrylase of *Entamoeba histolytica*. *PLoS ONE* 7(2): e30305.

Nakatani, T., Ohtsu, I., Nonaka, G., Wiriyathanawudhiwong, N., Morigasaki, S. and Takagi, H. (2012) Enhancement of thioredoxin/glutaredoxin-mediated L-cysteine synthesis from *S*-sulfocysteine increases L-cysteine production in *Escherichia coli*. *Microbial Cell Factories* 11:62.

Oda, Y., Mino, K., Ishikawa, K. and Ataka, M. (2005) Three-dimensional structure of a new enzyme, *O*-phosphoserine sulfhydrylase, involved in L-cysteine biosynthesis by a hyperthermophilic archaeon, *Aeropyrum pernix* K1, at 2.0 Å resolution. *Journal of Molecular Biology* 351, 334–344.

O'Leary, S.E., Jurgenson, C.T., Ealick, S.E. and Begley, T.P. (2008) *O*-phospho-L-serine and the thiocarboxylated sulfur carrier protein CysO-COSH are substrates for CysM, a cysteine synthase from *Mycobacterium tuberculosis*. *Biochemistry* 47, 11606–11615.

Paritala, H. and Carroll, K.S. (2013) New targets and inhibitors of mycobacterial sulfur metabolism. *Infectious Disorders – Drug Targets* 13, 85–115.

Poyraz, O., Jeankumar, V.U., Saxena, S., Schnell, R., Haraldsson, M., Yogeeswari, P., Sriram, D. and Schneider, G. (2013) Structure-guided design of novel thiazolidine inhibitors of *O*-acetyl serine sulfhydrylase from *Mycobacterium tuberculosis*. *Journal of Medicinal Chemistry* 56, 6457–6466.

Salsi, E., Bayden, A.S., Spyrakis, F., Amadasi, A., Campanini, B., Bettati, S., Dodatko, T., Cozzini, P., Kellogg, G.E., Cook, P.F. *et al.* (2010a) Design of *O*-acetylserine sulfhydrylase inhibitors by mimicking nature. *Journal of Medicinal Chemistry* 53, 345–356.

Salsi, E., Campanini, B., Bettati, S., Raboni, S., Roderick, S.L., Cook, P.F. and Mozzarelli, A. (2010b) A two-step process controls the formation of the bienzyme cysteine synthase complex. *The Journal of Biological Chemistry* 285, 12813–12822.

Schelle, M.W. and Bertozzi, C.R. (2006) Sulfate metabolism in mycobacteria. *ChemBioChem* 7, 1516–1524.

Schnell, R., Oehlmann, W., Singh, M. and Schneider, G. (2007) Structural insights into catalysis and inhibition of *O*-acetylserine sulfhydrylase from *Mycobacterium tuberculosis*. Crystal structures of the enzyme α-aminoacrylate intermediate and an enzyme-inhibitor complex. *The Journal of Biological Chemistry* 282, 23473–23481.

Schnell, R., Sriram, D. and Schneider, G. (2015) Pyridoxal-phosphate dependent mycobacterial cysteine synthases: structure, mechanism and potential as drug targets. *Biochimica et Biophysica Acta* 1854, 1175–1183.

Sekowska, A., Kung, H.F. and Danchin, A. (2000) Sulfur metabolism in *Escherichia coli* and related bacteria: facts and fiction. *Journal of Molecular Microbiology and Biotechnology* 2, 145–177.

Senaratne, R.H., De Silva, A.D., Williams, S.J., Mougous, J.D., Reader, J.R., Zhang, T., Chan, S., Sidders, B., Lee, D.H., Chan, J. *et al.* (2006) 5′-Adenosinephosphosulphate reductase (CysH) protects *Mycobacterium tuberculosis* against free radicals during chronic infection phase in mice. *Molecular Microbiology* 59, 1744–1753.

Shepard, W., Soutourina, O., Courtois, E., England, P., Haouz, A. and Martin-Verstraete, I. (2011) Insights into the Rrf2 repressor family – the structure of CymR, the global cysteine regulator of *Bacillus subtilis*. *FEBS Journal* 278, 2689–2701.

Singh, P., Brooks, J.F. 2nd, Ray, V.A., Mandel, M.J. and Visick, K.L. (2015) CysK plays a role in biofilm formation and colonization by *Vibrio fischeri*. *Applied and Environmental Microbiology* 81, 5223–5234.

Spyrakis, F., Felici, P., Bayden, A.S., Salsi, E., Miggiano, R., Kellogg, G.E., Cozzini, P., Cook, P.F., Mozzarelli, A. and Campanini, B. (2013a) Fine tuning of the active site modulates specificity in the interaction of *O*-acetylserine sulfhydrylase isozymes with serine acetyltransferase. *Biochimica et Biophysica Acta* 1834, 169–181.

Spyrakis, F., Singh, R., Cozzini, P., Campanini, B., Salsi, E., Felici, P., Raboni, S., Benedetti, P., Cruciani, G., Kellogg, G.E. *et al.* (2013b) Isozyme-specific ligands for *O*-acetylserine sulfhydrylase, a novel antibiotic target. *PLoS ONE* 8(10): e77558.

Tai, C.H., Nalabolu, S.R., Jacobson, T.M., Minter, D.E. and Cook, P.F. (1993) Kinetic mechanisms of the A and B isozymes of *O*-acetylserine sulfhydrylase from *Salmonella typhimurium* LT-2 using the natural and alternative reactants. *Biochemistry* 32, 6433–6442.

Tai, C.H., Nalabolu, S.R., Simmons, J.W. 3rd, Jacobson, T.M. and Cook, P.F. (1995) Acid–base chemical mechanism of *O*-acetylserine sulfhydrylases-A and -B from pH studies. *Biochemistry* 34, 12311–12322.

Tai, C.H., Burkhard, P., Gani, D., Jenn, T., Johnson, C. and Cook, P.F. (2001) Characterization of the allosteric anion-binding site of *O*-acetylserine sulfhydrylase. *Biochemistry* 40, 7446–7452.

Tanous, C., Soutourina, O., Raynal, B., Hullo, M.F., Mervelet, P., Gilles, A.M., Noirot, P., Danchin, A., England, P. and Martin-Verstraete, I. (2008) The CymR regulator in complex with the enzyme CysK controls cysteine metabolism in *Bacillus subtilis*. *The Journal of Biological Chemistry* 283, 35551–35560.

Tian, H., Guan, R., Salsi, E., Campanini, B., Bettati, S., Kumar, V.P., Karsten, W.E., Mozzarelli, A. and Cook, P.F. (2010) Identification of the structural determinants for the stability of substrate and aminoacrylate external Schiff bases in *O*-acetylserine sulfhydrylase-A. *Biochemistry* 49, 6093–6103.

Tyrrell, R., Verschueren, K.H., Dodson, E.J., Murshudov, G.N., Addy, C. and Wilkinson, A.J. (1997) The structure of the cofactor-binding fragment of the LysR family member, CysB: a familiar fold with a surprising subunit arrangement. *Structure* 5, 1017–1032.

Wirtz, M. and Hell, R. (2006) Functional analysis of the cysteine synthase protein complex from plants: structural, biochemical and regulatory properties. *Journal of Plant Physiology* 163, 273–286.

Wirtz, M., Berkowitz, O., Droux, M. and Hell, R. (2001) The cysteine synthase complex from plants. Mitochondrial serine acetyltransferase from *Arabidopsis thaliana* carries a bifunctional domain for catalysis and protein–protein interaction. *European Journal of Biochemistry* 268, 686–693.

Wirtz, M., Birke, H., Heeg, C., Muller, C., Hosp, F., Throm, C., Konig, S., Feldman-Salit, A., Rippe, K., Petersen, G. *et al.* (2010) Structure and function of the hetero-oligomeric cysteine synthase complex in plants. *The Journal of Biological Chemistry* 285, 32810–32817.

Woehl, E., Tai, C.-H., Dunn, M.F. and Cook, P.F. (1996) Formation of the α-aminoacrylate intermediate limits the overall reaction catalyzed by O-acetylserine sulfhydrylase. *Biochemistry* 35, 4776–4783.

Yi, H., Dey, S., Kumaran, S., Lee, S.G., Krishnan, H.B. and Jez, J.M. (2013) Structure of soybean serine acetyltransferase and formation of the cysteine regulatory complex as a molecular chaperone. *The Journal of Biological Chemistry* 288, 36463–36472.

Zhao, C., Moriga, Y., Feng, B., Kumada, Y., Imanaka, H., Imamura, K. and Nakanishi, K. (2006) On the interaction site of serine acetyltransferase in the cysteine synthase complex from *Escherichia coli*. *Biochemical and Biophysical Research Communications* 341, 911–916.

Zhu, X., Yamaguchi, T. and Masada, M. (1998) Complexes of serine acetyltransferase and isozymes of cysteine synthase in spinach leaves. *Bioscience, Biotechnology, and Biochemistry* 62, 947–952.

Zocher, G., Wiesand, U. and Schulz, G.E. (2007) High resolution structure and catalysis of *O*-acetylserine sulfhydrylase isozyme B from *Escherichia coli*. *FEBS Journal* 274, 5382–5389.

16 Metabolic Engineering of *Corynebacterium glutamicum* for L-Valine Production

X. Wang[1]* and P. J. Quinn[2]

[1]*State Key Laboratory of Food Science and Technology and Synergetic Innovation Center of Food Safety and Nutrition, Jiangnan University, Wuxi, China;* [2]*Department of Biochemistry, King's College London, UK*

16.1 Abstract

L-Valine is one of the three branched-chain amino acids that is essential for animal health and important in metabolism, so it is widely added to food, medicine and feed products. It is predominantly produced through microbial fermentation, and the production efficiency largely depends on the quality of the microorganisms used. In recent years, continuing efforts have been made to determine the mechanisms and means of regulation of L-valine biosynthesis in *C. glutamicum*, the bacterium that is most used for amino acid production. Metabolic engineering based on the mechanisms of biosynthesis and regulation of L-valine production in *C. glutamicum* provides an effective alternative to traditional breeding for strain development, and industrially competitive L-valine-producing *C. glutamicum* strains have now been constructed by genetically defined metabolic engineering. This article reviews the global metabolic and regulatory networks responsible for L-valine biosynthesis, the molecular mechanisms of regulation and the strategies employed in *C. glutamicum* strain engineering.

16.2 Introduction

The metabolic engineering of microorganisms is the process of altering the genetics of appropriate strains to modify their metabolic potential for the enhanced synthesis of specific products (Stephanopoulos, 1999). The identity of the gene products concerned in the pathways of synthesis of these products requires detailed knowledge of the metabolic fluxes and environmental factors that influence them (Imam *et al.*, 2015; Simeonidis and Price, 2015; Unthan *et al.*, 2015). This chapter considers the metabolic engineering of *C. glutamicum* to optimize production of L-valine.

L-Valine is one of the three branched-chain amino acids that are incorporated into proteins. It is widely used as a nutritional supplement and flavouring agent in the food industry and as an amino acid additive in the stock feed industry (Karau and Grayson, 2014). It is also important for the manufacture of composite amino acid injections or synthetic antibiotics in the pharmaceutical industry (Park *et al.*, 2007). Because of its widespread use, the supply of L-valine is invariably limiting. Historically, L-valine and other amino acids were obtained by extraction and

*Corresponding author. E-mail address: xwang@jiangnan.edu.cn

purification from a variety of protein sources (Eyal and Bressler, 1993; Pasupuleti, 2010). More recently, L-valine has been mainly produced by microbial fermentation, and its production largely depends on the quality of the microbial strains used. Therefore, the development of efficient bacterial strains to increase the production of L-valine is attracting considerable interest.

A number of microorganisms have been reported to produce L-valine by fermentation, including *Bacillus subtilis* (Chattopadhyay and Banerjee, 1978), *Serratia marcescens* (Khachatryan, 1986), *Brevibacterium flavum* (Xu *et al.*, 2010a,b) and *C. glutamicum* (Bartek *et al.*, 2010). *C. glutamicum* is among the most important organisms in industrial biotechnology and dominates the large-scale production of a variety of amino acids, including L-glutamate, L-lysine, L-isoleucine and L-valine (Yin *et al.*, 2012, 2013; Eikmanns and Blombach, 2014; Xu *et al.*, 2014). Most L-valine-producing strains of *C. glutamicum* have been developed by traditional breeding methods. Such methods are time-consuming and labour intensive, and strains that do survive multiple rounds of random mutagenesis involved in the improvement process are genetically undefined and susceptible to further changes. Consequently, the metabolic engineering of *C. glutamicum* strains with clearly defined genetic backgrounds is emerging as the most efficient solution to improve the production of L-valine and other amino acids and their precursors (Hao *et al.*, 2010; Wendisch, 2014). Because the genomes of several *C. glutamicum* strains have been sequenced and their genes annotated (Kalinowski *et al.*, 2003; Yukawa *et al.*, 2007), metabolic engineering through rational genetic manipulation has been used to construct L-valine-producing strains of *C. glutamicum* (Chen *et al.*, 2015) and *Bacillus licheniformis* (Liang *et al.*, 2015) as a means of improving L-valine production, yields and quality, reducing by-product formation and expanding the substrate spectra (Elišáková *et al.*, 2005; Blombach *et al.*, 2009; Oldiges *et al.*, 2014).

16.3 Biosynthetic Pathway of L-Valine in *C. glutamicum*

The biosynthesis of L-valine in *C. glutamicum* involves a series of reactions, among which the four most important are catalysed by the enzymes acetohydroxyacid synthase (AHAS), acetohydroxyacid isomeroreductase (AHAIR), dihydroxyacid dehydratase (DHAD) and transaminase B (TA) (Fig. 16.1). Together, these convert pyruvate to L-valine (Leyval *et al.*, 2003; McHardy *et al.*, 2003). AHAS condenses two molecules of pyruvate to 2-acetolactate. AHAS is a tetramer composed of two identical large subunits and two identical small subunits, and the large and small subunits are encoded by *ilvB* and *ilvN*, respectively. As the first common enzyme shared by the biosynthesis pathways of all three branched chain amino acids, AHAS is present in microorganisms and in plants but not in animals, so it is an attractive target for both drug and herbicide discovery (Gedi and Yoon, 2012). AHAS activity requires thiamine diphosphate, Mg^{2+} and flavin adenine dinucleotide (FAD). AHAIR is encoded by *ilvC* and catalyses both the isomerization and reduction of 2-acetolactate, forming dihydroxyisovalerate, and its activity requires the cofactor NADPH. DHAD is encoded by *ilvD* and catalyses the dehydration of dihydroxyisovalerate to form 2-ketoisovalerate. TA is encoded by *ilvE* and converts 2-ketoisovalerate into L-valine; in *C. glutamicum*, this reaction could be catalysed by the enzyme AvtA (valine-pyruvate aminotransferase) encoded by *avtA* as well (Marienhagen *et al.*, 2005). TA also catalyses the final reaction steps in the biosynthetic pathways of L-leucine and L-isoleucine in *C. glutamicum* (McHardy *et al.*, 2003).

The enzymes AHAS, AHAIR, DHAD and TA not only catalyse the four reactions in the biosynthetic pathway of L-isoleucine as outlined above, they can also use different substrates. For example, AHAS uses both pyruvate and 2-ketobutyrate as substrates when synthesizing L-isoleucine, but it uses only pyruvate when synthesizing L-valine. In *C. glutamicum*, the L-valine biosynthetic pathway is interwoven with the biosynthetic pathways of several major metabolites, as shown in Fig. 16.1. The key precursor, pyruvate, is also used for the biosynthesis of L-alanine, acetyl CoA, acetate, oxaloacetate and L-isoleucine; another key precursor, 2-ketoisovalerate, is also used for the biosynthesis of L-leucine and D-pantothenate. Therefore, strengthening the biosynthetic pathway of L-valine and weakening the other pathways that consume the key precursors is a strategy that can be designed to optimize L-valine production in *C. glutamicum*.

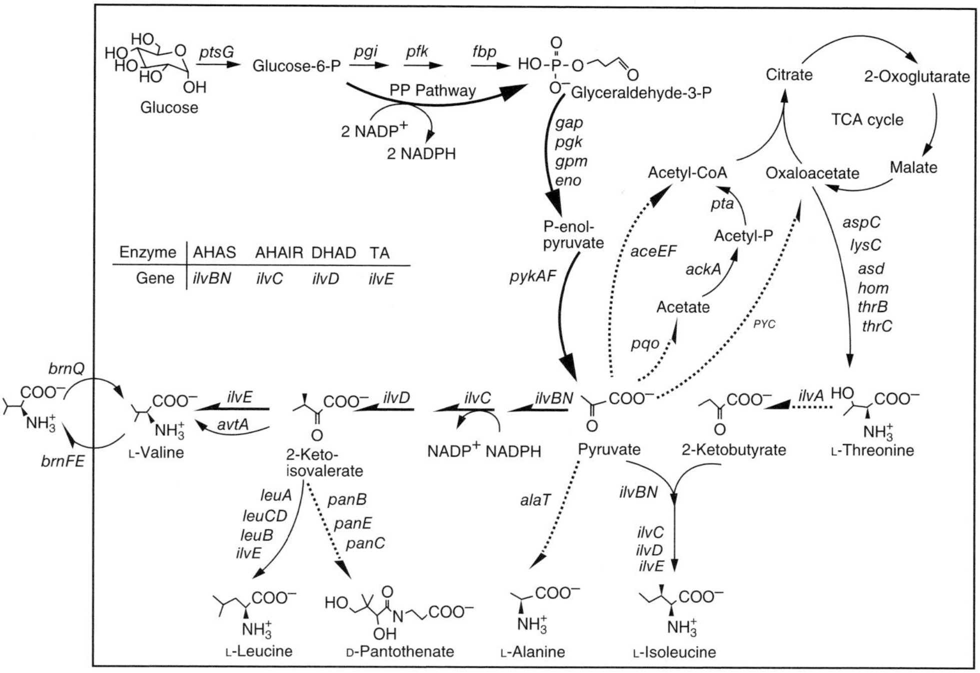

Fig. 16.1. The central metabolism and biosynthetic pathway of L-valine in *Corynebacterium glutamicum* and strategies for metabolic engineering to improve L-valine production. The frame indicates the cell wall. The thick solid arrow-headed pathways indicate the increased flux that can be achieved by directly overexpressing the corresponding genes in the pathway. The dotted arrow-headed pathways indicate the reactions that consume the precursors for L-valine production. The increased flux through the pentose phosphate (PP) pathway provides more ATP and NADPH, which are required for L-valine production. Enzyme abbreviations with their gene designations: AHAS, two-component acetohydroxyacid synthase (genes *ilvBN*); AHAIR, acetohydroxyacid isomeroreductase (gene *ilvC*); DHAD, dihydroxyacid dehydratase (gene *ilvD*); TA, transaminase B (gene ilvE). Other abbreviations and gene designations: *aceEF*, pyruvate dehydrogenase complex; *ackA*, acetate kinase; *alaT*, alanine aminotransferase; *asd*, aspartate-semialdehyde dehydrogenase; *aspC*, aspartate aminotransferase; *avtA*, alanine/valine-pyruvate aminotransferase; *brnFE*, a two-component branched chain amino acid export permease; *brnQ*, branched-chain amino acid transport system carrier protein; *eno*, enolase; *fbp*, fructose-1,6-bisphosphatase; *gap*, glyceraldehyde 3-phosphate dehydrogenase; *gpm*, phosphoglycerate mutase; *hom*, homoserine dehydrogenase; *ilvA*, L-threonine dehydratase; *leuA*, isopropylmalate synthase; *leuB*, 3-isopropylmalate dehydrogenase; *leuCD*, two-component isopropylmalate dehydratase; *lysC*, aspartokinase; *panB*, 3-methyl-2-oxobutanoate hydroxymethyltransferase; *panC*, pantothenate synthetase; *panE*, ketopantoate reductase (this activity encoded by *ilvC* in *C. glutamicum*; *pfk*, ATP-dependent 6-phosphofructokinase; *pgi*, glucose-6-phosphate isomerase; *pgk*, 3-phosphoglycerate kinase; *pqo*, pyruvate:quinone oxidoreductase; *pta*, phosphate acetyltransferase; *ptsG*, glucose-specific EII permease of the phosphotransferase system (PTS); *pykAF*, pyruvate kinases; TCA cycle, tricarboxylic acid cycle; *thrB*, homoserine kinase; *thrC*, threonine synthase.

The export of L-valine, as well as that of the other two branched-chain amino acids, X and Y, is controlled by the exporter BrnFE, a two-component export permease belonging to the LIV-E transporter family (Marin and Kramer, 2007), which consists of BrnF and BrnE, which are encoded by *brnF* and *brnE*, respectively (Kennerknecht *et al.*, 2002).

16.4 Regulation of L-Valine Biosynthesis in *C. glutamicum*

The biosynthesis and export of L-valine in *C. glutamicum* is strictly regulated by transcriptional repression or feedback inhibition. Understanding these regulatory mechanisms is important for the rational metabolic engineering of L-valine production.

16.4.1 Transcriptional repression

The transcription of the genes *ilvB* and *ilvN* that encode the first key enzyme in the L-valine biosynthetic pathway, AHAS, is regulated in *C. glutamicum*. On its chromosome, the genes *ilvB*, *ilvN* and *ilvC* are located in the operon *ilvBNC*. According to Northern blot experiments, the *ilvBNC* operon can be transcribed into three lengths of mRNAs, suggesting there are three promoters in the operon (Keilhauer *et al.*, 1993). After deleting the upstream sequences of *ilvB*, two different lengths of mRNAs were obtained, but after deleting the upstream sequences of *ilvN*, only one type of mRNA was obtained (Keilhauer *et al.*, 1993). This indicates that the three promoters in the *ilvBNC* operon are upstream of the genes *ilvB*, *ilvN* and *ilvC*, respectively, so that the *ilvC* gene can be transcribed three times, and so has the most efficient expression among the three genes. In addition, a DNA sequence that is 292 bp upstream of the *ilvBNC* operon encodes a leader peptide composed of 25 amino acids, including two L-isoleucines, three L-valines and two L-leucines, and its downstream sequence can form an mRNA that contains both a stem–loop structure and a transcription terminator structure (Morbach *et al.*, 2000). This leader peptide can regulate the transcription of the operon *ilvBNC*.

16.4.2 Feedback inhibition

The main object of the feedback inhibition by L-valine in *C. glutamicum* is the key enzyme AHAS (Fig. 16.1). The large and small subunits of AHAS are the catalytic and regulatory domains, respectively. AHAS activity can be inhibited by the binding of any of the three branched-chain amino acids (L-valine, L-isoleucine and L-leucine), and the binding site for these is usually in the regulatory domains (Elišáková *et al.*, 2005). The half-inhibitory concentration for L-valine, L-isoleucine and L-leucine are 0.9, 3.1 and 6.0 mM, respectively (Leyval *et al.*, 2003). At a concentration of 5 mM, neither a single branched-chain amino acid nor different combinations of any of the three can inhibit more than 57% of AHAS activity (Elišáková *et al.*, 2005), indicating that the binding sites of AHAS are the same for all three. The mechanism of inhibition is not clear, but several mutated AHASs that can release the inhibition have been reported (Elišáková *et al.*, 2005; Yin *et al.*, 2012), and the most efficient point mutations are the 20th, 21st and 22nd amino acid residues in the regulatory domains of AHAS (Elišáková *et al.*, 2005).

AHAS variants from five branched-chain producing strains of *C. glutamicum* have been characterized and amino acid substitutions found in both the catalytic and regulatory subunits in which reduced sensitivity to feedback inhibition by branched-chain amino acid products was observed (Guo *et al.*, 2015). In one variant, the accumulation of L-valine at a concentration of 9.5 g l^{-1} at an hourly rate of 0.13 g l^{-1} was recorded. Some strains also showed marked resistance to feedback inhibition such that only 13% of the original activity was lost in the presence of 10 mM branched-chain amino acid.

16.4.3 Regulation of transport

The accumulation of L-valine can repress the transcription of key genes or inhibit the activity of key enzymes in the biosynthetic pathway of L-valine in *C. glutamicum*. As a result, the extracellular transport of L-valine by BrnFE is as important as its intracellular biosynthesis (Kennerknecht *et al.*, 2002). The activity and expression of BrnFE are regulated by the global regulator Lrp, which is encoded by *lrp*. Lange *et al.* (2012) found that adding any of the three branched chain amino acids or L-methionine to the medium can stimulate the expression *brnFE*, but that this stimulation requires the presence of Lrp. According to microarray and DNAase footprint analysis, Lrp binds to the DNA between *lrp* and *brnF* (Lange *et al.*, 2012) and activates the expression of the *brnFE* operon. Because of its specific function, Lrp has been used to develop the biosensor for monitoring L-valine production in single cells of *C. glutamicum* strains (Mustafi *et al.*, 2012, 2014).

16.5 Metabolic Engineering of L-Valine Production in *C. glutamicum*

Most industrial L-valine producing *C. glutamicum* strains were obtained by traditional mutation breeding, but metabolic engineering is now being used to modify these strains (Bailey, 1991; Nielsen, 1998). In recent years, many tools for genetic manipulation in *C. glutamicum* have been developed and used for improving different amino acid production strains. These tools include various expression vectors (Radmacher *et al.*, 2002; Xu *et al.*, 2010a,b), which are typically shuttle vectors between *Escherichia coli* and *C. glutamicum*; the promoters used in these vectors are carefully chosen because the activity of a promoter may be significantly different in *E. coli* from that in *C. glutamicum* (Holátko *et al.*, 2009; Xu *et al.*, 2011). Various gene deletion systems in *C. glutamicum* have also been established (Okibe *et al.*, 2011; Tan *et al.*, 2012). Using these tools for the genetic manipulation of *C. glutamicum*, more L-valine producing strains have been developed.

16.5.1 Accumulating the key precursors in the biosynthetic pathway of L-valine

There are several key precursors in the biosynthetic pathway of L-valine, among which pyruvate and 2-keto-isovalerate are the most important for L-valine production (Fig. 16.1).

Pyruvate is an important intracellular metabolite in *C. glutamicum*; it can be consumed by several pathways to generate acety L-CoA, oxaloacetate, L-alanine and L-isoleucine, as well as L-valine (Fig. 16.1). Using pyruvate as substrate, the pyruvate dehydrogenase complex (PDHC), which is encoded by several genes, including *ace* (Eikmanns and Blombach, 2014) can produce acetyl-CoA, the pyruvate:quinine oxidoreductase encoded by *pqo* can produce acetate, the alanine carboxylase (PC) encoded by *pyc* can produce oxaloacetate, and alanine aminotransferase (AA) encoded by the genes *alaT* and *avtA* can generate L-alanine (Marienhagen and Eggeling, 2008). Deleting these genes could lead to the accumulation of pyruvate, thereby leading to increased L-valine production. For example, an *aceE* deletion mutant, *C. glutamicum* ATCC14067 ΔaceE, accumulated more L-valine than the wild type ATCC14067 (Hu *et al.*, 2013, 2014).

Pyruvate can also be catalysed by AHAS to synthesize L-isoleucine, as can another substrate, 2-ketobutyrate. However, AHAS has a much higher affinity for 2-ketobutyrate than for pyruvate, so when 2-ketobutyrate is present, more pyruvate will be used to synthesize L-isoleucine than to synthesize L-valine. Because L-isoleucine is difficult to separate from L-valine owing to their similar chemical structures, it is the key contaminant for L-valine production in *C. glutamicum*. To avoid the biosynthesis of L-isoleucine, the production of 2-ketobutyrate has to be controlled. As 2-ketobutyrate is produced from L-threonine by the action of L-threonine dehydratase, which is encoded by *ilvA* (Fig. 16.1), the deletion of *ilvA* could favourably increase the availability of pyruvate, so improving the production of L-valine and avoiding the accumulation of L-isoleucine (Sahm and Eggeling, 1999). The deletion of other genes in the L-isoleucine pathway, such as *ilvB*, *ilvN*, *ilvC*, *ilvD* and *ilvE*, is not feasible because these genes are also used by the L-valine biosynthetic pathway, and their deletion would block L-valine production. The *aceE* and *ilvA* deletion mutant *C. glutamicum* ATCC14067 ΔaceE ΔilvA accumulated more L-valine than ATCC14067 ΔaceE (Hu *et al.*, 2013), indicating that more L-valine can be produced when more pyruvate is accumulated in the cell.

At the end of biosynthetic pathway of L-valine, the substrate 2-ketoisovalerate is shared by the biosynthetic pathways of L-leucine and D-pantothenate. Isopropylmalate synthase, encoded by *leuA* is the first key enzyme in the L-leucine biosynthetic pathway (Fig. 16.1). Therefore, knocking out *leuA* can prevent the accumulation of L-leucine. Three enzymes encoded by *panB*, *panC* and *panE* are needed in the pathway from 2-ketoisovalerate to D-pantothenate. As the genes *panB* and *panC* are located adjacently on the chromosome, they can be knocked out together to prevent the biosynthesis of D-pantothenate – although as D-pantothenate is a constituent of coenzyme A, its reduced availability could also lead to a reduced flux from pyruvate to acetyl-CoA.

16.5.2 Strengthening the biosynthetic pathway of L-valine by overexpressing the key genes

After deletion of the key genes in the competing pathways to enable the accumulation of the key precursors of L-valine, the next step towards further improving the yield of L-valine in *C. glutamicum* is to increase the expression of key genes in its biosynthetic pathway. Starting from pyruvate, the biosynthesis of L-valine requires the four enzymes encoded by the five genes *ilvB*, *ilvN*, *ilvC*, *ilvD* and *ilvE*. On the chromosome, *ilvB*, *ilvN* and *ilvC* are next to each other and *ilvD* is only separated from *ilvB* by two genes, while *ilvE* is far away from the other genes. The next paragraph describes a number of examples of the metabolic engineering of these five genes in *C. glutamicum* (strain) ATCC13032 to increase the production of L-valine. Figure 16.2 gives details of these engineered strains and of their production of L-valine.

When overexpressed in *C. glutamicum* ATCC13032, the gene cluster *ilvBNCD* was able to accumulate 42 mM L-valine (Radmacher *et al.*, 2002). The same gene cluster overexpressed in an *ilvA* deletion mutant, ATCC13032 ΔilvA, increased the concentration of L-valine to 58 mM (Radmacher *et al.*, 2002). The gene clusters *ilvBNC* and *ilvBNCE* overexpressed in ATCC13032 ΔilvA could accumulate 48 and 54 mM L-valine, respectively. A further dramatic increase in L-valine production was obtained when *panBC* was deleted in ATCC13032 ΔilvA; the resulting *C. glutamicum* ATCC13032 ΔilvA ΔpanBC pJC1ilvBNCD could accumulate 92 mM (11 g l^{-1}) L-valine after being grown under limiting conditions for 48 h in flask fermentation (Radmacher *et al.*, 2002). Further deletion of *alaT* in *C. glutamicum* ATCC13032 ΔilvA ΔpanBC pJC1ilvBNCD (Radmacher *et al.*, 2002) did not obviously increase the yield of L-valine, but it significantly reduced the accumulation of L-alanine (Wada *et al.*, 2008). Although the yield of L-valine was not obviously increased by the deletion of *aceE* in *C. glutamicum* ATCC13032 ΔaceE, ATCC13032 ΔaceE pJC4ilvBNCE could produce 210 mM (25 g l^{-1}) of L-valine in feed-batch fermentation (Blombach *et al.*, 2007), suggesting that overexpression of the gene cluster ilvBNCE is a great strategy for improving L-valine production in *C. glutamicum*. Additional deletion of the *pqo* gene in *C. glutamicum* ATCC13032 ΔaceE pJC4ilvBNCE could further

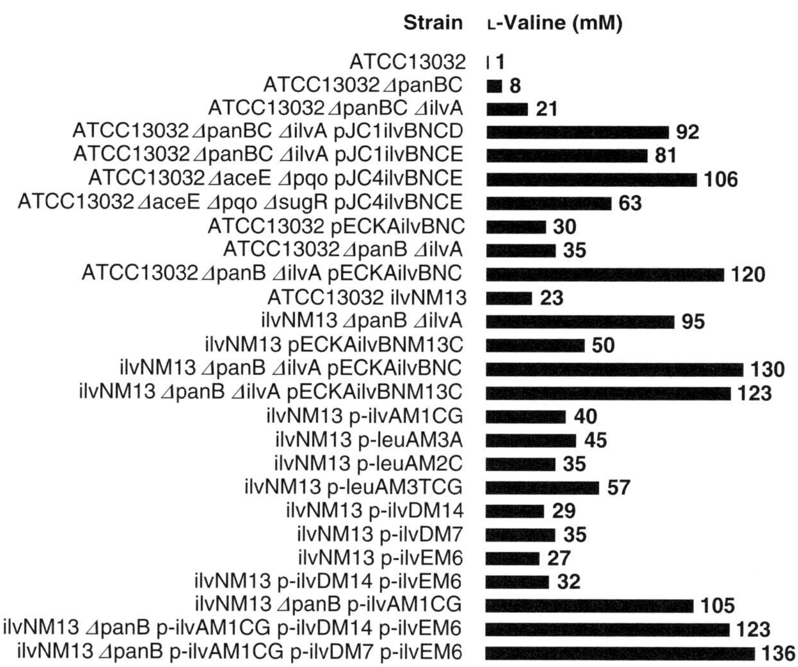

Fig. 16.2. Metabolic engineering of *Corynebacterium glutamicum* strain ATCC13032 to increase L-valine production (Radmacher *et al.*, 2002; Elišáková *et al.*, 2005; Blombach *et al.*, 2007, 2009; Wada *et al.*, 2008).

improve L-valine production. The growth of both ATCC13032 ΔaceE pJC4ilvBNCE and ATCC13032 ΔaceE Δpqo pJC4ilvBNCE requires acetic acid, which can affect glucose consumption, and thereby the biosynthesis of L-valine (Wendisch *et al.*, 2000). However, using ethanol instead of acetic acid can increase glucose consumption in these strains; ATCC13032 ΔaceE Δpqo pJC4ilvBNCE could accumulate 301 mM (35 g l^{-1}) of L-valine after 66 h of feed-batch fermentation (Blombach *et al.*, 2009). Additional deletion of *pgi*, which encodes phosphoglucose isomerase, and *pyc*, which encodes pyruvate carboxylase, in ATCC13032 ΔaceE Δpqo Δpgi pJC4ilvBNCE led to a further increase of L-valine production. ATCC13032 ΔaceE Δpqo Δpgi pJC4ilvBNCE could produce 412 mM (48 g l^{-1}) of L-valine in feed-batch fermentation (Blombach *et al.*, 2008).

16.5.3 Optimizing L-valine accumulation by chromosomal mutagenesis

Because L-valine is synthesized via a four-step biosynthetic pathway from pyruvate without the requirement of ATP, reducing the activity of the H^+-ATPase encoded by *atpG* can improve L-valine production through the accumulation of pyruvate (via increased glucose utilization) in *C. glutamicum* (Aoki *et al.*, 2005; Wada *et al.*, 2008; Sawada *et al.*, 2012). Mutation of *atpG* (T817C) on the chromosome of ATCC13032 led to the point mutation (Ser273Pro) of H^+-ATPase, generating *C. glutamicum* mutant strain A-1. Introduction of the plasmid pVKilvN53C into the strain A-1 resulted in the production of 96 mM (116 g l^{-1}) L-valine after 72 h fermentation (Wada *et al.*, 2008). The plasmid pVKilvN53C contains two genes, *ilvN53* and *ilvC*, wherein the gene *ilvN53* encodes the small subunit of AHAS without the C-terminal 53 amino acids, resulting in resistance to feedback inhibition of L-valine.

C. glutamicum mutant strain ilvNM13 was constructed from the wild type ATCC13032 by site-directed mutagenesis (Elišáková *et al.*, 2005) of chromosomal *ilvN*, in which amino acid residues 20–22, GlyIleIle, were changed to AspAspPhe, leading to a complete discharge of feedback inhibition by the three branched chain amino acids. After deleting the genes *panB* and *ilvA*, and overexpressing the plasmid pECKAilvBNC, the strain ilvNM13 ΔilvA ΔpanB pECKAilvBNC can produce 130 mM (15 g l^{-1}) of L-valine. In consideration of the instability of the expression plasmid, Holátko *et al.* (2009) deleted the gene *panB*, attenuated the promoter activity of the genes *leuA* and *ilvA*, and enhanced the promoter activity of the genes *ilvD* and *ilvE* on the chromosome of strain *ilvNM13*; the resulting plasmid-free strain, ilvNM13 ΔpanBP-ilvAM-1CG P-ilvDM7 P-ilvEM6, could produce 136 mM (16 g l^{-1}) L-valine within 48 h of fermentation. These data indicate that site-directed mutagenesis on the chromosome of *C. glutamicum* is feasible for enhancing the production of L-valine (Hou *et al.*, 2012).

Exchange of the native *C. glutamicum* promoter of the *aceE* gene with a mutated dihydrodipicolinate synthase gene *dapA* promoter and overexpressing the plasmid pJC4 ilvBNCE could increase L-valine production, and additional deletion of the *pqo* gene and the phosphoenolpyruvate carboxylase gene *ppc* could further improve the yield. For example, *C. glutamicum* aceE A16 Δpqo Δppc/pJC4ilvBNCE produced up to 738 mM (87 g l^{-1}) L-valine with an overall yield of 0.36 mole $mole^{-1}$ glucose in fed-batch fermentations (Buchholz *et al.*, 2013).

16.5.4 Balancing the cofactors to improve L-valine production

In *C. glutamicum*, the synthesis of one mole of L-valine from one mole of glucose generates two moles of NADH but consumes two moles of NADPH. The conversion of NADH to NADPH is difficult in this bacterium, and so it might be a rate-limiting factor in L-valine production. In order to solve this cofactor imbalance, one way is to increase the levels of NADPH; the other way is to convert the coenzyme requirement for L-valine synthesis from NADPH to NADH in *C. glutamicum*.

The pentose phosphate pathway (PP pathway), by converting glucose-6-phosphate to glyceraldehyde-3-phosphate, could provide more NADPH (see Fig. 16.1). Deleting the gene *pgi* could increase the carbon flow through the PP pathway to generate more NADPH (Bartek *et al.*, 2010).

The coenzyme requirement for L-valine synthesis was converted from NADPH to NADH via modification of AHAIR and the introduction of *Lysinibacillus sphaericus* NAD-specific leucine dehydrogenase in place of endogenous TA in *C. glutamicum* strain R (JCM Catalogue no. 18229) lacking the lactate dehydrogenase gene *ldhA*. The intracellular NADH/NAD(+) ratio significantly decreased, and glucose consumption and L-valine production drastically improved. Moreover, L-valine yield increased and succinate formation decreased concomitantly with the decreased intracellular redox state. These observations suggest that the intracellular NADH/NAD(+) ratio is the primary rate-limiting factor for L-valine production under oxygen deprivation conditions. The L-valine productivity and yield were even better and by-products derived from pyruvate further decreased as a result of a feedback resistance-inducing mutation in the acetohydroxy acid synthase encoded by *ilvBN*. The resultant strain produced 1470 mM L-valine after 24 h with a yield of 0.63 mole mole^{-1} glucose (Hasegawa *et al.*, 2012). None the less, succinate was still produced and acted to suppress the L-valine yield. By additionally disrupting *ppc*, succinate production was effectively suppressed, but both glucose consumption and L-valine production decreased significantly due to the severely elevated intracellular NADH/NAD(+) ratio. In contrast, this perturbed intracellular redox state was more than compensated for by the deletion of three genes associated with NADH-producing acetate synthesis and the overexpression of five glycolytic genes, including *gapA*, which encodes NADH-inhibited glyceraldehyde-3-phosphate dehydrogenase. Inserting the feedback-resistant mutant AHAS and NAD-preferring mutant AHAIR into the chromosome resulted in higher L-valine yield and productivity. Deleting the alanine transaminase gene *avtA* suppressed alanine production. The resultant strain produced 1280 mM L-valine at a yield of 88% mole mole^{-1} of glucose after 24 h under oxygen deprivation (Hasegawa *et al.*, 2013).

16.6 Outlook

Rational design is important in the metabolic engineering of *C. glutamicum*. An efficient *C. glutamicum* strain should not only maximize the accumulation of L-valine, but also minimize the accumulation of by-products, which can simplify the subsequent separation and purification processes, and therefore reduce the production costs. Currently, developing L-valine production *C. glutamicum* mainly focuses on the L-valine biosynthetic pathway, as this is a key factor to improving L-valine production, although not the only one. Oxygen transfer rates, for example, have a significant impact on bioprocess performance and need to be assessed for candidate strains considered for L-valine production (Kass, 2014). Furthermore, to construct the optimal strain of *C. glutamicum* for L-valine production, many other factors must also be taken into account. To obtain the global optimization strategy, future research should focus on the overall metabolic engineering of the metabolic networks in *C. glutamicum* cells, and combine the modern biotechnologies of transcriptomics, proteomics, metabolomics and computer modelling (Park *et al.*, 2007; Yu *et al.*, 2013), so that a comprehensive analysis is done of all the nodes that influence the production of L-valine.

Thorough understanding of the metabolic and regulatory mechanisms of the biosynthesis of L-valine is urgently needed for designing system-wide metabolic engineering strategies in *C. glutamicum*. Systems metabolic engineering has shown its superiority and effectiveness over simple engineering of the strain (Tan *et al.*, 2011; Xu *et al.*, 2010a,b), and the fast development of systems biology and synthetic biotechnology enables the genome-wide reprogramming of cells (Shinfuku *et al.*, 2009; Ravasi *et al.*, 2012). Thus, the engineering of *C. glutamicum* into platforms is promising for the production of various amino acids, and not only for L-valine.

Acknowledgements

This work was supported by the National Key Basic Research Program of China (2012CB725202) and the National Natural Science Foundation of China (31370131).

References

Aoki, R., Wada, M., Takesue, N., Tanaka, K. and Yokota, A. (2005) Enhanced glutamic acid production by a H^+-ATPase-defective mutant of *Corynebacterium glutamicum*. *Bioscience, Biotechnology, and Biochemistry* 69, 1466–1472.

Bailey, J.E. (1991) Towards a science of metabolic engineering. *Science* 252, 1668–1674.

Bartek, T., Blombach, B., Zönnchen, E., Makus, P., Lang, S., Eikmanns, B.J. and Oldiges, M. (2010) Importance of NADPH supply for improved L-valine formation in *Corynebacterium glutamicum*. *Biotechnology Progress* 26, 361–371.

Blombach, B., Schreiner, M.E., Holátko, J., Bartek, T., Oldiges, M. and Eikmanns, B.J. (2007) L-Valine production with pyruvate dehydrogenase complex-deficient *Corynebacterium glutamicum*. *Applied and Environmental Microbiology,* 73, 2079–2084.

Blombach, B., Schreiner, M.E., Bartek, T., Oldiges, M. and Eikmanns, B.J. (2008) *Corynebacterium glutamicum* tailored for high-yield L-valine production. *Applied Microbiology and Biotechnology*, 79, 471–479.

Blombach, B., Arndt, A., Auchter, M. and Eikmanns, B.J. (2009) L-valine production during growth of pyruvate dehydrogenase complex deficient *Corynetacterium glutamicum* in the presence of ethanol or by inactivation of the transcriptional regulator SugR. *Applied and Environmental Microbiology* 75, 1197–1200.

Buchholz, J., Schwentner, A., Brunnenkan, B., Gabris, C., Grimm, S., Gerstmeir, R., Takors, R., Eikmanns, B.J. and Blombach, B. (2013) Platform engineering of *Corynebacterium glutamicum* with reduced pyruvate dehydrogenase complex activity for improved production of L-lysine, L-valine, and 2-ketoisovalerate. *Applied and Environmental Microbiology*, 79, 5566–5575.

Chattopadhyay, S.P. and Banerjee, A.K. (1978) Production of valine by a *Bacillus* sp. *Zeitschrift für Allgemeine Mikrobiologie* 18, 243–254.

Chen, C., Li, Y., Hu, J., Dong, X. and Wang, X. (2015) Metabolic engineering of *Corynebacterium glutamicum* ATCC13869 for L-valine production. *Metabolic Engineering* 29, 66–75.

Eikmanns, B.J. and Blombach, B. (2014) The pyruvate dehydrogenase complex of *Corynebacterium glutamicum*: an attractive target for metabolic engineering. *Journal of Biotechnology* 192(Pt B), 339–345.

Elišáková, V., Pátek, M., Holátko, J., Nešvera, J., Leyval, D., Goergen, J.L. and Delaunay, S. (2005) Feedback-resistant acetohydroxy acid synthase increases valine production in *Corynebacterium glutamicum*. *Applied and Environmental Microbiology* 71, 207–213.

Eyal, A.M. and Bressler, E. (1993) Industrial separation of carboxylic and amino acids by liquid membranes: applicability, process considerations, and potential advantages. *Biotechnology and Bioengineering* 41, 287–295.

Gedi, V. and Yoon, M.Y. (2012) Bacterial acetohydroxyacid synthase and its inhibitors – a summary of their structure, biological activity and current status. *The FEBS Journal* 279, 946–963.

Guo, Y.F., Han, M., Xu, J.Z. and Zhang, W.G. (2015) Analysis of acetohydroxyacid synthase variants from branched-chain amino acids-producing strains and their effects on the synthesis of branched-chain amino acids in *Corynebacterium glutamicum*. *Protein Expression and Purification* 109, 106–112.

Hao, N., Yan, M., Zhou, H., Liu, H.M., Cai, P. and Ouyang, P.K. (2010) The effect of AmtR on growth and amino acids production in *Corynebacterium glutamicum*. *Applied Biochemistry and Microbiology* 46, 561–566.

Hasegawa, S., Uematsu, K., Natsuma, Y., Suda, M., Hiraga, K., Jojima, T., Inui, M. and Yukawa, H. (2012) Improvement of the redox balance increases L-valine production by *Corynebacterium glutamicum* under oxygen deprivation. *Applied and Environmental Microbiology* 78, 865–875.

Hasegawa, S., Suda, M., Uematsu, K., Natsuma, Y., Hiraga, K., Jojima, T., Inui, M. and Yukawa, H. (2013) Engineering of *Corynebacterium glutamicum* for high-yield L-valine production under oxygen deprivation conditions. *Applied and Environmental Microbiology* 79, 1250–1257.

Holátko, J., Elišáková, V., Prouza, M., Sobotka, M., Nešvera, J. and Pátek, M. (2009) Metabolic engineering of the L-valine biosynthesis pathway in *Corynebacterium glutamicum* using promoter activity modulation. *Journal of Biotechnology* 139, 203–210.

Hou, X., Chen, X., Zhang, Y., Qian, H. and Zhang, W. (2012) L-Valine production with minimization of by-products' synthesis in *Corynebacterium glutamicum* and *Brevibacterium flavum*. *Amino Acids* 43, 2301–2311.

Hu, J., Tan, Y., Li, Y., Hu, X., Xu, D. and Wang, X. (2013) Construction and application of an efficient multiple-gene-deletion system in *Corynebacterium glutamicum*. *Plasmid* 70, 303–313.

Hu, J.Y., Li, Y.Y., Zhang, H.L., Tan, Y.Z. and Wang, X.Y. (2014) Construction of a novel expression system for use in *Corynebacterium glutamicum*. *Plasmid* 75, 18–26.

Imam, S., Schaeuble, S., Brooks, A.N., Baliga, N.S. and Price, N.D. (2015) Data-driven integration of genome-scale regulatory and metabolic network. *Frontiers in Microbiology* 6, 1–10.

Kalinowski, J., Bathe, B., Bartels, D., Bischoff, N., Bott, M., Burkovski, A., Dusch, N., Eggeling, L., Eikmanns, B.J., Gaigalat, L. *et al.* (2003) The complete *Corynebacterium glutamicum* ATCC13032 genome sequence and its impact on the production of L-aspartate-derived amino acids and vitamins. *Journal of Biotechnology* 104, 5–25.

Karau, A. and Grayson, I. (2014) Amino acids in human and animal nutrition. In: Zorn, H. and Czermak, P. (eds) *Biotechnology of Food and Feed Additives. Advances in Biochemical Engineering/Biotechnology* 143, 189–228.

Kass, F., Prasad, A., Tillack, J., Moch, M., Giese, H., Buchs, V., Wiechert, W. and Oldiges, M. (2014) Rapid assessment of oxygen transfer impact for *Corynebacterium glutamicum*. *Bioprocess and Biosystems Engineering* 37, 2567–2577.

Keilhauer, C., Eggeling, L. and Sahm, H. (1993) Isoleucine synthesis in *Corynebacterium glutamicum*: molecular analysis of the *ilvB-ilvN-ilvC* operon. *Journal of Bacteriology* 175, 5595–5603.

Kennerknecht, N., Sahm, H., Yen, M.R., Pátek, M., Saier, M.-H. Jr and Eggeling, L. (2002) Export of L-isoleucine from *Corynebacterium glutamicum*: a two-gene-encoded member of a new translocator family. *Journal of Bacteriology* 184, 3947–3956.

Khachatryan, A.Z., Durgar'Yan, S.S. and Martirosov, S.M. (1986) Dependence of valine production by *Serratia marcescens* on the ion composition of the fermentation medium. *Prikladnaya Biokhimiya i Mikrobiologiya* 22, 554–556.

Lange, C., Mustafi, N., Frunzke, J., Kennerknecht, N., Wessel, M., Bott, M. and Wendisch, V.F. (2012) Lrp of *Corynebacterium glutamicum* controls expression of the *brnFE* operon encoding the export system for L-methionine and branched-chain amino acids. *Journal of Biotechnology* 158, 231–241.

Leyval, D., Uy, D., Delaunay, S., Goergen, J.L. and Engasser, J.M. (2003) Characterisation of the enzyme activities involved in the valine biosynthetic pathway in a valine-producing strain of *Corynebacterium glutamicum*. *Journal of Biotechnology* 104, 241–252.

Liang, C., Huo, Y., Qi, G., Wei, X., Wang, Q. and Chen, S. (2015) Enhancement of L-valine production in *Bacillus licheniformis* by blocking three branched pathways. *Biotechnology Letters* 37, 1243–1248.

Marienhagen, J. and Eggeling, L. (2008) Metabolic function of *Corynebacterium glutamicum* aminotransferases AlaT and AvtA and impact on L-valine production. *Applied and Environmental Microbiology* 74, 7457–7462.

Marienhagen, J., Kennerknecht, N., Sahm, H. and Eggeling, L. (2005) Functional analysis of all aminotransferase proteins inferred from the genome sequence of *Corynebacterium glutamicum*. *Journal of Bacteriology* 187, 7639–7646.

Marin, M. and Kramer, R. (2007) Amino acid transport systems in biotechnologically relevant bacteria. In: Wendisch, V.F. (ed.) *Amino Acid Biosynthesis – Pathways, Regulation and Metabolic Engineering*. Springer, Heidelberg, Germany, pp. 289–325.

McHardy, A.C., Tauch, A., Ruckert, C., Pühler, A. and Kalinowski, J. (2003) Genome-based analysis of biosynthetic aminotransferase genes of *Corynebacterium glutamicum*. *Journal of Biotechnology* 104, 229–240.

Morbach, S., Junger, C., Sahm, H. and Eggeling, L. (2000) Attenuation control of *ilvBNC* in *Corynebacterium glutamicum*: evidence of leader peptide formation without the presence of a ribosome binding site. *Journal of Bioscience and Bioengineering* 90, 501–507.

Mustafi, N., Grünberger, A., Kohlheyer, D., Bott, M. and Frunzke, J. (2012) The development and application of a single-cell biosensor for the detection of L-methionine and branched-chain amino acids. *Metabolic Engineering* 14, 449–457.

Mustafi, N., Grünberger, A., Mahr, R., Helfrich, S., Nöh, K., Blombach, B., Kohlheyer, D. and Frunzke, J. (2014) Application of a genetically encoded biosensor for live cell imaging of L-valine production in pyruvate dehydrogenase complex-deficient *Corynebacterium glutamicum* strains. *PLoS ONE* 9(1): e85731.

Nielsen, J. (1998) Metabolic engineering: techniques for analysis of targets for genetic manipulations. *Biotechnology and Bioengineering* 58, 125–132.

Okibe, N., Suzuki, N., Inui, M. and Yukawa, H. (2011) Efficient markerless gene replacement in *Corynebacterium glutamicum* using a new temperature-sensitive plasmid. *Journal of Microbiological Methods* 85, 155–163.

Oldiges, M., Eikmanns, B.J. and Blombach, B. (2014) Application of metabolic engineering for the biotechnological production of L-valine. *Applied Microbiology and Biotechnology* 98, 5859–5870.

Park, J.H., Lee, K.H., Kim, T.Y. and Lee, S.Y. (2007) Metabolic engineering of *Escherichia coli* for the production of L-valine based on transcriptome analysis and *in silico* gene knockout simulation. *Proceedings of the National Academy of Sciences of the United States of America* 104, 7797–7802.

Pasupuleti, V., Holmes, C. and Demain, A. (2010) Applications of protein hydrolysates in biotechnology. In: Pasupuleti, V.K. and Demain, A.L. (eds) *Protein Hydrolysates in Biotechnology*. Springer, Dordrecht, The Netherlands. pp. 1–9.

Radmacher, E., Vaitsikova, A., Burger, U., Krumbach, K., Sahm, H. and Eggeling, L. (2002) Linking central metabolism with increased pathway flux: L-valine accumulation by *Corynebacterium glutamicum*. *Applied and Environmental Microbiology* 68, 2246–2250.

Ravasi, P., Peiru, S., Gramajo, H. and Menzella, H.G. (2012) Design and testing of a synthetic biology framework for genetic engineering of *Corynebacterium glutamicum*. *Microbial Cell Factories* 11:147.

Sahm, H. and Eggeling, L. (1999) D-Pantothenate synthesis in *Corynebacterium glutamicum* and use of *panBC* and genes of L-valine synthesis for its overproduction. *Applied and Environmental Microbiology* 65, 1973–1979.

Sawada, K., Kato, Y., Imai, K., Li, L., Wada, M., Matsushita, K. and Yokota, A. (2012) Mechanism of increased respiration in an H^+-ATPase-defective mutant of *Corynebacterium glutamicum*. *Journal of Bioscience and Bioengineering* 113, 467–473.

Shinfuku, Y., Sorpitiporn, N., Sono, M., Furusawa, C., Hirasawa, T. and Shimizu, H. (2009) Development and experimental verification of a genome-scale metabolic model for *Corynebacterium glutamicum*. *Microbial Cell Factories* 8:43.

Simeonidis, E. and Price, N.D. (2015) Genome-scale modeling for metabolic engineering. *Journal of Industrial Microbiology and Biotechnology* 42, 327–338.

Stephanopoulos, G. (1999) Metabolic fluxes and metabolic engineering. *Metabolic Engineering* 1, 1–11.

Tan, Y., Xu, D., Li, Y. and Wang, X. (2012) Construction of a novel *sacB*-based system for marker-free gene deletion in *Corynebacterium glutamicum*. *Plasmid* 67, 44–52.

Unthan, S., Baumgart, M., Radek, A., Herbst, M., Siebert, D., Brühl, N., Bartsch, A., Bott, M., Wiechert, W., Marin, K. *et al.* (2015) Chassis organism from *Corynebacterium glutamicum* – a top-down approach to identify and delete irrelevant gene clusters. *Biotechnology Journal* 10, 290–301.

Wada, M., Hijikata, N., Aoki, R., Takesue, N. and Yokota, A.(2008) Enhanced valine production in *Corynebacterium glutamicum* with defective H^+-ATPase and *C*-terminal truncated acetohydroxyacid synthase. *Bioscience, Biotechnology, and Biochemistry* 72, 2959–2965.

Wendisch, V.F. (2014) Microbial production of amino acids and derived chemicals: synthetic biology approaches to strain development. *Current Opinion in Biotechnology* 30, 51–58.

Wendisch, V.F., de Graaf, A.A., Sahm, H. and Eikmanns, B.J. (2000) Quantitative determination of metabolic fluxes during co[-]utilization of two carbon sources: comparative analyses with *Corynebacterium glutamicum* during growth on acetate and/or glucose. *Journal of Bacteriology* 182, 3088–3096.

Xu, D., Tan, Y., Huan, X., Hu, X. and Wang, X. (2010a) Construction of a novel shuttle vector for use in *Brevibacterium flavum*, an industrial amino acid producer. *Journal of Microbiological Methods* 80, 86–92.

Xu, D., Tan, Y., Shi, F. and Wang, X. (2010b) An improved shuttle vector constructed for metabolic engineering research in *Corynebacterium glutamicum*. *Plasmid* 64, 85–91.

Xu, D., Tan, Y., Li, Y. and Wang, X. (2011) Construction of a novel promoter-probe vector and its application for screening strong promoter for *Brevibacterium flavum* metabolic engineering. *World Journal of Microbiological Biotechnology* 27, 961–968.

Xu, J.Z., Han, M., Zhang, V.Y., Guo, F., Qian, H. and Zhang, W.G. (2014) Improvement of L-lysine production combines with minimization of by-products synthesis in *Corynebacterium glutamicum*. *Journal of Chemical Technology and Biotechnology* 89, 1924–1933.

Yin, L., Hu, X., Xu, D., Ning, J., Chen, J. and Wang, X. (2012) Co-expression of feedback-resistant threonine dehydratase and acetohydroxy acid synthase increase L-isoleucine production in *Corynebacterium glutamicum*. *Metabolic Engineering* 14, 542–550.

Yin, L., Shi, F., Hu, X., Chen, C. and Wang, X. (2013) Increasing L-isoleucine production in *Corynebacterium glutamicum* by overexpressing global regulator Lrp and two-component export system BrnFE. *Journal of Applied Microbiology* 114, 1369–1377.

Yukawa, H., Omumasaba, C.A., Nonaka, H., Kós, P., Okai, N., Suzuki, N., Suda, M., Tsuge, Y., Watanabe, J., Ikeda, Y. *et al.* (2007) Comparative analysis of the *Corynebacterium glutamicum* group and complete genome sequence of strain R. *Microbiology* 153, 1042–1058.

Yu, X., Li, Y. and Wang, X. (2013) Molecular evolution of threonine dehydratase in bacteria. *PLoS ONE* 8(12): e80750.

17 Flavour Formation From Leucine by Lactic Acid Bacteria (LAB)

M.I. Afzal,[1]* S. Delaunay[2] and C. Cailliez-Grimal[3]

[1]*Department of Biosciences, COMSATS Institute of Information Technology, Islamabad, Pakistan;* [2]*Laboratoire Réactions et Génie des Procédés (LRGP), UMR 7274 du CNRS and Université de Lorraine, Vandoeuvre-lès-Nancy, France;* [3]*Laboratoire d'Ingénierie des Biomolécules, Université de Lorraine, Vandoeuvre-lès-Nancy, France*

17.1 Abstract

Leucine belongs to the branched-chain amino acids, which in lactic acid bacteria (LAB) can be intracellularly converted into different aromatic compounds through various enzymes and metabolic routes. The aldehyde 3-methylbutanal, the alcohol 3-methylbutanol and the carboxylic acid 3-methylbutyric acid that originate from leucine are associated as key aromatic compounds in some cheese types, but also linked to spoilage in meat and fish products. This chapter summarizes the overall flavour perception and desirability of the aromatic compounds resulting from leucine catabolism and their impact on different food products, along with the enzymes and metabolic pathways leading to their formation in lactic acid bacteria. Proposed strategies will also be highlighted for the control of desired concentrations of these aromatic compounds in different food products.

17.2 Introduction

LAB are equipped with various enzymes that efficiently catabolize amino acids to generate different aromatic compounds, which contribute to the specific flavour of food products. The aromatic compounds resulting from leucine catabolism include 3-methylbutanal (an aldehyde), 3-methylbutanol (an alcohol) and 3-methylbutyric acid (a carboxylic acid, which is also known as isovaleric acid/isovalerate or 3-methylbutanoic acid). The metabolic pathways involved in leucine catabolism that lead to these compounds are shown in Fig. 17.1. 3-Methylbutanal is a potent flavour compound which is generally perceived as a malty/chocolate-like flavour in hard cheese varieties, but as a feet/off-odour in meat and fish. 3-Methylbutanol is generally perceived as a fruity flavour, while 3-methylbutyric acid is perceived as a cheesy, sweaty and putrid off-odour in many food products. The flavour perception of the branched-chain alcohol and carboxylic acid always seem to be low compared with the aldehyde because their respective concentrations are always less than those of the corresponding aldehyde. 3-Methylbutanal is a potent flavour component that is responsible for milk spoilage and is caused by the malty flavour-producing strains *Streptococcus lactis* var. *maltigenes* and *Carnobacterium maltaromaticum* (formerly *Lactobacillus maltaromaticus*); various scientists have studied the possible consequences of this particular flavour component and also the microflora

*Corresponding author. E-mail address: inam.afzal@comsats.edu.pk

responsible for it in cheese, meat and fish products (Miller *et al.*, 1974; Afzal *et al.*, 2010).

There has been a debate among groups of scientists about the desirability and non-desirability of 3-methylbutanal in various food products. It is generally assumed that a balanced flavour perception depends upon numerous factors, including moisture content, texture and, more importantly, the internal balance of different odorous compounds and any slight imbalance between them that may impair the final flavour perception. Due the complexity of the large pool of internal aromatic compounds, it seems quite a complex and difficult task to obtain a desired flavour from food products.

The bacteria and yeasts responsible for the production of 3-methylbutanal, 3-methylbutanol and 3-methylbutyric acid in chocolate, cheese, meat and fish products include *Lactococcus*, *Lactobacillus*, *Streptococcus*, *Carnobacterium*, *Enterococcus*, *Staphylococcus*, *Pseudomonas*, *Brochothrix*, *Debaryomyces hansenii*, *Yarrowia lipolytica* and *Geotrichum candidum*. Details of the different species of these that are found in various cocoa, dairy, fish/shellfish and meat products are given in Table 17.1.

The biosynthesis of flavour compounds from leucine catabolism can take place in bacteria by one of two metabolic routes or by a combination of them both. The two routes are: the direct pathway, operated via the α-keto-acid decarboxylase (KADC) enzyme (the KADC pathway); and the indirect pathway, operated via the α-keto-acid dehydrogenase (KADH) enzyme

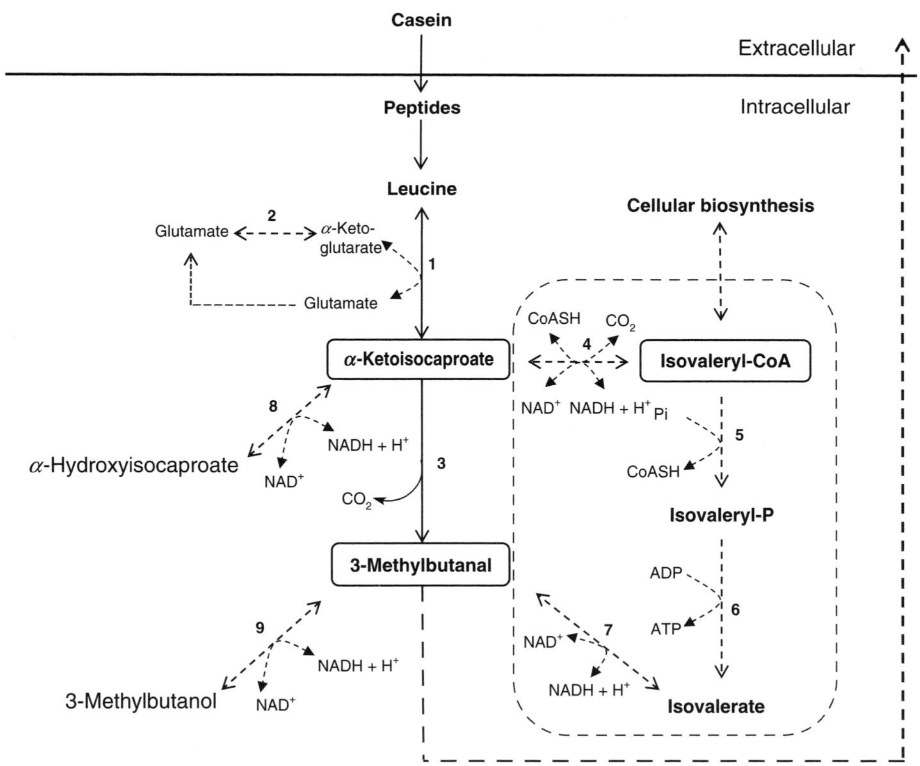

Fig. 17.1. The proposed intracellular metabolic pathways involved in leucine catabolism by lactic acid bacteria (LAB). The solid line indicates the direct pathway and dashed lines the indirect pathway. The enzymes indicated by bold numbers are: **1**, AT, aminotransferase (transaminase), i.e. L-leucine aminotransferase/L-branched chain amino acid aminotransferase; **2**, GDH, glutamate dehydrogenase; **3**, KADC, α-keto-acid decarboxylase; **4**, KADH, α-keto-acid dehydrogenase; **5**, PTA, phosphotransferase; **6**, ACK, acyl kinase; **7**, AldDH, aldehyde dehydrogenase; **8**, HADH, α-hydroxyacid dehydrogenase; and **9**, AlcDH, alcohol dehydrogenase.

along with several intermediate enzymes which include the phosphotransferase, acyl kinase and aldehyde dehydrogenase enzymes (the KADH pathway). A recent study by Afzal *et al.* (2012) has confirmed the presence and functionality of both catabolic routes in one strain of *C. maltaromaticum*.

The functionality of both metabolic routes depends upon the intracellular redox environment or the presence (and balance) of the oxidizing and reducing factors NAD^+/NADH and H^+. Indeed, high formation of 3-methylbutanal by bacteria has been attributed to the presence of oxygen in the medium and to the stimulation of KADC pathway, while the absence of oxygen or presence of a reducing environment was found to favour the formation of isovaleric acid or the stimulation of KADH pathway (Kieronczyk *et al.*, 2006; Afzal *et al.*, 2012, 2013).

As the desirability of the presence of 3-methylbutanal, 3-methylbutanol and 3-methylbutyric acid varies according to particular food type involved, different strategies have been developed to control the desired concentration of these flavour compounds. These strategies include addition of substrate, potential microbial adjuncts, bacteriocin-induced lysis or modification of the extracellular redox environment. This chapter highlights current knowledge on the catabolic pathways involved in leucine catabolism, the presence and implications of the presence of the different flavour compounds that have originated from leucine, flavour perception and desirability, and the development of various strategies to control the desired formation of these compounds in different food products.

17.3 Leucine Catabolic Pathways Among LAB

The first step involved in leucine catabolism among LAB is catalysed by an aminotransferase (transaminase) enzyme (L-leucine aminotransferase or to the more general L-branched chain amino acid aminotransferase) that yields α-ketoisocaproic acid. This is considered to be a central metabolite and the rate-controlling step towards the formation of the aromatic compounds being considered in this chapter, which include 3-methylbutanal, 3-methylbutanol and 3-methylbutyric acid (isovaleric acid/isovalerate or 3-methylbutanoic acid) (see Fig. 17.1). During this reaction, the amino group of leucine is transferred to α-ketoglutarate in a reaction catalysed by glutamate dehydrogenase, resulting in the formation of glutamic acid, a step that is linked to the first step in leucine catabolism yielding α-ketoisocaproic acid. Both aminotransferase and glutamate dehydrogenase activities have been well documented in *Lactococcus*, *Lactobacillus*, *Streptococcus*, *Staphylococcus* and *Carnobacterium* species. A vast diversity of aminotransferase activity has also been found to be present in a large number of dairy- and meat-related LAB. Moreover, aromatic aminotransferases have also found to be involved in leucine catabolism in many studies (e.g. Yvon *et al.*, 1997; Rijnen *et al.*, 2003). In order to achieve an intense aroma in food products, attention has been focused on the glutamate dehydrogenase enzyme, which, as already noted, is linked to aminotransferase activity. Furthermore, bacterial strains possessing high glutamate dehydrogenase activities also show high formation of aromatic compounds (Tanous *et al.*, 2002; Kieronczyk *et al.*, 2006).

The α-ketoisocaproic acid resulting from the transamination of leucine can be converted into 3-methylbutanal, 3-methylbutanol and 3-methybutyric acid via two metabolic routes. The first, a direct pathway, involves the formation of 3-methylbutanal by non-oxidative decarboxylation catalysed by the enzyme α-keto-acid decarboxylase (KADC). The aldehyde 3-methylbutanal can then quickly be reduced to 3-methylbutanol by alcohol dehydrogenase. This pathway is designated the KADC pathway. The indirect pathway comprises a group of several intermediate enzymes, starting with α-keto-acid dehydrogenase (KADH), which catalyses an oxidative decarboxylation. During this reaction, α-ketoisocaproic acid is converted to isovaleryl-CoA (also known as 3-methylbutryl-CoA and 3-methylbutanoyl-CoA). The other enzymes involved in this pathway (which is designated the KADH pathway) include phosphotransferase, acyl kinase and aldehyde dehydrogenase.

At present, only *Carnobacterium* species have been found to possess the functionality of both of these catabolic routes for leucine. The α-keto-acid decarboxylase enzyme of the direct pathway has been well studied, although only strains of *Lactococcus lactis* and *C. maltaromaticum* have been found to possess this particular enzyme activity (Smit *et al.*, 2004; Afzal *et al.*, 2012). In *L. lactis*, it was observed that most of the 3-methylbutanal and 3-methylbutanol was formed via the direct (KADC) route, so this activity has been proposed to be the rate-controlling step in aldehyde and alcohol formation from α-ketoisocaproic acid. The gene encoding the KADC enzyme has been well characterized in *L. lactis* (de la Plaza *et al.*, 2004; Smit *et al.*, 2005).

The KADH pathway has been found in *C. maltaromaticum*, *Propionibacterium freudenreichii*, *Lactobacillus* (*Lb.*) *helveticus* and *Enterococcus faecalis* (Ward *et al.*, 1999; Thierry *et al.*, 2002; Helinck *et al.*, 2004; Afzal *et al.*, 2012). It is interesting to note the 3-methylbutanal formed by the direct (KADC) route can be converted to 3-methylbutyric (isovaleric) acid by aldehyde dehydrogenase. KADH activity, along with the genes encoding the KADH complex (*bkdA*, *bkdB*, *bkdC* and *bkdD*) have been identified in *C. maltaromaticum* (Afzal *et al.*, 2012). Moreover, this activity was found to be dependent on both NAD^+ and $NADP^+$ in this bacterium.

The functionality of both the KADC and the KADH metabolic routes depends upon many factors, including the concentration of dissolved oxygen in the medium, the intracellular redox environment and the presence of oxidizing and reducing factors (NAD^+/NADH, H^+), as has been observed for *C. maltaromaticum*, *Lb. helveticus* and *L. lactis* (Helinck *et al.*, 2004; Kieronczyk *et al.*, 2004; Afzal *et al.*, 2012). During a study on the physiology of leucine catabolism by *C. maltaromaticum* in bioreactor under different dissolved oxygen states, higher concentrations of 3-methylbutanal and 3-methylbutanol were attributed to the presence of oxygen or the stimulation of the KADC pathway by KADC activity (Afzal *et al.*, 2013). Similar results were observed in *L. lactis* and *Proteus vulgaris* (Kieronczyk *et al.*, 2006; Deetae *et al.*, 2011).

17.4 Leucine Catabolic Activities and Their Effects on the Sensory Characteristics of Foods

The 3-methylbutanal (aldehyde), 3-methylbutanol (alcohol) and 3-methylbutyric acid (carboxylic acid) that originate from leucine catabolism are key odorant compounds in many food products, as mentioned above and summarized in Table 17.1. In this section, we review the flavour perception and desirability of these flavour compounds according to the different food types in which they are found. Also discussed is the spoilage potential of the different food products linked to these odorant compounds and the microorganisms that are responsible.

Recently, the sensory quality of chocolate has been determined by the qualitative and quantitative composition of the volatile compounds produced by mixed starter cultures composed of *Lb. fermentum* L18, *Acetobacter pasteurianus* A149, *Pichia kluyveri* and *Kluyveromyces marxianus* (Crafack *et al.*, 2014). 3-Methylbutanal, 3-methylbutanol and 3-methylbutyric acid were detected in the sensory profiles. The flavour perceptions of 3-methylbutanal, 3-methylbutanol and 3-methylbutyric acid were perceived, respectively, as giving a malty, chocolate like and cocoa flavour, a fruity flavour and a sweaty odour, and all were considered desirable. No spoilage defect was detected or related to the presence of these odorant compounds.

On the contrary, the formation of 3-methylbutanal and 3-methylbutanol by *C. maltaromaticum* (formerly *Lb. maltaromaticus*) in milk was perceived as giving a malty off-odour, although milk spoilage was not clearly reported (Miller *et al.*, 1974). Many researchers linked the presence of 3-methylbutanal and 3-methylbutanol as a key odorant compounds in certain cheese varieties, especially hard cheeses. Avsar *et al.* (2004) and Ayad *et al.* (2003) investigated the sensory profiles of Cheddar and Proosdij-type/Gouda cheese, respectively. They associated the nutty, chocolate-like flavour perception in these cheeses to the presence of 3-methylbutanal and declared it to be desirable. The flavour of the corresponding alcohol and carboxylic acid were perceived as fruity and cheesy like, but their desirability was not clearly reported. However, no product defect/spoilage was observed due to the presence of these flavour compounds. Some researchers reported the flavour perception of the aldehyde and alcohol as unclean, roasted or off-flavour in Egyptian Ras and ewes' raw milk cheeses; these were considered undesirable though there was no associated product defect or spoilage (Ayad *et al.*,

Table 17.1. Flavour perception and desirability of aromatic compounds originating from leucine catabolism in different food products.

Food product type	Food product variety	Bacteria/yeasts involved	Flavour compounds: 3-Methylbutanal (aldehyde) Flavour/ perception	Desirability[a]	3-Methylbutanol (alcohol) Flavour/ perception	Desirability[a]	3-Methylbutyric acid (organic acid) Flavour/ perception	Desirability[a]	Spoilage[a]	Reference(s)
Cocoa	Chocolate	*Lactobacillus* (*Lb.*) *fermentum* L18, *Acetobacter pasteurianus* A149, *Pichia kluyveri*, *Kluyveromyces marxianus*	Malty, chocolate, cocoa	D	Fruity	D	Sweat	D	NS	Crafack *et al.*, 2014
Dairy	Skimmed-milk	*Carnobacterium maltaromaticum*	Malty, off-odour	NR	Off-odour	NR	NE	NR	NR	Miller *et al.*, 1974
	Cheddar cheese	*Lactococcus* (*L.*) *lactis* subsp. *lactis*, *L. lactis* subsp. *cremoris*	Chocolate/ nutty	D	Fruity	NR	Cheesy	NR	NS	Banks *et al.*, 2001; Avsar *et al.*, 2004; Ayad *et al.*, 2004
	Egyptian Ras cheese	Without the addition of starter cultures	Unclean	ND	Unclean	ND	NE	NR	NS	
	Proosdij type/ Gouda cheese	Mixed strain mesophilic starter culture Bos, mixed strain thermophilic starter culture APS, *L. lactis* subsp. *lactis* B851	Chocolate-like	D	NE	NR	NE	NR	NS	Ayad *et al.*, 2003
	Ewes' raw milk cheese	*L. lactis* (CSC and wild strains)	Roasted hazelnuts	ND	Off-flavour	ND	NE	NR	NS	Morales *et al.*, 2003
	Teleme cheese	*L. lactis* subsp. *lactis*, *L. lactis* subsp. *cremoris*, *Lb. delbrueckii* subsp. *bulgaricus*, *Streptococcus thermophiles*	NE	NR	Fruity	NR	NE	NR	NS	Massouras *et al.*, 2006

	Camembert-type cheese	*Proteus vulgaris* 1M10, *Debaryomyces hansenii* 304	Fruity	NR	Fruity	NR	NE	NR	NS	Deetae *et al.*, 2009
	Swiss cheese	*Propionibacterium freudenreichii* subsp. *shermanii* TL 34 and ITGP23	Malty	NR	Fruity	NR	Rancid, putrid	NR	NS	Thierry *et al.*, 2002
Fish and shellfish	Salmon	*C. divergens*, *C. maltaromaticum*	Cheese, feet, off-odour	ND	Cheese, feet, off-odour	ND	Cheese, feet, off-odour	ND	Weakly S	Brillet *et al.*, 2005
	Shrimp	*Brochothrix thermosphacta*, *Serratia liquefaciens* like, *C. maltaromaticum*	Cheese, feet, sour, fermented, butter, buttermilk like, sour, nauseous off-odour	ND	Cheese, feet, sour, fermented, butter, buttermilk like, sour, nauseous off-odour	ND	Off-odour	ND	S	Jaffrès *et al.*, 2009
Meat	Beef	*Hafnia alvei*, *Enterobacter agglomerans*, *S. liquefaciens*, *Alteromonas putrefaciens*, *Aeomonas hydrophila*, *Pseudomonas fragi*, *B. thermosphacta*	Off-odour	ND	Fruity, off-odour	ND	Strong cheese, off-odour	ND	S	Dainty *et al.*, 1989; Casaburi *et al.*, 2015
	Ham	*C. piscicola* [now *C. maltaromaticum*], *Leuconostoc* (*Ln.*) *camosum*, *Ln. mesenteroides*	Off-odour	ND	Off-odour	ND	Off-odour	ND	S	Budde *et al.*, 2003

Continued

Table 17.1. Continued.

Food product type	Food product variety	Bacteria/yeasts involved	Flavour compounds						Spoilage[a]	Reference(s)
			3-Methylbutanal (aldehyde)		3-Methylbutanol (alcohol)		3-Methylbutyric acid (organic acid)			
			Flavour/ perception	Desirability[a]	Flavour/ perception	Desirability[a]	Flavour/ perception	Desirability[a]		
	Sausage	*Staphylococcus carnosus* 833, *C. piscicola* [now *C. maltaromaticum*] strain 545, *Lb. plantarum* 4045, *S. carnosus*, *S. xylosus*	Sausage/ fatty	NR	Whisky like, sausage/ fatty	NR	Sausage/ fatty	NR	NS	Masson *et al.*, 1996; Larrouture-Thiveyrat *et al.*, 2003; Herranz *et al.*, 2006
	Chicken	*Pseudomonas* spp., *S. liquefaciens*, *Shewanella*, *B. thermosphacta*	Off-odour, putrid	ND	Off-odour, putrid	ND	Off-odour, putrid	ND	S	Nychas *et al.*, 1998; Casaburi *et al.*, 2015

[a]Abbreviations: CSC, commercial starter cultures; D, desirable; ND, non-desirable; NE, not examined; NR, not reported; NS, not spoiled; S, spoiled.

2003; Morales *et al.*, 2003). The branched chain aldehyde, the alcohol and the carboxylic acid were all considered to be potent odorous compounds in soft cheese varieties like Camembert- and Swiss-type cheeses (Thierry *et al.*, 2002; Deetae *et al.*, 2009). The microorganisms that are responsible for the production of these flavour compounds are listed in Table 17.1.

In meat products, including beef, ham and chicken fillets, the flavour perception of the aldehyde, alcohol and carboxylic acid from leucine catabolism was perceived as off-odour, putrid, fruity and strongly cheese like, and considered to be undesirable and related to product spoilage (Dainty *et al.*, 1989; Nychas *et al.*, 1998; Budde *et al.*, 2003; Casaburi *et al.*, 2015). The desirability of 3-methylbutanal, 3-methylbutanol and 3-methylbutyric acid was not reported for sausage production, but the sausages were not considered to be spoiled; the flavour perception of these odorous compounds was designated as sausage/fatty or whisky like (Masson *et al.*, 1996; Larrouture-Thiveyrat *et al.*, 2003; Herranz *et al.*, 2006).

The flavour perception of odorant compounds in fish and shellfish, including salmon and shrimps, has also been investigated (Brillet *et al.*, 2005; Jaffrès *et al.*, 2009). Like meat products, product defect/spoilage in salmon and shrimps was related to the presence of 3-methylbutanal, 3-methylbutanol and 3-methylbutyric acid, and perceived as cheesy, feet like, buttermilk like, sour and nauseous off-odour; this was considered to be highly undesirable.

According to the particular food type, numerous strategies have been developed by many researchers to control the desired concentrations of these odorant compounds in the final food product. These strategies include the following: the addition of amino acids to the milk medium for hard cheesemaking (Wallace and Fox, 1997); the use of culture adjuncts (Whetstine *et al.*, 2006); bacteriocin-induced lysis in the cheesemaking medium (de la Plaza *et al.*, 2004; Martínez-Cuesta *et al.*, 2006); the use of strains possessing enzymes complementing the metabolic pathways for hard cheesemaking (Ayad *et al.*, 2001; Amárita *et al.*, 2006); environmental modification by the use of shaking and static conditions (Deetae *et al.*, 2011); and the addition of oxidizing and reducing agents in models of cheesemaking (Kieronczyk *et al.*, 2006).

17.5 Conclusions

The aldehyde, alcohol and carboxylic acid resulting from leucine catabolism are associated as key flavour compounds in many hard cheese varieties and chocolate, where they are considered to be highly desirable. None the less, in the case of meat and fish, these compounds are related to product defect/spoilage and are considered to be highly undesirable. The flavour compounds originate from leucine either via a direct metabolic route in which the key enzyme is an α-keto-acid decarboxylase enzyme or via indirect route in which the key enzyme is an α-keto-acid dehydrogenase enzyme complex. Numerous strategies have been proposed to control the formation of these aromatic compounds to a desired amount and composition in specific food products. Further knowledge on the flavour-forming pathways and the use of potential microbial cultures to achieve the desired flavours will help us to better understand the role of these flavour compounds in particular food types before proposing their application at the industrial scale.

References

Afzal, M.I., Jacquet, T., Delaunay, S., Borges, F., Milliere, J.B., Revol-Junelles, A.M. and Cailliez-Grimal, C. (2010) *Carnobacterium maltaromaticum*: identification, isolation tools, ecology and technological aspects in dairy products. *Food Microbiology* 27, 573–579.

Afzal, M.I., Delaunay, S., Paris, C., Borges, F., Revol-Junelles, A.M. and Cailliez-Grimal, C. (2012) Identification of metabolic pathways involved in the biosynthesis of flavor compound 3-methylbutanal from leucine catabolism by *Carnobacterium maltaromaticum* LMA 28. *International Journal of Food Microbiology* 157, 332–339.

Afzal, M.I., Boulahya, K.A., Paris, C., Delaunay, S. and Cailliez-Grimal, C. (2013) Effect of oxygen on the biosynthesis of flavor compound 3-methylbutanal from leucine catabolism during batch culture in *Carnobacterium maltaromaticum* LMA 28. *Journal of Dairy Science* 96, 352–359.

Amárita, F., de la Plaza, M., Fernández de Palencia, P., Requena, T. and Peláez, C. (2006) Cooperation between wild lactococcal strains for cheese aroma formation. *Food Chemistry* 94, 240–246.

Avsar, Y.K., Karagul-Yuceer, Y., Drake, M.A., Singh, T.K., Yoon, Y. and Cadwallader, K.R. (2004) Characterization of nutty flavor in Cheddar cheese. *Journal of Dairy Science* 87, 1999–2010.

Ayad, E.H.E., Verheul, A., Engels, W.J., Wouters, J.T. and Smit, G. (2001) Enhanced flavour formation by combination of selected lactococci from industrial and artisanal origin with focus on completion of a metabolic pathway. *Journal of Applied Microbiology* 90, 59–67.

Ayad, E.H.E., Verheul, A., Bruinenberg, P., Wouters, J.T.M. and Smit, G. (2003) Starter culture development for improving the flavour of Proosdij-type cheese. *International Dairy Journal* 13, 159–168.

Ayad, E.H.E., Awad, S., El Attar, A., de Jong, C. and El-Soda, M. (2004) Characterisation of Egyptian Ras cheese. 2. Flavour formation. *Food Chemistry* 86, 553–561.

Banks, J.M., Yvon, M., Gripon, J.C., de la Fuente, M.A., Brechany, E.Y., Williams, A.G. and Muir, D.D. (2001) Enhancement of amino acid catabolism in Cheddar cheese using α-ketoglutarate: amino acid degradation in relation to volatile compounds and aroma character. *International Dairy Journal* 11, 235–243.

Brillet, A., Pilet, M.F., Prevost, H., Cardinal, M. and Leroi, F. (2005) Effect of inoculation of *Carnobacterium divergens* V41, a bio-preservative strain against *Listeria monocytogenes* risk, on the microbiological, chemical and sensory quality of cold-smoked salmon. *International Journal of Food Microbiology* 104, 309–324.

Budde, B.B., Hornbaek, T., Jacobsen, T., Barkholt, V. and Koch, A.G. (2003) *Leuconostoc carnosum* 4010 has the potential for use as a protective culture for vacuum-packed meats: culture isolation, bacteriocin identification, and meat application experiments. *International Journal of Food Microbiology* 83, 171–184.

Casaburi, A., Piombino, P., Nychas, G.J., Villani, F. and Ercolini, D. (2015) Bacterial populations and the volatilome associated to meat spoilage. *Food Microbiology* 45, 83–102.

Crafack, M., Keul, H., Eskildsen, C.E., Petersen, M.A., Saerens, S., Blennow, A., Skovmand-Larsen, M., Swiegers, J.H., Petersen, G.B., Heimdal, H. and Nielsen, D.S. (2014) Impact of starter cultures and fermentation techniques on the volatile aroma and sensory profile of chocolate. *Food Research International* 63, 306–316.

Dainty, R.H., Edwards, R.A., Hibbard, C.M. and Marnewick, J.J. (1989) Volatile compounds associated with microbial growth on normal and high pH beef stored at chill temperatures. *Journal of Applied Bacteriology* 66, 281–289.

de la Plaza, M., Fernández de Palencia, P., Peláez, C. and Requena, T. (2004) Biochemical and molecular characterization of α-ketoisovalerate decarboxylase, an enzyme involved in the formation of aldehydes from amino acids by *Lactococcus lactis*. *FEMS Microbiology Letters* 238, 367–374.

Deetae, P., Spinnler, H.E., Bonnarme, P. and Helinck, S. (2009) Growth and aroma contribution of *Microbacterium foliorum*, *Proteus vulgaris* and *Psychrobacter* sp. during ripening in a cheese model medium. *Applied Microbiology and Biotechnology* 82, 169–177.

Deetae, P., Saint-Eve, A., Spinnler, H.E. and Helinck, S. (2011) Critical effect of oxygen on aroma compound production by *Proteus vulgaris*. *Food Chemistry* 126, 134–139.

Helinck, S., Le Bars, D., Moreau, D. and Yvon, M. (2004) Ability of thermophilic lactic acid bacteria to produce aroma compounds from amino acids. *Applied and Environmental Microbiology* 70, 3855–3861.

Herranz, B., Fernandez, M., de la Hoz, L. and Ordonez, J.A. (2006) Use of bacterial extracts to enhance amino acid breakdown in dry fermented sausages. *Meat Science* 72, 318–325.

Jaffrès, E., Sohier, D., Leroi, F., Pilet, M.F., Prévost, H., Joffraud, J.J. and Dousset, X. (2009) Study of the bacterial ecosystem in tropical cooked and peeled shrimps using a polyphasic approach. *International Journal of Food Microbiology* 131, 20–29.

Kieronczyk, A., Skeie, S., Langsrud, T., Le Bars, D. and Yvon, M. (2004) The nature of aroma compounds produced in a cheese model by glutamate dehydrogenase positive *Lactobacillus* INF15D depends on its relative aminotransferase activities towards the different amino acids. *International Dairy Journal* 14, 227–235.

Kieronczyk, A., Cachon, R., Feron, G. and Yvon, M. (2006) Addition of oxidizing or reducing agents to the reaction medium influences amino acid conversion to aroma compounds by *Lactococcus lactis*. *Journal of Applied Microbiology* 101, 1114–1122.

Larrouture-Thiveyrat, C., Pepin, M., Leroy-Sétrin, S. and Montel, M.C. (2003) Effect of *Carnobacterium piscicola* on aroma formation in sausage mince. *Meat Science* 63, 423–426.

Martínez-Cuesta, M.C., Requena, T. and Peláez, C. (2006) Cell membrane damage induced by lacticin 3147 enhances aldehyde formation in *Lactococcus lactis* IFPL730. *International Journal of Food Microbiology* 109, 198–204.

Masson, F., Talon, R. and Montel, M.C. (1996) Histamine and tyramine production by bacteria from meat products. *International Journal of Food Microbiology* 32, 199–207.

Massouras, T., Pappa, E.C. and Mallatou, H. (2006) Headspace analysis of volatile flavour compounds of teleme cheese made from sheep and goat milk. *International Journal of Dairy Technology* 59, 250–256.

Miller, A., Morgan, M.E. and Libbey, L.M. (1974) *Lactobacillus maltaromicus*, a new species producing a malty aroma. *International Journal of Systematic Bacteriology* 24, 346–354.

Morales, P., Fernández-Garcıía, E., Gaya, P. and Nuñez, M. (2003) Formation of volatile compounds by wild *Lactococcus lactis* strains isolated from raw ewes' milk cheese. *International Dairy Journal* 13, 201–209.

Nychas, G.J.E., Drosinos, E. and Board, R.G. (1998) The microbiology of meat and poultry. In: Board, R.G. and Davies, A.R. (eds) *Chemical Changes in Stored Meat*. Blackie, London, pp. 288–326.

Rijnen, L., Yvon, M., van Kranenburg, R., Courtin, P., Verheul, A., Chambellon, E. and Smit, G. (2003) Lactococcal aminotransferases AraT and BcaT are key enzymes for the formation of aroma compounds from amino acids in cheese. *International Dairy Journal* 13, 805–812.

Smit, B.A., Engels, W.J., Wouters, J.T. and Smit, G. (2004) Diversity of L-leucine catabolism in various microorganisms involved in dairy fermentations, and identification of the rate-controlling step in the formation of the potent flavour component 3-methylbutanal. *Applied Microbiology and Biotechnology* 64, 396–402.

Smit, B.A., van Hylckama Vlieg, J.E., Engels, W.J., Meijer, L., Wouters, J.T. and Smit, G. (2005) Identification, cloning, and characterization of a *Lactococcus lactis* branched-chain α-keto acid decarboxylase involved in flavor formation. *Applied and Environmental Microbiology* 71, 303–311.

Tanous, C., Kieronczyk, A., Helinck, S., Chambellon, E. and Yvon, M. (2002) Glutamate dehydrogenase activity: a major criterion for the selection of flavour-producing lactic acid bacteria strains. *Antonie Van Leeuwenhoek* 82, 271–278.

Thierry, A., Maillard, M.B. and Yvon, M. (2002) Conversion of L-leucine to isovaleric acid by *Propionibacterium freudenreichii* TL 34 and ITGP23 *Applied and Environmental Microbiology* 68, 608–615.

Wallace, J.M. and Fox, P.F. (1997) Effect of adding free amino acids to Cheddar cheese curd on proteolysis, flavour and texture development. *International Dairy Journal* 7, 157–167.

Ward, D.E., Ross, R.P., van der Weijden, C.C., Snoep, J.L. and Claiborne, A. (1999) Catabolism of branched-chain α-keto acids in *Enterococcus faecalis*: the *bkd* gene cluster, enzymes, and metabolic route. *Journal of Bacteriology* 181, 5433–5442.

Whetstine, M.E., Drake, M.A., Broadbent, J.R. and McMahon, D. (2006) Enhanced nutty flavor formation in Cheddar cheese made with a malty *Lactococcus lactis* adjunct culture. *Journal of Dairy Science* 89, 3277–3284.

Yvon, M., Thirouin, S., Rijnen, L., Fromentier, D. and Gripon, J.C. (1997) An aminotransferase from *Lactococcus lactis* initiates conversion of amino acids to cheese flavor compounds. *Applied and Environmental Microbiology* 63, 414–419.

18 Microbial Degradation of Phenolic Amino Acids

D.E. Holmes[1]* and J.A. Smith[2]

[1]*Department of Physical and Biological Sciences, Western New England University, Springfield, Massachusetts, USA;* [2]*Department of Biology, American International College, Springfield, Massachusetts, USA*

18.1 Abstract

Ferroglobus placidus is a hyperthermophilic archaeon from the family Archaeoglobaceae that is able to completely oxidize the phenolic amino acids tyrosine and phenylalanine to CO_2 under strict anaerobic conditions with ferric iron acting as its sole electron acceptor. The compounds phenylacetate and 4-hydroxyphenylacetate are intermediates formed during the degradation of these amino acids and, as in several anaerobic mesophilic bacteria, the benzoyl-CoA and glutaryl-CoA reductase pathways are involved in the complete degradation process. It has been suggested that the general strategies for the anaerobic degradation of phenolic amino acids are highly conserved among bacteria and archaea living in both mesophilic and hyperthermophilic environments.

18.2 Introduction

Amino acids can comprise as much as 50% of the organic carbon and 100% of the organic nitrogen in marine sediments (Lee, 1988). Up to 4% of the total amino acid pool in these sediments is phenylalanine, and other amino acids, such as glutamate, aspartate and glycine, are even more abundant (Lee, 1988; Dauwe and Middelburg, 1998). This abundance of amino acids in sedimentary environments has resulted in the evolution of anaerobic degradation pathways that allow prokaryotes to utilize these compounds as carbon and energy sources (Barker, 1981).

To date, the majority of studies on aromatic amino acid catabolism have focused on the degradation of aromatic amino acids in the presence of oxygen, with anaerobic studies being mostly limited to partial oxidation (Barker, 1981; Baena *et al.*, 1998, 1999, 2000; Díaz *et al.*, 2007; Fonknechten *et al.*, 2010; Russell *et al.*, 2013; Wöhlbrand *et al.*, 2013). Some studies have shown that mesophilic denitrifying and sulfate-reducing bacteria are able to couple the complete oxidation of aromatic amino acids with anaerobic respiration (Bak and Widdel, 1986; Kuever *et al.*, 2001). However, the occurrence of anaerobic degradation of these aromatic amino acids in hot (>80°C) environments is mostly unexplored. While it has been shown that *Archaeoglobus fulgidus* and *Pyrococcus furiosus* can catabolize phenylalanine with sulfate or elemental sulfur serving as electron acceptors for growth, both of these archaea utilize partial oxidation pathways rather than the complete oxidation pathways observed in their mesophilic bacterial counterparts (Mai and Adams, 1994; Parthasarathy *et al.*, 2013).

*Corresponding author: E-mail address: dholmes@bio.umass.edu

Although few studies of amino acid catabolism in hot environments have been conducted, this metabolism is evolutionarily significant if one considers the fact that amino acids are thought to have first evolved in hydrothermal vents (Miller, 1953; Parker *et al.*, 2011). It has been proposed that the chemically reducing atmosphere of early Earth allowed the formation of certain basic compounds of life, such as amino acids and nucleic acids (Miller, 1953; Parker *et al.*, 2011). Some of the only environments today that are reminiscent of early Earth are those found in hydrothermal vents. Therefore, biochemical studies of organisms isolated from hydrothermal vent sediments, such as *Ferroglobus placidus*, can provide considerable insight into amino acid metabolism on early Earth.

This chapter focuses on the anaerobic degradation of phenolic amino acids by the hyperthermophilic archaeon *F. placidus*. This organism is unusual among hyperthermophilic archaea in that it can completely oxidize tyrosine and phenylalanine to carbon dioxide and transfer the electrons so produced to insoluble iron (ferric) oxide (Aklujkar *et al.*, 2014), which is commonly present as the insoluble iron hydroxide (oxyhydroxide) in aqueous environments.

18.3 Phenylalanine Metabolism

F. placidus can grow with phenylalanine provided as the sole electron donor and insoluble iron (ferric) oxyhydroxide provided as the electron acceptor. It can completely oxidize phenylalanine and transfer its electrons to iron according to the following stoichiometry:

$$C_9H_{11}NO_2 + 40Fe(III) + 16H_2O \rightarrow 9CO_2 + NH_3 + 40Fe(II) + 40H^+ \quad (18.1)$$

Here, *F. placidus* utilizes a pathway similar to that of the nitrate-reducing bacterium *Thauera aromatica* (Schneider *et al.*, 1997; Debnar-Daumler *et al.*, 2014). Like *T. aromatica*, phenylalanine is first oxidized to phenylacetate and then to benzoyl-CoA, though *F. placidus* uses iron as the electron acceptor rather than nitrate.

The first step in this pathway involves the formation of a 2-oxoacid, phenylpyruvate, by a pyridoxal-phosphate-dependent aminotransferase (Fig. 18.1). Phenylpyruvate is then decarboxylated to phenylacetaldehyde by phenylpyruvate carboxylase. Next, phenylacetaldehyde is oxidized to phenylacetate. In *T. aromatica*, an NAD-dependent aldehyde dehydrogenase catalyses this step. However, this enzyme is not present in the *F. placidus* genome; rather, it appears to use a tungsten-containing aldehyde:ferredoxin oxidoreductase (AOR) homologous to an enzyme found in another nitrate-reducing bacterium (*Aromatoleum aromaticum*) capable of phenylalanine degradation to carry out this step (Debnar-Daumler *et al.*, 2014). Although the AOR found in *A. aromaticum* can indeed catalyse the conversion of phenylacetaldehyde to phenylacetate, unlike *F. placidus*, AOR is not the primary enzyme that it uses for this reaction. Rather, an NAD-dependent dehydrogenase protein homologous to that found in *T. aromatica* is the main enzyme and AOR is thought to function as a detoxifying enzyme when high aldehyde concentrations accumulate in the cell's cytoplasm.

The metabolism of phenylacetate to benzoyl-CoA also appears to proceed in a manner similar to that in *T. aromatica* (Schneider *et al.*, 1997; Heider *et al.*, 1998). Phenylacetate is first ligated to CoA, forming phenylacetyl-CoA. In the following steps, phenylacetyl-CoA is oxidized to benzoyl-CoA by a phenylacetyl-CoA dehydrogenase, a hydrolase and a 2-oxoacid-ferredoxin oxidoreductase protein. Once benzoyl-CoA has been synthesized, it enters the benzoyl-CoA reduction pathway for further degradation (Holmes *et al.*, 2011, 2012).

18.4 Tyrosine Metabolism

F. placidus can also grow when tyrosine is provided as the sole electron donor with insoluble iron oxyhydroxide provided as an electron acceptor according to the following stoichiometry:

$$C_9H_{11}NO_3 + 38Fe(III) + 15H_2O \rightarrow 9CO_2 + NH_3 + 38Fe(II) + 38H^+ \quad (18.2)$$

The anaerobic catabolism of tyrosine by *F. placidus* has not been as carefully studied as that of phenylalanine, but the structural similarity between the two amino acids suggests that their

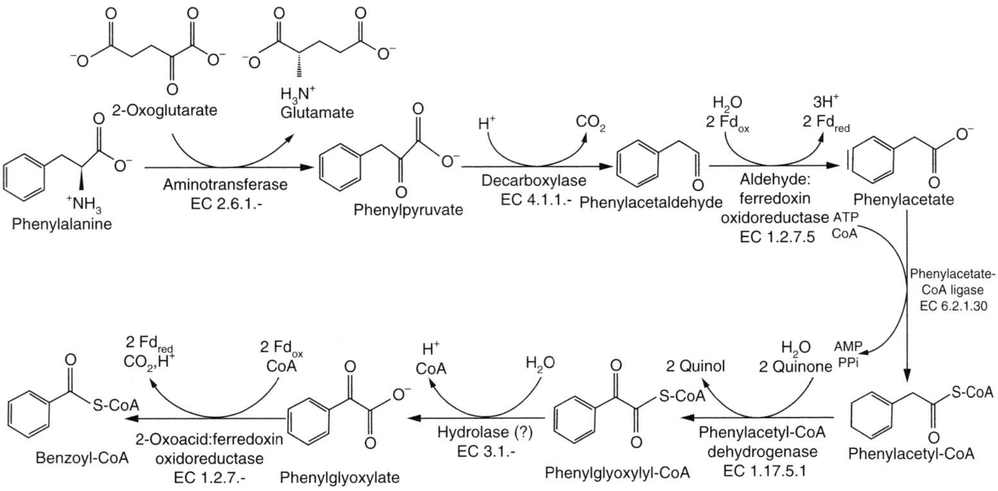

Fig. 18.1. Proposed pathway utilized by *Ferroglobus placidus* to oxidize phenylalanine to benzoyl-CoA.

catabolic pathways are likely to be mechanistically similar. As in a methanogenic consortium (Balba and Evans, 1980; Evans and Fuchs, 1988), the intermediate 4-hydroxyphenylacetate was detected in cultures of *F. placidus* grown with tyrosine provided as the electron donor. Neither phenylacetate nor phenol were detected as intermediates during growth with tyrosine, and genes coding for a tyrosine phenol-lyase (Kumagai *et al.*, 1970) are not present in the *F. placidus* genome.

It is likely that the first step in tyrosine degradation involves transfer of the amino group from tyrosine to 2-oxoglutarate by an aminotransferase protein to form 4-hydroxyphenylpyruvate and glutamate (Fig. 18.2). In the following step, it is likely that hydroxyphenylacetaldehyde is formed by a decarboxylase protein, and that this is then oxidized by an aldehyde ferredoxin oxidoreductase to 4-hydroxyphenylacetate. Next, it has been proposed that CoA is ligated to 4-hydroxyphenylacetate to form 4-hydroxyphenylacetyl-CoA in a manner similar to that of a denitrifying *Pseudomonas* species (Mohamed *et al.*, 1993). It is then likely that a molybdopterin-binding oxidoreductase catalyses the formation of 4-hydroxyphenylglyoxylyl-CoA, followed by hydrolysis to 4-hydroxyphenylglyoxylate and oxidation to 4-hydroxybenzoyl-CoA by enzymes that are similar to those proposed in the phenylacetate degradation pathway. While it is known that *F. placidus* is capable of carrying out the dehydroxylation of 4-hydroxybenzoyl-CoA, the enzyme responsible for this activity is not clear (Holmes *et al.*, 2012).

18.5 Benzoyl-CoA Reduction Pathway

Both the tyrosine and phenylalanine degradation pathways converge at benzoyl-CoA – a key metabolite formed during the anaerobic degradation of all known benzene ring-containing compounds. Once benzoyl-CoA is formed, it is further metabolized to glutaryl-CoA by the benzoyl-CoA reductase pathway. Like all anaerobic benzoate-degrading bacteria that have been studied to date (Carmona *et al.*, 2009), *F. placidus* utilizes the enzyme benzoate-CoA ligase to catalyse the conversion of benzoate to benzoyl-CoA (Fig. 18.3).

The next step involves reduction of the aromatic ring to cyclohex-1,5-diene-1-carboxyl-CoA by benzoyl-CoA reductase (BCR). Two different types of BCR have been described. Class I BCR complexes accept electrons from a reduced ferredoxin protein. The reaction catalysed by this enzyme is ATP dependent and irreversible (Boll and Fuchs, 1995). The best studied class I BCRs are those from the facultative denitrifying species *T. aromatica* (Breese *et al.*, 1998), *Azoarcus* spp. (López Barragán *et al.*,

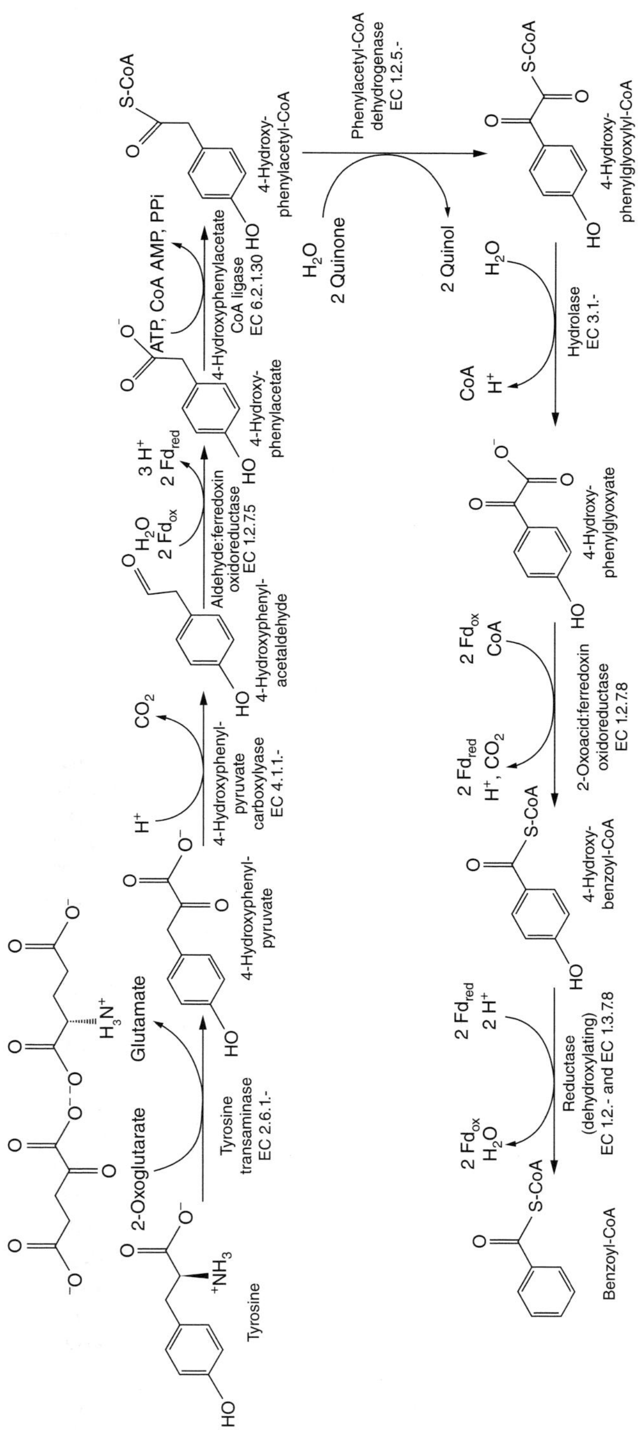

Fig. 18.2. Proposed pathway utilized by *Ferroglobus placidus* to oxidize tyrosine to benzoyl-CoA.

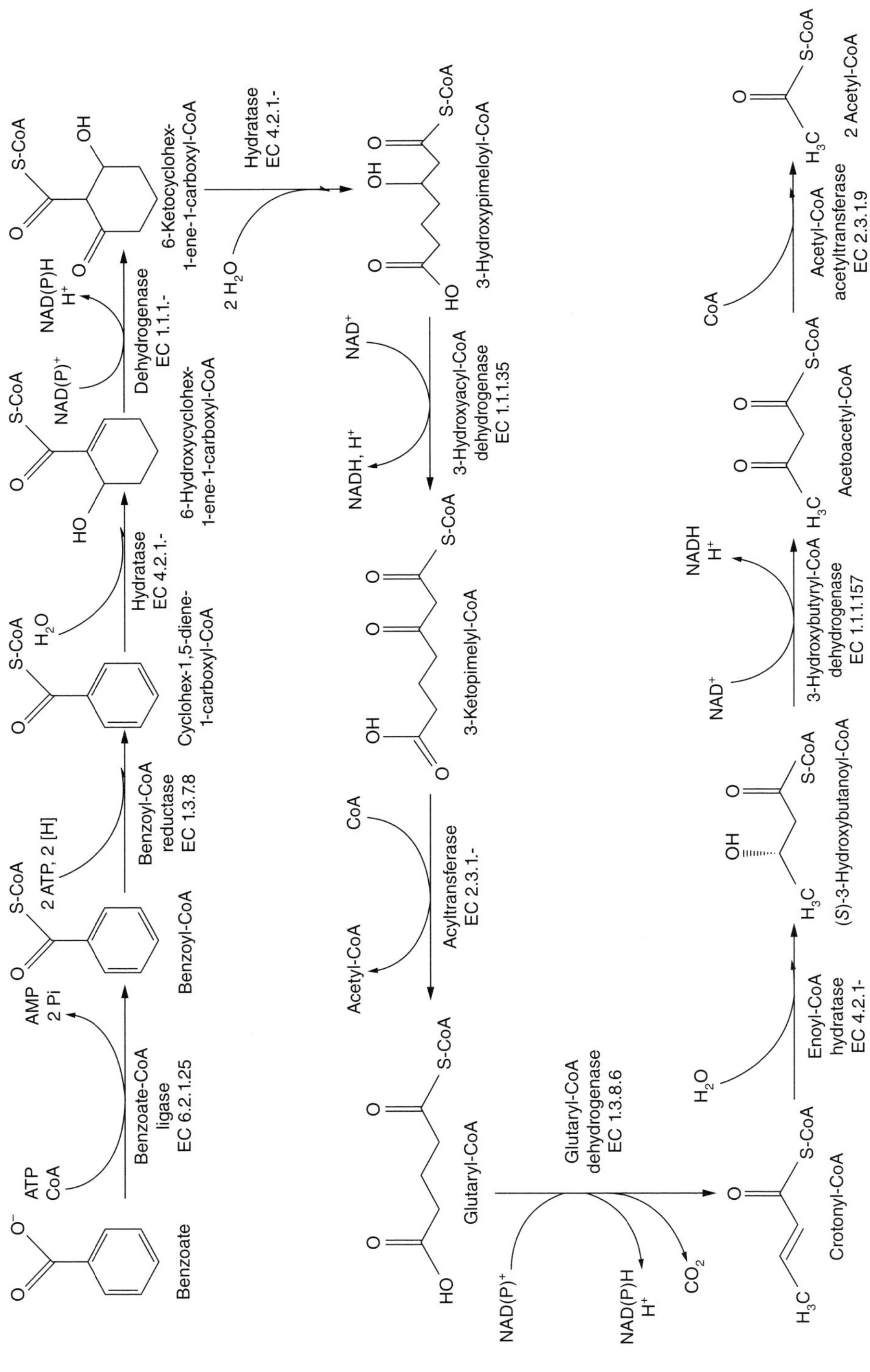

Fig. 18.3. Proposed pathway utilized by *Ferroglobus placidus* to oxidize benzoyl-CoA to glutaryl-CoA and then to acetyl-CoA.

2004) and the phototroph *Rhodopseudomonas palustris* (Egland *et al.*, 1997).

A diversity of strict anaerobic microorganisms such as *Geobacter* species (Butler *et al.*, 2007; Wischgoll *et al.*, 2005), *Syntrophus aciditrophicus* (Schocke and Schink, 1999) and *Desulfococcus multivorans* (Peters *et al.*, 2004) have a class II BCR. The reaction catalysed by this complex is reversible and ATP independent (Kung *et al.*, 2009, 2010; Löffler *et al.*, 2011). It has been proposed that strict anaerobes must use class II BCRs because the amount of energy available from benzoate oxidation coupled to the reduction of iron, sulfate or protons is not sufficient to support the substantial energetic requirement of the ATP-dependent class I BCR reaction (Kung *et al.*, 2009, 2010; Löffler *et al.*, 2011). *F. placidus*, however, is completely unique among all anaerobic prokaryotes in that it has an ATP-consuming class I BCR complex that catalyses the conversion of benzoyl-CoA to cyclohex-1,5-diene-1-carboxyl-CoA (Schmid *et al.*, 2015).

In the following steps of the benzoyl-CoA reduction pathway, cyclohex-1,5-diene-1-carboxyl-CoA is converted to the C7 dicarboxylic acid derivative 3-hydroxypimeloyl-CoA. Two slightly different versions of these reactions have been described: the BadK-BadH-BadI pathway found in the phototroph *Rhodopseudomonas palustris* (Harwood *et al.*, 1998; Gibson and Harwood, 2002) and the *Thauera–Azoarcus* type pathway (Breese *et al.*, 1998) that has been found in all other anaerobic benzoate-degrading organisms studied to date (Carmona *et al.*, 2009). It appears that *F. placidus* catalyses these steps in a manner similar to the *Thauera–Azoarcus* pathway (Schmid *et al.*, 2015). First, a hydratase protein hydrolyses cyclohex-1,5-diene-1-carboxyl-CoA to 6-hydroxycyclohex-1-ene-1-carboxyl-CoA; this is then oxidized to 6-ketocyclohex-1-ene-1-carboxyl-CoA. Although homologues for the three genes coding for the hydratase and dehydrogenase enzymes involved in these steps have not yet been identified, hydratase and dehydrogenase activities have been detected in cell extracts (Schmid *et al.*, 2015). In the next step, two water molecules are added, forming 3-hydroxypimeloyl-CoA; 3-ketopimelyl-CoA is then formed by another oxidation reaction and this substrate is next converted to glutaryl-CoA by an acyltransferase protein. Thereafter glutaryl-CoA is converted to acetyl-CoA (Section 18.6).

18.6 Glutaryl-CoA Pathway

It has been proposed that *F. placidus* uses a decarboxylating glutaryl-CoA dehydrogenase homologous to that of *Pseudomonas putida* (Hartel *et al.*, 1993; Aklujkar *et al.*, 2014) to produce crotonyl-CoA directly (Fig. 18.3). In the next step, a water molecule is added to form (*S*)-3-hydroxybutanoyl-CoA. Electrons are then transferred by 3-hydroxybutyryl-CoA dehydrogenase to NAD^+ to form acetoacetyl-CoA. Finally, an acetyl group is transferred by acetoacetyl-CoA thiolase to form two molecules of acetyl-CoA, which can then enter the citric acid cycle or the Wood–Ljungdahl pathway for further degradation.

18.7 Homology with Mesophilic Bacterial Proteins

Many of the genes involved in the degradation of aromatics by *F. placidus* have homologues in such mesophilic bacteria as *T. aromatica*, *Geobacter metallireducens*, *Aromatoleum aromaticum* and *Rhodopseudomonas palustris* (Table 18.1). Genes coding for the enzymes involved in the degradation of these compounds are arranged in clusters within the *F. placidus* genome – another feature that it shares with mesophilic aromatic-degrading bacteria (Schneider *et al.*, 1997; Larimer *et al.*, 2004; Rabus *et al.*, 2005; DiDonato *et al.*, 2010; Fuchs *et al.*, 2011; Ebenau-Jehle *et al.*, 2012; Holmes *et al.*, 2012; Jiang *et al.*, 2012).

F. placidus is the only hyperthermophilic archaeon that has been shown to utilize aromatic compounds as electron donors. While one other species from the family Archaeoglobaceae – *Geoglobus acetivorans* – has homologues for many of the degradation genes for aromatics found in *F. placidus*, growth on aromatic compounds such as benzoate, phenylalanine and tyrosine has not been observed in this organism (Mardanov *et al.*, 2015). Like *F. placidus*, many of the putative degradation genes for aromatics found in the genome of *G. acetivorans* are homologous to those of mesophilic bacteria.

The functional conservation between aromatics degradation genes from the above two hyperthermophilic archaea and mesophilic bacteria is intriguing. It is unlikely that *F. placidus* or *G. acetivorans* acquired these genes from bacteria

Table 18.1. Homologous degradation genes for aromatics found in mesophilic bacteria.

Gene ID	Gene designation	Gene name	Best hit in known aromatic degraders	% Identity/ Similarity
Ferp_1031	Didehydro-pimeloyl-CoA hydratase	*dph*	*Aromatoleum aromaticum*	44/61
Ferp_1035	Cyclohex-1-ene-1-carboxyl-CoA hydratase	*badK1*	*Rhodopseudomonas palustris*	38/59
Ferp_1036	Thiolase family protein	*thl1*	*R. palustris*	31/47
Ferp_1040	2-Ketocyclohexanecarboxyl-CoA hydrolase	*badI*	*R. palustris*	32/50
Ferp_1041	2-Hydroxy-glutaryl-CoA reductase, component A	*hgdC*	*Geobacter bemidjiensis*	44/60
Ferp_1042	2-Hydroxy-glutaryl-CoA reductase, component D	*hgdA*	*G. bemidjiensis*	27/45
Ferp_1043	2-Hydroxy-glutaryl-CoA reductase, component D	*hgdB*	*G. bemidjiensis*	23/43
Ferp_1044	AMP-dependent synthetase and ligase	*bcl1*	*R. palustris*	36/52
Ferp_1180	Ferredoxin	*fdx*	*A. aromaticum*	51/63
Ferp_1184	Benzoyl-CoA reductase, gamma subunit	*bzdN*	*A. aromaticum*	44/66
Ferp_1185	Benzoyl-CoA reductase, beta subunit	*bzdO*	*A. aromaticum*	49/66
Ferp_1186	Benzoyl-CoA reductase, delta subunit	*bzdP*	*A. aromaticum*	42/63
Ferp_1187	Benzoyl-CoA reductase, alpha subunit	*bzdQ*	*A. aromaticum*	55/72
Ferp_1188	Putative reductase	*badX*	*A. aromaticum*	40/58
Ferp_1189	Thiolase family protein	*thl2*	*G. metallireducens*	36/53
Ferp_2180	Benzoate-CoA ligase	*bcl2*	*A. aromaticum*	33/51
Ferp_1233	2-Hydroxycyclohexanecarboxyl-CoA dehydrogenase	*badH*	*R. palustris*	40/52
Ferp_1566	Pimeloyl-CoA dehydrogenase (large subunit)	*pimC*	*R. palustris*	25/43
Ferp_1579	Pimeloyl-CoA dehydrogenase (small subunit)	*pimD*	*R. palustris*	28/47
Ferp_1942	Cyclohex-1-ene-1-carboxyl-CoA hydratase	*badK2*	*R. palustris*	41/57
Ferp_1422	Aldehyde:ferredoxin oxidoreductase	*aor*	*A. aromaticum*	28/45

via horizontal gene transfer, as the guanine–cytosine (GC) content of the aromatics pathway genes is similar to the mean GC content of both genomes. The mean GC content of *F. placidus* and *G. acetivorans* is 44 and 47%, respectively, and all of the genes from the aromatics degradation pathways fall within 7% of this percentage indicating a likely origin by vertical transmission. In addition, relatively few mobile genetic elements or proviral sequences are present in either the *F. placidus* or *G. acetivorans* genomes (Anderson *et al.*, 2011; Mardanov *et al.*, 2015).

It is also extremely unlikely that these two types of organisms (hyperthermophilic archaea and mesophilic bacteria) were ever in direct contact as they have completely different physiologies and were isolated from entirely different habitats. Both *F. placidus* and *G. acetivorans* were isolated from marine hydrothermal vent sediments (Hafenbradl *et al.*, 1996; Slobodkina *et al.*, 2009), whereas many of the bacterial species with homologous enzymes could never have endured hyperthermophilic conditions and were isolated from freshwater environments far from the domains of *F. placidus* and *G. acetivorans* (Lovley and Phillips, 1986; Anders *et al.*, 1995; Larimer *et al.*, 2004; Rabus and Widdel, 1995).

The conservation of these pathways across different domains of life is probably the result of convergent evolution. Evidence for this comes from the fact that several of the predicted proteins involved in the degradation of aromatic compounds are structurally different. For example, although many of the mesophilic oxidoreductase proteins use $NAD(P)^+$ as electron carriers, $NAD(P)^+$ is rarely seen in the reactions catalysed by *F. placidus* that are described here. Rather, many of the oxidoreductase proteins of *F. placidus* transfer their electrons to a ferredoxin protein (see Table 18.2). Adenine dinucleotide coenzymes such as $NAD(P)^+$ are structurally complex (Huang *et al.*, 2000; Gazzaniga *et al.*, 2009), while 4Fe-4S clusters, as in ferredoxin, are among the simplest electron transfer groups, and are thought to have been among the earliest redox-active molecules to have emerged on the primitive Earth (Kim *et al.*, 2012). These proteins are also extremely thermostable and can withstand high temperatures such as those found in environments where organisms like *F. placidus* thrive.

Another striking feature of *F. placidus* is the fact that candidate genes of the aromatic amino acid degradation pathways in this archaeon appear to encode both molybdoenzymes and tungstoenzymes; in contrast, most of the mesophilic bacteria use only molybdoenzymes. Tungstoenzymes are not very common among mesophilic bacterial species, but are frequently associated with hyperthermophilic archaea (Bevers *et al.*, 2009).

18.8 Industrial Applications

In addition to providing an insight into the origins of life, studies of the thermally stable enzymes of hyperthermophilic organisms that are involved in the degradation of aromatic amino acids can have broad industrial applications. For example, intermediates formed during the anaerobic degradation of tyrosine and phenylalanine are used to add scent to perfumes, to add flavour to cheese, to synthesize penicillin and other pharmaceutical products, and to dye fabric (Ferrero *et al.*, 1991; Gummalla and Broadbent, 2001; Luengo *et al.*, 2001; Letizia *et al.*, 2003; Barden, 2011; Barel, 2014).

Currently, many of these intermediates are being synthesized by industry in ways that generate toxic side products such as ethylbenzene and benzaldehyde. During mass production, it is likely that many of these toxic compounds make their way into the environment so causing harm to humans and wildlife. Therefore, it is important to develop a technique that can allow the production of large quantities of aromatic amino acid intermediates without the utilization or production of hazardous chemicals.

One sustainable approach that would minimize exposure to harmful chemicals involves the formation of intermediates such as phenylacetate by microorganisms. If the enzymatic mechanisms involved in the formation of these intermediates come to be fully understood, it will be possible to genetically engineer microorganisms and enzymes that are extremely efficient in the production of these compounds. Thermally stable enzymes like those produced by *F. placidus* are particularly desirable, as they can be employed for use in harsh industrial conditions but still retain their specific catalytic activity.

18.9 Conclusions

F. placidus represents a group of prokaryotes that can couple the complete oxidation of an aromatic amino acid with anaerobic respiration. This organism's genome contains genes that code for enzymes from aromatic amino acid degradation pathways associated with mesophilic, nitrate-reducing, sulfate-reducing and phototrophic bacteria (Schneider *et al.*, 1997; Larimer *et al.*, 2004; Rabus *et al.*, 2005; DiDonato *et al.*, 2010; Fuchs *et al.*, 2011; Ebenau-Jehle *et al.*, 2012; Jiang *et al.*, 2012).

Not only are many of the enzymes involved in the first stages of aromatic amino acid degradation in *F. placidus* homologous to those in mesophilic bacteria, benzoyl-CoA is an important intermediate formed by all of these organisms during the anaerobic degradation process (Carmona *et al.*, 2009). Benzoyl-CoA is also an intermediate that is formed during the anaerobic degradation of such aromatic compounds as ethylbenzene, benzoate and phenol (Fuchs

Table 18.2. Comparison of electron donors associated with oxidoreductase proteins from the phenylalanine pathway found in various prokaryotic species.

Reaction	Enzyme name	Organism	Gene ID	Electron donor
Phenylacetaldehyde → Phenylacetate	Phenylacetaldehyde oxidoreductase	*Aromatoleum aromaticum*	ebA4954	NAD(P)H
		Thauera aminoaromatica	AIO06093	NAD(P)H
		Rhodopseudomonas palustris	WP_011473920	NAD(P)H
		Geobacter metallireducens	WP_004513074	NAD(P)H
		Delta NaphS2[a]	WP_006423128	NAD(P)H
		Ferroglobus placidus	Ferp_1422	Ferredoxin
Phenylglyoxylate → Benzoyl-CoA	Phenylglyoxylate oxidoreductase	*A. aromaticum*	ebA5400	NAD(P)H
		T. aminoaromatica	Tmz1t_2965	NAD(P)H
		R. palustris	WP_011159320	NAD(P)H
		G. metallireducens	WP_004513641	NAD(P)H
		Delta NaphS2[a]	WP_006425705	NAD(P)H
		F. placidus	Ferp_1033	Ferredoxin

[a]Deltaproteobacterium strain NaphS2.

et al., 2011), and many of the organisms that are able to anaerobically degrade aromatic amino acids can also gain energy from the oxidation of these aromatic compounds.

It does not seem too surprising that such a diversity of microorganisms have acquired the ability to degrade aromatic amino acids anaerobically. These compounds are abundant carbon and nitrogen sources in many oxygen-limiting environments, and organisms that can degrade these compounds will have a competitive advantage. *F. placidus* is the only hyperthermophilic archaeon known to date that has been shown to oxidize any of the aromatic amino acids completely. It seems likely that organisms such as *F. placidus* occupy an important niche within hyperthermophilic environments and that they warrant further investigation.

References

Aklujkar, M., Risso, C., Smith, J.A., Beaulieu, D., Dubay, R., Giloteaux, L., DiBurro, K. and Holmes, D. (2014) Anaerobic degradation of aromatic amino acids by the hyperthermophilic archaeon, *Ferroglobus placidus*. *Microbiology* 160, 2694–709.

Anders, H. J., Kaetzke, A., Kämpfer, P., Ludwig, W. and Fuchs, G. (1995) Taxonomic position of aromatic-degrading denitrifying pseudomonad strains K-172 and Kb-740 and their description as new members of the genera *Thauera*, as *Thauera aromatica* sp. nov., and *Azoarcus*, as *Azoarcus evansii* sp. nov, respectively, members of the beta-subclass of the *Proteobacteria*. *International Journal of Systematic Bacteriology* 45, 327–333.

Anderson, I., Risso, C., Holmes, D., Lucas, S., Copeland, A., Lapidus, A., Cheng, J.F., Bruce, D., Goodwin, L., Pitluck, S. *et al.* (2011) Complete genome sequence of *Ferroglobus placidus* AEDII12DO. *Standards in Genomic Sciences* 5, 50–60.

Baena, S., Fardeau, M.L., Labat, M., Ollivier, B., Thomas, P., Garcia, J.L. and Patel, B.K.C. (1998) *Aminobacterium colombiense* gen. nov. sp. nov., an amino acid-degrading anaerobe isolated from anaerobic sludge. *Anaerobe* 4, 241–250.

Baena, S., Fardeau, M.L., Ollivier, B., Labat, M., Thomas, P., Garcia, J.L. and Patel, B.K. (1999) *Aminomonas paucivorans* gen. nov., sp. nov., a mesophilic, anaerobic, amino-acid-utilizing bacterium. *International Journal of Systematic Bacteriology* 49, 975–982.

Baena, S., Fardeau, M.L., Labat, M., Ollivier, B., Garcia, J.L. and Patel, B.K. (2000) *Aminobacterium mobile* sp. nov., a new anaerobic amino-acid-degrading bacterium. *International Journal of Systematic and Evolutionary Microbiology* 50, 259–264.

Bak, F. and Widdel, F. (1986) Anaerobic degradation of phenol and phenol derivatives by *Desulfobacterium phenolicum* sp. nov. *Archives of Microbiology* 146, 177–180.

Balba, M.T. and Evans, W.C. (1980) Methanogenic fermentation of the naturally occurring aromatic amino acids by a microbial consortium. *Biochemical Society Transactions* 8, 625–627.

Barden, T.C. (2011) Indoles: industrial, agricultural and over-the-counter uses. *Topics in Heterocyclic Chemistry* 26, 31–46.

Barel, A.O. (2014) *Handbook of Cosmetic Science and Technology*, 4th edn. CRC Press (Taylor and Francis Group), Boca Raton, Florida.

Barker, H.A. (1981) Amino acid degradation by anaerobic bacteria. *Annual Review of Biochemistry* 50, 23–40.

Bevers, L.E., Hagedoorn, P.L. and Hagen, W.R. (2009) The bioinorganic chemistry of tungsten. *Coordination Chemistry Reviews* 253, 269–290.

Boll, M. and Fuchs, G. (1995) Benzoyl-coenzyme A reductase (dearomatizing), a key enzyme of anaerobic aromatic metabolism – ATP dependence of the reaction, purification and some properties of the enzyme from *Thauera aromatica* strain K172. *European Journal of Biochemistry* 234, 921–933.

Breese, K., Boll, M., Alt-Morbe, J., Schagger, H. and Fuchs, G. (1998) Genes coding for the benzoyl-CoA pathway of anaerobic aromatic metabolism in the bacterium *Thauera aromatica*. *European Journal of Biochemistry* 256, 148–154.

Butler, J.E., He, Q., Nevin, K.P., He, Z., Zhou, J. and Lovley, D.R. (2007) Genomic and microarray analysis of aromatics degradation in *Geobacter metallireducens* and comparison to a *Geobacter* isolate from a contaminated field site. *BMC Genomics* 8:180.

Carmona, M., Zamarro, M.T., Blázquez, B., Durante-Rodríguez, G., Juárez, J.F., Valderrama, J.A., Barragán, M.J., García, J.L. and Díaz, E. (2009) Anaerobic catabolism of aromatic compounds: a genetic and genomic view. *Microbiology and Molecular Biology Reviews* 73, 71–133.

Dauwe, B. and Middelburg, J.J. (1998) Amino acids and hexosamines as indicators of organic matter degradation in North Sea sediments. *Limnology and Oceanography* 43, 782–798.

Debnar-Daumler, C., Seubert, A., Schmitt, G. and Heider, J. (2014) Simultaneous involvement of a tungsten-containing aldehyde:ferredoxin oxidoreductase and a phenylacetaldehyde dehydrogenase in anaerobic phenylalanine metabolism. *Journal of Bacteriology* 196, 483–492.

Díaz, C., Baena, S., Fardeau, M.L. and Patel, B.K. (2007) *Aminiphilus circumscriptus* gen. nov., sp. nov., an anaerobic amino-acid-degrading bacterium from an upflow anaerobic sludge reactor. *International Journal of Systematic and Evolutionary Microbiology* 57, 1914–1918.

DiDonato, R.J. Jr, Young, N.D., Butler, J.E., Chin, K.J., Hixson, K.K., Mouser, P., Lipton, M.S., DeBoy, R. and Methe, B.A. (2010) Genome sequence of the deltaproteobacterial strain NaphS2 and analysis of differential gene expression during anaerobic growth on naphthalene. *PLOS ONE* 5(11): e14072.

Ebenau-Jehle, C., Thomas, M., Scharf, G., Kockelkorn, D., Knapp, B., Schuhle, K., Heider, J. and Fuchs, G. (2012) Anaerobic metabolism of indoleacetate. *Journal of Bacteriology* 194, 2894–2903.

Egland, P.G., Pelletier, D.A., Dispensa, M., Gibson, J. and Harwood, C.S. (1997) A cluster of bacterial genes for anaerobic benzene ring biodegradation. *Proceedings of the National Academy of Sciences of the United States of America* 94, 6484–6489.

Evans, W.C. and Fuchs, G. (1988) Anaerobic degradation of aromatic compounds. *Annual Review of Microbiology* 42, 289–317.

Ferrero, M. A., Reglero, A., Martinezblanco, H., Fernandezvalverde, M. and Luengo, J.M. (1991) *In vitro* enzymatic-synthesis of new penicillins containing keto acids as side-chains. *Antimicrobial Agents and Chemotherapy* 35, 1931–1932.

Fonknechten, N., Chaussonnerie, S., Tricot, S., Lajus, A., Andreesen, J.R., Perchat, N., Pelletier, E., Gouyvenoux, M., Barbe, V., Salanoubat, M. *et al.* (2010) *Clostridium sticklandii*, a specialist in amino acid degradation: revisiting its metabolism through its genome sequence. *BMC Genomics* 11:555.

Fuchs, G., Boll, M. and Heider, J. (2011) Microbial degradation of aromatic compounds – from one strategy to four. *Nature Reviews Microbiology* 9, 803–816.

Gazzaniga, F., Stebbins, R., Chang, S.Z., McPeek, M.A. and Brenner, C. (2009) Microbial NAD metabolism: lessons from comparative genomics. *Microbiology and Molecular Biology Reviews* 73, 529–541.

Gibson, J. and Harwood, C.S. (2002) Metabolic diversity in aromatic compound utilization by anaerobic microbes. *Annual Review of Microbiology* 56, 345–369.

Gummalla, S. and Broadbent, J.R. (2001) Tyrosine and phenylalanine catabolism by *Lactobacillus* cheese flavor adjuncts. *Journal of Dairy Science* 84, 1011–1019.

Hafenbradl, D., Keller, M., Dirmeier, R., Rachel, R., Rossnagel, P., Burggraf, S., Huber, H. and Stetter, K.O. (1996) *Ferroglobus placidus* gen nov, sp nov, a novel hyperthermophilic archaeum that oxidizes Fe2+ at neutral pH under anoxic conditions. *Archives of Microbiology* 166, 308–314.

Hartel, U., Eckel, E., Koch, J., Fuchs, G., Linder, D. and Buckel, W. (1993) Purification of glutaryl-CoA dehydrogenase from *Pseudomonas* sp., an enzyme involved in the anaerobic degradation of benzoate. *Archives of Microbiology* 159, 174–181.

Harwood, C.S., Burchhardt, G., Herrmann, H. and Fuchs, G. (1998) Anaerobic metabolism of aromatic compounds via the benzoyl-CoA pathway. *FEMS Microbiology Reviews* 22, 439–458.

Heider, J., Boll, M., Breese, K., Breinig, S., Ebenau-Jehle, C., Feil, U., Gad'on, N., Laempe, D., Leuthner, B., Mohamed, M.E., Schneider, S., Burchhardt, G. and Fuchs G. (1998) Differential induction of enzymes involved in anaerobic metabolism of aromatic compounds in the denitrifying bacterium *Thauera aromatica*. *Archives of Microbiology* 170, 120–131.

Holmes, D.E., Risso, C., Smith, J.A. and Lovley, D.R. (2011) Anaerobic oxidation of benzene by the hyperthermophilic archaeon *Ferroglobus placidus*. *Applied and Environmental Microbiology* 77, 5926–5933.

Holmes, D.E., Risso, C., Smith, J.A. and Lovley, D.R. (2012) Genome-scale analysis of anaerobic benzoate and phenol metabolism in the hyperthermophilic archaeon *Ferroglobus placidus*. *The ISME Journal* 6, 146–157.

Huang, F., Bugg, C.W. and Yarus, M. (2000) RNA-catalyzed CoA, NAD, and FAD synthesis from phosphopantetheine, NMN, and FMN. *Biochemistry* 39, 15548–15555.

Jiang, K., Sanseverino, J., Chauhan, A., Lucas, S., Copeland, A., Lapidus, A., Del Rio, T.G., Dalin, E., Tice, H., Bruce, D. *et al*. (2012) Complete genome sequence of *Thauera aminoaromatica* strain MZ1T. *Standards in Genomic Sciences* 6, 325–335.

Kim, J.D., Rodriguez-Granillo, A., Case, D.A., Nanda, V. and Falkowski, P.G. (2012) Energetic selection of topology in ferredoxins. *PLOS Computational Biology* 8(4): e1002463.

Kuever, J., Konneke, M., Galushko, A. and Drzyzga, O. (2001) Reclassification of *Desulfobacterium phenolicum* as *Desulfobacula phenolica* comb. nov and description of strain Sax(T) as *Desulfotignum balticum* gen. nov., sp nov. *International Journal of Systematic and Evolutionary Microbiology* 51, 171–177.

Kumagai, H., Yamada, H., Matsui, H., Ohkishi, H. and Ogata, K. (1970) Tyrosine phenol lyase. II. Cofactor requrements. *The Journal of Biological Chemistry* 245, 1773–1777.

Kung, J.W., Löffler, C., Dorner, K., Heintz, D., Gallien, S., Van Dorsselaer, A., Friedrich, T. and Boll, M. (2009) Identification and characterization of the tungsten-containing class of benzoyl-coenzyme A reductases. *Proceedings of the National Academy of Sciences of the United States of America* 106, 17687–17692.

Kung, J.W., Baumann, S., von Bergen, M., Müller, M., Hagedoorn, P.L., Hagen, W.R. and Boll, M. (2010) Reversible biological Birch reduction at an extremely low redox potential. *Journal of the American Chemical Society* 132, 9850–9856.

Larimer, F.W., Chain, P., Hauser, L., Lamerdin, J., Malfatti, S., Do, L., Land, M.L., Pelletier, D.A., Beatty, J.T., Lang, A.S. *et al*. (2004) Complete genome sequence of the metabolically versatile photosynthetic bacterium *Rhodopseudomonas palustris*. *Nature Biotechnology* 22, 55–61.

Lee, C. (1988) Amino acids and amine biogeochemistry in marine particulate material and sediments. In: Blackburn, T.H. and Sorensen, J. (eds) *Nitrogen Cycling in Coastal Marine Environments*. Wiley, Chicester, UK, pp. 125–141.

Letizia, C.S., Cocchiara, J., Lalko, J. and Api, A.M. (2003) Fragrance material review on linalyl phenylacetate. *Food and Chemical Toxicology* 41, 1017–1021.

Löffler, C., Kuntze, K., Vazquez, J.R., Rugor, A., Kung, J.W., Böttcher, A. and Boll, M. (2011) Occurrence, genes and expression of the W/Se-containing class II benzoyl-coenzyme A reductases in anaerobic bacteria. *Environmental Microbiology* 13, 696–709.

López Barragán, M.J., Carmona, M., Zamarro, M.T., Thiele, B., Boll, M., Fuchs, G., García, J.L. and Díaz, E. (2004) The *bzd* gene cluster, coding for anaerobic benzoate catabolism, in *Azoarcus* sp. strain CIB. *Journal of Bacteriology* 186, 5762–5774.

Lovley, D.R. and Phillips, E.J.P. (1986) Availability of ferric iron for microbial reduction in bottom sediments of the fresh-water tidal Potomac River. *Applied and Environmental Microbiology* 52, 751–757.

Luengo, J.M., García, J.L. and Olivera, E.R. (2001) The phenylacetyl-CoA catabolon: a complex catabolic unit with broad biotechnological applications. *Molecular Microbiology* 39, 1434–1442.

Mai, X.H. and Adams, M.W.W. (1994) Indolepyruvate ferredoxin oxidoreductase from the hyperthermophilic archaeon *Pyrococcus furiosus* – a new enzyme involved in peptide fermentation. *The Journal of Biological Chemistry* 269, 16726–16732.

Mardanov, A.V., Slododkina, G.B., Slobodkin, A.I., Beletsky, A.V., Gavrilov, S.N., Kublanov, I.V., Bonch-Osmolovskaya, E.A., Skryabin, K.G. and Ravin, N.V. (2015) The *Geoglobus acetivorans* genome: Fe(III) reduction, acetate utilization, autotrophic growth, and degradation of aromatic compounds in a hyperthermophilic archaeon. *Applied and Environmental Microbiology* 81, 1003–1012.

Miller, S.L. (1953) A production of amino acids under possible primitive earth conditions. *Science* 117, 528–529.

Mohamed, M.E.-S., Seyfried, B., Tschech, A. and Fuchs, G. (1993) Anaerobic oxidation of phenylacetate and 4-hydroxyphenylacetate to benzoyl-Coenzyme-A and CO_2 in denitrifying *Pseudomonas sp* - evidence for an α-oxidation mechanism. *Archives of Microbiology* 159, 563–573.

Parker, E.T., Cleaves, H.J., Dworkin, J.P., Glavin, D.P., Callahan, M., Aubrey, A., Lazcano, A. and Bada, J.L. (2011) Primordial synthesis of amines and amino acids in a 1958 Miller H_2S-rich spark discharge experiment. *Proceedings of the National Academy of Sciences of the United States of America* 108, 5526–5531.

Parthasarathy, A., Kahnt, J., Chowdhury, N.P. and Buckel, W. (2013) Phenylalanine catabolism in *Archaeoglobus fulgidus* VC-16. *Archives of Microbiology* 195, 781–797.

Peters, F., Rother, M. and Boll, M. (2004) Selenocysteine-containing proteins in anaerobic benzoate metabolism of *Desulfococcus multivorans*. *Journal of Bacteriology* 186, 2156–2163.

Rabus, R. and Widdel, F. (1995) Anaerobic degradation of ethylbenzene and other aromatic hydrocarbons by new denitrifying bacteria. *Archives of Microbiology* 163, 96–103.

Rabus, R., Kube, M., Heider, J., Beck, A., Heitmann, K., Widdel, F. and Reinhardt, R. (2005) The genome sequence of an anaerobic aromatic-degrading denitrifying bacterium, strain EbN1. *Archives of Microbiology* 183, 27–36.

Russell, W.R., Duncan, S.H., Scobbie, L., Duncan, G., Cantlay, L., Calder, A.G., Anderson, S.E. and Flint, H.J. (2013) Major phenylpropanoid-derived metabolites in the human gut can arise from microbial fermentation of protein. *Molecular Nutrition and Food Research* 57, 523–535.

Schmid, G., Rene, S.B. and Boll, M. (2015) Enzymes of the benzoyl-coenzyme A degradation pathway in the hyperthermophilic archaeon *Ferroglobus placidus*. *Environmental Microbiology* 17, 3289–3300.

Schneider, S., Mohamed, M.E. and Fuchs, G. (1997) Anaerobic metabolism of L-phenylalanine via benzoyl-CoA in the denitrifying bacterium *Thauera aromatica*. *Archives of Microbiology* 168, 310–320.

Schocke, L. and Schink, B. (1999) Energetics and biochemistry of fermentative benzoate degradation by *Syntrophus gentianae*. *Archives of Microbiology* 171, 331–337.

Slobodkina, G.B., Kolganova, T.V., Querellou, J., Bonch-Osmolovskaya, E.A. and Slobodkin, A.I. (2009) *Geoglobus acetivorans* sp. nov., an iron(III)-reducing archaeon from a deep-sea hydrothermal vent. *International Journal of Systematic and Evolutionary Microbiology* 59, 2880–2883.

Wischgoll, S., Heintz, D., Peters, F., Erxleben, A., Sarnighausen, E., Reski, R., Van Dorsselaer, A. and Boll, M. (2005) Gene clusters involved in anaerobic benzoate degradation of *Geobacter metallireducens*. *Molecular Microbiology* 58, 1238–1252.

Wöhlbrand, L., Jacob, J.H., Kube, M., Mussmann, M., Jarling, R., Beck, A., Amann, R., Wilkes, H., Reinhardt, R. and Rabus, R. (2013) Complete genome, catabolic sub-proteomes and key-metabolites of *Desulfobacula toluolica* Tol2, a marine, aromatic compound-degrading, sulfate-reducing bacterium. *Environmental Microbiology* 15, 1334–1355.

19 The Biosynthesis of Tryptophan

E.J. Parker*

Maurice Wilkins Centre for Molecular Biodiscovery, Biomolecular Interaction Centre and Department of Chemistry, University of Canterbury, Christchurch, New Zealand

19.1 Abstract

In microorganisms and plants, L-tryptophan is biosynthesized from chorismate by a series of seven enzyme-catalysed reactions. The enzymes pathway has attracted attention both as a source of therapeutic targets and because of its evolutionary details. This chapter describes the enzymatic steps responsible for the biosynthesis of tryptophan. The seven enzyme-catalysed reactions take chorismate, generated from the shikimate pathway, through to L-tryptophan. The details of many of these reactions are now known, and accompanied by the structures that are now available for the enzymes, which means that the features of the biosynthetic pathway are now well understood.

19.2 Introduction

L-Tryptophan (Trp), the most chemically complex and the least abundant of the 20 common proteinogenic amino acids, plays an important role in metabolism. In addition to its role as a component of proteins, Trp is the biosynthetic precursor to a large number of complex microbial natural products. Here, we review the biosynthetic pathway by which this important aromatic amino acid is synthesized. The pathway for Trp is only found in plants and microorganisms and thus it has attracted significant attention for the development of new antimicrobial therapies.

There has been a particular interest recently in this pathway in *Mycobacterium tuberculosis*, the pathogen responsible for tuberculosis (Xiong *et al.*, 2013). Auxotrophic mutants of *M. tuberculosis* containing gene knockouts within the tryptophan biosynthetic pathway have been shown to be unable to replicate in cultured murine macrophages, while also showing an attenuation of virulence in mice (Zhang *et al.*, 2013). These results indicate that tryptophan is not bioavailable within the murine macrophage.

The evolutionary relationships between the proteins responsible for Trp biosynthesis have also attracted considerable attention. Three out of the seven enzymes that comprise the pathway have a $(\beta\alpha)_8$-barrel structure, which is the most common fold among single-domain enzymes.

19.3 Overview of Tryptophan Biosynthesis

The series of seven enzyme-catalysed reactions of the tryptophan pathway have been identified and characterized in several organisms. The starting point for the biosynthesis of Trp is

*E-mail address: emily.parker@canterbury.ac.nz

chorismate, which is biosynthesized through the seven steps of the shikimate pathway; chorismate is also the precursor to a range of other aromatic metabolites, including the other aromatic amino acids phenylalanine and tyrosine (Fig. 19.1).

The first committed step in the biosynthesis of tryptophan from chorismate is catalysed by anthranilate synthase (Morollo and Eck, 2001; Tang *et al.*, 2001; Lin *et al.*, 2009). This step is a 1,4-nucleophilic substitution by ammonia followed by the elimination of pyruvate. The ammonia used in the reaction is generated *in situ* by a glutamine amidotransferase. Anthranilate synthase acts as the key control point in partitioning chorismate towards Trp biosynthesis and, as such, it is found to be allosterically inhibited by Trp.

The indole ring of tryptophan is formed by the next three reactions of the pathway (Fig. 19.2). A ribose moiety is linked to anthranilate, in a reaction catalysed by phosphoribosyl anthranilate synthase (Kim *et al.*, 2002; Schlee *et al.*, 2009). Following the isomerization of phosphoribosyl anthranilate, the indole ring is constructed by an intramolecular electrophilic substitution, in a reaction catalysed by indole glycerol phosphate synthase. This reaction chemistry is analogous to that observed in histidine biosynthesis, as is described below (Sterner *et al.*, 1995, 2001). The final steps in Trp production are catalysed by the tryptophan synthase complex (Raboni *et al.*, 2009). In these reactions, indole is firstly released and then transferred to serine to introduce the amino acid functionality.

19.4 Anthranilate Synthase

This enzyme is one of a group of chorismate-utilizing enzymes that are predicted to share

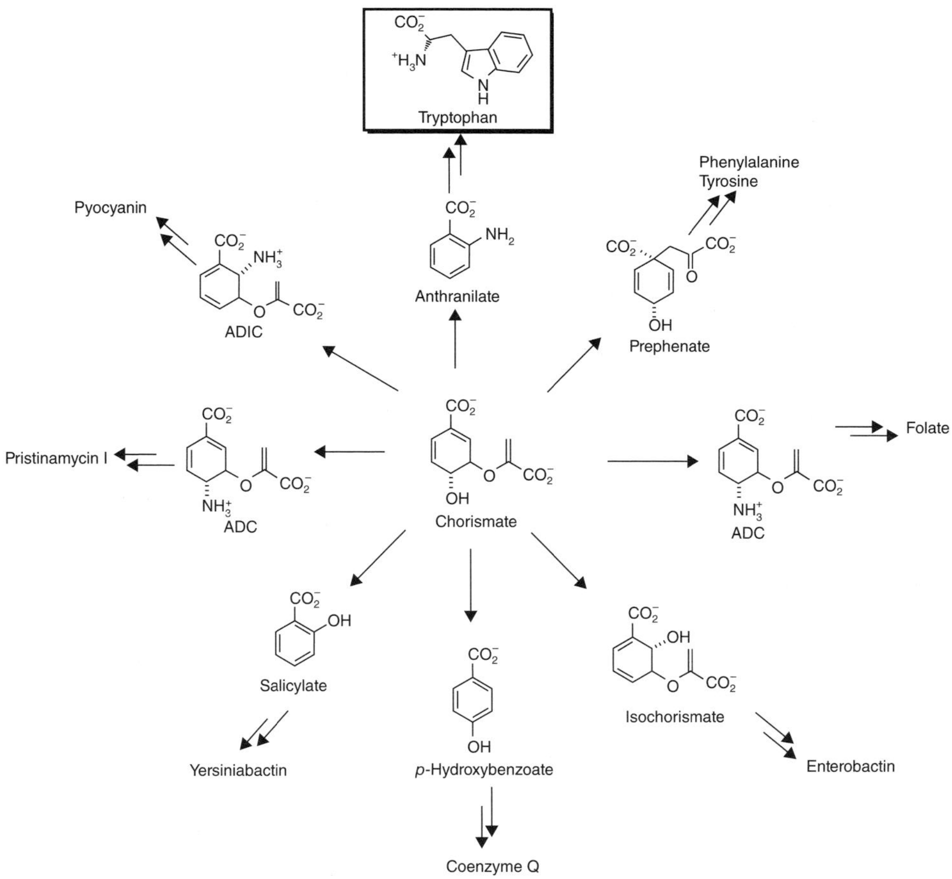

Fig. 19.1. The multiple end products generated from chorismate, which itself is derived from the shikimate pathway.

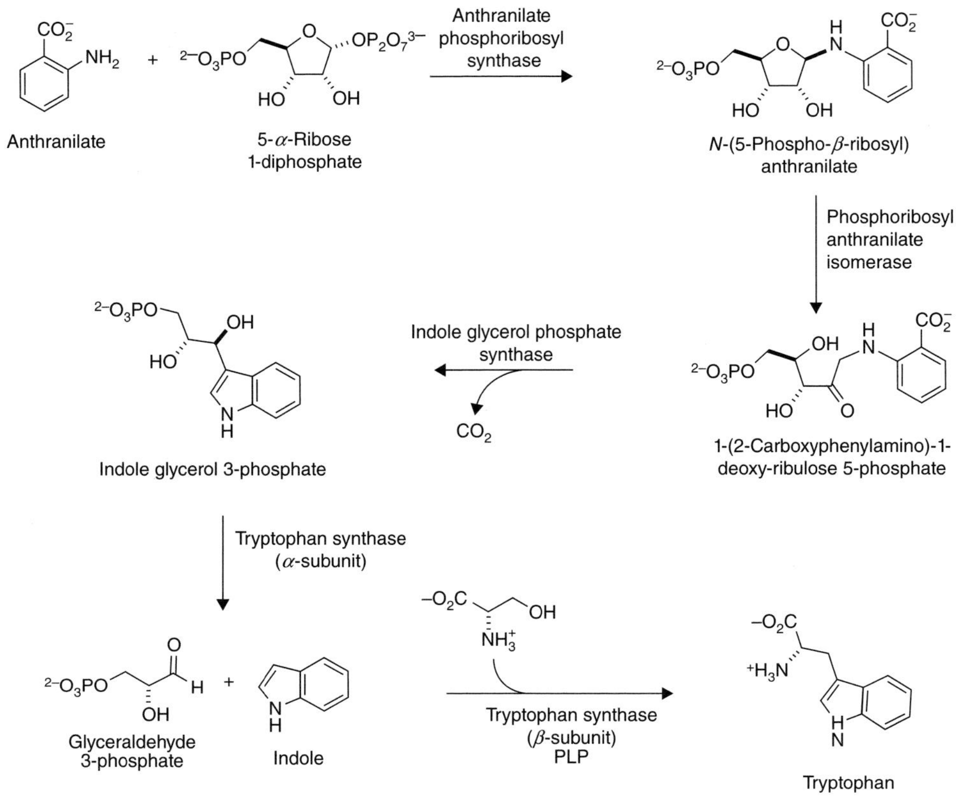

Fig. 19.2. Overview of the biosynthetic route to tryptophan from anthranilate.

many mechanistic features and are likely to have common evolutionary origins (Fig. 19.3) (Bulloch *et al.*, 2004; He *et al.*, 2004; He and Toney, 2006). Anthranilate synthase (AS) comprises two distinct functional components: AS-I (TrpE) and AS-II (TrpG). AS-I catalyses the production of anthranilate using both chorismate and ammonia as substrates through a 2-amino-2-deoxyisochorismate (ADIC) intermediate (Morollo *et al.*, 1993). The ammonia is provided by the amidotransferase activity of AS-II from glutamine (Mouilleron and Golinelli-Pimpaneau, 2007).

Significantly, different organizations of the subunits AS-I and AS-II have been observed from different bacteria, with hetero-oligomers (dimers or tetramers) or a single fused polypeptide having been reported (Romero *et al.*, 1995; Ashenafi *et al.*, 2015). Structures have been reported for the heterotetramers (2:2 ratio of AS-I and AS-II) from *Sulfolobus solfataricus* (Knochel *et al.*, 1999), *Salmonella typhimurium* and *Serratia marcescens* (Spraggon *et al.*, 2001), although the oligomeric associations observed for these proteins are different, with the enzymes from the mesophilic *S. typhimurium* and *S. marcescens* showing association through AS-I subunit, whereas the thermophilic *S. solfataricus* enzyme shows association through the AS-II subunit. The structure of the *M. tuberculosis* AS-I has recently been reported in complex with a competitive inhibitor (Bashiri *et al.*, 2015). This enzyme crystallized as a homodimer, even in the presence of its partner AS-II; the protein showed a significantly different dimer interface than has been observed for the *S. typhimurium* and *S. marcescens* AS-I interactions. The predicted binding site for the allosteric inhibitor Trp is found near the interface region (Fig. 19.4).

Fusion enzymes of AS-I and AS-II have been observed for some bacteria, including *Bacillus subtilis* (Patel *et al.*, 1974), *Acinetobacter calcoaceticus* (Ashenafi *et al.*, 2015) and *Pseudomonas putida* (Queener *et al.*, 1973). These enzymes function as dimers. More recently, the fused AS from *Streptomyces venezuelae* has been reported to be monomeric, and unlike other AS complexes,

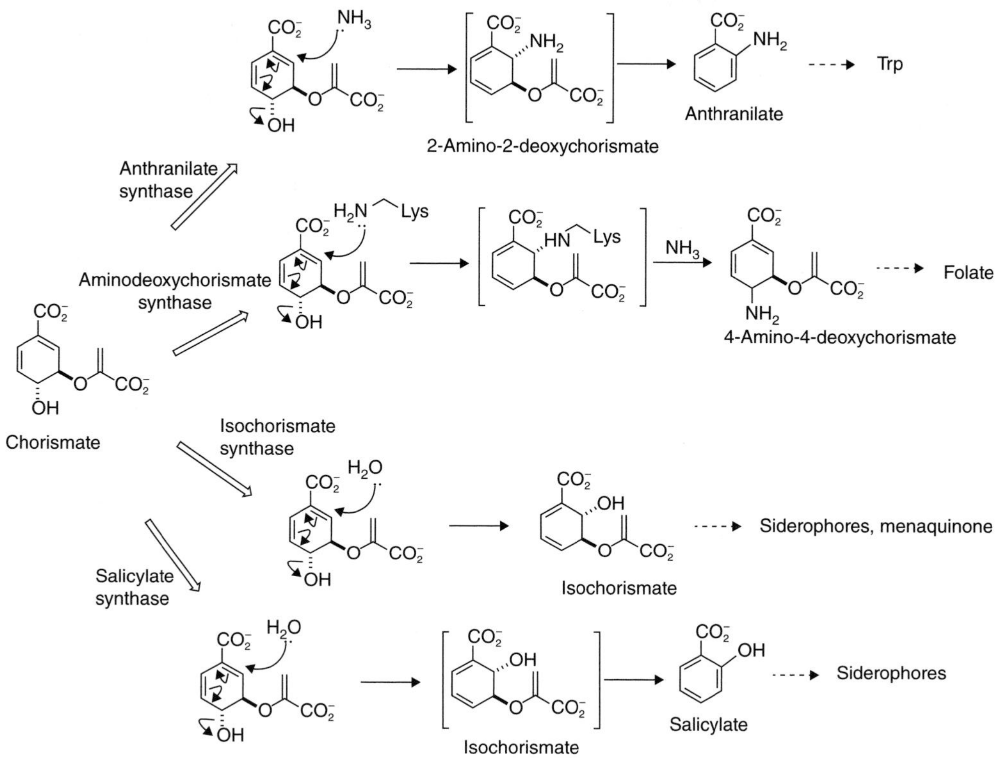

Fig. 19.3. Related enzymatic reactions using chorismate as a substrate.

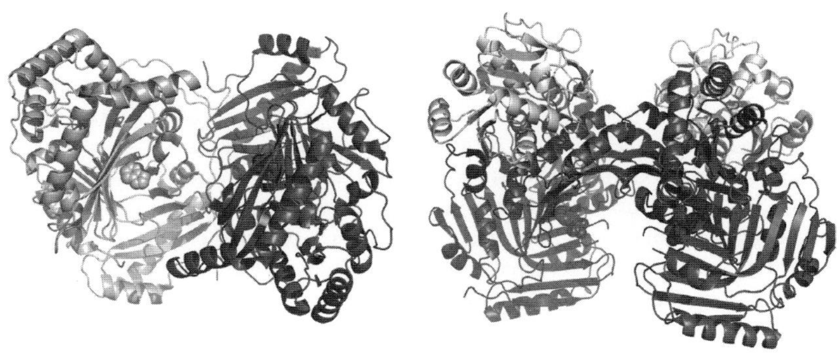

Fig. 19.4. Structures of (left) AS-I homodimer of anthranilate synthase (AS) from *Mycobacterium tuberculosis* (PDB code 5CWA) (Bashiri *et al.*, 2015) and (right) the AS-I:AS-II heterotetramer of AS from *Serratia marcescens* (PDB code 1I7S).

does not use ammonia as an alternative to glutamine (Ashenafi *et al.*, 2015).

As noted above, chorismate is also a substrate for other enzymes leading into other pathways. For AS activity, ammonia (generated from glutamine by the action of AS-II) acts nucleophilically to deliver first the intermediate ADIC. The analogous reaction in which water is a nucleophile is catalysed by isochorismate synthase (ICS) to give isochorismate. This reaction is also the first step of the salicylate synthase-catalysed reaction (Fig. 19.3). Chorismate-utilizing enzymes share a common fold and similarities in their reaction mechanisms suggest that they have originated from a common ancestor. Very recently, it was demonstrated that the nucleophilic specificity of the *S. typhimurium* AS could be extended to accommodate water rather than ammonia as a nucleophile by two residue substitutions, thereby giving a bifunctional AS/ICS (Plach *et al.*, 2015). Interestingly, these findings demonstrated that new activity could be straightforwardly established on an enzyme of primary metabolism. Similarly, an aminodeoxychorismate synthase (see Fig. 19.3), an enzyme that does not catalyse the elimination of pyruvate, has also been altered to give AS activity using Janus-predicted mutations (Culbertson *et al.*, 2015).

19.5 Anthranilate Phosphoribosyltransferase

Anthranilate phosphoribosyltransferase (AnPRT, also referred to as TrpD) catalyses the second step of Trp synthesis from chorismate, transferring a phosphoribosyl unit from phosphoribosyl pyrophosphate (PRPP) to anthranilate forming phosphoribosylanthranilate (PRA). AnPRT is a member of the family of phosphoribosyltransferase (PRT) enzymes (Musick and Nyhan, 1981; Sinha and Smith, 2001). These enzymes play roles in purine and nucleotide biosynthesis and salvage pathways as well as in amino acid metabolism. AnPRT represents a type III category enzyme and has a novel fold which distinguishes it from the type I, type II and type IV groups of PRT enzymes (Ivens *et al.*, 2001; Lee *et al.*, 2006). AnPRT monomers comprise a smaller *N*-terminal α-helical domain and a larger *C*-terminal α/β domain, which are joined by a hinge region (Fig. 19.5). This fold has a similar architecture to that of some pyrimidine nucleoside phosphorylases, which catalyse the near reverse reaction to that of AnPRT.

Structures of AnPRT are available from a number of microbial organisms, including *S. solfataricus* (Ivens *et al.*, 2001), *Pectobacterium carotovorum* (Kim *et al.*, 2002), *Acinetobacter bayliyi*, *Xanthomonas campestris* and *M. tuberculosis* (Ivens *et al.*, 2001; Kim *et al.*, 2002; Lee *et al.*, 2006; Marino *et al.*, 2006). All of the structures are homodimers with the chains interacting via the smaller *N*-terminal domain. An engineered monomeric form of the *S. solfataricus* enzyme has been generated with similar activity to that of the wild-type enzyme but lower thermostability (Schwab *et al.*, 2008).

The structures of alternative substrates and of various mutants of the *M. tuberculosis* AnPRT have shed light on an unusual channel that may play a role in the delivery of anthranilate to the active site (Castell *et al.*, 2013; Cookson *et al.*, 2014, 2015). Multiple anthranilate binding sites within this channel have been observed. This mechanism of delivery of anthranilate to the active site may account for the substrate inhibition by anthranilate that is observed for this enzyme. It has been proposed that the movement of anthranilate through this channel to the active site is facilitated by the conformational mobility of a key active site loop (Cookson *et al.*, 2015). The role of the channel may be to deliver anthranilate to PRPP following the binding of this reactive substrate to ensure that it is attached to anthranilate rather than water. The multiple anthranilate binding sites have also been investigated as possible binding sites for inhibitors of AnPRT (Evans *et al.*, 2014). Two anthranilate binding sites were also observed for the *S. solfataricus* enzyme (Ivens *et al.*, 2001).

19.6 Phosphoribosyl Anthranilate Isomerase

PRA isomerase (PRAI, TrpF) catalyses the third step in the biosynthesis of Trp in which PRA is rearranged to give 1-(2-carboxyphenylamino)-1-deoxy-ribulose 5-phosphate (CdRP). This Amadori rearrangement has attracted significant attention because of the relationship between PRAI and the analogous reaction in histidine biosynthesis (Henn-Sax *et al.*, 2002). The catalysis requires a number of proton transfer steps and leads to the enolamine as the first product (see Fig. 19.6). This enolamine is thought to tautomerize to give the ketoamine (CdRP), which is the substrate of the next enzyme of the pathway.

The structural organization of the PRAI enzyme differs widely in microorganisms. In *Eschericia coli*, TrpF is fused to the *C*-terminus of the next enzyme in the pathway (Wilmanns *et al.*, 1992). The two components of this bifunctional protein can be separately expressed and are fully

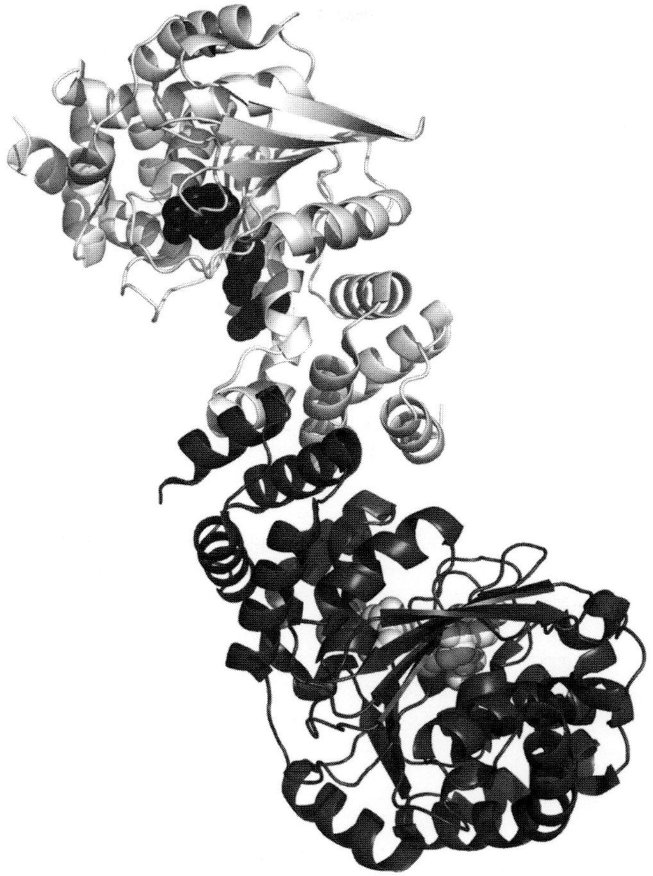

Fig. 19.5. The structure of anthranilate phosphoribosyltransferase (AnPRT) from *Sulfolobus solfataricus*. The two domains are shown in different shades; the substrate-binding sites are indicated by contrastingly shaded spheres within each domain.

active (Hommel *et al.*, 1995). The structure of TrpF has been determined and this protein is one of the three in the Trp biosynthetic pathway to have a $(\beta\alpha)_8$-barrel structure. In mesophilic organisms, PRA isomerase is monomeric, whereas in the hyperthermophile *Thermotoga maritima*, the enzyme is dimeric. The dimer could be disrupted by mutation to give a thermolabile monomer, suggesting that dimerization delivers thermostability (Thoma *et al.*, 2000).

An analogous Amadori rearrangement is catalysed by *N′*-[(5′-phosphoribosyl)formimino]-5-aminoimidazole-4-carboxamide-ribonucleotide (ProFAR) isomerase, which operates within the histidine biosynthetic pathway (Henn-Sax *et al.*, 2002). Structurally, this enzyme is similar to PRA isomerase, and the evolutionary relationships and substrate specificities of these enzymes have been studied in considerable detail (Leopoldseder *et al.*, 2004). It was noted that *trpF*, the gene encoding PRA isomerase, was missing in the genome of several members of the *Actinobacteria* phylum, and it was proposed that a bisubstrate-specific protein was able to operate in both Trp and histidine biosynthesis. The existence of a dual substrate isomerase, named PriA, has been confirmed in *Streptomyces coelicolor* and *M. tuberculosis* (Due *et al.*, 2011). The sequence and structural basis of the dual specificity has been investigated in considerable detail (List *et al.*, 2011).

19.7 Indole Glycerol Phosphate Synthase

Indole 3-glycerol phosphate (IGP) synthase catalyses the ring closure of CdRP to give indole 3-glycerol phosphate (Fig. 19.7). This transformation consists of a sequence of condensation, decarboxylation and dehydration steps, and the

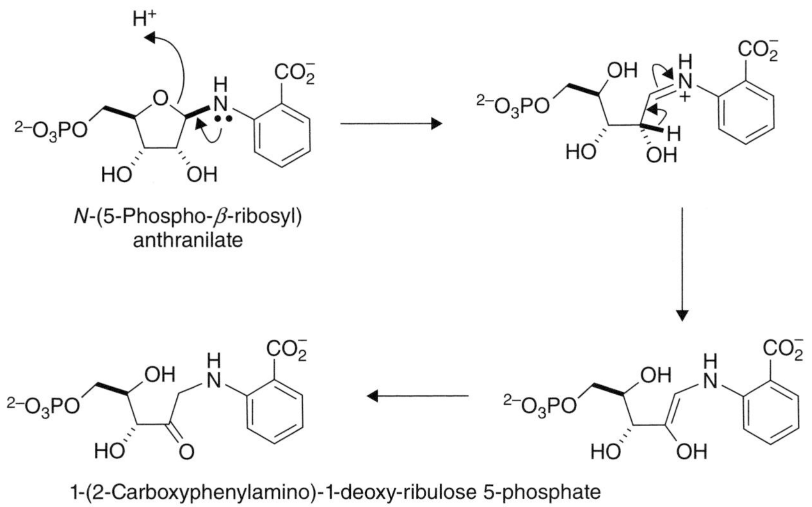

Fig. 19.6. The Amadori rearrangement catalysed by phosphoribosylanthranilate isomerase (PRAI), which converts phosphoribosylanthranilate (PRA) to 1-(2-carboxyphenylamino)-1-deoxy-ribulose 5-phosphate (CdRP).

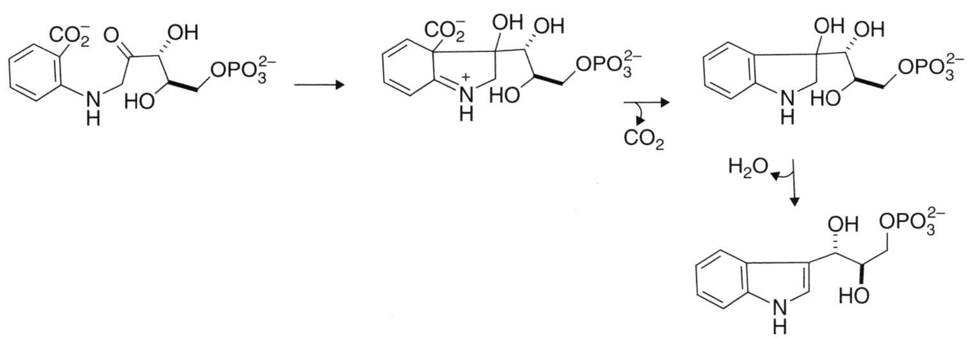

Fig. 19.7. The reaction catalysed by indole glycerol phosphate synthase, which catalyses the ring closure of 1-(2-carboxyphenylamino)-1-deoxy-ribulose 5-phosphate (CdRP) to give indole 3-glycerol phosphate.

kinetic mechanism of the sequence has been recently determined (Schlee *et al.*, 2013). The crystal structures of the monofunctional IGP synthase proteins have been determined from the hyperthermophiles *S. solfataricus* and *T. maritima*. In addition, the structure of the bifunctional PRA isomerase:IGP synthase has been determined from *E. coli* (Wilmanns *et al.*, 1992). The enzyme has a typical $(\beta\alpha)_8$ barrel structure and three highly conserved residues, Lys53, Lys110 and Glu159 (*S. solfataricus* numbering), have been identified as critical for catalysis. Molecular dynamics studies suggest that the mobility of active site loops is important, with both the $\beta 1 \alpha 1$ loop and the $\beta 2 \alpha 2$ loop playing important roles in protein function (Shen *et al.*, 2008; Schlee *et al.*, 2013; Zaccardi *et al.*, 2014). The temperature dependency of the reaction has also been examined in detail for the *S. solfataricus* enzyme, and has demonstrated that the rate-determining step changes as a function of temperature, and that even at adaptive temperatures, mesophilic and thermophilic IGP synthases have different overall rate profiles (Zaccardi *et al.*, 2012).

19.8 Tryptophan Synthase

This last enzyme in the tryptophan biosynthetic pathway has attracted a great deal of attention and been extensively reviewed as the one of the first enzymes to be discovered that catalyses two distinct reactions at independent active sites (in the α- and β-domains), which are connected via an internal tunnel (Leopoldseder *et al.*, 2006; Dunn, 2012). The enzyme was the first well-documented example of substrate channelling in amino acid biosynthesis. The overall reaction chemistry is relatively straightforward (see Fig. 19.8); glyceraldehyde 3-phosphate is eliminated in the α-Trp synthase reaction, producing indole, and this is channelled to the second active site. In the pyridoxal 5′-phosphate (PLP)-dependent β-reaction, serine undergoes the nucleophilic displacement of water by indole via an elimination–addition mechanism that is typical of the β-substitution reactions of PLP. The net result is the generation of Trp.

Tryptophan synthase is heterotetrameric with an $\alpha_2\beta_2$-configuration, and while the chemical transformations that generate first indole, and then the reaction to form Trp, take place in the active sites of the α- and β-units, respectively, it is clear that there is an intimate relationship between these two functions, with both substrate channelling between the reaction centres and complex allostery giving the overall functional response. Many of the details of the interactions between the active sites and their substrates and intermediates have been elucidated by a series of

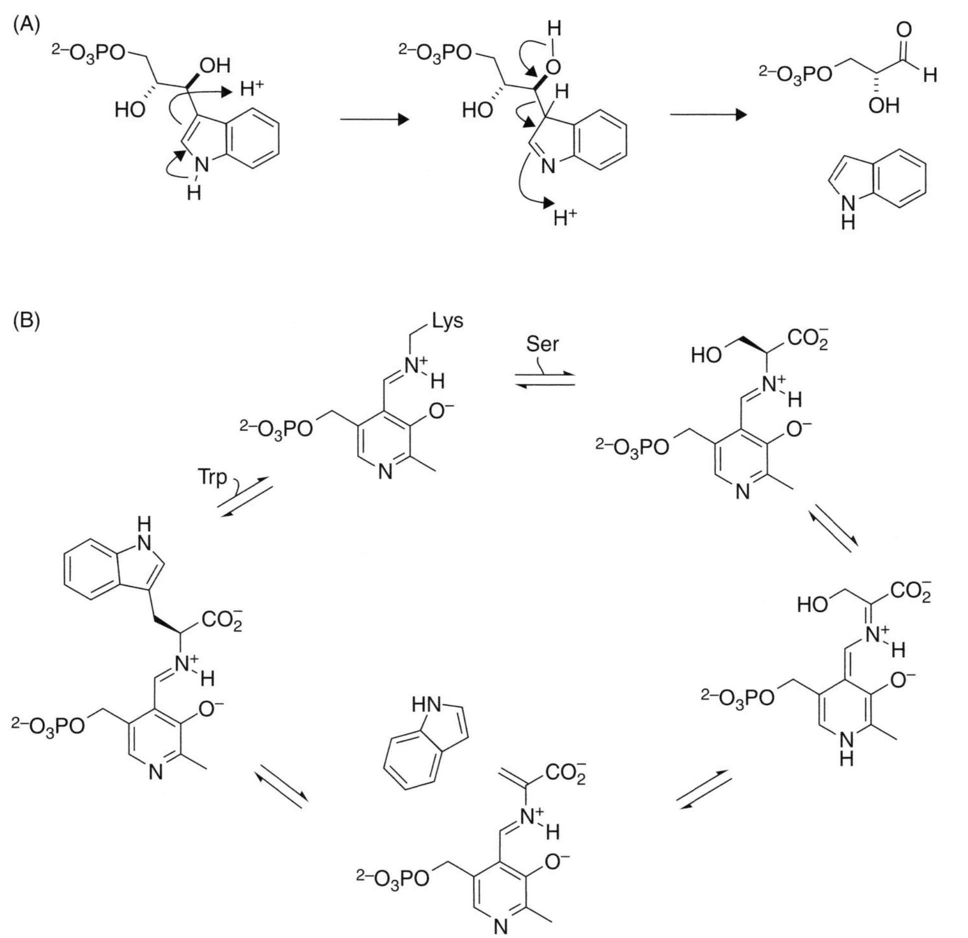

Fig. 19.8. (A) The generation of indole via a retro-aldol reaction catalysed by the α-subunit of tryptophan synthase; (B) The formation of tryptophan (Trp) from indole and serine (Ser) via a pyridoxal 5′-phosphate (PLP)-dependent reaction.

structures in complex with various analogues. Open and closed conformations have been shown and the whole complex has been shown to set up for the efficient utilization of the substrates IGP and serine though a system of allosteric interactions to switch on and off the activity of the α-site according the stage of catalysis of the β-site. This system also prevents escape of intermediate indole to bulk solvent by the channel interconnecting the two sites (Dunn, 2012).

The functional interdependence of the α- and β-units of tryptophan synthase is illustrated by their lower catalytic activity in the absence of their binding partner. Recently, the Arnold group carried out directed evolution studies to enhance the stand-alone function of the β-subunit. The mutations that enhanced activity of the β-subunit alone were found to reproduce the allosteric enhancement that interaction with the α-subunit provided (Buller *et al.*, 2015). Complexation with the β-subunit also alters the activity of the α-subunit, and the dynamic network connecting these phenomena has been exposed by NMR studies (Axe and Boehr, 2013).

19.9 Conclusions

Tryptophan is the largest of the 20 proteinogenic amino acids and the least abundant, both in terms of its frequency in protein sequences and its cellular abundance. It plays a role as a building block for chemical diversity and is incorporated into a number of diverse natural products in bacteria; it therefore undergoes significant modification, with a variety of different modifications observed (Alkhalaf and Ryan, 2015). Its role as both an important component of proteins and a substrate for numerous other products means that the route for tryptophan biosynthesis is of considerable importance in microorganisms. The close evolutionary relationships between some of the enzymes and the potential for their inhibition in the development of new therapeutics have meant that this pathway has attracted considerable interest (see Chapter 20).

References

Alkhalaf, L.M. and Ryan, K.S. (2015) Biosynthetic manipulation of tryptophan in bacteria: pathways and mechanisms. *Chemistry and Biology* 22, 317–328.

Ashenafi, M., Reddy, P.T., Parsons, J.F. and Byrnes, W.M. (2015) The fused anthranilate synthase from *Streptomyces venezuelae* functions as a monomer. *Molecular and Cellular Biochemistry* 400, 9–15.

Axe, J.M. and Boehr, D.D. (2013) Long-range interactions in the alpha subunit of tryptophan synthase help to coordinate ligand binding, catalysis, and substrate channeling. *Journal of Molecular Biology* 425, 1527–1545.

Bashiri, G., Johnston, J.M., Evans, G.L., Bulloch, E.M.M., Goldstone, D.C., Jirgis, E.N.M., Kleinboelting, S., Castell, A., Ramsay, R.J., Manos-Turvey, A. *et al.* (2015) Structure and inhibition of subunit I of the anthranilate synthase complex of Mycobacterium tuberculosis and expression of the active complex. *Acta Crystallographica Section D, Biological Crystallography* 71, 2297–2308.

Buller, A.R., Brinkmann-Chen, S., Romney, D.K., Herger, M., Murciano-Calles, J. and Arnold, F.H. (2015) Directed evolution of the tryptophan synthase β-subunit for stand-alone function recapitulates allosteric activation. *Proceedings of the National Academy of Sciences of the United States of America* 112, 14599–14604.

Bulloch, E.M.M., Jones, M.A., Parker, E.J., Osborne, A.P., Stephens, E., Davies, G.M., Coggins, J.R. and Abell, C. (2004) Identification of 4-amino-4-deoxychorismate synthase as the molecular target for the antimicrobial action of (6*S*)-6-fluoroshikimate. *Journal of the American Chemical Society* 126, 9912–9913.

Castell, A., Short, F.L., Evans, G.L., Cookson, T.V.M., Bulloch, E.M.M., Joseph, D.D.A., Lee, C.E., Parker, E.J., Baker, E.N. and Lott, J.S. (2013) Substrate capture mechanism of *Mycobacterium tuberculosis* anthranilate phosphoribosyltransferase provides a mode for inhibition. *Biochemistry* 52, 1776–1787.

Cookson, T.V.M., Castell, A., Bulloch, E.M.M., Evans, G.L., Short, F.L., Baker, E.N., Lott, J.S. and Parker, E.J. (2014) Alternative substrates reveal catalytic cycle and key binding events in the reaction catalysed by anthranilate phosphoribosyltransferase from *Mycobacterium tuberculosis*. *Biochemical Journal* 461, 87–98.

Cookson, T.V.M., Evans, G.L., Castell, A., Baker, E.N., Lott, J.S. and Parker, E.J. (2015) Structures of *Mycobacterium tuberculosis* anthranilate phosphoribosyltransferase variants reveal the conformational changes that facilitate delivery of the substrate to the active site. *Biochemistry* 54, 6082–6092.

Culbertson, J.E., Chung, D.H., Ziebart, K.T., Espiritu, E. and Toney, M.D. (2015) Conversion of aminodeoxychorismate synthase into anthranilate synthase with Janus mutations: mechanism of pyruvate elimination catalyzed by chorismate enzymes. *Biochemistry* 54, 2372–2384.

Due, A.V., Kuper, J., Geerlof, A., von Kries, J.P. and Wilmanns, M. (2011) Bisubstrate specificity in histidine/tryptophan biosynthesis isomerase from *Mycobacterium tuberculosis* by active site metamorphosis. *Proceedings of the National Academy of Sciences of the United States of America* 108, 3554–3559.

Dunn, M.F. (2012) Allosteric regulation of substrate channeling and catalysis in the tryptophan synthase bienzyme complex. *Archives of Biochemistry and Biophysics* 519, 154–166.

Evans, G.L., Gamage, S.A., Bulloch, E.M.M., Baker, E.N., Denny, W.A. and Lott, J.S. (2014) Repurposing the chemical scaffold of the anti-arthritic drug lobenzarit to target tryptophan biosynthesis in *Mycobacterium tuberculosis*. *ChemBioChem* 15, 852–864.

He, Z. and Toney, M.D. (2006) Direct detection and kinetic analysis of covalent intermediate formation in the 4-amino-4-deoxychorismate synthase catalyzed reaction. *Biochemistry* 45, 5019–5028.

He, Z., Lavoie, K.D.S., Bartlett, P.A. and Toney, M.D. (2004) Conservation of mechanism in three chorismate-utilizing enzymes. *Journal of the American Chemical Society* 126, 2378–2385.

Henn-Sax, M., Thoma, R., Schmidt, S., Hennig, M., Kirschner, K. and Sterner, R. (2002) Two $(\beta\alpha)_8$-barrel enzymes of histidine and tryptophan biosynthesis have similar reaction mechanisms and common strategies for protecting their labile substrates. *Biochemistry* 41, 12032–12042.

Hommel, U., Eberhard, M. and Kirschner, K. (1995) Phosphoribosyl anthranilate isomerase catalyzes a reversible Amadori reaction. *Biochemistry* 34, 5429–5439.

Ivens, A., Mayans, O., Szadkowski, H., Wilmanns, M. and Kirschner, K. (2001) Purification, characterization and crystallization of thermostable anthranilate phosphoribosyltransferase from *Sulfolobus solfataricus*. *European Journal of Biochemistry* 268, 2246–2252.

Kim, C., Xuong, N.-H., Edwards, S., Madhusudan, Yee, M.-C., Spraggon, G. and Mills, S.E. (2002) The crystal structure of anthranilate phosphoribosyltransferase from the enterobacterium *Pectobacterium carotovorum*. *FEBS Letters* 523, 239–246.

Knochel, T., Ivens, A., Hester, G., Gonzalez, A., Bauerle, R., Wilmanns, M., Kirschner, K. and Jansonius, J.N. (1999) The crystal structure of anthranilate synthase from *Sulfolobus solfataricus*: functional implications. *Proceedings of the National Academy of Sciences of the United States of America* 96, 9479–9484.

Lee, C.E., Goodfellow, C., Javid-Majd, F., Baker, E.N. and Lott, S.J. (2006) The crystal structure of TrpD, a metabolic enzyme essential for lung colonization by *Mycobacterium tuberculosis*, in complex with its substrate phosphoribosylpyrophosphate. *Journal of Molecular Biology* 355, 784–797.

Leopoldseder, S., Claren, J., Jurgens, C. and Sterner, R. (2004) Interconverting the catalytic activities of $(\beta\alpha)_8$-barrel enzymes from different metabolic pathways: sequence requirements and molecular analysis. *Journal of Molecular Biology* 337, 871–879.

Leopoldseder, S., Hettwer, S. and Sterner, R. (2006) Evolution of multi-enzyme complexes: the case of tryptophan synthase. *Biochemistry* 45, 14111–14119.

Lin, X., Xu, S., Yang, Y., Wu, J., Wang, H., Shen, H. and Wang, H. (2009) Purification and characterization of anthranilate synthase component I (TrpE) from *Mycobacterium tuberculosis* H37Rv. *Protein Expression and Purification* 64, 8–15.

List, F., Sterner, R. and Wilmanns, M. (2011) Related $(\beta\alpha)$-barrel proteins in histidine and tryptophan biosynthesis: a paradigm to study enzyme evolution. *ChemBioChem* 12, 1487–1494.

Marino, M., Deuss, M., Svergun, D.I., Konarev, P.V., Sterner, R. and Mayans, O. (2006) Structural and mutational analysis of substrate complexation by anthranilate phosphoribosyltransferase from *Sulfolobus solfataricus*. *The Journal of Biological Chemistry* 281, 21410–21421.

Morollo, A.A. and Eck, M.J. (2001) Structure of the cooperative allosteric anthranilate synthase from *Salmonella typhimurium*. *Nature Structural Biology* 8, 243–247.

Morollo, A.A., Finn, M.G. and Bauerle, R. (1993) Isolation and structure determination of 2-amino-2-deoxyisochorismate – an intermediate in the biosynthesis of anthranilate. *Journal of the American Chemical Society* 115, 816–817.

Mouilleron, S. and Golinelli-Pimpaneau, B. (2007) Conformational changes in ammonia-channeling glutamine amidotransferases. *Current Opinion in Structural Biology* 17, 653–664.

Musick, W.D.L. and Nyhan, W.L. (1981) Structural features of the phosphoribosyl-transferases and their relationship to the human deficiency disorders of purine and pyrimidine metabolism. *Critical Reviews in Biochemistry and Molecular Biology* 11, 1–34.

Patel, N., Holmes, W.M. and Kane, J.F. (1974) Homologous and hybrid complexes of anthranilate synthase from *Bacillus* species. *Journal of Bacteriology* 119, 220–227.

Plach, M.G., Loeffler, P., Merkl, R. and Sterner, R. (2015) Conversion of anthranilate synthase into isochorismate synthase: implications for the evolution of chorismate-utilizing enzymes. *Angewandte Chemie – International Edition* 54, 11270–11274.

Queener, S.W., Queener, S.F., Meeks, J.R. and Gunsalus, I.C. (1973) Anthranilate synthase from *Pseudomonas putida* – purification and properties of a 2-component enzyme. *The Journal of Biological Chemistry* 248, 151–161.

Raboni, S., Bettati, S. and Mozzarelli, A. (2009) Tryptophan synthase: a mine for enzymologists. *Cellular and Molecular Life Sciences* 66, 2391–2403.

Romero, R.M., Roberts, M.F. and Phillipson, J.D. (1995) Anthranilate synthase in microorganisms and plants. *Phytochemistry* 39, 263–276.

Schlee, S., Deuss, M., Bruning, M., Ivens, A., Schwab, T., Hellmann, N., Mayans, O. and Sterner, R. (2009) Activation of anthranilate phosphoribosyltransferase from *Sulfolobus solfataricus* by removal of magnesium inhibition and acceleration of product release. *Biochemistry* 48, 5199–5209.

Schlee, S., Dietrich, S., Kurcon, T., Delaney, P., Goodey, N.M. and Sterner, R. (2013) Kinetic mechanism of indole-3-glycerol phosphate synthase. *Biochemistry* 52, 132–142.

Schwab, T., Skegro, D., Mayans, O. and Sterner, R. (2008) A rationally designed monomeric variant of anthranilate phosphoribosyltransferase from *Sulfolobus solfataricus* is as active as the dimeric wild-type enzyme but less thermostable. *Journal of Molecular Biology* 376, 506–516.

Shen, H.B., Xu, F., Hu, H.R., Wang, F.F., Wu, Q., Huang, Q. and Wang, H.H. (2008) Coevolving residues of $(\beta\alpha)_8$-barrel proteins play roles in stabilizing active site architecture and coordinating protein dynamics. *Journal of Structural Biology* 164, 281–292.

Sinha, S.C. and Smith, J.L. (2001) The PRT protein family. *Current Opinion in Structural Biology* 11, 733–739.

Spraggon, G., Kim, C., Nguyen-Huu, X., Yee, M.C., Yanofsky, C. and Mills, S.E. (2001) The structures of anthranilate synthase of *Serratia marcescens* crystallized in the presence of (i) its substrates, chorismate and glutamine, and a product, glutamate, and (ii) its end-product inhibitor, L-tryptophan. *Proceedings of the National Academy of Sciences of the United States of America* 98, 6021–6026.

Sterner, R., Dahm, A., Darimont, B., Ivens, A., Liebl, W. and Kirschner, K. (1995) $(\beta\alpha)_8$-Barrel proteins of tryptophan biosynthesis in the hyperthermophile *Thermotoga maritima*. *The EMBO Journal* 14, 4395–4402.

Sterner, R., Merz, A., Thoma, R. and Kirschner, K. (2001) Phosphoribosylanthranilate isomerase and indoleglycerol-phosphate synthase: tryptophan biosynthetic enzymes from *Thermotoga maritima*. *Methods in Enzymology* 331, 270–280.

Tang, X.-F., Ezaki, S., Atomi, H. and Imanaka, T. (2001) Anthranilate synthase without an LLES motif from a hyperthermophilic archaeon is inhibited by tryptophan. *Biochemical and Biophysical Research Communications* 281, 858–865.

Thoma, R., Hennig, M., Sterner, R. and Kirschner, K. (2000) Structure and function of mutationally generated monomers of dimeric phosphoribosylanthranilate isomerase from *Thermotoga maritima*. *Structure* 8, 265–276.

Wilmanns, M., Priestle, J.P., Niermann, T. and Jansonius, J.N. (1992) 3-Dimensional structure of the bifunctional enzyme phosphoribosylanthranilate isomerase – indoleglycerolphosphate synthase from *Escherichia coli* refined at 2.0 Å resolution. *Journal of Molecular Biology* 223, 477–507.

Xiong, X., Xu, Z., Yang, Z., Liu, Y., Wang, D., Dong, M., Parker, E.J. and Zhu, W. (2013) Key targets and relevant inhibitors for the drug discovery of tuberculosis. *Current Drug Targets* 14, 676–699.

Zaccardi, M.J., Mannweiler, O. and Boehr, D.D. (2012) Differences in the catalytic mechanisms of mesophilic and thermophilic indole-3-glycerol phosphate synthase enzymes at their adaptive temperatures. *Biochemical and Biophysical Research Communications* 418, 324–329.

Zaccardi, M.J., O'Rourke, K.F., Yezdimer, E.M., Loggia, L.J., Woldt, S. and Boehr, D.D. (2014) Loop–loop interactions govern multiple steps in indole-3-glycerol phosphate synthase catalysis. *Protein Science* 23, 302–311.

Zhang, Y.J., Reddy, M.C., Ioerger, T.R., Rothchild, A.C., Dartois, V., Schuster, B.M., Trauner, A., Wallis, D., Galaviz, S., Huttenhower, C. *et al.* (2013) Tryptophan biosynthesis protects Mycobacteria from CD4 T-cell-mediated killing. *Cell* 155, 1296–1308.

20 Tryptophan Biosynthesis in Bacteria: Drug Targets and Immunology

J.S. Lott*

School of Biological Sciences, University of Auckland, Auckland, New Zealand

20.1 Abstract

Tryptophan is an energetically expensive amino acid to manufacture, which makes it an important point of contest between pathogenic bacteria and their hosts. Further, the mammalian innate immune response depletes the amount of tryptophan available for uptake by bacteria. This is an effective mammalian response against natural auxotrophs but, potentially, it could also be exploited as a novel therapeutic strategy against bacteria that are prototrophic for tryptophan synthesis.

20.2 Introduction

In many bacteria, tryptophan (Trp) is synthesised from chorismate, as is described in more detail in Chapter 19, and outlined briefly here in Section 20.3. In contrast, a number of pathogenic microorganisms, including *Toxoplasma gondii*, *Chlamydia* species and *Leishmania* parasites are auxotrophic for tryptophan production. This biosynthetic deficit is exploited by the mammalian immune system, which uses tryptophan starvation, mediated by the tryptophan-degrading enzyme indoleamine 2,3-dioxygenase (IDO), as a defence mechanism against intracellular pathogens. This chapter will describe how this innate immune response could potentially be utilized as a new therapeutic approach against other pathogenic bacteria that are naturally competent for tryptophan biosynthesis, i.e. prototrophs, such as *Mycobacterium tuberculosis.*

20.3 The Tryptophan Biosynthetic Pathway

The tryptophan biosynthetic pathway (Fig. 20.1) in most bacteria comprises six catalytic steps, beginning with chorismate. The seven enzymes responsible and the genes that encode them are usually represented by the letters A–G (hence TrpA–TrpG), as originally assigned based on the gene order in *Escherichia coli* rather than on their place in the reaction pathway (Crawford, 1975, 1989). The first committed step in the biosynthesis is the conversion of chorismate to anthranilate, which is catalysed by the enzyme anthranilate synthase (AS). AS is usually a heterodimeric enzyme, and is composed of two functional components: AS-I (TrpE), which catalyses the production of anthranilate from chorismate and ammonia, and AS-II (TrpG), a glutamine amidotransferase (GAT), which provides the required ammonia by converting glutamine

*E-mail address: s.lott@auckland.ac.nz

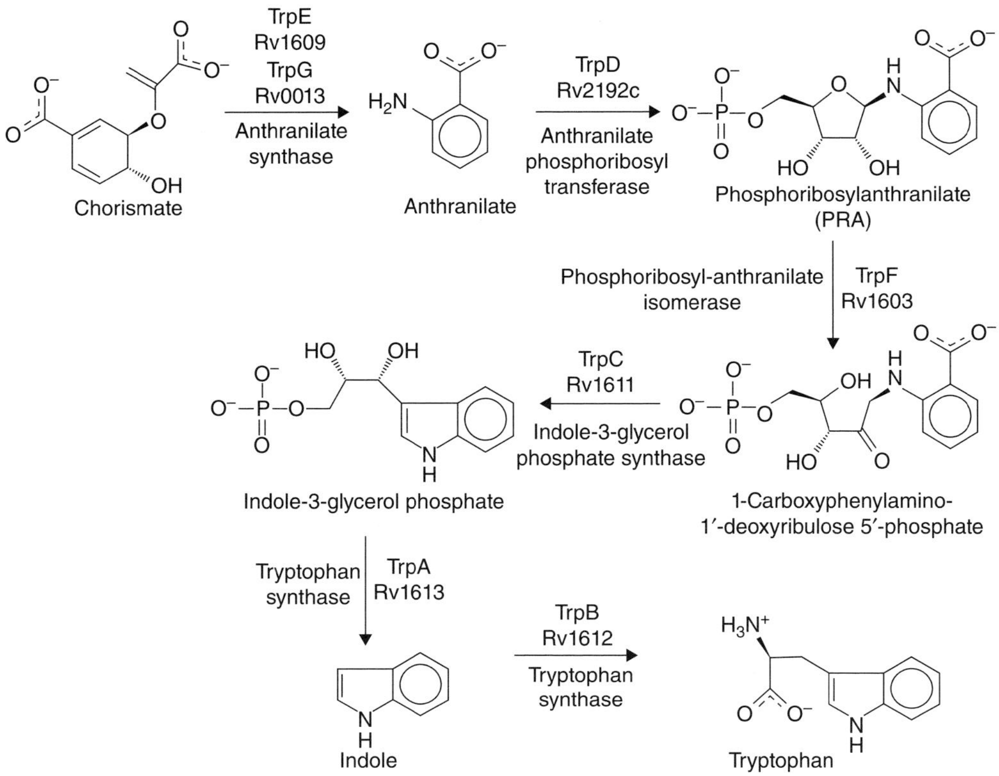

Fig. 20.1. The tryptophan biosynthetic pathway, indicating the compounds and enzymes involved, as further described in the text. The names of the enzyme proteins (TrpA–G) and the open reading frames that encode them in *M. Tuberculosis* (given as Rv numbers) are also shown.

to glutamate. Anthranilate phosphoribosyl transferase (AnPRT; TrpD) then transfers a 5′-phosphoribose unit from phosphoribosyl pyrophosphate (PRPP) on to the amino group of anthranilate to produce phosphoribosylanthranilate (PRA). The ribose ring of PRA is opened by PRA isomerase (PRAI; TrpF) to produce the isomer 1-carboxyphenylamino-1′-deoxyribulose-5′-phosphate. A ring closure reaction catalysed by indole-3-glycerol phosphate synthase (IGPS; TrpC) forms the characteristic indole heterocyclic ring system. Finally, the heterotetrameric enzyme tryptophan synthase (TrpAB) cleaves the indole ring from the glycerol phosphate backbone and forms the complete amino acid by condensation of the indole moiety with serine. Allosteric regulation of the pathway is provided by feedback inhibition by tryptophan, which binds to a regulatory site in the TrpE protein (Morollo and Eck, 2001; Spraggon *et al.*, 2001). In some genomes, the genes encoding adjacent steps in the pathway are found fused into single open reading frames; for example, in *Salmonella typhimurium*, *trpE* and *trpG* are fused, and in *E. coli*, *trpD* and *trpG* are fused, as are *trpC* and *trpF*.

20.4 Tryptophan Depletion is a Response of the Innate Immune System

The first evidence of tryptophan degradation being a strategy employed by the immune system was the observation that treatment of cultured human fibroblasts with interferon-γ (IFN-γ) blocked the growth of the obligate intracellular protozoal parasite *Toxoplasma gondii* (Pfefferkorn and Guyre, 1984). The action of IFN-γ was independent of whether or not the fibroblasts were themselves infected, but its effectiveness was dependent on the amount of tryptophan available: a higher concentration of available tryptophan required a greater dose of IFN-γ for effective growth inhibition (Pfefferkorn, 1984). Subsequent work showed that the IFN-γ-induced depletion of tryptophan was mediated by the induction of a specific isoform of the mammalian enzyme indoleamine 2,3-dioxygenase (IDO-1). IDO-1 converts tryptophan to *N*-formylkynurenine, and this is then further metabolized to kynurenine by the enzyme *N*-formylkynurenine formamidase (Fig. 20.2); the latter enzyme is not induced by IFN-γ (Pfefferkorn *et al.*, 1986).

Other intracellular protozoal parasites such as *Leishmania donovani* were subsequently shown to be susceptible to IDO-1-mediated tryptophan starvation (Murray *et al.*, 1989). When strongly induced by IFN-γ, IDO-1 is capable of depleting almost all of the available tryptophan in lung tissue (Fujigaki *et al.*, 2002). This observation is in concordance with the observation that tryptophan biosynthesis in *Francisella tularensis* is required for bacterial growth and survival in a lung-specific fashion. *F. tularensis* infection induces lung-specific IDO-1 expression, and a knockout of IDO-1 in the host permits partial rescue of auxotrophic bacteria (Peng and Monack, 2010). Furthermore, tryptophan starvation is effective against intracellular bacteria such as *Chlamydophila psittaci* (Byrne *et al.*, 1986; Murray *et al.*, 1989), *Chlamydia trachomatis* (Thomas *et al.*, 1993) and *Streptococcus agalactiae* (MacKenzie *et al.*, 1998), which are natural auxotrophs for tryptophan biosynthesis. Kynurenines are also potent negative regulators of inflammation and T cell activity, and dendritic cells expressing IDO-1 can inhibit the proliferation of active T cells due to both tryptophan depletion and the accumulation of kynurenine, and induce the expansion of regulatory T cells, so resulting in an immune tolerance that may allow chronic bacterial infection (Mellor, 2005; Zelante *et al.*, 2009).

20.5 *Mycobacterium tuberculosis* is a Globally Significant Human Pathogen

Tuberculosis (TB) remains one of the most significant causes of death worldwide, ranking alongside the human immunodeficiency virus (HIV) as the deadliest communicable disease. The World Health Organization (WHO) estimates that around 1.5 million people died from TB in 2014 (WHO, 2015). Drug-sensitive forms of the disease are curable with the currently available four-drug therapy (isoniazid, rifampicin, pyrazinamide and ethambutol) and the mortality rate has been slowly declining in recent years. For instance, it is estimated that TB treatment has saved 35 million lives since the turn of the millennium. However, there is substantial concern that the emergence of multidrug-resistant (MDR – defined as resistant to isoniazid and rifampicin) and extensively drug-resistant (XDR – defined as MDR-TB plus resistance to at least one fluoroquinolone and a second-line injectable antibiotic) strains of *Mycobacterium tuberculosis*, the causative agent of the disease, may reverse this progress. Almost half a million people developed MDR-TB in 2014, and over 46,000 people developed XDR-TB. The success rate for treatment of MDR-TB is less than 50%, and for XDR-TB it is only 26%, leading to an estimated total death toll of 190,000 from the two types of the disease in 2014. There is then a pressing global need for the development of new anti-TB therapies, especially antibiotics that function via new modes of action, as these will be effective against infections that are resistant to current drugs (Ma *et al.*, 2010).

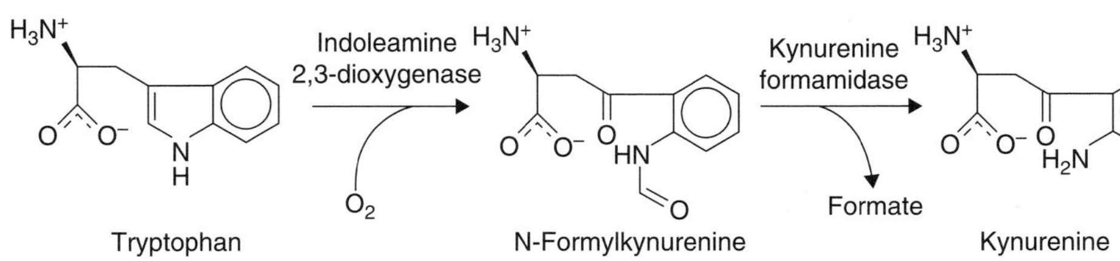

Fig. 20.2. The inducible mammalian tryptophan breakdown pathway, indicating the compounds and enzymes involved, as further described in the text.

20.6 The Tryptophan Biosynthetic Pathway in *Mycobacterium tuberculosis*

The complete genome sequence of *M. tuberculosis* was published in 1998 (Cole *et al.*, 1998), thus enabling the open reading frames (ORFs) that encode the enzymes of the tryptophan biosynthetic pathway to be identified by sequence comparison. However, not all of the enzymes in the pathway could be unambiguously deciphered from sequence information alone and, unlike the situation in *E. coli*, the ORFs are not arranged in a single operon. Consequently, the subsequent biochemical confirmation of which ORFs encode the pathway has only recently been completed.

Two ORFs (Rv1609 and Rv2386c) were initially identified as candidate TrpE genes encoding the major subunit of anthranilate synthase, AS-I. (The ORFs in *M. tuberculosis* are designated by an 'Rv' number because the first sequenced genome was from the virulent H37Rv laboratory strain of this bacterium. Rv numbers suffixed with 'c' indicate that the ORF is encoded on the complementary strand of the genome, relative to the designated start point for ORF numbering, which is the gene *dnaA*.) Biochemical and structural analysis showed that Rv2386c in fact encodes a chorismate-utilizing enzyme that is similar to AS-I in both sequence and active site chemistry; this is a salicylate synthase, MbtI, which catalyses the first committed step of mycobactin synthesis (Harrison *et al.*, 2006). Rv1609 was therefore confirmed as the true TrpE. The identity of TrpG has proven more elusive. The genome encodes four potential GATs (Rv0013, Rv0788, Rv1602 and Rv2604c), but only Rv0013 forms a functional AS complex with TrpE (Bashiri *et al.*, 2015), though the interaction between the two subunits is weak, and it was not possible to crystallize the complex. This implies that Rv0013 may also provide GAT activity to other biosynthetic pathways, as is the case in *Bacillus subtilis*, where the same GAT functions in the synthesis of both tryptophan and *p*-aminobenzoate synthesis (Slock *et al.*, 1990).

Complementation studies have shown that the ORF Rv1603 shows both PRAI (TrpF) activity and *N*′-[(5′-phosphoribosyl)formimino]-5-aminoimidazole-4-carboxamide ribonucleotide (ProFAR) isomerase (HisA) activity, the latter being a related chemical transformation that is a key catalytic step in histidine biosynthesis (Barona-Gómez and Hodgson, 2003). Specifically, both reactions are Amadori rearrangements of amino-aldoses to their respective amino-ketoses. Hence, Rv1603 is thought to be bifunctional and to function as both TrpF and HisA in the respective biosynthetic pathways. The predicted TrpB (Rv1612), TrpC (Rv1611) and TrpD (Rv2192c) proteins have all been heterologously expressed and characterized, and demonstrate the expected biochemical activities *in vitro* (Lee *et al.*, 2006; Czekster *et al.*, 2009; Shen *et al.*, 2009b).

M. tuberculosis was the subject of a consortium approach to the large-scale determination of protein structures – the TB Structural Genomics Consortium (TBSGC) – which resulted in a large number of proteins from the bacterium being structurally analysed in a concerted fashion (Murillo *et al.*, 2007; Chim *et al.*, 2011). As part of this effort, a number of proteins in the *M. tuberculosis* tryptophan biosynthesis pathway have been structurally characterized. TrpE (PDBID: 5CWA) and TrpB (PDBID: 2O2E and 2O2J) have been determined as ligand-free structures. The structure of TrpC has been determined both in ligand-free form (PDBID: 3QJA and 4FB7) and in complex with some potential inhibitors. The structure of TrpD has been determined in *apo* form (PDBID: 2BPQ and 1ZVW), in complex with natural substrates (PDBID: 3QR9), and in complex with numerous inhibitors and alternate substrates.

20.7 Tryptophan Biosynthesis is Essential for Host Colonization by *M. tuberculosis*

As described in the previous section, *M. tuberculosis* is fully competent for tryptophan biosynthesis, a trait that it shares with many pathogenic bacteria. None the less, its critical dependence on this biosynthetic capability during infection was revealed when knockout strains of *M. tuberculosis* H37Rv auxotrophic for proline or tryptophan were investigated for their potential as vaccine candidates (Smith *et al.*, 2001). An *M. tuberculosis* strain with a defined mutation in the *trpD* gene, which effectively disrupted tryptophan biosynthesis, failed to cause disease in mice. It appears then that tryptophan biosynthesis is required for TB disease formation. The essentiality of the biosynthetic pathways for some other amino acids, for example leucine (Bange *et al.*, 1996; Hondalus *et al.*, 2000), lysine (Pavelka *et al.*, 2003) and methionine (Berney *et al.*, 2015), for *in vivo* growth and disease formation by *M. tuberculosis* has also been demonstrated. This implies that the bacterium has to survive in niches with limited concentrations of amino acids, presumably including the phagosomes of macrophages in the lungs, during the establishment of an infection.

The tryptophan auxotrophic strain of *M. tuberculosis* described by Smith *et al.* (2001) was strikingly avirulent, even in severe, combined immune-deficient (SCID) mice, which lack a functional adaptive immune system. Not only did the mice survive infection, they also cleared the *trpD* mutant bacteria to below detectable levels in the lung. This implies that if a suitable inhibitor of tryptophan biosynthesis can be identified, it may act as an effective antibiotic, at least during the initiation of infection. Nevertheless, it remains unproven from these data whether an established infection by biosynthetically intact bacteria could be effectively killed via this approach. The possibility exists that even though tryptophan biosynthesis is required to establish infection, it may not be critical to the maintenance of an existing infection.

20.8 The Role of Indoleamine 2,3-Dioxygenase 1 (IDO-1) in *M. tuberculosis* Infection

Increased IDO-1 activity and decreased tryptophan concentrations have been found both in serum (Suzuki *et al.*, 2012) and pleural fluid (Suzuki *et al.*, 2013) from the lungs of patients with pulmonary TB, suggesting that infection with *M. tuberculosis* also triggers this innate immune response, possibly in a lung-specific fashion. TB patients also showed significant increases in serum kynurenine concentration compared with control subjects. This implies that if the bacterial tryptophan biosynthetic pathway were to be inhibited, the increased activity of IDO-1 will contribute to *M. tuberculosis* cell death through tryptophan starvation.

However, non-survivors of TB infection also had significantly higher kynurenine concentrations and significantly lower tryptophan concentrations than survivors, indicating that elevated IDO-1 activity was actually a predictor of death from *M. tuberculosis* infection. A caveat on these studies is that they are drawn from a high-mortality cohort of older patients with a high incidence of other co-morbidities; none the less, it appears possible that the immunomodulatory effects of kynurenine actually serve to increase the severity of pulmonary infection, at least in these cases.

20.9 Inhibitors of Tryptophan Biosynthesis in *M. tuberculosis*

One important point in favour of the choice of tryptophan biosynthesis as a potential anti-tuberculosis therapeutic target is that humans do not have a tryptophan biosynthetic pathway, thus reducing the chance that any inhibitors of this pathway will have negative effects on the human host. A number of inhibitors targeted at various enzymes in the pathway have been described, but while there are promising leads, none has yet been described that constitute viable therapeutics.

Inhibitors of plant tryptophan synthase that were designed as potential herbicides, with sub-micromolar potency (Sachpatzidis *et al.*, 1999) have been predicted, but not tested, to be functional against *M. tuberculosis* TrpAB (Dias *et al.*, 2006). A sub-micromolar inhibitor of *M. tuberculosis* TrpC has been identified by library screening. This has whole-cell activity against the bacterium *in vitro* (Shen *et al.*, 2009a), although its antibacterial activity was not reduced in the presence of exogenous tryptophan, indicating that its mode of cell killing may be through a mechanism other than tryptophan starvation. TrpE is a chorismate-utilizing enzyme, a class of enzymes for which many inhibitors have been designed, though almost none of these have achieved better than micromolar potency against their targets. It has also been found that inhibitors designed against one family member often function against other family members (Kozlowski *et al.*, 1995). Consistent with this, isochorismate-like inhibitors designed to act against the salicylate synthase MbtI (Manos-Turvey *et al.*, 2010, 2012; Chi *et al.*, 2012) were also active against TrpE, albeit with reduced potency (Bashiri *et al.*, 2015). There was also evidence of species specificity, as the best inhibitor of the series was more effective against TrpE from *Serratia marcescens* than against either enzyme from *M. tuberculosis*.

A series of inhibitors of *M. tuberculosis* TrpD have been identified by targeted library screening combined with structure-guided design and synthesis (Castell *et al.*, 2013; Cookson *et al.*, 2014; Evans *et al.*, 2014; Cookson *et al.*, 2015). TrpD has a surprisingly complex substrate-binding pattern, with PRPP binding in the base of the active site cleft, and two molecules of anthranilate binding nearer the mouth of the cleft, one of which is in a catalytically productive orientation, with the second thought to have a role in the substrate capture mechanism of the enzyme (Lee *et al.*, 2006; Castell *et al.*, 2013). The most effective of the designed inhibitors exploit the ability of the enzyme to bind anthranilate in multiple places in the active site cleft, but are still only micromolar inhibitors of the enzyme with modest activity against mycobacterial cells *in vitro* (Evans *et al.*, 2014).

The most promising evidence that tryptophan biosynthesis is a viable *in vivo* target for therapeutic compounds comes from the work of Zhang *et al.* (2013), which used transposon insertion mutagenesis to identify that an intact tryptophan biosynthetic pathway is necessary for bacterial survival in immunocompetent mice, but not in mice lacking $CD4^+$ T cells. These authors subsequently showed that TrpE knockout bacteria were susceptible to tryptophan starvation both in a defined medium *in vitro* and by IDO-mediated tryptophan depletion in IFN-γ stimulated macrophages. Most strikingly, they showed that two derivatized anthranilates, fluorinated at the 5- or 6- positions on the ring, are effective antibacterial agents, both in axenic culture and, with more modest effect, when administered intravenously in a mouse model of TB disease. The genome of an escape mutant showed that it had a mutation in the *trpE* gene that encodes a hypermorphic version of the enzyme that has lost its usual allosteric feedback inhibition (Bashiri *et al.*, 2015), so confirming tryptophan biosynthesis as the cellular target.

Nevertheless, it was found that none of the four possible fluoroanthranilates tested acted as inhibitors, but as alternate substrates for TrpD *in vitro*, producing fluorinated versions of PRA, which are more stable than the parental compound, and are able to act as substrates for *E. coli* TrpFC *in vitro*, resulting in the production of fluoro-indole-3-glycerol phosphate (Cookson *et al.*, 2014). It is unknown whether fluoro-indole-3-glycerol phosphate is an effective substrate for TrpA. Zhang *et al.* (2013) were unable to directly assay whether the fluoroanthranilates were direct inhibitors of TrpE, so it remains possible that rather than acting as inhibitors of the pathway, they are instead incorporated to produce fluorinated tryptophan: 5-fluoroanthranilate would produce 5-fluorotryptophan, and 6-fluoroanthranilate would produce 4-fluorotryptophan (Fig. 20.3). Therefore, two possible mechanisms for the antibacterial action of the fluoroanthranilates can be suggested: the inhibition of

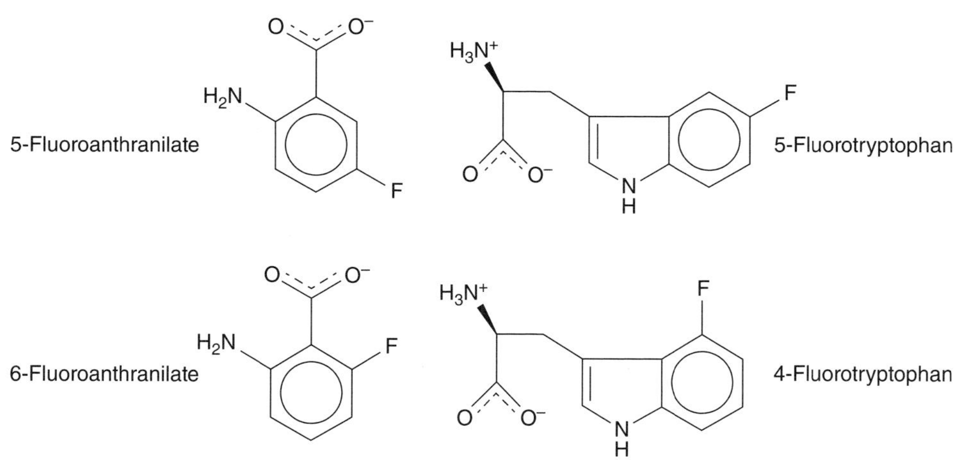

Fig. 20.3. Fluoroanthranilates and their cognate fluorotryptophans.

tryptophan biosynthesis downstream of TrpD by a fluorinated intermediate, leading to tryptophan starvation; or the production of fluorinated tryptophan. Tryptophan fluorinated at the 4-, 5- or 6-position has generally cytotoxic cellular effects in *E. coli* due to its incorporation into proteins (Browne *et al.*, 1970).

20.10 Conclusions and Future Prospects

The tryptophan biosynthetic pathway in bacteria has been well explored from a biochemical point of view, but has not generally been considered to be a valid target for antibiotic development, presumably because most bacteria are able to import tryptophan from their environment. However, even bacteria that are competent to synthesize tryptophan often obtain it from their environment when it is available there (Mellor, 2005), presumably because tryptophan is the most energy-consuming of the amino acids to synthesize. This gives a hint about the importance of tryptophan to pathogenic bacteria, which are often balancing nutritional needs with immune evasion or immune modulation during the course of an infection. Most antibiotics are effective against bacteria grown *in vitro*, but compounds that rely on some special feature of the host environment, and target processes that only become critical for bacterial survival when the bacteria are in the host, should effectively synergize with the immune response, thereby resulting in substantial antibacterial efficacy.

The depletion of available tryptophan by induced IDO-1 is an innate immune reaction that highlights the special situation of tryptophan, though the modulation of T cell activity by kynurenine shows that this immunomodulation can be a double-edged sword. In the case of mycobacterial infection, there is evidence to suggest that IDO-1 induction may synergize with conventional antibiotics (Hayashi *et al.*, 2001) but, conversely, that strong IDO-1 induction is a poor prognosis for the eventual outcome of the infection (Suzuki *et al.*, 2012). The fluoroanthranilates are the best candidate compounds identified to date that target tryptophan biosynthesis, but more work is required to understand their mode of action; they may induce tryptophan starvation, or they may actually function as pro-drugs for the *in situ* production of fluorotryptophan.

The identification of conditionally dependent targets for new therapeutics, such as tryptophan biosynthesis, represents a conceptual shift toward a holistic approach to drug discovery which favours antibiotics that function by synergizing with the host response to the bacterial infection. It is to be hoped that the pursuit of such approaches will yield new drugs that act via entirely new modes of action, and which may therefore help to address the pressing issue of antibiotic resistance.

References

Bange, F.C., Brown, A.M. and Jacobs, W.R. Jr (1996) Leucine auxotrophy restricts growth of *Mycobacterium bovis* BCG in macrophages. *Infection and Immunity* 64, 1794–1799.

Barona-Gómez, F. and Hodgson, D.A. (2003) Occurrence of a putative ancient-like isomerase involved in histidine and tryptophan biosynthesis. *EMBO Reports* 4, 296–300.

Bashiri, G., Johnston, J.M., Evans, G.L., Bulloch, E.M., Goldstone, D.C., Jirgis, E.N., Kleinboelting, S., Castell, A., Ramsay, R.J., Manos-Turvey, A. *et al.* (2015) Structure and inhibition of subunit I of the anthranilate synthase complex of *Mycobacterium tuberculosis* and expression of the active complex. *Acta Crystallographica Section D, Biological Crystallography* 71, 2297–2308.

Berney, M., Berney-Meyer, L., Wong, K.-W., Chen, B., Chen, M., Kim, J., Wang, J., Harris, D., Parkhill, J., Chan, J. *et al.* (2015) Essential roles of methionine and *S*-adenosylmethionine in the autarkic lifestyle of *Mycobacterium tuberculosis*. *Proceedings of the National Academy of Sciences of the United States of America* 112, 10008–10013.

Browne, D.T., Kenyon, G.L. and Hegeman, G.D. (1970) Incorporation of monofluorotryptophans into protein during the growth of *Escherichia coli*. *Biochemical and Biophysical Research Communications* 39, 13–19.

Byrne, G.I., Lehmann, L.K. and Landry, G.J. (1986) Induction of tryptophan catabolism is the mechanism for gamma-interferon-mediated inhibition of intracellular *Chlamydia psittaci* replication in T24 cells. *Infection and Immunity* 53, 347–351.

Castell, A., Short, F.L., Evans, G.L., Cookson, T.V., Bulloch, E.M., Joseph, D.D., Lee, C.E., Parker, E.J., Baker, E.N. and Lott, J.S. (2013) The substrate capture mechanism of *Mycobacterium tuberculosis* anthranilate phosphoribosyltransferase provides a mode for inhibition. *Biochemistry* 52, 1776–1787.

Chi, G., Manos-Turvey, A., O'Connor, P.D., Johnston, J.M., Evans, G.L., Baker, E.N., Payne, R.J., Lott, J.S. and Bulloch, E.M. (2012) Implications of binding mode and active site flexibility for inhibitor potency against the salicylate synthase from *Mycobacterium tuberculosis*. *Biochemistry* 51, 4868–4879.

Chim, N., Habel, J.E., Johnston, J.M., Krieger, I., Miallau, L., Sankaranarayanan, R., Morse, R.P., Bruning, J., Swanson, S., Kim, H. *et al.* (2011) The TB Structural Genomics Consortium: a decade of progress. *Tuberculosis* 91, 155–172.

Cole, S.T., Brosch, R., Parkhill, J., Garnier, T., Churcher, C., Harris, D., Gordon, S.V., Eiglmeier, K., Gas, S., Barry, C.E. 3rd *et al.* (1998) Deciphering the biology of *Mycobacterium tuberculosis* from the complete genome sequence. *Nature* 393, 537–544.

Cookson, T., Castell, A., Bulloch, E.M.M., Evans, G.L., Short, F.L., Baker, E.N., Lott, J.S. and Parker, E.J. (2014) Alternative substrates reveal catalytic cycle and key binding events in the reaction catalysed by anthranilate phosphoribosyltransferase from *Mycobacterium tuberculosis*. *The Biochemical Journal* 461, 87–98.

Cookson, T.V.M., Evans, G.L., Castell, A., Baker, E.N., Lott, J.S. and Parker, E.J. (2015) Structures of *Mycobacterium tuberculosis* anthranilate phosphoribosyltransferase variants reveal the conformational changes that facilitate delivery of the substrate to the active site. *Biochemistry* 54, 6082–6092.

Crawford, I.P. (1975) Gene rearrangements in the evolution of the tryptophan synthetic pathway. *Bacteriological Reviews* 39, 87–120.

Crawford, I.P. (1989) Evolution of a biosynthetic pathway: the tryptophan paradigm. *Annual Review of Microbiology* 43, 567–600.

Czekster, C.M., Neto, B.A., Lapis, A.A., Dupont, J., Santos, D.S. and Basso, L.A. (2009) Steady-state kinetics of indole-3-glycerol phosphate synthase from *Mycobacterium tuberculosis*. *Archives of Biochemistry and Biophysics* 486, 19–26.

Dias, M., Canduri, F., da Silveira, N.J.F, Czekster, C.M., Basso, L.A., Palma, M.S., Santos, D.S. and de Azevedo, W.F. Jr (2006) Molecular models of tryptophan synthase from *Mycobacterium tuberculosis* complexed with inhibitors. *Cell Biochemistry and Biophysics* 44, 375–384.

Evans, G.L., Gamage, S.A., Bulloch, E.M., Baker, E.N., Denny, W.A. and Lott, J.S. (2014) Repurposing the chemical scaffold of the anti-arthritic drug lobenzarit to target tryptophan biosynthesis in *Mycobacterium tuberculosis*. *ChemBioChem* 15, 852–864.

Fujigaki, S., Saito, K., Takemura, M., Maekawa, N., Yamada, Y., Wada, H. and Seishima, M. (2002) L-Tryptophan-L-kynurenine pathway metabolism accelerated by *Toxoplasma gondii* infection is abolished in gamma interferon-gene-deficient mice: cross-regulation between inducible nitric oxide synthase and indoleamine-2,3-dioxygenase. *Infection and Immunity* 70, 779–786.

Harrison, A.J., Yu, M., Gårdenborg, T., Middleditch, M., Ramsay, R.J., Baker, E.N. and Lott, J.S. (2006) The structure of MbtI from *Mycobacterium tuberculosis*, the first enzyme in the biosynthesis of the siderophore mycobactin, reveals it to be a salicylate synthase. *Journal of Bacteriology* 188, 6081–6091.

Hayashi, T., Rao, S.P., Takabayashi, K., Van Uden, J.H., Kornbluth, R.S., Baird, S.M., Taylor, M.W., Carson, D.A., Catanzaro, A. and Raz, E. (2001) Enhancement of innate immunity against *Mycobacterium avium* infection by immunostimulatory DNA is mediated by indoleamine 2,3-dioxygenase. *Infection and Immunity* 69, 6156–6164.

Hondalus, M.K., Bardarov, S., Russell, R., Chan, J., Jacobs, W.R. Jr and Bloom, B.R. (2000) Attenuation of and protection induced by a leucine auxotroph of *Mycobacterium tuberculosis*. *Infection and Immunity* 68, 2888–2898.

Kozlowski, M., Tom, N.J., Seto, C.T., Sefler, A.M. and Bartlett, P.A. (1995) Chorismate-utilizing enzymes isochorismate synthase, anthranilate synthase, and *p*-aminobenzoate synthase: mechanistic insight through inhibitor design. *Journal of the American Chemical Society* 117, 2128–2140.

Lee, C.E. Goodfellow, C., Javid-Majd, F., Baker, E.N. and Lott, S.J. (2006) The crystal structure of TrpD, a metabolic enzyme essential for lung colonization by *Mycobacterium tuberculosis*, in complex with its substrate phosphoribosylpyrophosphate. *Journal of Molecular Biology* 355, 784–797.

Ma, Z., Lienhardt, C., McIlleron, H., Nunn, A.J. and Wang, X. (2010) Global tuberculosis drug development pipeline: the need and the reality. *The Lancet* 375, 2100–2109.

MacKenzie, C.R., Hadding, U. and Däubener, W. (1998) Interferon-γ-induced activation of indoleamine 2, 3-dioxygenase in cord blood monocyte-derived macrophages inhibits the growth of group B streptococci. *Journal of Infectious Diseases* 178, 875–878.

Manos-Turvey, A., Bulloch, E.M., Rutledge, P.J., Baker, E.N., Lott, J.S. and Payne, R.J. (2010) Inhibition studies of *Mycobacterium tuberculosis* salicylate synthase (Mbtl). *ChemMedChem* 5, 1067–1079.

Manos-Turvey, A. Cergol, K.M., Salam, N.K., Bulloch, E.M.M., Chi, G., Pang, A., Britton, W.J., West, N.P., Baker, E.N., Lott, J.S. and Payne, R.J. (2012) Synthesis and evaluation of *M. tuberculosis* salicylate synthase (Mbtl) inhibitors designed to probe plasticity in the active site. *Organic and Biomolecular Chemistry* 10, 9223–9236.

Mellor, A. (2005) Indoleamine 2,3 dioxygenase and regulation of T cell immunity. *Biochemical and Biophysical Research Communications* 338, 20–24.

Morollo, A.A. and Eck, M.J. (2001) Structure of the cooperative allosteric anthranilate synthase from *Salmonella typhimurium*. *Nature Structural and Molecular Biology* 8, 243–247.

Murillo, A.C., Li, H.Y., Alber, T., Baker, E.N., Berger, J.M., Cherney, L.T., Cherney, M.M., Cho, Y.S., Eisenberg, D., Garen, C.R. *et al.* (2007) High throughput crystallography of TB drug targets. *Infectious Disorders – Drug Targets* 7, 127–139.

Murray, H.W., Szuro-Sudol, A., Wellner, D., Oca, M.J., Granger, A.M., Libby, D.M., Rothermel, C.D. and Rubin, B.Y. (1989) Role of tryptophan degradation in respiratory burst-independent antimicrobial activity of gamma interferon-stimulated human macrophages. *Infection and Immunity* 57, 845–849.

Pavelka, M.S. Jr, Chen, B., Kelley, C.L., Collins, F.M. and Jacobs, W.R. Jr (2003) Vaccine efficacy of a lysine auxotroph of *Mycobacterium tuberculosis*. *Infection and Immunity* 71, 4190–4192.

Peng, K. and Monack, D.M. (2010) Indoleamine 2,3-dioxygenase 1 is a lung-specific innate immune defense mechanism that inhibits growth of *Francisella tularensis* tryptophan auxotrophs. *Infection and Immunity* 78, 2723–2733.

Pfefferkorn, E.R. (1984) Interferon gamma blocks the growth of *Toxoplasma gondii* in human fibroblasts by inducing the host cells to degrade tryptophan. *Proceedings of the National Academy of Sciences of the United States of America* 81, 908–912.

Pfefferkorn, E.R. and Guyre, P.M. (1984) Inhibition of growth of *Toxoplasma gondii* in cultured fibroblasts by human recombinant gamma interferon. *Infection and Immunity* 44, 211–216.

Pfefferkorn, E.R., Rebhun, S. and Eckel, M. (1986) Characterization of an indoleamine 2, 3-dioxygenase induced by gamma-interferon in cultured human fibroblasts. *Journal of Interferon Research* 6, 267–279.

Sachpatzidis, A., Dealwis, C., Lubetsky, J.B., Liang, P.H., Anderson, K.S. and Lolis, E. (1999) Crystallographic studies of phosphonate-based α-reaction transition-state analogues complexed to tryptophan synthase. *Biochemistry* 38, 12665–12674.

Shen, H., Wang, F. Zhang, Y., Huang, Q., Xu, S., Hu, H., Yue, J. and Wang, H. (2009a) A novel inhibitor of indole-3-glycerol phosphate synthase with activity against multidrug-resistant *Mycobacterium tuberculosis*. *The FEBS Journal* 276, 144–154.

Shen, H., Yang, Y., Wang, F., Zhang, Y., Ye, N., Xu, S. and Wang, H. (2009b) Characterization of the putative tryptophan synthase β-subunit from *Mycobacterium tuberculosis*. *Acta Biochimica et Biophysica Sinica* 41, 379–388.

Slock, J., Stahly, D.P., Han, C.Y., Six, E.W. and Crawford, I.P. (1990) An apparent *Bacillus subtilis* folic acid biosynthetic operon containing *pab*, an amphibolic *trpG* gene, a third gene required for synthesis of *para*-aminobenzoic acid, and the dihydropteroate synthase gene. *Journal of Bacteriology* 172, 7211–7226.

Smith, D.A., Parish, T., Stoker, N.G. and Bancroft, G.J. (2001) Characterization of auxotrophic mutants of *Mycobacterium tuberculosis* and their potential as vaccine candidates. *Infection and Immunity* 69, 1142–1150.

Spraggon, G., Kim, C., Nguyen-Huu, X., Yee, M.C., Yanofsky, C. and Mills, S.E. (2001) The structures of anthranilate synthase of *Serratia marcescens* crystallized in the presence of (i) its substrates, chorismate and glutamine, and a product glutamate, and (ii) its end-product inhibitor, L-tryptophan. *Proceedings of the National Academy of Sciences of the United States of America* 98, 6021–6026.

Suzuki, Y., Suda, T., Asada, K., Miwa, S., Suzuki, M., Fujie, M., Furuhashi, K., Nakamura, Y., Inui, N., Shirai, T. *et al.* (2012) Serum indoleamine 2,3-dioxygenase activity predicts prognosis of pulmonary tuberculosis. *Clinical and Vaccine Immunology* 19, 436–442.

Suzuki, Y., Miwa, S., Akamatsu, T., Suzuki, M., Fujie, M., Nakamura, Y., Inui, N., Hayakawa, H., Chida, K. and Suda, T. (2013) Indoleamine 2,3-dioxygenase in the pathogenesis of tuberculous pleurisy. *International Journal of Tuberculosis and Lung Disease* 17, 1501–1506.

Thomas, S.M., Garrity, L.F., Brandt, C.R., Schobert, C.S., Feng, G.S., Taylor, M.W., Carlin, J.M. and Byrne, G.I. (1993) IFN-gamma-mediated antimicrobial response. Indoleamine 2, 3-dioxygenase-deficient mutant host cells no longer inhibit intracellular *Chlamydia* spp. or *Toxoplasma* growth. *The Journal of Immunology* 150, 5529–5534.

WHO (2015) *Global Tuberculosis Report 2015*. World Health Organization, Geneva, Switzerland. Available at: http://www.who.int/tb/publications/global_report/en/ (accessed 29 August 2016).

Zelante, T., Fallarino, F., Bistoni, F., Puccetti, P. and Romani, L. (2009) Indoleamine 2,3-dioxygenase in infection: the paradox of an evasive strategy that benefits the host. *Microbes and Infection* 11, 133–141.

Zhang, Y.J., Reddy, M.C., Ioerger, T.R., Rothchild, A.C., Dartois, V., Schuster, B.M., Trauner, A., Wallis, D., Galaviz, S. and Huttenhower, C. *et al.* (2013) Tryptophan biosynthesis protects mycobacteria from CD4 T-cell-mediated killing. *Cell* 155, 1296–1308.

21 The Kynurenine Pathway of Tryptophan Metabolism in Microorganisms

R. S. Phillips*

Departments of Chemistry and of Biochemistry and Molecular Biology, University of Georgia, Athens, Georgia, USA

21.1 Abstract

In this chapter, the kynurenine pathway in bacteria, archaea, unicellular fungi and protists is reviewed. The kynurenine pathway is the major catabolic pathway for L-tryptophan metabolism in prokaryotes and eukaryotes. There are two different kynurenine pathways, sometimes called 'inducible' and 'constitutive', in microorganisms. The former of these pathways results in the 'complete' catabolism of L-tryptophan, while the latter leads to the biosynthesis of NAD(P). The enzymes in the first pathway are tryptophan dioxygenase, kynurenine formamidase and kynureninase. The second pathway requires two additional enzymes, kynurenine 3-monooxygenase and 3-hydroxyanthranilate dioxygenase, to produce quinolinate and thereby lead to the NAD(P)$^+$ biosynthetic pathway. Catabolism via the latter pathway requires three further enzymes, 2-amino-3-carboxymuconate semialdehyde decarboxylase, 2-aminomuconate semialdehyde dehydrogenase and 2-aminomuconate deaminase. In addition, in some bacteria, metabolites of the kynurenine pathway are used for the biosynthesis of secondary metabolites.

21.2 Introduction

L-Tryptophan (L-Trp) was named by Neumeister (1890), even before the amino acid was isolated, as the unknown source of indole, which had been found earlier by Bopp (1849) in faeces and decaying meat. L-Trp was subsequently isolated from enzyme-hydrolysed casein in 1903 (Hopkins and Cole, 1903) and the structure of L-Trp was proven by Ellinger (1906) to be β-(3-indolyl)-L-alanine. Kynurenine was first discovered as a metabolite in the urine of animals fed L-Trp; it was also discovered in *Bacillus subtilis* by Kotake (1933). The catabolic pathway of L-Trp, proceeding through the intermediate kynurenine in animals, was subsequently elucidated by Kotake and co-workers, and others; because kynurenine was the first intermediate to be identified, it was dubbed 'the kynurenine pathway'. In animals, another metabolite, kynurenic acid, was also found to arise from kynurenine by transamination. Most of the early interest in the kynurenine pathway focused on its function in mammalian metabolism, and it was found that L-Trp metabolism via the kynurenine pathway is also a *de novo* pathway leading to the biosynthesis of NAD(P)$^+$ (Beadle *et al.*, 1947), in addition to

*E-mail address: plp@uga.edu

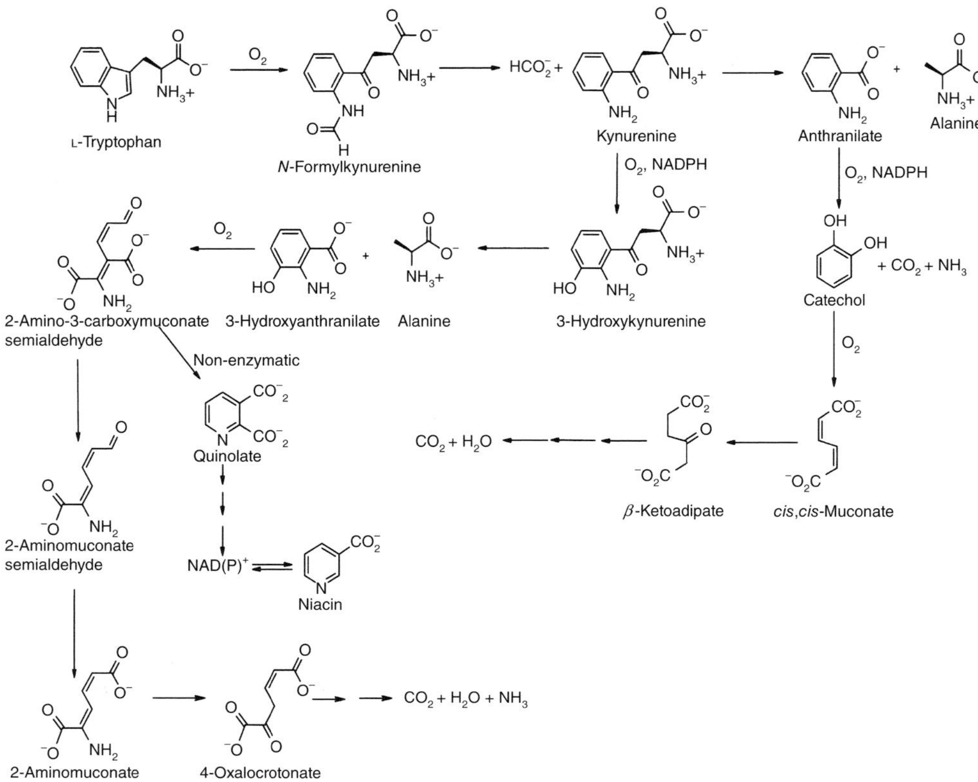

Fig. 21.1. Outline of the kynurenine pathway of L-tryptophan metabolism in eukaryotes and prokaryotes. The sequence of reactions on the top and right represent the 'inducible' pathway to the complete catabolism of L-tryptophan to CO_2 and H_2O. The sequence on top leading down to the left represents the 'constitutive' pathway (the hydroxykynurenine branch of the kynurenine pathway), which leads to $NAD(P)^+$ synthesis via quinolate or to complete catabolism via 2-aminomuconate, which is subsequently deaminated. In *Burkholderia cepacia*, 2-aminomuconate is deaminated to 4-oxalocrotonate (as shown here), while in eukaryotes, 2-aminomuconate is deaminated to 2-ketoadipate (not shown).

providing energy through complete catabolism to CO_2 and H_2O (Fig. 21.1). Thus, the nutritional deficiency disease, pellagra, which is caused by a lack of nicotinic acid or its precursor tryptophan, could be treated in rats with either niacin or L-Trp (Krehl, 1949). It has been estimated that 90–95% of dietary L-Trp in humans is catabolized through the kynurenine pathway, primarily in the liver. It has also been found recently that dysfunction of the kynurenine pathway in humans is involved in a variety of serious pathologies, including Huntington's disease (Pearson and Reynolds, 1992), HIV-related dementia (Sardar and Reynolds, 1995), Alzheimer's disease (Ting *et al.*, 2007) and cancer (Opitz *et al.*, 2011).

After the Second World War, Osamu Hayaishi was a young assistant professor at Osaka University Medical School in Japan in search of a research project (Hayaishi, 2008). Professor Kotake gave Hayaishi a bottle of L-Trp which had miraculously survived the war intact, and Hayaishi isolated a bacterium by enrichment from local soil that would grow on L-Trp as the sole carbon and nitrogen source. Hayaishi identified the bacterium as a strain of *Pseudomonas fluorescens* and, in a series of studies, he characterized the enzymes responsible for L-Trp metabolism (Stanier *et al.*, 1951). He subsequently spent much of his career studying enzymes of the kynurenine pathway in prokaryotes and eukaryotes (Hayaishi, 2008), and also demonstrated that both of the new oxygen atoms in the *N*-formylkynurenine obtained by the oxidation of L-Trp using extracts of tryptophan-adapted *Pseudomonas* are derived from O_2 rather than water, so reporting the first example of an oxygenase (Hayaishi *et al.*, 1957a). The kynurenine pathway in pseudomonads was found to differ from the mammalian kynurenine pathway in that it produces anthranilate, rather than hydroxyanthranilate, as an intermediate, and does not result in the production of NAD(P)$^+$ (see Fig. 21.1).

In both the prokaryotic and eukaryotic pathways, L-Trp is initially cleaved by a haemoprotein dioxygenase to give *N*-formyl-L-kynurenine (*N*-formylkynurenine). However, there are two distinct enzymes that can perform this reaction, tryptophan dioxygenase and indoleamine dioxygenase (IDO), which show minimal sequence homology, although they both contain a haem group. In mammals, tryptophan dioxygenase is expressed mainly in the liver, while IDO is expressed in other tissues, including intestine, lung and brain tissue. In humans, there are two isoenzymes of IDO, IDO-1 and IDO-2. *N*-formylkynurenine undergoes subsequent hydrolysis by a formamidase, resulting in the formation of L-kynurenine (subsequently given as plain kynurenine). In the eukaryotic pathway, kynurenine undergoes subsequent hydroxylation by a mitochondrial flavoenzyme, kynurenine monooxygenase, before the action of the pyridoxal-5′-phosphate (PLP)-dependent kynureninase cleaves it to produce 3-hydroxyanthranilate and L-alanine. In contrast, Hayaishi found that the *P. fluorescens* kynureninase preferentially cleaves kynurenine to give anthranilate and L-alanine (Hayaishi and Stanier, 1952).

In eukaryotes and also in some prokaryotes, 3-hydroxyanthranilate goes on to react with O_2, in a reaction catalysed by a non-haem iron dioxygenase, which results in 2-amino-3-carboxymuconate semialdehyde (Ichiyama *et al.*, 1965; Zhang *et al.*, 2005); this can either undergo spontaneous (i.e. non-enzymatic) cyclization to quinolinate (Colabroy and Begley, 2005a), or undergo enzymatic decarboxylation by 2-amino-3-carboxymuconate semialdehyde decarboxylase (Martynowski *et al.*, 2006), thereby committing the molecule to catabolic destruction. In a reaction catalysed by quinolinate phosphoribosyl transferase, the quinolinate is then converted to give nicotinate mononucleotide, so ultimately leading to the synthesis of NAD(P)$^+$. In contrast, in *P. fluorescens*, anthranilate from the kynurenine pathway is subsequently converted to catechol, CO_2 and NH_3 in a reaction catalysed by a non-haem iron anthranilate dioxygenase (Taniuchi *et al.*, 1964). The catechol product in then further converted to *cis,cis*-muconate in a reaction catalysed by another non-haem iron dioxygenase, pyrocatechase (Hayaishi *et al.*, 1957b). The *cis,cis*-muconate is then taken into the muconate pathway for complete catabolism to CO_2 and H_2O via β-ketoadipate (3-ketoadipate), which undergoes cleavage to acetate and succinate (Ornston and Stanier, 1966).

21.3 Distribution of the Kynurenine Pathway in Microorganisms

After Hayaishi elucidated the kynurenine pathway in *P. fluorescens*, a similar pathway was discovered in a few other species of bacteria, such as *Xanthomonas* sp. and *Bacillus megaterium* (Bouknight and Sadoff, 1975). It was generally assumed, though, that the kynurenine pathway was restricted to only a few species of bacteria that are able to grow in the laboratory on L-Trp as the sole carbon and nitrogen source. The kynurenine pathway was also identified in several eukaryotic microorganisms, particularly fungi. It is especially interesting that Shetty and Gaertner (1973) found that the common bread mould, *Neurospora crassa*, contains two different kynureninases, one of which is expressed in the presence of L-Trp, the other one of which is expressed constitutively, hence implying two separate pathways. In contrast, the common brewer's and baker's yeast, *Saccharomyces cerevisiae*, was found to make only the constitutive kynureninase (Shetty and Gaertner, 1973). Consequently, these two kynurenine pathways were termed 'inducible' and 'constitutive', as one was induced by L-Trp and the other one was not. The inducible form of kynureninase shows a preferential reaction with kynurenine, similar to the *Pseudomonas* enzyme, and the constitutive form reacts preferentially with 3-hydroxy-L-kynurenine, like the mammalian enzyme. Thus, it was suggested that these fungi can use excess L-Trp to produce energy through the inducible catabolic pathway, but can make essential NAD(P)$^+$ from L-Trp through the constitutive pathway. With the advent of genome sequencing, the kynurenine pathway was, surprisingly, discovered to be much more widely distributed in microorganisms than previously thought. Orthologues of tryptophan dioxygenase, indoleamine dioxygenase, kynurenine formamidase, kynureninase, kynurenine monooxygenase, 3-hydroxyanthranilate dioxygenase and 2-amino-3-carboxymuconate semialdehyde decarboxylase are found in the genomes of dozens of bacteria, archaea, protists and fungi.

21.4 Regulation of the Kynurenine Pathway in Microorganisms

As mentioned above, the catabolic kynurenine pathway is inducible. It was found in early studies that there is basal expression of tryptophan dioxygenase and kynurenine formamidase in *P. fluorescens* in the absence of kynurenine, and that their expression is further induced by kynurenine rather than L-Trp (Palleroni and Stanier, 1964). In contrast, kynureninase does not show basal expression in the absence of kynurenine, but is induced by kynurenine. This result implies that expression of kynureninase is controlled by a different promoter than that used for tryptophan dioxygenase and kynurenine formamidase. The kynurenine pathway in *P. aeruginosa* was recently shown to be regulated by *kyn*R, a transcriptional regulator of the Lrp/AsnC family (Knoten *et al.*, 2011). Homologues of *kyn*R are found in other Gram-negative bacteria that contain the kynurenine pathway. As in *P. fluorescens*, the genes for *kyn*A (tryptophan dioxygenase) and *kyn*B (kynurenine formamidase) in *P. aeruginosa* are expressed in the presence of kynurenine. However, *kyn*A is located distant from *kyn*B and *kyn*U (kynureninase) in the genome. Interestingly, *kyn*R only binds the promoter of *kyn*A in the presence of kynurenine, but it binds to the *kyn*BU promoter both in the absence and presence of kynurenine. A mutant of *kyn*R was found to no longer induce transcription in response to kynurenine. A homologous transcriptional regulator in the *kyn* operon of *Burkholderia xenovorans* was found to bind kynurenine with a K_d of 82 μM (Hall *et al.*, 2013), and the binding of kynurenine increased the affinity of the transcription factor for the promoter region of the operon. In contrast, tryptophan dioxygenase, kynurenine formamidase and kynureninase are induced together by L-Trp, but not by kynurenine, in *Xanthomonas pruni* (Brown and Wagner, 1970), which makes $NAD(P)^+$ using the kynurenine pathway. In several fungi – *Neurospora crassa*, *Penicillium roqueforti*, *Aspergillus niger* and *Rhizopus stolonifer* – the inducible kynureninases were found to be expressed in the presence of L-Trp (Shetty and Gaertner, 1973, 1975).

21.5 Tryptophan/Indoleamine Dioxygenase

Tryptophan dioxygenase and indoleamine dioxygenase catalyse the oxidative cleavage of the 2,3-double bond in the pyrrole ring of L-Trp to give *N*-formylkynurenine (see Fig. 21.1). They are both haemoproteins, and are the only known haem-containing dioxygenases, though there is no significant sequence homology between them. Tryptophan dioxygenase (*kyn*A) is often found contiguous with kynureninase (*kyn*U) in the genomes of bacteria that contain the inducible kynurenine pathway, suggesting co-regulation and co-expression, as discussed above. Fungi also can contain orthologues of both tryptophan dioxygenase and indoleamine dioxygenase in their genomes. It is generally thought that the inducible tryptophan dioxygenase functions in catabolism, whereas indoleamine dioxygenase is constitutive and plays a role in NAD biosynthesis.

There are two isoenzymes of indoleamine dioxygenase in humans (Ball *et al.*, 2009), but three indoleamine dioxygenase isoenzymes, IDOα, IDOβ and IDOγ, have been found in fungi (Yuasa and Ball, 2012). *Aspergillus oryzae* contains all three IDO isoenzymes in the genome, while *S. cerevisiae* has only one (Fig. 21.2). Some other fungi also have all three isoenzymes. IDOα and IDOβ show relatively high activity, comparable to that of mammalian indoleamine dioxygenase, but IDOγ has very low activity and probably does not function as an indoleamine dioxygenase *in vivo*, even though it is widely distributed in fungi (Yuasa and Ball, 2013). IDOα and IDOβ show high sequence homology to each other, but IDOγ has more homology with bacterial indoleamine dioxygenases. One group of bacterial indoleamine dioxygenase homologues has low activity and may not be functional *in vivo*, but there is another group of bacterial indoleamine dioxygenase orthologues that have activity comparable to that of the mammalian enzymes (Yuasa *et al.*, 2011). It should be noted that a paralogue of indoleamine dioxygenase, *Prn*B, catalyses the rearrangement and decarboxylation of 7-Cl-L-Trp, resulting in monodechloroaminopyrrolnitrin in the pyrrolnitrin biosynthetic pathway (Kirner *et al.*, 1998). Some of the bacterial genes annotated as indoleamine dioxygenase show high homology with *Prn*B (see Fig. 21.2), and so they may actually be *Prn*B.

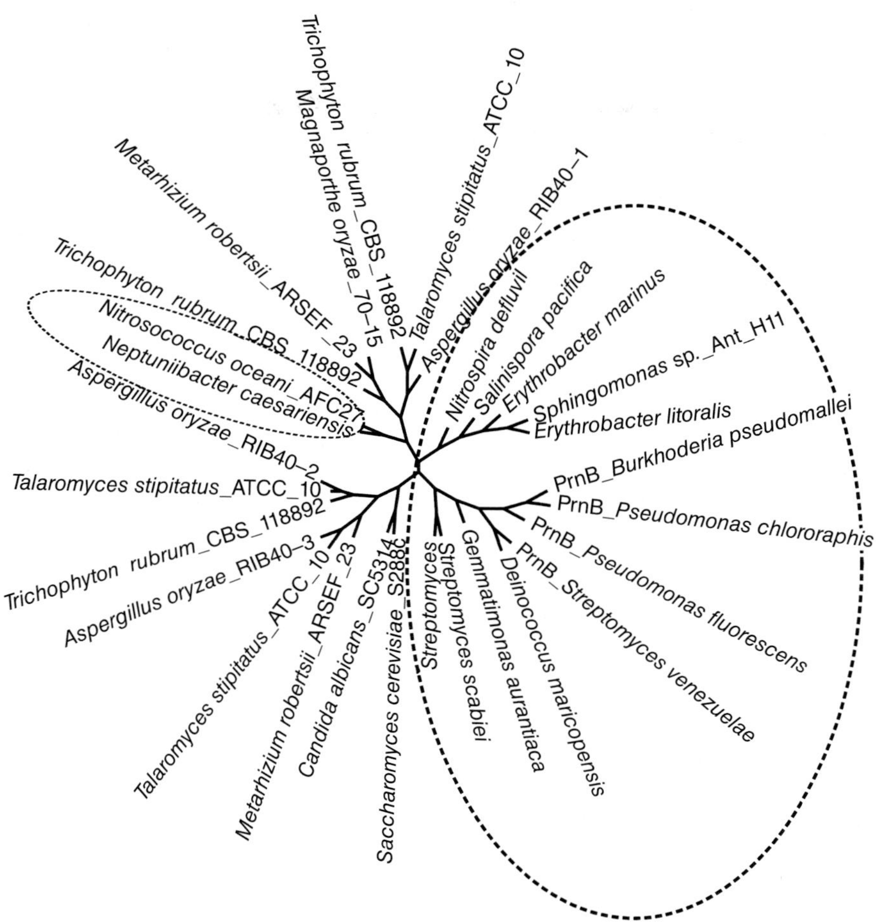

Fig. 21.2. Phylogenetic tree of indoleamine dioxygenases from bacteria (circled) and fungi. Note that the paralogue indoleamine dioxygenases (PrnBs) from bacteria are also very closely related. Orthologous genes are indicated by the numbers that have been assigned to them.

21.6 Kynurenine Formamidase

Kynurenine formamidase catalyses the hydrolysis of the formyl group of *N*-formylkynurenine to give kynurenine (see Fig. 21.1). Remarkably, although tryptophan dioxygenase, indoleamine dioxygenase, kynurenine monooxygenase and kynureninase all show a high degree of homology throughout biology, the kynurenine formamidases from prokaryotes and eukaryotes are completely unrelated proteins in both sequence and structure (Kurnasov *et al.*, 2003a). Bacterial kynurenine formamidase contains a binuclear zinc site with a nucleophilic bridging hydroxide (Díaz-Sáez, 2014), and thus resembles other binuclear zinc hydrolases, such as leucine aminopeptidase. In contrast, kynurenine formamidase from fungi and other eukaryotes is an α/β fold hydrolase, related to carboxyesterases, with a catalytic Asp/His/Ser triad in the active site and no metal (Han *et al.*, 2012).

21.7 Kynurenine Monooxygenase

Kynurenine monooxygenase is a flavoenzyme, containing a tightly but non-covalently bound FAD, and it catalyses the hydroxylation of kynurenine using O_2 and NADPH (see Fig. 21.1). The presence of this enzyme in a genome is suggestive of a constitutive pathway leading to $NAD(P)^+$ from L-Trp. Eukaryotic kynurenine monooxygenase is a mitochondrial enzyme which binds to the external side of the outer mitochondrial membrane with a *C*-terminal membrane anchor sequence of about 50 residues (Hirai *et al.*, 2010). The sequence of prokaryotic kynurenine monooxygenase lacks the *C*-terminal membrane anchor, but may be still membrane associated, as with the kynurenine monooxygenase from *Cytophaga hutchisonii* (Kurnasov *et al.*, 2003b), because a detergent, Triton X-100, is required to solubilize the recombinant enzyme. However, a soluble kynurenine monooxygenase was reported from *P. fluorescens* (Crozier and Moran, 2007). The kynurenine monooxygenase sequences shown in Fig. 21.3 fall into two clades, corresponding to the prokaryotes and eukaryotes. In general, the bacteria that have kynurenine monooxygenase are found in the lower clade of

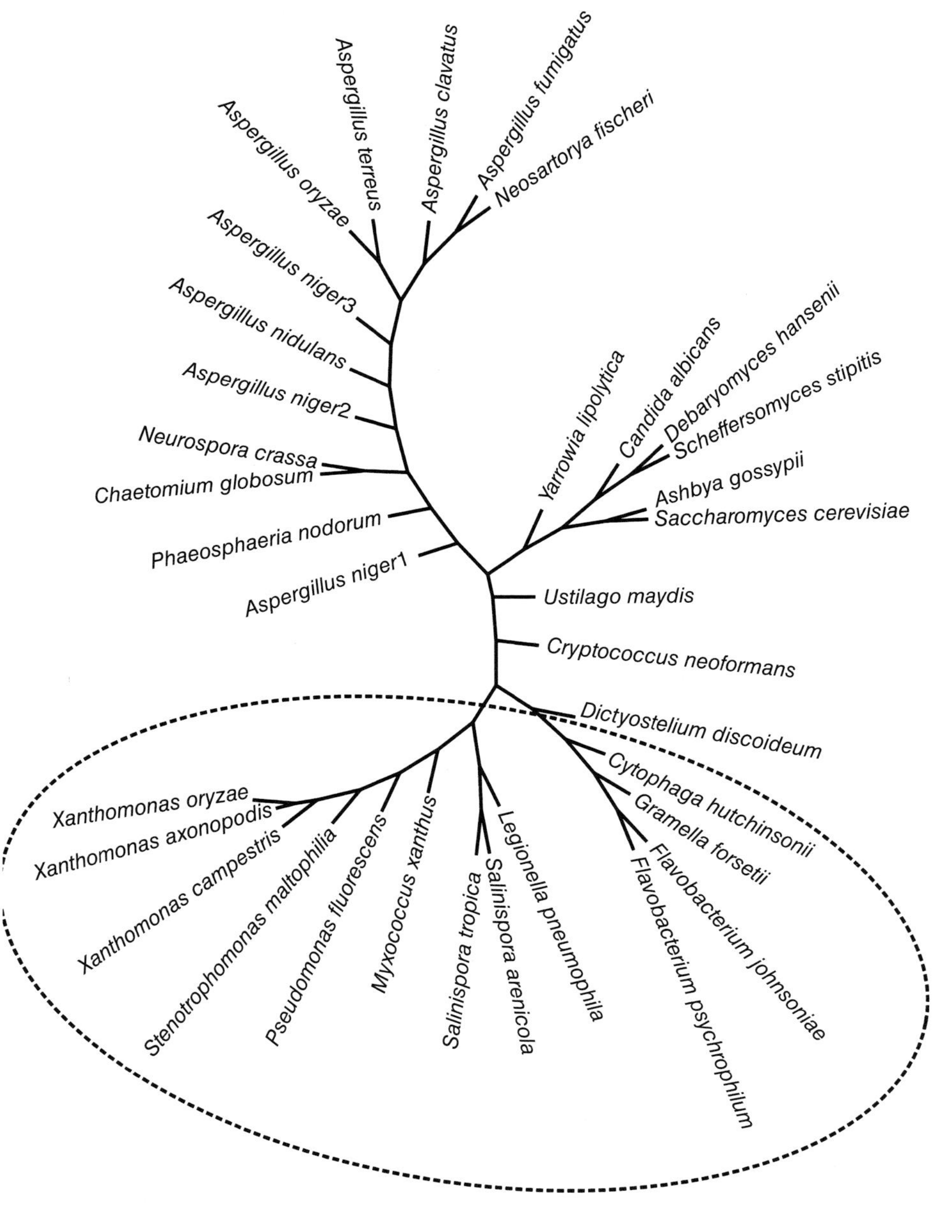

Fig. 21.3. Phylogenetic tree of kynureninase monooxygenases in prokaryotic (circled) and eukaryotic microorganisms. Orthologous genes are indicated by the numbers that have been assigned to them.

kynureninases, corresponding to the biosynthetic pathway for NAD(P)$^+$. Surprisingly, the kynurenine monooxygenase of *Dictyostelium*, a slime mould, is found within the bacterial clade, rather than the clade with other eukaryotes, suggesting a possible recent gene transfer. No archaea are known at this time to have kynurenine monooxygenase in their genomes.

21.8 Kynureninase

Kynureninase is a PLP-dependent enzyme that catalyses the hydrolytic cleavage of kynurenine to give alanine and anthranilate (see Fig. 21.1). The kynureninase sequences of bacteria, archaea and fungi belong to well defined clades (Fig. 21.4). The top part of Fig. 21.4 (the top clade) consists of bacteria and archaea that contain the inducible catabolic pathway, and include *P. fluorescens*. It is interesting that the only archaea found to have kynureninase are halophiles, not thermophiles. The left-hand side of the top clade in Fig. 21.4 consists of bacteria that can use the kynurenine pathway for NAD(P)$^+$ biosynthesis, and include *C. hutchisonii*. The bottom clade contains eukaryotes, mostly moulds and fungi, but also includes the slime mould *Dictyostelium* and *Monosiga*, a choanoflagellate. *S. cerevisiae* has only the constitutive kynureninase. As expected from the

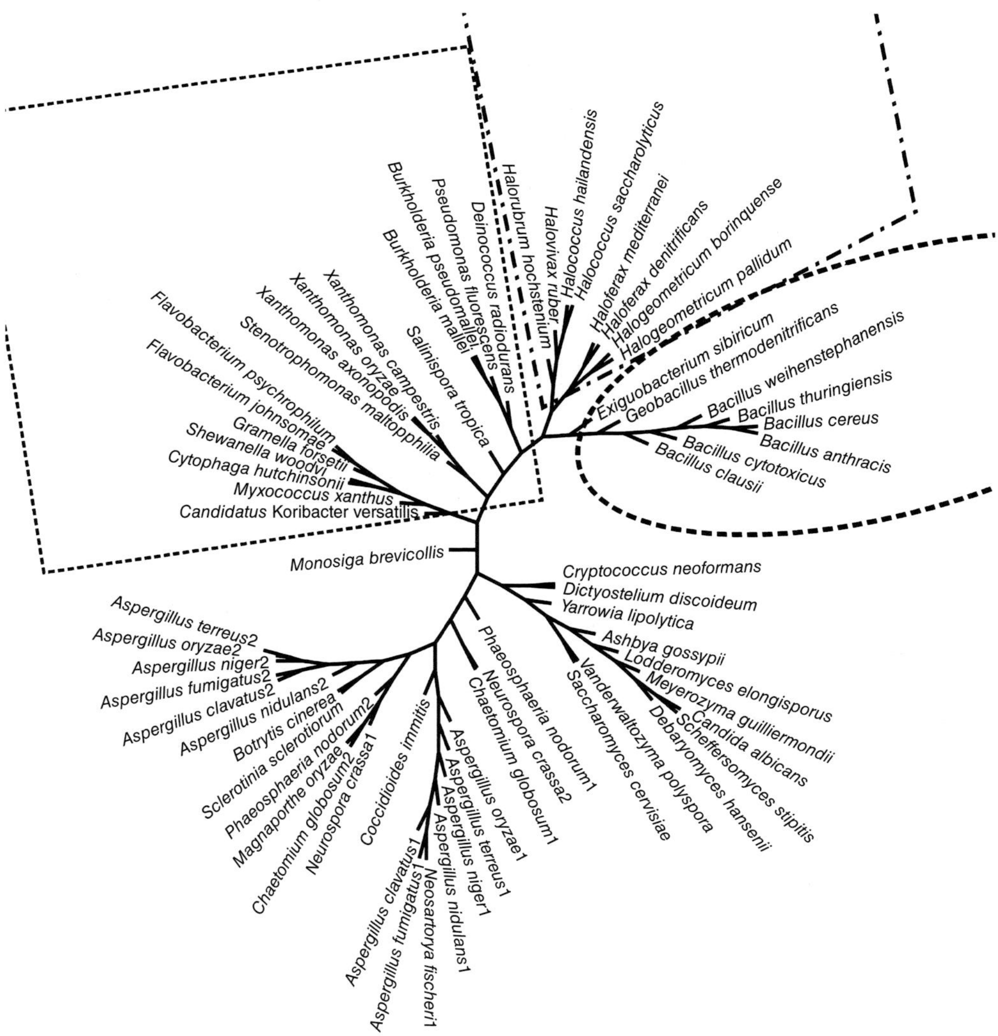

Fig. 21.4. Phylogenetic tree of kynureninases in eukaryotic and prokaryotic microorganisms. Bacteria are shown inside dashed lines (the square and the circle) and archaea inside dotted–dashed lines.

biochemical studies discussed above, *Neurospora* has two separate genes for kynureninase, but various species of *Aspergillus*, *Chaetomium* and *Phaeospheria* also have two kynureninases, suggesting that they too have both the constitutive and the inducible kynurenine pathways. The two isoenzymes of fungal kynureninase are found in separate subclades.

The active sites of kynureninase from *P. fluorescens* and *Homo sapiens* are remarkably similar, differing in only three amino acids, Trp64/His102, Gly281/Ser332 and Thr282/Asn333, respectively, yet they show different substrate preferences (Momany *et al.*, 2004; Lima *et al.*, 2007, 2009a) (Fig. 21.5). The latter two amino acids are involved in binding of the PLP cofactor; Gly282/Ser332 donates a hydrogen bond to the PLP phosphate from the peptide backbone NH. Thr282/Asn333 donates a hydrogen bond to the PLP phosphate from the side chain OH and NH_2, respectively. However, the carbonyl O of Asn333 in the human enzyme is also located in a position to accept a hydrogen bond from the 3-OH of the substrate (Lima *et al.*, 2009a). Additionally, the Trp64/His102 is located within 3 Å of the substrate aromatic ring, and may exert steric restraints. Mutation of the human enzyme to the residues of the *Pseudomonas* enzyme resulted in a loss of activity with 3-hydroxykynurenine, but activity with kynurenine was retained (Lima *et al.*, 2009a). This suggested that these changes in the active site are at least partially responsible for the difference in substrate specificity, but a comparison of sequences of the constitutive and inducible kynureninases in fungi shows that they both have the residues corresponding to the human sequence (Fig. 21.5). Thus, the catabolic kynurenine pathway in fungi must have evolved separately from that in bacteria, and there must be an alternative structural strategy for changing the substrate specificity of kynureninase. Furthermore, a kynureninase orthologue in *Streptosporangium sibiricum* exhibits the kynurenine signature (W, G, T, i.e. Trp, Gly, Thr) but has high activity with 4-methyl-3-hydroxykynurenine, its physiological substrate (Giessen *et al.*, 2011), while another orthologue from *Streptomyces chrysomallus* has a mixed motif (W, S, T – Trp, Ser, Thr) but has comparable activity with both kynurenine and 3-hydroxykynurenine (Keller *et al.*, 2010).

21.9 3-Hydroxyanthranilate 3,4-Dioxygenase

3-Hydroxyanthranilate 3,4-dioxygenase is a non-haem Fe^{2+}-containing *meta*-cleavage dioxygenase with a cupin barrel fold (Zhang *et al.*, 2005) that produces 2-amino-3-carboxymuconate semialdehyde (see Fig. 21.1) in the penultimate step in the formation of quinolinate through the kynurenine pathway. The enzyme is found in those bacteria that use the kynurenine pathway for $NAD(P)^+$ biosynthesis, as well as in fungi and protists, but has not been found in archaea. It is interesting that some fungi, such as *Aspergillus oryzae*, contain two homologues of hydroxyanthranilate dioxygenase in their genomes (Fig. 21.6).

21.10 2-Amino-3-carboxymuconate Semialdehyde Decarboxylase

2-Amino-3-carboxymuconate semialdehyde decarboxylase is part of the catabolic pathway of L-Trp in higher eukaryotes (see Fig. 21.1). It is a Zn-containing metalloenzyme belonging to the amidohydrolase superfamily (Li *et al.*, 2006). As the major catabolic pathway for L-Trp in bacteria and fungi proceeds via anthranilate (see Fig. 21.1), this enzyme is relatively rare in bacteria. It has been reported as part of the L-Trp catabolic pathway in *Burkholderia cepacia* (Colabroy and Begley, 2005b), whereas the enzyme in *P. fluorescens* is in the pathway for 2-nitrobenzoate metabolism via 3-hydroxyanthranilate, rather than in the kynurenine pathway (Muraki *et al.*, 2003). Similarly, there are only a few fungal examples of this enzyme, as this branch of the kynurenine pathway is primarily used for $NAD(P)^+$ biosynthesis.

21.11 2-Aminomuconate Semialdehyde Dehydrogenase and 2-Aminomuconate Deaminase

To complete the catabolism in the hydroxykynurenine branch of the kynurenine pathway, 2-aminomuconate semialdehyde must undergo oxidation to 2-aminomuconate catalysed by a dehydrogenase, followed by deamination (see Fig. 21.1). The enzymes that do this have been rarely reported in microorganisms, but they

```
Haloferax_medit   .DGELYMDGN SLGLFSV.DA EDALEHVIDE WRELAIRGWT DADP.....D
Halorubrum_sacc   .DGVLYMDGN SLGPASD.AA LASLDRVVDE WRDRLIAGWT DADP.....P
Bacillus_cereus   .EGTIYLDGN SLGLLSK.RA EKSLLTLLDS WKEYGIDGWT EGEH.....P
Dictyostelium     DKPVIYLTGN SLGLQPKEIE KQLVCNYLND WRKYGVEGHH KGDH.....P
Homo sapiens      DENAIYFLGN SLGLQPK.MV KTYLEEELDK WAKIAAYGHE VGKR.....P
Aspergillus_fum   SEPCIYFCGN SLGIQPK.AT QKYLEAQLDT WSSIGVCGHF TKIEDSPLKE
Aspergillus_ory   SDPCIYFCGN SLGIQPK.AT AKYLEAQLDT WSSIGVSGHF VDLEGSPLKQ
Aspergillus_nid   .NPCIYFCGN SLGLQPK.AT AKYMEAHLDT WASIGVNGHF TKIEDSPLDP
Neurospora_cras   DTPCIYFVGN SLGAQPK.AV RHYLEAQLET WASIGVNGHF SSLSNSPLTP
Aspergillus_fum   DEKCIYLCGN SLGLQPR.ST RKYIDHYLRT WATKGVTGHF VPHDDQLLPP
Aspergillus_ory   DPRSIYLCGN SLGLQPR.NT RKYLEHYLRT WAIKGVTGHF TPHDDQLLPP
Aspergillus_nid   DARCIYLCGN SLGLQPR.NT KKYINYYLRT WAIKGVTGHF THHDDELLPP
Neurospora_cras   DAKCIYLCGN SLGVQPK.RT VTRLNQYLTT WATQGVQGHF KPLEESPLPT
Xanthomonas       ...QTYLVGN SLGLQPR.AA RAMVDEVLDQ WGTLGVEGHF TGPT.....Q
Burkholderia      ......LDGN SLGAQPRASA ARAQQVIGAE WGEGLIRSWN T.......AG
Burkholderia_m    ......LDGN SLGAQPRAAA ARAQQVIGAE WGEGLIRSWN T.......AG
Pseudomonas_flu   ......LDGN SLGARPVAAL ARAQAVIAEE WGNGLIRSWN S.......AG
Alcaligenaceae    ......LDGN SLGAKPRAAD ARAQAVVGEQ WGQGLIRSWN S.......AG
Acinetobacter_c   ......LDGN SLGARPKKSL EVAQHIIAKE WGEDLINSWN K.......AD
Streptomyces_an   ......LDGN SLGRLPAATP ELLRGVVEQG WGDGLVRSW. ........ES
Streptosporangi   ......LNGN SLGPPAAAVP IALSNAVRQ. WSEHLIRGWW D.......DG

Haloferax_medit   YDPAHSAAAF QTGTISILAT APLDGALDIV EDAGGVAALR EKSLALTDFL
Halorubrum_sacc   YTPADSAGAW QIGTPPLLAA APLEGSVELL REAG.IERLR EKSLALTDFL
Bacillus_cereus   LTAADHAGAY QIGTPHVLST APLIGSLEIF KDAG.IERLR EKSLHITRYM
Dictyostelium     FVAEDGALGF RMSNPSVADC TALRASLSVF EKAGGIKKLA EKSTIITGYL
Homo sapiens      LQLIPGVCGF RISNPPILLV CSLHASLEIF KQAT.MKALR KKSVLLTGYL
Aspergillus_fum   FKPIPGAGGY QISNPSAIDL ACLCAALSVF DETS.IAELR KKSVLMTAYL
Aspergillus_ory   FKPIPGAGGW QLSNPSAIDL ACLCASLSVF DETS.MAELR KKSVMLTAYL
Aspergillus_nid   FRPIPGAGGF QLSTSSATDL TCLNASLSIF DQTS.ISELR RKSVQLTAYL
Neurospora_cras   FQPTPGAQGF QVSNPSAIDL TSLAAALSVF NKTS.MKDLR SKALVLTAYT
Aspergillus_fum   FVPQQGAAGF QLSNPSVLDM NAVVASLELF NQTS.MAEIR KKSLNLTGYL
Aspergillus_ory   FVPQPGAAGF QLSNPSVLDM NAVAASLEIF NRTS.MAEIR KKSLDLTGYL
Aspergillus_nid   FVPQSGAAGY QLSNPSVLDM NAVVASLELF NRTS.MAEIR QKSLNLTGYL
Neurospora_cras   FHPVAGAAGF QLSNPSILDI TSLSASLEIF QEAGGMEALR SKSLKLTSFL
Xanthomonas       FVPSPGAEGW QLSNPPVLAL APLRASLTLF DQAG.MAALR AKSERLTGHL
Burkholderia      FAPDPGIARF LCGTQPIVSM SMVECGLDVF LQTD.MDAIR RKSLALTDAF
Burkholderia_m    FEPDPGIARF LCGTQPIVSM SMVECGLDVF AQTD.MHAIR RKSLALTDAF
Pseudomonas_flu   YEPSNGIARY LCGTQPITSL AMVECGLDVF AQTD.MASLR RKSLALTDLF
Alcaligenaceae    YEPAEGIRRY LCGTQPITSL ALVECGLDIA AQAD.MAEVR RKSLALGDLF
Acinetobacter_c   YEPANSIRRY LCGTQPIISM SLIECGIDIF LQAD.MQQIR VKSLKLTDLF
Streptomyces_an   YDPVPSIERF LVSTPPLLSL AAIDPALALV EEAG.VERIR AKGLLLGQLV
Streptosporangi   FVPADGITRA RTGTPQILSL LALEAALEPL EAAG.MAAVR AKSLRLADYL
```

Fig. 21.5. Partial sequence alignment of kynureninases from different species of eukaryotes and prokaryotes, showing the residues involved in substrate specificity denoted by the standard one-letter amino acid designations.

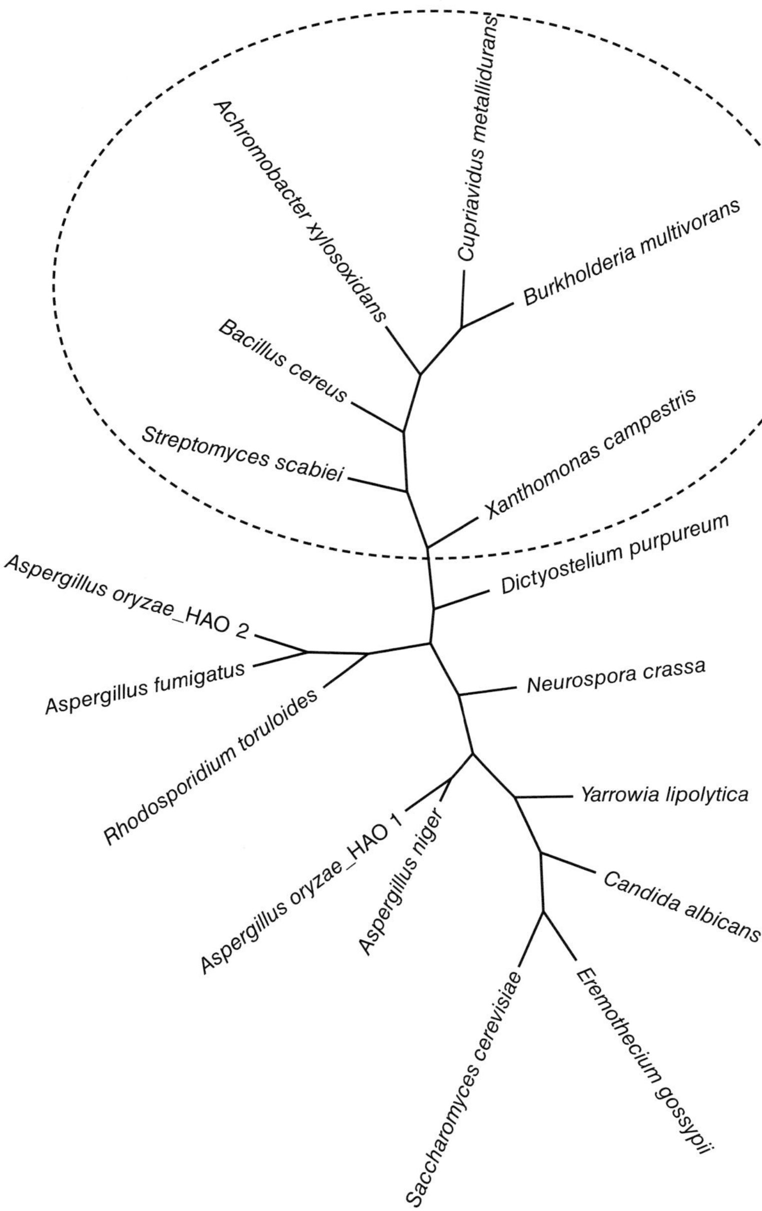

Fig. 21.6. Phylogenetic tree of 3-hydroxyanthranilate 3,4-dioxygenases in bacteria (circled), protists and fungi. Orthologous genes are indicated by the numbers that have been assigned to them.

have been found in *B. cepacia* (Colabroy and Begley, 2005b). It is interesting that the deamination product of 2-aminomuconate in *B. cepacia* is 4-oxalocrotonate, in contrast to that in eukaryotes, which in instead make 2-ketoadipate (α-ketoadipate) as the end point of the hydroxykynurenine branch of the pathway in L-Trp catabolism (Colabroy and Begley, 2005b).

21.12 Roles of the Kynurenine Pathway in Microorganisms

For most bacteria, archaea and some fungi, the primary function of the inducible kynurenine pathway is energy-yielding catabolism, as discussed above. As we have seen, in some bacteria, and in fungi, the constitutive kynurenine pathway can also be a source of metabolically essential NAD(P)$^+$. There is evidence that this NAD(P)$^+$ biosynthetic pathway in bacteria has arisen by lateral gene transfer (Lima *et al.*, 2009b), although most bacteria use the alternative pathway for NAD(P)$^+$ synthesis from aspartate. The presence of a specific pathway utilizing L-Trp for catabolism is surprising, because L-Trp is the rarest of the 20 common amino acids, and it is unlikely to be found in concentrations sufficient to support growth outside the laboratory. None the less, the broad distribution of the inducible pathway in bacteria and fungi suggests that it can be utilized for catabolism, but may also be used to support secondary metabolism.

P. aeruginosa produces two quinolone quorum-sensing signalling molecules, the *Pseudomonas* quinolone signal (PQS, 2-heptyl-3-hydroxy-4(1*H*)-quinolone) and its precursor, 2-heptyl-4(1*H*)-quinolone (HHQ), which are important for virulence. These are biosynthesized from anthranilate (see Fig. 21.7), which itself can be made via several different pathways, including the L-Trp biosynthetic pathway (Dulcey *et al.*, 2013). However, it has been found that the anthranilate used to make PQS comes preferentially from the kynurenine pathway when the cells are grown in rich media (Farrow and Pesci, 2007). Moreover, a number of microbial antibiotics are biosythesized from L-Trp using kynurenine pathway metabolites. Sibirimycin, a pyrrolo[1,4]benzodiazepine, is made by *S. sibiricum* from L-Trp by methylation of 3-hydroxykynurenine in the 4-position, followed by kynureninase-catalysed hydrolysis to give 4-methyl-3-hydroxyanthranilate, which undergoes oxidative cyclization (Giessen *et al.*, 2011). Actinomycin is made from 4-methyl-3-hydroxyanthranilic acid in *S. chrysomallus* by a biosynthetic pathway with genes for tryptophan dioxygenase, kynurenine formamidase, and kynureninase that are separate from those in the normal kynurenine pathway (Keller *et al.*, 2010). There are two kynureninase orthologues located in the 66 kb actinomycin biosynthetic gene cluster, AcmH and AcmK. AcmH is full length, but AcmK is shorter, lacking the first 121 amino acids in the sequence, and is unlikely to be active, because part of the cofactor and substrate binding site is composed of residues within this range of missing amino acids. The quinoxaline and quinoline rings in the quinomycin antitumour antibiotics are made from L-Trp by a novel pathway involving β-hydroxy-L-Trp, which is oxidized to β-hydroxy-L-kynurenine by a tryptophan dioxygenase (Hirose *et al.*, 2011).

Remarkably, in a strain of *Chlamydia psittaci*, the *trpE* gene for anthranilate synthase in the *trp* operon is replaced by *kyn*U, which codes for kynureninase (Wood *et al.*, 2004). This presumably allows the pathogenic organism to overcome the immune response of the host, which depletes L-Trp due to activation of the kynurenine pathway.

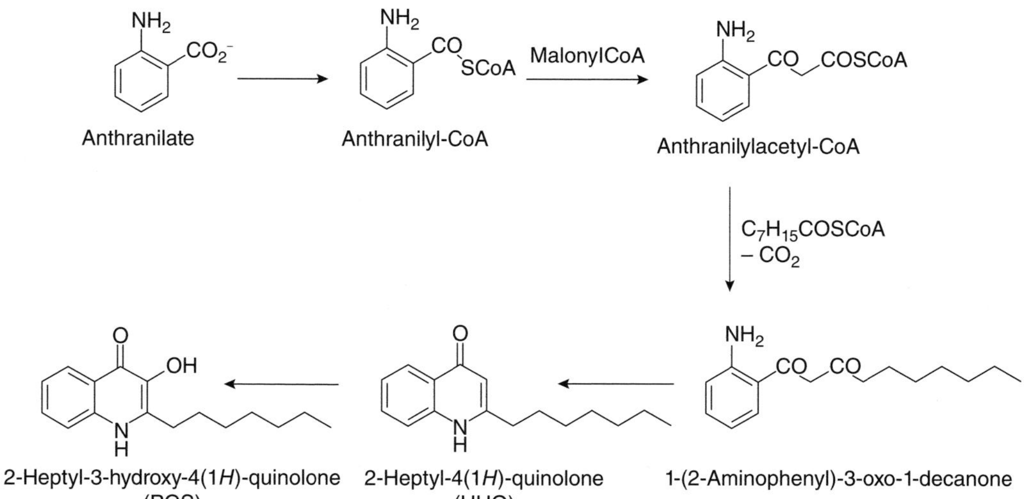

Fig. 21.7. Biosynthesis of the *Pseudomonas* quinolone signal (PQS) molecule 2-heptyl-3-hydroxy-4(1H)-quinolone and its precursor, 2-heptyl-4(1H)-quinolone (HHQ).

21.13 Conclusions

The kynurenine pathway is widely distributed in microorganisms. While its major role is catabolic, it is also often used as an alternative pathway for the biosynthesis of NAD(P)$^+$. Furthermore, products of the kynurenine pathway may be used to form secondary metabolites.

References

Ball, H.J., Yuasa, H.J., Austin, C.J.D., Weiser, S. and Hunt, N.H. (2009) Indoleamine 2,3-dioxygenase-2; a new enzyme in the kynurenine pathway. *The International Journal of Biochemistry and Cell Biology* 41, 467–471.

Beadle, G.W., Mitchell, H.K. and Nyc, J.F. (1947) Kynurenine as an intermediate in the formation of nicotinic acid from tryptophane by *Neurospora*. *Proceedings of the National Academy of Sciences of the United States of America* 33, 155–158.

Bopp, F. (1849) Einiges über Albumin, Caseïn und Fibrin. *Justus Liebigs Annalen der Chemie* 69, 16–37.

Bouknight, R.R. and Sadoff, (1975) H.L. Tryptophan catabolism in *Bacillus megaterium*. *Journal of Bacteriology* 121, 70–76.

Brown, A.T. and Wagner, C. (1970) Regulation of enzymes involved in the conversion of tryptophan to nicotinamide adenine dinucleotide in a colorless strain of *Xanthomonas pruni*. *Journal of Bacteriology* 101, 456–463.

Colabroy, K.L. and Begley, T.P. (2005a) The pyridine ring of NAD is formed by a nonenzymatic pericyclic reaction. *Journal of the American Chemical Society* 127, 840–841.

Colabroy, K.L. and Begley, T.P. (2005b) Tryptophan catabolism: identification and characterization of a new degradative pathway. *Journal of Bacteriology* 187, 7866–7869.

Crozier, K.R. and Moran, G.R. (2007) Heterologous expression and purification of kynurenine-3-monooxygenase from *Pseudomonas fluorescens* strain 17400. *Protein Expression and Purification* 51, 324–333.

Díaz-Sáez, L., Srikannathasan, V., Zoltner, M. and Hunter, W.N. (2014) Structures of bacterial kynurenine formamidase reveal a crowded binuclear zinc catalytic site primed to generate a potent nucleophile. *Biochemical Journal* 462, 581–589.

Dulcey, C.E., Dekimpe, V., Fauvelle, D.-A., Milot, S., Groleau, M.-C., Doucet, N., Rahme, L.G., Lépine, F. and Déziel, E. (2013) The end of an old hypothesis: the *Pseudomonas* signaling molecules 4-hydroxy-2-alkylquinolines derive from fatty acids, not 3-ketofatty acids. *Chemistry and Biology* 20, 1481–1491.

Ellinger, A. (1906) Über die Constitution der Indolgruppe im Eiweiss. III. Mittheilung: Oxydation des Tryptophans zu *β*-Indolaldehyd. *Berichte der Deutschen Chemischen Gesellschaft* 39, 2515–2522.

Farrow, J.M., and Pesci, E.C. (2007) Two distinct pathways supply anthranilate as a precursor of the *Pseudomonas* quinolone signal. *Journal of Bacteriology* 189, 3425–3433.

Giessen, T.W., Kraas, F.I. and Marahiel, M.A. (2011) A four-enzyme pathway for 3,5-dihydroxy-4-methylanthranilic acid formation and incorporation into the antitumor antibiotic sibiromycin. *Biochemistry* 50, 5680–5692.

Hall, R.S., Martí-Arbona, R., Hennelly, S.P. Maity, T.S. Mu, F. Dunbar, J.M. Unkefer, C.J. and Unkefer, P.J. (2013) *In-vitro* characterization of an L-kynurenine-responsive transcription regulator of the oxidative tryptophan degradation pathway in *Burkholderia xenovorans*. *Journal of Molecular Biology Research* 3, 55–67.

Han, Q., Robinson, H. and Li, J. (2012) Biochemical identification and crystal structure of kynurenine formamidase from *Drosophila melanogaster*. *Biochemical Journal* 446, 253–260.

Hayaishi, O. (2008) From oxygenase to sleep. *Journal of Biological Chemistry* 283, 19165–19175.

Hayaishi, O. and Stanier, R.Y. (1952) The kynureninase of *Pseudomonas fluorescens*. *The Journal of Biological Chemistry* 195, 735–740.

Hayaishi, O., Rothberg, S., Mehler, A.H. and Saito, Y. (1957a) Studies on oxygenases. Enzymatic formation of kynurenine from tryptophan. *The Journal of Biological Chemistry* 229, 889–896.

Hayaishi, O., Katagiri, M. and Rothberg, S. (1957b) Studies on oxygenases. Pyrocatechase. *The Journal of Biological Chemistry* 229, 905–920.

Hirai, K., Kuroyanagi, H., Tatebayashi, Y., Hayashi, Y., Hirabayashi-Takahashi, K., Saito, K., Haga, S., Uemura, T. and Izumi, S. (2010) Dual role of the carboxyl-terminal region of pig liver L-kynurenine 3-monooxygenase: mitochondrial-targeting signal and enzymatic activity. *Journal of Biochemistry* 148, 639–650.

Hirose, Y., Watanabe, K., Minami, A., Nakamura, T., Oguri, H. and Oikawa, H. (2011) Involvement of common intermediate 3-hydroxy-L-kynurenine in chromophore biosynthesis of quinomycin family antibiotics. *Journal of Antibiotics* 64, 117–122.

Hopkins, F.G., and Cole, S.W. (1903) A contribution to the chemistry of proteids: Part II. The constitution of tryptophane, and the action of bacteria upon it. *The Journal of Physiology* 29, 451–466.

Ichiyama, A., Nakamura, S., Kawai, H., Honjo, T., Nishizuka, Y., Hayaishi, O. and Senoh, S. (1965) Studies on the metabolism of the benzene ring of tryptophan in mammalian tissues. II. Enzymic formation of alpha-aminomuconic acid from 3-hydroxyanthranilic acid. *The Journal of Biological Chemistry* 240, 740–749.

Keller, U., Lang, M., Crnovcic, I., Pfennig, F. and Schauwecker, F. (2010) The actinomycin biosynthetic gene cluster of *Streptomyces chrysomallus*: a genetic hall of mirrors for synthesis of a molecule with mirror symmetry. *Journal of Bacteriology* 192, 2583–2595.

Kirner, S., Hammer, P.E. Hill, D.S., Altmann, A., Fischer, I., Weislo, L.J., Lanahan, M., van Pée, K.-H. and Ligon, J.M. (1998) Functions encoded by pyrrolnitrin biosynthetic genes from *Pseudomonas fluorescens*. *Journal of Bacteriology* 180, 1939–1943.

Knoten, C.A., Hudson, L.L., Coleman, J.P., Farrow, J.M. and Pesci, E.C. (2011) KynR, a Lrp/AsnC-type transcriptional regulator, directly controls the kynurenine pathway in *Pseudomonas aeruginosa*. *Journal of Bacteriology* 193, 6567–6575.

Kotake, Y. (1933) Studien über den Intermediären Stoffwechsel des Tryptophans. *Hoppe-Seyler's Zeitschrift für Physiologische Chemie* 214, 1–32.

Krehl, W.A. (1949) Niacin in amino acid metabolism. *Vitamins and Hormones* 7, 111–146.

Kurnasov, O., Jablonski, L., Polanuyer, B., Dorrestein, P., Begley, T. and Osterman, A. (2003a) Aerobic tryptophan degradation pathway in bacteria: novel kynurenine formamidase. *FEMS Microbiology Letters* 227, 219–227.

Kurnasov, O., Goral, V., Colabroy, K., Gerdes, S., Anantha, S., Osterman, A. and Begley, T.P. (2003b) NAD biosynthesis: identification of the tryptophan to quinolinate pathway in bacteria. *Cell Chemical Biology* 10, 1195–1204.

Li, T., Iwaki, H., Fu, R., Hasegawa, Y., Zhang, H. and Liu, A. (2006) α-Amino-β-carboxymuconic-ε-semialdehyde decarboxylase (ACMSD) is a new member of the amidohydrolase superfamily. *Biochemistry* 45, 6628–6634.

Lima, S., Khristoforov, R., Momany, C. and Phillips, R.S. (2007) Crystal structure of *Homo sapiens* kynureninase. *Biochemistry* 46, 2735–2744.

Lima, S., Kumar, S., Gawandi, V., Momany, C. and Phillips, R.S. (2009a) Crystal structure of the *Homo sapiens* kynureninase-3-hydroxyhippuric acid inhibitor complex: insights into the molecular basis of kynureninase substrate specificity. *Journal of Medicinal Chemistry* 52, 389–396.

Lima, W.C., Varani, A.M. and Menck, C.F.M. (2009b) NAD biosynthesis evolution in bacteria: lateral gene transfer of kynurenine pathway in *Xanthomonadales* and *Flavobacteriales*. *Molecular Biology and Evolution* 26, 399–406.

Martynowski, D., Eyobo, Y., Li, T., Yang, K., Liu, A. and Zhang, H. (2006) Crystal structure of α-amino-β-carboxymuconate-ε-semialdehyde decarboxylase: insight into the active site and catalytic mechanism of a novel decarboxylation reaction. *Biochemistry* 45, 10412–10421.

Momany, C., Levdikov, V., Blagova, L., Lima, S. and Phillips R.S. (2004) Three-dimensional structure of kynureninase from *Pseudomonas fluorescens*. *Biochemistry* 43, 1193–1203.

Muraki, T., Taki, M., Hasegawa, Y., Iwaki, H. and Lau, P.C.K. (2003) Prokaryotic homologs of the eukaryotic 3-hydroxyanthranilate 3,4-dioxygenase and 2-amino-3-carboxymuconate-6-semialdehyde decarboxylase in the 2-nitrobenzoate degradation pathway of *Pseudomonas fluorescens* strain KU-7. *Applied and Environmental Microbiology* 69, 1564–1572.

Neumeister, R. (1890) Zur Physiologie der Eiweissresorption und zur Lehre von den Peptonen. *Zeitschrift für Biologic* 20, 309–373.

Opitz, C.A., Litzenburger, U.M., Sahm, F., Ott, M., Tritschler, I., Trump, S., Schumacher, T., Jestaedt, L., Schrenk, D., Weller, M. and Jugold, M. (2011) An endogenous tumour-promoting ligand of the human aryl hydrocarbon receptor. *Nature* 478, 197–203.

Ornston, L.N. and Stanier, R.Y. (1966) The conversion of catechol and protocatechuate to β-ketoadipate by *Pseudomonas putida*. *The Journal of Biological Chemistry* 241, 3776–3786.

Palleroni, N.J., and Stanier, R.Y. (1964) Regulatory mechanisms governing synthesis of the enzymes for tryptophan oxidation by *Pseudomonas fluorescens*. *Journal of General Microbiology* 35, 319–334.

Pearson, S.J. and Reynolds, G.P. (1992) Increased brain concentrations of a neurotoxin, 3-hydroxykynurenine, in Huntington's disease. *Neuroscience Letters* 144, 199–201.

Sardar, A.M. and Reynolds, G.P. (1995) Frontal cortex indoleamine-2,3-dioxygenase activity is increased in HIV-1-associated dementia. *Neuroscience Letters* 187, 9–12.

Shetty, A.S. and Gaertner, F.H. (1973) Distinct kynureninase and hydroxykynureninase activities in microorganisms: occurrence and properties of a single physiologically discrete enzyme in yeast. *Journal of Bacteriology* 113, 1127–1133.

Shetty, A.S. and Gaertner, F.H. (1975) Kynureninase-type enzymes of *Penicillium roqueforti*, *Aspergillus niger*, *Rhizopus stolonifer*, and *Pseudomonas fluorescens*: further evidence for distinct kynureninase and hydroxykynureninase activities. *Journal of Bacteriology* 122, 235–244.

Stanier, R.Y., Hayaishi, O. and Tsuchida, M. (1951) The bacterial oxidation of tryptophan. I. A general survey of the pathways. *Journal of Bacteriology* 62, 355–366.

Taniuchi, H., Hatanaka, M., Kuno, S., Hayaishi, O., Nakajima, M. and Kurihara, N. (1964) Enzymatic formation of catechol from anthranilic acid. *The Journal of Biological Chemistry* 239, 2204–2211.

Ting, K.K., Brew, B. and Guillemin, G. (2007) The involvement of astrocytes and kynurenine pathway in Alzheimer's disease. *Neurotoxicity Research* 12, 247–262.

Wood, H., Roshick, C. and McClarty, G. (2004) Tryptophan recycling is responsible for the interferon-γ resistance of *Chlamydia psittaci* GPIC in indoleamine dioxygenase-expressing host cells. *Molecular Microbiology* 52, 903–916.

Yuasa, H.J. and Ball, H.J. (2012) The evolution of three types of indoleamine 2,3 dioxygenases in fungi with distinct molecular and biochemical characteristics. *Gene* 504, 64–74.

Yuasa, H.J., and Ball H.J. (2013) Indoleamine 2,3-dioxygenases with very low catalytic activity are well conserved across kingdoms: IDOs of Basidiomycota. *Fungal Genetics and Biology* 56, 98–106.

Yuasa, H.J., Ushigoe, A. and Ball, H.J. (2011) Molecular evolution of bacterial indoleamine 2,3-dioxygenase. *Gene* 485, 22–31.

Zhang, Y., Colabroy, K.L., Begley, T.P. and Ealick, S.E. (2005) Structural studies on 3-hydroxyanthranilate-3,4-dioxygenase: the catalytic mechanism of a complex oxidation involved in NAD biosynthesis. *Biochemistry* 44, 7632–7643.

22 Histidine Degradation in Bacteria

A.J. Nieuwkoop[1] and R.A. Bender[2*]
[1]*Biomedical Sciences Department, Grand Valley State University, Allendale, Michigan, USA;* [2]*Molecular, Cellular and Developmental Biology, University of Michigan, Ann Arbor Michigan, USA*

22.1 Abstract

The ability to degrade histidine is both broadly distributed and highly conserved within the bacterial kingdom. In many cases, the complete catabolism of histidine to glutamate, ammonia and either formate or formamide allows the use of histidine as a carbon source, nitrogen source, or both. In some cases, the cells use intermediates and components of the pathway to generate important signalling molecules that affect properties such as virulence or biofilm formation. The strong conservation of the enzymes that are involved and the genetic clustering of the genes of the histidine utilization (*Hut*) pathway suggests horizontal transfer over evolutionary time.

22.2 The Histidine Utilization (Hut) Pathway

The ability to degrade histidine is both common and highly conserved within the bacterial kingdom, though and relatively rare in the archaea and in eukaryotes (Tabor, 1954; Magasanik, 1978; Itoh *et al.*, 2007; Bender, 2012). The histidine catabolic pathway (Fig. 22.1) begins with the release of an ammonia from histidine to form urocanate (Tabor *et al.*, 1952; Magasanik and Bowser, 1955). In step two, the urocanate is hydrated to form imidazolone propionate (IP) (Revel and Magasanik, 1958; Matherly and Phillips, 1983). In step three, a ring cleavage occurs to create formiminoglutamate (FIG) (Revel and Magasanik, 1958). At this point, there are two different ways that FIG can be degraded (Bender, 2012). In one case (seen in *Klebsiella* and *Bacillus*), FIG undergoes hydrolysis to create glutamate and formamide as a waste product (Magasanik and Bowser, 1955); this is a four-step pathway. In the other, seen in both *Pseudomonas* and *Streptomyces*, the imino group of FIG is first hydrolysed to yield ammonia and formylglutamate (FG). The FG is subsequently hydrolysed to yield a mixture of formate and glutamate (Tabor and Hayaishi, 1952); this is a five-step pathway.

22.2.1 Histidase

Histidase, the first enzyme in the pathway, is a homotetramer with about 500 amino acids per monomer (Hernandez and Phillips, 1993) that is encoded by the gene *hutH*. The deamination of histidine by histidase differs from pattern commonly found for amino acids. For most amino acids, the amino group is removed by transamination or oxidative deamination, leaving

* Corresponding author. E-mail address: rbender@umich.edu

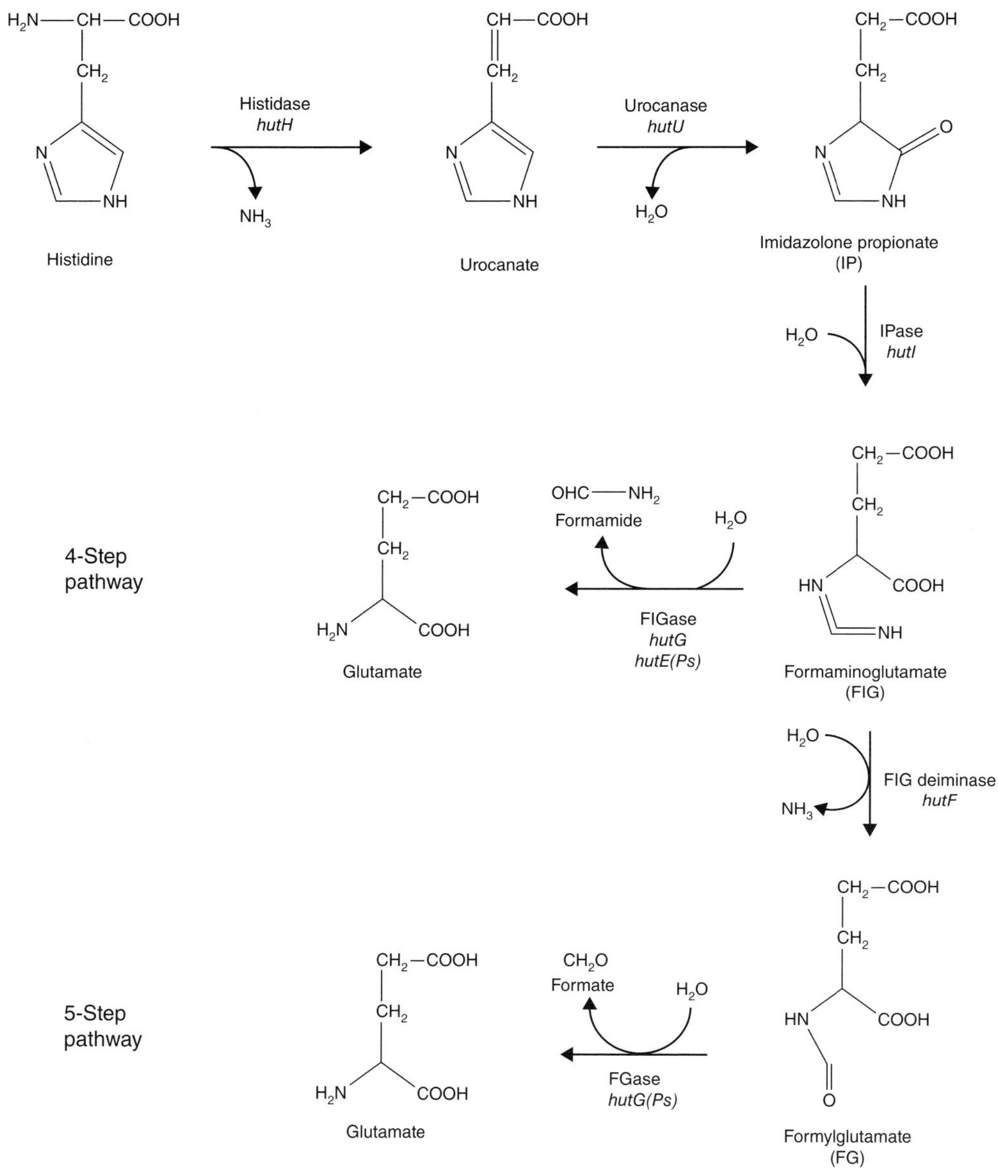

Fig. 22.1. Two variants of the conserved histidine utilization (Hut) pathway: a four-step pathway seen in *Klebsiella* and *Bacillus*; and a five-step pathway, seen in *Pseudomonas* and *Streptomyces*. The designations for the enzymes catalysing each step are shown in italics.

the cognate keto acid (McFall and Newman, 1996). However, histidase functions in a non-oxidative manner that removes the amino group, resulting in the formation of urocanate (Tabor *et al.*, 1952; Magasanik and Bowser, 1955). To accomplish this, a strong electrophile is required within the active site. Examination of the three-dimensional structure of the *Pseudomonas* histidase revealed a novel imidazole structure, 4-methylidene-imidazole-5-one (MIO) generated from the cyclization of residues 142 to 144 (Schwede *et al.*, 1999). The MIO structure (Fig. 22.2) allows the enzyme to remove a non-acidic proton from the β-carbon atom of histidine while keeping the protonation of the nearby amino group on the adjacent carbon (Schwede

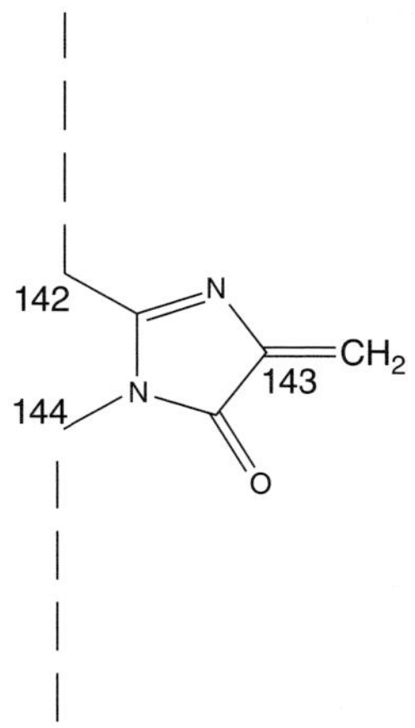

Fig. 22.2. 4-Methylidene-imidazole-5-one (MIO), a novel imidazole structure formed at the active site of histidase by the cyclization of residues 142 to 144. The dashed line represents the polypeptide chain of histidase. The numbers indicate the asymmetric carbon atoms derived from three amino acids in the chain, alanine, serine and glycine: Ala142, Ser143 and Gly144.

et al., 1999).The product of this *β*-elimination reaction is *trans*-urocanate, a very strongly UV-absorbing compound. Bacteria that lack histidase have no discernible phenotype other than the inability to use histidine as a carbon source.

22.2.2 Urocanase

Urocanase, the next enzyme in the Hut pathway, is a homodimer containing 557 amino acids per subunit (Lynch and Phillips, 1972; Kessler *et al.*, 2004); it is encoded by the gene *hutU*. Associated with each monomer of urocanase there is one molecule of tightly bound NAD^+, which is necessary for its activity (Egan and Phillips, 1977; Retey, 1994; Kessler *et al.*, 2004). The crystal structure of urocanase (Kessler *et al.*, 2004) has confirmed the unusual role of this cofactor, in that it serves as an electrophile rather than as a redox centre. The electrophilic attack on the carbon–carbon double bond of the imidazole ring results in the hydration of urocanate to yield hydroxyimidazole propionate. A spontaneous tautomerization of this enol to the keto form leaves imidazolone propionate (Revel and Magasanik, 1958; Matherly and Phillips, 1983). Mutants that lack urocanase can use histidine as a nitrogen source, but not as a carbon source.

22.2.3 Imidazolone propionate hydrolase (IPase)

Imidazolone propionate hydrolase (IPase), the third enzyme in the Hut pathway, is a homodimeric metalloenzyme encoded by the gene *hutI*. It ranges in size from 400 to 450 amino acids per monomer, depending on the species. Imidazolone propionate (IP) is quite unstable, with a half-life of less than 25 min (Revel and Magasanik, 1958; Rao and Greenberg, 1961), so the detailed characterization of IPase has taken longer than that of the other enzymes of the pathway. IPase was the last of the Hut enzymes to have its crystal structure described (Tyagi *et al.*, 2008), and even today the details of the enzyme and its active site metal are not precisely known. Nevertheless, it is clear that IPase is a hydrolase that cleaves the imidazolone ring to yield formiminoglutamate (FIG) (Revel and Magasanik, 1958). This is a completely different reaction from the non-enzymatic cleavage of IP, which can result in the formation of either formylisoglutamine (Feinberg and Greenberg, 1959) or 4-oxoglutaramic acid (4-ketoglutaramic acid) (Hassall and Greenberg, 1963).

Mutants lacking IPase will die if grown in the presence of histidine if both histidase and urocanase are present and active (Hagen *et al.*, 1975; Boylan and Bender, 1984). The exact reason for this phenomenon is unknown, but it could be poisoning from the build-up of formylisoglutamine or 4-ketoglutaramic acid derived from the non-enzymatic breakdown of IP (Revel and Magasanik, 1958; Matherly and Phillips, 1980).

These first three steps in the pathway of histidine utilization are identical in all bacteria that have a Hut pathway. However, there are two different ways that the FIG resulting from the action of IPase is catabolized (see Fig. 22.1).

22.2.4 Formiminoglutamate hydrolase (FIGase)

The final enzyme in the four-step pathway, formiminoglutamate hydrolase (FIGase), is a poorly characterized enzyme that cleaves FIG to formamide (a waste product) and glutamate, and is encoded by the gene *hutG* (see Fig. 22.1; also Section 22.6, which describes the gene *hutE(Ps)*). The FIGase examined from *Bacillus subtilis* is estimated to be about 220 kDa by sedimentation velocity analysis (Kaminskas *et al.*, 1970). The DNA sequence of the *hutG* gene indicates that the subunit mass is approximately 35 kDa, suggesting that FIGase is composed of six identical subunits. Both the *Bacillus* and *Klebsiella* FIGase enzymes require a manganese cofactor. The enzyme has a surprisingly poor affinity for its substrate, with a K_m of approximately 40 or 50 mM (Lund and Magasanik, 1965; Kaminskas *et al.*, 1970).

22.2.5 Formiminoglutamate (FIG) deiminase

In the five-step pathway, FIG is first hydrolysed to formylglutamate (FG) and ammonia, freeing a second usable molecule of ammonia from each molecule of histidine (Wickner and Tabor, 1972) in a reaction catalysed by FIG deiminase. This enzyme is encoded by the *hutF* gene from *Pseudomonas* (Hu and Phillips, 1988). The FIG deiminase in *Pseudomonas* (the HutF enzyme) is a homodimer with about 450 amino acids per monomer (Wickner and Tabor, 1972; Martí-Arbona and Raushel, 2006). The enzyme requires an atom of zinc per monomer in order to be active (Martí-Arbona and Raushel, 2006). FIG deiminase has a much higher affinity for FIG than does the FIGase of the four-step pathway: the K_m for FIG is about 220 μM for FIG deiminase (Martí-Arbona and Raushel, 2006) vs 40–50 mM for FIGase (Lund and Magasanik, 1965; Kaminskas *et al.*, 1970). Perhaps the HutF enzyme found in the five-step pathway is better adapted to its role in histidine utilization than the HutG enzyme used in the four-step pathway.

22.2.6 Formylglutamate hydrolase (FGase)

The final reaction in the five-step pathway is the splitting of FG to glutamate and formate, both of which can be utilized for biosynthetic purposes (Tabor and Mehler, 1954). This requires the action of formylglutamate hydrolase (FGase), which is encoded by the *hutG* gene from *Pseudomonas* (Hu and Phillips, 1988). As the use of this gene name can lead to confusion between the *hutG* gene of *Pseudomonas* for FGase and the unrelated *hutG* gene of *Klebsiella* encoding FIGase in the four-step pathway, the *hutG* gene and HutG enzyme from the five-step pathway are designated as *hutG(Ps)* and HutG(Ps). In *Pseudomonas*, this enzyme is a monomer of about 50 kDa (about 460 amino acids) and is activated by a divalent metal cation, either Co^{2+} or Fe^{2+} (Hu *et al.*, 1987). The enzyme has a relatively high K_m for its substrate, at approximately 12 mM. It may be that this enzyme was not originally for this pathway, but was adopted from elsewhere and modified to permit the cell to obtain usable glutamate from histidine degradation.

22.2.7 Histidine and urocanate permeases

There is considerable variation in the presence and specificity of transport enzymes associated with the Hut pathway.

22.2.7.1 Urocanate permease in enteric bacteria

Salmonella typhimurium [*S. enterica* Typhimurium, known as *S.* Typhimurium] has a variety of functional permeases capable of transporting histidine from the environment, and these vary in affinity, capacity and regulation (Ames, 1964). However, there is no permease gene within the *hut* operons of *S.* Typhimurium (Fig. 22.3) and the regulation of the histidine permeases of *S.* Typhimurium is independent of the *hut* system (Ames, 1964). *S.* Typhimurium cannot grow with urocanate as sole source of carbon or nitrogen, even if all the Hut enzymes are present (Brill and Magasanik, 1969; Meiss *et al.*, 1969), so no urocanate transporter is present in *S.* Typhimurium.

In contrast, the *K. pneumoniae hut* cluster contains a transport gene, *hutT*, immediately downstream of *hutH* as part of the *hutUHT* operon (Fig. 22.3). A Hut-regulated transporter for urocanate was characterized that carries urocanate or the non-metabolizable analogue, imidazole propionate, but not histidine (Schlesinger and Magasanik, 1965), and it appears that *hutT* encodes this urocanate-specific transporter.

22.2.7.2 Histidine and urocanate permeases in pseudomonads

In contrast to the enteric bacteria, the pseudomonads appear to lack significant *hut*-independent

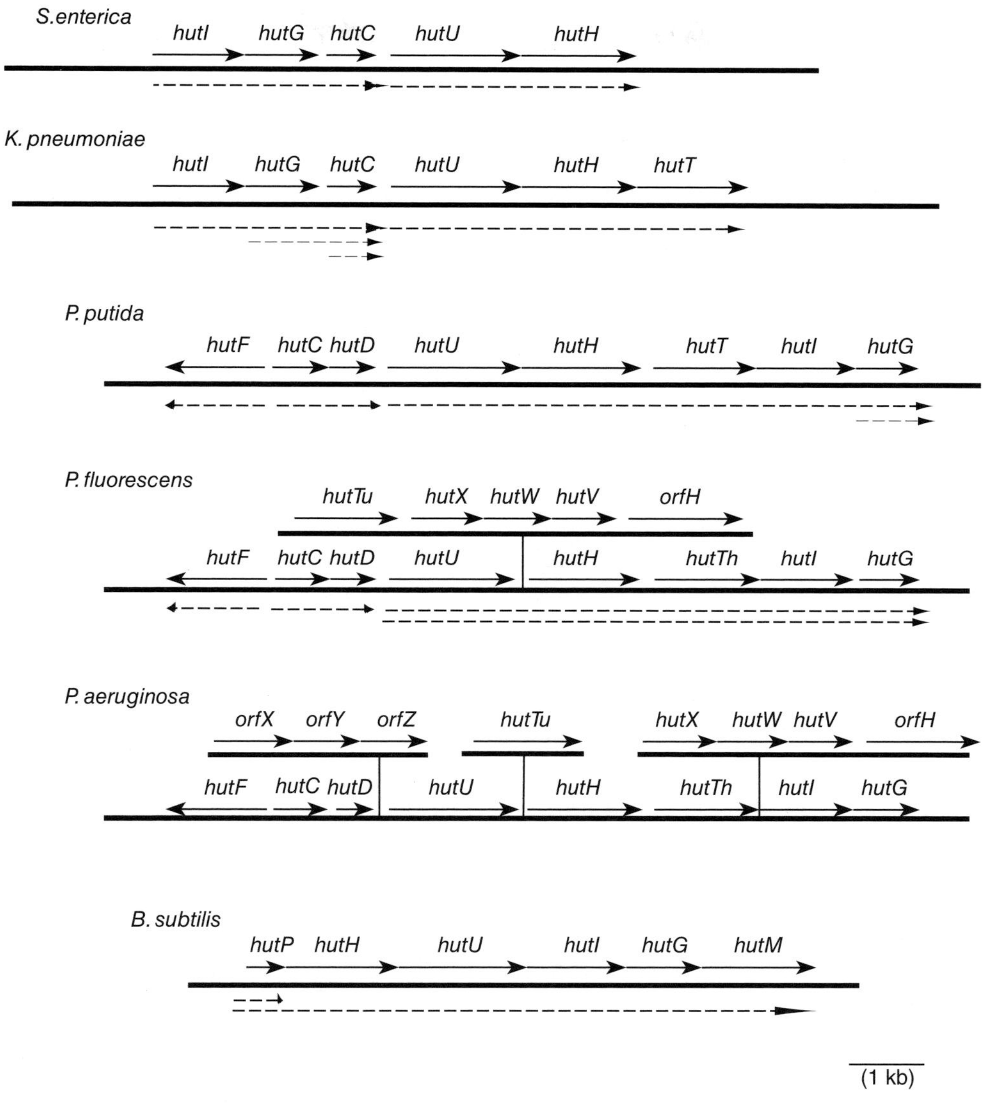

Fig. 22.3. Gene arrangements of the *hut* operons in several well-studied bacteria: *Salmonella enterica*, *Klebsiella pneumoniae*, the *Pseudomonas* spp. *P. putida*, *P. fluoresens* and *P. aeruginosa*, and *Bacillus subtilis*. The solid lines with arrowheads represent the length and direction of the genes. The broken lines with arrowheads indicate the transcriptional units. The heavy black lines represent the chromosomes of the bacteria. The genes are aligned by the *hutU* (urocanase) gene and the apparent insertions of extra genes within the operons are positioned above the main line. Genes with *hut* designations have been shown to be part of the Hut system. Those with *orf* (open reading frame) designations are part of the *hut* transcription units, but their functions in histidine degradation (if any) are unknown. Note that *hutG* in the *Pseudomonas* spp. is not an orthologue of *hutG* in the other bacteria as described in the text, and is usually denoted by *hutG(Ps)*. See text for the gene designations.

histidine permeases (Zhang *et al.*, 2012). Instead, the *Pseudomonas fluorescens hut* cluster (see Fig. 22.3) contains a *hutTh* gene, which encodes a permease that transports histidine but not urocanate (Zhang *et al.*, 2012). Clones from *P. fluorescens* that contain the entire *hut* region confer the ability to degrade histidine on *Escherichia coli*. Clones from which *hutT* has been removed do not, even though all of the enzymes of the pathway are present and expressed

(Zhang *et al.*, 2012). In a similar fashion, insertion mutations within the *hutT* gene of *P. aeruginosa* abolish the ability to grow with histidine as a carbon source (Rietsch *et al.*, 2004). Furthermore, the *P. fluorescens* and *P. aeruginosa hut* gene clusters (but not that found in *P. putida*) also contain a urocanate-specific permease gene, *hutTu*, that supports growth on urocanate but not histidine. They also contain a cluster of three genes, *hutWXV* (see Fig.22.3), that encode an ABC-type transport system that also weakly supports growth on histidine (Zhang *et al.*, 2012).

22.2.7.3 The Hut-specific permease of Bacillus subtilis

The product of the last gene in the *hut* operon of *B. subtilis*, HutM, is 36% identical with and 56% similar to the HutT enzyme of *P. fluorescens*. Although this degree of similarity is comparable to that seen with many other amino acid permeases, the location of the *hutM* gene argues that this is also a histidine transporter gene. It is not known whether *B. subtilis* has a Hut-specific urocanate transporter.

22.3 Hut Operon Structure and Conservation

The *hut* genes are almost always tightly clustered at a single site, arranged in two or more operons. The genetic arrangement of these genes is shown in Fig. 22.3.

22.3.1 The *hut* operons of enteric bacteria

The arrangement of *hut* genes is similar in many common genera of enteric bacteria. For example, *Klebsiella*, *Salmonella*, *Citrobacter* and *Enterobacter* have a *hut* cluster located next to the biotin synthesis (*bio*) genes. *Vibrio* has a similar cluster, but it is not close to *bio*. *Erwinia* and *Serratia* also have *hut* genes, but they encode the five-step pathway typical of the pseudomonads. This difference, along with the almost universal clustering of the *hut* genes, argues for widespread lateral transfer of the Hut system in evolutionary time. However, *hut* is far from universal, even within the enterics – *Escherichia*, *Edwardsiella*, *Shigella* and *Proteus* all lack an orthologue of *hut*.

The *hut* cluster of *K. pneumoniae* contains six genes: four enzyme-encoding genes, the gene for the Hut repressor HutC, and also a gene for the transport of urocanate, HutT (see Fig. 22.3). Genetic and cloning experiments (Goldberg and Magasanik, 1975; Boylan and Bender, 1984) have established the order of the genes to be *hutIGCUH*, the same as the order that is found in *S. enterica* (Smith *et al.*, 1971). The *hutT* gene for the likely urocanate transporter of *K. pneumoniae* is located at the end of the cluster, generating the order *hutIGCUHT-bio*. The urocanate transporter gene *hutT* is absent from the *hut* clusters of all of the other enterics, including *S. enterica*.

The *hutUH* genes of *K. pneumoniae* form an operon (Schwacha *et al.*, 1990). *The hutUH* operon is transcribed from a promoter located between *hutC* and *hutU* (Nieuwkoop and Bender, 1988). This is similar to the arrangement determined by genetic analysis for the *S. enterica hutUH* genes (Smith *et al.*, 1971). The *hutT* gene lies 98 bp downstream from *hutH* in *K. pneumoniae* and it is not known whether it is part of a *hutUHT* operon or is transcribed from its own promoter, or both. The position of *hutT* so close to *hutUH* suggests a *hutUHT* operon, especially as HutT activity is co-regulated with HutH (Schlesinger and Magasanik, 1965). The *hutG* gene is not coordinately regulated with *hutH* and *hutU* (Magasanik *et al.*, 1965) and therefore is not part of the *hutUHT* operon. In *S. enterica*, the *hutI*, *hut-G* and *hut-C* genes are transcribed as a single *hutIGC* operon from a promoter left of *hutI* (Smith and Magasanik, 1971; Magasanik, 1978), but the *hutC* gene of *K. pneumoniae* is transcribed independently (Schwacha *et al.*, 1990).

22.3.2 The *hut* operons of pseudomonads

The arrangement of the *hut* operons varies somewhat from species to species within the pseudomonads, with *P. putida* having the simplest arrangement. *P. fluorescens* and *P. aeruginosa* appear to have additional *hut* genes inserted into a framework similar to that of *P. putida*, as depicted in Fig. 22.3.

22.3.2.1 *The* hut *operons of* Pseudomonas putida

The *hut* genes of *P. putida* had originally been mapped by both genetic and physical methods (Hu and Phillips, 1988). The gene order was found to be *hutF-hutC-hutU-hutH-hutI-hutG*, with *hutF* transcribed right to left and the remaining genes transcribed left to right. The first studies determined three transcription units inducible by urocanate: *hutF, hutC* and *hutU-G*. A fourth transcription unit, induced by both urocanate and FG, transcribes only the *hutG* gene (Hu and Phillips, 1988). This configuration is very much like that found in *K. pneumoniae*. Analysis and annotation of the complete genome sequence within *P. putida* confirmed these initial findings and added two additional genes to the cluster: *hutD* (Allison and Phillips, 1990), a gene involved in regulation (see below); and *hutT*, a transporter, whose specificity is unknown, but is probably for either histidine or urocanate. This arrangement is shown in Fig. 22.3. Based on the closeness of *hutG* to *hutI*, (only11 bp separation) and their induction profiles (Hu and Phillips, 1988), it is very likely that *hutG* is transcribed both as part of a *hutUHTIG* operon (*hutU–G* operon) and also independently as a separate transcript.

22.3.2.2 *The* hut *operons of* P. fluorescens

The *hut* operons of *P. fluorescens* strain SBW25 are similar to those of *P. putida*, with the exception of an insertion of five genes between *hutU* and *hutH* (Fig. 22.3). The first of these five genes encodes a urocanate permease (Zhang *et al.*, 2012). The next three genes encode an ABC-type transporter that can transport histidine, albeit inefficiently (Zhang *et al.*, 2012). The fifth gene encodes a protein of unknown function, whose amino acid sequence is 36% identical to that of HutH (Zhang and Rainey, 2007), although it lacks the three amino acids necessary for forming the MIO at the active site of histidase and thus has no enzymatic activity. Accordingly, its role is unknown.

The entire stretch of genes in the *P. fluorescens hut* cluster from *hutU* through to *hutG* forms a single operon, again suggesting the role of most of them as part of the Hut system. Insertions in the *P. fluorescens hutU* gene abolish *hutG* expression completely (Zhang and Rainey, 2007). This arrangement is in stark contrast to the arrangement found in *P. putida*, where *hutG* appears to have its own promoter (Hu and Phillips, 1988). The termination codon of *hutC* has overlapping bases with the initiation codon of *hutD*. The close apposition seen here and in other *Pseudomonas* spp. strongly suggests that *hutC* and *hutD* form a single *hutCD* transcription unit (Zhang and Rainey, 2007). It is interesting to note that like the Hut systems from enteric bacteria (Macaluso *et al.*, 1990; Goss and Bender, 1995), all of the *hut* operons of *P. fluorescens* (and possibly of other pseudomonads) (Hendrickson *et al.*, 2001; Janiyani and Ray, 2002) can be transcribed by the 'housekeeping' RNA polymerase, which utilizes sigma-70 as its promoter recognition subunit (Zhang and Rainey, 2007, 2008). However, the *hutU–G* operon of *P. fluorescens* has been found to be transcribed from a promoter that is recognized by the unusual RNA polymerase that carries sigma-54 (also known as sigma-N) as its promoter recognition subunit (Zhang and Rainey, 2007, 2008).

22.3.2.3 *The* hut *operons of* P. aeruginosa

The hut operons of *P. aeruginosa* contain the same five genes as *P. fluorescens* inserted downstream of *hutU*, except that their order is rearranged (Fig. 22.3). An additional insertion is also noted of a possible operon with three open reading frames (ORFs) of unknown function located between the *hutCD* operon and the *hutU–G* operon of *P. aeruginosa*. To date, little is known about the transcriptional units of *P. aeruginosa* or other pseudomonads, though the arrangement of their operons suggests that they are probably similar to those of both *P. putida* and *P. fluorescens*. Overall, a common organizational pattern is clear for the Gram-negative organisms: the *hut* genes are clustered, with a similar arrangement of the core enzyme genes, with evidence of duplication (and speciation) of the transport gene.

22.3.3 The *hut* operon of *Bacillus subtilis*

The *hut* genes of the Gram-positive organism *B. subtilis* are also clustered. Early genetic analysis established the order *hutHUIG*, with regulatory elements that are both tightly linked and located close to *hutH* (Kimhi and Magasanik, 1970). This same study showed that the expression of the *hutH*, *hutU*, *hutI* and *hutG* gene products was co-regulated, strongly suggesting that they constitute a single *hutHUIG* operon. Regulatory elements (originally named *hutR* and *hutC*) are now understood to include a gene (*hutP*) that encodes a protein involved in the anti-termination of a single *hutPHUIGM* transcript, as well as a binding site where HutP acts, located between *hutP* and *hutH* (Oda *et al.*, 1988; Wray and Fisher, 1994; Yoshida *et al.*, 1995). The *hut* genes of *B. subtilis* are contained within a single operon, in contrast to the multiple operon schemes found in the Gram-negative organisms. DNA sequencing analysis has revealed some interesting features of this system. A 3 bp overlap is noted between *hutH* and *hutU*, and a 7 bp overlap is found between *hutI* and *hutG*. Additionally, there is a very short (13 bp) spacer region between *hutU* and *hutI*, and a longer region (76 bp) between *hutG* and *hutM*, which encodes the histidine transporter. These three groupings strongly resemble the groupings in other organisms (see above). There are obvious similarities between the Hut enzymes of *B. subtilis* and those of the Gram-negative bacteria, although their regulation is significantly different. For example, the inducer of the *hut* operon in *B. subtilis* is histidine and not urocanate (Chasin and Magasanik, 1968), and regulation is by HutP-mediated anti-termination rather than by a traditional HutC-mediated repressor protein (Oda *et al.*, 1988; Wray and Fisher, 1994; Yoshida *et al.*, 1995).

22.4 Regulation of Hut Expression

A histidine molecule requires a large amount of energy to synthesize and is one of the most expensive of the amino acids in the cell, requiring an equivalent of 20 high-energy phosphate bonds for its synthesis (Akashi and Gojobori, 2002). Therefore, a pathway for degrading it is tightly controlled. In addition, a portion of the histidine biosynthetic pathway is shared with the purine biosynthetic pathway, and so a futile cycle of histidine synthesis and degradation would be extremely damaging. It should be noted that histidine degradation (forming glutamate) is not the reverse of histidine synthesis (starting with ribose phosphate). The regulation of the Hut pathway has been well studied in three bacterial groups: the enteric bacteria (*K. pneumoniae* and *S. enterica*), several *Pseudomonas* spp. (*P. putida*, *P. aeruginosa* and *P. fluorescens*) and the Gram-positive *B. subtilis*. The details of the regulatory mechanisms vary among these three groups and have recently been described elsewhere (Bender, 2012). However, there are a few common regulatory features that can be mentioned here:

1. The Hut enzymes are not synthesized unless exogenous histidine is present at concentrations that exceed the normal internal pools generated by histidine synthesis.
2. The relative affinities of the degradative enzymes and the tRNA synthetases are such that internal pools of histidine will not be depleted below a level where protein synthesis can continue.
3. The Hut enzymes are not usually formed at maximal rates unless the cells are also limited in some essential requirement (such as a carbon source) that can be met by the degradation of histidine.

22.5 Theme and Variation Within the Hut Paradigm

The Hut system has also been studied in a number of organisms other than the ones so far discussed. While these studies are not as completely detailed, they demonstrate a number of the shared topics that have already been mentioned.

22.5.1 The hut system of *Streptomyces* spp.

The streptomycetes use the five-step pathway like the pseudomonads (Kendrick and Wheelis, 1982; Kroening and Kendrick, 1987). Like the pathway of the pseudomonads, the entire pathway is inducible (Kendrick and Wheelis, 1982), with the true inducer being urocanate and not histidine (Kroening and Kendrick, 1987), but it differs in that Hut expression is not repressed by either glucose or ammonia (Kendrick and Wheelis, 1982). The identity of the *hutH* gene has been determined; it is unlinked to any other *hut* gene and is transcribed as a single-gene operon (Wu *et al.*, 1992, 1995). It appears that *hutH* is transcriptionally regulated by a repressor (Wu *et al.*, 1995). Annotation studies suggest that the remaining *hut* genes are unlinked to *hutH*, yet may form a single operon elsewhere on the *S. coelicolor* chromosome. The regulation of this putative *hut* operon has not been examined beyond noting that it is inducible. The importance of this genetic separation between histidase and the remaining pathway is unknown, but might suggest that the system is mainly a urocanate-degrading pathway.

The histidase of the streptomycetes varies from the other systems that have been studied in that it is post-translationally modified. The nature of the reversible modification is not known (Kroening and Kendrick, 1989). Inactivation has been shown to occur upon the induction of sporulation, nutritional downshift or phosphate limitation (Kroening and Kendrick, 1987). Urocanate may also play a role in this regulation, but its role is unclear as there is disagreement about whether it is necessary at all for the transcription of *hutH* (Kroening and Kendrick, 1989; Wu *et al.*, 1995). Thus, the streptomycetes model presents two variations from the general themes noted so far: separation of *hutH* from the rest of the *hut* genes and post-translational regulation of the activity of the histidase enzyme.

22.5.2 The hut system of *Ralstonia eutropha*

The *hut* cluster of *R. eutropha* also resembles that of the pseudomonads, with two exceptions: the *hutF* gene is part of a single *hutHUDIFG* operon that includes all of the enzyme-encoding genes; and the *hutD* gene (which acts as a regulator of Hut expression in pseudomonads) is co-transcribed with the other enzymatic genes rather than as a regulatory gene. The Hut enzymes of *R. eutropha* have been shown to be induced by urocanate and only weakly repressed by good carbon sources such as succinate or pyruvate (Schlesier and Friedrich, 1982a). Regulation by ammonium does not appear to be present, although ammonium does appear to repress the transport of histidine (but not of urocanate) into the cell (Schlesier and Friedrich, 1982a).

R. eutropha is a facultative chemolithoautotroph and can use either organic compounds or molecular hydrogen as its energy source (Schlegel and Eberhardt, 1972). Growth of *R. eutropha* with a hydrogen–oxygen gas mixture results in a repression of histidase formation that is stronger than the carbon catabolite repression demonstrated by a good carbon/energy source such as pyruvate (Schlesier and Friedrich, 1982b). This repression can be relieved by the presence of carbon dioxide, which appears to allow the excess energy generated by hydrogen oxidation to be used by the energy-demanding process of autotrophic carbon fixation (Schlesier and Friedrich, 1982b). Therefore, the repressive effect of hydrogen seems to result from an imbalance between energy generation and the need for energy in biosynthesis.

22.5.3 The hut system of *Caulobacter crescentus*

While growth yield is a strong indicator of which pathway is used for histidine degradation (discerned from the number of usable nitrogens), it is not always correct. *C. crescentus* degrades histidine with the production of formate rather than formamide as an end product (Ferber *et al.*, 1988), suggesting that it utilizes the five-step pathway. The *hut* cluster of *C. crescentus* contains orthologues of the five-step pathway-specific genes *hutG(Ps)* and *hutF*, again pointing towards the five-step pathway, but, growth yields suggest that only two nitrogen atoms are assimilated from histidine, rather than the three expected from the five-step pathway (Ferber *et al.*, 1988). This is unexpected, because *C. crescentus* assimilates ammonia by the efficient glutamine synthetase/glutamate synthase pathway and grows well with glutamate as the sole nitrogen source (Ely *et al.*, 1978). However, histidine (like many of the other amino acids) inhibits the growth of *C. crescentus* in minimal medium and it appears that some metabolic process must occur to detoxify it before growth can occur (Ferber and Ely, 1982). This sensitivity to amino acids is an interesting subject on its own, but it serves here as a caution against only using growth tests to determine which *hut* pathway is in use by an organism.

22.6 Unanswered Questions

A careful evolutionary study is beyond the scope of this review. However, a few comments are worth making and are probably best left as questions rather than conclusions. It has been suggested that the four-step pathway (see Fig. 22.1) is primitive and that the HutF and HutG(Ps) enzymes are adaptations that allowed an extra nitrogen to be extracted, and perhaps an extra carbon as well (Hu and Phillips, 1988; Itoh *et al.*, 2007). An extra carbon, in the form of formate, might be especially advantageous for autotrophs with Hut, e.g. *R. eutropha*. Consistent with this evolutionary argument, the HutF enzyme of the five-step pathway has a higher affinity for FIG than the HutG enzyme of the four-step pathway. Thus, if an organism acquired a HutF–HutG(Ps) pair, its original HutG enzyme would become irrelevant and would be lost.

Most organisms have either a four or a five-step pathway, but there are exceptions. *P. aeruginosa* appears to have both pathways, the five-step pathway represented by the standard *hut* operon structure (Fig. 22.3), and an unlinked FIGase, encoded by hutE(Ps) (see Fig. 22.1), whose regulation is different from that of the *hut* operon (Gerth *et al.*, 2012). Curiously, competition assays between variants lacking one or the other of the FIG-degrading enzymes show that the four-step pathways appears to confer an advantage during growth with histidine as sole carbon source. Unexpectedly, neither pathway has an advantage when histidine is the sole nitrogen source, when the five-step pathway would be expected to be far superior (Gerth *et al.*, 2012). Clearly, there are regulatory phenomena that are not fully understood at work here.

It is also curious that many organisms (e.g. *P. fluorescens* and *R. eutropha*) with the five-step pathway (yielding formate rather than formamide) have an active formamidase that would liberate the extra ammonia if formamide were an end product. In contrast, organisms (e.g. *K. pneumoniae* and *B. subtilis*) that have the four-step pathway (which does yield formamide) do not appear to have an active formamidase.

The presence of the *orfH* gene within the *hut* cluster of several species raises the question of what this HutH orthologue does. The HutH orthologue (*orfH* in Fig. 22.3) is clearly not a histidase and cannot replace *hutH* (Zhang and Rainey, 2008). None the less, its presence within the *hut* operons of many organisms suggests that it has something to do with histidine catabolism. The fact that *orfH* is usually accompanied by transporter genes (*hutTu* and *hutXWV*) seems to argue that these genes function as a unit to introduce a substrate related to histidine into the Hut pathway. Curiously, *K. pneumoniae* appears to have an orthologue of *hutD* and is linked to orthologues of *orfH* and the four amino acid transport genes. Obviously, the roles of these genes and their physiological and evolutionary connection to histidine catabolism remain open questions.

The nitrogen regulation of Hut in the pseudomonads and the oxygen regulation of Hut in *K. pneumoniae* are both poorly characterized relative to other systems. In addition, the evolutionary arguments are ripe for a careful study. Until recently, Hut was studied as an isolated unit, with no function other than the conversion of histidine to glutamate, but the intermediates in the pathway are quite reactive and have interesting regulatory effects on other systems such as those described for biofilm production or pathogenicity (Bender, 2012). In short, Hut provides an excellent vehicle to allow the probing of regulatory and metabolic issues: it is dispensable, its enzymes are easily measured, its chemistry is filled with novelty, its intermediates are potentially functional as effectors, and there is a long history of unexplained observations dating back more than 6 decades. In a word, the potential of Hut has not yet been exhausted.

References

Akashi, H. and Gojobori, T. (2002) Metabolic efficiency and amino acid composition in the proteomes of *Escherichia coli* and *Bacillus subtilis. Procedings of the National Academy of Sciences of the United States of America* 99, 3695–3700.

Allison, S. and Phillips, A. (1990) Nucleotide sequence of the gene encoding the repressor for the histidine utilization genes of *Pseudomonas putida. Journal of Bacteriology* 172, 5470–5476.

Ames, G. (1964) Uptake of amino acids by *Salmonella typhimurium*. *Archives of Biochemistry and Biophysics* 104, 1–18.

Bender, R.A. (2012) Regulation of the histidine utilization (*hut*) system in bacteria. *Microbiology and Molecular Biology Reviews* 76, 565–584.

Boylan, S. and Bender, R. (1984) Genetic and physical maps of *Klebsiella aerogenes* genes for histidine utilization (*hut*). *Molecular and General Genetics* 193, 99–103.

Brill, W. and Magasanik, B. (1969) Genetic and metabolic control of histidase and urocanase in *Salmonella typhimurium*, strain 15-59. *The Journal of Biological Chemistry* 244, 5392–5402.

Chasin, L. and Magasanik, B. (1968) Induction and repression of the histidine-degrading enzymes of *Bacillus subtilis*. *The Journal of Biological Chemistry* 243, 5165–5178.

Egan, R. and Phillips, A. (1977) Presence of tightly bound NAD+ in urocanase of *Pseudomonas putida*. *The Journal of Biological Chemistry* 252, 5701–5707.

Ely, B., Amarasinghe, A. and Bender, R. (1978) Ammonia assimilation and glutamate formation in *Caulobacter crescentus*. *Journal of Bacteriology* 133, 225–230.

Feinberg, R.H. and Greenberg, D.M. (1959) Studies on the enzymic decomposition of urocanic acid. *The Journal of Biological Chemistry* 234, 2670–2676.

Ferber, D. and Ely, B. (1982) Resistance to amino acid inhibition in *Caulobacter crescentus*. *Molecular and General Genetics* 187, 446–452.

Ferber, D., Khambaty, F. and Ely, B. (1988) Utilization of histidine by *Caulobacter crescentus*. *Journal of General Microbiology* 134, 2149–2154.

Gerth, M.L., Ferla, M.P. and Rainey, P.B. (2012) The origin and ecological significance of multiple branches for histidine utilization in *Pseudomonas aeruginosa* PAO1. *Environmental Microbiology* 14, 1929–1940.

Goldberg, R. and Magasanik, B. (1975) Gene order of the histidine utilization (*hut*) operons in *Klebsiella aerogenes*. *Journal of Bacteriology* 122, 1025–1031.

Goss, T. and Bender, R. (1995) The nitrogen assimilation control protein, NAC, is a DNA binding transcription activator in *Klebsiella aerogenes*. *Journal of Bacteriology* 177, 3546–3555.

Hagen, D., Gerson, S. and Magasanik, B. (1975) Isolation of super-repressor mutants in the histidine utilization system of *Salmonella typhimurium*. *Journal of Bacteriology* 121, 583–593.

Hassall, H. and Greenberg, D.M. (1963) Studies on the enzymic decomposition of urocanic acid. V. The formation of 4-oxoglutaramic acid, a nonenzymic oxidation product of 4(5)-imidazolone-5(4)-propionic acid. *The Journal of Biological Chemistry* 238, 1423–1431.

Hendrickson, E., Plotnikova, J., Mahajan-Miklos, S., Rahme, L.G. and Ausubel, S.M. (2001) Differential roles of the *Pseudomonas aeruginosa* PA14 *rpoN* gene in pathogenicity in plants, nematodes, insects, and mice. *Journal of Bacteriology* 183, 7126–7134.

Hernandez, D. and Phillips, A. (1993) Purification and characterization of *Pseudomonas putida* histidine ammonia-lyase expressed in *Escherichia coli*. *Protein Expression and Purification* 4, 473–478.

Hu, L., Mulfinger, L. and Phillips, A. (1987) Purification and properties of formylglutamate amidohydrolase from *Pseudomonas putida*. *Journal of Bacteriology* 169, 4696–4702.

Hu, L. and Phillips, A. (1988) Organization and multiple regulation of histidine utilization genes in *Pseudomonas putida*. *Journal of Bacteriology* 170, 4272–4279.

Itoh, Y., Nishijyo, T. and Nakada, Y. (2007) Histidine catabolism and catabolite regulation. In: Ramos, J.-L. and Filloux, A. (eds) *Pseudomonas, Volume 5. A Model System in Biology.* Springer, Dordrecht, The Netherlands, pp. 371–395.

Janiyani, K. and Ray, M. (2002) Cloning, sequencing, and expression of the cold-inducible *hutU* gene from the antarctic psychrotrophic bacterium *Pseudomonas syringae*. *Applied and Environmental Microbiology* 68, 1–10.

Kaminskas, E., Kimhi, Y. and Magasanik, B. (1970) Urocanase and *N*-formimino-L-glutamate formiminohydrolase of *Bacillus subtilis*, two enzymes of the histidine degradation pathway. *The Journal of Biological Chemistry* 245, 3536–3544.

Kendrick, K. and Wheelis, M. (1982) Histidine dissimilation in *Streptomyces coelicolor*. *Journal of General Microbiology* 128, 2029–2040.

Kessler, D., Retey, J. and Schulz, G. (2004) Structure and action of urocanase. *Journal of Molecular Biology* 342, 183–194.

Kimhi, Y. and Magasanik, B. (1970) Genetic basis of histidine degradation in *Bacillus subtilis*. *The Journal of Biological Chemistry* 245, 3545–3548.

Kroening, T. and Kendrick, K. (1987) *In vivo* regulation of histidine ammonia-lyase activity from *Streptomyces griseus*. *Journal of Bacteriology* 169, 823–829.

Kroening, T. and Kendrick, K. (1989) Cascading regulation of histidase activity in *Streptomyces griseus. Journal of Bacteriology* 171, 1100–1105.

Lund, P. and Magasanik, B. (1965) *N*-formimino-L-glutamate formiminohydrolase of *Aerobacter aerogenes. The Journal of Biological Chemistry* 240, 4316–4319.

Lynch, M. and Phillips, A. (1972) Urocanase of *Pseudomonas putida.* Subunit structure and origin of enzyme-bound α-ketobutyrate. *The Journal of Biological Chemistry* 247, 7799–7805.

Macaluso, A., Best, E. and Bender, R. (1990) Role of the nac gene product in the nitrogen regulation of some NTR-regulated operons of *Klebsiella aerogenes. Journal of Bacteriology* 172, 7249–7255.

Magasanik, B. (1978) Regulation in the *hut* system. In: Miller J and Reznikoff W (eds) *The Operon.* Cold Spring Harbor Laboratory, Cold Spring Harbor, New York, pp. 373–387.

Magasanik, B. and Bowser, H. (1955) The degradation of histidine by *Aerobacter aerogenes. The Journal of Biological Chemistry* 213, 571–580.

Magasanik, B., Lund, P., Neidhardt F. and Schwartz, D.T. (1965) Induction and repression of the histidine-degrading enzymes in *Aerobacter aerogenes. The Journal of Biological Chemistry* 240, 4320–4324.

Martí-Arbona, R. and Raushel, F. (2006) Mechanistic characterization of *N*-formimino-L-glutamate iminohydrolase from *Pseudomonas aeruginosa. Biochemistry* 45, 14256–14262.

Matherly, L. and Phillips, A. (1980) Substrate-mediated inctivation of urocanase from *Pseudomonas putida*: evidence for an essential sulfhydryl group. *Biochemistry* 19, 5814–5818.

Matherly, L. and Phillips, A. (1983) Identification of the enol tautomer of imidazolone propionate as the urocanase reaction product. *Archives of Biochemistry and Biophysics* 220, 314–317.

McFall, E. and Newman, E. (1996) Amino acids as carbon sources. In: Neidhardt, F.C., Ingraham, J.L., Lin, E.C.C., Low, K.B., Magasanik, B., Reznikoff, W.S., Riley, M., Schaechter, M. and Umbarger, H.E (eds) Escherichia coli *and* Salmonella*: Cellular and Molecular Biology,* 2nd edn. ASM Press, Washington, DC, pp. 358–379.

Meiss, H., Brill, W. and Magasanik, B. (1969) Genetic control of histidine degradation in *Salmonella typhimurium*, strain LT-2. *The Journal of Biological Chemistry* 244, 5382–5391.

Nieuwkoop, A. and Bender, R. (1988) RNA polymerase as a repressor of transcription in the *hut(P)* region of mutant *Klebsiella aerogenes* histidine utilization operons. *Journal of Bacteriology* 170, 4986–4990.

Oda, M., Sugishita, A. and Furukawa, K. (1988) Cloning and nucleotide sequences of histidase and regulatory genes in the *Bacillus subtilis hut* operon and positive regulation of the operon. *Journal of Bacteriology* 170, 3199–3205.

Rao, D. and Greenberg, D. (1961) Studies on the enzymic decomposition of urocanic acid. IV. Purification and properties of 4(5)-imidazolone-5(4)-propionic acid hydrolase. *The Journal of Biological Chemistry* 236, 1758–1763.

Retey, J. (1994) The urocanase story: a novel role of NAD^+ as electrophile. *Archives of Biochemistry and Biophysics* 314, 1–16.

Revel, H. and Magasanik, B. (1958) The enzymatic degradation of urocanic acid. *The Journal of Biological Chemistry* 233, 930–935.

Rietsch, A., Wolfgang, M. and Mekalanos, J. (2004) Effect of metabolic imbalance on expression of type III secretion genes in *Pseudomonas aeruginosa. Infection and Immunity* 72, 1383–1390.

Schlegel, H. and Eberhardt, U. (1972) Regulatory phenomena in the metabolism of *Knallgasbacteria. Advances in Microbial Physiology* 7, 205–242.

Schlesier, M. and Friedrich, B. (1982a) Histidine utilization by *Alcaligenes eutrophus*: regulation of histidase formation under heterotrophic conditions of growth. *Archives of Microbiology* 132, 254–259.

Schlesier, M. and Friedrich, B. (1982b) Effect of molecular hydrogen on histidine utilization by *Alcaligenes eutrophus. Archives of Microbiology* 132, 260–265.

Schlesinger, S. and Magasanik, B. (1965) Imidazolepropionate, a nonmetabolizable inducer for the histidine-degrading enzymes in *Aerobacter aerogenes. The Journal of Biological Chemistry* 240, 4325–4330.

Schwacha, A., Cohen, J., Gehring, K., and Bender, R.A. (1990) Tn1000-mediated insertion mutagenesis of the histidine utilization (*hut*) gene cluster from *Klebsiella aerogenes:* genetic analysis of *hut* and unusual target specificity of Tn1000. *Journal of Bacteriology* 172, 5991–5998.

Schwede, T., Retey, J. and Schulz, G. (1999) Crystal structure of histidine ammonia-lyase revealing a novel polypeptide modification as the catalytic electrophile. *Biochemistry* 38, 5355–5361.

Smith, G. and Magasanik, B. (1971) Nature and self-regulated synthesis of the repressor of the *hut* operons in *Salmonella typhimurium. Procedings of the National Academy of Sciences of the United States of America* 68, 1493–1497.

Smith, G., Halpern, Y. and Magasanik, B. (1971) Genetic and metabolic control of enzymes responsible for histidine degradation in *Salmonella typhimurium.* 4-imidazolone-5-propionate amidohydrolase and *N*-formimino-L-glutamate formiminohydrolase. *The Journal of Biological Chemistry* 246, 3320–3329.

Tabor, H. (1954) Metabolic studies on histidine, histamine, and related imidazoles. *Pharmacological Reviews* 6, 299–343.

Tabor, H. and Hayaishi, O. (1952) The enzymatic conversion of histidine to glutamic acid. *The Journal of Biological Chemistry* 194, 171–175.

Tabor, H. and Mehler, A. (1954) Isolation of *N*-formyl-L-glutamic acid as an intermediate in the enzymatic degradation of L-histidine. *The Journal of Biological Chemistry* 210, 559–568.

Tabor, H., Mehler, A., Hayaishi, O. and White, J. (1952) Urocanic acid as an intermediate in the enzymatic conversion of histidine to glutamic and formic acids. *The Journal of Biological Chemistry* 196, 121–128.

Tyagi, R., Eswaramoorthy, S., Burley, S., Raushel, F.M. and Swaminathan, S. (2008) A common catalytic mechanism for proteins of the HutI family. *Biochemistry* 47, 5608–5615.

Wickner, R. and Tabor, H. (1972) *N*-formimino-L-glutamate iminohydrolase from histidine-adapted *Pseudomonas.* Purification and properties. *The Journal of Biological Chemistry* 247, 1605–1609.

Wray, L.J. and Fisher, S. (1994) Analysis of *Bacillus subtilis hut* operon expression indicates that histidine-dependent induction is mediated primarily by transcriptional antitermination and that amino acid repression is mediated by two mechanisms: regulation of transcription initiation and inhibition of histidine transport. *Journal of Bacteriology* 176, 5466–5473.

Wu, P., Kroening, T., White, P. and Kendrick, K.E. (1992) Purification of histidase from *Streptomyces griseus* and nucleotide sequence of the *hutH* structural gene. *Journal of Bacteriology* 174, 1647–1655.

Wu, P., Srinivasan, K. and Kendrick, K. (1995) Regulated expression of the histidase structural gene in *Streptomyces griseus. Journal of Bacteriology* 177, 854–857.

Yoshida, K., Sano, H., Seki, S., Oda, M., Fujimura, M. and Fujita, Y. (1995) Cloning and sequencing of a 29 kb region of the *Bacillus subtilis* genome containing the *hut* and *wapA* loci. *Microbiology* 141, 337–343.

Zhang, X. and Rainey, P. (2007) Genetic analysis of the histidine utilization (*hut*) genes in *Pseudomonas fluorescens* SBW25. *Genetics* 176, 2165–2176.

Zhang, X. and Rainey, P. (2008) Dual involvement of CbrAB and NtrBC in the regulation of histidine utilization in *Pseudomonas fluorescens* SBW25. *Genetics* 178, 185–195.

Zhang, X.X., Chang, H., Tran, S.L., Gauntlett, J.C., Cook, G.M. and Rainey, P.B. (2012) Variation in transport explains polymorphism of histidine and urocanate utilization in a natural *Pseudomonas* population. *Environmental Microbiology* 14, 1941–1951.

23 The Histidine Phosphatase Superfamily in Pathogenic Bacteria

O.O. Coker[1] and P. Palittapongarnpim[2]*

[1]Department of Medicine and Therapeutics, Faculty of Medicine, Chinese University of Hong Kong, Hong Kong; [2]Department of Microbiology, Faculty of Science, Mahidol University, Thailand

23.1 Abstract

The histidine phosphatase superfamily is a large group of proteins whose functions are based on the phosphorylation of a histidine residue during catalysis. Apart from a few exceptions, they share a signature arginine, histidine, glycine (RHG) near the *N*-terminal. They are found in both eukaryotes and bacteria and may function as cofactor-dependent phosphoglycerate mutases or histidine acid phosphatases. They can be grouped into two large branches based on the primary structure of the proteins, both of which are functionally diverse. Proteins belonging to this superfamily have been characterized to function in biosynthesis, signalling and scavenging pathways. Some have also been linked to pathogenesis in bacteria such as *Francisella tularensis* and *Mycobacterium tuberculosis*. Orthologues of such proteins are identified in diverse species of bacteria, suggesting that they may also play roles in the pathogenesis of other bacteria. The general properties, subgrouping, structure, mechanism of catalysis and functions of these proteins are described in this chapter, with emphasis on pathogenic bacteria.

23.2 Introduction

Histidine is a fairly unusual amino acid that performs important functions in many proteins. It is classified as a basic amino acid. However, unlike lysine and arginine, which have typical pK_a values of 10 and 12, respectively, histidine has a pK_a of only about 6.5. This makes the imidazole ring at the end of its side chain either protonated or deprotonated depending on its microenvironment. The imidazole ring can function as a ligand for metal ions in metalloenzymes, which need a set of ligands that coordinate the metal ions and hold them in place in the active sites. The nitrogen atoms of the imidazole ring of histidine are good Lewis bases that can donate electrons to the metal ions. Histidine is also found frequently in many other enzymes because its imidazole side chain can perform many different roles in catalysis. When protonated, histidine can act as a general acid during catalysis and donate its proton to the substrate; when it is deprotonated, it can act as a general base. Some enzymes, such as the ribonucleases, contain a pair of histidines, one protonated and the other not.

Phosphatases play important roles in the physiology of cells. They provide a means of

*Corresponding author. E-mail address: prasit.pal@mahidol.ac.th

scavenging phosphate groups from the environment. They are involved in the formation and degradation of several cellular structures. Together with kinases and phosphorylases, they provide a reversible system to phosphorylate and dephosphorylate many kinds of molecules, including proteins. These provide the basis of many regulatory mechanisms of cellular processes. Dephosphorylation is required in various environments of different pH, and both acid and alkaline phosphatases are found widely in nature. Phosphatases that work in acidic or alkaline conditions have to work through different mechanisms and, therefore, they have very different structures.

The phosphatases that function optimally in an acidic environment include enzymes from three major groups: the metal-ion dependent phosphatases, tyrosine acid phosphatases and histidine acid phosphatases. With the advent of next-generation sequencing technology, the whole genomes of many bacteria have been sequenced at a rapid rate, and many of the phosphatase genes are now annotated computationally, mainly by sequence homology and via the presence of characteristic signature sequences, although some are annotated as hypothetical proteins. Only a small fraction of the phosphatases have been characterized functionally and/or structurally, and only a few have had their substrate specificities characterized. Because off this, the physiological functions of each particular phosphatase are rarely known. These difficulties are complicated by the fact that several paralogues of phosphatases are usually present in an organism and the functions of these paralogous phosphatases may overlap.

In this chapter, we summarize the recent advances in the understanding of histidine acid phosphatases. Eukaryotic histidine acid phosphatases are relatively well characterized and recently reviewed (Rigden, 2008). As such, we focus on bacterial histidine phosphatases especially in pathogens.

23.3 The Histidine Phosphatase Superfamily

23.3.1 Definition and general properties

The histidine phosphatase superfamily is a large group of proteins that share a catalytic core with a nucleophilic histidine that is phosphorylated during the catalytic cycle. The proteins usually contain a signature sequence of RHG (arginine, histidine, glycine) or RHA (arginine, histidine, alanine) near the *N*-terminus (Fig. 23.1). They include the cofactor-dependent phosphoglycerate mutases (dPGMs) in addition to histidine acid phosphatases (HAPs) and have a variety of substrates. The dPGMs are the first characterized and the most extensively studied members of this family. This creates a bias in available automatic annotations of the biochemical functions of the superfamily members. Many of them have been annotated to be dPGMs but subsequently found to be acid phosphatases. Some enzymes may be able to catalyse both reactions, usually with one dominant activity.

Members of the phosphatase superfamily have a variety of biochemical and physiological functions and, therefore, usually have limited sequence similarity. They are also found in a variety of organisms, including bacteria and eukaryotes. However, HAPs are rarely found in archaea, in which the genes are speculated to have been horizontally transferred from bacteria or eukaryotes.

23.3.2 Subgrouping

Based primarily on amino acid sequence similarity, there are two major branches in the phosphatase superfamily. Members of the first branch include the dPGMs and phosphatases of various functions, such as the SixA of *Escherichia coli*, a HAP, or the PhoE of *Bacillus stearothermophilus*, a dPGM (Rigden *et al.*, 2003). Most known members of the first branch are reported from bacteria, and they are found mainly in cytosols and rarely contain disulfide bonds. The second branch of the phosphatase superfamily comprises acid phosphatases and phytases, many of which are of broad specificities. Most known members of this branch are from eukaryotic organisms. They usually contain secretory signals and are likely to have extracytoplasmic localization, and most of them contain disulfide bonds. There is no predilection for the positions of the disulfide bonds, suggesting that they have general stabilization function, which conforms to the more exposed cellular localization of these phosphatases.

```
FtHap        .......... .......... .......... .......... ..........
EcoliSixA    .......... .......... .......... .......... ..........
EcPhytase    .......... .......... .......... .......... ..........
SfNeo13      .......... .......... .......... .......... ..........
STCobC       .......... .......... .......... .......... ..........
EcolidPGM    .......... .......... .......... .......... ..........
MtbdPGM      .......... .......... .......... .......... ..........
BsPhoE       .......... .......... .......... .......... ..........
MtbRv2419c   .......... .......... .......... .......... ..........
MtbRv2228c   MKVVIEADGG SRGNPGPAGY GAVVWTADHS TVLAESKQAI GRATNNVAEY
MtbRv2135c   .......... .......... .......... .......... ..........
MtbRv3214    .......... .......... .......... .......... ..........

FtHap        .......... .......... .......... .......... ..........
EcoliSixA    .......... .......... .......... .......... ..........
EcPhytase    .......... .......... .......... .......... ..........
SfNeo13      .......... .......... .......... .......... ..........
STCobC       .......... .......... .......... .......... ..........
EcolidPGM    .......... .......... .......... .......... ..........
MtbdPGM      .......... .......... .......... .......... ..........
BsPhoE       .......... .......... .......... .......... ..........
MtbRv2419c   .......... .......... .......... .......... ..........
MtbRv2228c   RGLIAGLDDA VKLGATEAAV LMDSKLVVEQ MSGRWKVKHP DLLKLYVQAQ
MtbRv2135c   .......... .......... .......... .......... ..........
MtbRv3214    .......... .......... .......... .......... ..........

FtHap        .......... .......... .......... .......... ..........
EcoliSixA    .......... .......... .......... .......... ..........
EcPhytase    .......... .......... .......... .......... ..........
SfNeo13      .......... .......... .......... .......... ..........
STCobC       .......... .......... .......... ..MPPAKKRY FSEKNVAQSG
EcolidPGM    .......... .......... .......... .......... ..........
MtbdPGM      .......... .......... .......... .......... ..........
BsPhoE       .......... .......... .......... .......... ..........
MtbRv2419c   .......... .......... .......... .......... ..........
MtbRv2228c   ALASQFRRIN YEWVPRARNT YADRLANDAM DAAAQSAAAD ADPAKIVATE
MtbRv2135c   .......... .......... .......... .......... ..........
MtbRv3214    .......... .......... .......... .......... ..........

FtHap        .....MVGYS SKLIFVSMIT RHG....... .......DRA PFANIENANY
EcoliSixA    .......... ....MQVFIM RHG....... ........DA ALDAAS....
EcPhytase    .....QSEPE LKLESVVIVS RHGVRAPTKA TQLMQDVTPD AWPTWP...V
SfNeo13      .......... MTAAQRFVLV RHA....... ........EA SLNVLRDDDV
STCobC       ALSCFCAFNL MRNTMRLWLV RHG....... ........ET EANVAG...L
EcolidPGM    .......... ..AVTKLVLV RHG....... ........ES QWNKEN...R
MtbdPGM      .......... MANTGSLVLL RHG....... ........ES DWNALN...L
BsPhoE       .......... ...ATTLYLT RHG....... ........ET KWNVER...R
MtbRv2419c   .......... .MRARRLVML RHG....... ........QT DYNVGS...R
MtbRv2228c   SPTSPGWTGA RGTPTRLLLL RHG....... ........QT ELSEQR...R
MtbRv2135c   .......... ....MTVILL RHA....... ........RS TSNTAG...V
MtbRv3214    ....QGAMAM GVRNHRLLLL RHG....... ........ET AWSTLG...R
```

Fig. 23.1. Sequence alignment of some histidine phosphatase superfamily proteins from pathogenic bacteria. The standard one-letter abbreviations are used to indicate amino acid residues. Bold indicates residues that are conserved in most of the sequences and play roles in catalysis. The alignments were generated by ClustalX version 2.1. Abbreviations: BsPhoE, acid phosphatase (PhoE) from *Bacillus stearothermophilus*; EcolidPGM, cofactor dependent phosphoglycerate mutase from *Escherichia coli*; EcoliSixA, phosphohistidine phosphatase from *E. coli*; EcPhytase, phytase from *E. coli*; FtHap, histidine acid phosphatase from *Francisella tularensis*; MtbdPGM, phosphoglycerate mutase (Rv0489) from *Mycobacterium tuberculosis*; MtbRv2135c: acid phosphatase, Rv2135c from *M. tuberculosis*; MtbRv2228c, Rv2228c from *M. tuberculosis*; MtbRv2419c, glucosyl-3-phosphoglycerate phosphatase (Rv2419c) from *M. tuberculosis*; MtbRv3214, acid phosphatase (Rv3214) from *M. tuberculosis*; SfNeo13, histidine phosphatase Neo13 from *Streptomyces fradiae*; STCobC, histidine phosphatase CobC from *Salmonella enterica* serotype Typhimurium.

FtHap	SWGTELSE..	....	.........**L**	TPIGMNQEYN	LGLQL.....
EcoliSixA	....DSVRP.	....	.........**L**	TTNGCDESRL	MANWL.....
EcPhytase	KLGWLTPRGG	ELIAYLGHYQ	RQRLVADGL**L**	AKKGCPQSGQ	VAIIADVDER
SfNeo13	MPGFDPHAG.	....	.........**L**	TPLGERQARA	LAEHLPGDPA
STCobC	YSGHAPTP..	....	.........**L**	TEKGIGQAKT	LHTLL.....
EcolidPGM	FTGWYDVD..	....	.........**L**	SEKGVSEAKA	AGKLL.....
MtbdPGM	FTGWVDVG..	....	.........**L**	TDKGQAEAVR	SGELI.....
BsPhoE	MQGWQDSP..	....	.........**L**	TEKGRQDAMR	LGKRL.....
MtbRv2419c	MQGQLDTE..	....	.........**L**	SELGRTQAVA	AAEVL.....
MtbRv2228c	YSGRGNPG..	....	.........**L**	NEVGWRQVGA	AAGYL.....
MtbRv2135c	LAGRSGVD..	....	.........**L**	DEKGREQATG	LIDRI.....
MtbRv3214	HTGGTEVE..	....	.........**L**	TDTGRTQAEL	AGQLL.....

FtHap	..RKRYIDKF	GLLPEHYVDQ	SIYVLSSHTN	**R**TVVSAQSLL	MGLYPAGTGP
EcoliSixA	KGQKVEIER.	....	...VLVSPFL	**R**A...EQTLE	EVGDCLNLPS
EcPhytase	TRKTGEAFAA	GLAPDCAITV	HTQTDTSSPD	**P**LFNPLKTGV	CQLDNANVTD
SfNeo13	VGPPGPAPAP	RPGPGSGPRP	AVRVYSSPQR	**R**A...ARTAE	VLAAALGVPV
STCobC	..RHAHLTG.	....	..CYRASWSA	**R**....AHTAR	LVLEGRDVPQ
EcolidPGM	.KEEGYSFDF	....	...AYTSVLK	**R**AIHTLWNVL	DELDQAWLPV
MtbdPGM	.AEHDLLPDV	....	...LYTSLLR	**R**AITTAHLAL	DSADRLWIPV
BsPhoE	..EAVELAA.	....	...IYTSTSG	**R**A...LETAE	IVRGGRLIPI
MtbRv2419c	.GKRQPLL..	....	...IVSSDLR	**R**A...YDTAV	KLGERTGLVV
MtbRv2228c	.ARRGGIAA.	....	...VVSSPLQ	**R**A...YDTAV	TAARALALDV
MtbRv2135c	....GDLPIR	A.........	...VASSPML	**R**C...QRTVE	PLAEALCLEP
MtbRv3214	....GELELD	DPI.......	...VICSPRR	**R**T...LDTAK	LAGLTVNEVT

FtHap	LIGDGDPAIK	DRFQPIPI**M**T	LSADSRLI..	..QFPYEQYL	AVLKKYVYNS
EcoliSixA	SA........	E.........	....	..VLP.....	......ELTP
EcPhytase	AI...LSRAG	G.SIADFT**G**H	RQTAFRELER	VLNFP.QSNL	CLKREKQDES
SfNeo13	LA...DGRLA	E..LRAPR**A**F	PRPMTVR...	..QWDAVLED	RLRRPAAEPA
STCobC	HI...LPELN	EMYFGDWE**M**R	HHRDLT....	..HEDAESYA	AWCTDWQNAV
EcolidPGM	EK...SWKLN	ERHYGALQ**G**L	NKAETAE...	..KYGDEQVK	QWRRGFAVTP
MtbdPGM	RR...SWRLN	ERHYGALQ**G**L	DKAETKA...	..RYGEEQFM	AWRRSYDTPP
BsPhoE	YQ...DERLR	EIHLGDWE**G**K	THDEIRQ...	..MDPIAFDH	FWQAPHLYAP
MtbRv2419c	RV...DTRLR	ETHLGDWQ**G**L	THAQIDA...	..DAP.GARL	AWREDATWAP
MtbRv2228c	VV...DDDLV	ETDFGAWE**G**L	TFAEAAE...	..RDP.ELHR	RWLQDTSITP
MtbRv2135c	LI...DDRFS	EVDYGEWT**G**R	KIGDLV....	..DEP.....	LWRVVQAHPS
MtbRv3214	GL......LA	EWDYGSYE**G**L	TTPQIRE...	..SEP.....	DWLV.WTHGC

FtHap	PEWQNKTKEA	APNFAKWQQI	LGN...RISG	LNDVITVG**D**V	LIVAQAHGKP
EcoliSixA	C.........	....	.GD.......	.VGLVS..**A**Y	LQALTNEGV.
EcPhytase	CSLTQALPSE	LKVSADNVSL	TGA.....VS	LASMLT..**E**I	FLLQQAQGMP
SfNeo13	....	....	.GV.....ES	WAAQRR..**R**V	REFLDSRCPV
STCobC	P.........	....	TNG.....EG	FQAFTR..**R**V	ERFISRLDAF
EcolidPGM	PELTKDDERY	PGHDPRYAKL	SEKELPLTES	LALTID..**R**V	IPYWNETILP
MtbdPGM	PPIERGSQFS	QDADPRYADI	GGG..PLTEC	LADVVA..**R**F	LPYFTDVIVG
BsPhoE	Q.........	....	.RG.....ER	FCDVQQ..**R**A	LEAVQSIVDR
MtbRv2419c	H.........	....	.GG.....ES	RVDVAA..**R**S	RPLVAELVAS
MtbRv2228c	P.........	....	.GG.....ES	FDDVLR..**R**V	RRGRDRIIVG
MtbRv2135c	A.........	.......AVF	PGG.....EG	LAQVQT..**R**A	VAAVREHDRR
MtbRv3214	P.........	....	.AG.....ES	VAQVND..**R**A	DSAVALALEH

Fig. 23.1. Continued.

23.3.3 Structure and mechanism

The structural characteristics of the histidine phosphatase superfamily include their core α/β domains, with β-sheets flanked by α-chains on both sides ($\alpha/\beta/\alpha$-units). This domain contributes to the catalytic activity of the enzyme by using the histidine residue in the first β-chain to

```
FtHap       LPKGLSQEDA DQIIALTDWG LAQQFKSQKV SYI..MGGKL TNRMIEDLNN
EcoliSixA   .......... .......... .......ASV LVI..SHLPL VGYLVAELCP
EcPhytase   EPGWGRITDS HQWNTLLSLH NAQFYLLQRT PEVARSRATP LLDLIKTALT
SfNeo13     GADG...... .......... .......ARW VLV..SHSET LQAVLFELLG
STCobC      SDC....... .......... .......QNL LIV..SHQGV LSLLIARLLT
EcolidPGM   RMKSG..... .......... .......ERV IIA..AHGNS LRALVKYLDN
MtbdPGM     DLRVG..... .......... .......KTV LIV..AHGNS LRALVKHLDQ
BsPhoE      HEG....... .......... .......ETV LIV..THGVV LKTLMAAFKD
MtbRv2419c  EPEWGGADEP D......... .......RPV VLV..AHGGL IAALSAALLK
MtbRv2228c  YEG....... .......... .......ATV LVV..SHVTP IKMLLRLALD
MtbRv2135c  LADQHG.... .......... .......HDV LWLACTHGDV IKAVIADAFG
MtbRv3214   MSS....... .......... .......RDV LFV..SHGHF SRAVITRWVQ

FtHap       AV.NGKSKYK MTYYSGHALT LLEVMGTLGV PLDTAPGYAS NLEMELYKDG
EcoliSixA   GE........ .......... .......... ...TPPM... ......FTTS
EcPhytase   PHPPQKQAYG VTLPTSVLFI AGHDTNLANL GGALELNWTL PGQPDNTPPG
SfNeo13     LE........ .......... .......... ...DALLHRT RFK...ISNT
STCobC      MP........ .......... .......... ...AASLWHF R.....VEQG
EcolidPGM   MS........ .......... .......... ...EEEILEL N.....IPTG
MtbdPGM     MS........ .......... .......... ...DDEIVGL N.....IPTG
BsPhoE      TP........ .......... .......... ...LDHLWSP PY....MYGT
MtbRv2419c  LP........ .......... .......... .......... ........VA
MtbRv2228c  AG........ .......... .......... ...SGVLYRL H.....LDLA
MtbRv2135c  MH........ .......... .......... ...LDSFQRI T.....ADPG
MtbRv3214   LP........ .......... .......... ...LAEGSRF A.....MPTA

FtHap       DIYTVKLRYN .......... .......... ....GKYVKL ..PIMDKNNS
EcoliSixA   AIASVTLDES .......... .......... ....GNG... ..TFNWQMSP
EcPhytase   GELVFERWRR LSDNSQWIQV SLVFQTLQQM RDKTPLSLNT PPGEVKLTLA
SfNeo13     GVFIVDRTED .......... .......... ....GCA... ..SLVVANSK
STCobC      CWSAIDICE. .......... .......... ....GFA... ..TLKVLNSR
EcolidPGM   VPLVYEFDEN FKPLKRYYL. .......... ....GNA... ..DEIAAKAA
MtbdPGM     IPLRYDLDSA MRPLVRG... .......... ....GTYLDP ..EAAAAGAA
BsPhoE      SVTIIEVDG. .......... .......... ....GTF... ..HVAVEGDV
MtbRv2419c  NWPALG.... .......... .......... ....GMG... ..NASWTQLS
MtbRv2228c  SLSIAEFYAD .......... .......... ....GAS... ..SVRLVNQT
MtbRv2135c  SVSVVRYTQL RPFVLHVNHT .......... ....GARLAP ALQAAASAQG
MtbRv3214   SIGICGFEH. .......... .......... ....GVR... ..QLAVLGLT

FtHap       CSLDALNKYM QSINEKFQKH HHHHH.....
EcoliSixA   CNLKMAKAI. .......... ..........
EcPhytase   GCEERNAQGM CSLAGFTQIV NEARIPACSL
SfNeo13     THLARTA... .......... ..........
STCobC ACWRPE.... .......... ..........
EcolidPGM   AVANQGKAK. .......... ..........
MtbdPGM     AVAGQGRGGV PRGAAALEHH HHHH......
BsPhoE      SHIEEVKEV. .......... ..........
MtbRv2419c  GHWAPGSDFE SIRWRLDVWN ASAQVSSDVL
MtbRv2228c  GYL....... .......... ..........
MtbRv2135c  ASPEPNAAVP PGDAVIGGST D.........
MtbRv3214   GHPQPIAAG. .......... ..........
```

Fig. 23.1. Continued.

remove the phosphate group from the substrate. The primary structure of the histidine phosphatase superfamily is characterized by the signature motif RHG, usually near the *N*-terminus, as shown in Fig. 23.1. Some histidine phosphatase superfamily members also have another α-domain, which contributes to specific substrate binding.

In addition to the nucleophile histidine, the active site consistently contains two arginine residues, including the ones in the signature motif, and another histidine residue which provides a positively charged environment to stabilize the negatively charged phosphate group. The active sites of most of the branch two enzymes and some of the branch one enzymes also contain a third arginine residue, which resides near the RHG signature. Therefore, many of the members of the histidine phosphatase superfamily have the signature of RHGXRXP (where X is an unspecified or unknown amino acid residue). The enzymes also contain a glutamic acid residue (most of the branch one enzymes) or an aspartic acid residue (many branch two enzymes, including phytases; Kumar *et al.*, 2012) to provide a proton for the leaving dephosphorylated substrate. The carbonyl group of the highly conserved glycine in the signature motif, RHG, makes a hydrogen bond with the nucleophilic histidine side chain, holding it in an appropriate conformation. This residue may occasionally be alanine.

While the active sites have a relatively consistent structure, the binding sites to the substrate vary enormously to accommodate the diversity of the substrates of histidine phosphatase superfamily and their different degrees of specificity. When the substrate is a macromolecule, the active site may be rather exposed, while the active site of an enzyme that has a small substrate may be more restricted, which is sometimes facilitated by the presence of an extra α-domain.

The catalysis of the histidine phosphatase superfamily has a two-step mechanism. First, the phosphate group of the substrate is attacked by the nucleophilic histidine, resulting in the formation of a phosphohistidine intermediate and the release of an alcohol molecule (the dephosphorylated substrate). The second and rate-limiting step is the hydrolysis of the phosphohistidine intermediate, assisted by the glutamic or aspartic acid residue in the active site.

23.4 Functions of the Histidine Phosphatase Superfamily in Bacteria

The diverse functions of the histidine phosphatase superfamily are reflected by the diversity of the primary structures of their proteins, which have evolved to catalyse the dephosphorylation of various substrates.

23.4.1 Biosynthetic pathways

The most studied function of the histidine phosphatase superfamily is, paradoxically, the mutase function of the dPGMs, which may sometimes also have phosphatase activity. Phosphoglycerate mutases (PGMs) (EC 5.4.2.1) interconvert 3- and 2-phosphoglyceric acid (3-PGA and 2-PGA), which is an essential step in glycolysis or gluconeogenesis. Most organisms, both prokaryotic and eukaryotic, contain one of the two types of PGM. dPGMs require a cofactor, 2,3-bisphosphoglyceric acid, while the other type of PGM, the iPGM, is cofactor independent. iPGMs are present in higher plants, while dPGMs are present in a variety of organisms, such as bacteria, yeasts and vertebrates. In the bacterial superkingdom, both dPGMs and iPGMs exhibit a patchy distribution, and species in the same genus may possess different kinds of PGMs. Some organisms, such as *E. coli*, contain both types of PGM (Foster *et al.*, 2010). 3D-structures of their dPGMs have been solved for a number of bacteria, including *E. coli* (Bond *et al.*, 2001, 2002), *M. tuberculosis* (Müller *et al.*, 2005), *Bacillus stearothermophilus* (Rigden *et al.*, 2002) and more recently, *Burkholderia pseudomallei* (Davies *et al.*, 2011) and *Staphylococcus aureus* (Roychowdhury *et al.*, 2014).

HAPs are is involved in the biosynthesis of serine. One of the serine biosynthetic pathways in microorganisms and higher plants has three steps, involving 3-phosphoglycerate dehydrogenase, 3-phosphoserine aminotransferase and then phosphoserine phosphatase (PSP). Most known PSPs are Mg^{2+}-dependent. However, two novel metal-independent PSPs have been identified in *Hydrogenobacter thermophilus*. Their primary structures are similar to those of the dPGMs, and they are also similar to proteins in other bacteria that are annotated as dPGMs. The fact that some of the bacteria can synthesize serine but lack a metal-dependent PSP suggests that some of annotated dPGMs may actually function as metal-independent PSPs (Chiba *et al.*, 2012). This hypothesis remains to be confirmed.

HAPs play roles in diverse pathways of metabolism of bacteria, including cobalamin synthesis by CobC (an α-ribazole-phosphate phosphatase that removes the 5′-O-P from adenosylcobalamin-5′-phosphate), which has been studied most extensively in *Salmonella enterica* serotype Typhimurium (O'Toole *et al.*, 1994; Zayas and

Escalante-Semerena, 2007) and neomycin synthesis by Neo13 (5′-phosphoribostamycin phosphatase) from *Streptomyces fradiae* (Huang *et al.*, 2005). Considering the difficulty in ascertaining the natural substrate of a phosphatase, many more metabolic roles of HAPs most likely remain to be discovered.

23.4.2 Scavenging functions

Another obvious function of HAPs is scavenging for phosphate that is essential for growth. This is in agreement with the fact that many HAPs, particularly the branch two enzymes, including phytases, are secretory proteins.

The primary roles of most phytase branch two enzymes are to release phosphate from the indigestible phytic acid that is found in grains and oil seeds. Phytases are important additives for animal feed. Most commercial phytases are produced from fungi, particularly *Aspergillus*, but many bacteria from diverse environments, including members of the family *Enterobacteriaceae*, such as *E. coli* and *Klebsiella* (Böhm et al., 2010), also produce phytases (Jorquera *et al.*, 2008), presumably also for scavenging phosphates from the environment. A line of transgenic pigs expressing *E. coli* phytase under control of the mouse parotid gland secretory protein promoter has been constructed. The transgene cassette directs the secretion of the phytase in the saliva (Forsberg *et al.*, 2013), and enables more efficient digestion of plant phosphate and availability of calcium and other major electrolyte macrominerals (Forsberg *et al.*, 2014).

Glucose-1-phosphatases are branch two enzymes that may also play a role in scavenging phosphate and/or glucose. The enzymes usually also have phytase and glucose-6-phosphatase activity (Herter *et al.*, 2006; Kim *et al.*, 2009). *E. coli* glucose-1-phosphatase allowed the bacterium to grow in a medium containing glucose-1-phosphate as the only carbon source (Lee *et al.*, 2003).

One of the most interesting environments in which the scavenging functions of phosphatases may play an important role for pathogenic bacteria is the acidic environment in the phagosome of phagocytic cells. HAPs are potentially important for the survival of intracellular bacteria in such environments.

23.4.3 Signalling

Common signal transduction systems in prokaryotes include the histidine to aspartate (His-to-Asp) phosphorelay. *E. coli* utilizes about 30 distinct His-to-Asp pathways that respond to various environmental stimuli (Mizuno, 1997). Some HAPs play roles in this signalling system. One of the most studied systems is *E. coli* SixA phospho-histidine phosphatase, which modulates the complex ArcB-to-ArcA phosphorelay system (Matsubara and Mizuno, 2000; Hamada *et al.*, 2005) that allows *E. coli* to sense certain aerobic and anaerobic growth conditions. SixA exhibits an activity towards the phosphorylated His-717 of ArcB, a tripartite sensor kinase, which normally donates its phosphate group to the Asp-54 in ArcA, a response regulator. Phospho-ArcA functions as a transcriptional repressor for the *sdh*CDAB (succinate dehydrogenase) operon. SixA, therefore, plays a role in downregulating the phosphorelay under certain anaerobic conditions and may help to derepress the *sdh* operon. It is possible that HAPs, as well as other phosphatases, are involved in other two-component or phosphorelay systems of *E. coli* and other bacteria.

23.4.4 Pathogenesis

Apart from cell signalling, which is important for pathogenic bacteria to orchestrate responses to the host environment and defence mechanisms, HAPs also play some other roles.

Acid phosphatases play roles in the survival of *F. tularensis*, the cause of tularaemia (Mohapatra *et al.*, 2013), in macrophages. There are four acid phosphatase genes identified in highly virulent type A strains, *acpA*, *acpB*, *acpC* and *hap*, the last being a HAP gene. In *F. tularensis* subsp. *novicida*, the expression of *acpA* and *hap* increased 219- and 10-fold, respectively, when the bacteria were in macrophages. All mice infected with a strain having mutations in all four genes survived within macrophages for longer than 9 weeks, while a mutant in the *acpA*, *acpC* and *hap* genes was defective for survival in macrophages and caused delayed deaths in a mouse model of tularaemia. However, Child *et al.* (2010) reported contradictory evidence on the role of acid phosphatases on the pathogenesis of type A

F. tularensis. The 3D-structure of the *F. tularensis* HAP has been solved (Singh *et al.*, 2009), and the structure of this 37 KDa periplasmic HAP is similar to that of human prostatic acid phosphatase, in exhibiting a two-domain fold, an α/β core domain and a cap domain that caps the core. The core domain contains the catalytic histidine residue (His17), the phosphoryl binding site and an aspartate residue (Asp261) that protonates the leaving group. The cap domain also contributes to the recognition of substrates. The structure conforms to the role of phosphate scavenging from small molecules available in host macrophages.

23.5 Role of Histidine Acid Phosphatase Superfamily Proteins in the Pathogenicity of *Mycobacterium tuberculosis*

Several HAPs have been identified and characterized in *M. tuberculosis*, including the dPGM Rv0489 (Müller *et al.*, 2005; Coker *et al.*, 2013), the glucosyl-3-phosphoglycerate phosphatase Rv2419c (Mendes *et al.*, 2011; Zheng *et al.*, 2014) and two other HAPs of unknown substrate specificity (Rv2135c and Rv3214) (Watkins *et al.*, 2005; Watkins and Baker, 2006; Coker *et al.*, 2013). The HAP domain appears to fuse with the RNase H domain in Rv2228c, which has both RNase and α-ribazole phosphatase (CobC) activity (Watkins and Baker, 2008, 2010). There is still another uncharacterized protein, Rv3837c, with the signature RHG near the *N*-terminus. Although none of the mycobacterial histidine acid phosphatases that have been characterized have been directly linked to the pathogenesis of *M. tuberculosis*, some play roles that are indirectly linked.

The metabolism of phosphate is important in the pathogenesis of *M. tuberculosis*. It is essential for the survival and virulence of the bacteria after ingestion by alveolar macrophages. To enable infecting mycobacteria to survive within the phosphate-deficient and acidic environment of the macrophages, which is aimed at killing them, mycobacterial acid phosphatases may play roles by recycling phosphate inside the cell or by liberating phosphate from host molecules (Watkins and Baker, 2006).

The pathogenicity of *M. tuberculosis* has been partly attributed to its cell envelope. The cell envelope is made up of a complex mixture of polysaccharides and lipids, some of which are unique to mycobacteria. Mycolic acids are fatty acids that are unique to mycobacteria and form an integral part of the cell wall. The synthesis of these acids is regulated by methylglucose lipopolysaccharides (MGLPs), and the biosynthesis of MGLPs requires the concerted action of several enzymes, including phosphatases. Rv2419c is a HAP identified as catalysing the second step of the pathway leading to the biosynthesis of MGLPs (Mendes *et al.*, 2011), so it performs an essential function in the synthesis of the mycobacterial cell wall and is a potential drug target in the treatment of mycobacterial infections.

Rv2135c is another HAP found to be associated with the membrane fractions of mycobacteria (Malen *et al.*, 2010). It is predicted to be in the same operon as Rv2136c, the only homologue of undecaprenyl pyrophosphate phosphatase, which plays an important role in the transfer of building blocks of cell wall synthesis from the cytoplasmic space to the extracellular space in *M. tuberculosis* (Coker and Palittapongarnpim, 2011; Darby *et al.*, 2011). Despite the apparently essential role of Rv2136c, it has found actually been found to be non-essential (Sassetti *et al.*, 2003), indicating that there are additional proteins that can perform the same function. Rv2135c is one the possible candidates that may complement the functions of Rv2136c, because both of them, and their homologues, are predicted to share the same operon in several species of mycobacteria and *Actinomycetales* (Coker and Palittapongarnpim, 2011). An alternative clue to the function of Rv2135c is the fact that the novel type of phosphoserine phosphatase of *Hydrogenobacter thermophiles* was identified as a very similar protein with known crystallographic structural data (Coker *et al.*, 2013). Rv2135c is, therefore, likely to play some role in virulence and pathogenesis, though the details remain to be elucidated.

Rv0489 has been confirmed to be the dPGM of *M. tuberculosis* (Müller *et al.*, 2005; Coker *et al.*, 2013). It was purified from the phagosome of infected murine bone marrow macrophages infected with mycobacteria but was not found in mycobacteria in broth culture (Mattow *et al.*, 2006). This suggests a likely involvement of the Rv0489 gene in the pathogenesis of *M. tuberculosis*.

The *N*-terminal domain of Rv2228c is similar to that of RNase H and can degrade both a DNA/RNA hybrid and double-stranded RNA. The coupling with HAP at the *C*-terminal suggests a related function of the two domains. It is hypothesized that the protein can provide an economic means to generate inorganic phosphate from a hybrid RNA/DNA substrate or dsRNA and thereby improve survival in the phosphate-deprived environment inside the macrophage (Watkins and Baker, 2008, 2010). Homologues of Rv2228c appear in some other pathogenic mycobacteria, such as *M. leprae*.

Another candidate phosphatase that may play a role in the macrophage environment is Rv3214, which is a broad-substrate member of the HAPs. It has been identified by mass spectrometry in the culture filtrate, the membrane protein fraction and whole cell lysates of *M. tuberculosis* (Malen *et al.*, 2010; de Souza *et al.*, 2011). The presence of the protein in the culture filtrate suggests that it may be secreted and potentially involved in scavenging phosphate in infected macrophages.

23.6 Conclusions

The histidine phosphatase superfamily plays an essential role in phosphate metabolism, which is particularly important for bacterial pathogens in the acidic environment inside macrophages. The low level of sequence similarity between members of the family makes the prediction of their functions unreliable. These range from biosynthesis to scavenging to regulatory functions. At present, the functions of relatively small fractions of the family have been experimentally determined. The determination of more 3D-structures will provide important clues to the further characterization of this superfamily.

References

Böhm, K., Herter, T., Müller, J.J., Borriss, R. and Heinemann, U. (2010) Crystal structure of *Klebsiella* sp. ASR1 phytase suggests substrate binding to a preformed active site that meets the requirements of a plant rhizosphere enzyme. *The FEBS Journal* 277, 1284–1296.

Bond, C.S., White, M.F. and Hunter, W.N. (2001) High resolution structure of the phosphohistidine-activated form of *Escherichia coli* cofactor-dependent phosphoglycerate mutase. *The Journal of Biological Chemistry* 276, 3247–3253.

Bond, C.S., White, M.F. and Hunter, W.N. (2002) Mechanistic implications for *Escherichia coli* cofactor-dependent phosphoglycerate mutase based on the high-resolution crystal structure of a vanadate complex. *Journal of Molecular Biology* 316, 1071–1081.

Chiba, Y., Oshima, K., Arai, H., Ishii, M. and Igarashi, Y. (2012) Discovery and analysis of cofactor-dependent phosphoglycerate mutase homologs as novel phosphoserine phosphatases in *Hydrogenobacter thermophilus*. *The Journal of Biological Chemistry* 287, 11934–11941.

Child, R., Wehrly, T.D., Rockx-Brouwer, D., Dorward, D.W. and Celli, J. (2010) Acid phosphatases do not contribute to the pathogenesis of type A *Francisella tularensis*. *Infection and Immunity* 78, 59–67.

Coker, O.O. and Palittapongarnpim, P. (2011) Current understanding of *de novo* synthesis of bacterial lipid carrier (undecaprenyl phosphate): more enzymes to be discovered. *African Journal of Microbiology Research* 5, 2555–2565.

Coker, O.O., Warit, S., Rukseree, K., Summpunn, P., Prammananan, T. and Palittapongarnpim, P. (2013) Functional characterization of two members of histidine phosphatase superfamily in *Mycobacterium tuberculosis*. *BMC Microbiology* 13:292.

Darby, C.M., Venugopal, A., Ehrt, S. and Nathan, C.F. (2011) *Mycobacterium tuberculosis* gene *Rv2136c* is dispensable for acid resistance and virulence in mice. *Tuberculosis* 91, 343–347.

Davies, D.R., Staker, B.L., Abendroth, J.A, Edwards, T.E., Hartley, R., Leonard, J., Kim, H., Rychel, A.L., Hewitt, S.N., Myler, P.J. and Stewart, L.J. (2011) An ensemble of structures of *Burkholderia pseudomallei* 2,3-bisphosphoglycerate-dependent phosphoglycerate mutase. *Acta Crystallographica Section F, Structural Biology and Crystallization Communications* 67, 1044–1050.

de Souza, G.A., Leversen, N.A., Malen, H. and Wiker, H.G. (2011) Bacterial proteins with cleaved or uncleaved signal peptides of the general secretory pathway. *Journal of Proteomics* 75, 502–510.

Forsberg, C.W., Meidinger, R.G., Liu, M., Cottrill, M., Golovan S. and Phillips, J.P. (2013) Integration, stability and expression of the *E. coli* phytase transgene in the Cassie line of Yorkshire Enviropig. *Transgenic Research* 22, 379–389.

Forsberg, C.W., Meidinger, R.G., Ajakaiye, A., Murray, D., Fan, M.Z., Mandell, I.B. and Phillips, J.P. (2014) Comparative carcass and tissue nutrient composition of transgenic Yorkshire pigs expressing phytase in the saliva and conventional Yorkshire pigs. *Journal of Animal Science* 92, 4417–4439.

Foster, J.M., Davis, P.J., Raverdy, S., Sibley, M.H., Raleigh, E.A., Kumar, S. and Carlow, C.K. (2010) Evolution of bacterial phosphoglycerate mutases: non-homologous isofunctional enzymes undergoing gene losses, gains and lateral transfers. *PloS ONE* 5(10): e13576.

Hamada, K., Kato, M., Shimizu, T., Ihara, K., Mizuno, T. and Hakoshima, T. (2005) Crystal structure of the protein histidine phosphatase SixA in the multistep His-Asp phosphorelay. *Genes to Cells* 10, 1–11.

Herter, T., Berezina, O.V., Zinin, N.V., Velikodvorskaya, G.A., Greiner, R. and Borriss, R. (2006) Glucose-1-phosphatase (AgpE) from *Enterobacter cloacae* displays enhanced phytase activity. *Applied Microbiology and Biotechnology* 70, 60–64.

Huang, F., Haydock, S.F., Mironenko, T., Spiteller, D., Li, Y. and Spencer, J.B. (2005) The neomycin biosynthetic gene cluster of *Streptomyces fradiae* NCIMB 8233: characterisation of an aminotransferase involved in the formation of 2-deoxystreptamine. *Organic and Biomolecular Chemistry* 3, 1410–1418.

Jorquera, M., Martinez, O., Maruyama, F., Marschner, P. and de la Luz Mora, M. (2008) Current and future biotechnological applications of bacterial phytases and phytase-producing bacteria. *Microbes and Environments* 23, 182–191.

Kim, Y.O, Kim, H.W., Park, I.S., Lee, J.H., Lee, S.J. and Kim, K.K. (2009) Purification, characterization, and gene cloning of glucose-1-phosphatase from *Citrobacter braakii*. *The Journal of General and Applied Microbiology* 55, 345–350.

Kumar, V., Singh, G., Verma, A.K. and Agrawal, S. (2012) *In silico* characterization of histidine acid phytase sequences. *Enzyme Research* 2012: 845465.

Lee, D.C., Cottrill, M.A., Forsberg, C.W. and Jia, Z. (2003) Functional insights revealed by the crystal structures of *Escherichia coli* glucose-1-phosphatase. *The Journal of Biological Chemistry* 278, 31412–31418.

Malen, H., Pathak, S., Softeland, T., de Souza, G.A. and Wiker, H.G. (2010) Definition of novel cell envelope associated proteins in Triton X-114 extracts of *Mycobacterium tuberculosis* H37Rv. *BMC Microbiology* 10:132.

Matsubara, M. and Mizuno, T. (2000) The SixA phospho-histidine phosphatase modulates the ArcB phosphorelay signal transduction in *Escherichia coli*. *FEBS Letters* 470, 118–124.

Mattow, J., Siejak, F., Hagens, K., Becher, D., Albrecht, D., Krah, A., Schmidt, F., Jungblut, P.R., Kaufmann, S.H. and Schaible, U.E. (2006) Proteins unique to intraphagosomally grown *Mycobacterium tuberculosis*. *Proteomics* 6, 2485–2494.

Mendes, V., Maranha, A., Alarico, S., da Costa, M.S. and Empadinhas, N. (2011) *Mycobacterium tuberculosis* Rv2419c, the missing glucosyl-3-phosphoglycerate phosphatase for the second step in methylglucose lipopolysaccharide biosynthesis. *Scientific Reports* 1: 177.

Mizuno, T. (1997) Compilation of all genes encoding two-component phosphotransfer signal transducers in the genome of *Escherichia coli*. *DNA Research* 4, 161–168.

Mohapatra, N.P, Soni, S., Rajaram, M.V., Strandberg, K.L. and Gunn, J.S. (2013) Type A *Francisella tularensis* acid phosphatases contribute to pathogenesis. *PloS ONE* 8(2): e56834.

Müller, P., Sawaya, M.R., Pashkov, I., Chan, S., Nguyen, C., Wu, Y., Perry, L.J. and Eisenberg, D. (2005) The 1.70 Å X-ray crystal structure of *Mycobacterium tuberculosis* phosphoglycerate mutase. *Acta Crystallographica Section D, Biological Crystallography* 61, 309–315.

O'Toole, G.A, Trzebiatowski, J.R. and Escalante-Semerena, J.C. (1994) The *cobC* gene of *Salmonella typhimurium* codes for a novel phosphatase involved in the assembly of the nucleotide loop of cobalamin. *The Journal of Biological Chemistry* 269, 26503–26511.

Rigden, D.J. (2008) The histidine phosphatase superfamily: structure and function. *Biochemical Journal* 409, 333–348.

Rigden, D.J., Mello, L.V., Setlow, P. and Jedrzejas, M.J. (2002) Structure and mechanism of action of a cofactor-dependent phosphoglycerate mutase homolog from *Bacillus stearothermophilus* with broad specificity phosphatase activity. *Journal of Molecular Biology* 315, 1129–1143.

Rigden, D.J., Littlejohn, J.E., Henderson K. and Jedrzejas, M.J. (2003) Structures of phosphate and trivanadate complexes of *Bacillus stearothermophilus* phosphatase PhoE: structural and functional analysis in the cofactor-dependent phosphoglycerate mutase superfamily. *Journal of Molecular Biology* 325, 411–420.

Roychowdhury, A., Kundu, A., Gujar, A., Bose, M. and Das, A.K. (2014) Expression, purification, crystallization and preliminary X-ray diffraction studies of phosphoglycerate mutase from *Staphylococcus aureus* NCTC8325. *Acta Crystallographica Section F, Structural and Biological Communications* 70, 53–56.

Sassetti, C.M., Boyd, D.H. and Rubin, E.J. (2003) Genes required for mycobacterial growth defined by high density mutagenesis. *Molecular Microbiology* 48, 77–84.

Singh, H., Felts, R.L., Schuermann, J.P., Reilly, T.J. and Tanner, J.J. (2009) Crystal structures of the histidine acid phosphatase from *Francisella tularensis* provide insight into substrate recognition. *Journal of Molecular Biology* 394, 893–904.

Watkins, H.A and Baker, E.N. (2006) Structural and functional analysis of Rv3214 from *Mycobacterium tuberculosis*, a protein with conflicting functional annotations, leads to its characterization as a phosphatase. *Journal of Bacteriology* 188, 3589–3599.

Watkins, H.A. and Baker, E.N. (2008) Cloning, expression, purification and preliminary crystallographic analysis of the RNase HI domain of the *Mycobacterium tuberculosis* protein Rv2228c as a maltose-binding protein fusion. *Acta Crystallographica Section F, Structural and Biological Crystallization Communications* 64, 746–749.

Watkins, H.A. and Baker, E.N. (2010) Structural and functional characterization of an RNase HI domain from the bifunctional protein Rv2228c from *Mycobacterium tuberculosis*. *Journal of Bacteriology* 192, 2878–2886.

Watkins, H.A., Yu, M. and Baker, E.N. (2005) Cloning, expression, purification and preliminary crystallographic data for Rv3214 (EntD), a predicted cofactor-dependent phosphoglycerate mutase from *Mycobacterium tuberculosis*. *Acta Crystallographica Section F, Structural and Biological Crystallization Communications* 61, 753–755.

Zayas, C.L. and Escalante-Semerena, J.C. (2007) Reassessment of the late steps of coenzyme B_{12} synthesis in *Salmonella enterica*: evidence that dephosphorylation of adenosylcobalamin-5′-phosphate by the CobC phosphatase is the last step of the pathway. *Journal of Bacteriology* 189, 2210–2218.

Zheng, Q., Jiang, D., Zhang, W., Zhang, Q., Zhao, Q., Jin, J., Li, X., Yang, H., Bartlam, M., Shaw, N., Zhou, W. and Rao, Z. (2014) Mechanism of dephosphorylation of glucosyl-3-phosphoglycerate by a histidine phosphatase. *The Journal of Biological Chemistry* 289, 21242–21251.

24 Functions and Metabolism of D-Amino Acids in Microorganisms

S. Takahashi,[1]* K. Abe,[1] K. Shibata[2] and Y. Kera[1]

[1]Department of Bioengineering, Nagaoka University of Technology, Nagaoka, Niigata, Japan; [2]Department of Chemistry and Biochemistry, Fukushima National College of Technology, Iwaki, Fukushima, Japan

24.1 Abstract

All naturally occurring amino acids, except glycine, have two chiral forms, L (laevorotatory) and D (dextrorotatory). L-Amino acids occur predominantly in nature, but certain D-amino acids have also been found in various organisms. Microorganisms are a major source of D-amino acids in nature, and in microorganisms, D-amino acids have been known for a long time to serve as essential components of bacterial cell wall peptidoglycan and peptide antibiotics. However, other types of free D-amino acids have also been recently shown to be produced by a wide range of microorganisms and to play important roles in various cellular processes. Many enzymes participate in D-amino acid metabolism in microorganisms, and increasing genome sequence information has further identified unique enzymes and metabolic routes for D-amino acids. These findings indicate that the range of D-amino acids in microorganisms is more extensive than previously expected. In this chapter, we describe the physiological functions of D-amino acids in microorganisms and the enzymes that are involved in D-amino-acid metabolism.

24.2 Introduction

Natural amino acids, except for glycine, have a chiral carbon atom at their α-carbon positions, so they have two stereoisomeric forms, the L- and D-forms. In nature, L-amino acids are far more predominant than D-amino acids and are one of the vital compounds for organisms as a component of proteins synthesized by the ribosomes. In contrast, D-amino acids are considerably less observed in organisms and for a long time they have therefore received less attention from the scientific community than L-amino acids. Significant amounts of various D-amino acids have recently been found not only in bacteria, but also plants and animals, including humans, where they are suggested to play a role in a various biological processes (Friedman, 2010). These findings indicate that greater attention should be paid to studies not only on the distribution and functions of D-amino acids but also on D-amino acid-metabolizing enzymes in various organisms.

Although D-amino acids can be generated by chemical reactions with the help of extreme conditions, such as higher temperature, the main suppliers of D-amino acids in nature may well be microorganisms, particularly bacteria.

*Corresponding author. E-mail address: shoutaka@vos.nagaokaut.ac.jp

Hence, large amounts of D-amino acids are mainly found in environments that are abundant with microorganisms, such as soils, fermented foods, and animal excrement. Bacteria have been known for a long time to produce two particular D-amino acids, D-Ala and D-Glu, which are components of the bacterial cell wall peptidoglycan (PG) and have been termed 'canonical D-amino acids'. Recently, various other D-amino acids previously not found in PG have been shown to be produced as a free form by various bacterial species (Cava *et al.*, 2011a). These newly found D-amino acids are often called 'non-canonical D-amino acids', and have been shown to be involved in certain cellular processes, such as biofilm disassembly and cell wall remodelling (Cava *et al.*, 2011a).

In microorganisms, D-amino acids can be produced by several enzymatic routes, but they are mainly synthesized by the amino acid racemases. This enzyme family is divided into two groups on the basis of whether or not pyridoxal-5′-phosphate (PLP) is required as a cofactor for the reaction. PLP-dependent amino acid racemases are further classified into five distinct structural fold types, I–V. Microbial amino acid racemases are classified into the fold type I, II or III (Hernández and Cava, 2015). Several types of enzymes are responsible for the degradation of D-amino acids in order to utilize them for cell growth or to detoxify them; these include enzymes catalysing the oxidation, dehydrogenation and dehydration of D-amino acids.

24.3 Functions of D-Amino Acids

24.3.1 Functions of D-amino acids in prokaryotic microorganisms

D-Amino acids have been known for a long time to be a major component of the bacterial cell wall PG that provides a protective barrier for the cell, enabling cell survival under variable physico chemical conditions. The major D-amino acids of PG are D-Ala and D-Glu, although others, such as D-Asp, D-Ser, D-Asn and D-Lys, are also observed in the PG of certain bacteria.

D-Asp is found in the interpeptide bridge of PG layer of lactic acid bacteria, such as *Lactobacillus lactis* and *Enterococcus* spp. (Bellais *et al.*, 2006). In these bacterial PGs, the β-carboxy group of D-Asp is linked to the ε-amino group of L-Lys by D-Asp ligase, and the α-carboxy group of the D-Asp is partially amidated to form D-*iso*-Asn in the mature PG.

D-Ser is found in the terminal position of PG precursors in vancomycin-resistant *Enterococcus* spp. (Meziane-Cherif *et al.*, 2014). Vancomycin is an antibiotic that can inhibit PG synthesis by binding to the peptidoglycan monomer, preventing the formation of the peptide cross bridge, and D-Ser can decrease the effect of vancomycin by lowering the binding affinity of vancomycin to the PG precursor.

D-Lys is found in PG of the hyperthermophilic bacterium *Thermotoga maritima* (Boniface *et al.*, 2009). This D-amino acid is incorporated by MurE synthetase into position 3 of the PG stem instead of *meso*-diaminopimelic acid or L-Lys in the nascent PG. However, it is unknown whether the presence of the unusual D-amino acid is related to the thermophilic phenotype.

Another popular function of the D-amino acids in prokaryotic organisms is as a component of the extracellular polymer poly-γ-glutamic acid (PGA). This polymer is produced by various Gram-positive bacteria, primarily *Bacillus* spp., and also by some archaea (Ogunleye *et al.*, 2015). PGA can be composed of only D- or L-Glu or of both enantiomers via γ-amide linkages, and is synthesized in a ribosome-independent manner. The pathogenic *B. anthracis* produces PGA composed of only D-Glu, which enables this bacterium to evade the host immune system, to protect it from phage infections and to prevent antibodies from gaining access into the bacterium. *Staphylococcus epidermidis* produces PGA containing both Glu enantiomers which protects it against antimicrobial peptides (Kocianova *et al.*, 2005). PGA allows some bacteria and an archaea to survive at high salt concentrations and can serve as a carbon and nitrogen source under nutrient starvation conditions.

Teichoic acids are extracellular anionic molecules comprising polymeric glycerol phosphate moieties containing D-Ala. These acids play a role as a component of the cell wall envelope of many Gram-positive bacteria (Neuhaus and Baddiley, 2003).

D-Amino acids also function as components of various bacterially produced peptide antibiotics, including gramicidins, bacitracins and daptomycins (Fig. 24.1) (Martínez-Rodríguez *et al.*, 2010). Gramicidins are produced by *Brevibaillus brevis*

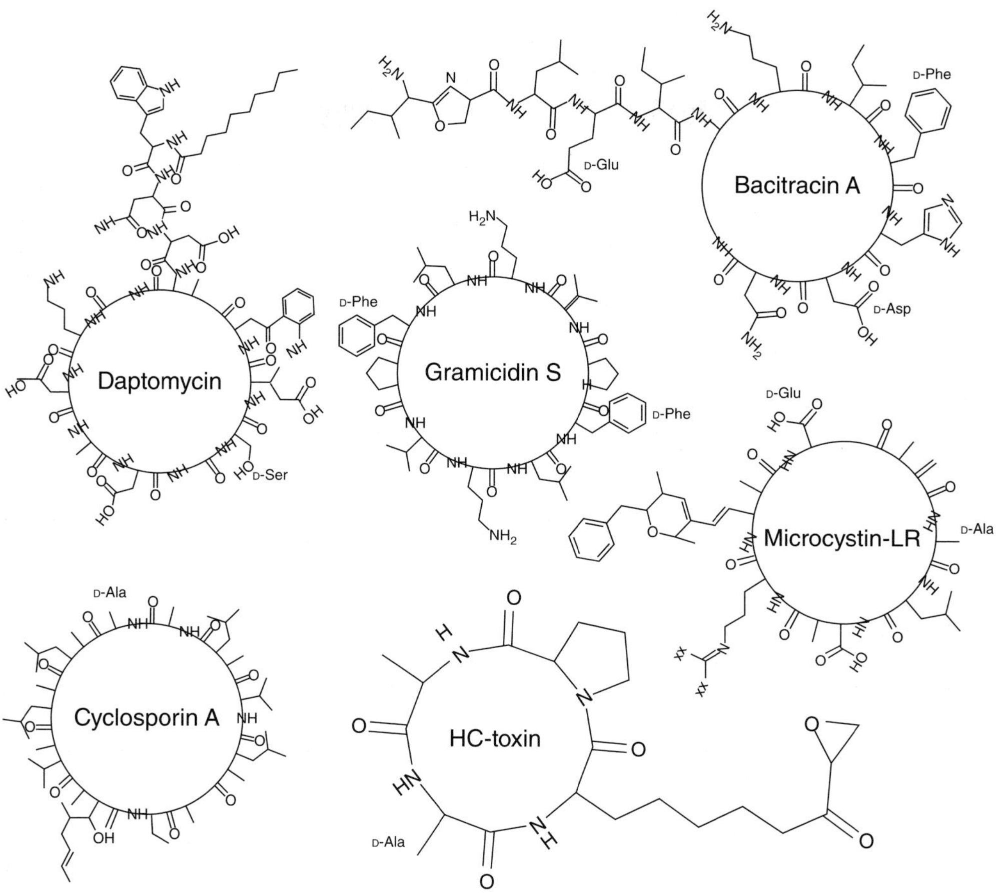

Fig. 24.1. The structures of antibiotic peptides containing D-amino acids produced by microorganisms.

and contain D-Phe, D-Leu or D-Val (Wallace, 1986). Bacitracins are produced by *Bacillus subtilis* and *B. licheniformis* and contains D-Phe, D-Asp and D-Glu (Toscano and Storm, 1982). Daptomycin is produced by *Streptomyces roseosporus* and contains D-Ala and D-Ser (Debono *et al.*, 1987). These bacterial peptide antibiotics are synthesized mainly by non-ribosomal peptide synthetases. D-Amino acids are also found in ribosome-synthesized peptides, such as lantibiotics (Ltns) including lactocin S produced by *Lactobacillus sakei* and Ltn-α and Ltn-β produced by *Lactococcus lactis* (Cotter *et al.*, 2005). These peptides contain some D-Ala residues, which are generated by the conversion of L-Ser during their maturation.

In the anaerobic bacteria *Clostridium* spp., D-Pro plays a role in a unique amino acid fermentation called the Stickland reaction, which involves the coupled oxidation and reduction of amino acids used for ATP production (Fig. 24.2) (Bouillaut *et al.*, 2013). In this reaction, D-Pro is reduced by D-Pro reductase as a Stickland acceptor, which is coupled with the oxidation of other L-amino acids, such as L-Ala, L-Ile and L-Leu, as Stickland donors.

Until recently, the role of the non-canonical D-amino acids produced in various bacteria was not really known, but it has now been shown that they play a role in cell wall remodelling or biofilm dispersion in certain bacterial species (Cava *et al.*, 2011b). In *Vibrio cholerae*, the non-canonical D-amino acids, D-Met and D-Leu, that it produced were incorporated into its cell wall PG to edit its composition and structure. Because the production of non-canonical D-amino acids in this bacterium depends on the stress-associated

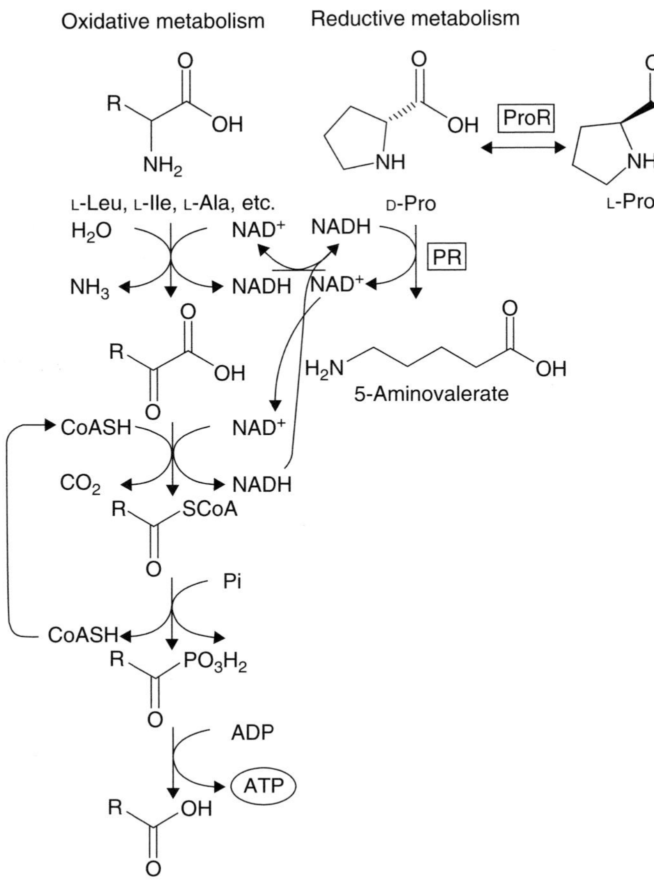

Fig. 24.2. The Stickland reaction with L-amino acids as electron donors and D-Pro as an electron acceptor. The reaction involves the coupled oxidation and reduction of amino acids used for ATP production. Abbreviations: Pi, inorganic phosphate; PR, D-proline reductase; ProR, D-proline racemase.

sigma factor RpoS, the change in PG structure caused by the incorporation of D-amino acids is considered to be an adaptation to various environmental stresses (Cava *et al.*, 2011b).

The non-canonical D-amino acids D-Leu, D-Met, D-Trp and D-Tyr produced by *B. subtilis* are proposed to be involved in biofilm disassembly (Kolodkin-Gal *et al.*, 2010). These D-amino acids also prevented biofilm formation by the pathogens *Staphylococcus aureus* and *Pseudomonas aeruginosa*. In *B. subtilis*, this disassembly was initially believed to be caused by the release of TasA fibres (amyloid-like protein fibres that normally form part of the matrix around the cell) from cells caused by the incorporation of the D-amino acids into the cell wall, but it has recently been shown to be caused by interference in protein synthesis caused by the D-amino acids in the certain strains carrying a mutation in the *dtd* gene that encodes D-aminoacyl-tRNA deacylase, which is responsible for removing D-amino acids from D-aminoacyl-tRNAs (Leiman *et al.*, 2013).

24.3.2 Functions of D-amino acids in eukaryotic microorganisms

In eukaryotic microorganisms, D-amino acids have been shown to be a component of peptides, such as cyclosporin A and HC-toxin (Fig. 24.1) (Martínez-Rodríguez *et al.*, 2010). Some fungi, such as *Tolypocladium inflatum*, produce a cyclic peptide cyclosporin A containing one D-Ala. Cyclosporin A exhibits various biological activities, including immunosuppressive, antiinflammatory, antifungal and antiparasitic properties (Survase *et al.*, 2011). In addition, the fungus *Cochliobolus carbonum* produces the peptide HC-toxin, which is a cyclic tetrapeptide with the structure cyclo(D-Pro-L-Ala-D-Ala-L-Aeo), where Aeo is 2-amino-9,10-epoxi-8-oxodecanoic acid. HC-toxin causes the inhibition of maize root growth (Walton, 2006). The ascospore walls of the yeast *Saccharomyces cerevisiae* contain D,L-dityrosine, which has been proposed to be important for protecting the spores from proteases (Briza *et al.*, 1990).

24.4 D-Amino Acid Synthesis

24.4.1 PLP-dependent amino acid racemase

24.4.1.1 *Alanine racemase*

Alanine racemase (Alr; EC 5.1.1.1) is a PLP-dependent amino acid racemase and is ubiquitous among bacteria; it also occurs in a few fungi. In bacterial species, the enzyme belongs to the fold type III group of PLP-dependent enzymes and is responsible for the production of D-Ala, an essential component of the bacterial cell wall. Alr, therefore, is ubiquitous among bacteria and is an essential enzyme in most of them. The disruption of this gene results in impaired growth in the absence of D-Ala in the growth medium. Gram-negative bacteria possess two *alr* genes, whereas most Gram-positive bacteria, except for certain *Bacillus* spp., possess only one *alr* gene.

The two alanine racemases of *Escherichia coli* encoded by the *alr* and *dadX* genes share a moderate amino acid similarity. Alr encoded by *alr* is named biosynthetic racemase and is constitutively expressed to produce D-Ala to form the cell wall PG. The Alr encoded by the *dadX* gene is named catabolic racemase and is inducibly expressed in the presence of L-Ala and repressed by glucose. This catabolic racemase is essential for the catabolism of L-Ala to provide D-Ala to D-Ala dehydrogenase; it can also provide D-Ala for cell wall biosynthesis (Awasthy *et al.*, 2012). *B. subtilis* also possesses two Alrs; these are encoded by the *dal* and *yncD* genes, and they have different optimal pHs and temperatures (Pierce *et al.*, 2008). The physiological role(s) of the Alrs in *B. subtilis* remains to be elucidated.

Alr is also found in the fungi *Tolypocladium niveum*, *C. carbonum* and *Schizosaccharomyces pombe*. Most fungal Alr belongs to the fold type I group of PLP-dependent enzymes, whereas the yeast *S. pombe* has the bacterial type (fold type III) of PLP-dependent Alr, which is considered to be derived from *Gammaproteobacteria* by horizontal gene transfer (Uo *et al.*, 2001). This enzyme of the fungi *T. niveum* and *C. carbonum* is responsible for providing D-Ala for production of the cyclic D-amino acid-containing peptides cyclosporin A and HC-toxin, respectively (see Fig. 24.1) (Hoffmann, 1994; Cheng, 2000). In *S. pombe*, this enzyme is involved in the utilization of D-Ala for cell growth and possibly decreases the toxicity of the D-amino acid (Uo *et al.*, 2001). The enzyme is also found to produce D-Ala in the parasitic protozoan *Leishmania amazonensis* and the microalgae *Chlamydomonas reinhardtii* and *Thalassiosira* spp. (Nishimura *et al.*, 2006; Panizzutti *et al.*, 2006), although the physiological significance of the enzyme and of the role of D-Ala in the protozoa and the microalgae remains unknown.

24.4.1.2 *Serine racemase*

Serine racemase (Srr; EC 5.1.1.18) is a PLP-dependent enzyme found in both bacteria and eukaryotic microorganisms. In bacteria, Srr is found in the vancomycin-resistant bacterium *Enterococcus gallinarum* BM4174 (Arias, 2003). This Srr is encoded by the *vanT* gene and is unique in terms of being a membrane-bound amino acid racemase. The transmembrane domain is suggested to be responsible for the transport of L-Ser, whereas the C-terminal domain has a significant sequence identity with bacterial alanine racemase, thus also exhibiting a significant alanine racemase activity (18% of the serine racemase activity) (Arias, 2003). This bacterial enzyme is responsible for providing D-Ser for the synthesis of PG precursors terminating in D-Ser instead of D-Ala, which bestows vancomycin resistance.

In eukaryotic microorganisms, Srr is identified in the yeast *S. pombe* and the slime mould *Dictyostelium discoideum* (Goto *et al.*, 2009; Ito *et al.*, 2012b). The primary structures of these eukaryotic Srrs are quite different from those of bacterial Srrs, but are similar to those of mammalian Srrs, which belong to the fold type II group of PLP-dependent enzymes, which include serine/threonine dehydratase. These eukaryotic Srrs also show a significant dehydratase activity toward both Ser enantiomers. In addition, similar to mammalian Srrs, their activities are enhanced by the presence of divalent cations, such as Mg^{2+} and Ca^{2+}; more interestingly, the activity of the mould Srr is also activated by Na^{+}. However, the physiological significance of these Srrs remains unknown.

24.4.1.3 Arginine/lysine racemase

Arginine racemase (ArgR; EC 5.1.1.9) activity was first identified in an extract of *Pseudomonas graveolens* IFO 3460 (later named *P. taetrolens* NBRC34609) (Soda *et al.*, 1967). Later on, this racemase was shown to contain PLP as a cofactor and to act on not only Arg but also Lys, to a similar extent; it was also weakly active against Met and non-proteinogenic amino acids, such as ornithine and ethionine. Consequently, this enzyme is now being called an arginine racemase with broad substrate specificity (Matsui and Oikawa, 2010). The racemase is a homodimeric protein with a subunit molecular mass of 41.5 kDa for the recombinant enzyme produced in *E. coli*. It belongs to the fold type III group of PLP-dependent enzymes and is unique among PLP-dependent amino acid racemases in its localization to the periplasmic space and in having a disulfide bond (Matsui and Oikawa, 2010). The enzyme plays an essential role in the catabolism of D-Arg and D-Lys.

24.4.1.4 Broad substrate specificity amino acid racemase

Broad substrate specificity amino acid racemase (Bsr; EC 5.1.1.10) is a PLP-dependent racemase acting on various amino acids. Most of this type of bacterial racemase belongs to the fold type III group of PLP-dependent enzymes. It has been identified in various bacteria, such as the Gram-negative *P. putida* strains, *Aeromonas caviae* (previously classified as *A. punctate* subsp. *caviae*) and *V. cholerae*, and the Gram-positive *Lactobacillus otakiensis* JCM15040. In addition, this type of amino acid racemase has also recently been identified in the hyperthermophilic archaeon *Pyrococcus horikoshii* OT-3.

Bsr was first reported in the *P. putida* strain IFO12996 (previously classified as *P. striata*) (Soda and Osumi, 1969), and this Bsr acts on L-Leu, L-α-aminobutyric acid, L-Ala, L-norvaline (L-norVal) and L-norleucine (L-norLeu), with higher activities towards L-ethionine, D-Lys, L-Met and D-Arg. The Bsr from *P. putida* strain SCRC-744 has broad-spectrum activity towards various amino acids, with D-Gln as the best substrate; it is not active towards D-Glu, D-Asp, D-Thr, D-Ile, D-Val and D-Pro (Asano and Endo, 1988). The enzyme from *P. putida* strain ATCC17642 isolated using D-Thr as the sole nitrogen source catalyses not only the racemization of amino acids (except for the acidic and aromatic amino acids), with L-Lys as the best substrate, followed by L-ornithine, L-ethionine, L-Arg, L-Gln and L-Met, but also the epimerization of Thr (D- to L- and L- to D-) at the α-position (Lim *et al.*, 1993).

The Bsr from *A. caviae* isolated using D-Ser as the sole nitrogen source acts on various amino acids, excepting the acidic and aromatic amino acids. L-Lys is the most preferred substrate, and the basic and sulfur amino acids, as well as L-Gln, are good substrates (Inagaki *et al.*, 1987).

The Bsr of *V. cholerae* attracts attention for the production of non-canonical D-amino acids, such as D-Met and D-Leu, which have recently been shown to be involved in the modulation of cell wall structure (Espaillat *et al.*, 2014). This Bsr is localized in the periplasmic space and has a high sequence homology with the alanine racemase of this bacterium. The enzyme can catalyse the racemization of 12 of the 19 natural chiral amino acids (excepting the acidic and aromatic amino acids) but exhibits only a minimal activity toward Ile, Val and Thr. The best substrate is Lys, followed by D-Arg and D-Met. The high production of D-Met and D-Leu by this bacterium *in vivo* is considered to be due to the relatively high availability of these L-amino acids in this organism.

Only one example of Bsr in Gram-positive bacteria has been reported, in the lactic acid bacterium *L. otakiensis* JCM15040 (Mutaguchi *et al.*, 2013). This enzyme was initially annotated as a putative γ-aminobutyrate aminotransferase in the genome sequence. However, the recombinant enzyme expressed in *E. coli* displayed no γ-aminobutyrate aminotransferase activity but preferentially catalysed the epimerization of L-Ile to D-*allo*-Ile and vice versa. In addition, it catalyses the racemization of many non-polar amino acids, including Leu, Val, Phe, Met, Ser and Ala, though the activities towards Ser and Ala are very low.

Bsr has also recently been found in the hyperthermophilic archaeon *P. horikoshii* OT-3 (Kawakami *et al.*, 2015). The enzyme was initially annotated as γ-aminobutyrate aminotransferase in the genome sequence, but the recombinant enzyme displays racemase activity towards Met, Leu, Phe, Ala, Ser, Ile, Val, Trp and Tyr. Phe is the most preferred substrate, followed by Leu and then Met. This enzyme is the first example of a PLP-dependent amino acid racemase with broad substrate specificity in a hyperthermophilic archaean and it is considered to be responsible for the utilization of D-amino acids for cell growth.

24.4.2 PLP-independent amino acid racemase

24.4.2.1 Aspartate racemase

The aspartate racemase (AspR; EC 5.1.1.13) of prokaryotic organisms is independent of PLP for catalysis, although only a PLP-dependent AspR activity was suggested to be in the crude extract of the acidothermophilic archaeon *Thermoplasma acidophilum* HO-62 (Long *et al.*, 2001). Microbial AspR is mainly distributed in lactic acid bacteria, such as *Streptococcus* spp. and *Lactobacillus* spp., and bifidobacteria, such as *Bifidobacterium bifidum* NBRC14252. The enzyme is also identified in cyanobacteria, such as *Microcystis aeruginosa* strain PCC 7806, and in several hyperthermophilic archaea, including *Desulfurococcus* sp. strain SY, *P. horikoshii* OT3 and *Thermococcus* sp. strain KS-8. It is also suggested that AspR exists ubiquitously in species of the archaea *Thermococcus* and *Pyrococcus*.

The AspR of *Streptococcus thermophiles* is a homodimeric enzyme and specifically catalyses the racemization of Asp but not of Ala and Glu (Okada *et al.*, 1991). The AspR of *B. bifidum* NBRC14252 AspR is a monomeric enzyme, which is unique because other microbial AspRs are dimers (Yamashita *et al.*, 2004). This enzyme acts exclusively on Asp and is inactive towards other amino acids, including Glu. Its thermal stability increases significantly in the presence of substrates, particularly the D-enantiomer. The enzyme is responsible for providing D-Asp to construct the cell wall PG.

The AspR of the cyanobacterium *M. aeruginosa* PCC 7806 is a member of the microcystin biosynthetic gene cluster (Sielaff *et al.*, 2003). Microcystins are hepatotoxic cyclic heptapeptides containing D-Asp, D-Glu and D-Ala, which are produced by various freshwater cyanobacteria (see Fig. 24.1) (Tillett *et al.*, 2000). This bacterial AspR can racemize Asp but not Glu and is responsible for providing D-Asp for microcystin synthesis.

The AspR of the hyperthermophilic archaean *Desulfurococcus* sp. strain SY has a significant primary structure homology with that of the bacterium *S. thermophiles* (32.5% identity) (Yohda *et al.*, 1996). This enzyme is highly specific for Asp, and its activity increases with increase in temperature from 37°C to 90°C by approximately 500-fold. Its role in hyperthermophilic archaea remains to be elucidated, but it is thought that AspR is involved in the utilization of D-Asp formed at high temperature or in the production of D-Asp as a cell component (Yohda *et al.*, 1996).

The crystal structure of AspR was first determined for the enzyme from *P. horikoshii* (Liu *et al.*, 2002). It has a dimeric structure, and two catalytic cysteine residues are located on both sides of a cleft between the two domains.

24.4.2.2 Glutamate racemase

Glutamate racemase (MurI, GluR; EC 5.1.1.3) has been identified in a wide range of bacterial species. Many GluRs are homodimers but a few are monomers or exist in equilibrium between monomeric and dimeric states depending on the presence or absence of substrates (Potrykus *et al.*, 2009). The reaction catalysed by this enzyme is a rate-limiting step in the bacterial cell wall biosynthetic pathway, and inhibition of the enzyme reaction can restrict bacterial growth.

Several bacterial species, including *E. coli* and *Mycobacterium tuberculosis*, possess one copy of the GluR gene (*murI*), which is therefore essential for cell growth in these bacteria. In contrast, *Bacillus* spp. and *Staphylococcus* spp. contain two GluR genes. The GluRs of *B. subtilis*, encoded by the *racE* and *yrpC* genes, have similar optimum pHs and substrate specificity, but different thermal stabilities and enzyme kinetic parameters (Ashiuchi *et al.*, 1999). Early studies suggested different roles of the two GluRs; RacE appeared to be responsible for providing D-Glu for PG synthesis and YrpC for poly-γ-glutamic acid synthesis. However, it was later shown that both GluRs are involved in PG synthesis (Oh *et al.*, 2015). In several bacteria, GluR can act as an inhibitor of DNA gyrase by inhibiting the DNA binding of gyrase (Ashiuchi *et al.*, 2002). This inhibition is independent and not linked with the racemase activity. In *E. coli*, the inhibitory effect is only observed in the presence of the PG precursor. The physiological significance of this gyrase inhibition is as yet unknown, but it is speculated that it may be involved in the coordination of the processes of cell wall biosynthesis and DNA replication at the time of cell division.

24.4.2.3 Proline racemase

Proline racemase (ProR; EC 5.1.1.4) has been identified in the pathogenic *Clostridium* spp. and in the eukaryotic pathogenic parasites *Trypanosoma* spp. The clostridial enzyme (PrdF) is important for energy metabolism via the Stickland pathway, in which the oxidation of the Stickland donor amino acid, such as L-Ile, L-Leu,and L-Ala, is coupled to the reduction of the Stickland acceptor amino acid, D-Pro (see Fig. 24.2) (Bouillaut *et al.*, 2013). This pathway is unique to clostridia and serves as the primary energy source for these species.

T. cruzi has two paralogous ProR genes per haploid genome, which are differentially expressed during cell development (Chamond *et al.*, 2003). The parasite produces the intracellular and extracellular isoforms of ProR at different developmental stages. These isoforms exhibit distinct kinetic properties, and the form that is secreted has been shown to be an effective mitogen for host B cells, and to contribute to immune evasion and persistence.

Recently, ProR has also been identified in the hyperthermophilic archaean *Thermococcus litoralis* (Watanabe *et al.*, 2015). This archaeal enzyme catalyses not only the racemization of Pro but also the epimerization of 3- and 4-hydroxy-Pro with similar kinetic constants. The enzyme is the first example of a bifunctional ProR and hydroxyproline epimerase (HypE) from archaea. Its role remains to be elucidated.

24.4.2.4 Phenylalanine racemase (gramicidin S synthetase A)

Phenylalanine racemase (gramicidin S synthetase A, GrsA; EC 5.1.1.11) is responsible for providing D-Phe, the first amino acid of gramicidin S, which is a cyclic decapeptide antibiotic produced by certain strains of *B. brevis* (see Fig. 24.1) (Sun *et al.*, 2014). This enzyme reaction is completely different from that of other amino acid racemases in its requirement for ATP and 4′-phosphopantetheine. In the reaction, D- or L-Phe is initially activated by adenylation with ATP in the adenylation domain, and then racemization occurs at the stage of thioester bond formation with the free thiol group of 4′-phosphopantetheine in the peptidyl-carrier protein (PCP) domain by the action of the epimerization domain (Fig. 24.3). Only the D-isomer is then transferred to gramicidin S synthetase B (GrsB) that is responsible for the synthesis of a pentapeptide of gramicidin S.

24.4.3 D-Amino acid aminotransferase

D-Amino acid aminotransferase (D-AAT; EC 2.6.1.21) is a PLP-dependent enzyme that catalyses reversible transamination between various D-amino acids and α-keto acids. The enzyme belongs to the fold type IV group of PLP-dependent enzymes. D-AAT has been found in various Gram-positive bacteria, including *Rhodospirillum rubrum*, *Bacillus* spp., *Geobacillus* spp., *Bradyrhizobium* (*Rhizobium*) *japonicum*, *Lactobacillus salivarius* and *Staphyloccus haemolyticus*.

The enzymes of *Bacillus* spp. can utilize various D-amino acids as amino donors, including D-Ala, D-α-aminobutyrate, D-theanine (γ-glutamylethylamine), D-Met, D-Asn, D-Asp, D-Glu and D-Gln, and some keto acids as amino acceptors, including α-ketoglutarate, pyruvate and α-ketobutyrate (Yonaha *et al.*, 1975). However, D-AAT from the thermophilic *Bacillus* sp. YM-1 has a narrow substrate specificity compared with that of the mesophilic *B. sphaericus* (Tanizawa *et al.*, 1989). The optimum temperature of *L. salivarius* D-AAT is comparable to that of the D-AATs of *Bacillus* spp. and *Geobacillus* spp. (60°C), but its thermal stability is lower (Kobayashi *et al.*, 2013). The substrate specificity of the *L. salivarius* enzyme greatly differs from those of other D-AATs; D-*allo*-Ile is a more preferred amino donor than D-Ala, and α-ketobutylate, glyoxylate and indole-3-pyruvate are more preferred amino acceptors than α-ketoglutarate. The *Geobacillus* spp. enzyme is highly thermostable, but its catalytic activity is approximately one third of that of the highly active *Bacillus* D-AATs (Lee *et al.*, 2006).

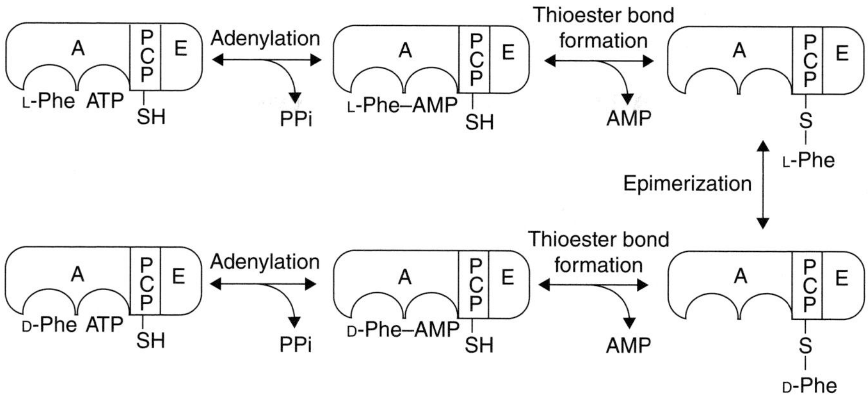

Fig. 24.3. The stages of the racemization reaction of Phe by phenylalanine racemase (gramicidin S synthetase A). Abbreviations: A, the adenylation domain; E, the epimerization domain; PCP, the thiolation domain; PPi, inorganic pyrophosphate.

Most *Bacillus* spp. and *S. haemolyticus* possess both D-AAT and GluR (Fotheringham *et al.*, 1998). D-AAT can synthesize D-Glu using α-ketoglutarate as the amino acceptor and D-Ala, produced by Alr, as the amino donor; this pathway provides an important route for the synthesis of the bacterial cell wall D-Glu. For this reason, in addition to GluR, this enzyme is also a target for the development of new antimicrobial agents. In addition, it has been also proposed that the D-AAT activity of *Bacillus* spp. and other species may prevent racemase-mediated D-Glu overflow and toxicity (Oh *et al.*, 2015).

24.4.4 Enzyme side reactions

D-Amino acids can also be produced as side reaction products in just PLP or in PLP-containing enzymes. The tryptophan synthase of *E. coli*, a PLP-dependent enzyme responsible for the synthesis of L-Trp, can also catalyse the racemization of Trp (Miles *et al.*, 1986). The aspartate aminotransferase of *E. coli*, a PLP-dependent enzyme that catalyses the reversible transamination of acidic L-amino acids and the corresponding α-keto acids, can catalyse the racemization of Asp, Glu and Ala (Vacca *et al.*, 1997). The methionyl-tRNA synthetase of *E. coli* can weakly catalyse the α-proton exchange reaction of L-Met *in vitro*, suggesting the possible conversion of L-Met to D-Met (Williams and Rosevear, 1991). In addition, the cystathionine β-lyase (MetC) of *E. coli*, a PLP-enzyme that catalyses the hydrolysis of L-cystathionine, has been shown to exhibit a significant alanine racemase activity. This can support the cell growth of *E. coli* strains lacking two different alanine racemases (encoded by the *alr* and *dadX* genes) and with a methionine repressor gene (*metJ* gene) mutation (Kang *et al.*, 2011). In addition, the crystalline tyrosine phenol-lyase of *Escherichia intermedia*, a PLP-dependent enzyme that catalyses the reversible hydrolytic cleavage of L-Tyr, can racemize Ala (Kumagai *et al.*, 1970).

24.5 D-Amino Acid Degradation

24.5.1 D-Amino acid oxidase

D-Amino acid oxidase (DAO; EC 1.4.3.3) catalyses the oxidative deamination of neutral and basic D-amino acids to give the corresponding α-keto acids along with the production of ammonia and hydrogen peroxide. DAO contains a non-covalently bound flavin adenine dinucleotide (FAD) as a cofactor and localizes in peroxisomes in eukaryotic organisms (Pollegioni *et al.*, 2007). In microorganisms, this enzyme was first identified in the fungus *Neurospora crassa* (Horowitz, 1944), and then mainly in various other fungi, such as the yeasts *Rhodotorula gracilis* (syn. *Rhodosporidium toruloides*) (Pilone *et al.*, 1989) and *Trigonopsis variabilis* (Kubicek-Pranz and Röhr, 1985). The expression of fungal DAOs is generally induced by the presence of D-amino acids in the culture medium (Takahashi *et al.*, 2012). D-Ala is usually the best inducer for gene expression in many fungi. These fungal enzymes have a relatively broad substrate specificity, with a higher activity toward neutral D-amino acids than towards basic D-amino acids and higher catalytic efficiency than animal enzymes; they are responsible for the utilization of D-amino acids for cell growth (Ohnishi *et al.*, 1962; Pollegioni *et al.*, 2007). This enzyme also plays a protective role against D-amino acid toxicity (Ohnishi *et al.*, 1962).

Recently, DAO has been identified in some *Actinobacteria*, including *Arthrobacter protophormiae* (Geueke *et al.*, 2007), *Streptomyces coelicolor* A3(2) (Saito *et al.*, 2013) and the thermophilic actinobacterium *Rubrobacter xylanophilus* (Takahashi, 2014). Moreover, genome sequence information revealed that homologous genes for DAO are widely distributed in various Gram-positive actinobacteria with a few Gram-negative bacteria (Takahashi *et al.*, 2014). These bacterial DAOs have markedly different enzymatic properties. The DAO from *A. protophormiae* has a higher catalytic activity towards both basic and neutral D-amino acids; this substrate specificity is unique among DAOs. Other bacterial DAOs exhibit a higher activity towards branched chain D-amino acids and much lower catalytic efficiency than the *A. protophormiae* enzyme. The *R. xylanophilus* DAO has a much higher thermal stability than other bacterial DAOs due to a tight cofactor binding (Fig. 24.4). The physiological role of bacterial DAOs remains to be elucidated, but in *S. coelicolor* A3(2) it is suggested that they are involved in the degradation of non-canonical D-amino acids released from the cell wall (Saito *et al.*, 2013).

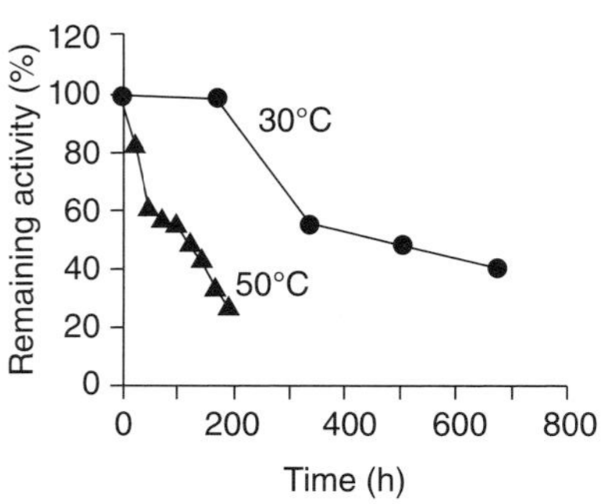

Fig. 24.4. The high thermal stability of the D-amino acid oxidase of *Rubrobacter xylanophilus*.

24.5.2 D-Aspartate oxidase

D-Aspartate oxidase (DDO, DASOX, DASPO; EC 1.4.3.1) catalyses the oxidative deamination of acidic D-amino acids. This enzyme contains FAD as a cofactor, is localized in the peroxisome, and shares homology with DAO in its primary structure, suggesting that the two enzymes have a common ancestral protein. Most DDOs exhibit a higher activity not only towards D-Asp but also to D-Glu, although a few of them show a higher activity towards D-Glu than D-Asp, and are often called D-Glu oxidase (Fukunaga *et al.*, 1998).

In microorganisms, DDO was first identified in the fungus *Fusarium sacchari* (Wakayama *et al.*, 1994) and then in the yeasts *Cryptococcus humicola* UJ1 (Yamada *et al.*, 1996), *Candida boidinii* 2201 (Fukunaga *et al.*, 1998) and *Trichoderma harzianum* SKW-36 (Yano *et al.*, 2012). The primary structure of microbial DDO has only been identified for the *C. humicola* enzyme and this has a higher homology with fungal DAOs than with animal DDOs, suggesting its close relationship to fungal DAOs but not to animal DDOs (Takahashi *et al.*, 2004). These microbial DDOs, however, show relatively different substrate specificities; the *C. humicola* DDO exhibits a much higher activity towards D-Asp than towards D-Glu. In contrast, the enzyme of *C. boidinii* has a higher catalytic efficiency towards D-Glu, namely its D-Glu oxidase activity. The *T. harzianum* enzyme catalyses not only the oxidation of acidic D-amino acids but also, weakly, that of some neutral D-amino acids, such as D-Asn and D-Gln. The expression of *C. humicola* DDO is significantly induced by the presence of D-Asp in the culture medium, with peroxisome proliferation, but not by the presence of D-Glu in the culture medium; the expression of the *C. boidinii* enzyme is induced by D-Glu but not by D-Asp. In contrast, the *T. harzianum* enzyme is not induced by any D-amino acids. In *C. humicola*, DDO plays an essential role in the assimilation of D-Asp as a carbon or nitrogen source, or both, for cell growth (Takahashi *et al.*, 2005). In addition, it may be responsible for the detoxification of acidic D-amino acids.

24.5.3 D-Amino acid dehydrogenase

D-Amino acid dehydrogenase (DAD; EC 1.4.99.1) is a membrane-bound FAD enzyme that catalyses the dehydrogenation of free D-amino acids in the presence of artificial electron acceptors. This enzyme is widely distributed in Gram-negative bacteria such as *E. coli*, *Pseudomonas* spp., *Salmonella typhimurium* [*S. eneterica* serovar Typhimurium], the thermophilic bacterium *Rhodothermus marinus* and the pathogenic bacterium *Helicobacter pylori* (Tanigawa *et al.*, 2010; Satomura *et al.*, 2015). The enzyme has also been found in the archaean *Pyrobaculum islandicum* (Satomura *et al.*, 2002). Most bacterial DADs prefer D-Ala as a substrate, while the enzyme of *H. pylori* is highly active towards D-Pro, and the enzyme of *R. marinus* shows higher activity towards D-Phe, D-Pro and *cis*-4-hydroxy-D-Pro. The enzyme of the archaeon *P. islandicum* also exhibits higher activity towards D-Pro and *allo*-4-hydroxy-D-Pro and has a low similarity of primary structure with those of the mesophilic bacterial D-Ala dehydrogenases, and thus it is classified as a D-Pro dehydrogenase (EC 1.5.99.13).

Most Gram-negative bacterial DADs are considered to be mainly responsible for L-Ala catabolism (Wild and Klopotowski, 1981), and these genes constitute an operon (the *dad* operon) with a catabolic alanine racemase gene. The alanine racemase converts L-Ala to D-Ala, which is then converted by DAD to pyruvate, ammonium and hydrogen, subsequently yielding protons and electrons. It has been shown that in *E. coli*, the electrons generated by this enzyme reaction are transferred to a cytochrome, suggesting that the enzyme plays a role in the utilization of D-amino acids as an energy substrate. *E. coli* DAD is inducibly expressed in the presence of either L-Ala or D-Ala in the culture medium, but L-Ala is the actual activator (Wild *et al.*, 1985). Expression of the enzyme is repressed by leucine-responsive regulatory protein (Lrp), a global regulator that controls the expression of various genes, including the genes involved in amino acid metabolism.

The *DAD* gene (*dadh* gene) of the thermophilic bacterium *R. marinus* constitutes an operon with the 4-hydroxyproline 2-epimerase gene (*hypE*) instead of the alanine racemase gene (Satomura *et al.*, 2015). The expressions of both the *dadh* and *hypE* genes are extensively induced by the presence of *trans*-4-hydroxy-L-Pro, suggesting their involvement in the catabolism of *trans*-4-hydroxy-L-Pro. This function of DAD, called D-hydroxyproline dehydrogenase, is also reported in the bacteria *P. aeruginosa* and *P. putida* (Watanabe *et al.*, 2012).

P. aeruginosa PAO1 also possesses another type of DAD named DauA (Li *et al.*, 2010). DauA has higher catalytic efficiency towards D-Arg and D-Lys; its expression is induced by these basic D-amino acids, and gene disruption causes inability of growth on only the basic D-amino acids. Therefore, DauA is named anabolic D-Arg dehydrogenase. The *dauA* gene constitutes an operon with the catabolic L-Arg dehydrogenase gene (*dauB*). The coupled reaction of the two dehydrogenases converts D-Arg to the L-isomer to utilize the D-amino acid through the L-Arg catabolic pathway (Li *et al.*, 2010). Further, *P. aeruginosa* PAO1 possesses another DAD named DguA (He *et al.*, 2014). The expression of the *dguA* gene is specifically induced by exogenous D-Glu or D-Gln, and is regulated by the transcriptional activator DguR. DguA is responsible for the utilization of D-Glu and D-Gln as the sole source of carbon and/or nitrogen.

24.5.4 D-Serine/threonine dehydratase

D-Serine dehydratase (DSD; EC 4.2.1.18) is a PLP-dependent enzyme that catalyses the dehydration of D-Ser and D-Thr to yield pyruvate and 2-oxobutanoate, respectively, and ammonia. This enzyme can be divided into two evolutionarily distinct types. One type of DSD belongs to the fold type II group of the PLP-dependent enzymes and includes most bacterial DSDs, such as the enzymes of *E. coli* and *Salmonella typhimurium* [*S. eneterica* serovar Typhimurium] (Bharath *et al.*, 2011). Although the majority of the fold type II PLP-dependent enzymes are dimers, these bacterial DSDs are monomers. The other type of DSD enzyme belongs to the fold type III group of PLP-dependent enzymes and includes the enzymes of the bacterium *Burkholderia cepacia* (formerly *Pseudomonas cepacia*) and the yeast *S. cerevisiae* (Wong *et al.*, 1980; Ito *et al.*, 2012a). The enzymatic and biochemical properties of *B. cepacia* DSD are currently poorly understood.

In *E. coli*, the DSD encoded by the *dsdA* gene is inducibly expressed by the presence of D-Ser (Palchaudhuri *et al.*, 1988). This induction is regulated by the DSD activator gene (*dsdC*) and the cyclic AMP-binding protein. Bacterial strains express DSD to utilize D-Ser as a carbon, nitrogen and energy source. In addition, the enzyme plays a role in the detoxification of D-Ser (Korte-Berwanger *et al.*, 2013).

The enzyme of *S. cerevisiae* (DsdSC) is a dimeric protein and requires not only PLP but also Zn^{2+} for maximum activity (Ito *et al.*, 2012a). The Zn^{2+} is located in the active site and is indispensable for both α-hydrogen abstraction and hydroxyl group elimination from D-Ser. In this yeast species, the enzyme contributes minimally to the detoxification of D-Ser.

24.5.5 D-Proline reductase

D-Proline reductase (PR; EC 1.21.4.1) catalyses the reductive cleavage of the D-Pro ring to yield 5-aminovalerate; it is also involved in the amino acid metabolism of several *Clostridium* spp. (see Fig. 24.2) (Kabisch *et al.*, 1999). PR is a membrane-bound protein consisting of a 26 kDa subunit containing a catalytically active selenocysteine (PrdB) and 23 and 45 kDa subunits derived from the proprotein PrdA. The 23 kDa subunit contains a substrate-activating pyruvoyl group at its *N*-terminus. The reduction of D-Pro by PR is coupled to the oxidation of other amino acids to obtain energy via a typical Stickland reaction.

24.5.6 D-Threonine aldolase

D-Threonine aldolase (DTA; EC 4.1.2.42) is a PLP-enzyme that catalyses the reversible cleavage of D-Thr into Gly and acetaldehyde. The enzyme belongs to the fold type III PLP-dependent group of enzymes and structurally resembles bacterial alanine racemases. DTA activity has been found in several genera of bacteria, including the Gram-positive *Arthrobacter* sp. DK-38 and the Gram-negative *Alcaligenes xylosoxidans* IFO12669, *Xanthomonas oryzae* IAM1657 and *Pseudomonas* sp. DK-2, but not in fungi (Kataoka *et al.*, 1997). The physiological significance of DTA in microorganisms remains to be elucidated.

The DTA of the Gram-positive *Arthrobacter* sp. DK-38 can act on not only D-Thr but also D-*allo*Thr and several other D-β-hydroxy-α-amino acids with comparable catalytic efficiency; it has been designated a low-specificity DTA (Kataoka *et al.*, 1997). The enzyme also requires a divalent cation such as Co^{2+}, Ni^{2+}, Mn^{2+} or Mg^{2+} for both its catalytic activity and its thermal stability. The enzyme of the Gram-negative *A. xylosoxidans* IFO12669 shows significant identity with the *Arthrobacter* sp. DTA (54%) in its primary structure (Uhl *et al.*, 2015). This bacterial DTA is also a low-specificity DTA and requires a divalent cation for its catalytic activity; its crystal structure has been determined recently.

24.6 Conclusions

For a long time, D-amino acids were considered to have limited functions in microorganisms. However, the recent identification of various free D-amino acids in a wide variety of microorganisms and newly identified important functions of D-amino acids have shed light on earlier studies on D-amino acids in microorganisms. Microorganisms are the major providers of D-amino acids in nature and have been shown to possess a wide range of D-amino acid metabolic systems to utilize these amino acids, to regulate their functions or to detoxify their toxic effects to in order to survive under different environmental conditions. An increasing amount of genomic information helps us to isolate or identify novel enzymes, and the metabolic routes and functions of D-amino acids, and is beginning to reveal the characteristics of the extensive group of D-amino acids that is found in microorganisms.

References

Arias, C.A. (2003) Role of the transmembrane domain of the VanT serine racemase in resistance to vancomycin in *Enterococcus gallinarum* BM4174. *Journal of Antimicrobial Chemotherapy* 51, 557–564.

Asano, Y. and Endo, K. (1988) Amino acid racemase with broad substrate specificity, its properties and use in phenylalanine racemization. *Applied Microbiology and Microbiology* 29, 523–527.

Ashiuchi, M., Soda, K. and Misono, H. (1999) Characterization of *yrpC* gene product of *Bacillus subtilis* IFO 3336 as glutamate racemase isozyme. *Bioscience, Biotechnology, and Biochemistry* 63, 792–798.

Ashiuchi, M., Kuwana, E., Yamamoto, T., Komatsu, K., Soda, K. and Misono, H. (2002) Glutamate racemase is an endogenous DNA gyrase inhibitor. *The Journal of Biological Chemistry* 277, 39070–39073.

Awasthy, D., Bharath, S., Subbulakshmi, V. and Sharma, U. (2012) Alanine racemase mutants of *Mycobacterium tuberculosis* require D-alanine for growth and are defective for survival in macrophages and mice. *Microbiology* 158, 319–327.

Bellais, S., Arthur, M., Dubost, L., Hugonnet, J.E., Gutmann, L., van Heijenoort, J., Legrand, R., Brouard, J.P., Rice, L. and Mainardi, J.L. (2006) Aslfm, the D-aspartate ligase responsible for the addition of D-aspartic acid onto the peptidoglycan precursor of *Enterococcus faecium*. *The Journal of Biological Chemistry* 281, 11586–11594.

Bharath, S.R., Bisht, S., Savithri, H.S. and Murthy, M.R. (2011) Crystal structures of open and closed forms of D-serine deaminase from *Salmonella typhimurium* – implications on substrate specificity and catalysis. *The FEBS Journal* 278, 2879–2891.

Boniface, A., Parquet, C., Arthur, M., Mengin-Lecreulx, D. and Blanot, D. (2009) The elucidation of the structure of *Thermotoga maritima* peptidoglycan reveals two novel types of cross-link. *The Journal of Biological Chemistry* 284, 21856–21862.

Bouillaut, L., Self, W.T. and Sonenshein, A.L. (2013) Proline-dependent regulation of *Clostridium difficile* Stickland metabolism. *Journal of Bacteriology* 195, 844–854.

Briza, P., Ellinger, A., Winkler, G. and Breitenbach, M. (1990) Characterization of a DL-dityrosine-containing macromolecule from yeast ascospore walls. *The Journal of Biological Chemistry* 265, 15118–15123.

Cava, F., Lam, H., de Pedro, M.A. and Waldor, M.K. (2011a) Emerging knowledge of regulatory roles of D-amino acids in bacteria. *Cellular and Molecular Life Sciences* 68, 817–831.

Cava, F., de Pedro, M.A., Lam, H., Davis, B.M. and Waldor, M.K. (2011b) Distinct pathways for modification of the bacterial cell wall by non-canonical D-amino acids. *The EMBO Journal* 30, 3442–3453.

Chamond, N., Gregoire, C., Coatnoan, N., Rougeot, C., Freitas-Junior, L.H., da Silveira, J.F., Degrave, W.M. and Minoprio, P. (2003) Biochemical characterization of proline racemases from the human protozoan parasite *Trypanosoma cruzi* and definition of putative protein signatures. *The Journal of Biological Chemistry* 278, 15484–15494.

Cheng, Y.Q. (2000) A eukaryotic alanine racemase gene involved in cyclic peptide biosynthesis. *The Journal of Biological Chemistry* 275, 4906–4911.

Cotter, P.D., O'Connor, P.M., Draper, L.A., Lawton, E.M., Deegan, L.H., Hill, C. and Ross, R.P. (2005) Posttranslational conversion of L-serines to D-alanines is vital for optimal production and activity of the lantibiotic lacticin 3147. *Proceedings of the National Academy of Sciences of the United States of America* 102, 18584–18589.

Debono, M., Barnhart, M., Carrell, C.B., Hoffmann, J.A., Occolowitz, J.L., Abbott, B.J., Fukuda, D.S., Hamill, R.L., Biemann, K. and Herlihy, W.C. (1987) A21978C, a complex of new acidic peptide antibiotics: isolation, chemistry, and mass spectral structure elucidation. *The Journal of Antibiotics (Tokyo)* 40, 761–777.

Espaillat, A., Carrasco-Lopez, C., Bernardo-Garcia, N., Pietrosemoli, N., Otero, L.H., Alvarez, L., de Pedro, M.A., Pazos, F., Davis, B.M., Waldor, M.K. *et al.* (2014) Structural basis for the broad specificity of a new family of amino-acid racemases. *Acta Crystallographica Section D, Biological Crystallography* 70, 79–90.

Fotheringham, I.G., Bledig, S.A. and Taylor, P.P. (1998) Characterization of the genes encoding D-amino acid transaminase and glutamate racemase, two D-glutamate biosynthetic enzymes of *Bacillus sphaericus* ATCC 10208. *Journal of Bacteriology* 180, 4319–4323.

Friedman, M. (2010) Origin, microbiology, nutrition, and pharmacology of D-amino acids. *Chemistry and Biodiversity* 7, 1491–1530.

Fukunaga, S., Yuno, S., Takahashi, M., Taguchi, S., Kera, Y., Odani, S. and Yamada, R.-H. (1998) Purification and properties of D-glutamate oxidase from *Candida boidinii* 2201. *Journal of Fermentation and Bioengineering* 85, 579–583.

Geueke, B., Weckbecker, A. and Hummel, W. (2007) Overproduction and characterization of a recombinant D-amino acid oxidase from *Arthrobacter protophormiae*. *Applied Microbiology and Microbiology* 74, 1240–1247.

Goto, M., Yamauchi, T., Kamiya, N., Miyahara, I., Yoshimura, T., Mihara, H., Kurihara, T., Hirotsu, K. and Esaki, N. (2009) Crystal structure of a homolog of mammalian serine racemase from *Schizosaccharomyces pombe*. *The Journal of Biological Chemistry* 284, 25944–25952.

He, W., Li, G., Yang, C.K. and Lu, C.D. (2014) Functional characterization of the *dguRABC* locus for D-Glu and D-Gln utilization in *Pseudomonas aeruginosa* PAO1. *Microbiology* 160, 2331–2340.

Hernández, S.B. and Cava, F. (2015) Environmental roles of microbial amino acid racemases. *Environmental Microbiology* 18, 1673–1685.

Hoffmann, K., Schneider-Scherzer, E., Kleinkauf, H. and Zocher, R. (1994) Purification and characterization of eucaryotic [sic] alanine racemase acting as key enzyme in cyclosporin biosynthesis. *The Journal of Biological Chemistry* 269, 12710–12714.

Horowitz, N.H. (1944) The D-amino acid oxidase of *Neurospora*. *The Journal of Biological Chemistry* 154, 141–149.

Inagaki, K., Tanizawa, K., Tanaka, H. and Soda, K. (1987) Purification and characterization of amino acid racemase with very broad substrate specificity from *Aeromonas caviae*. *Agricultural and Biological Chemistry* 51, 173–180.

Ito, T., Koga, K., Hemmi, H. and Yoshimura, T. (2012a) Role of zinc ion for catalytic activity in D-serine dehydratase from *Saccharomyces cerevisiae*. *The FEBS Journal* 279, 612–624.

Ito, T., Murase, H., Maekawa, M., Goto, M., Hayashi, S., Saito, H., Maki, M., Hemmi, H. and Yoshimura, T. (2012b) Metal ion dependency of serine racemase from *Dictyostelium discoideum*. *Amino Acids* 43, 1567–1576.

Kabisch, U.C., Grantzdorffer, A., Schierhorn, A., Rucknagel, K.P., Andreesen, J.R. and Pich, A. (1999) Identification of D-proline reductase from *Clostridium sticklandii* as a selenoenzyme and indications for a catalytically active pyruvoyl group derived from a cysteine residue by cleavage of a proprotein. *The Journal of Biological Chemistry* 274, 8445–8454.

Kang, L., Shaw, A.C., Xu, D., Xia, W., Zhang, J., Deng, J., Woldike, H.F., Liu, Y. and Su, J. (2011) Upregulation of MetC is essential for D-alanine-independent growth of an *alr*/*dadX*-deficient *Escherichia coli* strain. *Journal of Bacteriology* 193, 1098–1106.

Kataoka, M., Ikemi, M., Morikawa, T., Miyoshi, T., Nishi, K., Wada, M., Yamada, H. and Shimizu, S. (1997) Isolation and characterization of D-threonine aldolase, a pyridoxal-5'-phosphate-dependent enzyme from *Arthrobacter* sp. DK-38. *European Journal of Biochemistry* 248, 385–393.

Kawakami, R., Ohmori, T., Sakuraba, H. and Ohshima, T. (2015) Identification of a novel amino acid racemase from a hyperthermophilic archaeon *Pyrococcus horikoshii* OT-3 induced by D-amino acids. *Amino Acids* 47, 1579–1587.

Kobayashi, J., Shimizu, Y., Mutaguchi, Y., Doi, K. and Ohshima, T. (2013) Characterization of D-amino acid aminotransferase from *Lactobacillus salivarius*. *Journal of Molecular Catalysis B: Enzymatic* 94, 15–22.

Kocianova, S., Vuong, C., Yao, Y., Voyich, J.M., Fischer, E.R., DeLeo, F.R. and Otto, M. (2005) Key role of poly-γ-DL-glutamic acid in immune evasion and virulence of *Staphylococcus epidermidis*. *The Journal of Clinical Investigation* 115, 688–694.

Kolodkin-Gal, I., Romero, D., Cao, S., Clardy, J., Kolter, R. and Losick, R. (2010) D-Amino acids trigger biofilm disassembly. *Science* 328, 627–629.

Korte-Berwanger, M., Sakinc, T., Kline, K., Nielsen, H.V., Hultgren, S. and Gatermann, S.G. (2013) Significance of the D-serine-deaminase and D-serine metabolism of *Staphylococcus saprophyticus* for virulence. *Infection and Immunity* 81, 4525–4533.

Kubicek-Pranz, E.M. and Röhr, M. (1985) Formation of D-amino-acid oxidase in the yeast *Trigonopsis variabilis*. *Canadian Journal of Microbiology* 31, 625–628.

Kumagai, H., Kashima, N. and Yamada, H. (1970) Racemization of D- or L-alanine by crystalline tyrosine phenol-lyase from *Escherichia intermedia*. *Biochemical and Biophysical Research Communications* 39, 796–801.

Lee, S.G., Hong, S.P., Song, J.J., Kim, S.J., Kwak, M.S. and Sung, M.H. (2006) Functional and structural characterization of thermostable D-amino acid aminotransferases from *Geobacillus* spp. *Applied and Environmental Microbiology* 72, 1588–1594.

Leiman, S.A., May, J.M., Lebar, M.D., Kahne, D., Kolter, R. and Losick, R. (2013) D-Amino acids indirectly inhibit biofilm formation in *Bacillus subtilis* by interfering with protein synthesis. *Journal of Bacteriology* 195, 5391–5395.

Li, C., Yao, X. and Lu, C.D. (2010) Regulation of the *dauBAR* operon and characterization of D-amino acid dehydrogenase DauA in arginine and lysine catabolism of *Pseudomonas aeruginosa* PAO1. *Microbiology* 156, 60–71.

Lim, Y.H., Yokoigawa, K., Esaki, N. and Soda, K. (1993) A new amino acid racemase with threonine α-epimerase activity from *Pseudomonas putida*: purification and characterization. *Journal of Bacteriology* 175, 4213–4217.

Liu, L., Iwata, K., Kita, A., Kawarabayasi, Y., Yohda, M. and Miki, K. (2002) Crystal structure of aspartate racemase from *Pyrococcus horikoshii* OT3 and its implications for molecular mechanism of PLP-independent racemization. *Journal of Molecular Biology* 319, 479–489.

Long, Z., Lee, J.A., Okamoto, T., Sekine, M., Nimura, N., Imai, K., Yohda, M., Maruyama, T., Sumi, M., Kamo, N. *et al.* (2001) Occurrence of D-amino acids and a pyridoxal 5′-phosphate-dependent aspartate racemase in the acidothermophilic archaeon, *Thermoplasma acidophilum*. *Biochemical and Biophysical Research Communications* 281, 317–321.

Martínez-Rodríguez, S., Martínez-Gómez, A.I., Rodríguez-Vico, F., Clemente-Jiménez, J.M. and Las Heras-Vázquez, F.J. (2010) Natural occurrence and industrial applications of D-amino acids: an overview. *Chemistry and Biodiversity* 7, 1531–1548.

Matsui, D. and Oikawa, T. (2010) Detection and function of the intramolecular disulfide bond in arginine racemase: an enzyme with broad substrate specificity. *Chemistry and Biodiversity* 7, 1591–1602.

Meziane-Cherif, D., Stogios, P.J., Evdokimova, E., Savchenko, A. and Courvalin, P. (2014) Structural basis for the evolution of vancomycin resistance D,D-peptidases. *Proceedings of the National Academy of Sciences of the United States of America* 111, 5872–5877.

Miles, E.W., Phillips, R.S., Yeh, H.J.C. and Cohen, L.A. (1986) Isomerization of (3*S*)-2,3-dihydro-5-fluoro-L-tryptophan and of 5-fluoro-L-tryptophan catalyzed by tryptophan synthase: studies using fluorine-19 nuclear magnetic resonance and difference spectroscopy. *Biochemistry* 25, 4240–4249.

Mutaguchi, Y., Ohmori, T., Wakamatsu, T., Doi, K. and Ohshima, T. (2013) Identification, purification, and characterization of a novel amino acid racemase, isoleucine 2-epimerase, from *Lactobacillus species*. *Journal of Bacteriology* 195, 5207–5215.

Neuhaus, F.C. and Baddiley, J. (2003) A continuum of anionic charge: structures and functions of D-alanyl-teichoic acids in Gram-positive bacteria. *Microbiology and Molecular Biology Reviews* 67, 686–723.

Nishimura, K., Tomoda, Y., Nakamoto, Y., Kawada, T., Ishii, Y. and Nagata, Y. (2006) Alanine racemase from the green alga *Chlamydomonas reinhardtii*. *Amino Acids* 32, 59–62.

Ogunleye, A., Bhat, A., Irorere, V.U., Hill, D., Williams, C. and Radecka, I. (2015) Poly-γ-glutamic acid: production, properties and applications. *Microbiology* 161, 1–17.

Oh, S.Y., Richter, S.G., Missiakas, D.M. and Schneewind, O. (2015) Glutamate racemase mutants of *Bacillus anthracis*. *Journal of Bacteriology* 197, 1854–1861.

Ohnishi, E., Macleod, H. and Horowitz, N.H. (1962) Mutants of *Neurospora* deficient in D-amino acid oxidase. *The Journal of Biological Chemistry* 237, 138–142.

Okada, H., Yohda, M., Giga-Hama, Y., Ueno, Y., Ohdo, S. and Kumagai, H. (1991) Distribution and purification of aspartate racemase in lactic acid bacteria. *Biochimica et Biophysica Acta, Protein Structure and Molecular Enzymology* 1078, 377–382.

Palchaudhuri, S., Patel, V. and McFall, E. (1988) DNA sequence of the D-serine deaminase activator gene *dsdC*. *Journal of Bacteriology* 170, 330–334.

Panizzutti, R., de Souza Leite, M., Pinheiro, C.M. and Meyer-Fernandes, J.R. (2006) The occurrence of free D-alanine and an alanine racemase activity in *Leishmania amazonensis*. *FEMS Microbiology Letters* 256, 16–21.

Pierce, K.J., Salifu, S.P. and Tangney, M. (2008) Gene cloning and characterization of a second alanine racemase from *Bacillus subtilis* encoded by *yncD*. *FEMS Microbiology Letters* 283, 69–74.

Pilone M.S., Pollegioni, L., Casalin, P., Curti, B. and Ronchi, S. (1989) Properties of D-amino-acid oxidase from *Rhodotorula gracilis*. *European Journal of Biochemistry* 180, 199–204.

Pollegioni, L., Piubelli, L., Sacchi, S., Pilone, M.S. and Molla, G. (2007) Physiological functions of D-amino acid oxidases: from yeast to humans. *Cellular and Molecular Life Sciences* 64, 1373–1394.

Potrykus, J., Flemming, J. and Bearne, S.L. (2009) Kinetic characterization and quaternary structure of glutamate racemase from the periodontal anaerobe *Fusobacterium nucleatum*. *Archives of Biochemistry and Biophysics* 491, 16–24.

Saito, Y., Takahashi, S., Kobayashi, M., Abe, K. and Kera, Y. (2013) D-Amino acid oxidase of *Streptomyces coelicolor* and the effect of D-amino acids on the bacterium. *Annals of Microbiology* 64, 1167–1177.

Satomura, T., Kawakami, R., Sakuraba, H. and Ohshima, T. (2002) Dye-linked D-proline dehydrogenase from hyperthermophilic archaeon *Pyrobaculum islandicum* is a novel FAD-dependent amino acid dehydrogenase. *The Journal of Biological Chemistry* 277, 12861–12867.

Satomura, T., Ishikura, M., Koyanagi, T., Sakuraba, H., Ohshima, T. and Suye, S. (2015) Dye-linked D-amino acid dehydrogenase from the thermophilic bacterium *Rhodothermus marinus* JCM9785: characteristics and role in *trans*-4-hydroxy-L-proline catabolism. *Applied Microbiology and Microbiology* 99, 4265–4275.

Sielaff, H., Dittmann, E., Tandeau De Marsac, N., Bouchier, C., Von Dohren, H., Borner, T. and Schwecke, T. (2003) The *mcyF* gene of the microcystin biosynthetic gene cluster from *Microcystis aeruginosa* encodes an aspartate racemase. *Biochemical Journal* 373, 909–916.

Soda, K. and Osumi, T. (1969) Crystalline amino acid racemase with low substrate specificity. *Biochemical and Biophysical Research Communications* 35, 363–368.

Soda, K., Yorifuji, T. and Ogata, K. (1967) Occurrence of arginine racemase in bacterial extract. *Biochimica et Biophysica Acta, Enzymology* 146, 606–608.

Sun, X., Li, H., Alfermann, J., Mootz, H.D. and Yang, H. (2014) Kinetics profiling of gramicidin S synthetase A, a member of nonribosomal peptide synthetases. *Biochemistry* 53, 7983–7989.

Survase, S.A., Kagliwal, L.D., Annapure, U.S. and Singhal, R.S. (2011) Cyclosporin A – a review on fermentative production, downstream processing and pharmacological applications. *Biotechnology Advances* 29, 418–435.

Takahashi, S., Takahashi, T., Kera, Y., Matsunaga, R., Shibuya, H. and Yamada, R.H. (2004) Cloning and expression in *Escherichia coli* of the D-aspartate oxidase gene from the yeast *Cryptococcus humicola* and characterization of the recombinant enzyme. *The Journal of Biochemistry* 135, 533–540.

Takahashi, S., Kakuichi, T., Fujii, K., Kera, Y. and Yamada, R.H. (2005) Physiological role of D-aspartate oxidase in the assimilation and detoxification of D-aspartate in the yeast *Cryptococcus humicola*. *Yeast* 22, 1203–1212.

Takahashi, S., Okada, H., Abe, K. and Kera, Y. (2012) D-Amino acid-induced expression of D-amino acid oxidase in the yeast *Schizosaccharomyces pombe*. *Current Microbiology* 65, 764–769.

Takahashi, S., Furukawara, M., Omae, K., Tadokoro, N., Saito, Y., Abe, K. and Kera, Y. (2014) A highly stable D-amino acid oxidase of the thermophilic bacterium *Rubrobacter xylanophilus*. *Applied and Environmental Microbiology* 80, 7219–7229.

Tanigawa, M., Shinohara, T., Saito, M., Nishimura, K., Hasegawa, Y., Wakabayashi, S., Ishizuka, M. and Nagata, Y. (2010) D-Amino acid dehydrogenase from *Helicobacter pylori* NCTC 11637. *Amino Acids* 38, 247–255.

Tanizawa, K., Masu, Y., Asano, S., Tanaka, H. and Soda, K. (1989) Thermostable D-amino acid aminotransferase from a thermophilic *Bacillus* species. Purification, characterization, and active site sequence determination. *The Journal of Biological Chemistry* 264, 2445–2449.

Tillett, D., Dittmann, E., Erhard, M., von Döhren, H., Börner, T. and Neilan, B.A. (2000) Structural organization of microcystin biosynthesis in *Microcystis aeruginosa* PCC7806: an integrated peptide–polyketide synthetase system. *Chemistry and Biology* 7, 753–764.

Toscano, W.A. Jr and Storm, D.R. (1982) Bacitracin. *Pharmacology and Therapeutics* 16, 199–210.

Uhl, M.K., Oberdorfer, G., Steinkellner, G., Riegler-Berket, L., Mink, D., van Assema, F., Schurmann, M. and Gruber, K. (2015) The crystal structure of D-threonine aldolase from *Alcaligenes xylosoxidans* provides insight into a metal ion assisted PLP-dependent mechanism. *PLoS ONE* 10(4): e0124056.

Uo, T., Yoshimura, T., Tanaka, N., Takegawa, K. and Esaki, N. (2001) Functional characterization of alanine racemase from *Schizosaccharomyces pombe*: a eucaryotic [sic] counterpart to bacterial alanine racemase. *Journal of Bacteriology* 183, 2226–2233.

Vacca, R.A., Giannattasio, S., Graber, R., Sandmeier, E., Marra, E. and Christen, P. (1997) Active-site Arg → Lys substitutions alter reaction and substrate specificity of aspartate aminotransferase. *The Journal of Biological Chemistry* 272, 21932–21937.

Wakayama, M., Nakashima, S., Sakai, K. and Moriguchi, M. (1994) Isolation, enzyme production and characterization of D-aspartate oxidase from *Fusarium sacchari* var. *elongatum* Y-105. *Journal of Fermentation and Bioengineering* 78, 377–379.

Wallace, B.A. (1986) Structure of gramicidin A. *Biophysical Journal* 49, 295–306.

Walton, J.D. (2006) HC-toxin. *Phytochemistry* 67, 1406–1413.

Watanabe, S., Morimoto, D., Fukumori, F., Shinomiya, H., Nishiwaki, H., Kawano-Kawada, M., Sasai, Y., Tozawa, Y. and Watanabe, Y. (2012) Identification and characterization of D-hydroxyproline dehydrogenase and Δ^1-pyrroline-4-hydroxy-2-carboxylate deaminase involved in novel L-hydroxyproline metabolism of bacteria: metabolic convergent evolution. *The Journal of Biological Chemistry* 287, 32674–32688.

Watanabe, S., Tanimoto, Y., Nishiwaki, H. and Watanabe, Y. (2015) Identification and characterization of bifunctional proline racemase/hydroxyproline epimerase from archaea: discrimination of substrates and molecular evolution. *PLoS ONE* 10(3): e0120349.

Wild, J. and Klopotowski, T. (1981) D-Amino acid dehydrogenase of *Escherichia coli* K12: positive selection of mutants defective in enzyme activity and localization of the structural gene. *Molecular and General Genetics* 181, 373–378.

Wild, J., Hennig, J., Lobocka, M., Walczak, W. and Klopotowski, T. (1985) Identification of the *dadX* gene coding for the predominant isozyme of alanine racemase in *Escherichia coli* K12. *Molecular and General Genetics* 198, 315–322.

Williams, J.S. and Rosevear, P.R. (1991) A novel α-proton exchange reaction catalyzed by *Escherichia coli* methionyl-tRNA synthetase. *Biochemistry* 30, 6412–6416.

Wong, H.C., Allenza, P. and Lessie, T.G. (1980) Hydroxyamino acid utilization and α-ketobutyrate toxicity in *Pseudomonas cepacia*. *Journal of Bacteriology* 144, 441–443.

Yamada, R., Ujiie, H., Kera, Y., Nakase, T., Kitagawa, K., Imasaka, T., Arimoto, K., Takahashi, M. and Matsumura, Y. (1996) Purification and properties of D-aspartate oxidase from *Cryptococcus humicolus* UJ1. *Biochimica et Biophysica Acta* 1294, 153–158.

Yamashita, T., Ashiuchi, M., Ohnishi, K., Kato, S., Nagata, S. and Misono, H. (2004) Molecular identification of monomeric aspartate racemase from *Bifidobacterium bifidum*. *European Journal of Biochemistry* 271, 4798–4803.

Yano, S., Ozaki, H., Matsuo, S., Ito, M. and Wakayama, M. (2012) Production, purification and characterization of D-aspartate oxidase from the fungus *Trichoderma harzianum* SKW-36. *Advances in Bioscience and Biotechnology* 3, 7–13.

Yohda, M., Endo, I., Abe, Y., Ohta, T., Iida, T., Maruyama, T. and Kagawa, Y. (1996) Gene for aspartate racemase from the sulfur-dependent hyperthermophilic archaeum, *Desulfurococcus* strain SY, *The Journal of Biological Chemistry* 271, 22017–22021.

Yonaha, K., Misono, H., Yamamoto, T. and Soda, K. (1975) D-Amino acid aminotransferase of *Bacillus sphaericus*. Enzymologic and spectrometric properties. *The Journal of Biological Chemistry* 250, 6983–6989.

25 Pathways of Utilization of D-Amino Acids in Higher Organisms

J.P.F. D'Mello*

Formerly of SAC, University of Edinburgh King's Buildings Campus, Edinburgh, UK

25.1 Abstract

Although an integral part of the peptidoglycan structure of bacteria, D-amino acids are ubiquitous in higher organisms by virtue of endogenous synthesis and symbiotic microbial activity. The occurrence of D-alanine, D-aspartate and D-glutamate in higher organisms is now well documented. Distinct pathways of synthesis, utilization and salvage of D-amino acids have been elucidated at generic and specific levels, as well as in terms of metabolic and nutritional efficacy. Thus, the most significant discovery of recent times confirmed that D-serine serves as a signalling molecule in both plants and animals. In particular, D-serine is the endogenous ligand for the 'glycine' binding site of the *N*-methyl-D-aspartate subtype of glutamate receptor in the central nervous system (CNS) and retina, where serine racemase is also localized. Furthermore, on the basis of the anatomical localization, neuronal biosynthesis, uptake, degradation and activation of L-glutamate receptors, D-aspartate is also emerging as an amino acid of particular interest. During homeostatic regulation, D-amino acids that are surplus to immediate utilization are degraded by D-amino acid oxidases (DAAOs), with the resulting keto acid potentially available for further catabolism or transamination. D-Amino acids may also arise during microbial activity in the digestive tract of animals, particularly ruminants, and in the fermentation processes involved in the preparation of dairy products, vinegar and wine. In animal nutrition, pure DL-methionine is an effective replacement for its L-isomer but D-lysine is not utilized. Toxicity, however, is not an inevitable consequence of the occurrence of D-amino acids in higher organisms, as there are highly effective endogenous pathways for oxidation and salvage. Moreover, the intervention of microbes is not essential for the avoidance of toxicity by higher organisms. It is concluded that D-serine and D-aspartate are essential signalling and modulatory molecules in higher organisms. In addition, given its ubiquitous distribution, there may be justification to investigate a possible physiological role for D-glutamate, reflecting the critically important roles of its L-isomer in neurotransmission. The prospects for D-amino acid utilization in diagnostic and clinical applications may well define the pathway for future research. In vertebrate nutrition, the availability of D-amino acids in heat-processed foods and animal feed also deserves further investigation. Nevertheless, it appears that under normal conditions, toxicity of D-amino acids in higher organisms is the exception rather than the rule.

*E-mail address: jpfdmello@hotmail.co.uk

25.2 Introduction

A fundamental requirement in the ribosomal biosynthesis of proteins is that the full complement of the 20 to 21 canonical amino acids should be simultaneously available for activation and formation of the peptide bond. Of equal importance is that the configuration of the amino acids supplied should be exclusively in the form of L-isomers. This chiral prerequisite is clearly exemplified in the requirement of higher animals for the essential amino acids. Thus, the diet of adult humans must supply L-lysine, L-threonine, L-methionine, L-leucine, L-isoleucine, L-valine, L-tryptophan and L-phenylalanine, with L-cysteine and L-tyrosine partially contributing to the requirement of methionine and phenylalanine, respectively (Pillai and Kurpad, 2012). The non-essential amino acids (NEAA) are also required in the form of L-isomers. In the context of human nutrition, the principal NEAAs include L-glutamate, L-arginine, L-aspartate, L-alanine and L-proline. The requirements for amino acids are generally satisfied via the consumption of food proteins, but in animal nutrition, any dietary deficiencies are rectified by supplements of, for example, pure L-lysine (for pigs) and DL-methionine (for poultry). DL-Methionine is synthesized by chemical procedures, but L-lysine is obtained via established fermentative technology that is used for a wide range of L-amino acids (Baker, 1994).

One of the characteristic features of bacterial metabolism is associated with the biosynthesis of D-amino acids, which constitute a significant component of the peptidoglycan structure of the cell wall envelope (see Chapter 24). In addition, it has long been known that D-amino acids occur in other microorganisms, such as yeasts and fungi (Yoshimura and Esak, 2003). Significant advances in biotechnology, chromatographic techniques and immunodetection have provided fresh impetus for D-amino acid research in a comprehensive perspective of comparative biology. As a result, there is now considerable evidence to substantiate the occurrence and role of D-amino acids in higher organisms. For example, germinating pea (*Pisum sativum*) seedlings contain relatively high concentrations of D-alanine in conjugated forms (Fukuda *et al.*, 1973). The occurrence of D-serine in rice (*Oryza sativa*) was established by Gogami *et al.* (2009). More recently, Gordes *et al.* (2011) demonstrated the uptake and conversion of D-amino acids in *Arabidopsis thaliana*. These observations are consistent with the general concept of an active system of biosynthesis, utilization and oxidation of D-amino acids. Thus, the occurrence of serine racemases in barley (*Hordeum vulgare*) and other plant species was highlighted by Fujitani *et al.* (2007). In addition, there is overwhelming evidence of a specific role for D-serine and possibly of other D-amino acids (for example D-aspartate) in vertebrate animals that involves *de novo* synthesis by well-established endogenous pathways (Hashimoto *et al.*, 1993a,b; Ota *et al.*, 2012; Smith *et al.*, 2012). New evidence demonstrating that D-serine and serine racemase occur in retinal neurons as well as in the brain has been presented by Smith *et al.* (2012). The activity of D-amino acid oxidases is integrated within homeostatic mechanisms controlling D-amino acid metabolism.

By virtue of extensive pre-and post-gastric microbial fermentation, cattle and sheep may also be expected to assimilate those D-amino acids normally associated with peptidoglycans. As might be expected, quantitative data is sparse, without any citation of publications in the extensive reviews by Atasoglu and Wallace (2003) and Fox and Tedeschi (2003) with regard to *de novo* synthesis and metabolism of D-amino acids by rumen microbes and the prediction of dietary amino acid adequacy for ruminants. Yet, Schieber *et al.* (1999) had previously demonstrated the presence of D-alanine, D-glutamate and D-aspartate in five species of rumen bacteria, including *Fibrobacter succinogenes*, *Streptococcus bovis*, *Selenomonas ruminantium*, *Prevotella ruminicola* and *Anaerovibrio lipolytica*.

It is to be expected that foods and beverages prepared by fermentation technologies should contain D-amino acids emanating from microbial activity and/or subsequent processing procedures. This is borne out by extensive data (Bruckner and Hausch, 1989, 1990; Kato *et al.*, 1995; Abe *et al.*, 1999; Sardella *et al.*, 2013) relating to milk, dairy products and alcoholic beverages. There is now convincing evidence of amino acid racemization during the processing of raw materials used in foods and animal feedstuffs. The extent of racemization appears to be proportionate to the severity of the processing conditions (Horak *et al.*, 2014).

This brief overview of the existing evidence indicates that D-amino acids and their associated enzymes are widely distributed in nature and not just confined to microorganisms. As a general rule, it may be stated that the biosynthesis of D-amino acids in higher organisms is not always accompanied by adverse effects and in at least one case may contribute to essential signalling functions.

Nevertheless, the foregoing review of the literature should be set against renewed claims of the adverse effects of D-amino acids (Zhang and Sun, 2014). It should be acknowledged that almost every amino acid, irrespective of chirality, has been associated with the potential and capacity to precipitate deleterious effects in higher organisms, and this expectation should be considered in conjunction with the evidence.

25.3 Occurrence of D-Amino Acids in Higher Organisms

D-Amino acids are generally associated with the peptidoglycans of bacteria, but it has been patently clear for some considerable time that they are ubiquitous in higher organisms, including plants, and particularly in the physiological fluids and specific organs of vertebrate animals (Table 25.1).

The distribution of D-amino acids in higher organisms is the net effect of endogenous synthesis, microbial contribution during symbiotic relationships, cellular utilization and degradation by host enzymes. As shown in Table 25.1, D-amino acids occur in plants and animals, the latter group including invertebrates, amphibians, fish, avian species and mammals. D-Amino acids have been detected in body fluids, but the significant observation relates to their localization in the CNS of higher animals. D-Serine and D-aspartate have been associated with the functional metabolism of the brain, although D-alanine and D-glutamate are also present in the CNS. The existence of a blood–brain barrier implies that these D-amino acids arise as a result of endogenous metabolism, precluding any microbial intervention.

25.3.1 Functions

An overriding feature of recent research has been the role of at least two D-amino acids as signalling molecules in both plants and animals, though details of the mechanisms involved are still emerging. Specifically, the occurrence of D-serine has been established in *Arabidopsis* (Fujitani *et al.*, 2006), barley (Fujitani *et al.*, 2007) and rice (Gogami *et al.*, 2009). In plants, glutamate receptor-like genes regulated by D-serine have been reported, leading to the conclusion that there is a novel plant signal transduction cascade similar to the amino acid-mediated neural communication normally associated with animal systems (Michard *et al.*, 2011). In particular, reference has been made to the localization of D-serine regulation in the pistil. It may be of relevance to recall evidence of D-amino acid enhancement of ethylene in seed tissues of *Xanthium pennsylvanicum* Wallr. (Satoh and Esashi, 1980). The marked stimulation induced by D-phenylalanine, D-valine, D-leucine, D-threonine and D-methionine was contrasted with the ineffectiveness of the corresponding L-amino acids.

The overwhelming evidence of signalling functions relates to the CNS and the neuroendocrine systems of animals (D'Aniello, 2007; Ota *et al.*, 2012). In animals, D-serine is the endogenous ligand for the 'glycine' binding site of the *N*-methyl-D-aspartate subtype of glutamate receptor in the brain and retina (Smith *et al.*, 2012; Bardaweel *et al.*, 2014). Ota *et al.* (2012) also presents the case for D-aspartate in some detail, on the basis of its anatomical localization, neuronal biosynthesis and uptake and degradation, as well as the activation of L-glutamate receptors by aspartate.

25.4 Endogenous Synthesis and Degradation of D-Amino Acids in Higher Organisms

Both molecular and feeding studies have provided convincing evidence that higher organisms are endowed with the enzymatic systems for the synthesis and breakdown of D-amino acids. Biosynthesis is facilitated by the amino acid racemases, while degradation is effected by D-amino acid oxidases.

Table 25.1. Occurrence of free D-amino acids in higher organisms. Evidence is presented in chronological order separately for plants and animals.

Organism	D-Amino acids	Reference(s)
Pea	Large quantities of D-Ala found in germinating pea seedlings attributed to the action of D-amino acid aminotransferase	Fukuda *et al.* (1973); Ogawa *et al.* (1973)
Gymnosperms and mono- and dicotyledonous angiosperms	D-Asp, D-Asn, D-Gln, D-Ser and D-Ala in most plants examined. D-Pro, D-Val, D-Leu and D-Lys in certain plants	Bruckner and Westhauser (2003)
Rice	D-Ser concentrations maximized 6 days after germination	Gogami *et al.* (2009)
Human	D-Asp and D-Ala in normal and Alzheimer's brain	Fisher *et al.* (1991)
Rat	Large quantities of D-Ser in brain with traces of D-Asp and D-Ala	Hashimoto *et al.* (1992)
Amphibians	First report of D-Asp and D-aspartate oxidase in amphibian tissues	Kera *et al.* (1992)
Rat	Widespread distribution of D-Asp in rat periphery	Hashimoto *et al.* (1993b)
Human	Increases in D-amino acids in chronic renal failure	Young *et al.* (1994)
Rat	Presence of free D-Glu and D-Asp in liver, kidney and brain. D-Glu concentrations higher than values for D-Asp	Kera *et al.* (1995)
Chicken and pigeon	D-Glu and D-Asp in tissues	Kera *et al.* (1996)
Mammals	D-Asp and D-Ser in brain and periphery	Hashimoto and Oka (1997)
Fish	D-Asp and D-Glu in several tissues of five fish species under investigation	Kera *et al.* (1998)
Human	D-Ala, D-Ser, D-Thr, D-Phe, D-Tyr, D-Lys and D-Pro in blood. D-Ala and D-Ser in urine	Bruckner and Schieber (2001)
Mice	D-Pro and D-Leu in brain of mutants	Hamase *et al.* (2001)
Avians, amphibians and fishes	D-Asp and/or D-Glu found in various tissues with sex differences for several tissues of avians and amphibians	Kera *et al.* (2001)
Mammals	Substantial quantities of D-amino acids present in mammalian tissues	Hamase *et al.* (2002)
Rat	D-Ala in central nervous system and periphery	Morikawa *et al.* (2003)
Human	Occurrence of D-Asp in seminal plasma and spermatozoa	D'Aniello *et al.* (2005)
Mammals	D-Ser and D-Asp present in the central nervous system	Fuchs *et al.* (2005)
Human	D-Asp detected in various tissues, including eye lens, brain, skin, bone and teeth in samples from elderly individuals	Fujii (2005)
Bivalve	Large quantities of D-Asp, comparable to those of free L-Asp, in muscle tissues	Watanabe *et al.* (2005)
Human	D-Ala and D-Pro in saliva emanating from oral epithelial cells. D-Asp and D-Ala in submandibular glands. Influence of ingested food and oral bacteria suggested to be insignificant contributors to salivary D-amino acids	Nagata *et al.* (2006)
Rat	D-Asp detected in serum during time-course investigation	Ohgusu *et al.* (2006)
Bivalve species	Occurrence and functions of D-Asp in bivalves	Yamada *et al.* (2006)

Continued

Table 25.1. Continued.

Organism	D-Amino acids	Reference(s)
Animals (in general)	D-Asp in neuroendocrine systems	D'Aniello (2007)
Rat	D-Glu in brain may be involved in several physiological mechanisms	Mangas *et al.* (2007)
Mice	D-Asp and D-Glu in midbrain of Parkinson's disease mouse model	Xin *et al.* (2007)
Mammals	Asp racemase generates neuronal D-Asp and regulates adult neurogenesis	Kim *et al.* (2010)
Rat	Significant D-Asp values recorded in brain and peripheral tissues with low D-Glu levels	Han *et al.* (2011)
Mammals	D-Asp is an important bioactive substance (review)	Katane and Homma (2011)
Prawns	D-Ala in all tissues. D-Asp in nervous tissue and eye. D-Glu in testis	Yoshikawa *et al.* (2011)
Mammals	D-Ser and D Ala in brain and periphery. Appreciable quantities of D-amino acids in mammals	Miyoshi *et al.* (2011; 2012)
Animals (in general)	D-Asp in nervous and endocrine systems	Ota *et al.* (2012)
Mammals	D-Ser in neurotransmission and neuromodulation	Bardaweel *et al.* (2014)
Vertebrates	D-Asp in glandular tissues	Di Fiore *et al.* (2014)
Rat	D-Ala in the islets of Langerhans of the pancreas	Ota *et al.* (2014)
Mammals	Biosynthesis of D-Asp using rat and human homologues of mouse Asp racemase	Matsuda *et al.* (2015)

25.4.1 Amino acid racemases

The role of amino acid racemases in living organisms has long been established, particularly in bacteria (Cohen and Sallach, 1961). The distribution of these enzymes in *Streptococcus faecalis*, *Lactobacillus arabinosus*, *Escherichia coli* and *Clostridium sticklandii* was reviewed. However, the notion that animals lack these enzymes, as maintained by Cohen and Sallach (1961), was quickly dispelled by a series of subsequent studies. Moreover, it was established that plants also possess at least two specific amino acid racemases.

The functions and mechanisms of action of amino acid racemases in living organisms were extensively reviewed by Yoshimura and Esak (2003), but subsequent evidence has altered prevailing perceptions in comparative biochemistry and physiology. In particular, an intriguing observation relates to the occurrence and activity of serine racemase (Fig. 25.1), which is responsible for converting L-serine to its D-isomer, in both plants (Gogami *et al.*, 2009; Hepler *et al.*, 2012) and animals (Smith *et al.*, 2012). Specifically, the occurrence of D-serine and the characterization of its racemase have been established in *Arabidopsis* (Fujitani *et al.*, 2006), barley (Fujitani *et al.*, 2007) and rice (Gogami *et al.*, 2009). In plants, glutamate receptor-like genes regulated by D-serine have been reported, leading to the conclusion that there is a novel plant signal transduction cascade similar to amino acid-mediated neural communication normally associated with animal systems (Michard *et al.*, 2011).

The presence of alanine racemase in plants has also been confirmed by Ono *et al.* (2006) in studies with lucerne seedlings. This enzyme resembles bacterial alanine racemase in terms of its catalytic requirements and activity.

In animals, D-serine is the endogenous ligand for the 'glycine' binding site of the *N*-methyl-D-aspartate subtype of glutamate receptor in the brain and retina (Smith *et al.*, 2012). Data supporting the location of serine racemase and D-serine in retinal Müller glial cells has been reviewed and there is new evidence that serine racemase occurs in retinal neurons, including the ganglion cells. As D-aspartate is abundant in the brain (Table 25.1), the expression of aspartate racemase might be anticipated. This expectation was confirmed by Yamada *et al.* (2006) in studies aimed at purifying and cloning the

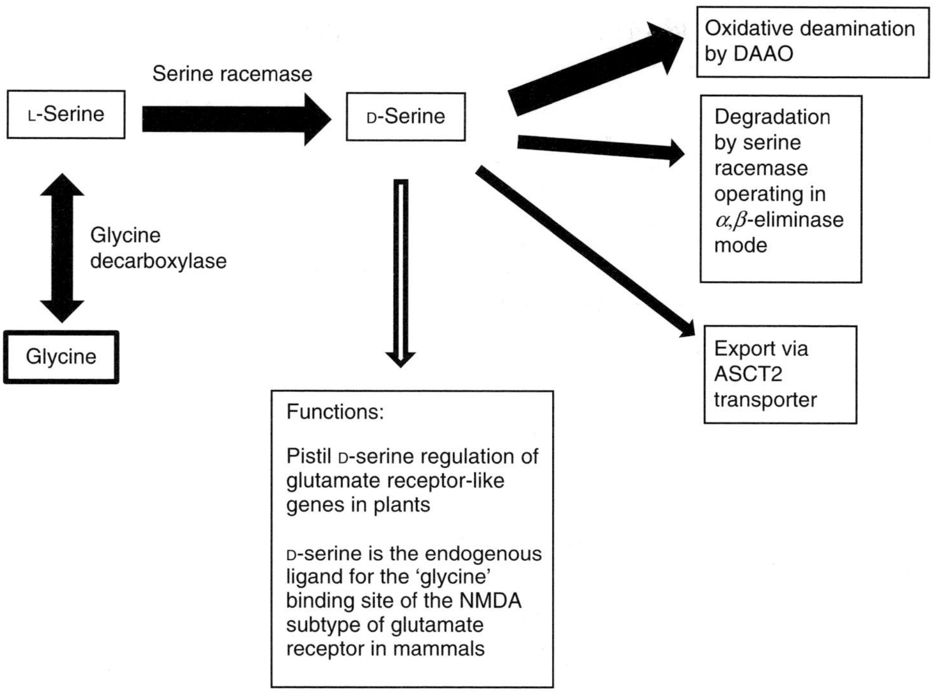

Fig. 25.1. Synthesis and metabolic fate of D-serine in higher organisms. Abbreviations: ASCT2, alanine-serine-cysteine transporter 2; DAAO, D-amino acid oxidase; NMDA, *N*-methyl-D-aspartate.

enzyme from a bivalve species. Kim *et al.* (2010) also identified and cloned aspartate racemase, which converts L-aspartate to its D-isomer in the mammalian brain. This enzyme co-localizes with D-aspartate in the CNS and neuroendocrine tissues. Yet Matsuda *et al.* (2015) indicated that the rat and human homologues of mouse aspartate racemase are not responsible for the biosynthesis of D-aspartate and that this issue awaits resolution.

25.4.2 D-Amino acid oxidases

The degradation of D-amino acids is catalysed by the D-amino acid oxidases (DAAO), as illustrated in Fig. 25.1 for serine. Considerable work has been published as evidence accumulated to demonstrate the critical role of DAAO, particularly in animal metabolism. The initial studies appeared to be somewhat tentative, focusing on establishing the distribution and substrate specificity of DAAOs along conventional lines, with the aim of evaluating species, inter-organ and sex differences (Table 25.2). With the advent of mutant DAAO-deficient rodents and sophisticated chromatographic hardware, the emphasis shifted to elucidating the role of DAAOs in D-serine elimination and their relationship to brain function in health and in disorders such as schizophrenia, as well as in novel clinical applications (Table 25.3).

Recent advances have revealed the importance of aspartate as a bioactive compound in mammals (Katane and Homma, 2011). D-Aspartate oxidase is now emerging as a key enzyme in regulatory processes governing its specific substrate, and is widely expressed in mammalian and amphibian tissues, consistent with the ubiquitous distribution of D-aspartate (Table 25.1). It is localized in peroxisomes in the mammalian liver (Van Veldhoven *et al.*, 1991), and like other DAAOs, is attributed with a detoxification role (D'Aniello *et al.*, 1993).

25.4.3 Homeostasis

The maintenance of equilibrium of D-amino acids in the metabolism of higher organisms is effected by the actions of racemases, DAAOs,

Table 25.2. Endogenous D-Amino acid oxidases (DAAOs): a historical perspective highlighting the progressive establishment of fundamental features and universal distribution in animals.

Source of enzyme	Observations	Reference
Animal tissues	A review of previous data	Blanchard *et al.* (1944)
Rat liver	DAAO levels measured in conjunction with nicotinic acid, riboflavin and arginase status in rats offered a protein-free diet	Seifter *et al.* (1948)
Sheep kidney, cobra venom and *Neurospora crassa*	Oxidation of various synthetic amino acids by mammalian DAAO, L-amino acid oxidase of cobra venom and the L-oxidase and DAAO of *Neurospora crassa*	Bender and Krebs (1950)
Mollusc liver	Mollusc liver DAAO similar to that of vertebrates	Blaschko and Hawkins (1952)
Mammalian liver and kidney	A summary of the specificity, mechanism of action and inhibition of DAAO; comparisons with fungal sources of DAAO	Cohen and Sallach (1961)
Rat liver	Combined biochemical and morphological study of particulate fractions from hepatic tissue. Analysis of urate oxidase, DAAO and catalase	Baudhuin *et al.* (1965)
Bovine and human nervous tissue	Distribution of DAAO in nervous tissues. Concluded that the physiological role of DAAO was elusive/enigmatic	Neims *et al.* (1966)
Rat brain	Distribution of DAAO confined to hindbrain, distributed almost equally between cerebellum and medulla	Gaunt and de Duve (1976)
Kidney of various animals	Activity and substrate specificity of DAAO	Konno and Yasumura (1981)
Kidney homogenate	Mouse mutant deficient in DAAO activity. Consistent sex differences in DAAO activity reported	Konno and Yasumura (1983)

substrate availability and other factors, such as transport (Pilleriin *et al.*, 2010). The case for D-serine is illustrated in Fig. 25.1. The role of glycine decarboxylase in the provision of L-serine may be a critical factor, given that glycine is also an activator of the *N*-methyl-D-aspartate (NMDA) subtype of glutamate receptor in the CNS (Smith *et al.*, 2012). The dual role of serine racemase has been proposed as an additional factor in serine homeostasis (Pilleriin *et al.*, 2010).

25.5 Food Composition and Safety

D-Amino acids are ubiquitous in a diverse range of common foods, particularly in animal and fermented products (Table 25.4). This distribution is to be expected given the extensive synthesis and salvage of D-amino acids in animals, and the application of fermentation processes in food biotechnology. The dynamics of metabolic pathways in the ruminant animal, as proposed later in this chapter, should account for the occurrence of D-amino acids in milk, but the extended use of bacteria in starter cultures will account for the origin of D-amino acids in fermented foods and beverages (Bruckner *et al.*, 1993). As shown in Table 25.4, milk, cheese, fish sauces, vinegar and wine are all associated with the presence of D-alanine, D-aspartate and D-glutamate, but the range of D-amino acids is greater for vinegar and wine. Thermal processing is generally associated with the racemization of L-amino acids to the corresponding D-enantiomers. It is unlikely that the occurrence of D-amino acids in common foods is detrimental to human health. Indeed, Chung and Reed (2015) suggest that D-aspartate, in combination with D-glutamate, may confer potential benefits in the immunoglobulin E binding to peanut allergens. However, racemization may reduce availability of L-lysine and other limiting L-amino acids in animal feed.

Table 25.3. Endogenous D-amino acid oxidase (DAAO): recent advances, including significant implications for human biochemistry and health. The emergence of neurochemical aspects of DAAOs is emphasized.

Source and/or assumed site of enzyme	Observations	Reference(s)
Mutant mouse lacking DAAO	Free D-amino acids significantly higher in various organs of mutant mice compared with controls. Suggested that DAAO is involved in the elimination of free D-amino acids	Nagata *et al.* (1989)
Mouse brain	Involvement of DAAO in elimination of D-serine in mouse brain	Nagata (1992)
Mutant mouse kidney and serum	Occurrence of free D-isomers of alanine, proline and serine in mutant mouse lacking DAAO	Nagata *et al.* (1992)
Mammals, including humans	*In vivo* role of DAAO and D-aspartate oxidase in detoxification of D-amino acids that may have accumulated during ageing	D'Aniello *et al.* (1993)
Mouse CNS and serum	Enhanced D-serine levels in mutant mice lacking DAAO activity	Hashimoto *et al.* (1993b)
Rat brain	DAAO is confined to the lower brainstem and cerebellum in rat brain. Concluded that localization of DAAO inversely correlated with distribution of free D-serine in mammalian brain	Horiike *et al.* (1994)
Brain	A review. The detoxifying action of DAAO in the metabolism of exogenous D-amino acids is emphasized. Modulation of D-serine levels in the brain also proposed	Pilone (2000)
Mutant mouse brain	Determination of free D-proline and D-leucine in the brain of mutant mice lacking DAAO	Hamase *et al.* (2001)
Mutant mouse brain	Determination of free D-amino acids in the brain of mutant mice lacking DAAO activity	Morikawa *et al.* (2001)
Central nervous system (CNS)	Genetic and physiological evidence for the DAAO gene in schizophrenia	Chumakov *et al.* (2002)
Mouse tissues	Spatiotemporal relationships in tissue distribution of D-serine, serine racemase and DAAO during postnatal development	Wang and Zhu (2003)
Human	Characterization of human D-amino acid oxidase. Recombinant human DAAO was expressed in *Escherichia coli*	Molla *et al.* (2006)
Human brain	Increased brain DAAO activity in schizophrenia measured in human brain cortex post-mortem	Madeira *et al.* (2008)
CNS	Co-administration of a D-amino acid oxidase inhibitor potentiates the efficacy of D-serine as a therapeutic agent for schizophrenia	Hashimoto *et al.* (2009)
Mutant rat brain and periphery	Determination of free D-serine and D-alanine in the brain and periphery of mutant rats lacking DAAO	Miyoshi *et al.* (2011)
Rodent brain	Modulation of NMDA (*N*-methyl-D-aspartate) receptor function by inhibition of DAAO in rodent brain	Strick *et al.* (2011)
Mutant rodent brain	D-Amino acids in the brain of mutant rodents lacking DAAO linked to behavioural changes. Implications for amyotrophic lateral sclerosis and schizophrenia	Yamanaka *et al.* (2012)
Nanoparticle system	A DAAO-nanoparticle system in experimental cancer enzymatic therapy for humans	Bava *et al.* (2013)
Rat spinal astrocytes	DAAO contributes to mechanical pain hypersensitivity induced by sleep deprivation in the rat	Wei *et al.* (2013)

Continued

Table 25.3. Continued.

Source and/or assumed site of enzyme	Observations	Reference(s)
CNS	DAAO inhibition with sodium benzoate proposed as a novel approach for early phase Alzheimer's disease	Lin *et al.* (2014)
CNS	Activity of DAAO is widespread in the human CNS	Sasabe *et al.* (2014)
Model liposomal system	Selective oxidation of D-amino acids catalysed by oligolamellar liposomes intercalated with DAAO	Yoshimoto *et al.* (2014)

25.6 Feeding and Nutrition

The exogenous provision of purified crystalline D-amino acids can result in their utilization in higher organisms. Thus, Gordes *et al.* (2011) reported that D-tryptophan, D-phenylalanine and D-histidine exogenously applied to *A. thaliana* resulted in the accumulation of the corresponding L-amino acids. In almost all cases, such applications were accompanied by accumulation of D-alanine and D-glutamate as well. It was concluded that microbial D-amino acids in the rhizosphere may be absorbed and metabolized via central metabolic mechanisms. Based on acquisition and assimilation studies of D-amino acids in wheat (*Triticum aestivum*), Hill *et al.* (2011) suggested that it may be necessary to reconsider the forms of soil nitrogen (N) that are important in the terrestrial N cycle. The occurrence of D-amino acid aminotransferases in pea seedlings is consistent with the existence of at least one component of an appropriate mechanism of assimilation (Ogawa *et al.*, 1973).

Since animals are endowed with DAAOs and certain specific L-amino acid aminotransferases, it is expected that D-amino acid utilization should occur according to the sequence shown in Fig. 25.2. However, the absence of certain specific L-aminotransferases means that the relevant D-amino acid cannot be converted into its L-isomer. The results of feeding studies with pure crystalline D-amino acids confirm variable degrees of utilization by animals (Table 25.5), although there is some uniformity in the responses observed. On the one hand, D-arginine, D-lysine and D-threonine, even from limited evidence, are recognized to be inactive in animals as precursors of their respective L-isomers. On the other, D-methionine is utilized with 100% efficiency in the dog and 90% in the chick and rat, but only 30% in the human, according to Baker (1994). In between, there are examples of other D-amino acids with highly variable levels of nutritional efficacy. For example, D-tryptophan is utilized with 100% efficiency by the rat, but has significantly lower levels of effectiveness in the chick and mouse, and zero bioavailability for the human. The efficacy of the L-and DL-forms of methionine and tryptophan for growth may be influenced by the presence of other D-amino acids in the diet. This premise was tested by D'Mello (1987) using young chicks fed dried bacterial cells derived from methanol. It was assumed that the D-amino acids in the peptidoglycan fraction of the bacterial cells would reduce the utilization of the two forms of methionine and tryptophan that were fed. However, it was observed that the efficacy of utilization of both the L-and DL-isomers of methionine and tryptophan for chick growth was not affected by dietary inclusion of the bacterial protein.

The relatively poor utilization of D-methionine by the human (Table 25.5), even in parenteral nutrition (Printen *et al.*, 1979), deserves further elaboration as the calculated efficacy depends on the assessment criteria. Even though there is a wide discrepancy in efficacy between the rat and human, when other measures are employed, there may be a greater degree of consistency in the responses to D-methionine. In the rat, for example, this isomer provides effective protection towards ototoxicity induced by the chemotherapeutic agent, cisplatin (Campbell *et al.*, 1996). This efficacy

Table 25.4. Occurrence of D-amino acids in foods and fermented products, including alcoholic beverages. It is generally accepted that certain types of fermentation may influence D-amino acid distribution and concentrations (Bruckner *et al.*, 1993; Abe *et al.*, 1999). In addition, thermal processing may induce the racemization of L-amino acids to the corresponding D-isomers (see Friedman, 1999).

Food	Observations	Reference
Milk	Significant quantities of D-Ala, D-Asp and D-Glu in raw cow's milk but not in human milk. D-Ala content increased after cold storage of milk	Gandolfi *et al.* (1992)
Milk	Subclinical mastitis in cows enhanced D-amino acid concentrations in milk	Casapó *et al.* (1995)
Milk, fermented milk, cheese and acid curd cheese	Significant quantities of D-amino acids detected in raw and pasteurized milk, yogurt, curdled milk, goat milk and fresh cheese. Occurrence attributed to microbial activity	Bruckner and Hausch (1990)
Milk and cheese	Presence of D-Ala, D-Asp and D-Glu attributed to microbial activity	Bruckner *et al.* (1992)
Cheese	D-Ala, D-Asp and D-Glu in commercially available Spanish samples	Sardella *et al.* (2013)
Casein	*In vitro* digestibility of alkali-treated casein reduced owing to racemization of amino acids	Bunjapamai *et al.* (1982)
Dairy products, raw ham, roasted coffee	Occurrence of D-Ala, D-Asp and D-Glu attributed to thermal racemization or to microbial activity	Palla *et al.* (1989)
Fermented foods	Significant quantities of D-amino acids found in sour milk, cheese, lactic acid-fermented juices, beer, wine and fermented black beans	Bruckner and Hausch (1989)
Yogurt	D-Ala, D-Asp and D-Glu; ratio of D- to L-amino acid highest for Ala	Jin *et al.* (1998)
Infant formula	D-Ala, D-Asp and D-Glu in reconstituted infant formula; no difference between conventional and microwave heat treatments	Marchelli *et al.* (1992)
Ham	D-Ala, D-Asp, D-Ser, D-Val and D-Leu in cooked ham samples	Gil-Díaz *et al.* (2009)
Fish sauces	Occurrence of D-amino acids in fish sauces and other fermented fish products. D-Ala most abundant, present with D-Asp and D-Glu. Fish sauce from Myanmar contained more of these D-amino acids than samples from other countries. D-Ala and D-Asp highest in fish sauce fermented for 22 months	Abe *et al.* (1999)
Sour dough	D-Ala, D-Glu and traces of other D-isomers detected in sour doughs fermented with *Lactobacillus brevis* and *Saccharomyces cerevisiae*. D-Amino acids decreased by more than 44% after baking	Gobbetti *et al.* (1994)
Vinegar	Grain vinegar samples contained D-Ala, D-Asp, D-Glu, D-Ser, D-Arg, D-Asn and D-Phe. Lactic acid fermentation favoured D-amino acid formation	Mutaguchi *et al.* (2013)
Wine	D-Ala, D-Asp, D-Glu, D-Ile and D-Leu found. In proportion to the L-isomer, D-Ala was the most prominent D-amino acid	Kato *et al.* (1995)

underlines the current optimism for the therapeutic uses of D-methionine in humans, not only for hearing loss (Campbell *et al.*, 2011), but also in other clinical circumstances, such as protection against radiation- or chemoradiation-induced mucositis in patients undergoing treatment for squamous cell cancer (Vuyyuri *et al.*, 2008).

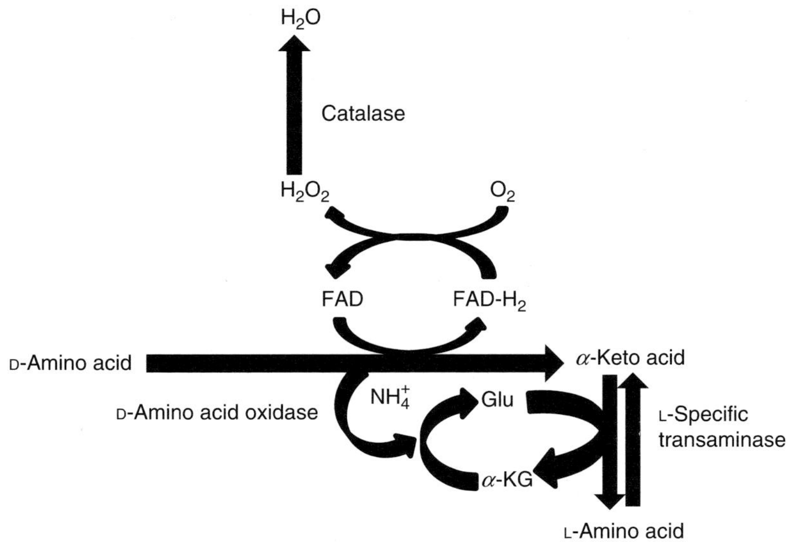

Fig. 25.2. The two-step conversion of D-amino acids to their respective L-isomers, based on studies with higher organisms, particularly animals. Abbreviations: FAD, flavin adenine dinucleotide; Glu, glutamate; α-KG, α-ketoglutarate. Adapted from Baker, 1994.

Table 25.5. Nutritional value of pure D-amino acids (relative to that of the corresponding L-isomer) for vertebrate animals, including humans. Based on the review of Baker (1994) and updated as appropriate in the text, particularly with respect to alternative criteria of efficacy.

D-Amino acid	Nutritional efficacy as determined in feeding studies
D-Arginine	Not utilized by the chick or rat. Assumed to be ineffective for other vertebrates, including humans
D-Lysine	Not utilized by the chick, rat, mouse or human
D-Threonine	Not utilized by the chick, rat, mouse or human
D-Isoleucine	Not utilized by the chick. Assumed to be ineffective for other animals, including humans
D-Histidine	10% utilization in the chick and mouse. Not utilized by the rat and assumed to be ineffective for other animals, including humans
D-Valine	70% utilization by the chick, but 15% and 5% in the rat and mouse, respectively. No data for the human
D-Leucine	Fully utilized by the chick; 50 and 15% utilization in the rat and mouse, respectively. No data for humans
D-Tryptophan	100% for the rat, 80% for the pig; 35, 30 and 20% for the dog, mouse and chick, respectively. Ineffective for the human
D-Phenylalanine	75 and 70% utilized by the chick and rat, respectively. No data presented for other species
D-Tyrosine	100% utilization in the chick and rat. No data for other species
D-Methionine	100% efficiency in the dog and pig, 90% in the chick and rat, 75% in the mouse and 30% in the human

25.6.1 Case study: D-amino acid kinetics in the ruminant animal

Pathways of utilization in the ruminant animal constitute a salutary model which exemplifies the complex facets of D-amino acid dynamics in higher organisms, particularly in the context of specialized symbiotic microbial ecology (Fig. 25.3). The central issues relate to dietary sources, microbial activity and endogenous

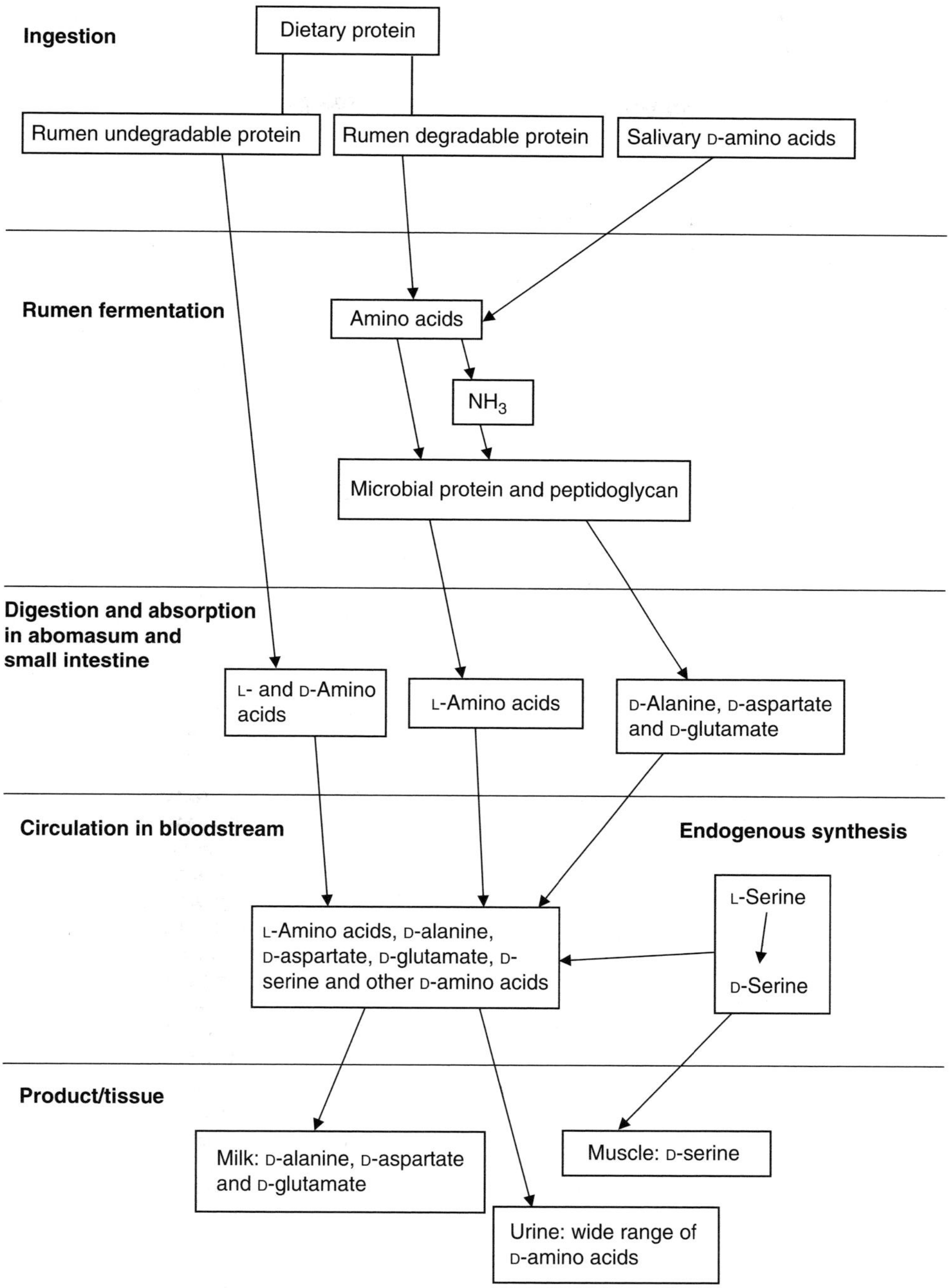

Fig. 25.3. Pathways of D-amino acid synthesis and distribution in the ruminant animal. A composite model based on discrete elements of data observed in other species. See text for explanation and sources of data.

synthesis as potential mechanisms for the supply of D-amino acids to the ruminant animal. Protein in the diet is generally regarded as being composed of two fractions: that which is degraded by the rumen microbes and that which escapes into the abomasum and small intestine to be digested by host enzymes. The former fraction is termed 'rumen degradable protein' (RDP) and the latter 'undegradable dietary protein' (UDP). The breakdown of RDP results in the production of microbial protein which contributes substantially to the amino acid requirements of the host animal. Ruminants produce copious amounts of saliva, which may contain D-amino acids (Nagata *et al.*, 2006), and these would also be fermented in the rumen. The peptidoglycan moiety of rumen microbes also supplies D-amino acids to the animal, which include D-alanine, D-aspartate and D-glutamate (Kamio *et al.*, 1981; Bruckner *et al.*, 1992). If the ruminant is fed on silage, there would be inputs of D-alanine and D-glutamate (Rooke *et al.*, 1984) and these would be exposed to fermentation by the rumen microorganisms. The diet may also contain UDP fractions present in heat- or alkali-treated ingredients, in which case a wide range of D-amino acids would be consumed and absorbed into the systems of the ruminant. Friedman (1999) concluded that the exposure of food proteins to certain processing treatments induces racemization of all L-amino acids to D-enantiomers. The relationship between amino acid racemization and processing conditions for cereal and oilseed by-products continues to attract attention (Horak *et al.*, 2014). A substantial number of the resulting D-isomers, particularly lysine and threonine, are not utilized by animals and others are of relatively low bioavailability (Table 25.5). The D-amino acids absorbed are transported in the bloodstream and metabolized in peripheral tissues, while a wide range may be excreted as D-amino acids in the urine, judging by analyses of human samples (Bruckner and Schieber, 2001).

25.7 Toxicology

It is instructive to recall that, given the appropriate conditions, several of the essential L-amino acids required by animals can precipitate adverse effects to varying degrees of severity (D'Mello and Lewis, 1970). Furthermore, the non-protein L-amino acids present in plants have also been associated with deleterious effects in animals (D'Mello, 2015). So toxicity is not an exclusive or inevitable consequence of the occurrence of D-amino acids in higher organisms; there are ubiquitous and highly effective endogenous pathways for oxidation and salvage (Figs 25.1 and 25.2). The intervention of microbes is not essential for the avoidance of ill effects. Nevertheless, it is possible to contrive conditions that might result in adverse effects. In this respect, there is considerable consistency in the responses to L- and D-amino acids. Thus, under particular circumstances, the term 'excitotoxicity' may be applied to both L-glutamate (Flores *et al.*, 2012) and D-serine (Smith *et al.*, 2012). Flores *et al.* (2012) referred to neurological and neurodegenerative disorders associated with L-glutamate perturbation of receptors and neurotransmission. It is instructive to note that amyotrophic lateral sclerosis, schizophrenia and Alzheimer's disease have all been linked with DAAO expression, but that the studies have involved the use of mutant DAAO-deficient animals (Hashimoto *et al.*, 2009; Yamanaka *et al.*, 2012; Tables 25.1 and 25.3). Smith *et al.* (2012) investigated the effect of the interaction between L-glutamate and D-serine toxicity on primary ganglion cells (Fig. 25.4), but pointed out that glycine status may be an additional factor to consider in excitotoxic conditions.

As with other toxicological disciplines, there are situations where it is difficult to separate cause-and-effect issues. For example, increases in D-amino acids during chronic renal failure have been attributed to depletion of the DAAO status (Young *et al.*, 1994). However, dietary intake of D-amino acids (Table 25.4), use of peptide antibiotics and DAAO-inhibiting drugs may also be contributory factors. Alternatively, these increases in D-amino acids may be entirely coincidental to the aetiology of renal failure. Environmental pollutants may also compromise D-amino acid homeostasis. For instance, Yu *et al.* (2014) suggested that chronic lead exposure during brain development might enhance D-serine degradation and contribute to learning and memory deficits in mammals. This raises an intriguing scenario in respect of other pollutants such as mercury and pesticides. Another aspect awaiting elucidation is the relationship between

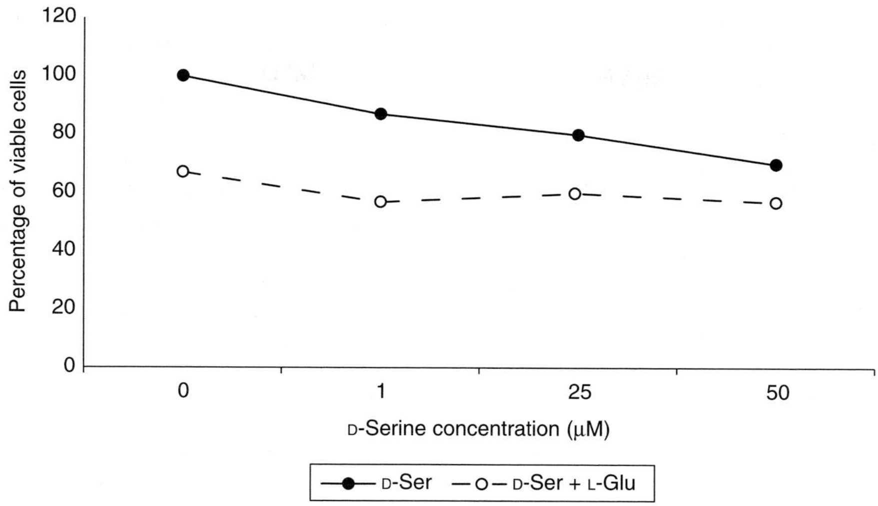

Fig. 25.4. Effects of D-serine or D-serine + L-glutamate on retinal ganglion cell viability. Exposure to D-serine decreased the viability of ganglion cells, particularly at 25 and 50 μM concentrations. Cell death was increased further in the presence of L-glutamate at a constant concentration of 20 μM. Abbreviations: Glu, glutamate; Ser, serine. Adapted from Smith *et al.*, 2012.

D-amino acid intake and immune responses, including allergies (Sela and Zisman, 1997).

25.8 Clinical Applications: Emerging Potential

The utilization of D-amino acids in free form or as components of medicinal compounds is an emerging feature that is likely to attract considerable attention in the future. Optimism has been expressed for the therapeutic uses of D-methionine in humans for cisplatin ototoxicity (Campbell *et al.*, 2011) and in other clinical circumstances, such as protection against radiation- or chemoradiation-induced mucositis in patients undergoing treatment for squamous cell cancer (Vuyyuri *et al.*, 2008). Hydrogen sulfide acts as a signalling molecule and cell protectant. Shibuya *et al.* (2013) reported the production of hydrogen sulfide from D-cysteine, pointing to the potential of this pathway in delivering hydrogen sulfide to specific tissues for therapeutic purposes.

Other pharmaceutical applications may emerge, and there is increasing interest in the incorporation of an expanded set of amino acid analogues for the synthesis of unnatural peptides. Connor and Tirrell (2007) suggested that, for the protein chemist, the 20 canonical L-amino acids are much too restrictive for macromolecular formulations. Effective technologies are now available to insert artificial (non-canonical) amino acids into polymers for novel applications. Currently, there is interest in the activity of D-amino acids as components of antimicrobial peptides. Thus, Silva *et al.* (2014) reported enhanced efficacy of lactoferricin peptides containing arginine and D-amino acid molecules towards *Mycobacterium avium*, a pathogen causing respiratory disease in susceptible individuals.

Arguably, the association of D-serine with CNS function and dysfunction provides the most promising opportunity for clinical intervention. For example, Hashimoto *et al.* (2009) concluded that the co-administration of D-serine with a DAAO inhibitor might be effective for the treatment of schizophrenia. Others suggest that benzoate inhibition of DAAO offers a promising approach for the treatment of early-phase Alzheimer's disease (Lin *et al.*, 2014).

25.9 Conclusions

The notion that the presence of D-amino acids in higher organisms is merely a coincidental observation inconsistent with the established rules of biochemistry is no longer tenable.

On the contrary, both plants and animals are endowed with amino acid racemases, D-amino acid oxidases (DAAOs) and transaminases that ensure homeostasis. D-Alanine, D-aspartate, D-glutamate and L-serine are regular constituents of the body fluids, tissues and organs of mammals and other species, depending on factors such as genetics, health and age.

The utilization of D-amino acids by higher organisms has been established with two distinct pathways. First, studies at the molecular and biochemical levels have unequivocally demonstrated that D-serine forms an integral part of the endogenous ligand for the 'glycine' binding site of the NMDA subtype of glutamate receptor in the brain and retina. On the basis of anatomical localization, neuronal biosynthesis, the uptake and degradation of aspartate and the activation of L-glutamate receptors, the case has also been advanced for D-aspartate in neuroendocrine functioning and signalling. The second approach has involved feeding studies with both plants and animals at the whole organism level. For example, exogenous application of D-tryptophan, D-phenylalanine and D-histidine to *A. thaliana* resulted in their conversion to the corresponding L-amino acids. The results of feeding studies with pure crystalline D-amino acids confirm variable degrees of utilization by animals, depending upon species and the D-amino acid in question. On the one hand, D-arginine, D-lysine and D-threonine are recognized to be inactive in animals as precursors of the respective L-isomers. On the other, D-methionine is utilized with 100% efficiency in the dog and 90% in the chick and rat but only 30% in the human. In between, there are examples of other D-amino acids with highly variable levels of nutritional efficacy. For example, D-tryptophan is utilized with 100% efficiency by the rat but with 0% bioavailability for the human. D-Amino acids may also arise from microbial activity in the digestive tract of animals, particularly ruminants, and in the fermentation processes involved in the manufacture of dairy products, vinegar and wine. The occurrence of D-amino acids in milk may serve as a marker of undesirable microbial proliferation in dairy products.

In the absence of overt or substantive evidence, it has to be assumed that the consumption of D-amino acids via dairy products and fermented beverages is unlikely to cause hazardous reactions in humans. The concept that D-amino acid synthesis or exogenous uptake is routinely accompanied by toxicity has to be revised in the light of recent developments. Nevertheless, it is prudent to avoid complacency, particularly in regard to immune responses. In addition, it is recognized that failure of homeostatic mechanisms may predispose individuals to deleterious effects. Thus, DAAO insufficiency may be an underlying feature of disorders such as schizophrenia or exposure to lead and other pollutants. In addition, the racemization of L-amino acids to the corresponding D-enantiomers during thermal processing may reduce metabolic L-amino acid availability, particularly in animal feed.

A number of issues remain unresolved at the present time. For example, the ubiquitous distribution of D-glutamate in higher organisms raises questions about a possible biochemical or physiological role, equivalent to that of its L-isomer. Currently, there is interest in the utilization of D-amino acids as components of antimicrobial peptides. The enhanced efficacy of lactoferricin peptides containing arginine and D-amino acid molecules towards *M. avium*, a pathogen causing respiratory disease in susceptible individuals, is one example of potential clinical application. The association of D-serine with brain function and dysfunction provides the most promising opportunity for intervention in the treatment of CNS disorders. The co-administration of D-serine with a DAAO inhibitor for the treatment of schizophrenia and the inhibition of DAAO by benzoate for the treatment of early-phase Alzheimer's disease are worthy cases for further research.

References

Abe, H., Park, J-N., Fukumoto, Y., Fujita, E., Tanaka, T., Washio, T., Otsuka, S., Shimizu, T. and Watanabe, K. (1999) Occurrence of D-amino acids in fish sauces and other fermented fish products. *Fisheries Science* 65, 637–641.

Atasoglu, C. and Wallace, R.J. (2003) Metabolism and *de novo* synthesis of amino acids by rumen microbes. In: D'Mello, J.P.F. (ed.) *Amino Acids in Animal Nutrition*. CAB International, Wallingford, UK, pp.265–290.

Baker, D.H. (1994) Utilization of precursors for L-amino acids. In: D'Mello, J.P.F. (ed.) *Amino Acids in Farm Animal Nutrition*. CAB International, Wallingford, UK, pp. 37–61

Bardaweel, S.K., Alzweiri, M. and Ishaqat, A.A. (2014) D-Serine in neurobiology: CNS neurotransmission and neuromodulation. *Canadian Journal of Neurological Sciences* 41, 164–176.

Baudhuin, P., Beufay, H. and de Duve, C. (1965) Combined biochemical and morphological study of particulate fractions from rat liver: analysis of preparations enriched in lysosomes or in particles containing urate oxidase, D-amino acid oxidase and catalase. *The Journal of Cell Biology* 26, 219–243.

Bava, A., Gornati, R., Cappellini, F., Caldinelli, L., Pollegioni, L. and Bernardini, G. (2013) D-Amino acid oxidase-nanoparticle system: a potential novel approach for cancer therapy. *Nanomedicine* 8, 1797–1806.

Bender, A.E. and Krebs, H.A. (1950) The oxidation of various synthetic α-amino-acids by mammalian D-amino-acid oxidase, L-amino-acid oxidase of cobra venom and the L- and D-amino-acid oxidases of *Neurospora crassa*. *Biochemical Journal* 46, 210–219.

Blanchard, M., Green, D.E., Nocito, V. and Ratner, S. (1944) L-Amino acid oxidase of animal tissues. *The Journal of Biological Chemistry* 155, 421–440.

Blaschko, H. and Hawkins, J. (1952) D-Amino-acid oxidase in the molluscan liver. *Biochemical Journal* 52, 306–310.

Bruckner, H. and Hausch, M. (1989) Gas chromatographic detection of D-amino acids as common constituents of fermented foods. *Chromatographia* 28, 487–492.

Bruckner, H. and Hausch, M. (1990) D-Amino acids in dairy products: detection, origin and nutritional aspects. I. Milk, fermented milk, fresh cheese and acid curd cheese. *Milchwissenschaft* 45, 357–360.

Bruckner, H. and Schieber, A. (2001) Determination of amino acid enantiomers in human urine and blood serum by gas chromatography-mass spectrometry. *Biomedical Chromatography* 15, 166–172.

Bruckner, H. and Westhauser, T. (2003) Chromatographic determination of L- and D-amino acids in plants. *Amino Acids* 24, 43–55.

Bruckner, H., Jack, P., Langer, M. and Godel, H. (1992) Liquid chromatographic determination of D-amino acids in cheese and cow milk. Implication of starter cultures, amino acid racemases and rumen microorganisms on formation, and nutritional considerations. *Amino Acids* 2, 271–284.

Bruckner, H., Becker, D. and Lupke, M. (1993) Chirality of amino acids of microorganisms used in food biotechnology. *Chirality* 5, 385–392.

Bunjapamai, S., Mahoney, R.R. and Fagerson, I.S. (1982) Determination of D-amino acids in some processed foods and effects of racemization on *in vitro* digestibility of casein. *Journal of Food Science* 47, 1229–1234.

Campbell, K.C.M., Rybak, L.P., Meech, R.P. and Hughes, L. (1996) D-Methionine provides excellent protection from cisplatin ototoxicity in the rat. *Hearing Research* 102, 90–98.

Campbell, K.C.M., Claussen, A., Meech, R.P., Verhulst, S., Fox, D. and Hughes, L. (2011) D-Methionine significantly rescues noise-induced hearing loss: timing studies. *Hearing Research* 282, 138–144.

Casapó, J., Csapó-Kiss, Z., Stefler, J., Martin, T.G. and Némethy, S. (1995) Influence of mastitis on D-amino acid content of milk. *Journal of Dairy Science* 78, 2375–2381.

Chumakov, I., Blumenfeld, M., Guerassimenko, O., Cavarec, L., Palicio, M., Abderrahim, H., Bougueleret, L., Barry, C., Tanaka, H., La Rosa, P. *et al.* (2002) Genetic and physiological data implicating the new human gene G72 and the gene for D-amino acid oxidase in schizophrenia. *Proceedings of the National Academy of Sciences of the United States of America* 99, 13365–13367.

Chung, S-Y. and Reed, S. (2015) IgE binding to peanut allergens is inhibited by combined D-aspartic acid and D-glutamic acids. *Food Chemistry* 166, 248–253.

Cohen, P.P. and Sallach, H.J. (1961) Nitrogen metabolism of amino acids. In: Greenberg, D.M (ed.) *Metabolic Pathways: Volume II*. Academic Press, New York and London, pp. 1–78.

Connor, R.E. and Tirrell, D.A. (2007) Non-canonical amino acids in protein polymer design. *Polymer Reviews* 47, 9–28.

D'Aniello, A. (2007) D-Aspartic acid: an endogenous amino acid with an important neuroendocrine role. *Brain Research Reviews* 53, 215–234.

D'Aniello, A., D'Onofrio, G., Pischetola, M., Vetere, A., Petrucelli, L. and Fisher, G.H. (1993) Biological role of D-amino acid oxidase and D-aspartate oxidase. Effects of D-amino acids. *The Journal of Biological Chemistry* 268, 26941–26949.

D'Aniello, G., Ronsini, S., Guida, F., Spinelli, P. and D'Aniello, A. (2005) Occurrence of D-aspartic acid in human seminal plasma and spermatozoa: possible role in reproduction. *Fertility and Sterility* 84, 1444–1449.

Di Fiore, M.M., Santillo, A. and Baccari, G.C. (2014) Current knowledge of D-aspartate in glandular tissues. *Amino Acids* 46, 1805–1818.

D'Mello, J.P.F. (1987) Utilisation of L- and DL-amino acids by broiler chicks fed diets containing Pruteen. *Animal Feed Science and Technology* 17, 1–13.
D'Mello, J.P.F. (2015) Toxicology of non-protein amino acids. In: D'Mello, J.P.F. (ed.) *Amino Acids in Higher Plants*. CAB International, Wallingford, UK, pp. 507–537.
D'Mello, J.P.F. and Lewis, D. (1970) Amino acids in chick nutrition. 3. Interdependence in amino acid requirements. *British Poultry Science* 11, 367–385.
Fisher, G.H., D'Aniello, A., Vetere, A., Padula, L., Cusano, G.P. and Man, E.H. (1991) Free D-aspartate and D-alanine in normal and Alzheimer brain. *Brain Research Bulletin* 26, 983–985.
Flores, G., Negrete-Díaz, J.V., Carrión, M., Andrade-Talavera, Y., Bello, S.A., Sihra, T.S. and Rodríguez-Moreno, A. (2012) Excitatory amino acids in neurological and neurodegenerative disorders. In: D'Mello, J.P.F. (ed.) *Amino Acids in Human Nutrition and Health*. CAB International, Wallingford, UK, pp. 427–453.
Fox, D.G. and Tedeschi, L.O. (2003) Predicting dietary amino acid adequacy for ruminants. In: D'Mello, J.P.F. (ed.) *Amino Acids in Animal Nutrition*. CAB International, Wallingford, UK, pp.389–410.
Friedman, M. (1999) Chemistry, nutrition and microbiology of D-amino acids. *Journal of Agricultural and Food Chemistry* 47, 3457–3479.
Fuchs, S.A., Berger, R., Klomp, L.W.J. and de Koning, T.J. (2005) D-Amino acids in the central nervous system in health and disease. *Molecular Genetics and Metabolism* 85, 168–180.
Fujii, N. (2005) D-Amino acids in elderly tissues. *Biological and Pharmaceutical Bulletin* 28, 1585–1589.
Fujitani, Y., Nakajima, N., Ishihara, K. and Sugimoto, M. (2006) Molecular and biochemical characterization of a serine racemase from *Arabidopsis thaliana*. *Phytochemistry* 67, 668–674.
Fujitani, Y., Horiuchi, T., Ito, K. and Sugimoto, M. (2007) Serine racemases from barley, *Hordeum vulgare* L., and other plant species represent a distinct eukaryotic group: gene cloning and recombinant protein characterization. *Phytochemistry* 68, 1530–1536.
Fukuda, M., Tokumura, A. and Ogawa, T. (1973) D-Alanine in germinating *Pisum sativum* seedlings. *Phytochemistry* 12, 2593–2596.
Gandolfi, L., Palla, G., Delprato, L., De Nisco, F., Marchelli, R. and Salvadori, C. (1992) D-Amino acids in milk as related to heat treatments and bacterial activity. *Journal of Food Science* 57, 377–379.
Gaunt, G.L. and de Duve, C. (1976) Subcellular distribution of D-amino acid oxidase and catalase in rat brain. *Journal of Neurochemistry* 26, 749–759.
Gil-Díaz, M., Santos-Delgado, M.J., Rubio-Barroso, S. and Polo-Díez, L.M. (2009) Free D-amino acid determination in ready-to-eat cooked ham irradiated with electron-beam by indirect chiral HPLC. *Meat Science* 82, 24–29.
Gobbetti, M., Simonetti, M.S., Rossi, J., Cossignani, L., Corsetti, A. and Damiani, P. (1994) Free D- and L-amino acid evolution during sour dough fermentation and baking. *Journal of Food Science* 59, 881–884.
Gogami, Y., Ito, K., Matsushima, Y. and Oikawa, T. (2009) Occurrence of D-serine in rice and characterization of rice serine racemase. *Phytochemistry* 70, 380–387.
Gordes, D., Kolukisaoglu, U. and Thurow, K. (2011) Uptake and conversion of D-amino acids in *Arabidopsis thaliana*. *Amino Acids* 40, 553–563.
Hamase, K., Inoue, T., Morikawa, A., Konno, R. and Zaitsu, K. (2001) Determination of free D-proline and D-leucine in the brains of mutant mice lacking D-amino acid oxidase activity. *Analytical Biochemistry* 298, 253–258.
Hamase, K., Morikawa, A. and Zaitsu, K. (2002) D-Amino acids in mammals and their diagnostic value. *Journal of Chromatography B: Biomedical Sciences and Applications* 781, 73–91.
Han, H., Miyoshi, Y., Ueno, K., Okamura, C., Mita, M., Lindner, W., Zaitsu, K. and Hamase, K. (2011) Simultaneous determination of D-aspartic acid and D-glutamic acid in rat tissues and physiological fluids using a multi-loop two-dimensional HPLC procedure. *Journal of Chromatography B: Biomedical Sciences and Applications* 879, 3196–3202.
Hashimoto, A. and Oka, T. (1997) Free D-aspartate and D-serine in the mammalian brain and periphery. *Progress in Neurobiology* 52, 325–353.
Hashimoto, A., Nishikawa, T., Oka, T., Takahashi, K. and Hayashi, T. (1992) Determination of free amino acid enantiomers in rat brain and serum by high-performance liquid chromatography after derivatization with *N-tert.*-butyloxycarbonyl-L-cysteine and *o*-phthaldialdehyde. *Journal of Chromatography B: Analytical Technologies in the Biomedical and Life Sciences* 582, 41–48.
Hashimoto, A., Nishikawa, T., Konno, R., Niwa, A., Yasumura, Y., Oka, T. and Takahashi, K. (1993a) Free D-serine and D-alanine in central nervous system and serum in mutant mice lacking D-amino acid oxidase. *Neuroscience Letters* 152, 33–36.
Hashimoto, A., Nishikawa, T., Oka, T., Hayashi, T. and Takahashi, K. (1993b) Widespread distribution of free D-aspartate in rat periphery. *FEBS Letters* 331, 4–8.

Hashimoto, K., Fujita, Y., Horio, M., Kunitachi, S., Iyo, M., Ferraris, D. and Tsukamoto, T. (2009) Co-administration of a D-amino acids oxidase inhibitor potentiates the efficacy of D-serine in attenuating prepulse inhibition deficits after administration of dizocilpine. *Biological Psychiatry* 65, 1103–1106.

Hepler, P.K., Kunkel, J.G., Rounds, C.M. and Winship, L.J. (2012) Calcium entry into pollen tubes. *Trends in Plant Science* 17, 32–38.

Hill, P.W., Quilliam, R.S., DeLuca, T.H., Farrar, J., Farrell, M., Roberts, P., Newsham, K.K., Hopkins, D.W., Bardgett, R.D. and Jones, D.L. (2011) Acquisition and assimilation of nitrogen as peptide-bound and D-enantiomers of amino acids by wheat. *PLoS ONE* 6(4): e19220.

Horak, J., Gerhardt, H., Theiner, J. and Lindner, W. (2014) Correlation between amino acid racemization and processing conditions for various wheat products, oil seed press cakes and lignin samples. *Food and Bioproducts Processing* 92, 355–368.

Horiike, K., Tojo, H., Arai, R., Nozaki, M. and Maeda, T. (1994) D-Amino acid oxidase is confined to the lower brain stem and cerebellum in rat brain: regional differentiation of astrocytes. *Brain Research* 652, 297–303.

Jin, D., Nagakura, K., Murofushi, S., Miyahara, T. and Toyo'oka, T. (1998) Total resolution of 17 DL-amino acids labelled with fluorescent chiral agent by high-performance liquid chromatography. *Journal of Chromatography A* 822, 215–224.

Kamio, Y., Itoh, Y. and Terawaki, Y. (1981) Chemical structure of peptidoglycan in *Selenomonas ruminantium*: cadaverine links to the D-glutamic acid residue of peptidoglycan. *Journal of Bacteriology* 146, 49–53.

Katane, M. and Homma, H. (2011) D-Aspartate – an important bioactive substance in mammals: a review from an analytical and biological point of view. *Journal of Chromatography B: Biomedical Sciences and Applications* 879, 3108–3121.

Kato, M., Fukushima, T., Santa, T., Homma, H. and Imai, K. (1995) Determination of D-amino acids, derivatized with 4-fluoro-7-nitro-2,3-benzoxadiazole (NBD-F), in wine samples by high-performance liquid chromatography. *Biomedical Chromatography* 9, 193–194.

Kera, Y., Nagasaki, H., Iwashima, A. and Yamada, R.-H. (1992) Presence of D-aspartate oxidase and free D-aspartate in amphibian (*Xenopus laevis, Cynops pyrrhogaster*) tissues. *Comparative Biochemistry and Physiology Part B: Comparative Biochemistry* 103, 345–348.

Kera, Y., Aoyama, H., Matsumura, H., Hasegawa, A., Nagasaki, H. and Yamada, R.-H. (1995) Presence of free D-glutamate and D-aspartate in rat tissues. *Biochimica et Biophysica Acta (BBA) – General Subjects* 1243, 282–286.

Kera, Y., Aoyama, H., Watanabe, N. and Yamada, R.-H. (1996) Distribution of D-aspartate oxidase and free D-glutamate and D-aspartate in chicken and pigeon tissues. *Comparative Biochemistry and Physiology Part B: Biochemistry and Molecular Biology* 115, 121–126.

Kera, Y., Hasegawa, S., Watanabe, T., Segawa, H. and Yamada, R.-H. (1998) D-Aspartate oxidase and free acidic D-amino acids in fish tissues. *Comparative Biochemistry and Physiology Part B: Biochemistry and Molecular Biology* 119, 95–100.

Kera, Y., Watanabe, T., Shibata, K., Nagasaki, H. and Yamada, R.-H. (2001) D-Aspartate oxidase and free acidic D-amino acids in lower vertebrates. *Journal of Molecular Catalysis B: Enzymatic* 12, 121–129.

Kim, P.M., Duan, X., Huang, A.S., Liu, C.Y., Ming, G-L., Song, H. and Snyder, S.H. (2010) Aspartate racemase, generating neuronal D-aspartate, regulates adult neurogenesis. *Proceedings of the National Academy of Sciences of the United States of America* 107, 3175–3179.

Konno, R. and Yasumura, Y. (1981) Activity and substrate specificity of D-amino acid oxidase in kidneys of various animals. *Doubutsugaku Zasshi* 90, 368–373.

Konno, R. and Yasumura, Y. (1983) Mouse mutant deficient in D-amino acid oxidase activity. *Genetics* 103, 277–285.

Lin, C.-H., Chen, P.-K., Chang, Y.-C., Chuo, L.-J., Chen, Y-S., Tsai, G.E. and Lane, H.-Y. (2014) Benzoate, a D-amino acid oxidase inhibitor, for the treatment of early-phase Alzheimer disease: a randomized, double-blind, placebo-controlled trial. *Biological Psychiatry* 75, 678–685.

Madeira, C., Freitas, M.E., Vargas-Lopes, C. and Panizzutti, R. (2008) Increased brain D-amino acid oxidase (DAAO) activity in schizophrenia. *Schizophrenia Research* 101, 76–83.

Mangas, A., Coveñas, R., Bodet, D., Geffard, M., Aguilar, L.A. and Yajeya, J. (2007) Immunocytochemical visualization of D-glutamate in the rat brain. *Neuroscience* 144, 654–664.

Marchelli, R., Dossena, A., Palla, G., Audhuy-Peaudecerf, M., Lefeuvre, S., Carnevali, P. and Freddi, M. (1992) D-Amino acids in reconstituted infant formula: a comparison between conventional and microwave heating. *Journal of the Science of Food and Agriculture* 59, 217–226.

Matsuda, S., Katane, M., Maeda, K., Kaneko, Y., Saitoh, Y., Miyamoto, T., Sekine, M. and Homma, H. (2015) Biosynthesis of D-aspartate in mammals: the rat and human homologs of mouse aspartate racemase are not responsible for the biosynthesis of D-aspartate. *Amino Acids* 47, 975–985.

Michard, E., Lima, P.T., Borges, F., Silva, A.C., Portes, M.T., Cavalho, J.E., Gilliham, M., Liu, L.H., Obermeyer, G. and Feijó, J.A. (2011) Glutamate receptor-like genes form Ca^{2+} channels in pollen tubes and are regulated by pistil D-serine. *Science* 332, 434–437.

Miyoshi, Y., Hamase, K., Okamura, T., Konno, R., Kasai, N., Tojo, Y. and Zaitsu, K. (2011) Simultaneous two-dimensional HPLC determination of free D-serine and D-alanine in the brain and periphery of mutant rats lacking D-amino-acid oxidase. *Journal of Chromatography B: Biomedical Sciences and Applications* 879, 3184–3189.

Miyoshi, Y., Koga, R., Oyama, T., Han, H., Ueno, K., Masuyama, K., Itoh, Y. and Hamase, K. (2012) HPLC analysis of naturally occurring free D-amino acids in mammals. *Journal of Pharmaceutical and Biomedical Analysis* 69, 42–49.

Molla, G., Sacchi, S., Bernasconi, M., Pilone, M.S., Fukui, K. and Pollegioni, L. (2006) Characterization of human D-amino acid oxidase. *FEBS Letters* 580, 2358–2364.

Morikawa, A., Hamase, K., Inoue, T., Konno, R., Niwa, A. and Zaitsu, K. (2001) Determination of free D-aspartic acid, D-serine and D-alanine in the brain of mutant mice lacking D-amino-acid oxidase activity. *Journal of Chromatography B: Analytical Technologies in the Biomedical and Life Sciences* 757, 119–125.

Morikawa, A., Hamase, K. and Zaitsu, K. (2003) Determination of D-alanine in the rat central nervous system and periphery using column-switching high-performance liquid chromatography. *Analytical Biochemistry* 312, 66–72.

Mutaguchi, Y., Ohmori, T., Akano, H., Doi, K. and Ohshima, T. (2013) Distribution of D-amino acids in vinegars and involvement of lactic acid bacteria in the production of D-amino acids. *Springerplus* 2: 691.

Nagata, Y. (1992) Involvement of D-amino acid oxidase in elimination of D-serine in mouse brain. *Experientia* 48, 753–755.

Nagata, Y., Konno, R., Yasumura, Y. and Akino, T. (1989) Involvement of D-amino acid oxidase in elimination of free D-amino acids in mice. *Biochemical Journal* 257, 291–292.

Nagata, Y., Yamamoto, K., Shimojo, T., Konno, R., Yasumura, Y. and Akino, T. (1992) The presence of free D-alanine, D-proline and D-serine in mice. *Biochimica et Biophysica Acta (BBA) – General Subjects* 1115, 208–211.

Nagata, Y., Higashi, M., Ishii, Y., Sano, H., Tanigawa, M., Nagata, K., Noguchi, K. and Urade, M. (2006) The presence of high concentrations of free D-amino acids in human saliva. *Life Sciences* 78, 1677–1681.

Neims, A.H., Zieverink, W.D. and Smilack, J.D. (1966) Distribution of D-amino acid oxidase in bovine and human nervous tissues. *Journal of Neurochemistry* 13, 163–168.

Ogawa, T., Fukuda, M. and Sasaoka, K. (1973) Occurrence of D-amino acid aminotransferase in pea seedlings. *Biochemical and Biophysical Research Communications* 52, 998–1002.

Ohgusu, T., Hamase, K., Tanaka, H., Shoyama, Y. and Zaitsu, K. (2006) High-throughput determination of free D-aspartic acid in mammals by enzyme immunoassay using specific monoclonal antibody. *Analytical Biochemistry* 357, 15–20.

Ono, K., Yanagida, K., Oikawa, T., Ogawa, T. and Soda, K. (2006) Alanine racemase of alfalfa seedlings (*Medicago sativa* L.): first evidence for the presence of an amino acid racemase in plants. *Phytochemistry* 67, 856–860.

Ota, N., Shi, T. and Sweedler, J.V. (2012) D-Aspartate acts as a signalling molecule in nervous and neuroendocrine systems. *Amino Acids* 43, 1873–1886.

Ota, N., Rubakhin, S.S. and Sweedler, J.V. (2014) D-Alanine in the islets of Langerhans of rat pancreas. *Biochemical and Biophysical Research Communications* 447, 328–333.

Palla, G., Marchelli, R., Dossena, A. and Casnati, G. (1989) Occurrence of D-amino acids in food. *Journal of Chromatography A* 475, 45–53.

Pillai, R.R. and Kurpad, A.V. (2012) Amino acid requirements: quantitative estimates. In: D'Mello, J.P.F. (ed.) *Amino Acids in Human Nutrition and Health*. CAB International, Wallingford, UK, pp.267–290.

Pilleriin, S., Walker, R. and Burnet, P.W.J. (2010) D-Serine metabolism in C6 glioma cells: involvement of alanine-serine-cysteine transporter (ASCT2) and serine racemase (SRR) but not D-amino acid oxidase (DAO). *Journal of Neuroscience Research* 88, 1829–1840.

Pilone, M.S. (2000) D-Amino acid oxidase: new findings. *Cellular and Molecular Life Sciences* 57, 1732–1747.

Printen, K.J., Brummel, M.C., Cho, E.S. and Stegink, L.D. (1979) Utilization of D-methionine during total parenteral nutrition in postsurgical patients. *The American Journal of Clinical Nutrition* 32, 1200–1205.

Rooke, J.A., Greife, H.A. and Armstrong, D.G. (1984) The effect of *in sacco* rumen incubation of a grass silage upon the total and D-amino acid composition of the residual silage dry matter. *The Journal of Agricultural Science* 102, 695–702.

Sardella, R., Lisanti, A., Marinozzi, M., Ianni, F., Natalini, B., Blanch, G.P. and del Castillo, M.L.R. (2013) Combined monodimensional chromatographic approaches to monitor the presence of D-amino acids in cheese. *Food Control* 34, 478–487.

Sasabe, J., Suzuki, M., Imanishi, N. and Aiso, S. (2014) Activity of D-amino acid oxidase is widespread in the human central nervous system. *Frontiers in Synaptic Neuroscience* 6: 14.

Satoh, S. and Esashi, Y. (1980) D-Amino-acid-stimulated ethylene production in seed tissues. *Planta* 149, 64–68.

Schieber, A., Bruckner, H. and Ling, J.R. (1999) GC-MS analysis of diaminopimelic acid stereoisomers and amino acid enantiomers in rumen bacteria. *Biomedical Chromatography* 13, 46–50.

Seifter, S., Harkness, D.M. and Rubin, L. (1948) The nicotinic acid, riboflavin, D-amino acid oxidase and arginase levels of the livers of rats on a protein-free diet. *The Journal of Biological Chemistry* 176, 1371–1381.

Sela, M. and Zisman, E. (1997) Different roles of D-amino acids in immune phenomena. *The FASEB Journal* 11, 449–456.

Shibuya, N., Koike, S., Tanaka, M., Ishigami-Yuasa, M., Kimura, Y., Ogasawara, Y., Fukui, K., Nagahara, N. and Kimura, H. (2013) A novel pathway for the production of hydrogen sulphide from D-cysteine in mammalian cells. *Nature Communications* 4: 1366.

Silva, T., Magalhães, B., Maia, S., Gomes, P., Nazmi, K., Bolscher, J.G., Rodrigues, P.N., Bastos, M. and Gomes, M.S. (2014) Killing of *Mycobacterium avium* by lactoferricin peptides: improved activity of arginine- and D-amino-acid-containing molecules. *Antimicrobial Agents and Chemotherapy* 58, 3461–3467.

Smith, S.B., Ha, Y. and Ganapathy, V. (2012) D-Serine and serine racemase in the retina. In: D'Mello, J.P.F. (ed.) *Amino Acids in Human Nutrition and Health*. CAB International, Wallingford, UK, pp.137–149.

Strick, C.A., Li, C., Scott, L., Harvey, B., James, L.C. and Duplantier, A.J., Hajós, M., Steyn, S.J., Piotrowski, M.A., James, L.C., Downs, J.T., Rago, B., Becker, S.L. *et al.* (2011) Modulation of NMDA receptor function by inhibition of D-amino acid oxidase in rodent brain. *Neuropharmacology* 61, 1001–1015.

Van Veldhoven, P.P., Brees, C. and Mannaerts, G.P. (1991) D-Aspartate oxidase, a peroxisomal enzyme in liver of rat and man. *Biochimica et Biophysica Acta (BBA) – General Subjects* 1073, 203–208.

Vuyyuri, S.B., Hamstra, D.A., Khanna, D., Hamilton, C.A., Markwart, S.M., Campbell, K.C., Sunkara, P., Ross, B.D. and Rehemtulla, A. (2008) Evaluation of D-methionine as a novel oral radiation protector for prevention of mucositis. *Clinical Cancer Research* 14, 2161.

Wang, L-Z. and Zhu, X-Z. (2003) Spatiotemporal relationships among D-serine, serine racemase and D-amino acid oxidase during mouse postnatal development. *Acta Pharmacologica Sinica* 24, 965–974.

Watanabe, T., Shibata, K., Kera, Y., Takahashi, S. and Yamada, R. (2005) Effects of hypoxic and osmotic stress on the free D-aspartate level in the muscle of blood shell *Scapharca broughtonii*. *Amino Acids* 28, 291–296.

Wei, H., Gong, N., Huang, J-L., Fan, H., Ma, A.-N., Li, X-Y., Wang, Y-X. and Pertovaara, A. (2013) Spinal D-amino acid oxidase contributes to mechanical pain hypersensitivity induced by sleep deprivation in the rat. *Pharmacology Biochemistry and Behavior* 111, 30–36.

Xin, L., Jie, L., Liu, C.-W. and Zhao, S-L. (2007) Determination of D-aspartic acid and D-glutamic acid in midbrain of Parkinson's disease mouse by reversed phase high performance liquid chromatography. *Chinese Journal of Analytical Chemistry* 35, 1151–1154.

Yamada, R.-H., Kera, Y. and Takahashi, S. (2006) Occurrence and functions of free D-aspartate and its metabolizing enzymes. *The Chemical Record* 6, 259–266.

Yamanaka, M., Miyoshi, Y., Ohide, H., Hamase, K. and Konno, R. (2012) D-Amino acids in the brain and mutant rodents lacking D-amino-acid oxidase activity. *Amino Acids* 43, 1811–1821.

Yoshikawa, N., Ashida, W., Hamase, K. and Abe, H. (2011) HPLC determination of the distribution of D-amino acids and effects of ecdysis on alanine racemase activity in kuruma prawn *Marsupenaeus japonicas*. *Journal of Chromatography B: Biomedical Sciences and Applications* 879, 3283–3288.

Yoshimoto, M., Okamoto, M., Ujihashi, K. and Okita, T. (2014) Selective oxidation of D-amino acids catalysed by oligolamellar liposomes intercalated with D-amino acid oxidase. *Langmuir* 30, 6180–6186.

Yoshimura, T. and Esak, N. (2003) Amino acid racemases: functions and mechanisms. *Journal of Bioscience and Bioengineering* 96, 103–109.

Young, G.A., Kendall, S. and Brownjohn, A.M. (1994) D-Amino acids in chronic renal failure and the effects of dialysis and urinary losses. *Amino Acids* 6, 283–293.

Yu, H., Li, T., Cui, Y., Liao, Y., Wang, F., Zhao, F. and Jin, Y. (2014) Effects of lead exposure on D-serine metabolism in the hippocampus of mice at the early developmental stages. *Toxicology* 325, 189–199.

Zhang, G. and Sun, H.J. (2014) Racemization in reverse: evidence that D-amino acid toxicity on earth is controlled by bacteria with racemases. *PLoS ONE* 9(3): e92101.

26 Rhizobial Amino Acid Metabolism: Polyamine Biosynthesis and Functions

M.F. Dunn*

Centro de Ciencias Genómicas, Universidad Nacional Autónoma de México, Cuernavaca, Morelos, Mexico

26.1 Abstract

Polyamines are ubiquitous polycationic molecules that have vital roles in many aspects of prokaryotic and eukaryotic cellular physiology. Most of the commonly occurring polyamines are synthesized from the basic L-amino acids arginine, ornithine and lysine. Polyamines produced by rhizobia, the nitrogen-fixing symbionts of leguminous plants, include the diamine putrescine and the triamines spermidine and homospermidine. The lucerne microsymbiont *Sinorhizobium meliloti* is able to synthesize an extended set of polyamines that includes norspermidine. Polyamines are implicated in a number of important processes in rhizobia, and this review summarizes their demonstrated or proposed roles in growth, motility, abiotic stress resistance, biofilm formation and symbiosis. The biosynthesis and catabolism of precursor amino acids and polyamines in rhizobia are discussed, along with a brief description of polyamine transport systems.

26.2 Introduction and Scope

A major source of reduced nitrogen for plant growth in agricultural and natural settings is derived from biological nitrogen fixation, a process performed strictly by archaebacteria (*Archaea*) and *Eubacteria* that have the nitrogenase and accessory enzyme systems necessary for the reduction of atmospheric nitrogen to ammonia. A significant quantity of fixed nitrogen is produced by rhizobia, a collective term for several genera of Gram-negative *Alphaproteobacteria*, including species of *Rhizobium*, *Bradyrhizobium*, *Mesorhizobium* and *Sinorhizobium*. Species in these genera establish nitrogen-fixing symbioses with compatible leguminous plants, where they induce the formation of specialized structures called nodules on the roots. The rhizobia within the plant cells of the nodule metabolically differentiate into their nitrogen-fixing forms, called bacteroids. The bacteroids receive dicarboxylic acids from the plant and use the energy derived from the catabolism of these acids to reduce atmospheric nitrogen to ammonia, which is exported to the plant as a nitrogen source (Dunn, 1998; Jones *et al.*, 2007; Udvardi and Poole 2013).

Polyamines are polycationic hydrocarbons with two or more amine groups. In a wide variety of organisms, polyamines are vital for diverse physiological functions, such as DNA replication, transcription, translation, abiotic stress resistance and bacterial pathogenesis (reviewed in Yoshida *et al.*, 2004; Igarashi and Kashiwagi 2006, 2015; Kusano *et al.*, 2008; Bitrián *et al.*, 2012; Di Martino *et al.*, 2013; Valdés-Santiago

*E-mail address: mike@ccg.unam.mx

and Ruiz-Herrera, 2014). The major habitats where rhizobia are found – bulk soil, the plant rhizosphere and inside the root nodules – all require different metabolic adaptations and survival strategies (Denison and Kiers, 2011). In most of the rhizobia–legume combinations so far examined, the ability of the microsymbiont to synthesize the amino acids L-arginine (Arg), L-ornithine (Orn) and L-lysine (Lys) is essential for an effective (nitrogen-fixing) symbiosis (reviewed in Randhawa and Hassani, 2002; Yadav, 2007; Dunn, 2015). These amino acids are the precursors of polyamines (see Section 26.6.2) Polyamines in rhizobia have so far been shown to be important in growth but, as described in Sections 26.7 and 26.8, they are also implicated in more specialized roles in free-living and symbiotically associated rhizobia.

This review deals with the metabolism of polyamines and their amino acid precursors in rhizobia. In Section 26.3, I discuss the chemical nature of polyamines and how they are analysed, and in Section 26.4, I survey their physiological functions in non-rhizobia. Section 26.5 deals with the rhizobial metabolism of polyamine precursors and Section 26.6 with polyamine biosynthesis, transport and degradation. Proposed or demonstrated functions of polyamines in free-living and symbiotically associated rhizobia are described in Sections 26.7 and 26.8, respectively.

26.3 Polyamine Chemistry, Detection and Analysis

Polyamines are low molecular weight organic bases most commonly occurring in the form of diamines, triamines and tetraamines, which have two, three and four amine groups, respectively. At physiological pH, all of the nitrogens in the polyamine have an approximate +1 charge, making them polycations (Brosnan and Brosnan, 1990). The synthesis of most polyamines involves the production of the diamine putrescine (Put; $H_2N(CH_2)_4NH_2$) directly from Orn (using Orn decarboxylase, ODC; or indirectly from Arg, using Arg decarboxylase, ADC and agmatinase), as described in more detail in Section 26.6.2 and in Chapter 8 of this volume. Triamines are made from Put by adding an aminopropyl or aminobutyl group to it, and tetraamines are formed from triamines by the addition of aminopropyl or (more rarely) aminobutyl groups (Burrell *et al.*, 2010).

Polyamines exist free in the cell cytoplasm or in association with a wide variety of oppositely charged macromolecules, including nucleic acids, proteins, cell wall components and lipids (Bachrach, 2005). The distinction between free polyamines and those bound to macromolecules is complicated by the possible redistribution of the two forms when cells are lysed prior to polyamine analysis (Tabor and Tabor, 1985). In practice, free polyamines are considered as those that are soluble in the dilute perchloric or trichloroacetic acid commonly used to lyse cells; the polyamines that are bound to macromolecules are precipitated under these conditions, and are obtained in their free forms by strong acid hydrolysis of the precipitated macromolecules (Bagni *et al.*, 1995; López-Gómez *et al.*, 2014b).

Methods commonly used for polyamine quantification include their derivatization with benzoyl chloride or dansyl chloride to produce UV-absorbing benzoyl polyamines and fluorescent dansyl polyamines. Either type of derivative can be analysed on a reverse-phase silica column by high-performance liquid chromatography (HPLC) using UV and fluorescence detection, respectively (Bagni *et al.*, 1995; Hunter, 1998; Morgan, 1998). HPLC analysis of non-derivatized polyamines using ion exchange has also been reported (Hamana *et al.*, 1990). Nuclear magnetic resonance (NMR) spectroscopy and HPLC- or gas chromatography-coupled mass spectrometry (GC-MS) have also been used for the identification of rhizobial polyamines (Smith, 1977; Fujihara and Harada, 1989; Fujihara *et al.*, 1994; Vauclare *et al.*, 2013; López-Gómez *et al.*, 2014a; Hernández *et al.*, 2015a). For dansyl polyamines, silica gel thin layer chromatography (TLC) with visualization under UV light has frequently been used (Smith, 1977; Bagni *et al.*, 1995; Madhubala, 1998; Hernández *et al.*, 2015a).

26.4 Physiological Functions of Polyamines in Non-rhizobia

In this section, I highlight some of the diverse functions of polyamines in non-rhizobia, as these are the organisms in which they have been most intensively studied. Polyamines and other simple organic molecules such as polyphosphate and poly-β-hydroxybutyrate are proposed to have played an essential role in the formation of protocells during life's early evolution (Norris *et al.*, 2014). In modern cells, polyamines are usually present in millimolar concentrations and, as already mentioned, exist both as free cytoplasmic forms and as forms bound to macromolecules. The many structural analogues of polyamines and their varied biosynthetic pathways are consistent with the discrete physiological functions of different polyamines (Michael, 2011). Polyamine physiology in plants, animals, fungi and bacteria has been recently reviewed by various authors (Igarashi and Kashiwagi, 2006, 2015; Wortham *et al.*, 2007; Shah and Swiatlo, 2008; Pegg and Casero, 2011; Shi and Chan, 2014; Valdés-Santiago and Ruiz-Herrera, 2014).

Polyamines influence or are essential for DNA replication, transcription, translation and growth under normal or stress conditions. Studies with mutants deficient in the production of one or more polyamines show that the polyamine requirement for growth differs significantly depending on the bacterial species and the specific polyamine(s) whose production is affected in the mutants. In general, polyamine-deficient mutants of bacteria have significantly reduced or zero growth in culture that can be chemically complemented by exogenously supplying the missing polyamine or its precursor (Cunningham-Rundles and Maas, 1975; Nakada and Itoh, 2003; Patel *et al.*, 2006; Chattopadhyay *et al.*, 2009; Hanfrey *et al.*, 2011; Shaw, 2011).

In *Escherichia coli*, polyamines modulate many aspects of metabolism by increasing the translational expression of selected transcriptional regulators that collectively form the polyamine modulon, thus allowing polyamines to have global effects on the growth and metabolism in this species (Yoshida *et al.*, 2004; Igarashi and Kashiwagi, 2006). More specific translational effects of polyamines on metabolism include increasing the synthesis of proteins required for biofilm formation in *E. coli* and *Yersinia pestis* (Wortham *et al.*, 2010; Sakamoto *et al.*, 2012). In several bacterial species, polyamine synthesis is absolutely required for biofilm formation (Patel *et al.*, 2006; Lee *et al.*, 2009; Burrell *et al.*, 2010). In *Y. pestis*, for example, studies with single mutants for the ODC and ADC genes (*odc* and *adc*) and with an *adc odc* double mutant showed that biofilm formation increased in proportion to the Put biosynthetic capacity of the mutants, with those unable to make Put being unable to form biofilms (Patel *et al.*, 2006). In contrast, Put synthesized by *Schwanella oneidensis* appears to favour biofilm dissociation rather than formation (Ding *et al.*, 2014).

In *Proteobacteria*, the ubiquitous second messenger cyclic dimeric guanosine 3′,5′-monophosphate (c-di-GMP) controls the production of exopolysaccharide (EPS), pili and the adhesive proteins required for biofilm formation (Gomelsky, 2011; Gao *et al.*, 2014). In *Vibrio cholerae*, polyamines control biofilm formation by interacting with a two-component sensory/response system that modulates levels of c-di-GMP which, in turn, control the expression of the genes that synthesize EPS (Pendergraft, 2012; Cockerell *et al.*, 2014; see also Section 26.7.3).

In *E. coli* and *Proteus mirabilis*, spermidine and Put, respectively, are required for the movement of cells on surfaces (swarming) and appear to act as signalling molecules to induce the swarming response (Sturgill and Rather, 2004; Kurihara *et al.*, 2009). However, at higher concentrations, polyamines inhibit *E. coli* swarming and chemotaxis by interacting with membrane porins to decrease the uptake of chemoattractants (Dela Vega and Delcour, 1996).

In *Pseudomonas aeruginosa*, quorum sensing systems are required for the induction of enzymes such as superoxide dismutase and catalase for resistance to oxidative stress. In quorum-sensing null mutants of *P. aeruginosa*, these enzymatic responses are not induced; instead, the increased synthesis of intracellular polyamines is thought to act in their place for coping with oxidative stress (Davenport *et al.*, 2015).

26.5 Synthesis and Degradation of Polyamine Precursors in Rhizobia

The synthesis of most polyamines begins with basic amino acid precursors, whose metabolism in rhizobia is described in the following two sections.

26.5.1 Lysine

In rhizobia and many other *Proteobacteria*, Lys is formed from L-aspartate in the nine-step pathway shown in Fig. 26.1. The biosynthesis of Lys and Arg share a common enzyme, ArgD, which is a bifunctional acetylornithine/succinyldiaminopimelate aminotransferase. Many other *Proteobacteria*, but not rhizobia, use a dedicated succinyldiaminopimelate aminotransferase (DapC) to catalyse this reaction during Lys biosynthesis (Liu *et al.*, 2010; Dunn, 2015). The ability of the microsymbiont to synthesize Lys is essential for symbiosis, as shown by *Rhizobium* sp. SMR8, *R. etli* CE3 and *Mesorhizobium ciceri* TAL620 Lys auxotrophs, which form ineffective (non-nitrogen fixing) interactions with their respective hosts (Yadav, 2007; Dunn, 2015). Lys catabolism in rhizobia probably occurs through the γ-glutamylation pathway described in Section 26.6.4.

26.5.2 Ornithine and arginine

A symbiotic requirement for Orn biosynthesis has been experimentally shown in the *S. meliloti*–lucerne interaction, while the ability to synthesize Arg is not required (Kerppola and Kahn, 1988; Kumar *et al.*, 2003). In *S. meliloti*, the

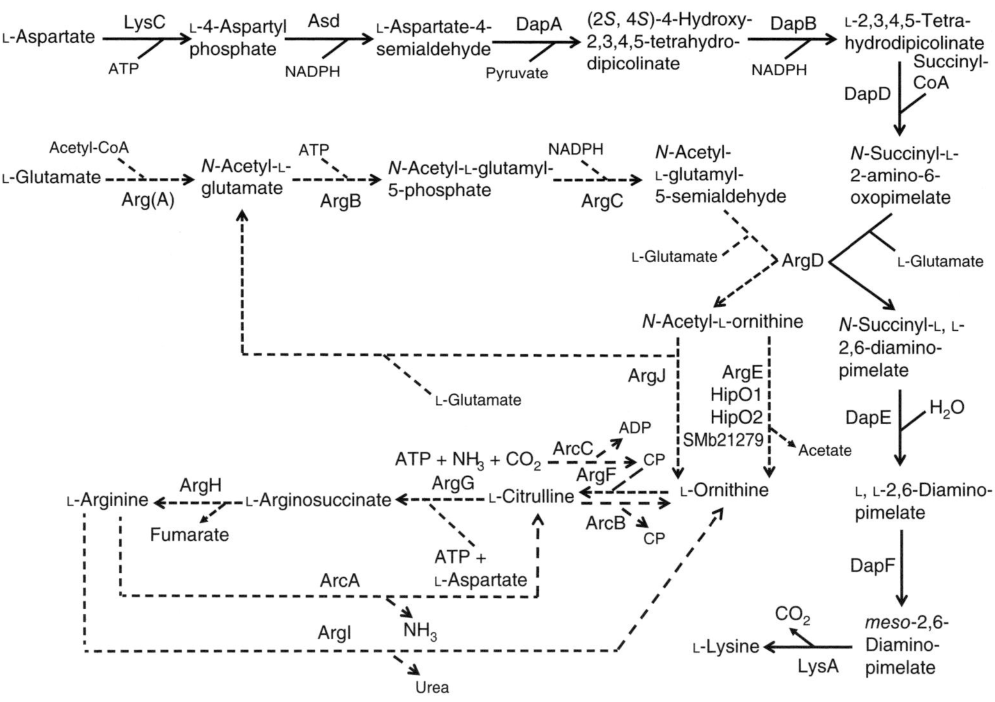

Fig. 26.1. Biosynthesis of the polyamine precursors ornithine, arginine and lysine in rhizobia. Enzymatic steps in pathways are depicted as follows: solid arrows, lysine biosynthesis; short-dashed arrows, ornithine and arginine biosynthesis; large-dashed arrows, interconversion of arginine and ornithine. Abbreviations: ArcA, arginine deiminase; ArcB, catabolic ornithine carbamoyltransferase; ArcC, carbamate kinase; Arg(A), short form of *N*-acetylglutamate synthase; ArgB, *N*-acetylglutamate kinase; ArgC, *N*-acetylglutamyl phosphate reductase; ArgD, acetylornithine/succinyldiaminopimelate aminotransferase; ArgE, *N*-acetylornithine deacetylase; ArgF, anabolic ornithine carbamoyltransferase; ArgG, arginosuccinate synthase; ArgH, arginosuccinate lyase; ArgI, arginase; Asd, aspartate-semialdehyde dehydrogenase; CP, carbamoyl phosphate; DapA, dihydrodipicolinate synthase; DapB, dihydrodipicolinate reductase; DapD, tetrahydrodipicolinate acylase; DapE, succinyl-diaminopimelate desuccinylase; DapF, diaminopimelate epimerase; HipO1, HipO2 and SMb21279, hippurate hydrolases; LysA, diaminopimelate decarboxylase; LysC, aspartate kinase.

synthesis of other Orn-derived products such as Orn-containing lipids is not required for a normal symbiosis (Dunn, 2015), making it most likely that the symbiotic defects of Orn auxotrophs stems from their inability to synthesize Put (Randhawa and Hassani, 2002; Dunn, 2014). This remains to be experimentally demonstrated.

Compared with Lys (Section 26.5.1), much more is known about Orn and Arg metabolism in rhizobia, and this section mostly concentrates on the pathways present in *S. meliloti* 1021. Because other rhizobia encode genes that are homologous to those for Arg biosynthesis in strain 1021, their biosynthetic pathways are probably similar. *S. meliloti* is, however, more complex than other rhizobia in terms of its enzymatic alternatives for catabolizing Orn and Arg (Dunn, 2015).

An interesting characteristic of rhizobial Arg pathways is the presence of both *N*-acetylornithine (NAO) deacetylase (ArgE) and glutamate *N*-acetyltransferase (ArgJ), two non-homologous enzymes for deacetylating NAO (Fig. 26.1). Surprisingly, an *S. meliloti* mutant in which both the genes for both enzymes (*argE* and *argJ*) were inactivated was not an Arg auxotroph. This is because *S. meliloti* encodes at least three enzymes with aminoacylase (Ama) activity that are able to deacetylate NAO, forming Orn, and compensate for the lack of ArgE and ArgJ activities in the double mutant (Hernández *et al.*, 2015b; Fig. 26.1). In strain 1021, ArgJ accounts for about 90% of the total NAO deacetylating activity, with the remaining activity equally split between ArgE and the aminoacylases (Hernández *et al.*, 2015b). The NAO deacetylating enzyme redundancies present in *S. meliloti* (and probably other rhizobia) contrast with *E. coli*, which has only ArgE to catalyse this reaction.

In strain 1021, Orn is also derived from Arg catabolism (Fig. 26.1). In *S. meliloti* 1021, arginase 1 (ArgI1; gene *argI1*) is encoded on the chromosome and arginase 2 (ArgI2; gene *argI2*) on symbiotic plasmid a (pSyma). Neither gene is located near others involved in Arg metabolism, and the ArgI2 amino acid sequence is only 28% identical to that of ArgI1, though ArgI1 has 47% sequence identity to the biochemically characterized arginase in *Agrobacterium tumefaciens*, a rhizobial phytopathogen (Cho *et al.*, 1996). In *S. meliloti* 1021, the transcription of *argI1*, but not of *argI2*, is highly induced by exogenous Arg (Hernández *et al.*, 2015a), and insertional inactivation of *argI1* significantly reduces growth with Arg as a sole nitrogen source (Dunn, 2014; A. Arteaga and M. Dunn, unpublished results). The transcriptional regulation of the *argI1* homologue in *S. meliloti* 2011 is at least partially dependent on the RNA chaperone Hfq (Torres-Quesada *et al.*, 2010).

In bacteroids, microaerobic cells in culture and possibly under oxidative stress, Orn is probably produced from Arg by the Arg deiminase pathway (ArcABC) (Santos *et al.*, 2010; Dunn, 2015). The Arg deiminase pathway consists of ArcA (Arg deiminase), ArcB (catabolic Orn carbamoyltransferase) and ArcC (carbamate kinase) (Fig. 26.1). In *R. etli* CNPAF512, free-living, aerobically grown cells produce the anabolic Orn carbamoyltransferase (ArgF of the biosynthetic pathway), whereas bacteroids contain the catabolic form (ArcB of the degradative pathway) (D'Hooghe *et al.*, 1997; Fig. 26.1).

In *S. meliloti* 1021, transcriptome analysis showed that the *arcA1BC* genes were all induced in microaerobiosis with the involvement of the FixJ and FixK transcriptional regulators. Inactivation of either *arcA1* or *arcB* in *S. meliloti* had no effect on its symbiosis with *Medicago sativa* (Becker *et al.*, 2004; Bobik *et al.*, 2006). Strain 1021 also encodes another Arg deiminase (*arcA2*, or *sma1670*) next to two contiguous *arcDs* (*sma1668* and *sma1667*) encoding possible Arg/Orn antiporters. In some bacteria, the *arcABC* genes are in an operon with *arcD*. Two basic amino acid decarboxylase genes (*sma0680* and *sma0682*) are encoded near the *arcA1BC* operon, and their sequences indicate that they encode ODC and ADC, respectively. Transcription of both of these decarboxylase genes increased 2–4-fold under low oxygen conditions and their expression was significantly lower in a *fixJ* or *fixK* mutant background. A possible amino acid transporter (SMa0678) with some similarity to the *S. meliloti* ArcD1 and ArcD2 is located upstream of *sma0680* and is regulated in the same manner as the decarboxylases (Bobik *et al.*, 2006). The proximity of the decarboxylases and *arcA1BC* and their apparent regulation by FixK suggests that the microaerobic, and possibly symbiotic, metabolism of Arg is geared to the production of Orn for Put synthesis (Fig. 26.1).

Thus, *S. meliloti* has multiple enzymes for Orn production by both biosynthetic and degradative routes. The clustering of genes such as basic amino acid decarboxylases and polyamine/basic amino acid antiporters indicates a common function of these protein products, perhaps to modulate polyamine synthesis for increased stress resistance (see Sections 26.7.4 and 26.8.2) or other functions.

26.6 Synthesis and Transport of Polyamines in Rhizobia

In rhizobia, polyamine research has focused mainly on the identification of the different polyamines produced during free-living growth and in bacteroids (see Section 26.6.1), and relatively little is known about the biochemical and genetic basis for their biosynthesis and transport (reviewed in Fujihara, 2009; Shaw, 2011; Dunn, 2015). Mechanisms used by bacteria to modulate their polyamine levels or activities include the control of biosynthetic and catabolic pathways, transport into and export from the cell, and the 'deactivation' of polyamines by chemical modification, usually acetylation (Tabor and Tabor, 1985). Acetylpolyamines appear to be absent in rhizobia (Hamana *et al.*, 1990), which is consistent with the lack of polyamine acetyltransferases encoded in their genomes.

26.6.1 Polyamines produced by rhizobia under free-living and symbiotic conditions

In rhizobia, polyamines in their free forms have been analysed in cells grown in culture (Hamana *et al.*, 1990, 2003), in whole nodule extracts (Chatterjee *et al.*, 1983; Lahiri *et al.*, 1992) and in nodules separated into their bacteroid and nodule cytoplasmic components (Fujihara and Harada, 1989; Fujihara *et al.*, 1994). A general review on the identity of free polyamines found in free-living and symbiotically associated rhizobia was presented by Fujihara (2009).

Rhizobia grown in minimal medium cultures commonly produce homospermidine (HSpd; $H_2N(CH_2)_4NH(CH_2)_4NH_2$), Put, cadaverine (Cad; $H_2N(CH_2)_5NH_2$), spermidine (Spd; $H_2N(CH_2)_3NH(CH_2)_4NH_2$) and aminobutrylhomospermidine (AbuHSpd; $H_2N(CH_2)_4NH(CH_2)_4NH(CH_2)_4NH_2$) (Hamana *et al.*, 1988, 1990, 2003; Fujihara and Harada, 1989; Fujihara *et al.*, 1995). The detection of Cad and Spd is somewhat variable in different species and AbuHSpd is restricted to fast-growing rhizobia (Fujihara and Harada, 1989). 4-Aminobutylcadaverine (AbuCad; $H_2N(CH_2)_5NH(CH_2)_4NH_2$) is synthesized by bacteroids of *R. tropici* in bean nodules but is absent from cells grown in culture (López-Gómez *et al.*, 2014b). An ion exchange HPLC analysis of free polyamines in a wide variety of culture-grown rhizobia failed to detect diaminopropane ($NH_2(CH_2)_3NH_2$), agmatine (Agm; $H_2N(CH_2)_4N(NH_2)_2$) and norspermidine (NSpd; $H_2N(CH_2)_3NH(CH_2)_3NH_2$) (Hamana *et al.*, 1990). The lack of diaminopropane and NSpd is consistent with the very limited occurrence in rhizobia of the genes encoding the enzymes required for their production (see Section 26.7), which are restricted principally to *S. meliloti* 1021 and related strains, and a few *Agrobacterium* and *Rhizobium* strains. In *S. meliloti*, NSpd is found only in association with macromolecules (Hernández *et al.*, 2015a), which explains why it has not been reported among the free polyamines produced by this species. Failure to detect Agm might result from this intermediate being very efficiently converted into Put by agmatinase, for which two or more orthologues are present in most rhizobia (Section 26.6.2). Most bacteria lack the ability to synthesize spermine (Spm; $H_2N(CH_3)_3NH(CH_2)_4NH(CH_2)_3NH_2$) (Shah and Swiatlo, 2008), and the genes encoding spermine synthase are absent in rhizobial genomes. Rhizobia and other bacteria can import spermine from the environment (e.g. rich culture media or the plant-produced spermine present in nodules) (Hamana *et al.*, 1988, 1990; Shah and Swiatlo, 2008).

The free polyamines identified in whole nodule extracts from a range of legumes include Put, Cad, Spd, HSpd and Spm, with HSpd usually present in highest concentration. Although some plants are able to produce uncommon polyamines like HSpd and NSpd (Rodríguez-Garay *et al.*, 1989), it appears that in whole nodule extracts, the HSpd is produced mainly by the bacteroids and the Spm by the plant, with the other polyamines being contributed in some proportion by both symbionts (Smith, 1977; Fujihara *et al.*, 1994, 1995; Yoneyama *et al.*, 1998; López-Gómez *et al.*, 2014a). In nodules formed on *Vigna mungo* (black gram), polyamine levels peak at the flowering stage, when nitrogenase activity at its highest level (Lahiri *et al.*, 1992). In bacteroids isolated from various symbiotic combinations, Put, Cad, HSpd and AbuHSpd have been reported (Fujihara *et al.*, 1995; Fujihara, 2009; López-Gómez *et al.*, 2014a).

26.6.2 Polyamine biosynthesis in rhizobia

Polyamine biosynthesis begins with the decarboxylation of the basic L-amino acids by pyridoxal 5′-phosphate (PLP)-dependent enzymes. Put is produced directly from Orn by the action of ODC or from Arg by the action of ADC and agmatinase (SpeB) (Fig. 26.2). Cad is produced by the decarboxylation of Lys by Lys decarboxylase (LDC). Enzymes designated L/ODC are able to decarboxylate both Lys and Orn (Lee *et al.*, 2007; Burrell *et al.*, 2010; Romano *et al.*, 2012). Sequence analysis of basic amino acid decarboxylases alone is not sufficient to determine which amino acid(s) they decarboxylate. Virtually all rhizobial genomes encode at least one basic amino acid decarboxylase and the HSpd synthase (Hss) required for synthesizing Put, HSpd and Cad, and the presence of these polyamines in most rhizobia has been confirmed by biochemical analysis (see Section 26.6.1). Four putative basic amino acid decarboxylases (ODC, ADC, LDC and L/ODC) are encoded in *S. meliloti* (Fig. 26.2), but definitively assigning their biochemical functions awaits their enzymatic characterization. In rhizobia and other bacteria that possess both ADC and ODC, the relative importance of each enzyme in producing Put is unclear (Burrell *et al.*, 2010; Dunn, 2014). Despite claims for the absence of ADC activity in rhizobia (Fujihara, 2009), and the fact that Agm has been assayed for but not detected in rhizobia (Hamana *et al.*, 1990), SpeB activity and the

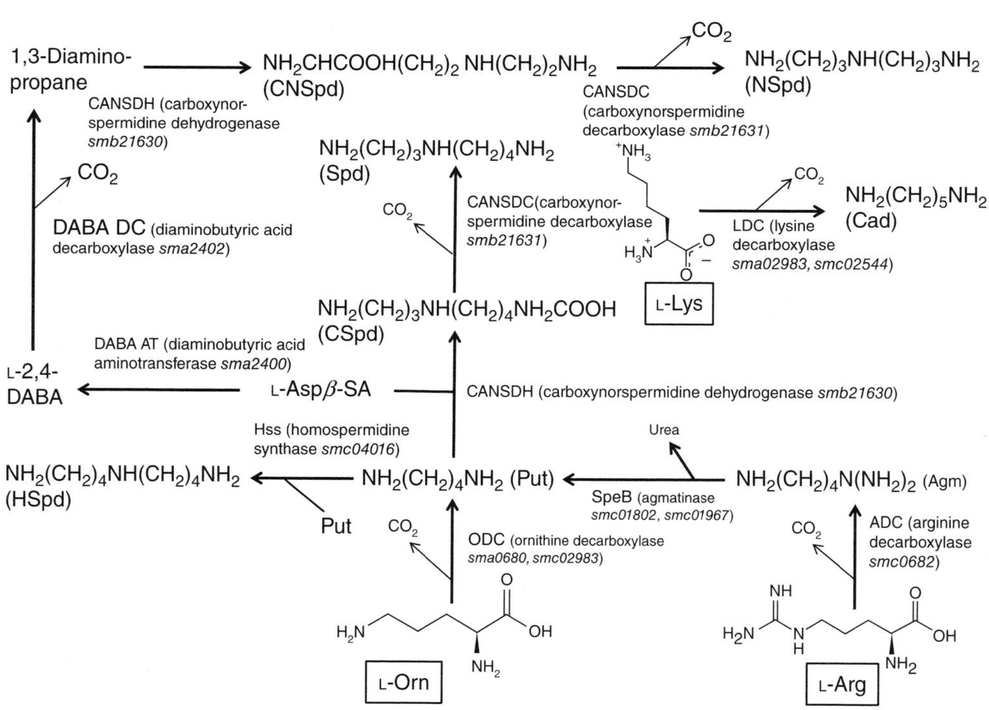

Fig. 26.2. Predicted pathways of polyamine biosynthesis in *Sinorhizobium meliloti* 1021. Accession numbers for genes encoding the enzymes are shown with their names. Abbreviations: Agm, agmatine; L-Arg, L-arginine; L-Asp β-SA, L-aspartate β-semialdehyde; Cad, cadaverine; CNSpd, carboxynorspermidine; CSpd, carboxyspermidine; HSpd, homospermidine; L-Lys, L-lysine; L-Orn, L-ornithine; L-2,4-DABA, L-2,4-diaminobutyric acid; NSpd, norspermidine; Put, putrescine; Spd, spermidine.

transcription of one of its encoding genes (*speB*) is measurable in *S. meliloti* 1021 (Díaz *et al.*, 2011), which is consistent with the functionality of the putative ADC encoded in its genome.

In contrast to other rhizobia, *S. meliloti* contains homologues of genes that encode enzymes in the 'alternative' polyamine biosynthesis pathway that has been described in *V. cholerae* (Lynch *et al.*, 2001; Lee *et al.*, 2009). Thus, *S. meliloti* probably makes Spd using the enzymes carboxynorspermidine (CNSpd) decarboxylase (Cansdc) and CNSpd dehydrogenase (Cansdh), and produces CNSpd using diaminobutyric acid (DABA) aminotransferase (DABA AT) and DABA decarboxylase (DABA DC) (Fig. 26.2). In *S. meliloti* 1021, DABA AT (RhbA) and DABA DC (RhbB) were identified as part of the rhizobactin synthesis pathway (Lynch *et al.*, 2001). This siderophore is synthesized from L-aspartate-β-semialdehyde in a pathway whose early steps are identical to the proposed *S. meliloti* polyamine pathway (Fig. 26.2). The inactivation of *rhbA* in *S. meliloti* 2011 abolished rhizobactin production, but the effect of the mutation on polyamine synthesis was not investigated (Lynch *et al.*, 2001). Consequently, we do not know if RhbA and RhbB function in both rhizobactin and polyamine biosynthesis.

26.6.3 Polyamine transport in rhizobia

Rhizobial genome sequences encode a variety of probable polyamine transporters, mostly arranged as contiguous genes encoding the components of ATP-binding cassette (ABC)-type systems. Two to five such systems occur in different strains and are usually annotated as the Spd/Put or Put ABC transporters. Four of these transporters occur in *S. meliloti* 1021, and the ability of this strain to transport Put, Cad, NSpd, Spd or Spm added to cultures (M. Dunn, unpublished results) indicates a relatively broad substrate specificity of at least one of these transporters.

Compared with those of cells grown in culture, the genes encoding an Spd/Put ABC transporter in *Sinorhizobium* sp. NGR234 are upregulated many fold in bacteroids of nodules formed on *Vigna unguiculata* and *Leucaena leucocephala* (Li *et al.*, 2013), which could perhaps increase the uptake of plant-produced polyamines (Section 26.6.1).

26.6.4 Polyamine degradation in rhizobia

Put catabolism by the γ-glutamylation pathway (Fig. 26.3) has been well studied in *E. coli* and *P. aeruginosa*. In *E. coli*, the enzymes comprising the pathway are PuuA (γ-glutamylputrescine synthetase), PuuB (γ-glutamylputrescine oxidoreductase), PuuC (γ-aminobutyraldehyde dehydrogenase) and PuuD (γ-glutamyl-γ-aminobutyrate hydrolase). Analogous enzymes designated PauA, B, C and D occur in *P. aeruginosa* but, in contrast to *E. coli*, this organism contains 2–6 orthologues of each enzyme, with the exception of PauC, which is present as a single copy (Chou, 2011). Cad degradation in *P. aeruginosa* follows a similar γ-glutamylation pathway, but yields glutarate as a final product (Fothergill and Guest, 1977). In this organism, some Pau orthologues act in both the degradation of Put and Cad (PauA1 and PauA5), while others are exclusive to just one of the pathways (PauA2 and PauA4, respectively). The control of polyamine catabolism in *E. coli* and *P. aeruginosa* is exerted through the PuuR and PauR transcriptional repressors, respectively (Chou, 2011; Nemoto *et al.*, 2012).

Based on the *P. aeruginosa* model (Chou, 2011), the predicted Put and Cad catabolic pathways in *S. meliloti* are shown in Fig. 26.3. The genes *smc01971*, *smc01972* and *smc01973* form an apparent operon encoding the PauC, PauB and PauA orthologues, with a putative PauD (SMc03943) encoded elsewhere in the genome (Dunn, 2015; see http://operondb.cbcb.umd.edu/cgi-bin/operondb/pairs.cgi?genome_id=847). Orthologues for PauA and PauB are also encoded by the contiguous genes *smc00762* and *smc00761*, which occur close to the genes encoding a probable Put ABC transporter. Further metabolism of the γ-aminobutyrate and aminovalerate intermediates from Put and Cad catabolism, respectively, requires GabT (4-aminobutyrate aminotransferase) and GabD (succinate semialdehyde dehydrogenase) (Chou, 2011). By amino acid similarity, the two likely *S. meliloti* GabDs are SMb21185 (whose encoding gene is next to *gabT*), and SMc02780. The other GabT orthologue from *P. aeruginosa*, GabT2, is most closely resembled by *S. meliloti* SMc04388,

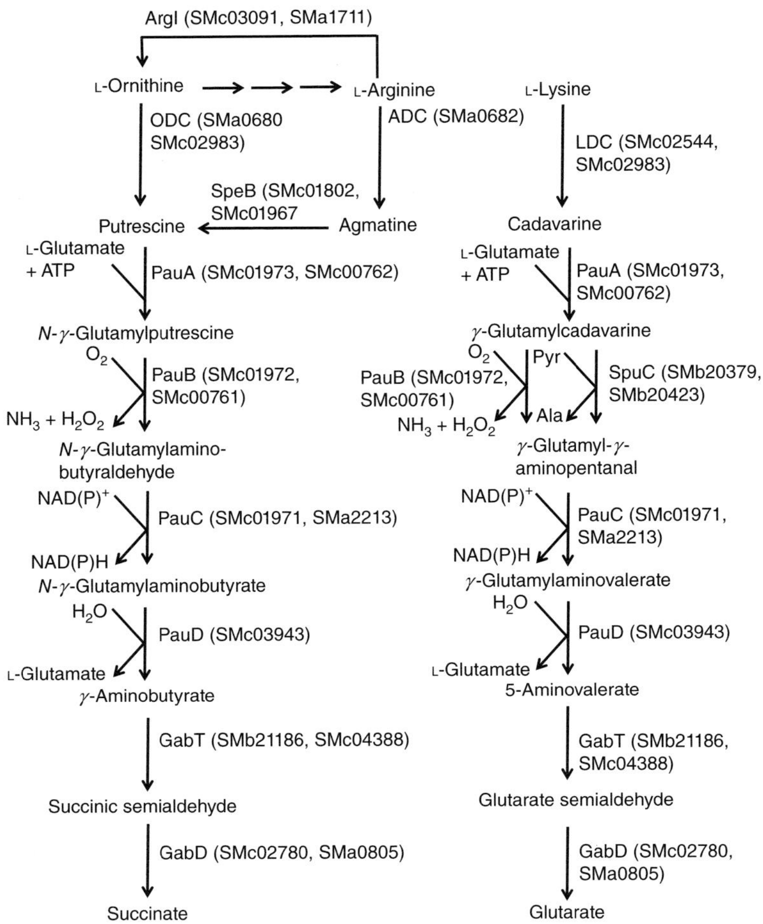

Fig. 26.3. Predicted pathways of cadavarine and putrescine degradation in *Sinorhizobium meliloti* 1021 based on orthology to enzymes occurring in *Pseudomonas aeruginosa* PAO1. In most cases, the two *S. meliloti* orthologues with the highest sequence identity to the *P. aeruginosa* enzyme are shown; it is not possible to predict which orthologues might act preferentially in putrescine or cadaverine catabolism. Abbreviations: ADC, arginine decarboxylase; Ala, L-alanine; ArgI, arginase; GabD, succinate semialdehyde/ glutarate semialdehyde dehydrogenase; GabT, 4-aminobutyrate/aminovalerate aminotransferase; LDC, lysine decarboxylase; ODC, ornithine decarboxylase; PauA, γ-glutamylcadaverine/putrescine synthetase; PauB, γ-glutamylcadaverine/putrescine oxidoreductase; PauC, γ-aminobutyraldehyde/aminopentanal dehydrogenase; PauD, γ-glutamyl-γ-aminobutyrate/aminovalerate hydrolase; Pyr, pyruvate; SpeB, agmatinase; SpuC, putrescine aminotransferase.

whose encoding gene lies next to one encoding an orthologue (SMc04387) of the PauR/PuuR negative transcriptional regulators (Nemoto *et al.*, 2012). If SMc04387 is indeed a PauR-type regulator, it could serve to control the intracellular levels of Cad and Put by regulating their rate of degradation.

Put as sole carbon source allows slow growth of *S. meliloti* 1021 in minimal medium containing ammonium, but does not allow growth when ammonium is used as a sole nitrogen source (Biondi *et al.*, 2009; A. Arteaga, unpublished results). Because the ammonium and glutamate liberated during Put degradation should provide utilizable nitrogen for the growth of strain 1021, the reason for its inability to use this polyamine as a nitrogen source is puzzling.

26.7 Functions of Polyamines in Free-living Rhizobia

Most published studies on polyamine function in rhizobia are correlative in nature and describe changes in free polyamine levels in cultures of rhizobia subjected to abiotic stress. These studies, some of which are described in more detail in this section (26.7) and in Section 26.8, are summarized in Table 26.1.

26.7.1 Requirement of polyamines for growth

Just a few studies have been done in which rhizobial polyamine synthesis has been disrupted by mutagenesis or the use of a basic amino acid decarboxylase inhibitor. The only demonstration of a requirement for polyamines for growth was shown by treating *Rhizobium leguminosarum* bv. viciae 3841 cultures with the ODC inhibitor difluoromethylornithine (DFMO) (Shaw, 2011). Cells grown in the presence of 5 mM DFMO had only 5% of the wild type levels of Put and HSpd and a drastically reduced growth rate and yield. Chemical complementation of the DFMO-treated cultures with 1 mM Put restored their growth and polyamine levels almost completely, indicating that the growth defect was indeed due to polyamine depletion. Whether the normal growth of strain 3841 requires Put itself and/or the HSpd synthesized from it is not known. The fact that a strain 3841 *hss* regulatory mutant that produced only 30% of the wild type level of HSpd (see pathway in Fig. 26.2) was able to grow normally in minimal medium argues for Put as the most important polyamine for growth. The growth of *R. leguminosarum* is presumably sensitive to inhibition by DFMO because it encodes only a single, alanine racemase-fold (DFMO-sensitive) ODC for Put synthesis (Shaw, 2011). In contrast, the growth and polyamine content of *S. meliloti* 1021 is not markedly affected by 5 mM DFMO, consistent with it having an aspartate aminotransferase-fold (DFMO-insensitive) ODC (SMa0680) in addition to its alanine racemase-fold type L/ODC (SMc02983) (V. Beccera and M. Dunn, unpublished results). In addition, *S. meliloti* could bypass the presumably DFMO-inhibited L/ODC using ADC and SpeB activities to synthesize Put (see Fig. 26.2).

A significant reduction in growth and polyamine levels in *A. tumefaciens* was caused by treatment of cultures with 1 mM DFMO. The growth inhibition could be reversed by the inclusion of 5 mM Put in the culture medium. Treatment with the ADC inhibitor difluoromethylarginine did not inhibit growth, indicating that ODC is the major or only enzyme for Put synthesis in this species (Ponappa *et al.*,1992).

26.7.2 Polyamines and motility

Rhizobia and other bacteria require polyamines to move on surfaces (swarming). In *S. meliloti* 2011, inactivation of the *rhbA* gene (encoding DABA AT, RhbA; see Fig. 26.2) resulted in an inability of the cells to swarm on agar plates. As mentioned in Section 26.6.2, RhbA participates in rhizobactin synthesis and the motility defect of the *rhbA* mutant was proposed to be due to its inability to produce this siderophore. However, neither the addition of iron nor that of purified rhizobactin to the medium used in the swarming assays restored the motility of the mutant (Nogales *et al.*, 2010). This raises the possibility that the inability of the mutant to synthesize CNSpd and NSpd (Fig. 26.3), rather than rhizobactin, could be the cause of its inability to swarm. In *R. etli*, an HSpd synthase (*hss*; see Fig. 26.2) mutant was unable to swarm, indicating that HSpd is required for surface motility in this species (Braeken *et al.*, 2008). Together, these results indicate a species-specific requirement for different polyamines for swarming.

26.7.3 Polyamines and biofilm formation

Polyamines modulate, and in some cases are required for, the formation of biofilms in diverse bacteria (Karatan and Michael, 2013; Section 26.4). Biofilms are surface-attached bacterial communities encased in a protective matrix of EPS, DNA and proteins. Numerous studies on gene expression in biofilms have revealed only a few genes that are exclusive to this response, while the expression of many genes that function in resistance to cellular stresses have been found. Thus, biofilm formation appears to be a kind of stress response that enhances the ability of bacteria to survive a variety of non-optimal

Table 26.1. A summary of studies reporting possible functions of polyamines in rhizobia.[a]

Observation[a]	Biological system	Explanation or proposed mechanism
Exogenous Put, Spd and Spm significantly increase nodulation and nitrogen fixation	*Galega orientalis–Rhizobium galagae* HAMBI 540 symbiosis	Increase binding of *R. galagae* to roots; stimulate malate uptake by bacteroids (Vassileva and Ignatov, 1999)
Exogenous polyamines stimulate malate uptake	*Bradyrhizobium japonicum* USDA 110 bacteroids	Direct interaction with the malate transporter (Whitehead *et al.*, 2001)
Bacteroid growth inhibited *in vitro* by Put and Spd	*B. japonicum* 138NR	Polyamines may limit bacterial multiplication in the nodule (Ozawa and Tsuji, 1993)
HSpd levels decrease and K^+ levels increase during response to salt stress in salt-resistant, but not salt-sensitive, strains	*Sinorhizobium fredii* P220 (salt resistant), *B. japonicum* A1017 (salt sensitive)	Inverse relation between HSpd (divalent) and K^+ (monovalent) maintains electroneutrality in cytoplasm (Fujihara and Yoneyama, 1993)
Salt stress modulates expression of genes for polyamine synthesis or transport, levels of HSpd decrease in salt-stressed bacteroids	*S. meliloti* strains 1021, A5 and 2011 grown in culture; *S. meliloti* 1021 bacteroids	Increased or decreased levels of specific polyamines important in osmotic stress resistance (e.g. by charge balancing) (Rüberg *et al.*, 2003; Domínguez-Ferreras *et al.*, 2006; Mrabet *et al.*, 2011; López-Gómez *et al.*, 2014a)
Salt-stressed bean nodules contain higher levels of polyamines bound to macromolecules	*R. tropici* CIAT 899	Decrease in nodule free polyamine content, possibly due to their ligation to macromolecules (López-Gómez *et al.*, 2014b)
HSpd levels high in acid-resistant strain at acidic pH and decline at higher pH. In pH sensitive strain, HSpd levels remain low and unchanged	*S. fredii* P220 (acid resistant), *B. japonicum* A1017 (acid sensitive)	High HSpd concentration at low pH buffers and maintains cytoplasmic pH, or protects against H^+-induced damage (Fujihara and Yoneyama, 1993)
daba at Mutant unable to swarm (move on surfaces)	*S. meliloti* 2011	Rhizobactin (Lynch *et al.*, 2001; Nogales *et al.*, 2010) or possibly CNSpd and/or NSpd required for swarming
hss Mutant unable to swarm	*Rhizobium etli* CNPAF512	HSpd required for swarming (Braeken *et al.*, 2008)
Reduced intracellular Put and HSpd levels caused by inhibition of ODC severely limit growth	*Rhizobium leguminosarum* bv. viciae 3841	Polyamines required for normal growth (Shaw, 2011)

[a]Abbreviations: CNSpd, carboxynorspermidine; *daba at*, diaminobutyric acid aminotransferase (DABA AT) gene; HSpd, homospermidine; *hss*, HSpd synthase gene; NSpd, norspermidine; ODC, ornithine decarboxylase; Put, putrescine; Spd, spermidine; Spm, spermine.

growth conditions (Landini, 2009). Although not yet demonstrated in rhizobia, the ability of polyamines to influence resistance to abiotic stress in some bacteria (see Section 26.4) derives from their ability to modulate biofilm formation (Burrell *et al.*, 2010; Chattopadhyay and Tabor, 2013; Karatan and Michael, 2013).

In *V. cholerae*, the expression of the *vps* genes required for biofilm formation increases in proportion to the intracellular concentration of c-di-GMP, which, in turn is controlled by NSpd and Spd. This occurs through a two-component system comprising NspS, a periplasmic protein that binds both Spd and NSpd, and MbaA. NspS has sequence similarity to a periplasmic binding protein for polyamines but does not function in transport. MbaA is a membrane-bound protein whose GGDEF and EAL domains are indicative of enzymes having, respectively, diguanylate cyclase (for c-di-GMP synthesis) and phosphodiesterase (for c-di-GMP degradation) activities. The effect of NSpd and Spd on biofilm formation is abolished in *nspS* or *mbaA* null mutants. The deletion of *mbaA* causes an increase in biofilm formation, EPS gene transcription, and EPS synthesis and transport, indicating that MbaA normally acts to decrease biofilm formation. These data are consistent with MbaA being a phosphodiesterase that degrades c-di-GMP. The experimental results obtained to date suggest that NspS interacts with and modulates the phosphodiesterase activity of MbaA as follows. NspS in combination with NSpd decreases MbaA phosphodiesterase activity, promoting higher c-di-GMP levels, *vps* gene expression and biofilm formation. In contrast, when NspS interacts with Spd higher MbaA phosphodiesterase activity is permitted, thereby lowering c-di-GMP levels, *vps* gene expression and biofilm formation. Therefore, by their interaction with the NspS/MbaA system, Spd and NSpd have opposite effects on biofilm formation in *V. cholerae* (Pendergraft, 2012; Cockerell *et al.*, 2014).

Polyamines may be involved in biofilm formation in rhizobia. The *S. meliloti* 1021 genome (but not the genomes of other rhizobia) encodes orthologues for the NspS/MbaA two-component polyamine sensing system described above. The *S. meliloti* NspS and MbaA encoding genes (*smc00991* and *smc00992*, respectively) are contiguous on the strain 1021 chromosome, as in *V. cholerae* (Cockerell *et al.*, 2014). SMc00991 is annotated as a putrescine periplasmic binding protein and SMc00992 as an integral membrane protein containing GGEDF and EAL domains (Gao *et al.*, 2014). Possibly, SMc00991 and SMc00992 regulate biofilm formation in *S. meliloti* by modulating c-di-GMP levels in response to specific polyamines.

26.7.4 Polyamines in abiotic stress resistance

Rhizobia, like other soil bacteria, are exposed to osmotic stress in naturally saline soils and, additionally, to stress caused by rainfall scarcity or excess, and freezing (Zahran, 1999; Fujihara, 2009). This stress may take the form of osmotic 'upshock' (a change from low to high salt conditions) or osmotic 'downshock' (a change from high to low salt concentrations).

In general, fast-growing *Rhizobium* and *Sinorhizobium* spp. are quite tolerant to high osmotic conditions, whereas the slow-growing *Bradyrhizobium japonicum* is quite sensitive (Fujihara, 2009). In salt-tolerant *S. fredii* P220, the osmotic solutes glutamate and K^+ accumulate in response to high-salt stress, while the concentrations of HSpd and Mg^{2+} decline. Because HSpd is not exported from the cells under these conditions, the decline in its intracellular level is probably due to catabolism. Strain P220 responds to osmotic downshock by accumulating Mg^{2+} and HSpd and reducing its intracellular concentrations of K^+ and glutamate. Under these conditions, the decreasing glutamate levels may reflect its consumption by the increased synthesis of Orn, a precursor of Put and HSpd. Mg^{2+} is an activator of ODC activity. Efflux of K^+ from the cell during osmotic downshock is accompanied by the import of H^+, and the increased levels of HSpd found under these conditions might serve as a sink for excess protons by protonation of its amine groups. The increased HSpd may also serve to compensate for the charge imbalance caused by K^+ efflux. In contrast to what occurs in strain P220, HSpd levels in salt-sensitive *B. japonicum* A1017 remain relatively constant during high-salt stress (Fujihara and Yoneyama, 1993; Fujihara, 2009).

The *S. meliloti* genome encodes several putative polyamine transporters (Dunn 2014; Section 26.6.3), and strains 1021 and A5 respond to high-salt stress by downregulating three of these systems (Domínguez-Ferreras *et al.*, 2006; Mrabet *et al.*, 2011), perhaps to decrease the importation of additional polyamines from the environment. The synthesis of Put (a precursor of HSpd) and/or Cad may also decrease under these conditions due to the downregulation of the putative L/ODC encoded by *smc02983* (Domínguez-Ferreras *et al.*, 2006). In *S. meliloti* 1021, the expression of *daba at* and *daba dc* were significantly downregulated during osmotic upshock (Rüberg *et al.*, 2003; Domínguez-Ferreras *et al.*, 2006), which could limit the synthesis of CNSpd and NSpd (see Fig. 26.2). Thus, *S. meliloti* may decrease its intracellular pool of polyamines as part of its osmoadaptive response in a manner similar to that of *S. fredii* P220, as described above.

RNA sequencing analysis showed that *odc* was downregulated, and the permease component of a Spd/Put transporter upregulated in *R. etli* CE3 subjected to heat or salt shock (López-Leal *et al.*, 2014). These expression changes could serve to modulate levels of polyamine by altering their synthesis or transport. Downregulation of the putative *odc*, the only such decarboxylase annotated in strain CE3, would limit Put synthesis to a corresponding degree. Lower Put levels brought about by repressing its synthesis are consistent with the decrease in Put and HSpd levels observed with salt shock in a variety of rhizobia (see Table 26.1).

Because plant roots excrete protons and organic acids, the rhizosphere can be 2 pH units lower than that of the surrounding soil. Likewise, the pH of the peribacteroid space is estimated to be 2 units lower than that of the plant cytosol. To survive this stress, free-living rhizobia and bacteroids employ several mechanisms (Rojas-Jiménez *et al.*, 2005; Muglia *et al.*, 2007), possibly including the modulation of polyamine levels.

Polyamines are essential for resistance to acid stress in enteric bacteria (Foster, 2004; Shah and Swiatio, 2008; Chattopadhyay and Tabor, 2013). *E. coli* contains both constitutively expressed and acid-inducible homologues of *ldc*, *adc* and *odc* (Michael, 2011). The acid-inducible ADC, for example, combats acid stress by protonating Arg to form Agm, thus consuming H^+ from the cytoplasm. Agm is then exported in exchange for Arg, thereby allowing the cycle of proton consumption to continue and raise the internal pH (Foster, 2004).

The *S. meliloti* genome encodes polyamine/basic amino acid antiporters (see Section 26.6.3) that could, along with the cognate basic amino acid decarboxylases responsible for polyamine biosynthesis, act to consume protons and maintain the intracellular pH during acid stress (Dunn, 2015). The putative *S. meliloti* ODC (SMa0680) is encoded adjacent to a predicted Put/Orn antiporter and its putative ADC gene (*sma0682*) is next to a predicted Agm/Arg antiporter.

A comparison of polyamine and ion levels in pH-resistant *S. fredii* P220 and pH-inhibited *B. japonicum* A1017 showed that both strains contained HSpd as the major polyamine (comprising about 80% of the total), as well as a minor amount of Put. In strain P220, the intracellular concentrations of both HSpd and Mg^{2+} significantly increased at acidic pHs, whereas in strain A1017 they remained relatively unchanged (Fujihara and Yoneyama, 1993). As mentioned above, the elevated Mg^{2+} concentration in strain P220 could promote HSpd synthesis because it activates ODC activity, which is rate limiting for HSpd synthesis. Mg^{2+} ions do not affect HSpd synthase activity (Fujihara, 2009). Neither strain excreted HSpd to the medium. The rise in HSpd content in the pH-resistant strain under acid stress might buffer the cytoplasmic pH or directly protect macromolecules against proton-induced damage (Fujihara and Yoneyama, 1993). Similarly, acid-sensitive *R. etli* UMR1632 under acidic stress produced up to three different, though unidentified, polyamines vs only one polyamine (possibly Spd) at neutral pH. In contrast, the acid-tolerant *R. tropici* UMR1899 made only one type of polyamine under acid stress (Graham *et al.*, 1994).

Rhizobial polyamines may also play a role in resistance to oxidative stress occurring as a normal part of the nodulation process (Chang *et al.*, 2009). The *S. meliloti* 1021 TolC is an outer membrane efflux protein whose inactivation increases the transcription of many genes whose products are involved in oxidative stress resistance. Genes relevant to polyamine metabolism that were upregulated in a strain 1021 *tolC* mutant included several for Lys and Arg synthesis, Arg conversion to Orn (*arcAB*), the putative *odc* and *adc* (*sma0680* and *sma0682*, respectively), and agmatinase (*speB*) (see Figs 26.1 and 26.2). Although polyamine levels were not measured in this work, their presumably higher levels in the mutant might act to scavenge free radicals produced by oxidative stress or to directly protect macromolecules from oxidative damage (Santos *et al.*, 2010).

Rhizobia in the soil or in nodules are subjected to heat stress in some climates (Zahran, 1999; Niste *et al.*, 2013). *R. tropici* PRF 81 is a very heat-tolerant strain whose prolonged growth at 35°C is not drastically reduced in comparison with cultures grown at its optimum temperature of 28°C. Proteome comparison of strain PRF 81 grown at these two temperatures showed that 59 proteins were significantly and reproducibly upregulated during growth at the higher temperature, with 26 of these being exclusive to this condition. Of the five amino acid biosynthesis proteins upregulated or exclusive to the high-temperature culture, three were involved in Lys and Arg biosynthesis, namely ArgB (*N*-acetylglutamate kinase), ArgD and Asd (aspartate-semialdehyde dehydrogenase) (Fig. 26.1). This indicates that growth under high heat conditions increases the synthesis of polyamine precursors, even though other genes involved in polyamine synthesis were not upregulated by heat stress (Gomes *et al.*, 2012).

26.8 Functions of Polyamines in Symbiotically-Associated Rhizobia

In this section, I summarize some of what is known about the possible or demonstrated roles of polyamines in the symbiosis between legumes and rhizobia under normal or abiotic stress conditions. Table 26.1 gives a summary of these studies.

26.8.1 Influence of polyamines on nodulation and nitrogen fixation

While polyamines are required for the normal growth of rhizobia (see Section 26.7.1), higher-than-physiological concentrations are inhibitory and could be important in controlling the proliferation of bacteroids *in nodulo*. One example comes from Goat's rue (*Galega orientalis*), which forms a symbiosis with *R. galagae. In vitro* assays showed that exogenous polyamines (Put, Spd and Spm) added at a concentration of 10 or 50 μM increased nodule number and biomass and nitrogenase activity up to several fold, but a 100 μM concentration lowered these parameters. Explanations proposed for the symbiotic enhancement seen with the lower polyamine concentrations, not yet all tested experimentally, include an increased binding of the rhizobia to roots by exogenous polyamines, the use of Put as an energy source by the microsymbiont, stimulation of malate uptake by bacteroids and modulation of the autoregulation of nodulation by the plant (Vassileva and Ignatov, 1999).

Differences in polyamine levels and the effect of exogenous polyamines on nitrogen fixation in a soybean wild type and a supernodulating cultivar indicate that differences in their respective polyamine metabolisms act to regulate nodule numbers. One interesting hypothesis is that hydrogen peroxide generated by nodule polyamine oxidase activity might trigger plant defence responses to constrain nodule numbers (Terakado-Yonooka and Fujihara, 2008). Bacteroid populations could also be affected by the growth inhibitory effects of specific polyamines. Mature soybean nodules contained concentrations of Put and Spd that were sufficient to inhibit the growth of isolated bacteroids of *B. japonicum* strain 138NR in an *in vitro* assay system. It was found that high concentrations of the polyamines were bactericidal, whereas more moderate concentrations were bacteriostatic (Ozawa and Tsuji, 1993).

26.8.2 Influence of polyamines on symbiosis under stress conditions

Relatively little has been reported on the possible role of rhizobial polyamines in abiotic stress resistance during symbiosis. As described in Section 26.7.4, polyamine levels in free-living rhizobia change in response to abiotic stress. López-Gómez *et al.* (2014a) studied the effect of salt stress on lucerne nodule polyamine content and symbiotic proficiency. Salt stress imposed on plants 5 weeks after inoculation with *S. meliloti* 1021 caused a significant decrease in nitrogen fixation and whole nodule extract Put levels for up to 6 weeks, while a decrease in HSpd was more transient. During salt stress, Put was detected only in the nodule cytosol but not in bacteroids, and HSpd had the opposite distribution (López-Gómez *et al.*, 2014a). The analysis of free polyamine levels in whole nodule extracts from salt-stressed bean nodules formed by *R. tropici* showed that, with the exception of Cad and AbuCad, they were lower than those of unstressed nodules. Interestingly, the content of Put and Spd bound to macromolecules increased in salt-stressed nodules, possibly explaining the reduction in their free-form levels (López-Gómez *et al.*, 2014b).

26.9 Concluding Remarks

Polyamines appear to be important in a number of processes in free-living and symbiotically associated rhizobia, including growth and motility and, possibly, stress resistance, biofilm formation and limiting bacteroid numbers *in planta*. Many of the studies described here rely on quantitating changes in polyamine levels in one or more wild type strains exposed to different growth or symbiotic conditions. Thus, the mechanisms responsible for any of the effects observed remain largely undetermined. The biochemical identification of polyamines produced by rhizobia, in combination with the analysis of their genome sequences, allows us to predict the probable pathways by which these compounds are synthesized. To confirm these predictions, much more work directed at studying the regulation and enzymology of these pathways is needed. Additionally, studies using defined polyamine biosynthetic mutants are required and should allow us to arrive at an understanding of the physiological functions of polyamines in rhizobia, some of which are probably specific to symbiosis.

Acknowledgements

Partial financial support for studies in the author's laboratory was provided by grants IN208811-3 and IN210114-3 from DGAPA-PAPIIT, Mexico.

References

Bachrach, U. (2005) Naturally occurring polyamines: interaction with macromolecules *Current Protein and Peptide Science* 6, 559–566.

Bagni, N., Mengoli, M. and Matsuzaki, S. (1995) Quantifying polyamines in *Agrobacterium rhizogenes* strains and in Ri plasmid transformed cells. In: Gartland, K.M.A and Davey, M.R. (eds) Agrobacterium Protocols. *Methods in Molecular Biology Volume* 44. Humana Press, Totowa, New Jersey, pp. 223–236.

Becker, A., Bergès, H., Krol, E., Bruand, C., Rüberg, S., Capela, D., Lauber, E., Meilhoc, E., Ampe, F., de Bruijn, F.J. *et al.* (2004) Global changes in gene expression in *Sinorhizobium meliloti* 1021 under microoxic and symbiotic conditions. *Molecular Plant–Microbe Interactions* 17, 292–303.

Biondi, E.G., Tatti, E., Comparini, D., Giuntini, E., Mocali, S., Giovannetti, L., Bazzicalupo, M., Mengoni, A. and Viti, C. (2009) Metabolic capacity of *Sinorhizobium* (*Ensifer*) *meliloti* strains as determined by phenotype microarray analysis. *Applied and Environmental Microbiology* 75, 5396–5404.

Bitrián, M., Zarza, X., Altabella, T., Tiburcio, A.F. and Alcázar, R. (2012) Polyamines under abiotic stress: metabolic crossroads and hormonal crosstalks in plants. *Metabolites* 2, 516–528.

Bobik, C., Meilhoc, E. and Batut, J. (2006) FixJ: a major regulator of the oxygen limitation response and late symbiotic functions of *Sinorhizobium meliloti*. *Journal of Bacteriology* 188, 4890–4902.

Braeken, K., Daniels, R., Vos, K., Fauvart, M., Bachaspatimayum, D., Vanderleyden, J. and Michiels, J. (2008) Genetic determinants of swarming in *Rhizobium etli*. *Microbial Ecology* 55, 54–64.

Brosnan, J.T. and Bronsan, M.E. (1990) Ammonium and organic cations. In: Pasternak, C.A. (ed.) *Monovalent Cations in Biological Systems*. CRC Press, Boca Raton, Florida, pp. 357–374.

Burrell, M., Hanfrey, C.C., Murray, E.J., Stanley-Wall, N.R. and Michael, A.J. (2010) Evolution and multiplicity of arginine decarboxylases in polyamine biosynthesis and essential role in *Bacillus subtilis* biofilm formation. *The Journal of Biological Chemistry* 285, 39224–39238.

Chang, C., Damiani, I., Puppo, A. and Frendo, P. (2009) Redox changes during the legume–*Rhizobium* symbiosis. *Molecular Plant* 2, 370–377.

Chatterjee, S., Choudhuri, M.M. and Ghosh, B. (1983) Changes in polyamine contents during root and nodule growth of *Phaseolus mungo*. *Phytochemistry* 22, 1553–1556.

Chattopadhyay, M.K. and Tabor, H. (2013) Polyamines are critical for the induction of the glutamate decarboxylase-dependent acid resistance system in *Escherichia coli*. *The Journal of Biological Chemistry* 288, 33559–33570.

Chattopadhyay, M.K., Tabor, C.W. and Tabor, H. (2009). Polyamines are not required for aerobic growth of a strain with deletions in all of the genes for polyamine biosynthesis. *Journal of Bacteriology* 191, 5549–5552.

Cho, K., Fuqua, C., Martin, B.S. and Winans, S.C. (1996) Identification of *Agrobacterium tumefaciens* genes that direct the complete catabolism of octopine. *Journal of Bacteriology* 178, 1872–1880.

Chou, H.T. (2011) L-Lysine decarboxylase and cadaverine gamma-glutamylation pathways in *Pseudomonas aeruginosa* PAO1. PhD thesis, Georgia State University, Atlanta, Georgia. Available at: http://scholarworks.gsu.edu/biology_diss/103 (accessed 6 September 2016).

Cockerell, S.R., Rutkovsky, A.C., Zayner, J.P., Cooper, R.E., Porter, L.R., Pendergraft, S.S., Parker, Z.M., McGinnis, M.W. and Karatan, E. (2014) *Vibrio cholerae* NspS, a homologue of ABC-type periplasmic solute binding proteins, facilitates transduction of polyamine signals independent of their transport. *Microbiology* 160, 832–843.

Cunningham-Rundles, S. and Maas, W.K. (1975) Isolation, characterization, and mapping of *Escherichia coli* mutants blocked in the synthesis of ornithine decarboxylase. *Journal of Bacteriology* 124, 791–799.

Davenport, P.W., Griffin, J.L. and Welch, M. (2015) Quorum sensing is accompanied by global metabolic changes in the opportunistic human pathogen *Pseudomonas aeruginosa*. *Journal of Bacteriology* 197, 2072–2082.

Dela Vega, A.L. and Delcour, A.H. (1996) Polyamines decrease *Escherichia coli* outer membrane permeability. *Journal of Bacteriology* 178, 3715–3721.

Denison, R.F. and Kiers, E.T. (2011) Life histories of symbiotic rhizobia and mycorrhizal fungi. *Current Biology* 21, R775–R785.

D'Hooghe, I., Vander Wauven, C., Michiels, J., Tricot, C., de Wilde, P., Vanderleyden, J. and Stalon, V. (1997) The arginine deiminase pathway in *Rhizobium etli*: DNA sequence analysis and functional study of the *arcABC* genes. *Journal of Bacteriology* 179, 7403–7409.

Díaz, R., Vargas-Lagunas, C., Villalobos, M.A., Peralta, H., Mora, Y., Encarnación, S., Girard, L. and Mora, J. (2011) *argC* orthologs from *Rhizobiales* show diverse profiles of transcriptional efficiency and functionality in *Sinorhizobium meliloti*. *Journal of Bacteriology* 193, 460–472.

Di Martino, M.L., Campilongo, R., Casalino, M., Micheli, G., Colonna, B. and Prosseda, G. (2013) Polyamines: emerging players in bacteria–host interactions. *International Journal of Medical Microbiology* 303, 484–491.

Ding, Y., Peng, N., Du, Y., Ji, L. and Cao, B. (2014) Disruption of putrescine biosynthesis in *Shewanella oneidensis* enhances biofilm cohesiveness and performance in Cr(VI) immobilization. *Applied and Environmental Microbiology* 80, 1498–1506.

Domínguez-Ferreras, A., Pérez-Arnedo, R., Becker, A., Olivares, J., Soto, M.J. and Sanjuán, J. (2006) Transcriptome profiling reveals the importance of plasmid pSymB for osmoadaptation of *Sinorhizobium meliloti*. *Journal of Bacteriology* 188, 7617–7625.

Dunn, M.F. (1998) Tricarboxylic acid cycle and anaplerotic enzymes in rhizobia. *FEMS Microbiology Reviews* 22, 105–123.

Dunn, M.F. (2015) Key roles of microsymbiont amino acid metabolism in rhizobia–legume interactions. *Critical Reviews in Microbiology* 41, 411–451.

Foster, J.W. (2004) *Escherichia coli* acid resistance: tales of an amateur acidophile. *Nature Reviews Microbiology* 2, 898–907.

Fothergill, J.C. and Guest, J.R. (1977) Catabolism of L-lysine by *Pseudomonas aeruginosa*. *Journal of General Microbiology* 99, 139–155.

Fujihara, S. (2009) Biogenic amines in rhizobia and legume root nodules. *Microbes and Environments* 24, 1–13.

Fujihara, S. and Harada, Y. (1989) Fast-growing root nodule bacteria produce a novel polyamine, aminobutylhomospermidine. *Biochemical and Biophysical Research Communications* 165, 659–666.

Fujihara, S. and Yoneyama, T. (1993) Effects of pH and osmotic stress on cellular polyamine contents in the soybean rhizobia *Rhizobium fredii* P220 and *Bradyrhizobium japonicum* A1017. *Applied and Environmental Microbiology* 59, 1104–1109.

Fujihara, S., Abe, H., Minakawa, Y., Akao, S. and Yoneyama, T. (1994) Polyamines in nodules from various plant–microbe symbiotic associations. *Plant and Cell Physiology* 35, 1127–1134.

Fujihara, S., Abe, H. and Yoneyama, T. (1995) A new polyamine 4-aminobutylcadaverine. Occurrence and its biosynthesis in root nodules of Adzuki bean plant *Vigna angularis*. *The Journal of Biological Chemistry* 270, 9932–9938.

Gao, S., Romdhane, S.B., Beullens, S., Kaever, V., Lambrichts, I., Fauvart, M. and Michiels, J. (2014) Genomic analysis of cyclic-di-GMP-related genes in rhizobial type strains and functional analysis in *Rhizobium etli*. *Applied Microbiology and Biotechnology* 98, 4589–4602.

Gomelsky, M. (2011) cAMP, c-di-GMP, c-di-AMP and now cGMP: bacteria use them all! *Molecular Microbiology* 79, 562–565.

Gomes, D.F., de Silvia Batista, J.S., Shiavon, A.L., Andrade, D.S. and Hungria, M. (2012) Proteomic profiling of *Rhizobium tropici* PRF 81: Identification of conserved and specific responses to heat stress. *BMC Microbiology* 12:84.

Graham, P.H., Draeger, K.J., Ferrey, M.L., Conroy, M.J., Hammer, B.E., Martinez, E., Aarons, S.R. and Quinto, C. (1994) Acid pH tolerance in strains of *Rhizobium* and *Bradyrhizobium*, and initial studies on the basis for acid tolerance of *Rhizobium tropici* UMR1899. *Canadian Journal of Microbiology* 40, 198–207.

Hamana, K., Matsuzaki, S. and Sakakibara, M. (1988) Distribution of *sym*-homospermidine in eubacteria, cyanobacteria, algae and ferns. *FEMS Microbiology Letters* 50, 11–16.

Hamana, K., Minamisawa, K. and Matsuzaki, S. (1990) Polyamines in *Rhizobium*, *Bradyrhizobium*, *Azorhizobium* and *Agrobacterium*. *FEMS Microbiology Letters* 71, 71–76.

Hamana, K., Sakamoto, A., Tachiyanagi, S. and Terauchi, E. (2003) Polyamine profiles of some members of the alpha subclass of the class *Proteobacteria*: polyamine analysis of twenty recently described genera. *Microbiology and Culture Collections* 19, 13–21.

Hanfrey, C.C., Pearson, B.M., Hazeldine, S., Lee, J., Gaskin, D.J., Woster, P.M., Phillips, M.A. and Michael, A.J. (2011) Alternative spermidine biosynthetic route is critical for growth of *Campylobacter jejuni* and is the dominant polyamine pathway in human gut microbiota. *The Journal of Biological Chemistry* 286, 43301–43312.

Hernández, V.M., Arteaga, A., Ortíz-Ortíz, C., Hernández-Lucas, I., Mora, J., Vázquez, A., Díaz, R., Thomas-Oates, J., Bergstrom, E., Girard, L. and Dunn, M.F. (2015a) Análisis de genómica funcional del metabolismo de la arginina y la biosíntesis de poliaminas en *Sinorhizobium meliloti*. In: *Abstracts, 39° Congreso Nacional de Microbiología, Querétaro 2015, Marzo 22–26*. Associación Mexicana de Microbiología, México. Available at: https://www.researchgate.net/publication/274375154_Analisis_de_genomica_funcional_del_metabolismo_de_la_arginina_y_la_biosintesis_de_poliaminas_en_Sinorhizobium (accessed 6 September 2016).

Hernández, V.M., Girard, L., Hernández-Lucas, I., Vázquez, A., Ortíz-Ortíz, C., Díaz, R. and Dunn, M.F. (2015b) Genetic and biochemical characterization of arginine biosynthesis in *Sinorhizobium meliloti* 1021. *Microbiology* 161, 1671–1682.

Hunter, K.J. (1998) A dansyl chloride-HPLC method for the determination of polyamines. In: Morgan, D.M.L. (ed.) *Polyamine Protocols. Methods in Molecular Biology, Volume 79*. Humana Press, Totowa, New Jersey, pp. 119–124.

Igarashi, K. and Kashiwagi, K. (2006) Polyamine modulon in *Escherichia coli*: genes involved in the stimulation of cell growth by polyamines. *Journal of Biochemistry* 139, 11–16.

Igarashi, K. and Kashiwagi, K. (2015) Modulation of protein synthesis by polyamines. *IUBMB Life* 67, 160–169.

Jones, K.M., Kobayashi, H, Davies, B.W., Taga, M.E. and Walker, G.C. (2007) How rhizobial symbionts invade plants: the *Sinorhizobium–Medicago* model. *Nature Reviews Microbiology* 5, 619–633.

Karatan, E. and Michael, A.J. (2013) A wider role for polyamines in biofilm formation. *Biotechnology Letters* 35, 1715–1717.

Kerppola, T.K. and Kahn, M.L. (1988) Symbiotic phenotypes of auxotrophic mutants of *Rhizobium meliloti* 104A14. *Journal of General Microbiology* 134, 913–919.

Kumar, A, Vij, N. and Randhawa, G.S. (2003) Isolation and symbiotic characterization of transposon *Tn*5-induced arginine auxotrophs of *Sinorhizobium meliloti*. *Indian Journal of Experimental Biology* 41, 1198–1294.

Kurihara, S., Suzuki, H., Tsuboi, Y. and Benno, Y. (2009) Dependence of swarming in *Escherichia coli* on spermidine and the spermidine importer. *FEMS Microbiology Letters* 294, 97–101.

Kusano, T., Berberich, T., Tateda, C. and Takahashi, Y. (2008) Polyamines: essential factors for growth and survival. *Planta* 228, 367–381.

Lahiri, K., Chattopadhyay, S., Chattopadhyay, S. and Ghosh, B. (1992) Polyamine metabolism in nodules of *Vigna mungo* during senescence. *Phytochemistry* 31, 4087–4090.

Landini, P. (2009) Cross-talk mechanisms in biofilm formation and responses to environmental and physiological stress in *Escherichia coli*. *Research in Microbiology* 160, 259–266.

Lee, J., Michael, A.J., Martynowski, D., Goldsmith, E.J. and Phillips, M.A. (2007) Phylogenetic diversity and the structural basis of substrate specificity in the β/α-barrel fold basic amino acid decarboxylases. *The Journal of Biological Chemistry* 282, 27115–27125.

Lee, J., Sperandio, V., Frantz, D.E., Longgood, J., Camilli, A., Phillips, M.A. and Michael, A.J. (2009) An alternative polyamine biosynthetic pathway is widespread in bacteria and essential for biofilm formation in *Vibrio cholerae*. *The Journal of Biological Chemistry* 284, 9899–9907.

Li, Y., Tian, C.F., Chen, W.F., Wang, L., Sui, X.H. and Chen, W.X. (2013) High-resolution transcriptomic analysis of *Sinorhizobium* sp. NGR234 bacteroids in determinate nodules of *Vigna unguiculata* and indeterminate nodules of *Leucaena leucocephala*. *PLoS ONE* 8(8): e70531.

Liu, Y., White, R.H. and Whitman, W.B. (2010) Methanocci use the diaminopimelate aminotransferase (DapL) pathway for lysine biosynthesis. *Journal of Bacteriology* 192, 3304–3310.

López-Gómez, M., Hildago-Castellanos, J., Iribarne, C. and Lluch, C. (2014a) Proline accumulation has prevalence over polyamines in nodules of *Medicago sativa* in symbiosis with *Sinorhizobium meliloti* during the initial response to salinity. *Plant and Soil* 374, 149–159.

López-Gómez, M., Cobos-Porras, L., Hidalgo-Castellanos, J. and Lluch, C. (2014b) Occurrence of polyamines in root nodules of *Phaseolus vulgaris* in symbiosis with *Rhizobium tropici* in response to salt stress. *Phytochemistry* 107, 32–41.

López-Leal, G., Tabche, M.L., Castillo-Ramírez, S., Mendoza-Vargas, A., Ramírez-Romero, M.A. and Dávila, G. (2014) RNA-Seq analysis of the multipartite genome of *Rhizobium etli* CE3 shows different replicon contributions under heat and saline shock. *BMC Genomics* 15:770.

Lynch, D., O'Brien, J., Welch, T., Clarke, P., Cuív, P.Ó., Crosa, J.H. and O'Connell, M. (2001) Genetic organization of the region encoding regulation, biosynthesis, and transport of Rhizobactin 1021, a siderophore produced by *Sinorhizobium meliloti*. *Journal of Bacteriology* 183, 2576–2585.

Madhubala, R. (1998) Thin-layer chromatographic method for assaying polyamines. In: Morgan, D.M.L. (ed.) *Polyamine Protocols. Methods in Molecular Biology, Volume 79*. Humana Press, Totowa, New Jersey, p. 131–136.

Michael, A.J. (2011) Exploring polyamine biosynthetic diversity through comparative and functional genomics. In: Pegg, A.E. and Casero, Jr., R.A. (eds) Polyamines: Methods and Protocols, Springer, pp. 39–50. DOI 10.1007/978-1-61779-034-8_2.

Morgan, D.M.L. (1998) Determination of polyamines as their benzoylated derivatives by HPLC. In: Morgan, D.M.L. (ed.) *Polyamine Protocols. Methods in Molecular Biology, Volume 79*. Humana Press, Totowa, New Jersey, pp. 111–118.

Mrabet, M., Zribi, K., Mhadhbi, H., Djébali, N., Mhamdi, R., Aouani, M.E. and Nakamura, K. (2011) Salt tolerance of a *Sinorhizobium meliloti* strain isolated from dry lands: growth capacity and protein profile changes. *Archives of Microbiology* 61, 361–369.

Muglia, C.I., Grasso, D.H. and Aguilar, M.O. (2007) *Rhizobium tropici* response to acidity involves activation of glutathione synthesis. *Microbiology* 153, 1286–1296.

Nakada, Y. and Itoh, Y. (2003) Identification of the putrescine biosynthetic genes in *Pseudomonas aeruginosa* and characterization of agmatine deiminase and *N*-carbamoylputrescine amidohydrolase of the arginine decarboxylase pathway. *Microbiology* 149, 707–714.

Nemoto, N., Kurihara, S., Kitahara, Y., Asada, K., Kato, K. and Suzuki, H. (2012) Mechanism for regulation of the putrescine utilization pathway by the transcription factor PuuR in *Escherichia coli* K-12. *Journal of Bacteriology* 194, 3437–3447.

Niste, M., Vidican, R., Pop, R. and Rotar, I. (2013) Stress factors affecting symbiosis activity and nitrogen fixation by *Rhizobium* cultured *in vitro*. *ProEnvironment* 6, 42–45.

Nogales, J., Domínguez-Ferreras, A., Amaya-Gómez, C.V., van Dillewijn, P., Cuéllar, V., Sanjuán, J., Olivares, J. and Soto, M.J. (2010) Transcriptome profiling of a *Sinorhizobium meliloti fadD* mutant reveals the role of rhizobactin 1021 biosynthesis and regulation genes in the control of swarming. *BMC Genomics* 11:157.

Norris, V., Reusch, R.N., Igarashi, K. and Root-Bernstein, R. (2014) Molecular complementarity between simple, universal molecules and ions limited phenotype space in the precursors of cells. *Biology Direct* 10:28.

Ozawa, T. and Tsuji, T. (1993) A possible role for polyamines in the repression of growth of *Bradyrhizobium japonicum* bacteroids in soybean nodules. *Plant and Cell Physiology* 34, 899–904.

Patel, C.N., Wortham, B.W., Lines, J.L., Fetherston, J.D., Perry, R.D. and Oliveira, M.A. (2006) Polyamines are essential for the formation of plague biofilm. *Journal of Bacteriology* 188, 2355–2363.

Pegg, A.E. and Casero, R.A. Jr (2011) Current status of the polyamine research field. In: Pegg, A.E. and Casero, R.A. Jr (eds) *Polyamines: Methods and Protocols. Methods in Molecular Biology, Volume 720*. Humana Press, Totowa, New Jersey, pp. 3–35.

Pendergraft, S.S. (2012) Characterization of protein–protein interaction within a polyamine responsive signaling system in *Vibrio cholerae*. MS thesis, Appalachian State University, Boone, North Carolina.

Ponappa, T., Lawson, D.R. and Miller, A.R. (1992) Inhibition of ornithine decarboxylase activity reduces polyamine levels and growth of *Agrobacterium tumefaciens*. *Current Microbiology* 24, 269–273.

Randhawa, G.S. and Hassani, R. (2002) Role of biosynthetic pathways of amino acids, nucleotide bases and vitamins in symbiosis. *Indian Journal of Experimental Biology* 40, 755–764.

Rodríguez-Garay, B., Phillips, G.C. and Kuehn, G.D. (1989) Detection of norspermidine and norspermine in *Medicago sativa* L. (alfalfa). *Plant Physiology* 89, 525–529.

Rojas-Jiménez, K., Sohlenkamp, C., Geiger, O., Martínez-Romero, E., Warner, D. and Vinuesa, P. (2005) A CIC chloride channel homolog and ornithine-containing membrane lipids of *Rhizobium tropici* CIAT899 are involved in symbiotic efficiency and acid tolerance. *Molecular Plant–Microbe Interactions* 18, 1175–1185.

Romano, A., Trip, H., Lonvaud-Funel, A., Lolkema, J.S. and Lucas, P.M. (2012) Evidence of two functionally distinct ornithine decarboxylation systems in lactic acid bacteria. *Applied and Environmental Microbiology* 78, 1953–1961.
Rüberg, S., Tian, Z.-X., Krol, E., Linke, B., Meyer, F., Wang, Y., Pühler, A., Weidner, S. and Becker, A (2003) Construction and validation of a *Sinorhizobium meliloti* whole genome DNA microarray: genome-wide profiling of osmoadaptive gene expression. *Journal of Biotechnology* 106, 255–268.
Sakamoto, A., Terui, Y., Yamamoto, T., Kashara, T., Nakamura, M., Tomitori, H., Yamamoto, K., Ishihama, A., Michael, A.J., Igarashi, K. and Kashiwagi, K. (2012) Enhanced biofilm formation and/or cell viability by polyamines through stimulation of response regulators UvrY and CpxR in the two-component signal transducing systems, and ribosome recycling factor. *The International Journal of Biochemistry and Cell Biology* 44, 1877–1886.
Santos, M.R., Cosme, A.M., Becker, J.D., Mederios, J.M.C., Mata, M.F. and Moreira, L.M. (2010) Absence of functional TolC protein caused increased stress response gene expression in *Sinorhizobium meliloti*. *BMC Microbiology* 10:80.
Shah, P. and Swiatlo, E. (2008) A multifaceted role for polyamines in bacterial pathogens. *Molecular Microbiology* 68, 4–16.
Shaw, F.L. (2011) From prediction to function: polyamine biosynthesis and formate metabolism in the α- and ε- proteobacteria. PhD thesis, University of East Anglia, Norwich, UK.
Shi, H. and Chan, Z. (2014) Improvement of plant abiotic stress tolerance through modulation of the polyamine pathway. *Journal of Integrative Plant Biology* 56, 114–121.
Smith, T.A. (1977) Homospermidine in *Rhizobium* and legume root nodules. *Phytochemistry* 16, 278–279.
Sturgill, G. and Rather, P.N. (2004) Evidence that putrescine acts as an extracellular signal required for swarming in *Proteus mirabilis*. *Molecular Microbiology* 51, 437–446.
Tabor, C.W. and Tabor, H. (1985) Polyamines in Microorganisms. *Microbiological Reviews* 49, 81–99.
Terakado-Tonooka, J. and Fijihara, S. (2008) Involvement of polyamines in the root nodule regulation of soybeans (*Glycine max*). *Plant Root* 2, 46–53.
Torres-Quesada, O., Oruezabal, R.I., Peregrina, A., Jofré, E., Lloret, J., Rivilla, R., Toro, N. and Jiménez-Zurdo, J.I. (2010) The *Sinorhizobium meliloti* RNA chaperone Hfq influences central carbon metabolism and the symbiotic interaction with alfalfa. *BMC Microbiology* 10:71.
Udvardi, M. and Poole, P.S. (2013) Transport and metabolism in legume–rhizobia symbioses. *Annual Review of Plant Biology* 64, 781–805.
Valdés-Santiago, L. and Ruiz-Herrera, J. (2014) Stress and polyamine metabolism in fungi. *Frontiers in Chemistry* 1: 42.
Vassileva, V. and Ignatov, G. (1999) Polyamine-induced changes in symbiotic parameters of the *Galega orientalis–Rhizobium galegae* nitrogen-fixing system. *Plant and Soil* 210, 83–91.
Vauclare, P., Bligny, R., Gout, E. and Widmer, F. (2013) An overview of the metabolic differences between *Bradyrhizobium japonicum* 110 bacteria and differentiated bacteroids from soybean (*Glycine max*) root nodules: an in vitro ^{13}C- and ^{31}P-nuclear magnetic resonance spectroscopy study. *FEMS Microbiology Letters* 343, 49–56.
Whitehead, L.F., Tyerman, S. and Day, D.A. (2001) Polyamines as potential regulators of nutrient exchange across the peribacteroid membrane in soybean root nodules. *Australian Journal of Plant Physiology* 28, 675–681.
Wortham, B.W., Patel, C.N. and Oliveira, M.A. (2007) Polyamines in bacteria: pleiotropic effects yet specific mechanisms. *Advances in Experimental Medical Microbiology* 603, 106–115.
Wortham, B.W., Oliveira, M.A., Fetherston, J.D. and Perry, R.D (2010) Polyamines are required for the expression of key Hms proteins important for *Yersinia pestis* biofilm formation. *Environmental Microbiology* 12, 2034–2047.
Yadav, A.S. (2007) Auxotrophy in rhizobia revisited. *Indian Journal of Microbiology* 47, 279–288.
Yoneyama, T., Fujihara, S. and Yagi, K. (1998) Natural abundance of ^{15}N in amino acids and polyamines from leguminous nodules: unique ^{15}N enrichment in homospermidine. *Journal of Experimental Botany* 49, 521–526.
Yoshida, M., Kashiwagi, K., Shigemasa, A., Taniguchi, S., Yamamoto, K., Makinoshima, H., Ishihama, A. and Igarash, K. (2004) A unifying model of the role of polyamines in bacterial cell growth, the polyamine modulon. *The Journal of Biological Chemistry* 279, 46008–46013.
Zahran, H.H. (1999) *Rhizobium*–legume symbiosis and nitrogen fixation under severe conditions and in an arid climate. *Microbiology and Molecular Biology Reviews* 63, 968–989.

27 Working Together: Amino Acid Biosynthesis in Endosymbiont-harbouring Trypanosomatidae

J.M.P. Alves*

Department of Parasitology, University of São Paulo, São Paulo, Brazil

27.1 Abstract

Endosymbiosis, i.e. a cell living inside another organism's cell, is a widespread phenomenon with important evolutionary and metabolic aspects. Among the many taxa presenting endosymbiosis are the genera *Strigomonas*, *Kentomonas* and *Angomonas* from the Trypanosomatidae family (phylum Euglenozoa). Protozoans from these three genera harbour in their cytoplasm one obligatory betaproteobacterial endosymbiont, *Candidatus* Kinetoplastibacterium spp., which affects important characteristics of the host's physiology and metabolism – in particular, the synthesis of amino acids, in which the two organisms collaborate closely in the completion of these highly important pathways. Although essential amino acids are those most heavily influenced by the presence of the bacterial endosymbiont, the interspecific collaboration also extends to other, non-essential amino acids. Across evolutionary time, a number of bacterial genes from these important pathways have been transferred to the trypanosomatid nucleus from both the endosymbiont and from other, long-gone bacteria.

27.2 Background

The Trypanosomatidae family comprises flagellate unicellular eukaryotes that are parasites of a large range of organisms from many taxonomic groups, including insects, vertebrates and plants (Vickerman, 1976, 1994). Special attention has been given to human parasites such as *Trypanosoma cruzi*, *T. brucei* and *Leishmania* spp., all of which cause significant mortality and morbidity across the developing world (Hoare, 1972). Taxonomically, the family belongs to the class Kinetoplastea – a group specially known for possessing a single, highly modified mitochondrion (the kinetoplast) per cell – and to the phylum Euglenozoa, along with the Euglenida and other, smaller classes (Adl *et al.*, 2012). The Euglenozoa phylum comprises organisms sporting many different lifestyles besides the aforementioned parasitic trypanosomatids, including free-living (in which case they can be osmotrophic, phagotrophic or photosynthetic), commensal and even endosymbiotic. The non-trypanosomatid Kinetoplastea can also present (ecto or endo) parasitic, commensal or free-living lifestyles, with the Trypanosomatidae being the only family of the Kinetoplastea that is exclusively parasitic. Therefore, this ancient group is especially interesting for genomic and metabolic evolutionary studies owing to the broad range of ecological niches that its members can occupy.

Parasites can be termed as being monoxenous or heteroxenous depending on whether they infect just one or more than one host, respectively. Trypanosomatids can belong to either of these

*E-mail address: jotajj@usp.br

categories, and much research has been published discussing which of these parasitic modes is most likely to be the ancestral one for the Trypanosomatidae (Vickerman, 1994; Lukeš *et al.*, 2014). Heteroxenous trypanosomatids alternate between an invertebrate (usually an insect, but also leeches and others) host and a vertebrate host (with the genus *Phytomonas*, which infects plants, being the most significant exception). Monoxenous trypanosomatids, in contrast, stay in an invertebrate host for their whole life cycle.

Due to their significant medical and veterinary interest, heteroxenous species from the genera *Trypanosoma* and *Leishmania*, specifically *T. cruzi*, *T. brucei* and *L. major*, are the most well studied Euglenozoa, having had their full genomes sequenced relatively early on in the genomic revolution of the past 15 years (Berriman *et al.*, 2005; El-Sayed *et al.*, 2005; Ivens *et al.*, 2005). However, biochemical and cell biology studies of monoxenous, insect-infecting trypanosomatids have also been performed for many decades, and have had an increase in attention in the recent past (Alves *et al.*, 2013b; Motta *et al.*, 2013; Runckel *et al.*, 2014; Kraeva *et al.*, 2015).

Among the insect-infecting Trypanosomatidae, the genera *Strigomonas*, *Angomonas* and *Kentomonas*, of the subfamily Strigomonadinae (Teixeira *et al.*, 2011; Votýpka *et al.*, 2014), are especially interesting because they harbour an obligate bacterial endosymbiont in their cytoplasm (de Souza and Motta, 1999), and are thus referred to as symbiont-harbouring trypanosomatids (SHTs). This bacterium (the trypanosomatid proteobacterial endosymbiont, or TPE) was acquired once by the common ancestor to all three genera and is named *Candidatus* Kinetoplastibacterium spp., and belongs to the *Alcaligenaceae* family of the *Betaproteobacteria*. Interestingly, the TPE is found in close association with the nucleus of the eukaryotic host, and the division of the endosymbiont is controlled by the trypanosomatid, with usually just one bacterium present per eukaryotic cell (Roitman and Camargo, 1985; de Souza and Motta, 1999; Motta *et al.*, 2010). A plethora of biochemical, cellular and now genomic and *in silico* studies have shown that the two organisms are dependent on each other, with neither able to survive on its own in the wild. The bacterium lacks a cell wall and cannot survive for more than a few hours outside its host; the aposymbiotic trypanosomatid (when the endosymbiont is killed by antibiotic treatment) can survive in culture when supplemented with many nutrients, but loses the capability of infecting insect epithelial gut cells. Many studies have clarified a large number of aspects in which the eukaryotic host is metabolically dependent on the endosymbiont, including, but not limited to, the acquisition of its essential haem (Lwoff, 1940; Newton, 1956; Chang *et al.*, 1975; Korený *et al.*, 2010; Alves *et al.*, 2011), vitamins (Roitman *et al.*, 1972; Klein *et al.*, 2013) and essential amino acids (Mundim *et al.*, 1974; Yoshida *et al.*, 1978; Alfieri and Camargo, 1982; Alves *et al.*, 2013b).

As shown recently, the complementation of the trypanosomatid's metabolism is done not only by the *Ca.* Kinetoplastibacterium spp. endosymbionts, but also by a large number of nuclear genes that have been acquired by horizontal gene transfer (HGT); in the specific case of amino acid biosynthesis, about half of the genes analysed have come from other bacterial sources into the genome of the common ancestor to many (and in some cases probably all) Trypanosomatidae (Alves *et al.*, 2013b).

It is, therefore, clear that there is an outstanding mutualistic symbiosis between these two lineages in which the bacterium might be on the way to becoming an organelle of the trypanosomatid cell. This makes the SHT biological system a very interesting model for the evolution of the eukaryotic cell.

27.3 *In silico* Metabolic Reconstruction

With the rise of high-throughput genome sequencing, powerful computational resources, metabolic information databases and comprehensive sequence databases containing a broad swath of annotated entries, purely computational prediction of an organism's metabolic potential became a reality. While such predictions should be considered preliminary in nature until direct biochemical evidence can be gathered, they are still very valuable in most instances. Besides speeding up biological discovery and guiding wet laboratory experiments, *in silico* predictions might often be the only metabolic information available for many organisms, be it due to experimental limitations (e.g. the ability to grow in culture) or simply the fact that the organism at hand has not attracted enough interest from the community (e.g. because of its lack of economic or medical significance).

The results described here were obtained by *in silico* prediction, and analysed in light of previously published cell biology and biochemical data when appropriate. The software used, ASGARD (Alves and Buck, 2007), was written in order to automate the search for and annotation of putatively present enzymes and pathways in newly sequenced genomes. It can use either the full set of predicted proteins or raw DNA sequence records, without the need for gene prediction, in order to yield a first glimpse into the metabolic capabilities of the genome. ASGARD takes advantage of the wealth of metabolic data present in the KEGG databases (Ogata *et al.*, 1999), plus the extensively annotated UniRef protein sequence database (Suzek *et al.*, 2007) and the many mappings of its entries to functional annotation categories such as Enzyme Commission (EC) or Gene Ontology (GO) numbers. It can also graphically display, in different user-selectable colours, up to five different metabolic reconstruction predictions in KEGG pathway maps, greatly aiding in comparative (or complementarity) analysis. In the particular case of the SHTs, this capability was especially useful, because it allowed the visualization of results from both the trypanosomatid host and the endosymbiotic bacterium in the same picture for each metabolic pathway. This made the finer details of the remarkable collaborative symbiosis between the two organisms immediately clear. The same maps also displayed data for up to three regular trypanosomatids, i.e. those not containing an endosymbiont, also enabling prompt comparison between the different categories of Trypanosomatidae under study.

While very useful and enabling, automatic annotation and functional prediction methods have their caveats and pitfalls, and must always be followed by manual curation of the results whenever the exact functional classification of a set of proteins is necessary for reaching a proper biological conclusion. For that reason, predicted enzymes for the pathways of interest were manually searched for against a number of protein domain databases in order to increase the certainty of their annotation as much as possible by using *in silico* methods. Any previously unpublished phylogenetic analyses presented here have been performed by the method of maximum likelihood, as implemented in the FastTree2 program (Price *et al.* 2010), after protein alignment using MUSCLE (Edgar, 2004).

27.4 Essential Amino Acid Biosynthesis

Previous nutritional data, obtained by studies involving techniques such as trace experiments with radioactive compounds, as well as experiments using wild or cured SHT cultures (for list of publications, see Alves *et al.*, 2013b), have shown that a number of amino acids seem to be essential to regular trypanosomatids (or RTs, i.e. those lineages not presenting a TPE) and must thus be provided in the culture medium – or, in nature, by the host. The list of amino acids that has been found to be needed in such cases comprises arginine/ornithine, cysteine, histidine, isoleucine, leucine, lysine, methionine, phenylalanine, threonine, tryptophan, tyrosine and valine. Aposymbiotic SHTs present this same list of essential amino acids; wild-type SHTs, in contrast, seem to require only the addition of methionine and/or tyrosine to the culture medium, indicating that the TPE is responsible for a large portion of the essential amino acid biosynthesis in these organisms. Below, we examine the details of the essential amino acid biosynthesis pathways in SHTs and their TPEs.

27.4.1 Arginine and ornithine

The biosynthesis of arginine and ornithine uses a pathway starting with glutamate and involving the urea cycle – of which these two amino acids are part, along with citrulline and argininosuccinate. Intermediates in this pathway are also part of the biosynthetic pathway for essential polyamines such as putrescine, spermidine and spermine. As can be seen in Fig. 27.1, the pathways are mosaics of eukaryotic and bacterial enzymes, with a significant contribution of genes that have been horizontally transferred to the trypanosomatid nucleus (Alves *et al.*, 2013b).

The first five steps of the pathway, going from glutamate to ornithine, can be performed

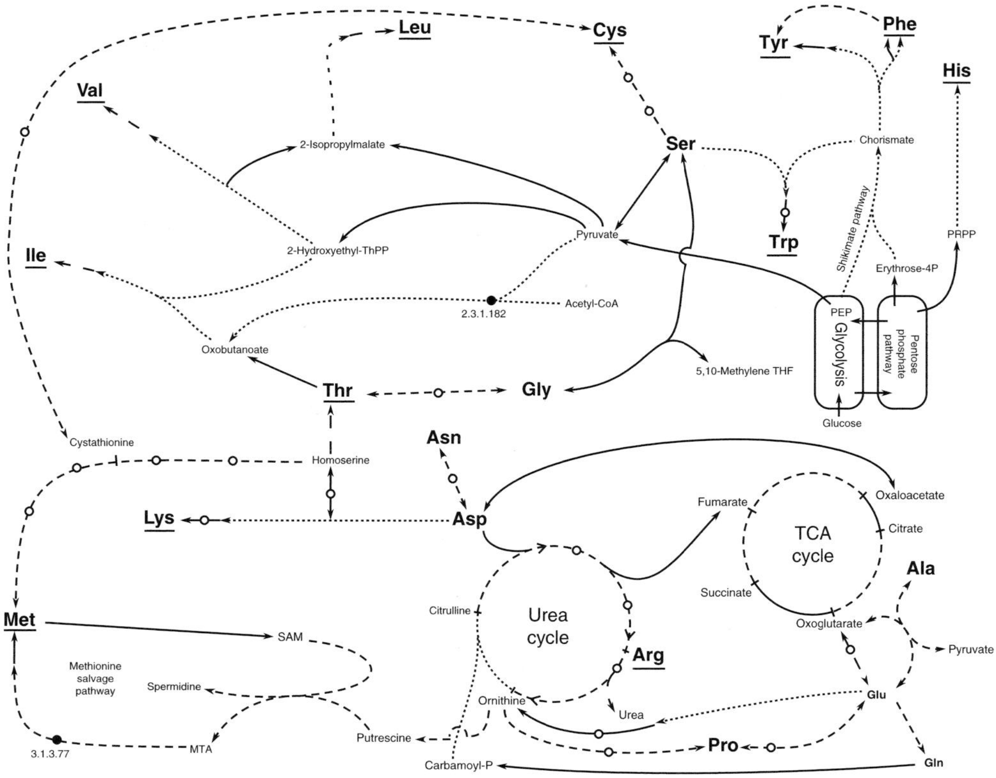

Fig. 27.1. Outline of amino acid biosynthesis in *Strigomonas culicis*, a symbiont-harbouring trypanosomatid species, as predicted from genomic data. The 20 main amino acids are represented by their standard three-letter abbreviations, and essential amino acids, as defined in the classic literature (see main text), are underlined. Dashed and dotted lines represent parts of metabolism that are performed solely by predicted proteins encoded by the trypanosomatid nuclear genome and by the bacterial endosymbiont, respectively; solid lines indicate that both organisms are capable of performing that part of the cellular metabolism. Arrow lengths are not proportional to the number of pathway steps taken in each case. Black circles and their accompanying Enzyme Commission (EC) numbers indicate enzymes that seem to be strictly necessary for their corresponding pathways, but whose genes have not been identified in the genomic analyses: EC 2.3.1.182, (*R*)-citramalate synthase; CimA; EC 3.1.3.77, acireductone synthase (E-1 enolase-phosphatase). Open circles indicate reactions that can be catalysed by one or more genes that are hypothesized to have been horizontally transferred to the trypanosomatid nucleus; bear in mind that not all genes from all pathways have been extensively analysed phylogenetically, therefore these candidates probably represent a lower bound for the actual number of horizontal gene transfer (HGT) events. Other abbreviations: MTA, 5′-methylthioadenosine; PEP, phosphoenolpyruvate; PRPP, 5′-phosphoribosyl diphosphate; SAM, *S*-adenosyl-methionine; THF, tetrahydrofolate; ThPP, thiamine diphosphate; TCA cycle, tricarboxylic acid cycle.

completely by enzymes from the TPE; alternatively, the last two of these five steps can also be performed by enzymes from the SHT. The final step is particularly interesting, as the two SHT enzymes that are capable of catalysing this reaction, the conversion of *N*-acetylornithine to ornithine, seem to have been transferred from other bacteria to the SHT genome (Alves *et al.*, 2013b). These two genes are also present in RT genomes, indicating that this is an ancient transfer that occurred before the establishment of endosymbiosis in the Strigomonadinae subfamily, and possibly originated from one or more gamma- and/or betaproteobacterial lineages.

The urea cycle steps, conversely, are catalysed by a mixture of TPE and SHT enzymes: the conversion of ornithine to citrulline by TPE enzymes, and all others by SHT enzymes. Once again, the influence of ancient HGT is of great importance in trypanosomatid metabolism, because two of the four urea cycle enzymes of the SHTs and RTs seem to have been transferred from a *Firmicutes* bacterium: argininosuccinate synthase (EC 6.3.4.5), which converts citrulline plus aspartate into argininosuccinate; and argininosuccinate lyase (EC 4.3.2.1), which then converts argininosuccinate into arginine plus fumarate. This has possibly happened in a single event, as the two genes are right next to each other in the genome, and the pair are surrounded by genes of eukaryotic origin.

Arginine can also be used directly in the SHTs as a substrate for the synthesis of polyamines, given the presence of arginine decarboxylase (EC 4.1.1.19) in the TPE genomes. Agmatine generated by this enzyme from arginine can then be converted by an SHT enzyme into putrescine.

It is, then, clear that the arginine and ornithine biosynthesis pathways are a patchwork of enzymes that originated not only from SHTs and TPE, but also from enzymes derived from other bacterial lineages that were transferred (most probably) to the common ancestor of all Trypanosomatidae.

27.4.2 Cysteine and methionine

The status of methionine as an essential amino acid for trypanosomatids is controversial; some studies include it in the list, but others do not. Our previous genomic studies (Alves *et al.*, 2013b) have shown that all genes necessary for the biosynthesis of methionine, with one unclear situation (detailed below), are present in the genomes of the Trypanosomatidae investigated. The conflicting conclusions regarding the essentiality of methionine are thus likely to be, as in the case of threonine detailed below, because these compounds are heavily used by the trypanosomatid, in which case external complementation (be it from the culture medium or the host environment) increases growth rate significantly.

All enzymes in the main methionine biosynthesis and salvage pathways (see Fig. 27.1) are encoded by genes from the nuclear genomes of SHTs (and RTs), with few present in the TPE genomes. Nevertheless, there is, once again, a significant bacterial contribution in the form of HGT events in the Met biosynthesis pathway, though not in the Met salvage pathway. Namely, both steps in the conversion of cystathionine to methionine can be performed by horizontally transferred enzymes; also, the synthesis of cystathionine itself, in the pathway starting from aspartate 4-semialdehyde, is catalysed in two steps that are also performed by enzymes that are hypothesized as having been transferred from ancient bacterial sources of diverse lineages. Conversely. the synthesis of aspartate 4-semialdehyde can be catalysed, in two steps by enzymes from the TPE genome, which could significantly help in the availability of Met for the SHT.

The other route of cystathionine biosynthesis, starting from serine or cysteine, presents an interesting case. At the time of our original genomic analyses, it was impossible to confidently ascertain by *in silico* means which of the predicted genes found was most likely to perform the transsulfuration step located between cysteine and cystathionine. This reversible step is catalysed by cystathionine gamma-lyase (EC 4.4.1.1), whose activity has long been known in trypanosomatids even though no gene had yet been identified for the enzyme. The difficulty in predicting the correct gene for this step comes from the similar protein domain compositions of the different sulfhydrolases involved in these pathways: cystathionine gamma-lyase, cystathionine beta-lyase (EC 4.4.1.8, in two different copies), cystathionine gamma-synthase (EC 2.5.1.48), and *O*-acetylhomoserine aminocarboxypropyl transferase (EC 2.5.1.49). As can be seen in the corresponding KEGG map, these enzymes are scattered in the cysteine and methionine pathways, and participate in many important functions. Previous annotations of the available RT genomes showed conflicting EC number assignments for genes that were clearly orthologous when analysed phylogenetically. As a result, we were unable to determine which gene was performing the central transsulfuration step. More interestingly, the EC 2.5.1.48, EC 2.5.1.49 and one of the two copies of EC 4.4.1.8 all seem to have originated from one or more HGT events; the other copy of EC 4.4.1.8 is of eukaryotic origin. Our main speculation at the time was whether one of these copies of EC 4.4.1.8 was actually the elusive enzyme responsible for the EC 4.4.1.1 activity.

However, a recent biochemical study (Giordana *et al.*, 2014) seems to have finally shed light on the identity of cystathionine gamma-lyase in these organisms. It turns out that the enzyme that we (and others) had identified as a cystathionine gamma-synthase (EC 2.5.1.48) is the one performing the EC 4.4.1.1 function. Although this experimental determination was conducted in *Leishmania*, it is clear from our phylogenetic analyses that this is the orthologous gene in other trypanosomatids, such as the SHTs.

Besides the pathways described above, cysteine can also be synthesized in two steps from serine that involve serine *O*-acetyltransferase (EC 2.3.1.30) and cysteine synthase (EC 2.5.1.47). While both of the genes for these enzymes are present in the SHT nucleus, and no copy is present in the TPE genome, there is some phylogenetic evidence for an HGT origin of both genes. Therefore, even though the SHT–TPE collaboration is not as central in the Cys/Met pathways as in those of other amino acids, there is still a great amount of bacterial input into the biosynthesis of these compounds in SHTs, as well in RTs.

27.4.3 Histidine

Histidine biosynthesis is the simplest case among those examined herein: all ten steps, from 5-phosphoribosyl diphosphate (from the pentose phosphate pathway) up to histidine itself, are performed by enzymes encoded in the endosymbiont genome, with no contribution whatsoever from the SHT nuclear genes. The only open question in this pathway is the identity of histidinol phosphate phosphatase (HPP, EC 3.1.3.15; preferred name histidinol-phosphatase), which could not be identified by similarity to previously known enzymes with this activity. Incidentally, this is the case not only for the TPEs, but also for all other *Betaproteobacteria* present in the KEGG databases at the time. Interestingly, there are currently several different gene families responsible for HPP activity (Mormann *et al.*, 2006; Petersen *et al.*, 2010), and one of these families involves inositol monophosphatase-like genes. Although the TPEs do not present any other inositol metabolism genes, two genes (BCUE_0333 and BCUE_0385) with inositol monophosphatase-like domains can be identified in the genome. It is possible to speculate that at least one of these genes could be a reasonable candidate for the 'missing' HPP in *Betaproteobacteria*.

27.4.4 Isoleucine, leucine and valine

The branched chain amino acids are essential amino acids that cannot be synthesized by most Trypanosomatidae, except for the SHTs. The biosynthesis of isoleucine, leucine and valine in SHTs is a case of labour division between the eukaryotic and bacterial components of the system. The initial steps, such as the conversion of threonine into oxobutanoate (by threonine ammonia-lyase, EC 4.3.1.19) or the synthesis of 2-(α-hydroxyethyl) thiamine diphosphate from pyruvate (by an acetyl-transferring pyruvate dehydrogenase, EC 1.2.4.1), can be catalysed by enzymes from either the SHT or TPE genomes. Given that the genes for these steps are also present in RT genomes, it is clear that these compounds are also being used in other metabolic processes unrelated to Ile, Leu and Val biosynthesis. Recent biochemical studies (Manchola *et al.*, 2015) have characterized branched chain amino acid transport into trypanosomatid cells, which is essential given the importance of these amino acids not only in protein synthesis but also in processes such as energy generation, the regulation of cellular differentiation and the biosynthesis of sterols and fatty acids.

All of the other steps in the synthesis of these amino acids except the very last can be catalysed only by enzymes whose genes are in the TPE genomes. The final enzyme, branched-chain amino acid aminotransferase (EC 2.6.1.42), can be detected only in the trypanosomatid and not in the TPE genome. Differently from the situation described above for histidinol-phosphate phosphatase, this enzyme can be identified, according to KEGG data, in most of the *Betaproteobacteria* examined. A significant exception is *Taylorella* spp., parasitic bacteria which are the closest known relatives of *Ca.* Kinetoplastibacterium (Alves *et al.*, 2013a). This could mean that both of these genera of bacteria either do not possess this gene, and depend on the host for that final step of Ile, Leu and Val synthesis, or do possess one or more novel genes that catalyse this step in the pathways that could not be detected by similarity searches.

The fact that the SHT expresses the final protein of the biosynthetic pathways of these amino acids is an indication of its role in controlling their production, and is reminiscent of the situation described in a very different endosymbiotic system, that of the aphid *Acyrthosiphon pisum* and its endosymbiont *Buchnera aphidicola* (which lives in special intestinal cells called bacteriocytes). In this case, as in the SHT–TPE system, the host possesses the final enzymes of the pathways, but the endosymbiont lacks them (Macdonald *et al.*, 2012). These enzymes have been shown to be upregulated in bacteriocytes, which would be a way for the host to control amino acid production (Hansen and Moran, 2011). Whether this also occurs in the case of the SHTs and their endosymbionts is something that can only be determined by future experimental data.

27.4.5 Lysine

The biosynthesis of lysine (see Fig. 27.1) is almost as simple a case as that of the histidine pathway, with almost the whole pathway being catalysed by enzymes encoded in the TPE genome. The eight steps from aspartate to *meso*-2,6-diaminopimelate (DAP) are performed solely by these bacterial enzymes, while the final step to generate lysine can be performed by either an enzyme from the TPE or from the SHT nucleus. Accordingly, earlier enzymatic and radioactive tracer experiments had already shown that there is no need to supplement the culture medium with lysine if DAP is present (Gutteridge, 1969). The gene responsible for this activity, diaminopimelate decarboxylase (EC 4.1.1.20), has some evidence of having been horizontally transferred from a bacterium into the genome of the trypanosomatid ancestor, as it is present both in SHTs and RTs.

27.4.6 Phenylalanine, tyrosine and tryptophan

Similar to the lysine pathway, the phenylalanine, tyrosine and tryptophan biosynthetic pathways comprise almost completely TPE enzymes, with only the enzymes for the final steps being present in the trypanosomatid genomes. Thus, the whole pathway from D-erythrose 4-phosphate and phosphoenolpyruvate (PEP) until chorismate (see Fig. 27.1), the common substrate for the synthesis of these three aromatic amino acids, is performed solely by TPE enzymes.

The next five steps of the tryptophan biosynthesis pathway are also catalysed by TPE enzymes, but the final step presents a more interesting situation. The corresponding enzyme, tryptophan synthase (EC 4.2.1.20), is actually a tetramer comprising two instances each of two different subunits (alpha and beta). The alpha subunit is present only in the TPE genome but not in any of the trypanosomatid genomes examined, whereas the beta subunit is present in all of the organism groups involved (TPEs, SHTs and at least one RT, *Herpetomonas muscarum*). Hence, the complete enzymatic complex is encoded in the TPE genomes, and only the second part (which condenses indole and serine to form tryptophan) is present in the SHT. The exact significance of this is not clear, although it might represent a mechanism for the trypanosomatid to regulate at least part of tryptophan production by controlling the final step of the pathway. Interestingly, the gene for the beta subunit present in the trypanosomatids is very similar to the corresponding enzyme from many *Bacteroidetes* bacteria, clustering robustly with this phylum in the phylogeny (Alves *et al.*, 2013b), so indicating a relatively recent transfer or, alternatively, a very conserved, negatively selected gene sequence.

The part of the pathway leading to Phe and Tyr mostly comprises TPE enzymes, again with the exception of some of the final steps in the pathway. Phenylalanine can actually be synthesized in two different ways, one of which involves only TPE enzymes – a generic aromatic amino acid aminotransferase (EC 2.6.1.57) and prephenate dehydratase (EC 4.2.1.51). Tyrosine can be readily obtained from Phe by the action of phenylalanine-4-hydroxylase (EC 1.14.16.1), an enzyme that is present in all Trypanosomatidae examined. It is, however, unclear whether Phe could supplement for Tyr in culture media, because no previous detailed nutritional or enzymatic data concerning these compounds are available.

27.4.7 Threonine

Threonine, like methionine, is another amino acid whose inclusion in the list of essential compounds for Trypanosomatidae was not formerly completely settled in the literature (see Alves *et al.*, 2013b, for a list of references). Also, as in the Met biosynthesis case, our genomic analysis data favours the possibility that threonine is not completely essential to trypanosomatids, but rather that its supplementation greatly improves the growth of these organisms in culture. Given the importance of threonine for many metabolic processes, including the synthesis of isoleucine, this is a reasonable hypothesis.

We have found two possible pathways for the biosynthesis of threonine in SHTs. One involves the presence of a gene encoding for threonine aldolase (EC 4.1.2.5), which directly interconverts glycine (plus acetaldehyde) and threonine. Besides being in SHTs, this gene can also be found in some RTs, such as *Leishmania major* and *Crithidia acanthocephali*, but not in *Herpetomonas muscarum* or *Trypanosoma* spp. In this case, the evolutionary history of the gene is different depending on the trypanosomatid group: the *Leishmania*/*Crithidia* orthologue clusters deep within *Firmicutes* bacteria, while the SHT orthologue seems to be basal to a large assemblage of both eukaryotic and bacterial lineages. In this case, we can safely draw a conclusion on the HGT status of the gene in the former group, but not in the latter group. This indicates a relatively recent transfer from a *Firmicutes* bacterium to the common ancestor of *Leishmania* and *Crithidia*, which are known to be closely related.

The other pathway for threonine synthesis is more intricate, utilizing part of the DAP pathway (of lysine biosynthesis) which, as mentioned above, is only present in the TPE genomes. The central enzyme in this pathway, which is also involved in a number of other amino acid biosynthetic pathways, is homoserine dehydrogenase (EC 1.1.1.3). This is present in all of the genomes examined and, in the Trypanosomatidae, it has originated in an HGT event from a *Firmicutes* bacterium – it clusters strongly with *Lysinibacillus* and *Solibacillus* (Alves *et al.*, 2013b). Conversely, the final two enzymes in this pathway, also present in all of the SHT and RT genomes examined, seem to be of eukaryotic origin.

Thus, though there is one pathway that could supply the trypanosomatid with threonine without the need for an endosymbiont, the synthesis of this amino acid by the SHT could be significantly increased by TPE activity, which produces a direct metabolic precursor (aspartate 4-semialdehyde) of the second possible pathway, configuring yet another instance of close collaboration between the trypanosomatid and its endosymbiont in the synthesis of amino acids.

27.5 Non-essential Amino Acid Biosynthesis

Non-essential amino acids in trypanosomatids are those that have been shown, in nutritional and enzymatic studies, not to be necessary in a defined culture medium for these organisms. There are many fewer non-essential amino acids in trypanosomatids than there are of the essential amino acids examined above. In this section, the metabolic interaction between the SHT and TPE genomes, as well as the genetic contribution of the HGT to the non-essential amino acid biosynthetic pathways of these organisms, are analysed in further detail.

27.5.1 Glycine and serine

Glycine (plus 5,10-methylenetetrahydrofolate) and serine (plus tetrahydrofolate) can be easily interconverted (see Fig. 27.1) by an enzyme, namely glycine hydroxymethyltransferase (EC 2.1.2.1) that is present in both the SHT and TPE genomes. Serine might also be synthesized directly from pyruvate, in a reaction catalysed by threonine ammonia-lyase (EC 4.3.1.19, mentioned above), which reportedly can also perform the conversion of pyruvate plus ammonia into serine. Given that all of these enzymes are present in both the SHT and TPE genomes, it seems clear that serine is easily synthesized.

27.5.2 Alanine, aspartate and asparagine

The synthesis of alanine is promptly achieved in Trypanosomatidae by the action of alanine transaminase (EC 2.6.1.2), the gene for which is present in the trypanosomatid but not the TPE genome. This enzyme catalyses the conversion of pyruvate (plus glutamate) directly into alanine (plus oxoglutarate), and vice versa, and is therefore responsible for the alanine (or Cahill) cycle.

Aspartate, which is an important substrate for the urea cycle (see Section 27.4.1), can be synthesized (together with α-ketoglutarate) by the action of aspartate transaminase (EC 2.6.1.1) from oxaloacetate (plus glutamate); the reaction is reversible. Just like alanine transaminase, the gene for aspartate transaminase is also present in trypanosomatid but not in TPE genomes.

In the case of both alanine and aspartate transaminases, similarity searches suggest that the genes are of eukaryotic origin, as they are very similar to the corresponding putative orthologues from many other eukaryotic lineages.

Another source of aspartate is the activity of the enzyme asparaginase (EC 3.5.1.1), which converts asparagine into aspartate. The gene encoding this enzyme is present in the trypanosomatid but not in the TPE genome, and the phylogeny shows that it is most likely a gene of eukaryotic origin.

Asparagine can be obtained in the reverse reaction, from aspartate (plus ammonia), by aspartate-ammonia ligase (EC 6.3.1.1), which is also present in trypanosomatid but not the TPE genome. Interestingly, this trypanosomatid gene seems to have been acquired in an ancient HGT event, because it is present in a wide range of trypanosomatids and is highly similar to the corresponding bacterial orthologues. In the phylogeny of this gene, the Trypanosomatidae group altogether in a pattern that generally reflects the family's currently accepted phylogeny. The trypanosomatid clade as a whole groups strongly with bacteria from several lineages, mainly large assemblies of *Gammaproteobacteria*.

In the case of the three amino acid biosynthetic pathways discussed here, it can be seen that the endosymbiont has retained only those genes necessary for its metabolic collaboration with the eukaryotic host, and has lost even the genes for non-essential, easy-to-synthesize amino acids such as Ala, Asp and Asn. The genetic contribution from the bacterial to the trypanosomatid metabolism is still far from negligible though, as both genes involved in the interconversion of asparagine and aspartate seem to have been acquired, by a distant ancestor to all Trypanosomatidae, from one or more gammaproteobacterial lineages.

27.5.3 Proline

Proline is a very important amino acid for trypanosomatids, and can be readily used (both as D-proline and as L-proline) as an energy source. It is also the most abundant amino acid in the tsetse fly midgut and haemolymph environments (Bursell *et al.*, 1973; Balogun, 1974). Although it does not involve the SHTs that are our focus here, it is interesting to notice that many species of the genus *Trypanosoma* possess a proline racemase gene (EC 5.1.1.4), which is able to interconvert D-proline and L-proline. This gene has been horizontally transferred from a *Firmicutes* bacterium to a common ancestor to all *Trypanosoma* spp. investigated (being later lost those from the *Trypanozoon* and *Nannomonas* subgenera), but not to other trypanosomatid genera (Caballero *et al.*, 2015). Not only does this once more show the important role of bacterial genes in amino acid metabolism in these organisms, but also, in this case, the borrowing of a tool used in host defence evasion by the *Trypanosoma* parasites, for when the enzyme proline racemase is released it causes indiscriminate B cell activation, which leads to immunosuppression and thus to survival of the parasite in the vertebrate host.

Proline can be synthesized in three enzymatic steps from glutamate using the following enzymes: glutamate 5-kinase (EC 2.7.2.11), glutamate-5-semialdehyde dehydrogenase (EC 1.2.1.41) and pyrroline-5-carboxylate reductase (EC 1.5.1.2), which is, as detailed below, derived from an HGT event. Genes for all three enzymes are present in the trypanosomatid nuclear genome, but there are none in the TPE genome.

Proline can also be obtained from ornithine, in one step, by the enzyme ornithine cyclodeaminase (EC 4.3.1.12). The gene for this enzyme is present exclusively in SHT nuclear genomes, being absent from both the TPE and RT genomes that have been examined to date. Its phylogeny indicates that this gene is a strong candidate for having been transferred from the TPE genome to the SHT nuclear genome as it clusters within the *Alcaligenaceae* family of the *Betaproteobacteria*.

27.5.4 Glutamine and glutamate

Glutamine can be obtained from glutamate (plus ammonia) in SHTs by the action of glutamine synthetase (EC 6.3.1.2), which is present in the SHT but not the TPE genome. The reverse – conversion of glutamine into glutamate – seems not to be possible, however, as neither organism possesses one of the enzymes catalysing the conversion of glutamine into glutamate plus ammonia – glutaminase (EC 3.5.1.2) or glutamin-(asparagin-)ase (EC 3.5.1.38; also known as glutaminase-asparaginase). The latter enzyme, if present, would also have been able to act in the conversion of asparagine into aspartate, mentioned above as performed by a different, more specific enzyme.

Glutamate can be synthesized from proline or from 2-oxoglutarate (from the citrate cycle). In the first case, three steps and two enzymes are involved: in the first, pyrroline-5-carboxylate reductase (EC 1.5.1.2) catalyses the conversion of proline into 1-pyrroline-5-carboxylate; the second step is the spontaneous conversion of this compound into glutamate 5-semialdehyde which, in the third step, is finally converted into glutamate by the action of 1-pyrroline-5-carboxylate dehydrogenase (EC 1.2.1.88). This last enzyme is also capable of using 1-pyrroline-5-carboxylate directly as a substrate to generate glutamate. The genes encoding each of these enzymes are present in the trypanosomatid but not the TPE genome. Hence, the bacterium is probably fully dependent on its host for its supply of both glutamate and glutamine.

Phylogenetically, the trypanosomatid gene for EC 1.5.1.2 seems to have originated by HGT from a bacterial group to a common ancestor of all trypanosomatids. The Trypanosomatidae clade in the phylogenetic tree generally agrees with the current consensus on the intra-familial relationships, with the clade itself clustering strongly within a group whose sister taxa are large assemblages of mainly *Firmicutes* and *Gammaproteobacteria*, on the opposite side of the tree from other Eukaryota. An interesting observation is the position of *Bodo saltans*, a free-living non-trypanosomatid kinetoplastid, which clusters very distantly from the Trypanosomatidae and robustly within a group of many other eukaryotes. The corresponding *Bodo saltans* genomic scaffold presents three different paralogues of this gene, organized in tandem, which form a monophyletic group in the tree.

These phylogenetic data indicate that there might have been a gene transfer of EC 1.5.1.2 from bacteria to the common ancestor of all Trypanosomatidae, while *Bodo saltans* retained the eukaryotic version that was probably present in the *Ca.* Kinetoplastea ancestor.

27.6 Conclusions

The study of amino acid biosynthesis in symbiont-harbouring trypanosomatids has a long history, involving methods ranging from biochemical and nutritional experiments all the way to genomic and *in silico* predictions, and back to confirmatory (or contradictory) wet laboratory data. It is clear from this quick exposition of the SHT–TPE metabolic interactions involving amino acid production that genes have been chiselled away by evolution from the TPE genomes, with only those necessary for the supplementation of the host's metabolic capabilities kept. So most of the amino acid biosynthetic genes still present in the TPE genomes participate in pathways for compounds that are essential for the SHT, with much less presence, if any, of genes for amino acids that can be readily produced by the trypanosomatid without TPE input.

It is also clear that the importance of the bacterial contribution to the metabolism of Trypanosomatidae as a whole goes way beyond the enzymatic steps provided by *Ca.* Kinetoplastibacterium to their Strigomonadinae hosts. Many genes have been transferred horizontally from probably several different bacterial lineages into the ancestors of some, or in some cases probably all, of the Trypanosomatidae. Besides previously published examples dealing with HGT in essential amino acid biosynthesis pathways, we have also seen here that at least a few enzymes participating in non-essential amino acid pathways seem to have also originated from HGT events. Given that a systematic analysis of all such genes has not yet been performed, the findings presented herein are very likely an underestimate of the actual amount of such genes. The bacterial origin of these genes is interesting not only from an evolutionary science viewpoint; it can also provide important hints into the rise and adaptation of parasites as well as significant targets for drug development, because HGT-derived enzymes that are present exclusively in the

Trypanosomatidae have a much greater chance of being mechanistically different from their eukaryotic counterparts. Thus, disruption of the activities of these enzymes might harm the parasite, but not the vertebrate host.

In short, these and many such genomic evolution studies only emphasize the famous quote from Theodosius Dobzhansky that 'nothing in Biology makes sense except in the light of evolution' (Dobzhansky, 1973).

References

Adl, S.M., Simpson, A.G.B., Lane, C.E., Lukeš, J., Bass, D. Bowser, S.S., Brown, M.W., Burki, F., Dunthorn, M., Hampl, V. *et al.* (2012) The revised classification of eukaryotes. *The Journal of Eukaryotic Microbiology* 59, 429–493.

Alfieri, S.C. and Camargo, E.P. (1982) Trypanosomatidae: isoleucine requirement and threonine deaminase in species with and without endosymbionts. *Experimental Parasitology* 53, 371–380.

Alves, J.M.P. and Buck, G.A. (2007) Automated system for gene annotation and metabolic pathway reconstruction using general sequence databases. *Chemistry and Biodiversity* 4, 2593–2602.

Alves, J.M.P., Voegtly, L., Matveyev, A.V., Lara, A.M., da Silva, F.M., Serrano, M.G., Buck, G.A., Teixeira, M.M.G. and Camargo, E.P. (2011) Identification and phylogenetic analysis of heme synthesis genes in trypanosomatids and their bacterial endosymbionts, *PLoS ONE* 6(8): e23518.

Alves, J.M.P., Serrano, M.G., da Silva, F.M., Voegtly, L.J., Matveyev, A.V., Teixeira, M.M., Camargo, E.P. and Buck, G.A. (2013a) Genome evolution and phylogenomic analysis of *Candidatus* Kinetoplastibacterium, the betaproteobacterial endosymbionts of *Strigomonas* and *Angomonas*. *Genome Biology and Evolution* 5, 338–350.

Alves, J.M.P., Klein, C.C., da Silva, F.M., Costa-Martins, A.G., Serrano, M.G., Buck, G.A., Vasconcelos, A.T.R., Sagot, M.-F., Teixeira, M.M.G., Motta, M.C.M. and Camargo, E.P. (2013b) Endosymbiosis in trypanosomatids: the genomic cooperation between bacterium and host in the synthesis of essential amino acids is heavily influenced by multiple horizontal gene transfers. *BMC Evolutionary Biology* 13:190.

Balogun, R.A. (1974) Studies on the amino acids of the tsetse fly, *Glossina morsitans*, maintained on *in vitro* and *in vivo* feeding systems. *Comparative Biochemistry and Physiology. A, Comparative Physiology* 49(2A), 215–222.

Berriman, M., Ghedin, E., Hertz-Fowler, C., Blandin, G., Renauld, H. *et al.* (2005) The genome of the African trypanosome *Trypanosoma brucei*. *Science* 309, 416–422.

Bursell, E., Billing, K.J., Hargrove, J.W., McCabe, C.T. and Slack, E. (1973) The supply of substrates to the flight muscle of tsetse flies. *Transactions of the Royal Society of Tropical Medicine and Hygiene* 67, 296.

Caballero, Z.C., Costa-Martins, A.G., Ferreira, R.C., Alves, J.M.P., Serrano, M.G., Camargo, E.P., Buck, G.A., Minoprio, P. and Teixeira, M.M.G. (2015) Phylogenetic and syntenic data support a single horizontal transference to a *Trypanosoma* ancestor of a prokaryotic proline racemase implicated in parasite evasion from host defences. *Parasites and Vectors* 8:222.

Chang, K.P., Chang, C.S. and Sassa, S. (1975) Heme biosynthesis in bacterium-protozoon symbioses: enzymic defects in host hemoflagellates and complemental role of their intracellular symbiotes. *Proceedings of the National Academy of Sciences of the United States of America* 72, 2979–2983.

de Souza, W. and Motta, M.C. (1999) Endosymbiosis in protozoa of the Trypanosomatidae family. *FEMS Microbiology Letters* 173, 1–8.

Dobzhansky, T. (1973) Nothing in biology makes sense except in the light of evolution. *The American Biology Teacher* 35, 125–129.

Edgar, R.C. (2004) MUSCLE: multiple sequence alignment with high accuracy and high throughput. *Nucleic Acids Research* 32, 1792–1797.

El-Sayed, N.M., Myler, P.J., Bartholomeu, D.C., Nilsson, D., Aggarwal, G. Tran, A.-N., Ghedin, E., Worthey, E.A., Delcher, A.L., Blandin, G, *et al.* (2005) The genome sequence of *Trypanosoma cruzi*, etiologic agent of Chagas disease. *Science* 309, 409–415.

Giordana, L., Mantilla, B.S., Santana, M., Silber, A.M. and Nowicki, C. (2014) Cystathionine γ-lyase, an enzyme related to the reverse transsulfuration pathway, is functional in *Leishmania* spp. *Journal of Eukaryotic Microbiology* 61, 204–213.

Gutteridge, W.E. (1969) Presence and properties of diaminopimelic acid decarboxylases in the genus *Crithidia*. *Biochimica et Biophysica Acta* 184, 366–373.

Hansen, A.K. and Moran, N.A. (2011) Aphid genome expression reveals host-symbiont cooperation in the production of amino acids. *Proceedings of the National Academy of Sciences of the United States of America* 108, 2849–2854.

Hoare, C.A. (1972) *Herpetosoma* from man and other mammals. In: Hoare, C.A. *The Trypanosomes of Mammals: A Zoological Monograph*. Blackwell Scientific, Oxford, UK pp. 288–314.

Ivens, A.C., Peacock, C.S., Worthey, E.A., Murphy, L., Aggarwal, G., Berriman, M., Sisk, E., Rajandream, M.-A., Adlem, E., Aert, R. *et al.* (2005) The genome of the kinetoplastid parasite, *Leishmania major*. *Science* 309, 436–442.

Klein, C.C., Alves, J.M.P., Serrano, M.G., Buck, G.A., Vasconcelos, A.T.R. Sagot, M.F., Teixeira, M.M., Camargo, E.P. and Motta, M.C. (2013) Biosynthesis of vitamins and cofactors in bacterium-harbouring trypanosomatids depends on the symbiotic association as revealed by genomic analyses. *PLoS ONE* 8(11): e79786.

Korený, L., Lukes, J. and Oborník, M. (2010) Evolution of the haem synthetic pathway in kinetoplastid flagellates: An essential pathway that is not essential after all? *International Journal for Parasitology* 40, 149–156.

Kraeva, N., Butenko, A., Hlaváčová, J., Kostygov, A., Myškova, J., Grybchuk, D., Leštinová, T., Votýpka, J., Volf, P., Opperdoes, F. *et al.* (2015) *Leptomonas seymouri*: adaptations to the dixenous life cycle analyzed by genome sequencing, transcriptome profiling and co-infection with *Leishmania donovani*. *PLoS Pathogens* 11(8): e1005127.

Lukeš, J., Skalický, T., Týč, J., Votýpka, J. and Yurchenko, V. (2014) Evolution of parasitism in kinetoplastid flagellates. *Molecular and Biochemical Parasitology* 195, 115–122.

Lwoff, M. (1940) *Recherches sur le Pouvoir de Synthèse des Flagellés Trypanosomides*. Masson et Cie, Paris.

Macdonald, S.J., Lin, G.G., Russell, C.W., Thomas, G.H. and Douglas, A.E. (2012) The central role of the host cell in symbiotic nitrogen metabolism. *Proceedings of the Royal Society B* 279, 2965–2973.

Manchola, N.C., Rapado, L.N., Barisón, M.J. and Silber, A.M. (2015) Biochemical characterization of branched chain amino acids uptake in *Trypanosoma cruzi*. *Journal of Eukaryotic Microbiology* 63, 299–308.

Mormann, S., Lömker, A., Rückert, C., Gaigalat, L., Tauch, A., Pühler, A. and Kalinowski, J. (2006) Random mutagenesis in *Corynebacterium glutamicum* ATCC 13032 using an IS6100-based transposon vector identified the last unknown gene in the histidine biosynthesis pathway. *BMC Genomics* 7:205.

Motta, M.C.M., Catta-Preta, C.M.C., Schenkman, S., de Azevedo Martins, A.C., Miranda, K.. de Souza, W. and Elias, M.C. (2010) The bacterium endosymbiont of *Crithidia deanei* undergoes coordinated division with the host cell nucleus. *PloS ONE* 5(8): e12415.

Motta, M.C.M., Martins, A.C. de A., de Souza, S.S., Catta-Preta, C.M.C., Silva, R., Klein, C.C., de Almeida, L.G.P., de Lima Cunha, O., Ciapina, L.P., Brocchi, M. *et al.* (2013) Predicting the proteins of *Angomonas deanei*, *Strigomonas culicis* and their respective endosymbionts reveals new aspects of the Trypanosomatidae family. *PLoS ONE* 8(4): e60209.

Mundim, M.H., Roitman, I., Hermans, M.A. and Kitajima, E.W. (1974) Simple nutrition of *Crithidia deanei*, a reduviid trypanosomatid with an endosymbiont. *The Journal of Protozoology* 21, 518–521.

Newton, B.A. (1956) A synthetic growth medium for the trypanosomid flagellate *Strigomonas (Herpetomonas) oncopelti*. *Nature* 177, 279–280.

Ogata, H., Goto, S., Sato, K., Fujibuchi, W., Bono, H. and Kanehisa, M. (1999) KEGG: Kyoto encyclopedia of genes and genomes. *Nucleic Acids Research* 27, 29–34.

Petersen, L.N., Marineo, S., Mandalà, S., Davids, F., Sewell, B.T. and Ingle, R.A. (2010) The missing link in plant histidine biosynthesis: *Arabidopsis* myoinositol monophosphatase-like2 encodes a functional histidinol-phosphate phosphatase. *Plant Physiology* 152, 1186–1196.

Price, M.N., Dehal, P.S. and Arkin, A.P. (2010) FastTree 2 – Approximately maximum-likelihood trees for large alignments. *PLoS ONE* 5(3): e9490.

Roitman, I. and Camargo, E.P. (1985) Endosymbionts of trypanosomatidae. *Parasitology Today* 1, 143–144.

Roitman, C., Roitman, I. and de Azevedo, H.P. (1972) Growth of an insect trypanosomatid at 37 C in a defined medium. *The Journal of Protozoology* 19, 346–349.

Runckel, C., Derisi, J. and Flenniken, M.L. (2014) A draft genome of the honey bee trypanosomatid parasite *Crithidia mellificae*. *PloS ONE* 9(4): e95057.

Suzek, B.E., Huang, H., McGarvey, P., Mazumder, R. and Wu, C.H. (2007) UniRef: comprehensive and non-redundant UniProt reference clusters. *Bioinformatics* 23, 1282–1288.

Teixeira, M.M.G., Borghesan, T.C., Ferreira, R.C., Santos, M.A., Takata, C.S.A., Campaner, M., Nunes, V.L., Milder, R.V., de Souza, W. and Camargo, E.P. (2011) Phylogenetic validation of the genera *Angomonas* and *Strigomonas* of trypanosomatids harboring bacterial endosymbionts with the description of new species of trypanosomatids and of proteobacterial symbionts. *Protist* 162, 503–524.

Vickerman, K. (1976) The diversity of the kinetoplastid flagellates. In: Lumsden, W.H.R. and Evans, D.A. (eds) *Biology of the Kinetoplastida*. Academic Press, London/New York/San Francisco, pp. 1–34.

Vickerman, K. (1994) The evolutionary expansion of the trypanosomatid flagellates. *International Journal for Parasitology* 24, 1317–1331.

Votýpka, J., Kostygov, A.Y., Kraeva, N., Grybchuk-Ieremenko, A., Tesařová, M. Grybchuk, D., Lukeš, J. and Yurchenko, V. (2014) *Kentomonas* gen. n., a new genus of endosymbiont-containing trypanosomatids of Strigomonadinae subfam. n. *Protist* 165, 825–838.

Yoshida, N., Jankevicius, J.V., Roitman, I. and Camargo, E.P. (1978) Enzymes of the ornithine-arginine metabolism of trypanosomatids of the genus *Herpetomonas*. *Journal of Eukaryotic Microbiology* 25, 550–555.

28 Amino Acid Metabolism in Helminths

H.V. Simpson[1]* and S. Umair[2]

[1]*Institute of Veterinary, Animal and Biomedical Sciences, Massey University, Palmerston North, New Zealand;* [2]*AgResearch Ltd, Palmerston North, New Zealand*

28.1 Abstract

Metabolism of amino acids is better known in nematodes than in other helminths, particularly in parasites of mammals, including humans. There are differences between species and also between life cycle stages, so the presence of genes encoding their proteins does not ensure functional enzymes, e.g. there are differences in the expression of glutamate synthase, arginine decarboxylase and Δ^1-piperideine-carboxylate reductase between the two closely related species or life cycle stages of sheep abomasal nematode parasites. The main differences in their metabolism of amino acids from that in other animals are in properties of individual enzymes, rather than the absence or presence of pathways. Unusual features are the lack a full ornithine-urea cycle in many helminths, the presence of creatinase activity, the synthesis of polyamines from either ornithine or agmatine and the incorporation of ammonia into glutamate.

28.2 Introduction

Cestodes (tapeworms) of the phylum Cestoda, trematodes (flukes and schistosomes) of the phylum Platyhelminthes and roundworms of the phylum Nematoda, although anatomically very different, are generally grouped as helminths. All cestodes are parasitic, but there are both parasitic and free-living species of trematodes and nematodes. Trematodes have both an absorptive tegument and an incomplete gut (Asch and Read, 1975; Hanna, 1980; Pappas, 1988), whereas nematodes have digestive tracts and absorption occurs across their outer cuticles, which are structurally unlike typical absorptive surfaces (Bird and Bird, 1981; Page, 2001), except in a few species in which there are microvilli on the cuticle (Riding, 1970). Adult cestodes consist of a head (the scolex) and segmental proglottids generated by the scolex, which mature as they move distally as new proglottids are formed and break off at the posterior end. As there is no digestive tract, absorption occurs only across the tegument, which resembles a brush border (Pappas *et al.*, 1973).

28.3 Overview

Since the extensive reviews of helminth nitrogen metabolism of McManus (1987) and Barrett (1981, 1983, 1991), most of the work on this group has been carried out on parasitic nematodes, with a focus on identifying novel anthelmintic targets. Biochemical studies of a limited range

*Corresponding author. E-mail address: H.V.Simpson@massey.ac.nz

of species, supplemented with annotated gene sequences in public databases, have resulted in only a partial picture of amino acid metabolism in helminths. Further, even closely related species have shown notable differences in enzyme properties and there are examples of enzymes not being expressed at some life cycle stages.

Proteolytic enzymes on the intestinal brush border and/or secreted by many parasitic nematodes of plants (Vanholme *et al.*, 2004) and animals (Sajid and McKerrow, 2002) break proteins down into peptides and amino acids. The uptake of nitrogenous compounds in helminths can be as ammonium, peptides (Meissner *et al.*, 2004) and amino acids (Isseroff *et al.*, 1976; Jeffs and Arme, 1985, 1987; Veljkovic *et al.*, 2004) and also through the γ-glutamyl cycle (Dass and Donahue, 1986; Abidi and Nizami, 1995). Excess nitrogen is excreted by free-living and parasitic nematodes as approximately 80% ammonia and 20% urea (Rogers, 1952; Wright, 1975a; Simpson *et al.*, 2009).

Amino acids are commonly oxidized (Grantham and Barrett; 1986a; Umair *et al.*, 2011c, 2012a) or decarboxylated (Singh *et al.*, 1983; Umair *et al.*, 2011b) by nematodes, and helminth aminotransferases are able to transaminate a large number of L-amino acids, usually using the glutamate to 2-oxoglutarate (α-ketoglutarate) system (Singh and Srivastava, 1983; Barrett, 1991; Walker and Barrett, 1991a,b). Amino acids contribute to nematode energy metabolism (Bruce *et al.*, 1972; Davies and Köhler, 1990) (Fig. 28.1) and some species are reported to survive *in vitro* with glutamine or alanine as the sole carbon source. The malate synthase–isocitrate lyase glyoxylate cycle in nematodes (Rothstein and Mayoh, 1964, 1966; Barrett *et al.*, 1970; Grantham and Barrett, 1986b) allows all amino acids to be glucogenic (see Fig. 28.1). Early studies of the interconversion of radiolabelled amino acids suggested there may be no essential amino acids for nematodes (Rothstein and Tomlinson, 1961; Rothstein, 1965); however, *in vitro* cultivation of helminths requires supplementation with the L-amino acids arginine, isoleucine, tryptophan, valine and histidine (Brockelman and Jackson, 1978; Hata, 1994), and also leucine, lysine, methionine, phenylalanine and threonine (Hata, 1994). The incorporation of ammonia via the glutamine synthetase (GS) and glutamate synthase (GOGAT) – the GO-GOGAT pathway (Umair *et al.*, 2011a) may reduce the number of amino acids that is essential for parasitic nematodes.

28.4 Glutamate

Glutamate is incorporated into proteins, is involved in the synthesis of other amino acids, is a

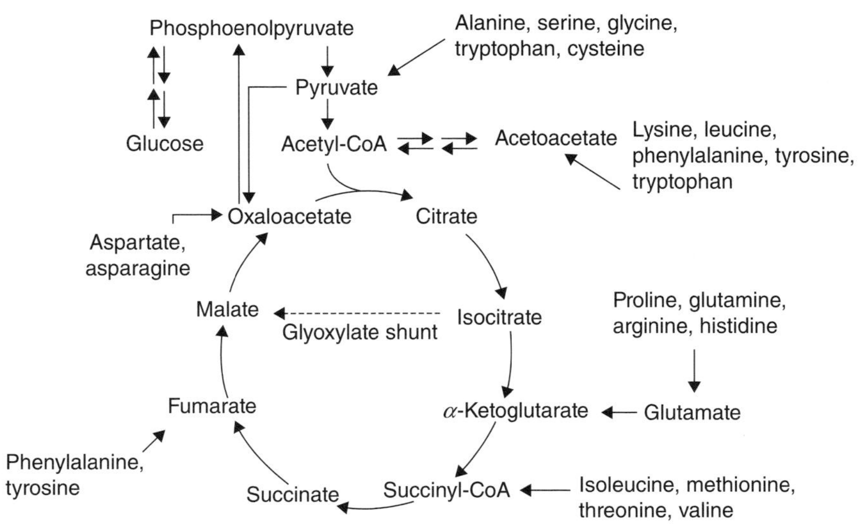

Fig. 28.1. Entry points of amino acids into energy metabolism. All amino acids are glucogenic, at least in nematodes, because of the presence of a functional glyoxylate cycle.

source of excreted ammonia and is also a link between nitrogen and energy metabolism through the interconversion of glutamate and 2-oxoglutarate which is catalysed by glutamate dehydrogenase (GDH) and a large group of transaminases (Barrett, 1991).

28.4.1 Glutamate dehydrogenase

GDH catalyses the reversible oxidative deamination of glutamate to ammonium and 2-oxoglutarate and can either function in ammonia assimilation or in the generation of ammonia and 2-oxoglutarate (Goldin and Frieden, 1971; Hudson and Daniel, 1993). GDHs appear to be universally present in helminths and many can use either NAD^+ or NADP, although not with equal efficiency (Grantham and Barrett, 1986a; Muhamad *et al.*, 2011, Umair *et al.*, 2011d). The allosteric regulation of *Haemonchus contortus* and *Teladorsagia circumcincta* GDH differed from that of mammalian GDH: ADP and ATP were more stimulatory, GTP was strongly inhibitory and ATP stimulated amination by *H. contortus* GDH more markedly than it did deamination (Umair *et al.*, 2012b).

28.4.2 Glutamine synthetase (GS)–glutamate synthase (GOGAT)

GS, a universal enzyme which catalyses the formation of glutamine from glutamate and ammonia, has been described in *Heligmosomoides polygyrus*, *Panagrellus redivivus* (Grantham and Barrett, 1988) and both L3 larvae and adult *T. circumcincta* (Muhamad, 2006). GOGAT, which catalyses the formation of two glutamate molecules from glutamine and 2-oxoglutarate (Vanoni and Curti, 1999), is present in plants, microorganisms and some insects (Hirayama *et al.*, 1998; Scaraffia *et al.*, 2005), in which it supports assimilation of ammonia if present at low concentrations and provided that ATP is available (Fisher and Sonenshein, 1991, Miflin and Habash, 2002). In contrast, GDH is able to assimilate ammonia only when it is present at higher concentrations (Helling, 1998). *In vitro*, there was reuptake by L3 *T. circumcincta* of excreted ammonia (Simpson *et al.*, 2009) and GOGAT activity was increased by the incubation of L3 with ammonia for a similar time period (Umair *et al.*, 2011a). GOGAT activity has been detected in homogenates of L3 and adult *T. circumcincta* (Muhamad, 2006; Umair *et al.*, 2011a) and in adult, but not L3, *H. contortus* (Umair *et al.*, 2011a), and databases contain GOGAT gene sequences for free-living and many parasitic nematodes.

28.4.3 Glutaminase

Glutaminase hydrolyses glutamine to glutamate and ammonia. High glutaminase activity was detected in *T. circumcincta* (Muhamad, 2006), low activity in *H. polygyrus* and *P. redivivus* (Grantham and Barrett, 1986a), but was undetectable in *Litomosoides carinii* (Davies and Köhler, 1990).

28.4.4 GABA (γ-aminobutyrate) shunt

The nematode neurotransmitter GABA (Feng *et al.*, 2002; Schuske *et al.*, 2004) is formed by the irreversible decarboxylation of L-glutamate by γ-aminobutyrate decarboxylase, the first step of the three-enzyme GABA shunt, which forms succinate from glutamate and bypasses part of the tricarboxylic acid (TCA) cycle (Balázs *et al.*, 1970). The other two enzymes are GABA transaminase (GABA-T) and succinic semialdehyde dehydrogenase. Although they have been demonstrated in specific tissues (Monteoliva *et al.*, 1965; Singh *et al.*, 1983), all enzymes of the shunt were detected in the homogenates of only some parasitic helminths, particularly larger species living in more anaerobic environments (Monteoliva *et al.*, 1965; Rasero *et al.*, 1968; MacKenzie *et al.*, 1989), suggesting that the GABA shunt may not be important in the metabolism of many helminths.

28.5 Proline

Proline required for the synthesis of collagen, an important component of the cuticle, can either be synthesized from ornithine or glutamate, or obtained directly from the host. A fully functional mammalian-type ornithine-proline-glutamate pathway has been demonstrated in many helminths (Grantham and Barrett, 1986b; Mohamed *et al.*, 2008; Umair *et al.*, 2011c).

28.6 Arginine

In helminths, arginine is a substrate for synthesis of polyamines, agmatine, nitric oxide, phosphoarginine, proline, glutamate and urea via the ornithine urea cycle. However, Umair *et al.* (2011b) found no evidence in sheep abomasal nematodes of the non-mammalian arginine deiminase or dihydrolase pathways, which generate ATP, ornithine, ammonia and CO_2 (Zúñiga *et al.*, 2002).

28.6.1 Arginase

Arginase, which irreversibly generates ornithine from arginine, has been described in numerous helminth species (Barrett, 1991). The *H. contortus* and *T. circumcincta* arginases were unusual in not requiring added Mn^{2+} or another bivalent metal ion for activity, in contrast to rat liver arginase (Kuhn *et al.*, 1991). Other divalent cations tested were inhibitory to the arginases of both species, as is also the case for *Fasciola hepatica* arginase (Mohamed *et al.*, 2005). The optimal pH of 8–8.5 for the arginases of the gastric nematode parasites (Muhamad, 2006; Umair *et al.*, 2011b) may relate to the acidity of their environment, as the arginase of the gastric bacterial pathogen *Helicobacter pylori* also has an exceptionally low optimal pH of 6.1 (McGee *et al.*, 2004).

28.6.2 Ornithine urea cycle

In ureotelic animals, excreted urea is formed by the ornithine urea cycle, though helminths generally lack a fully functional urea cycle (Janssens and Bryant, 1969) and urea is a minor excretory product (Rogers, 1952; Wright, 1975a). Although arginase is present, the other ornithine urea cycle enzymes, carbamoyl phosphate synthetase, ornithine transcarbamylase, argininosuccinate synthetase (ASS) and argininosuccinate lyase (ASL) have not been detected consistently. All ornithine urea cycle enzymes were detected in *P. redivivus* (Wright, 1975b), *Dicrocoelium lanceatum* (Rijavec and Kuralec, 1965) and *F. hepatica* (Rijavec and Kuralec, 1965; Mohamed *et al.*, 2005), but in *P. redivivus*, ASS and ASL activities were very low (Wright, 1975b).

28.6.3 Nitric oxide synthase (NOS)

Nitric oxide is a neurotransmitter in helminths, and NOS activity can be demonstrated histochemically in the central and peripheral nerves of cestodes and trematodes (Gustafsson *et al.*, 1996, 1998) and adult *H. contortus* (Umair *et al.*, 2011b). In contrast to its activity in neurons, NOS was undetectable in homogenates of whole adult or L3 *T. circumcincta* or *H. contortus* (Umair *et al.*, 2011b).

28.6.4 Agmatine

Agmatine is formed from arginine by arginine decarboxylase (ADC), which was detected in *H. contortus*, but not in *T. circumcincta* (Umair *et al.*, 2011b). In *H. contortus*, ADC appeared to be distinct from ornithine decarboxylase (ODC) (Umair *et al.*, 2013b). Agmatine is metabolized to putrescine and urea by agmatinase, an enzyme which is generally present in non-mammalian tissues (Tabor and Tabor, 1985), including both *H. contortus* and *T. circumcincta* (Umair *et al.*, 2011b). This was surprising for *T. circumcincta*, in which ADC was not found; however, this apparent activity may be due to arginase, which is capable of using either arginine or agmatine as substrate (Ahn *et al.*, 2004).

28.6.5 Polyamines

The polyamines putrescine, spermidine and spermine, which regulate cell growth, survival and differentiation, are universally synthesized from ornithine by the action of ODC (Pegg, 1986), but in microorganisms and plants, putrescine is also formed from agmatine by ADC (Slocum *et al.*, 1984). All three polyamines are actively transported in nematodes (Sharma *et al.*, 1989). Filaria appear to acquire putrescine from the host (Wittich *et al.*, 1987) and very low or negligible ODC activities have been reported in many other nematodes (Singh *et al.*, 1983; Wittich *et al.*, 1987; Sharma *et al.*, 1989). Unusually, putrescine was generated from arginine by both the ODC and ADC pathways in *H. contortus*, but not in *T. circumcincta* (Umair *et al.*, 2011b), and there was direct oxidation of spermine and spermidine in *Ascaris suum* (Müller and Walter, 1992). There were also unusual substrate specificities of the putrescine *N*-acetyltransferases of *A. suum* and *Onchocerca volvulus* (Wittich and Walter, 1989, 1990), and the ODCs of *Caenorhabditis elegans* (Schaeffer and Donatelli, 1990) and *H. contortus* (Klein *et al.*, 1997) were membrane bound, rather than cytosolic.

28.6.6 Arginine kinase

Phosphoarginine and arginine kinase are the usual phosphagen and phosphagen kinase in nematodes (Platzer *et al.*, 1995, 1999; Umair *et al.*, 2013c), but not in adult cestodes or digeneans (Barrett and Lloyd, 1981).

28.7 Glycine, Sarcosine, Serine and Threonine

28.7.1 Glycine

The metabolism of glycine, serine and threonine appear to be consistent with that in other animals. There are few reports on helminth glycine metabolism, apart from its interconversion to serine by L-serine hydroxymethyltransferase in *Nippostrongylus brasiliensis* (Walker and Barrett, 1991b), *Brugia pahangi* and *Dirofilaria immitis* (Barrett, 1983), and its production from sarcosine by the action of sarcosine oxidase (see Section 28.7.2).

28.7.2 Sarcosine

Sarcosine (*N*-methylglycine) is degraded to glycine either by sarcosine dehydrogenase (SarDH) or by sarcosine oxidase (SOX). Whereas in mammals, SarDH activity is essential to prevent sarcosinaemia (Reuber *et al.*, 1997), there was no SarDH activity in L3 and adult *H. contortus* and *T. circumcincta*, and sarcosine was demethylated by SOX. Genes encoding SOX have been annotated in *C. elegans*, *Ancylostoma caninum* and *H. contortus*.

Bacteria and plants generate sarcosine during the metabolism of creatinine by several pathways, including one using creatinase as the last enzyme (Wyss and Kaddurah-Daouk, 2000). Creatinase, which catalyses the conversion of creatine to sarcosine and urea, was demonstrated in adult, but not in L3, *H. contortus*, and in both L3 and adult *T. circumcincta* (Muhamad, 2006; Umair *et al.*, 2013a). The enzyme appears to be present in many nematodes, as genes encoding creatinase are reported from *H. contortus*, *A. suum*, *Toxocara canis*, *Brugia malayi*, *C. elegans* and *C. briggsae*.

28.7.3 Serine

L-serine is metabolized to phospholipids, glycine, taurine, cysteine and the mammalian neurotransmitter D-serine (Kalhan and Hanson, 2012), as well as providing methyl groups to tetrahydrofolate (THF), and subsequently to AdoMet (*S*-adenosylmethionine), for the methylation of proteins, DNA and RNA. Serine is obtained from the breakdown of protein, from glycine–serine recycling and by synthesis from pyruvate via oxaloacetate; it is metabolized to pyruvate by serine dehydratase, serine aminotransferase or serine hydroxymethyltransferase. L-Serine dehydratase, which deaminates serine to pyruvate, is active in *N. brasiliensis* (Walker and Barrett, 1992), *H. polygyrus* and *P. redivivus* (Grantham and Barrett, 1986a). L-Serine hydroxymethyltransferase, which catalyses the reversible conversion of serine and THF to glycine and 5,10-methyleneTHF (Schirch and Peterson, 1980), is present in *N. brasiliensis* (Walker and Barrett, 1991b), *B. pahangi* and *D. immitis* (Barrett, 1983). The gene is present in the *H. contortus* genome and in *C. elegans* is encoded by the *mel-32* gene, which it is maternally essential (Vatcher *et al.*, 1998). The reversible conversion of serine to cysteine is carried out by the helminth transsulfuration enzyme cystathionine β-synthase (Walker and Barrett, 1997).

28.7.4 Threonine

There is no evidence that threonine can be synthesized in helminths from aspartate, as it is in microorganisms and plants (Azevedo *et al.*, 1997), suggesting that its reported synthesis from labelled precursors by *Caenorhabditis briggsae* (Rothstein and Tomlinson, 1961) and *H. contortus* (Kapur and Sood, 1984) may be due to bacterial contamination or the presence of symbionts. Threonine is converted to pyruvate by threonine dehydrogenase, for which an encoding gene is annotated in the *H. contortus* genome. An intermediate in this pathway reacts with CoA to produce acetyl-CoA and glycine. There is an active threonine dehydratase in *N. brasiliensis* (Walker and Barrett, 1991b), *H. polygyrus* and *P. redivivus* (Grantham and Barrett, 1986a), and serine dehydratase can also metabolize threonine.

28.8 Methionine and Cysteine

Walker and Barrett (1997) comprehensively reviewed the metabolism of the sulfur amino acids by parasitic helminths and concluded that it generally resembled that of their hosts; the pathways investigated included the methionine cycle and AdoMet metabolism, transsulfuration, the transaminative catabolism of methionine, the oxidative catabolism of cysteine and glutathione synthesis. There were unusual properties of some nematode enzymes, such as the biochemical properties and amino acid structures of some methionine cycle enzymes, differences in AdoMet decarboxylases from the corresponding mammalian enzymes (Ndjonka *et al.*, 2003) and a cystathionine β-synthase in *N. brasiliensis* which catalysed the non-mammalian 'activated L-serine sulfhydrase' reaction (Walker and Barrett,1992).

28.8.1 Glutathione

The tripeptide glutathione (L-γ-glutamyl-L-cysteinylglycine) (GSH) is formed by successive additions to glutamate of cysteine and glycine by the enzymes γ-glutamylcysteine and GSH synthetases. It is degraded by four enzymes, together forming the γ-glutamyl cycle (Meister, 1981), which acts as an amino acid transporter. Glutathione has many cellular functions, particularly as an antioxidant, in leukotriene, steroid and prostaglandin metabolism, and as a conjugate of toxic compounds and xenobiotics (Meister, 1981). It is also a cellular detoxifying agent which reduces the efficiency of anthelmintic drugs.

Glutathione *S*-transferases (GSTs), which are responsible for conjugation of compounds to glutathione, have been identified and characterized or the encoding genes sequenced in many parasitic helminths. GSTs are secreted/excreted proteins (Moreno *et al.*, 2011) that are under investigation as targets for vaccines or the chemical treatment of parasitic trematodes and nematodes. Differences in nematode and host GSTs (Campbell *et al.*, 2001) have allowed vaccines to be developed to reduce parasite burdens and fecundity (Balloul *et al.*, 1987; Da Costa *et al.*, 1999).

28.9 Leucine, Isoleucine and Valine

P. redivivus and *H. polygyrus* catabolize branched chain amino acids (BCAAs) by pathways similar to those in rat liver (Grantham and Barrett, 1986b): ^{14}C-labelled leucine, isoleucine and valine were reduced to CO_2, and branched chain aminotransferases (BCATs) transaminated each to their corresponding keto acids. BCATs have been identified in *N. brasiliensis* (Walker and Barrett, 1991b), *F. hepatica* (Lee *et al.*, 1983) and *Ascaridia galli* (Singh and Srivastava, 1983), and Singh *et al.* (1983) detected decarboxylase activity for leucine and valine in the intestines, ovaries and cuticle of *A. galli*. The interaction of the three BCAAs seen in vertebrates did not appear to operate in *C. elegans* (Perelman and Lu, 2000).

28.10 Tyrosine, Phenylalanine and Tryptophan

The aromatic amino acids have been studied extensively as precursors of nematode neurotransmitters that are targeted by many chemical anthelmintics (Köhler, 2001). Dopamine, catecholamines, tyramine and octopamine are formed from tyrosine, serotonin from tryptophan and histamine from histidine (Blenau and Baumann,

2001). Octopamine is synthesized by the conversion of tyrosine to tyramine by tyrosine decarboxylase, then to octopamine by tyramine *β*-hydroxylase. Octopamine inhibits pharyngeal muscle pumping and egg laying, and is an antagonist to serotonin (Horvitz *et al.*, 1982). Tyramine also inhibits egg laying in *C. elegans* (Alkema *et al.*, 2005); also in *C. elegans*, phenylalanine hydroxylase is involved in synthesizing a melanin-like pigment (Calvo *et al.*, 2008).

Phenylalanine hydroxylase, tyrosine hydroxylase and tryptophan hydroxylase make up the aromatic amino acid hydroxylase family (Fitzpatrick, 2003). L-Phenylalanine is catabolized to L-tyrosine in *C. elegans* by phenylalanine hydroxylase (Calvo *et al.*, 2008). Tyrosine is catabolized in eukaryotes, including *C. elegans* (Fisher *et al.*, 2008), by a five-step pathway, the first step of which is the removal of the amino group by tyrosine aminotransferase; the following four steps produce fumarate and acetoacetate. The synthesis of tetrahydrobiopterin, an electron donor for these decarboxylases (Fitzpatrick, 2012), requires GTP-cyclohydrolase, the gene encoding of which has been sequenced and its expression monitored during development in *T. circumcincta* and *Dictyocaulus viviparus* (Baker *et al.*, 2011).

28.10.1 Tyrosinase

Tyrosinases have mono- and di-phenol oxidase activity and are involved in the synthesis of dihydroxyphenylalanine (DOPA) from tyrosine, melanin synthesis and protein crosslinking. Trematode tyrosinases have been extensively studied because of their role in eggshell sclerotinization and potential for chemotherapy or as vaccine candidates (Fitzpatrick *et al.*, 2007; He *et al.*, 2012; Bae *et al.*, 2015).

28.10.2 Chorismate mutase (CM)

CM, the last enzyme of the seven-step shikimate pathway, which synthesizes tyrosine and phenylalanine in plants and microorganisms, is expressed by plant parasitic nematodes of the genera *Meloidogyne* (Lambert *et al.*, 1999; Long *et al.*, 2006), *Globodera* (Jones *et al.*, 2003) and *Pratylenchus* (Haegeman *et al.*, 2011). There is no report of a complete shikimate pathway in nematodes and CM is suggested to have been acquired from microorganisms through horizontal gene transfer (Yan *et al.*, 1998). The expression of CM in the *Globodera pallida* suboesophageal gland (Jones *et al.*, 2003; Long *et al.*, 2006) supports the proposal that it is secreted to assist the nematodes in penetrating the root nodule through the development of feeding sites, although *Pratylenchus coffeae*, which also expresses the protein, does not form feeding sites. This enzyme may have a role in the suppression of the host response to the parasites (Curtis, 2007).

28.11 Alanine

Helminths very actively transaminate L-alanine reversibly to pyruvate using 2-oxoglutarate as the amino group donor, but D-alanine was transaminated at extremely low rates by *H. contortus* and *T. circumcincta* (Walker and Barrett, 1991a). Alanine may not be used as a substrate by other enzymes, such as alanine racemase, which was not detected in adult *N. brasiliensis* (Walker and Barrett, 1991a). Alanine transaminase (AlaAT) activity has been reported in all helminth groups (Barrett, 1991; Walker and Barrett, 1991a; Muhamad, 2006) in both cytosolic and mitochondrial forms; cytosolic AlaAT was responsible for 80% of the activity in adult *N. brasiliensis*, while in *H. contortus* 54% of activity was cytosolic and 22% was associated with the cell debris and cuticle fractions (Walker and Barrett, 1991a). *H. contortus* cytosolic AlaAT was less tolerant of temperatures of 45°C than the rat liver enzyme and responded differently to protective agents. Low host vitamin B_6 levels reduced AlaAT activity in *L. carinii* (Beg *et al.*, 1995) and *Hymenolepis diminuta* (Platzer and Roberts, 1970).

28.12 Aspartate and Asparagine

High aspartate transaminase (AspAT) activity has been reported in cestodes (Wertheim *et al.*, 1960; Rasero *et al.*, 1968), digeneans (Watts, 1970) and nematodes (Rasero *et al.*, 1968; Walker and Barrett, 1991b; Muhamad, 2006). The properties *T. circumcincta* AspAT were consistent with those of AspAT from other organisms (Muhamad, 2006).

Asparagine synthesis from aspartate by glutamine-dependent asparagine synthetase was identified in *P. redivivus* by Grantham and Barrett (1988). Asparagine can be catabolized by asparaginase or by transamination: asparaginase generates aspartate and ammonium, whereas transamination produces 2-oxosuccinamate. Asparagine transaminase activity was present in *H. diminuta, Hymenolepis citelli* (Wertheim *et al.*, 1960), *P. redivivus* and *H. polygyrus* (Grantham and Barrett, 1986a). Asparaginase activity was localized to the cuticle of *D. immitis* and an active recombinant protein was expressed (Tsuji *et al.*, 1999).

28.13 Lysine

L-Lysine cannot be synthesized in animal tissues, but is metabolized through either the pipecolate or saccharopine pathways (Broquist, 1991). A bifunctional enzyme (Gaziola *et al.*, 1997) with both lysine-ketoglutarate reductase (LKR) and saccharopine dehydrogenase (SacDH) activity (Galili *et al.*, 2001) forms the saccharopine pathway, the major route for lysine regulation and catabolism in plants and animals (Broquist, 1991; Tang *et al.*, 1997; Arruda *et al.*, 2000). This was the sole route for lysine catabolism in L3 abomasal nematodes, as Δ^1-piperideine-carboxylate reductase activity resulted in an incomplete pipecolate pathway, though this pathway was functional in adult worms (Umair *et al.*, 2012a). Nematode gene sequences deposited in databases are consistent with the existence of a bifunctional LKR-SacDH enzyme. Both LKR and SacDH were dual cofactor enzymes and not specific for either NAD^+ or $NADP^+$, in contrast to the case in other organisms, in which there is NADPH-specific LKR activity and NAD^+-specific SacDH activity (Tang *et al.*, 1997).

28.14 Conclusions

Helminth amino acid metabolism has been investigated in depth mainly in nematodes, particularly parasitic species, making it difficult to compare pathways over the many groups of helminths. As there are species and life cycle stage differences in the expression of enzymes, the presence of genes encoding the proteins does not ensure enzyme activity. The main differences in the metabolism of amino acids from that in other animals are in the properties of individual enzymes, rather than the absence or presence of pathways. Unusual features, at least in some helminths, are the lack a full ornithine-urea cycle, the presence of creatinase activity, the synthesis of polyamines from either ornithine or agmatine, and the incorporation of ammonia into glutamate through the GS-GOGAT pathway. There were differences between closely related species, such as common sheep abomasal nematode parasites, which were not identical in expressing glutamate synthase, arginine decarboxylase and Δ^1-piperideine-carboxylate reductase. Overall, helminth metabolism was more flexible than is generally seen in animals, particularly in the ability to synthesize amino acids from ammonia and the degradation of all amino acids to glucose because of the presence of a functional glyoxylate shunt.

References

Abidi, S.M.A. and Nizami, W.A. (1995) [^{3}H]-Amino acid uptake and metabolic studies on *Gigantocotyle explanatum* and *Gastrothylax crumenifer* (Digenea: Paramphistomidae). *International Journal for Parasitology* 25, 541–549.

Ahn, H.J., Kim, K.H., Lee, J., Ha, J.Y., Lee, H.H., Kim, D., Yoon, H.J., Kwon, A.R. and Suh, S.W. (2004) Crystal structure of agmatinase reveals structural conservation and inhibition mechanism of the ureohydrolase superfamily. *The Journal of Biological Chemistry* 279, 50505–50513.

Alkema, M.J., Hunter-Ensor, M., Ringstad, N. and Horvitz, H.R. (2005) Tyramine functions independently of octopamine in the *Caenorhabditis elegans* nervous system. *Neuron* 46, 247–260.

Arruda, P., Kemper, E.L., Papes, F. and Leite, A. (2000) Regulation of lysine catabolism in higher plants. *Trends in Plant Science* 5, 324–330.

Asch, H.L. and Read, C.P. (1975) Membrane transport in *Schistosoma mansoni*: transport of amino acids by adult males. *Experimental Parasitology* 38, 123–135.

Azevedo, R.A., Arruda, P., Turner, W.L. and Lea, P.J. (1997) The biosynthesis and metabolism of the aspartate derived amino acids in higher plants. *Phytochemistry* 46, 395–419.

Bae, Y.-A., Kim, S.-H., Ahn, C.-S., Kim, J.-G. and Kong, Y. (2015) Molecular and biochemical characterization of *Paragonimus westermani* tyrosinase. *Parasitology* 142, 807–815.

Baker, R.H., Buschbaum, S., Matthews, J.B., McKendrick, I.J., Schnieder, T., Strube, C. and Nisbet, A.J. (2011) GTP-cyclohydrolase and development in *Teladorsagia circumcincta* and *Dictyocaulus viviparus* (Nematoda: Strongylida). *Experimental Parasitology* 128, 309–317.

Balázs, R., Machiyama, Y., Hammond, B.J., Julian, T. and Richter, D. (1970) The operation of the γ-aminobutyrate bypath of the tricarboxylic acid cycle in brain tissue *in vitro*. *Biochemical Journal* 116, 445–461.

Balloul, J.-M., Grzych, J.-M., Pierce, R.M. and Capron, A. (1987) A purified 28,000 dalton protein from *Schistosoma mansoni* adult worms protects rats and mice against experimental schistosomiasis. *Journal of Immunology* 138, 3448–3453.

Barrett, J. (1981) *Biochemistry of Parasitic Helminths*. Macmillan, London.

Barrett, J. (1983) The biochemistry of filarial worms. *Helminthological Abstracts* 52A, 1–18.

Barrett, J. (1991) Amino acid metabolism in helminths. *Advances in Parasitology* 30, 39–105.

Barrett, J. and Lloyd, G.M. (1981) A novel phosphotransferase in the plerocercoids of *Shistocephalus solidus* (Cestoda: Pseudophyllidea). *Parasitology* 82, 11–16.

Barrett, J., Ward, C.W. and Fairbairn, D. (1970) The glyoxylate cycle and the conversion of triglycerides to carbohydrates in developing eggs of *Ascaris lumbricoides*. *Comparative Biochemistry and Physiology* 35, 577–586.

Beg, M.A., Fisten, J.L., Ingram, G.A. and Storey, D.M. (1995) Activities of glycogen phophorylase, alanine aminotransferase and aspartate aminotransferase in adult worms of *Litomosoides carinii* recovered from pyridoxine deficient cotton rats (*Sigmodon hispidus*). *Parasitology* 112, 227–232.

Bird, A.F. and Bird, J. (1991) *The Structure of Nematodes*, 2nd edn. Academic Press, San Diego, California.

Blenau, W. and Baumann, A. (2001) Molecular and pharmacological properties of insect biogenic amine receptors: Lessons from *Drosophila melanogaster* and *Apis mellifera*. *Archives of Insect Biochemistry and Physiology* 48, 13–38.

Brockelman, C.R. and Jackson, G.J. (1978) Amino acid, heme, and sterol requirements of the nematode, *Rhabditis maupasi*. *Journal of Parasitology* 64, 803–809.

Broquist, H.P. (1991) Lysine–pipecolic acid metabolic relationships in microbes and mammals. *Annual Review of Nutrition* 11, 435–448.

Bruce, J.I., Ruff, M.D., Belusko, R.J. and Werner, J.K. (1972) *Schistosoma mansoni* and *Schistosoma japonicum*: utilization of amino acids. *International Journal for Parasitology* 2, 425–430.

Calvo, A.C., Pey, A.L., Ying, M., Loer, C.M. and Martinez, A. (2008) Anabolic function of phenylalanine hydroxylase in *Caenorhabditis elegans*. *The FASEB Journal* 22, 3046–3058.

Campbell, A.M., Teesdale-Spittle, P.H., Barrett, J., Liebau, E., Jefferies, J.R. and Brophy, P.M. (2001) A common class of nematode glutathione *S*-transferase (GST) revealed by the theoretical proteome of the model organism *Caenorhabditis elegans*. *Comparative Biochemistry and Physiology Part B: Biochemistry and Molecular Biology* 128, 701–708.

Curtis, R.H.C. (2007) Do phytohormones influence nematode invasion and feeding site establishment? *Nematology* 9, 155–160.

Da Costa, A.V., Gaubert, S., Lafitte, S., Fontaine, J., Capron, A. and Grzych, J.M. (1999) Egg-hatching inhibition in mice immunized with recombinant *Schistosoma bovis* 28 kDa glutathione *S*-transferase. *Parasite Immunology* 21, 341–350.

Dass, P.D. and Donahue, M.J. (1986) γ-Glutamyl transpeptidase activity in *Ascaris suum*. *Molecular and Biochemical Parasitology* 20, 233–236.

Davies, K.P. and Köhler, P. (1990) The role of amino acids in the energy generating pathways of *Litomosoides carinii*. *Molecular and Biochemical Parasitology* 41, 115–124.

Feng, X.-P., Hayashi, J., Beech, R.N. and Prichard, R.K. (2002) Study of the nematode putative GABA type-A receptor subunits: evidence for modulation by ivermectin. *Journal of Neurochemistry* 83, 870–878.

Fisher, A.F., Page, K.E., Lithgow, G.J. and Nash, L. (2008) The *Caenorhabditis elegans* K10C2.4 gene encodes a member of the fumarylacetoacetate hydrolase family. A *Caenorhabditis elegans* model of type 1 tyrosinaemia. *The Journal of Biological Chemistry* 283, 9127–9135.

Fisher, S.H. and Sonenshein, A.L. (1991) Control of carbon and nitrogen metabolism in *Bacillus subtilis*. *Annual Review of Microbiology* 45, 107–135.

Fitzpatrick, J.M., Hirai, Y., Hirai, H. and Hoffmann, K.F. (2007) Schistosome egg production is dependent upon the activities of two developmentally regulated tyrosinases. *The FASEB Journal* 21, 823–835.

Fitzpatrick, P.F. (2003) Mechanism of aromatic amino acid hydroxylation. *Biochemistry* 42, 14083–14091.

Fitzpatrick, P.F. (2012) Allosteric regulation of phenylalanine hydroxylase. *Archives of Biochemistry and Biophysics* 519, 194–201.

Galili, G., Tang, G., Zhu, X. and Gakiere, B. (2001) Lysine catabolism: a stress and development super-regulated metabolic pathway. *Current Opinion in Plant Biology* 4, 261–266.

Gaziola, S.A., Teixeira, C.M., Lugli, J., Sodek, L. and Azevedo, R.A. (1997) The enzymology of lysine catabolism in rice seeds – isolation, characterization, and regulatory properties of a lysine 2-oxoglutarate reductase/saccharopine dehydrogenase bifunctional polypeptide. *European Journal of Biochemistry* 247, 364–371.

Goldin, B.R. and Frieden, C. (1971) L-glutamate dehydrogenases. *Current Topics in Cell Regulation* 4, 77–117.

Grantham, B.D. and Barrett, J. (1986a) Amino acid catabolism in the nematodes *Heligmosomoides polygyrus* and *Panagrellus redivivus*. 1. Removal of the amino group. *Parasitology* 93, 481–493.

Grantham, B.D. and Barrett, J. (1986b) Amino acid catabolism in the nematodes *Heligmosomoides polygyrus* and *Panagrellus redivivus*. 2. Metabolism of the carbon skeleton. *Parasitology* 93, 495–504.

Grantham, B.D. and Barrett, J. (1988) Glutamine and asparagine synthesis in the nematodes *Heligmosomoides polygyrus* and *Panagrellus redivivus*. *Journal of Parasitology* 74, 1052–1053.

Gustafsson, M.K.S., Lindholm, A.M., Terenina, N.B. and Reuter, M. (1996) NO nerves in a tapeworm. NADPH-diaphorase histochemistry in adult *Hymenolepis diminuta*. *Parasitology* 113, 559–565.

Gustafsson, M.K.S., Lindholm, A.M., Mäntylä, K., Reuter, M., Lundström, C.A. and Terenina, N. (1998) No news on the flatworm front! Nitric oxide synthase in parasitic and free-living flatworms. *Hydrobiologia* 383, 161–166.

Haegemann, A., Soumi, J. and Gheysen, G. (2011) Analysis of the transcriptome of the root lesion nematode *Pratylenchus coffeae* generated by 454 sequencing technology. *Molecular and Biochemical Parasitology* 178, 7–14.

Hanna, R.E.B. (1980) *Fasciola hepatica*: autoradiography of protein synthesis, transport, and secretion by the tegument. *Experimental Parasitology* 50, 297–304.

Hata, H. (1994) Essential amino acids and other essential components for development of *Angiostrongylus costaricensis* from third-stage larvae to young adults. *Journal of Parasitology* 80, 518–520.

He, Y., Cai, G., Ni, Y., Li, Y., Zong, H. and He, L. (2012) siRNA mediated knockdown of two tyrosinases genes from *Schistosoma japonicum* cultured *in vitro*. *Experimental Parasitology* 132, 394–402.

Helling, R.B. (1998) Pathway choice in glutamate synthesis in *Escherichia coli*. *Journal of Bacteriology* 180, 4571–4575.

Hirayama, C., Saito, H., Konno, K. and Shinbo, H. (1998) Purification and characterization of NADH-dependent glutamate synthase from the silkworm fat body (*Bombyx mori*). *Insect Biochemistry and Molecular Biology* 28, 473–482.

Horvitz, H.R., Chalfie, M., Trent, C., Sulston, J. E. and Evans, P.D. (1982) Serotonin and octopamine in the nematode *Caenorhabditis elegans*. *Science* 216, 1012–1014.

Hudson, R.C. and Daniel, R.M. (1993) L-glutamate dehydrogenases: distribution, properties and mechanism. *Comparative Biochemistry and Physiology Part B: Biochemistry and Molecular Biology* 106, 767–792.

Isseroff, H., Ertel, J.C. and Levy, M.G. (1976) Absorption of amino acids by *Schistosoma mansoni*. *Comparative Biochemistry and Physiology Part B: Biochemistry and Molecular Biology* 54, 125–133.

Janssens, P.A. and Bryant, C. (1969) The ornithine-urea cycle in some parasitic helminths. *Comparative Biochemistry and Physiology* 30, 261–272.

Jeffs, S.A. and Arme, C. (1985) *Hymenolepis diminuta*: characterization of the neutral amino acid transport loci of the metacestode. *Comparative Biochemistry and Physiology Part A: Physiology* 81, 387–390.

Jeffs, S.A. and Arme, C. (1987) *Echinococcus granulosus*: specificity of amino acid transport systems in protoscoleces. *Parasitology* 95, 71–78.

Jones, J.T., Furlanetto, C., Bakker, E., Banks, B., Blok, V., Chen, Q., Phillips, M. and Prior, A. (2003) Characterization of a chorismate mutase from the potato cyst nematode *Globodera pallida*. *Molecular Plant Pathology* 4, 43–50.

Kalhan, S.C. and Hanson, R.W. (2012) Resurgence of serine: an often neglected but indispensable amino acid. *The Journal of Biological Chemistry* 287, 19786–19791.

Kapur, J. and Sood, M.L. (1984) Amino acid biosynthesis in *Haemonchus contortus* from C^{14}-labelled precursors, *in vitro*. *Veterinary Parasitology* 15, 293–299.

Klein, R.D., Favreau, M.A., Alexander-Bowman, S.J., Nulf, S.C., Vanover, L., Winterrowd, C.A., Yarlett, N., Martinez, M., Keithly, J.S., Zantello, M.R. *et al.* (1997) *Haemonchus contortus*: cloning and functional expression of a cDNA encoding ornithine decarboxylase and development of a screen for inhibitors. *Experimental Parasitology* 87, 171–184.

Köhler, P. (2001) The biochemical basis of anthelmintic action and resistance. *International Journal for Parasitology* 31, 336–345.
Kuhn, N.J., Talbot, J. and Ward, S. (1991) pH-sensitive control of arginase by Mn(II) ions at submicromolar concentrations. *Archives of Biochemistry and Biophysics* 286, 217–221.
Lambert, K.N., Allen, K.D. and Sussex, I.M. (1999) Cloning and characterization of an esophageal-gland-specific chorismate mutase from the phytoparasitic nematode *Meloidogyne javanica*. *Molecular Plant–Microbe Interactions* 12, 328–336.
Lee, J.H., Lee, D.W., and Lee H.S. (1983) Purification and properties of branched chain amino acid aminotransferase from *Fasciola hepatica*. *Korean Journal of Parasitology* 21, 49–57.
Long, H., Wang, X., Xu, J.H. and Hu, Y.J. (2006) Isolation and characterization of another cDNA encoding a chorismate mutase from the phytoparasitic nematode *Meloidogyne arenaria*. *Experimental Parasitology* 113, 106–111.
MacKenzie, N.E., Van de Waa, E.A., Gooley, P.R., Williams, J.F., Bennett, J.L., Bjorge, S.M., Baille, T.A. and Geary, T.G. (1989) Comparison of glycolysis and glutaminolysis in *Onchocerca volvulus* and *Brugia pahangi* by ^{13}C nuclear magnetic resonance spectroscopy. *Parasitology* 99, 427–435.
McGee, D.J., Zabaleta, J., Viator, R.J., Testerman, T.L., Ochoa, A.C. and Mendz, G.L. (2004) Purification and characterization of *Helicobacter pylori* arginase, RocF: unique features among the arginase superfamily. *European Journal of Biochemistry* 271, 1952–1962.
McManus, D.P. (1987) Intermediary metabolism in parasitic helminths. *International Journal for Parasitology* 17, 79–95.
Meissner, B., Boll, M., Daniel, H. and Baumeister, R. (2004) Deletion of the intestinal peptide transporter affects insulin and TOR signaling in *Caenorhabditis elegans*. *The Journal of Biological Chemistry* 279, 36739–36745.
Meister, A. (1981) Metabolism and functions of glutathione. *Trends in Biochemical Sciences* 6, 231–234.
Miflin, B.J. and Habash, D.Z. (2002) The role of glutamine synthetase and glutamate dehydrogenase in nitrogen assimilation and possibilities for improvement in the nitrogen utilization of crops. *Journal of Experimental Botany* 53, 979–987.
Mohamed, S.A., Fahmy, A S., Mohamed, T.M. and Hamdy, S.M. (2005) Urea cycle of *Fasciola gigantica*: purification and characterization of arginase. *Comparative Biochemistry and Physiology Part B: Biochemistry and Molecular Biology* 142, 308–316.
Mohamed, S.A., Mohamed, T.M., Fahmy, A.S., El-Badry, M.O. and Abdel-Gany, S.S. (2008) *Fasciola gigantica*: enzymes of the ornithine-proline-glutamate pathway – characterization of Δ1-pyrroline-5-carboxylate dehydrogenase. *Experimental Parasitology* 118, 47–53.
Monteoliva, M., Rasero, F.S., Gorge, J.L. and Mayor, F. (1965) L-Glutamate-1-carboxylase in intestinal parasites. *Nature* 205, 1111–1112.
Moreno, Y., Gros, P.-P., Tam, M., Segura, M., Valanparambil, R., Geary, T.G. and Stevenson, M.M. (2011) Proteomic analysis of excretory–secretory products of *Heligmosomoides polygyrus* assessed with next-generation sequencing transcriptomic information. *PLoS Neglected Tropical Diseases* 10(5): e1370.
Muhamad, N. (2006) Nitrogen metabolism in *Ostertagia (Teladorsagia) circumcincta*. PhD thesis, Massey University, Palmerston North, New Zealand.
Muhamad, N., Simcock, D.C., Pedley, K.C., Simpson, H.V. and Brown, S. (2011) The kinetic properties of the glutamate dehydrogenase of *Teladorsagia circumcincta* and their significance for the lifestyle of the parasite. *Comparative Biochemistry and Physiology Part B: Biochemistry and Molecular Biology* 159, 71–77.
Müller, S. and Walter, R.D. (1992) Purification and characterization of polyamine oxidase from *Ascaris suum*. *Biochemical Journal* 283, 75–80.
Ndjonka, D., Zou, Y., Bi, X., Woster, P., Walter, R.D. and Lüersen, K. (2003) The activator-binding site of *Onchocerca volvulus S*-adenosylmethionine decarboxylase, a potential drug target. *Biological Chemistry* 384, 1195–1201.
Page, A.P. (2001) The nematode cuticle: synthesis, modification and mutants. In: Kennedy M.W and Harnett, W. (eds) *Parasitic Nematodes*. CAB International, Wallingford, UK, pp. 103–121.
Pappas, P.W. (1988) The relative role of the intestines and external surfaces in the nutrition of monogeneans and nematodes. *Parasitology* 96, S105–S121.
Pappas, P.W., Uglem, G.L. and Read, C.P. (1973) Mechanisms and specificity of amino acid transport in *Taenia crassiceps* larvae (Cestoda). *International Journal for Parasitology* 3, 641–651.
Pegg, A.E. (1986) Recent advances in the biochemistry of polyamines in eukaryotes. *Biochemical Journal* 234, 249–262.

Perelman, D. and Lu, N. (2000) Requirements for branched chain amino acids and their interactions in the nematode *Caenorhabditis elegans*. *Nematology* 2, 501–506.

Platzer, E.G. and Roberts, L.S. (1970) Developmental physiology of cestodes – Part VII Vitamin B6 and *Hymenolepis diminuta*: vitamin levels in the cestode and effects of deficiency of phosphorylase and transaminase activities. *Comparative Biochemistry and Physiology* 35, 535–552.

Platzer, E.G., Thompson, S.N., Borchardt, D.B. and Gamble, H.R. (1995) High energy phosphate metabolites observed by NMR in infective larvae of *Haemonchus contortus. Journal of Parasitology* 81, 434–438.

Platzer, E.G., Wang, W., Thompson, S.N. and Borchardt, D.B. (1999) Arginine kinase and phosphoarginine, a functional phosphagen, in the rhabditoid nematode *Steinernema carpocapsae*. *Journal of Parasitology* 85, 603–607.

Rasero, F.S., Monteoliva, M. and Mayor, F. (1968) Enzymes related to 4-aminobutyrate metabolism in intestinal parasites. *Comparative Biochemistry and Physiology* 25, 693–696.

Reuber, B.E., Karl, C., Reimann, S.A., Mihalik, S.J. and Dodt, G. (1997) Cloning and functional expression of a mammalian gene for a peroxisomal sarcosine oxidase. *The Journal of Biological Chemistry* 272, 6766–6776.

Riding, I. L. (1970) Microvilli on the outside of a nematode. *Nature* 226, 179–180.

Rijavec, M. and Kurelec, B. (1965) Harnstoffzyklus bei einigen Rinderparasiten (Helminthen). *Zeitschrift für Parasitenkunde* 26, 168–172.

Rogers, W.P. (1952) Nitrogen catabolism in nematode parasites. *Australian Journal of Scientific Research* B 5, 210–222.

Rothstein, M. (1965) Nematode biochemistry – V. Intermediary metabolism and amino acid interconversions in *Caenorhabditis briggsae*. *Comparative Biochemistry and Physiology* 14, 541–552.

Rothstein, M. and Mayoh, H. (1964) Nematode biochemistry IV. On isocitrate lyase in *Caenorhabditis briggsae*. *Archives of Biochemistry and Biophysics* 108, 134–142.

Rothstein, M. and Mayoh, H. (1966) Nematode biochemistry – VIII Malate synthase. *Comparative Biochemistry and Physiology* 17, 1181–1188.

Rothstein, M. and Tomlinson, G.A. (1961) Biosynthesis of amino acids by the nematode *Caenorhabditis briggsae*. *Biochimica et Biophysica Acta* 49, 625–627.

Sajid, M. and McKerrow, J.H. (2002) Cysteine proteases of parasitic organisms. *Molecular and Biochemical Parasitology* 120, 1–21.

Scaraffia, P.Y., Isoe, J., Murillo, A. and Wells, M.A. (2005) Ammonia metabolism in *Aedes aegypti*. *Insect Biochemistry and Molecular Biology* 35, 491–503.

Schaeffer, J.M. and Donatelli, M.R. (1990) Characterization of a high-affinity membrane-associated ornithine decarboxylase from the free-living nematode *Caenorhabditis elegans*. *Biochemical Journal* 270, 599–604.

Schirch, L. and Peterson, D. (1980) Purification and properties of mitochondrial serine hydroxymethyltransferase. *The Journal of Biological Chemistry* 255, 7801–7806.

Schuske, K., Beg, A.A. and Jorgensen, E.M. (2004) The GABA nervous system in *C. elegans*. *Trends in Neuroscience* 27, 407–414.

Sharma, V., Visen, P.K.S., Katiyar, J.C., Wittich, R.-M., Walter, R.D., Ghatak, S. and Shukla, O.P. (1989) Polyamine metabolism in *Ancylostoma ceylanicum* and *Nippostrongylus brasiliensis*. *International Journal for Parasitology* 19, 191–198.

Simpson, H.V., Muhamad, N., Walker, L.R., Simcock, D.C., Brown, S. and Pedley, K.C. (2009) Nitrogen excretion by the sheep abomasal parasite *Teladorsagia circumcincta*. *Experimental Parasitology* 123, 17–23.

Singh, G.N.A. and Srivastava, V.M.L. (1983) Metabolism of amino acids in *Ascaridia galli*: transamination. *Zeitschrift für Parasitenkunde* 69, 783–788.

Singh, G., Pampori, N.A. and Srivastava, V.M.L. (1983) Metabolism of amino acids in *Ascaridia galli*: decarboxylation reactions. *International Journal for Parasitology* 13, 305–307.

Slocum, R.D., Kaur-Sawhney, R. and Galston, A.W. (1984) The physiology and biochemistry of polyamines in plants. *Archives of Biochemistry and Biophysics* 235, 283–303.

Tabor, C.W. and Tabor, H. (1985) Polyamines in microorganisms. *Microbiological Reviews* 49, 81–99.

Tang, G., Miron, D., Zhu-Shimoni, J. X. and Galili, G. (1997) Regulation of lysine catabolism through lysine-ketoglutarate reductase and saccharopine dehydrogenase in *Arabidopsis*. *The Plant Cell* 9, 1305–1316.

Tsuji, N., Morales, T.H., Ozols, V., Carmody, A.B. and Chandrashekar, R. (1999) Identification of an asparagine amidohydrolase from the filarial parasite *Dirofilaria immitis*. *International Journal for Parasitology* 29, 1451–1455.

Umair, S., Bland, R.J. and Simpson, H.V. (2011a) Glutamate synthase, but not GABA shunt enzymes, contributes to nitrogen metabolism of the sheep abomasal nematode parasites *Haemonchus contortus* and *Teladorsagia circumcincta*. *Experimental Parasitology* 127, 9–13.
Umair, S., Patchett, M.L., Bland, R.J. and Simpson, H.V. (2011b) Arginine metabolism in the sheep abomasal nematode parasites *Haemonchus contortus* and *Teladorsagia circumcincta*. *Experimental Parasitology* 127, 506–514.
Umair, S., Leung, Y-M., Bland, R.J. and Simpson, H.V. (2011c) Enzymes of the ornithine-glutamate-proline pathway in the sheep abomasal nematode parasites *Haemonchus contortus* and *Teladorsagia circumcincta*. *Experimental Parasitology* 129, 115–119.
Umair, S., Knight, J.S., Patchett, M.L., Bland, R.J. and Simpson, H.V. (2011d) Molecular and biochemical characterisation of a *Teladorsagia circumcincta* glutamate dehydrogenase. *Experimental Parasitology* 129, 240–246.
Umair, S., Bland, R.J. and Simpson, H.V. (2012a) Lysine catabolism in *Haemonchus contortus* and *Teladorsagia circumcincta*. *Experimental Parasitology* 131, 101–106.
Umair, S., Knight, J.S., Patchett, M.L., Bland, R.J. and Simpson, H.V. (2012b) Nucleotide allosteric regulation of the glutamate dehydrogenases of *Teladorsagia circumcincta* and *Haemonchus contortus*. *Comparative Biochemistry and Physiology Part B: Biochemistry and Molecular Biology* 161, 255–260.
Umair, S., Knight, J.S. and Simpson, H.V. (2013a) Molecular and biochemical characterisation of ornithine decarboxylases in the sheep abomasal nematode parasites *Teladorsagia circumcincta* and *Haemonchus contortus*. *Comparative Biochemistry and Physiology Part B: Biochemistry and Molecular Biology* 165, 119–124.
Umair, S., Ria, C., Knight, J.S., Bland, R.J. and Simpson, H.V. (2013b) Sarcosine metabolism in *Haemonchus contortus* and *Teladorsagia circumcincta*. *Experimental Parasitology* 134, 1–6.
Umair, S., Knight, J.S., Bland, R.J. and Simpson, H.V. (2013c) Molecular and biochemical characterisation of arginine kinases in *Haemonchus contortus* and *Teladorsagia circumcincta*. *Experimental Parasitology* 134, 362–367.
Vanholme, B., De Meutter, J., Tytgat, T., Van Montagu, M., Coomans, A. and Gheysen, G. (2004) Secretions of plant-parasitic nematodes: a molecular update. *Gene* 332, 13–27.
Vanoni, M.A. and Curti, B. (1999) Glutamate synthase: a complex iron-sulfur flavoprotein. *Cellular and Molecular Life Sciences* 55, 617–638.
Vatcher, G.P. Thacker, C.M., Kaletta, T., Schnabel, H., Schnabel, R. and Baillie, D.L. (1998) Serine hydroxymethyltransferase is maternally essential in *Caenorhabditis elegans*. *The Journal of Biological Chemistry* 273, 6066–6073.
Veljkovic, E., Stasiuk, S., Skelly, P.J., Shoemaker, C.B. and Verrey, F. (2004) Functional characterization of *Caenorhabditis elegans* heteromeric amino acid transporters. *The Journal of Biological Chemistry* 279, 7655–7662.
Walker, J. and Barrett, J. (1991a) Studies on alanine aminotransferase in nematodes. *International Journal for Parasitology* 21, 377–380.
Walker, J. and Barrett, J. (1991b) Pyridoxal 5′-phosphate dependent enzymes in the nematode *Nippostrongylus brasiliensis*. *International Journal for Parasitology* 21, 641–649.
Walker, J. and Barrett, J. (1992) Biochemical characterisation of the enzyme responsible for activated L-serine sulphydrase activity in nematodes. *Experimental Parasitology* 74, 205–215.
Walker, J. and Barrett, J. (1997) Parasite sulphur amino acid metabolism. *International Journal for Parasitology* 27, 883–897.
Watts, S.D.M. (1970) Transamination in homogenates of rediae of *Cyptocotyle lingua* and of sporocysts of *Cercaria emasculans* Pelseneer, 1900. *Parasitology* 61, 499–504.
Wertheim, G., Zeledon, R. and Read, C.P. (1960) Transaminases of tapeworms. *Journal of Parasitology* 46, 497–499.
Wittich, R.-M. and Walter, R.D. (1989) A novel type of putrescine (diamine)-acetylating enzyme from the nematode *Ascaris suum*. *Biochemical Journal* 260, 265–269.
Wittich, R.-M. and Walter, R.D. (1990) Putrescine *N*-acetyltransferase in *Onchocerca volvulus* and *Ascaris suum*, an enzyme which is involved in polyamine degradation and release of *N*-acetylputrescine. *Molecular and Biochemical Parasitology* 38, 13–17.
Wittich, R.-M., Kilian, H.-D. and Walter, R.D. (1987) Polyamine metabolism in filarial worms. *Molecular and Biochemical Parasitology* 24, 155–162.
Wright, D.J. (1975a) Elimination of nitrogenous compounds by *Panagrellus redivivus*, Goodey, 1945 (Nematoda: Cephalobidae). *Comparative Biochemistry and Physiology Part B: Biochemistry and Molecular Biology* 52, 247–253.

Wright, D.J. (1975b) Studies on nitrogen catabolism in *Panagrellus redivivus*, Goodey, 1945 (Nematoda: Cephalobidae). *Comparative Biochemistry and Physiology Part B: Biochemistry and Molecular Biology* 52, 255–260.

Wyss, M. and Kaddurah-Daouk, R. (2000) Creatine and creatinine metabolism. *Physiological Reviews* 80, 1107–1213.

Yan, Y., Smant, G., Stokkermans, J., Qin, L., Helder, J., Baum, T., Schots, A. and Davis, E. (1998) Genomic organization of four β-1,4-endoglucanase genes in plant-parasitic cyst nematodes and its evolutionary implications. *Gene* 220, 61–70.

Zúñiga, M., Pérez, G. and González-Candelas, F. (2002) Evolution of arginine deiminase (ADI) pathway genes. *Molecular Phylogenetics and Evolution* 25, 429–444.

29 Microbial Degradation of Amino Acids in Anoxic Environments

A. Parthasarathy[1]* and N.P. Chowdhury[2]
[1]*Formerly of Department of Civil and Environmental Engineering, Stanford University, California, USA, currently of Science Academy of India, Chennai, India;* [2]*Max Planck Institute for Terrestrial Microbiology, Marburg, Germany*

29.1 Abstract

In the absence of terminal electron-acceptors such as sulfate, nitrate or molecular oxygen, only a few fermenting bacteria (*Firmicutes*) can catabolize amino acids. For every one of the 20 proteinogenic amino acids, at least one degradation pathway is known. Many pathways involve radical biochemistry and unique modes of energy conservation. This chapter summarizes the best studied pathways involving 3-methylaspartate, 4-hydroxybutyrate, 5-hydroxyvalerate, 2-hydroxyacids and 3,5-diaminohexanoate, and discusses the enzymatic reactions involved, bioenergetics and certain medical aspects.

29.2 Introduction

Protein degradation begins with hydrolysis, which produces peptides and amino acids, and contributes to the carbon and nitrogen cycles. In anoxic environments, where inorganic electron acceptors are available, bacteria and archaea couple the oxidation of amino acids and small peptides to the reduction of inorganic ions. However, this chapter deals with the catabolism of amino acids as sole carbon and energy substrates i.e. amino acid fermentation, which is biochemically more challenging. In bacterial fermentations, the substrate serves not only as an electron donor, but also as a terminal electron acceptor, because inorganic ions are absent (Thauer *et al.*, 1977). In soils, sewage, anoxic sediments and the guts of animals, the dominant amino acid fermenters are *Firmicutes* of the obligately anaerobic genera *Clostridium*, *Fusobacterium*, *Acidaminococcus*, *Peptostreptococcus* and *Megasphaera*. The major fermentation products are ammonia, short-chain fatty acids (SCFAs), acetate, CO_2 and molecular hydrogen. Minor products include H_2S, phenols and alcohols. SCFAs are further fermented into acetate, CO_2 and H_2. The latter two are the precursors of acetate in acetogenesis in clostridia and *Deltaproteobacteria*, and of methane in archaeal methanogenesis. H_2 consumption during methanogenesis lowers the partial pressure, making H_2 evolution by fermenters thermodynamically feasible. This relationship between fermenters and methanogens is a form of syntrophism.

Each of the 20 classical proteinogenic L-amino acids is fermented via at least one pathway; selenocysteine, the 21st, is inserted into proteins and recycled by dedicated systems. Selenocysteine fermentation has not been reported. The reductive branches of amino acid fermentations are especially interesting owing to the unique enzymes involved. In Stickland fermentation, one amino acid donates electrons while

*Corresponding author. E-mail address: harikuttanp@gmail.com

another accepts them, thereby generating ATP and reducing power.

Stickland electron donors include alanine, leucine, valine, isoleucine, methionine, phenylalanine, cysteine, serine, histidine, aspartate and glutamate; electron acceptors include glycine, proline, hydroxyproline, arginine, ornithine (derived from arginine) and tryptophan. Many proteinogenic amino acids can either be Stickland electron donors or electron acceptors, but isoleucine and valine are only degraded oxidatively. *Clostridium acetireducens* couples acetate reduction to the oxidation of isoleucine, valine, leucine or alanine (Örlygsson *et al.*, 1996), generating the corresponding 2-ketoacid. The 2-ketoacid is decarboxylated with the addition of CoASH to form the SCFA CoA-thioester. Substrate-level phosphorylation (SLP) generates ATP, with SCFA as the end product.

Lysine is fermented through β-lysine via the 3,5-diaminohexanoate pathway (Stadtman, 1973). Arginine is deiminated to citrulline, which is cleaved to ornithine and carbamyl phosphate. The latter is converted into CO_2 and ammonia, generating ATP. Ornithine-only fermentation by many clostridia yields ammonia, alanine, acetate and 5-aminovalerate (Dyer and Costilow, 1968). In *C. sticklandii*, ornithine is racemized, followed by an amino group migration from C5 to C4, forming D-*threo*-2,4-diaminopentanoate, which is oxidized to 2-amino-4-ketopentanoate and ammonia. 2-Amino-4-ketopentanoate is cleaved with CoASH, forming alanine and acetyl-CoA. The NADH produced during the oxidation of D-*threo*-2,4-aminopentanoate reduces D-proline, yielding 5-aminovalerate. Some strains of *C. botulinum*, *C. sporogenes* and *C. sticklandii* convert ornithine to L-proline by an NAD^+-bound cyclase-deaminase (Muth and Costilow, 1974). L-Proline is isomerized to D-proline, followed by reduction into 5-aminovalerate by a selenoenzyme-bound pyruvoyl-group (Kabisch *et al.*, 1999). 5-Aminovalerate is fermented by *C. viride* (Buckel *et al.*, 1994) via the 5-hydroxyvalerate pathway. Other *C. sticklandii* strains also couple ornithine oxidation to proline/lysine reduction.

Enteric bacteria decarboxylate lysine, ornithine and arginine to the primary amines cadaverine, putrescine and agmatine, respectively. Cadaverine (1,5-diaminopentane) is degraded syntrophically by co-cultures of the firmicute strain LC13R with either of the methanogens *Methanobrevibacter arboriphilus* or *Methanospirillum hungatei* M1h (Roeder and Schink, 2009). LC13R deaminates cadaverine to 5-aminovaleraldehyde, which is oxidized to 5-aminovalerate. 5-Aminovalerate is deaminated to glutaric semialdehyde, which is oxidized to glutarate. The two deaminations are catalysed by NAD^+-dependent glutamate dehydrogenase; the two aldehyde oxidations generate reduced ferredoxin, stimulating the release of H_2, which is consumed by the syntrophic partner. Glutaryl-CoA is oxidized to glutaconyl-CoA, which is decarboxylated to crotonyl-CoA. Crotonyl-CoA is reduced to butyrate and oxidized to acetate by SLP. Agmatine is deaminated by *Streptococcus faecalis* (Roon and Barker, 1972) to *N*-carbamylputrescine, which is cleaved into putrescine and carbamyl phosphate (which produces CO_2 and ammonia, and generates ATP). Putrescine fermentation by *Anaevorax odorimutans* starts with a transamination to 4-aminobutyraldehyde and a subsequent oxidation to 4-aminobutyrate (Matthies *et al.*, 2000), which is further degraded via the 4-hydroxybutyrate pathway.

Glycine fermentation occurs in *C. purinolyticum* (Dürre *et al.*, 1981), *Eubacterium acidaminophilum* and *C. litorale* (Fendrich *et al.*, 1990; Dietrichs *et al.*, 1991). Oxidation to CO_2 and ammonia is catalysed by the glycine decarboxylase complex. The reducing equivalents (3 NADPH) are transferred from the decarboxylase complex via thioredoxin to the glycine reductase complex. Reductive deamination yields acetyl phosphate (Cone *et al.*, 1976), which is converted to acetate by SLP. Deamination involves the selenocysteine-containing glycine reductase complex, whereby an unusual glycine-C2-selenoether intermediate is formed (Arkowitz and Abeles, 1990). Asparagine is deaminated into aspartate, and aspartate is deaminated directly to fumarate (desaturation) or converted via oxaloacetate and malate to fumarate. Fumarate is transformed via succinate and succinic semialdehyde to 4-hydroxybutyrate, which is degraded via the 4-hydroxybutyrate pathway (Takahashi *et al.*, 2000). A unique pyridoxal-5′-phosphate (PLP)-dependent reductive β-decarboxylation of aspartate to L-alanine occurs in *C. perfringens* (Meister *et al.*, 1951). L-Alanine is fermented via lactate in other *Firmicutes*.

Glutamine is deaminated directly to glutamate, whereas histidine is first deaminated to urocanate (Wickramasinghe and Fry, 1954). Following the oxidation, reduction and hydrolysis of urocanate, α-formamidino-glutarate is formed, which hydrolyses to formamide (Wachsman and Barker, 1954) and glutamate. Glutamate is transformed in non-Stickland fashion via five distinct pathways (Buckel, 2001). Glutamate, histidine, glutamine, leucine, alanine, cysteine, serine, threonine, methionine, phenylalanine, tyrosine and tryptophan are all fermented as single amino acids via the 2-hydroxyacid pathway that has been studied extensively by Buckel and coworkers (Kim *et al.*, 2004; Buckel *et al.*, 2012). *F. varium* catabolizes glutamate via the 2-hydroxyglutarate pathway when vitamin B_{12} is limiting, and via the 3-methylaspartate pathway when it is not limiting (Ramezani *et al.*, 2011).

29.3 Unusual Reactions in Amino Acid Fermentation

Direct deamination of most of the α-amino acids is difficult because the catalyst must abstract a non-acidic β-proton while the acidic hydrogen at the α-ammonium group remains, thus maintaining the ability of the leaving group. The pK_a ~40 of the β-hydrogen in α-amino acids is beyond the range of acid–base enzymatic catalysis (Buckel and Keese, 1995). Therefore, enzymes containing cofactors such as [4Fe-4S] clusters, flavins, *S*-adenosylmethionine (SAM), B_{12}-derivatives and/or PLP must activate the non-acidic hydrogens either of the amino acids, or of the 2-, 4- and 5-hydroxyacyl-CoA thioesters derived from them.

Carbon-skeleton mutases containing either B_{12} or SAM, as well as [4Fe-4S] clusters, mediate radical rearrangement of the initial amino acid into an isomer in which the hydrogen to be eliminated as a proton is in an activated position (Golding *et al.*, 1998). B_{12}-dependent glutamate mutase occurs in the 3-methylaspartate pathway (Barker, 1978; Buckel and Miller, 1987; Härtel and Buckel, 1996). Lysine fermentation involves rearrangements via SAM-[4Fe4S] cluster-PLP-dependent and B_{12}-PLP-dependent aminomutases, but the subsequent elimination of ammonia is oxidative (Frey *et al.*,1998). A glutamate aminomutase resembling the SAM-[4Fe-4S] cluster-PLP-dependent lysine aminomutase (Ruzicka and Frey, 2007) could be involved in degrading L-glutamate via β-glutamate and 3-oxoglutarate.

In the 2-hydroxyacid pathways (Buckel, 2001), the removal of non-acidic β-protons of 2-hydroxyacyl-CoA proceeds via radical chemistry initiated by [4Fe-4S] clusters (Buckel, 2001). In the 4-hydroxybutyrate pathway, radical chemistry via flavin and [4Fe-4S] clusters results in the elimination of an unactivated β-proton from 4-hydroxybutyryl-CoA. A non-radical FAD (flavin adenine dinucleotide)-catalysed oxidation introduces a double bond and activates the γ-hydrogen of 5-hydroxyvaleryl-CoA in the 5-hydroxyvalerate pathway (Eikmanns and Buckel, 1991a,b; Buckel *et al.*, 1994).

29.4 The 2-Hydroxyacid Pathway

Four variants of this pathway catabolize the following amino acids via the 2-hydroxyacids indicated; the major organisms using them are shown in parentheses:

1. Glutamate, histidine and glutamine via 2-hydroxyglutarate (*C. symbiosum*, *F. nucleatum*, *Acidaminococcus fermentans*, *C. sporosphaeroides*).
2. Leucine via 2-hydroxyisocaproate (*C. difficile*).
3. Alanine, cysteine and serine via lactate, threonine and methionine via 2-hydroxybutyrate, through a common set of enzymes (*C. propionicum*, *Megasphaera elsdenii*).
4. Phenylalanine via 3-phenyllactate, tyrosine via 4-hydroxy-3-phenyllactate and tryptophan via 3-indolelactate (collectively the 3-aryllactate pathways) by the action of a common set of enzymes (*C. sporogenes*).

29.4.1 Pathway and mechanism of dehydration

Serine and threonine are dehydrated to the corresponding α,β-unsaturated amino acids. Serine dehydratase is iron-sulfur dependent and threonine dehydratase is PLP dependent (Hofmeister *et al.*, 1993). The unsaturated amino acids tautomerize and rehydrate non-enzymatically to yield the 2-ketoacids, pyruvate and 2-oxobutyrate, respectively. Cysteine has been proposed to undergo the same reaction sequence, producing H_2S, ammonia and pyruvate (Lewis and Elsden, 1955). A PLP-dependent L-methioninase (Kreis and Hession, 1973) converts methionine into methanethiol, ammonia and 2-oxobutyrate. Other amino acids fermented via the 2-hydroxyisocaproate, lactate and 3-aryllactate pathways produce the corresponding 2-ketoacids via NAD^+-PLP-dependent transaminations. Every 2-ketoacid is oxidatively decarboxylated with CoA-thioesterification to the respective acyl-CoA one carbon atom shorter by a pathway-specific 2-ketoacid:ferredoxin-oxidoreductase. Subsequently, SLP occurs with the formation of the oxidized end products – acetate/SCFA/arylacetate. The 2-ketoacids are reduced by NADH-dependent *Re* (face)-stereospecific dehydrogenases (Berk *et al.*, 1996). The resulting (*R*)-2-hydroxyacids are thioesterified to the 2-hydroxyacyl-CoA by specific CoA-transferases, but the β-protons still remain unactivated.

In every 2-hydroxyacyl-CoA dehydratase (2-HADH), a substrate radical generated via [4Fe-4S] clusters causes *Umpolung* (charge reversal) at the keto group of the (*R*)-2-hydroxyacyl-CoA, as shown in Fig 29.1A (Buckel and Keese, 1995). The ketyl radical generated by one-electron reduction of the substrate eliminates water via proton transfer to a conserved glutamate residue (Knauer *et al.*, 2011). In the enoxy radical, the β-hydrogen (bold wedge, bottom formula, Fig 29.1A) has a $pK_a < 15$; binding with active-site residues reduces this value further. β-Proton abstraction by Fe-O^- from one of the [4Fe-4S] clusters generates the allylic ketyl radical, which yields the (*E*)-2-enoyl-CoA and recycles an electron. In *C. difficile*, the allylic ketyl radical was detected by electron paramagnetic resonance (EPR) spectroscopy of 2-hydroxyisocaproyl-CoA dehydratase turning over (*R*)-2-hydroxyisocaproyl-CoA (Kim *et al.*, 2008). Hence, the reversible α,β-*syn*-dehydration of (*R*)-2-hydroxyacyl-CoA into (*E*)-2-enoyl-CoA is a key reaction, making 2-HADH a key enzyme (Kim *et al.*, 2004) in all four variants of this pathway. The steps succeeding the dehydration are distinctive in each variant, based on differing kinetic and energy conservation constraints.

In the 3-aryllactate variant, 3-arylacrylates are reduced to 3-arylpropionates. In the other three variants, the reduction of a 2-enoyl-CoA to the corresponding SCFA CoA-thioester occurs. In every case except the 2-hydroxyglutarate variant, the reduced end products (SCFAs/3-arylpropionates) have the same chain length as the parent amino acid. The 2-hydroxyglutarate pathway differs from the other variants in several ways. It lacks the NAD^+-PLP-dependent transaminase; instead, an NAD^+-dependent glutamate dehydrogenase generates 2-oxoglutarate from glutamate. It also lacks a 2-ketoacid:ferredoxin-oxidoreductase. The 2-ketoacid reduction, CoA transfer and dehydration steps are exactly analogous to the other pathway variants. The dehydration of (*R*)-2-hydroxyglutaryl-CoA to (*E*)-glutaconyl-CoA is followed by a biotin-dependent decarboxylation yielding crotonyl-CoA (Buckel and Semmler, 1983). The hydration of crotonyl-CoA to 3-hydroxybutyryl-CoA leads via acetoacetyl-CoA to acetate with SLP (as in β-oxidation). Crotonyl-CoA is converted via butyryl-CoA to butyrate.

29.4.2 Activation of 2-hydroxyacyl-CoA dehydratase (2-HADH)

Most well-studied 2-HADHs are heterodimers; only the 2-hydroxyglutaryl-CoA dehydratase from *F. nucleatum* is a heterotrimer (Klees *et al.*, 1992). Every 2-HADH contains two [4Fe-4S] clusters and requires reduction by a [4Fe-4S] cluster-containing ATP-ase, the activator or 'archerase' (Bendrat *et al.*, 1993). Unlike other 2-HADHs, which interact only with their activators, the *C. sporogenes* aryllactyl-CoA dehydratase complex also contains the CoA-transferase (Dickert *et al.*, 2000). One-electron reduction of a 2-HADH by its activator involves a large conformational change of the helix-[4Fe-4S] cluster-helix motif of the latter coupled to the hydrolysis of 2 ATP molecules (Knauer *et al.*, 2011). An Mg-ATP-induced complex of 2-HADH and activator can be trapped with $[ADP\text{-}AlF_4]^-$. Once the electron transfer is complete, the activator dissociates from the 2-HADH, as deduced from chelation experiments (Kim *et al.*, 2008). Thus 'activated', the 2-HADH attains potentials of around −900 mV (Buckel *et al.*, 2012) and catalyses ~10^4 turnovers before inactivation by oxidation. Therefore, electron transfer involving [4Fe-4S] clusters as the sole cofactors in the 2-HADH-activator system facilitates radical dehydrations with minimal ATP hydrolysis.

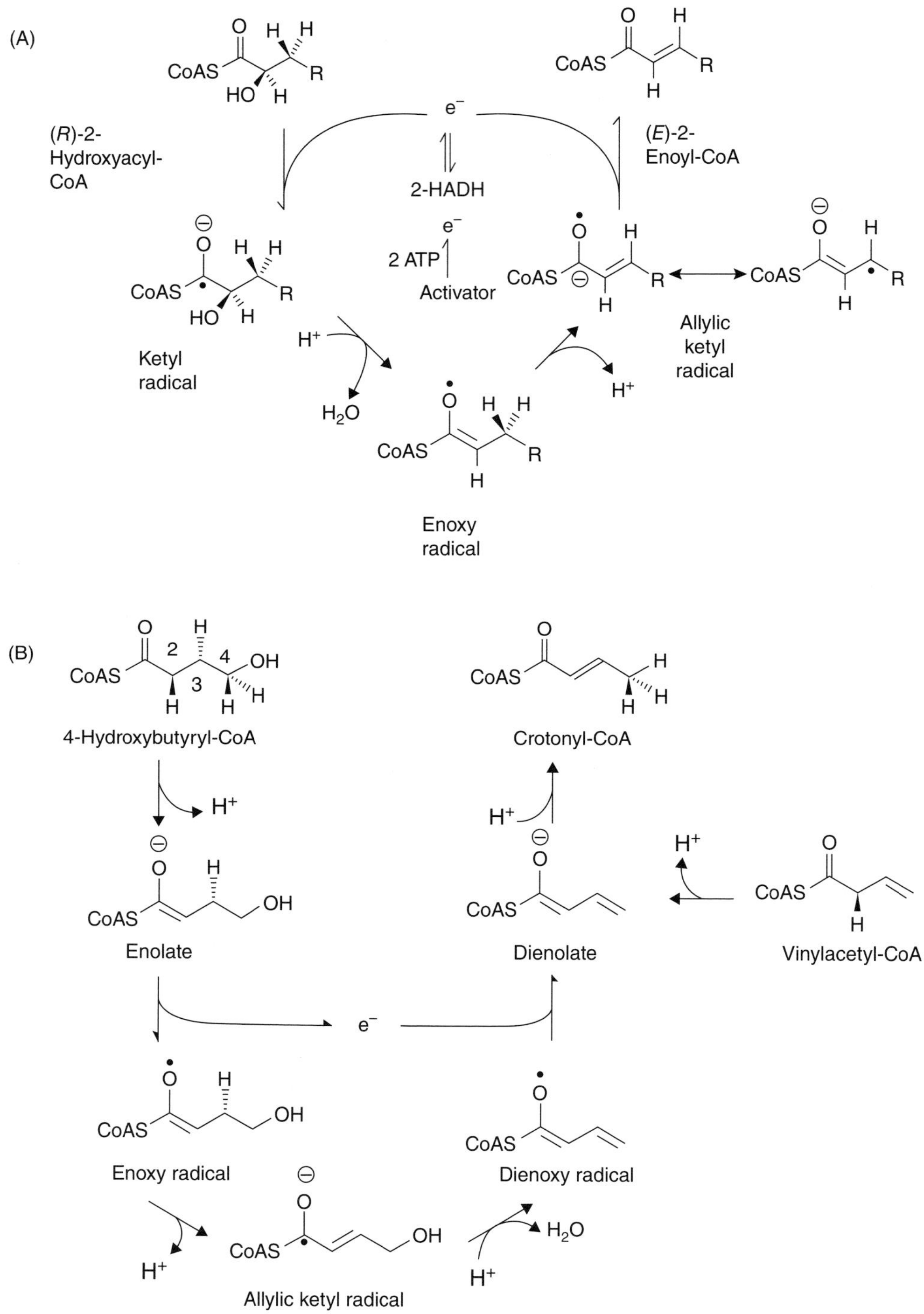

Fig. 29.1. (A) The mechanism of dehydration of (*R*)-2-hydroxyacyl-CoA into (*E*)-2-enoyl-CoA catalysed by 2-hydroxyacyl-CoA dehydratase (2-HADH) in the 2-hydroxyacid pathways. (B) The mechanism of dehydration of 4-hydroxybutytryl-CoA into (*E*)-crotonyl-CoA catalysed by 4-hydroxybutyryl-CoA dehydratase (4-HBDH) in the 4-hydroxybutyrate pathway.

29.4.3 Bioenergetics of the 2-hydroxyacid pathways

Reduced ferredoxins/flavodoxins (standard reduction potential, $E_0' = -400$ to -450 mV) are required (Thamer *et al.*, 2003) to generate the reduced activators. These highly negative redox potentials could be supplied by ferredoxin- or flavodoxin-oxidoreductases which oxidize 2-ketoacids coupled to the addition of CoA, yielding acyl-CoA/arylacyl-CoA. As the 2-hydroxyglutarate pathway does not involve 2-ketoacid:ferredoxin-oxidoreductases, ATP generation must be linked to other steps. Energy is conserved via the translocation of Na^+ to the outside of the cell by the membrane-bound glutaconyl-CoA decarboxylase subunits (Buckel and Semmler, 1983). ATP formation is coupled to the reduction of 3-arylacrylates in the 3-aryllactate pathways of *C. sporogenes* (Bader and Simon, 1983). Recent studies suggest that there is additional energy conservation in *C. sporogenes* via the translocation of Na^+ by a Rhodobacter nitrogen fixation protein (Rnf)-complex containing the NADH:ferredoxin-oxidoreductase (Li, 2014). Apart from SLP from CoA thioesters and H_2 production, some glutamate fermenters using the 2-hydroxyglutarate pathway also conserve energy via other mechanisms (see Section 29.7).

A summary is given for the 2-hydroxyglutarate pathway in the following equation:

$$5\ \text{Glutamate}^- + 6\ H_2O + 2\ H^+ \rightarrow 5\ NH_4^+ + 5\ CO_2 + 6\ \text{Acetate}^- + 2\ \text{Butyrate}^- + H_2$$

where $\Delta G^{o\prime} = -59$ kJ mol^{-1} glutamate; 70 kJ mol^{-1} ATP. (The $\Delta G^{o\prime}$ values do not include the energy conserved via electron-bifurcation.)

A summary is given for the alanine via 2-hydroxyacid (lactate) pathway in the following equation:

$$3\,(S)\text{-Alanine} + 2\ H_2O \rightarrow 3\ NH_4^+ + CO_2 + \text{Acetate}^- + 2\ \text{Propionate}^-$$

where $\Delta G^{o\prime} = -118$ kJ mol^{-1} acetate; 118 kJ mol^{-1} ATP.

In the 3-aryllactate pathways, the same equation holds with arylalanines, arylacetates and 3-arylpropionates instead of alanine, acetate and propionate, respectively.

29.4.4 Reaction constants of 2-HADH and physiological implications

Experimental and computational analysis of the equilibrium constants of the 2-HADH reactions (Parthasarathy *et al.*, 2010) revealed large substrate β-substituent effects on the kinetics. Consequently, enzyme function downstream of 2-HADH differs among the 2-hydroxyacid pathway variants. The dehydration of 2-hydroxyglutaryl-CoA is reversible and the subsequent decarboxylation drives the 2-hydroxyglutarate pathway towards completion. The dehydration of 2-hydroxyisocaproyl-CoA is essentially irreversible, and no subsequent decarboxylation occurs in the 2-hydroxyisocaproate pathway. The dehydration of lactyl-CoA to acrylyl-CoA is unfavourable in the forward direction. Physiological coupling of this reaction with the exergonic reduction of acrylyl-CoA to propionyl-CoA makes the pathway irreversible and quenches the toxic electrophile acrylyl-CoA. 3-Aryllactyl-CoA dehydrations are reversible, but favoured in the forward direction owing to stabilization by extended conjugation in the 3-arylacrylyl-CoA products (Li, 2014).

29.5 The 4-Hydroxybutyrate and 5-Hydroxyvalerate Pathways: Flavin-mediated Elimination of the Terminal Hydroxy Groups

29.5.1 The 4-hydroxybutyrate pathway

Aerobic and anaerobic bacteria produce 4-aminobutyrate/γ-aminobutyric acid (GABA) by several reactions, including the reduction of 4-aminobutyraldehyde (putrescine fermentation) and the decarboxylation of glutamate. *C. aminobutyricum* ferments 4-aminobutyrate to acetate and butyrate. The transamination of 4-aminobutyrate yields succinic semialdehyde, which is reduced to 4-hydroxybutyrate with NADH, followed by the thioesterification with CoA to 4-hydroxybutyryl-CoA catalysed by a specific CoA-transferase (Scherf and Buckel, 1991; Gerhardt *et al.*, 2000; Macieira *et al.*, 2009). The dehydration of 4-hydroxybuytryl-CoA to crotonyl-CoA entails removing the β-proton (pK_a ~ 40) and the γ-hydroxy group. 4-Hydroxybutyryl-CoA dehydratase (4-HBDH), the key enzyme, is a homotetramer with an FAD and a [4Fe–4S] cluster per subunit (Müh *et al.*, 1996). The dehydration of 4-hydroxybutyryl-CoA is reversible (Willadsen and Buckel, 1990), while the Δ-isomerization of vinylacetyl-CoA to crotonyl-CoA is irreversible (Scherf and Buckel, 1993). An FAD-based one-electron oxidation mechanism (Fig 29.1B) has been proposed for the dehydration. The substrate displaces active-site water molecules and induces electron transfer to the FAD and to the [4Fe-4S] cluster. Analysis of 4-HBDH mutants (Zhang *et al.*, 2015) indicates that a tyrosyl radical generates the one-electron acceptor $FADH^{\bullet}$ rather than the substrate itself. Following the loss of an acidic α-proton (bold wedge, C2, top left), the enolate formed is oxidized by $FADH^{\bullet}$ to yield an enoxy radical, in which the pK_a of the β-hydrogen is lowered by 26 units (Smith *et al.*, 2003). The β-proton (dotted wedge, C3, bottom left) is abstracted by the $FADH^{-}$ to yield the allylic ketyl radical (Buckel and Keese, 1995).

Alternatively, the enolate can be directly oxidized to the allylic ketyl radical via a proton-coupled electron-transfer (PCET). This radical expels the hydroxyl-group (Fig 29.1B, centre) aided by the [4Fe–4S] cluster and a catalytic histidine. The dienoxy radical formed is reduced to the dienolate, regenerating the $FADH^{\bullet}$. Protonation at C4 of the planar dienolate at the same side from which the OH leaves yields the product crotonyl-CoA (Buckel and Golding, 1998, 2006). Isotopic labelling studies have established that the 2-H*Re* and 3-H*Si* (where *Re* and *Si* designate the heterotropic faces) of 4-hydroxybutyryl-CoA exchanged with the solvent (Scott *et al.*, 2004; Friedrich *et al.*, 2008). The substitution of OH with H at C4 of 4-hydroxybutyryl-CoA with retention of the configuration was demonstrated by the chiral acetate method (Friedrich *et al.*, 2008). Crotonyl-CoA is reduced to butyrate via butyryl-CoA, and oxidized to acetate via 3-hydroxybutyryl-CoA and acetoacetyl-CoA, thus completing the pathway. SLP occurs from acetyl-CoA and butyryl-CoA.

A summary is given for the 4-hydroxybutyrate pathway in the following equation:

$$2\times \text{4-Aminobutyrate}^- + 2\,H_2O \rightarrow 2\,NH_4^+ + 2\,\text{Acetate}^- + \text{Butyrate}^- + H^+$$

where $\Delta G^{\circ\prime} \approx -37$ kJ mol^{-1} 4-aminobutyrate; 74 kJ mol^{-1} ATP.

29.5.2 The 5-hydroxyvalerate pathway

Three distinct microorganisms are needed to degrade glutamate via 5-aminovalerate. In *Bacillus subtilis*, osmotic stress induces the reduction of glutamate into the osmoprotectant proline (Bremer and Krämer, 2000). *C. sporogenes* reduces proline to 5-aminovalerate in a Stickland fermentation (Stickland, 1935; Barker, 1961). Finally, *C. viride* (Buckel *et al.*, 1994) ferments 5-aminovalerate (Fig 29.2A), which undergoes deamination and reduction into 5-hydroxyvalerate; the specific transfer of CoA yields 5-hydroxyvaleryl-CoA. The direct dehydration of 5-hydroxyvaleryl-CoA to 4-pentenoyl-CoA is precluded owing to the $pK_a \sim 40$ of the γ-hydrogen (Buckel and Keese, 1995).

An FAD-containing bifunctional acyl-CoA dehydrogenase/5-hydroxyvaleryl-CoA dehydratase (5-HVDH) converts 5-hydroxyvaleryl-CoA into 5-hydroxy-2-pentenoyl-CoA in an acyl-CoA-dehydrogenase reaction (a two-electron oxidation). The introduction of a double bond activates the γ-hydrogen, and 5-HVDH reversibly dehydrates 5-hydroxy-2-pentenoyl-CoA to 2,4-pentadienoyl-CoA (Eikmanns and Buckel, 1991a), which is reduced with NADH to 3-pentenoyl-CoA by the FAD-containing 2,4-pentadienoyl-CoA reductase (Eikmanns and Buckel, 1991b). 3-Pentenoyl-CoA is rearranged to 2-pentenoyl-CoA by 3-pentenoyl-CoA-Δ^3,Δ^2-isomerase. 2-Pentenoyl-CoA disproportionates into valerate, acetate and propionate via the respective CoA-thioesters (Barker *et al.*, 1987, Buckel *et al.*, 1994). SLP occurs from acetyl-CoA.

A summary is given for the 5-hydroxyvalerate pathway in the following equation:

$$2\times \text{5-Aminovalerate}^- + 2\,H_2O \rightarrow 2\,NH_4^+ + \text{Acetate}^- + \text{Propionate}^- + \text{Valerate}^- + H^+$$

where $\Delta G^{\circ\prime} \approx -37$ kJ mol^{-1} 5-aminovalerate; 74 kJ mol^{-1} ATP.

29.6 Pathways via Carbon Skeleton Rearrangements

29.6.1 Lysine fermentation via 3,5-diaminohexanoate

Lysine fermentation in *C. subterminale* and *C. sticklandii* (Fig 29.2B) begins with the rearrangement

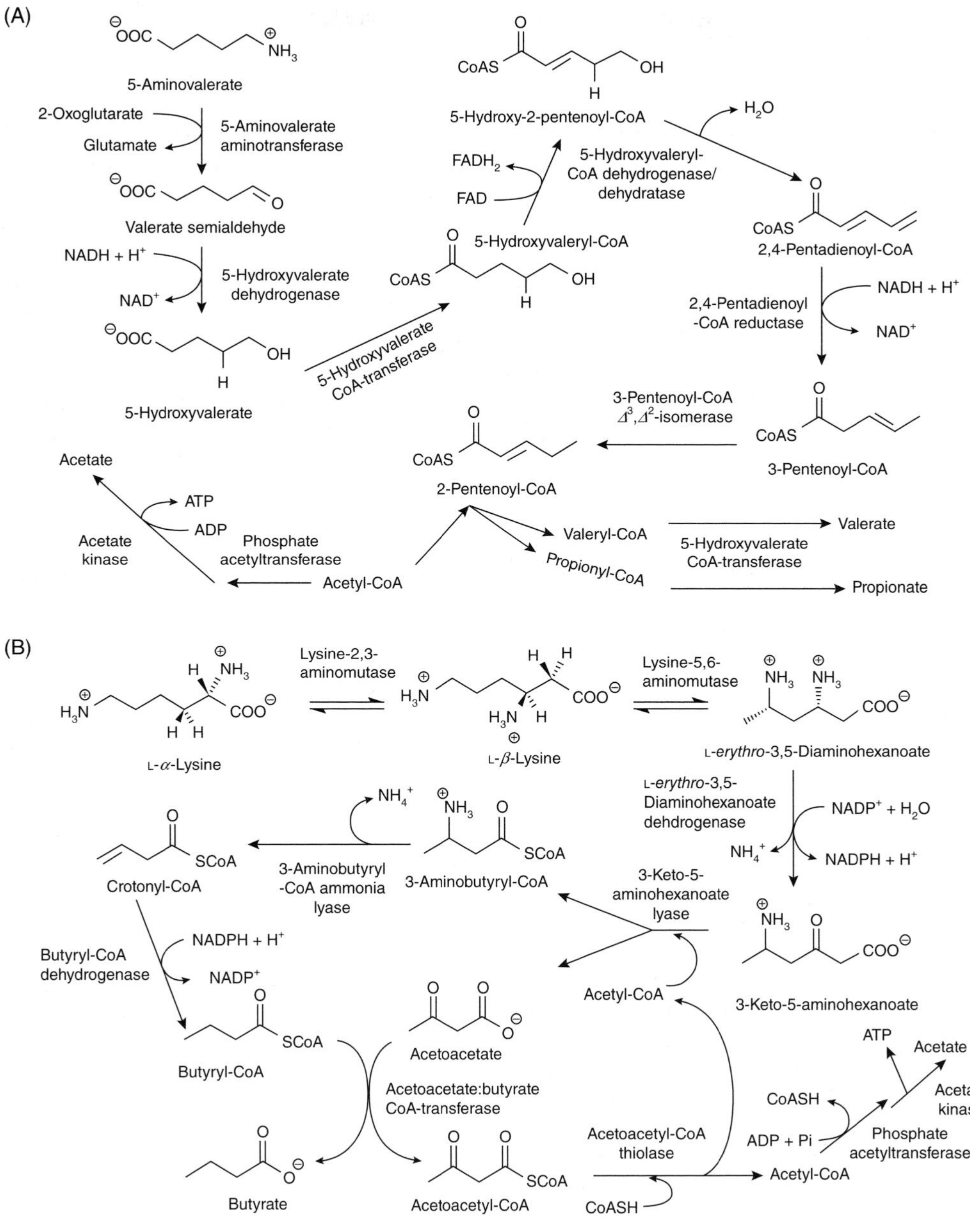

Fig. 29.2. (A) Fermentation of 5-aminovalerate (derived from glutamate) by *Clostridium viride*. (B) L-α-Lysine fermentation into butyrate and acetate via two aminomutases.

of L-α-lysine to L-β-lysine by lysine-2,3-aminomutase (Chirpich *et al.*, 1970; Stadtman, 1973). The PLP of lysine-2,3-aminomutase forms a Schiff base with L-α-lysine by removing a C3 hydrogen. One edge of the [4Fe-4S] cluster of the enzyme is coordinated by the termini of SAM-methionine. Upon reduction of the [4Fe-4S] cluster by reduced ferredoxin/flavodoxin, SAM generates the 5′-deoxyadenosine radical or Ado-CH2• (Frey *et al.*, 1998; Chen *et al.*, 2003). The L-α-lysine-3-yl radical rearranges to the L-β-lysine-2-yl radical. Ado-CH2• donates a methyl hydrogen back to the β-lysine-2-yl radical and β-lysine is formed.

β-Lysine is rearranged to L-*erythro*-3,5-diaminohexanoate by the PLP-B_{12}-dependent *β*-lysine-5,6-aminomutase (Stadtman and Renz, 1968). This reaction mechanism is nearly identical to that of lysine-2,3-aminomutase, except that coenzyme B_{12} generates the Ado-$CH_2^{\bullet}$ instead of SAM. The intermediate methylene radical in *β*-lysine-5,6-aminomutase is stabilized by interaction with cob(II)alamin, which is unnecessary in the case of *α*-lysine-2,3-aminomutase (Buckel *et al.*, 2005). The aminomutase reactions are followed by an $NADP^+$-dependent dehydrogenation of L-3,5-diaminohexanoate to 3-keto-5-aminohexanoate, which is cleaved to 3-aminobutyryl-CoA and acetoacetate via the addition of acetyl-CoA. An ammonia lyase deaminates 3-aminobutyryl-CoA to crotonyl-CoA. Other enzymes convert crotonyl-CoA to butyrate via butyryl-CoA and degrade acetoacetate to acetate with SLP (Kreimeyer *et al.*, 2007).

29.6.2 The 3-methylaspartate (mesaconate) pathway

This was the first glutamate fermentation pathway to be elucidated and occurs in *C. tetanomorphum* (Barker, 1978), *C. cochlearium*, *C. lentoputrescens*, *C. limosum*, *C. malenominatum* and *C. tetani* (Buckel and Barker, 1974; Buckel, 1980). L-Glutamate is converted by glutamate mutase into (2*S*,3*S*)-3-methylaspartate. Direct deamination of 3-methylaspartate via an ammonia lyase now occurs because the *α*-proton to be eliminated is acidic. The mesaconate formed by deamination is hydrated by mesaconase to (*S*)-citramalate, which is cleaved by citramalate lyase to acetate and pyruvate. This sequence of reactions is shown in the central part of Fig. 29.3. The pathway can then follow two different routes from pyruvate.

In the first variant, the final products are butyrate and CO_2 (Fig 29.3, right). The butyrate

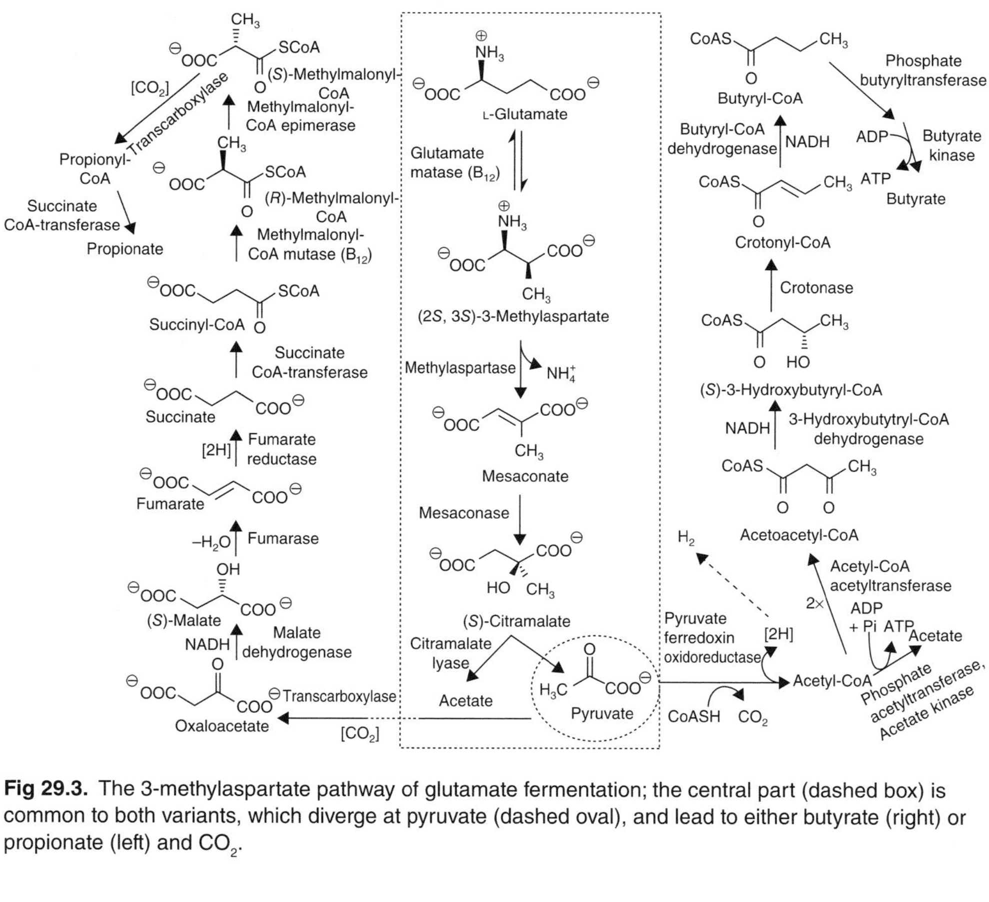

Fig 29.3. The 3-methylaspartate pathway of glutamate fermentation; the central part (dashed box) is common to both variants, which diverge at pyruvate (dashed oval), and lead to either butyrate (right) or propionate (left) and CO_2.

is formed from pyruvate via acetyl-CoA, acetoacetyl-CoA, (*S*)-3-hydroxybutyryl-CoA, crotonyl-CoA and butryryl-CoA, concomitant with the production of H_2 and additional acetate. The reduction of crotonyl-CoA to butyryl-CoA by an FAD-containing dehydrogenase occurs with energy conservation by electron bifurcation (see Section 29.7.3).

A summary of the 3-methylaspartate pathway to butyrate is given in the following equation:

$$5\ \text{Glutamate}^- + 6\ H_2O + 2\ H^+ \rightarrow 5\ NH_4^+ + 5\ CO_2 + 6\ \text{Acetate}^- + 2\ \text{Butyrate}^- + H_2$$

where $\Delta G^{\circ\prime} = -63.5$ kJ mol^{-1} glutamate; 106 kJ mol^{-1} ATP.

In a variation of the 3-methylaspartate pathway, the pyruvate formed from the cleavage of citramalate by citramalate lyase is degraded to acetate, CO_2 and propionate (Fig 29.3, left), accompanied by SLP. Additional energy is conserved in the pathway from pyruvate to propionate via fumarate (Lancaster *et al.*, 1999). The pathway leading to propionate has been found in *Selenomonas acidaminophila* (Nanninga *et al.*, 1987) and *Barkera propionica* (J. Sievers, W. Liesack, W. Buckel, unpublished).

A summary is given for this second variant of the 3-methylaspartate pathway in the following equation:

$$3\ \text{Glutamate}^- + 4\ H_2O \rightarrow 3\ NH_4^+ + 2\ CO_2 + 5\ \text{Acetate}^- + \text{Propionate}^-$$

where $\Delta G^{\circ\prime} = -69$ kJ mol^{-1} glutamate; ≥ 62 kJ mol^{-1} ATP.

29.6.2.1 *Glutamate mutase*

Studies on this first enzyme in the 3-methylaspartate (mesaconate) pathway, the coenzyme B_{12}-dependent glutamate mutase (GlmES, where E and S are the two subunits), revealed that the *glmS* and *glmE* genes are separated by *glmL*, which encodes a putative chaperone (Buckel *et al.*, 1999). In the X-ray diffraction (XRD) structure of *C. cochlearium* GlmES (Reitzer *et al.*, 1999), the B_{12} corrin ring is located between the subunits of the heterotetrameric enzyme. The 5,6-dimethylbenzimidazole ligand of coenzyme B_{12} is displaced from cobalt coordination by a conserved histidine from GlmS. The upper 5′-deoxyadenosine residue (derived by homolysis of the Co–C σ-bond of coenzyme B_{12}) protrudes into the active-site within GlmE. Substrate binding between one glutamate and three arginine residues (the 'arginine claw') in the GlmES-B_{12} complex induces homolytic cleavage of the carbon–cobalt bond of coenzyme B_{12}, producing cob(II)alamin and Ado-$CH_2^\bullet$. 4-*Si* hydrogen abstraction by Ado-$CH_2^\bullet$ yields 5′-deoxyadenosine and a 4-glutamyl radical detected by EPR spectroscopy (Bothe *et al.*, 1998). The 4-glutamyl radical fragments into acrylate and a glycine radical (Chih and Marsh, 2000), which recombine to the 3-methyleneaspartate radical. Finally, 3-methylaspartate is formed, thereby regenerating Ado-$CH_2^\bullet$ (Beatrix *et al.*, 1995).

29.7 Energy Conservation in Anaerobes and Electron Bifurcation

29.7.1 Anaerobic energy metabolism

Energy conservation in biology occurs via two major mechanisms, SLP and ion-gradient phosphorylation (IGP). IGP functions in tandem with an ATP-synthase driven by an electrochemical ion-gradient (H^+ or Na^+) across a membrane. In aerobes, energy is conserved by SLP in the oxidative branch of metabolism; in the reductive branch, ATP synthesis is coupled to the electron-transport chain (electron-transport phosphorylation, or ETP; Mitchell, 1975). In the absence of oxygen as the terminal electron acceptor, anaerobes use organic or inorganic compounds as electron sinks. Any anaerobic energy-conserving process that couples electron transport to phosphorylation is termed anaerobic respiration. In contrast, all anaerobic energy-conserving metabolic processes with SLP, but not ETP, are fermentations.

The accepted wisdom at one time was that fermentations usually occurred in the cytoplasm, and that anaerobic respiration involved an ion pump driven by electron transport in combination with an ATP synthase. Although the definition of fermentation excludes the participation of electron-transport chains, strict anaerobes such as *C. perfringens* (Ishimoto *et al.*, 1974; Hasan and Hall, 1975), *C. tertium* (Hasan and Hall, 1977) and *Propionibacterium* spp. (Kaneko and Ishimoto, 1977) contain subunits of the dissimilatory nitrate reductase, a potential electron sink for oxidative SLP. Fermentation is not entirely distinct from membrane electron transport because many redox processes cause the extrusion of H^+ or Na^+ from the cytoplasm, e.g. the decarboxylation of glutaconyl-CoA to crotonyl-CoA by biotin-dependent glutaconyl-CoA decarboxylase (see Section 29.4.3). This creates an electrochemical gradient that results in energy conservation through ion import via an ATP synthase. The discovery of a membrane-bound NADH:ferredoxin oxidoreductase capable of generating a Na^+ gradient, and its link to an F-type ATP-ase, has transformed our understanding of energy conservation in anaerobic bacteria (Boaingiu *et al.*, 2005; Imkamp *et al.*, 2007; Biegel and Müller, 2010).

29.7.2 Ferredoxins and anaerobic metabolism

Ferredoxins are important electron carriers in anaerobic energy metabolism in clostridia, acetogens and sulfate-reducing bacteria. Within bacterial cells, ferredoxins are over 90% reduced and function as electron donors in metabolic reactions with an E_0' of around −500 mV (Buckel and Thauer, 2013). Known ferredoxin-dependent reactions include the reduction of protons to H_2 (E_0' = −414 mV; Chen and Mortenson, 1974), and the reversible oxidation of pyruvate to acetyl-CoA and CO_2 (E_0' = −520 mV; Uyeda and Rabinowitz, 1971), which belongs to the category of 2-ketoacid oxidations yielding acyl-CoA/arylacyl-CoA (29.4.3). In most metabolic pathways, reduced ferredoxin has to be reoxidized by other means in order for its stoichiometry to remain balanced. However, in the case of *A. fermentans* growing on glutamate, H_2 was formed in a ferredoxin-dependent reaction with a soluble hydrogenase, but no reaction was known to generate reduced ferredoxin. Initial experiments on hydrogen formation (Härtel and Buckel, 1996) in *A. fermentans* showed that NADH generated by oxidising (*S*)-3-hydroxybutyryl-CoA to acetoacetyl-CoA could produce H_2 in an Na^+-dependent manner, though the observed hydrogen partial pressure could not be explained satisfactorily by an NADH-dehydrogenase model involving reverse electron transport from NADH to H_2.

Two novel types of ferredoxin-dependent catabolic enzymes could explain the enigma of ferredoxin reduction in anaerobic bacteria: (i) a soluble enzyme complex that couples the endergonic reduction of ferredoxin by NAD(P)H or H_2 to an exergonic reduction via flavin-based electron bifurcation (FBEB); and (ii) the membrane-associated Rnf A-G complex catalysing the oxidation of reduced ferredoxin with NAD^+ and H^+. The exergonic reduction of NAD^+ with reduced ferredoxin builds up an electrochemical H^+/Na^+ gradient (Biegel and Müller, 2010) that conserves energy in fermenters. These enzyme complexes have been found not only in glutamate fermenters, but also in acetogens and methanogenic archaea.

29.7.3 Flavin-based electron bifurcation (FBEB)

FBEB was first proposed in 2008 (Herrmann *et al.*, 2008) as a means whereby exergonic reduction of crotonyl-CoA with NADH could be coupled to endergonic reduction of ferredoxin (Fd) according to the reaction shown in Eqn 29.1:

$$\begin{aligned}&\mathrm{Fd_{oxidized}} + 2\ \mathrm{NADH} + \text{crotonyl-CoA}\\&\quad\rightarrow \mathrm{Fd_{reduced}}^{2-} + 2\ \mathrm{NAD^+} + \text{butyryl-CoA} \qquad (29.1)\end{aligned}$$

The proposed mechanism of this reaction was analogous to the ubiquinone-based electron bifurcation of mitochondria in which the exergonic one-electron oxidation of the fully reduced ubiquinol by the Rieske [2Fe-2S] cluster sustains the one-electron reduction of cytochrome *b* by the ubiquinol semiquinone radical (Brandt, 1996). FBEB was first discovered in a complex of butyryl-CoA dehydrogenase (Bcd) and electron-transferring flavoprotein (Etf) from the ethanol fermenter *C. kluyveri* (Li *et al.*, 2008), and later in the glutamate fermenters *C. tetanomorphum* (the 3-methylaspartate pathway) and *A. fermentans* (the 2-hydroxyglutarate pathway).

The Bcd-Etf complex (see Fig. 29.4A) catalysing the reaction shown in Eqn (29.1) contains FAD as the sole cofactor. In *C. tetanomorphum*, the Bcd/Etf subunits as determined by SDS-PAGE are α (Bcd) 40 kDa, β (EtfB) 36 kDa and γ (EtfC) 28 kDa. The apparent molecular mass of this whole complex has been determined as 574 kDa by size-exclusion chromatography, which is nearly twice that of the expected $\alpha_4\beta_2\gamma_2$ composition (288 kDa) found in *C. kluyveri* (Herrmann, 2008; Kölzer, 2008). Each Etf and Bcd subunit

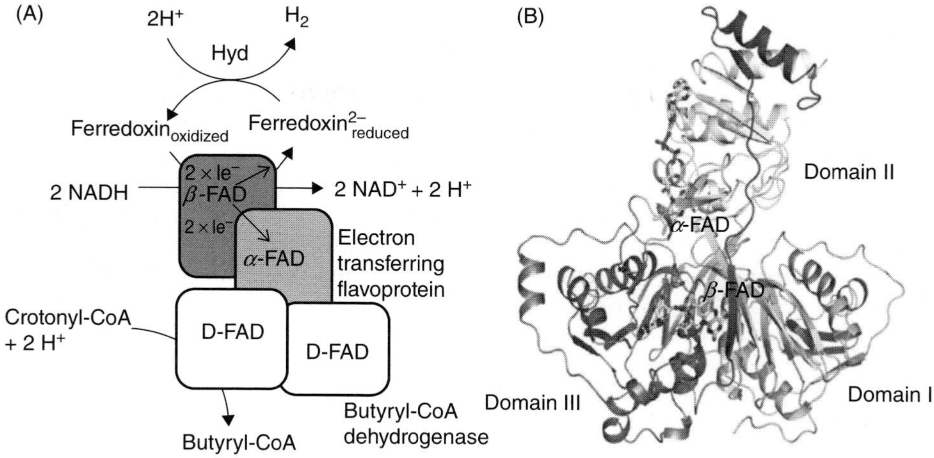

Fig. 29.4. (A) Schematic representation of the clostridial butyryl-CoA dehydrogenase–electron-transferring flavoprotein (Bcd-Etf) complex. (B) Structure of *Acidaminococcus fermentans* Etf; the protein consists of the α-subunit (domains I and II) and the β-subunit (domain III). Structure deposited in the Protein Data Base (PDB) as 4KPU. Abbreviations: α-, β-FAD, flavin adenine dinucleotides of the α- and β-subunits of the Etf protein; D-FAD, flavin adenine dinucleotide of butyryl-CoA dehydrogenase; Hyd: hydrogenase. From Chowdhury *et al.*, 2014.

contains an FAD, with a total of four FADs in the complex. During active bifurcation, NADH first binds the Etf β-subunit and completely reduces the β-FAD to β-FADH$^-$, which then becomes the bifurcating electron donor. The transfer of one electron to the α-FAD by an extensive domain movement of the α-subunit generates α-FAD$^{\bullet-}$. The resulting β-FADH$^{\bullet}$ is unstable and the remaining electron is immediately transferred to the interacting ferredoxin to form reduced ferredoxin (Fd$^-$). The stabilized semiquinone of the α-FAD$^{\bullet-}$ is stable and transfers one electron to the Bcd. In the next round, when another NADH is bound, the process repeats itself. Thus, Bcd receives two electrons which reduce crotonyl-CoA to butyryl-CoA, while the ferredoxin is reduced via Fd$^-$ to the completely reduced Fd^{2-}.

In the XRD structure of the electron-bifurcating *A. fermentans* Etf (see Fig 29.4B), the NADH-binding site is located at the β-subunit. However, in this organism, Etf and Bcd are separate proteins that do not form a tight complex (Chowdhury *et al.*, 2014). A docking model of Bcd-Etf has suggested the possible conformations of the bifurcating complex in *A. fermentans*, although the edge-to-edge distance (18 Å) between the FAD in Bcd and the α-FAD is too far for efficient electron transfer, suggesting that there is a reorientation of the domains during bifurcation.

Ferredoxin reduction under anaerobic conditions is tightly coupled to FBEB, but under aerobic conditions, the ferredoxin can be replaced by oxygen during electron bifurcation. Complete NADH oxidation by the dissociating Etf-Bcd from the lactate-fermenter *M. elsdenii* was observed under aerobic conditions with catalytic amounts of crotonyl-CoA or butyryl-CoA (Chowdhury *et al.*, 2015). In this case, the Etf-Bcd complex acts as an NADH oxidase, generating reactive oxygen species (ROS). FBEB is, therefore, regarded as the Achilles heel for anaerobes as the process produces toxic oxygen radicals and hydrogen peroxide when exposed to air. To counter this problem, anaerobes have evolved a cascade of oxidative stress response proteins, including catalase, NADH-peroxidase and rubrerythrin (Weinberg *et al.*, 2004; Mishra and Imlay, 2012 and 2013).

29.7.4 A membrane-associated Rhodobacter nitrogen fixation protein (Rnf)-complex for ferredoxin reoxidation and energy conservation

Fermentation in anaerobes occurs in an uninterrupted manner only under the continuous supply of oxidized ferredoxin and NADH. In clostridia, the mechanism of NAD$^+$ reduction by reduced ferredoxin (generated from pyruvate oxidation) to obtain NADH for butyrate synthesis remained unknown until recently. Membrane fractions of *C. tetanomorphum* revealed high NADH-dehydrogenase activities with ferricyanide as an electron acceptor, suggesting the presence of an enzyme whose natural substrates were NAD^+ and a low-potential electron donor. Reduced ferredoxin generated from electron bifurcation has a reduction potential that is more negative ($E_0' = -500$ mV) than that of the $NAD^+/NADH$ couple ($E_0' = -280$ mV) and the H^+/H_2 couple (−300 and −414 mV). The potential difference in ΔE_0 of around 220 mV corresponds to an exergonic reaction, so the reduction of NAD^+ by reduced ferredoxin results in the pumping of protons or Na^+ across the bacterial membrane (Tremblay *et al.*, 2012; Hess *et al.*, 2013), and this, in turn, drives the phosphorylation of ADP via a membrane-associated F_0F_1-type ATP synthase complex or an archaeal-type A_0A_1ATP synthase complex. Genome sequence analysis of many anaerobes revealed the presence of the genes for a membrane-associated RnfA-G complex (Biegel *et al.*, 2011) that catalyses the reversible oxidation of reduced ferredoxin with NAD^+, according to the reaction shown in Eqn 29.2:

$$\mathrm{Fd_{reduced}}^{2-} + \mathrm{H}^+ + \mathrm{NAD}^+ \leftrightarrow \mathrm{Fd_{oxidized}} + \mathrm{NADH} + \Delta\mu\mathrm{H}^+/\mathrm{Na}^+ \quad (29.2)$$

The *rnf* genes were first identified in *Rhodobacter capsulatus*, where they were required for nitrogen fixation. The deletion of *rnf* genes in *R. capsulatus*, did not affect the nitrogenase system; rather, the electron transport to the nitrogenase was abolished (Schmehl *et al.*, 1993). Homologous gene clusters were discovered in many aerobic and anaerobic bacteria (Biegel *et al.*, 2011) that coded for a membrane-bound NAD^+:ferredoxin-oxidoreductase. Inverted membrane vesicles of *C. tetanomorphum* catalysed NADH oxidation mediated by ferricyanide at rates of 10–25 U mg^{-1}. The protein concerned, when solubilized from the membranes and purified to homogeneity, revealed that the complex was composed of the six different subunits RnfABCDEG (Herrmann *et al.*, 2008). RnfC is soluble and predicted to interact with NADH. RnfA, RnfD and RnfE are composed of transmembrane α-helices, whereas RnfB and RnfG are membrane anchored. RnfD and RnfG contain covalently bound flavin mononucleotide (FMN).

Interestingly, four out of the six Rnf subunits are closely homologous to the Na^+-translocating NADH:quinone-oxidoreductase (Na^+-Nqr) from *Vibrio cholerae*, whose XRD structure was recently published (Steuber *et al.*, 2014). In the Nqr complex, electrons are proposed to flow from NADH via the subunits F, C, B and A to an external ubiquinone, causing a conformational change in NqrB, which triggers Na^+ translocation by NqrE and NqrD. Owing to the homology between the Nqr- and Rnf-subunits, the following path of electrons could be envisaged in Rnf, wherein the flow of electrons is in the opposite direction to that in Nqr:

$$\text{Ferredoxin} \rightarrow \text{RnfB} \rightarrow \text{RnfD (NqrB)} \rightarrow \text{RnfG (NqrC)} \rightarrow \text{RnfC (NqrF)} \rightarrow \text{NADH}$$

This induces a conformational change in RnfD, enabling RnfE and RnfA to pump Na^+ out of the cell, thereby generating a Na^+-gradient ($\Delta\mu Na^+$). Recent studies have confirmed the proposed mechanism in *Acetobacterium woodii* using Rnf-containing membrane vesicles, whereby the reduction of NAD^+ by reduced ferredoxin was coupled to the transport of radioactive [$^{22}Na^+$] within the vesicles (Hess *et al.*, 2013).

29.8 Biotechnological Applications of Amino Acid Fermentation

Potential applications of 2-HADH are the artificial fermentations of glucose to propionate via (*R*)-lactyl-CoA and acrylyl-CoA (Kandasamy *et al.*, 2013), and of glucose to (*E*)-glutaconate via (*R*)-2-hydroxyglutarate (Djurdjevic *et al.*, 2010, 2011), both in *Escherichia coli*. Glutaconate and its reduced form glutarate are useful for making polyesters and polyamides. Furthermore, the enzymes of the engineered pathway to glutarate could be used to produce adipate (Parthasarathy *et al.*, 2011), a constituent of nylon, as many of the enzymes of the 2-hydroxyglutarate pathway also accept the corresponding 6-carbon analogues as substrates. Synthesis of the preservative propionate via methylmalonyl-CoA is product inhibited in propionibacteria, but the acrylyl-CoA route based on lactyl-CoA dehydratase and acrylate reductase (Buckel *et al.*, 2012) in efficient lactate producers appears to be more promising for development. *Archaeoglobus fulgidus*, a hyperthermophile that degrades phenylalanine and tryptophan by a proposed pathway involving the reactions of the *C. sporogenes* pathway (Parthasarathy *et al.*, 2013) could be interesting for biotechnological applications.

4-HBDH could convert crotonyl-CoA to 4-hydroxybutyryl-CoA to generate poly-4-hydroxybutyrate in polyhydroxyalkanoate-producing bacteria; poly-4-hydroxybutyrate has properties different from those of the storage compound poly-3-hydroxybutyrate (Zhang *et al.*, 2009). 4-HBDH could also be applied to construct engineered pathways to produce chemicals from CO_2 rather than from glucose (Berg *et al.*, 2007). Glutamate mutase catalyses the rearrangement of L-2-hydroxyglutarate to L-*threo*-3-methylmalate (Roymoulik *et al.*, 2000), and offers a potential route to the latter starting from the precursor 2-oxoglutarate. The promiscuous 3-methylaspartase from the 3-methylaspartate pathway has been used to produce (*S*)-aspartate derivatives from readily available mesaconates and fumarates (Winkler and Williams, 1967; Kato and Asano, 1997).

29.9 Medical and Environmental Aspects of Amino Acid Fermentation

Amino acid fermentation occurs in human pathogens such as *C. difficile*, *C. tetani*, *C. perfringens*, *C. botulinum* and *F. nucleatum*, and in organisms important for the global elemental cycles. Some medical and environmental aspects of this are discussed below.

29.9.1 Medical aspects

An excess of branched-chain amino acids represses *C. difficile* codY, a global metabolic regulator of the sigma-factor *tcdR* for promoters of toxin genes (Dineen *et al.*, 2010). The addition of glycine and proline during growth on other amino acids decreases toxin yield (Karlsson *et al.*, 1999). Selenium is required for the growth of *C. difficile* in basal medium (Jackson *et al.*, 2006); the inhibitory effect of the selenium chelator auranofin (Jackson-Rosario *et al.*, 2009) suggests that the selenoenzymes proline reductase and glycine reductase play key roles in *C. difficile* physiology. The *C. botulinum* genome contains a collagenase (Sebaihia *et al.*, 2007); *C. difficile* lacks collagenase, but could degrade collagen by other means. Stickland fermentation of glycine, proline and hydroxy-proline derived from collagen may play a role in clostridial diseases.

In the polymicrobial disease periodontitis, butyrate, propionate and acetate produced by bacteria irritate the gums. *Porphyromonas gingivalis* degrades glutamyl-glutamate via glutamate, fermenting it through the 4-hydroxybutyrate pathway into butyrate and via methylmalonyl-CoA into propionate. Aspartyl-aspartate is hydrolysed to aspartate, which is degraded to butyrate and acetate (Takahashi *et al.*, 2000). Also implicated in periodontitis are *F. nucleatum*, which ferments glutamate via 2-hydroxyglutarate, and *Treponema denticola*, which utilizes serine, alanine, cysteine and glycine.

Clostridial butyrate production induces colonic regulatory T cells in the host (Atarashi *et al.*, 2011). *C. sporogenes* inhabits the human gut and converts tryptophan to 3-indolepyruvate, oxidizing it to 3-indoleacetate, and reducing it via (*R*)-3-indolelactate and (*E*)-3-indoleacrylate to 3-indolepropionate (IPA) (Li, 2014). IPA passes through the blood–brain barrier, scavenging ROS in the brain, and thereby protecting it from Alzheimer's disease (Chyan *et al.*, 1999); it could also potentially mitigate bowel inflammation in Crohn's disease; the interplay of gut microbial metabolites with the host central nervous system is termed the 'gut–brain axis' (Bercik *et al.*, 2012). The levels of phenylacetate, 3-phenylpropionate and 3-phenyllactate derived from phenylalanine fermentation and (4-hydroxyphenyl)-3-lactate from tyrosine fermentation are elevated in intestinal disease/sepsis (Fedotcheva *et al.*, 2008).

29.9.2 Environmental aspects

The 4-hydroxybutyrate pathway participates in the assimilation of CO_2 into organic material, which is quantitatively a very significant biosynthetic process, The Global Ocean Sampling database contains half as many 4-HBDH amino acid sequences as RubisCO (the CO_2-fixing enzyme) sequences. In the 3-hydroxypropionate/4-hydroxybutyrate cycle of autotrophs, one acetyl-CoA and two bicarbonate molecules are reductively converted via 3-hydroxypropionate to succinyl-CoA, which is reduced to 4-hydroxybutyrate and converted via crotonate to two acetyl-CoA molecules (Berg *et al.*, 2007). The key genes of this pathway are found in *Metallosphaera*, *Sulfolobus*, *Archaeoglobus* and *Cenarchaeum* species.

In *Igniococcus hospitalis* (*Desulfurococcales*), the carboxylations of acetyl-CoA to pyruvate, and of phosphoenol-pyruvate to oxaloacetate, lead to succinyl-CoA. Succinyl-CoA is converted via 4-hydroxybutyryl-CoA and crotonyl-CoA to two acetyl-CoA, which closes the cyclic pathway and leaves one acetyl-CoA for biosynthesis (Berg *et al.*, 2007). Thus, 4-HBDH is an enzymatic link between the Krebs cycle and fatty acid metabolism. The succinate-ethanol fermenter *Clostridium kluyveri* reduces succinyl-CoA to succinic semialdehyde, which enters the 4-hydroxybutyrate pathway leading to butyrate and acetate (Scherf *et al.*, 1994).

29.10 Conclusions

The preceding sections demonstrate that bacterial amino acid fermentation is a rich field for biochemical exploration that is full of fascinating enzymes, with interesting implications for biotechnology, medicine and the environment.

Acknowledgements

Prof. Wolfgang Buckel (MPI Marburg), our mentor and inspiration.

References

Arkowitz, R.A. and Abeles R.H. (1990) Isolation and characterization of a covalent selenocysteine intermediate in the glycine reductase system. *Journal of the American Chemical Society* 112, 870–872.

Atarashi, K., Tanoue, T., Shima, T., Imaoka, A., Kuwahara, T., Momose, Y., Cheng, G., Yamasaki, S., Saito, T., Ohba, Y. *et al.* (2011) Induction of colonic regulatory T cells by indigenous *Clostridium* species. *Science* 331, 337–341.

Bader, J., and Simon, H. (1983) ATP formation is coupled to the hydrogenation of 2-enoates in *Clostridium sporogenes. FEMS Microbiology Letters* 20, 171–175.

Barker, H.A. (1961) Fermentations of nitrogenous organic compounds. In: Gunsalus, G. (ed.) *The Bacteria*. Academic Press, New York, pp 151–207.

Barker, H.A. (1978) Explorations of bacterial metabolism. *Annual Reviews in Biochemistry* 47, 1–33.

Barker, H.A., D'Ari, L. and Kahn, J. (1987) Enzymatic reactions in the degradation of 5-aminovalerate by *Clostridium aminovalericum*. *The Journal of Biological Chemistry* 262, 8994–9003.

Beatrix, B., Zelder, O., Kroll, F., Örlygsson, G., Golding, B.T. and Buckel, W. (1995) Evidence for a mechanism involving transient fragmentation in coenzyme B_{12}-dependent carbon skeleton rearrangements. *Angewandte Chemie International Edition in English* 34, 2398–2401.

Bendrat, K., Müller, U., Klees, A.-G. and Buckel, W. (1993) Identification of the gene encoding the activator of (*R*)-2-hydroxyglutaryl-CoA dehydratase from *Acidaminococcus fermentans* by gene expression in *Escherichia coli. The FEBS Journal* 329, 329–331.

Bercik, P., Collins, S.M. and Verdu, E.F. (2012) Microbes and the gut–brain axis, *Neurogastroenterology and Motility* 24, 405–413.

Berg, I.A., Kockelkorn, D., Buckel, W. and Fuchs, G. (2007) A 3-hydroxypropionate/4-hydroxybutyrate autotrophic carbon dioxide assimilation pathway in archaea. *Science* 318, 1782–1786.

Berk, H., Buckel, W., Thauer, R.K. and Frey, P.A. (1996) *Re*-face stereospecificity at C4 of NAD(P) for alcohol dehydrogenase from *Methanogenium organophilum* and for (*R*)-2-hydroxyglutarate dehydrogenase from *Acidaminococcus fermentans* as determined by ^{1}H-NMR spectroscopy. *FEBS Letters* 399, 92–94.

Biegel, E. and Müller, V. (2010) Bacterial Na^+-translocating ferredoxin:NAD^+ oxidoreductase. *Proceedings of the National Academy of Sciences of the United States of America* 107, 18138–18142.

Biegel, E., Schmidt, S., Gonzalez, J.M. and Müller, V. (2011) Biochemistry, evolution and physiological function of the Rnf complex, a novel ion-motive electron transport complex in prokaryotes. *Cellular and Molecular Life Sciences: CMLS* 68, 613–634.

Boiangiu, C.D., Jayamani, E., Brügel, D., Herrmann, G., Kim, J., Forzi, L., Hedderich, R., Vgenopoulou, I., Pierik, A.J., Steuber, J. and Buckel, W. (2005) Sodium ion pumps and hydrogen production in glutamate fermenting anaerobic bacteria. *Journal of Molecular Microbiology and Biotechnology* 10, 105–119.

Bothe, H., Darley, D.J., Albracht, S.P., Gerfen, G.J., Golding, B.T. and Buckel, W. (1998) Identification of the 4-glutamyl radical as an intermediate in the carbon skeleton rearrangement catalyzed by coenzyme B_{12}-dependent glutamate mutase from *Clostridium cochlearium*. *Biochemistry* 37, 4105–4113.

Brandt, U. (1996) Bifurcated ubihydroquinone oxidation in the cytochrome bc_1 complex by proton-gated charge transfer. *FEBS Letters* 387, 1–6.

Bremer, E. and Krämer, R. (2000) Coping with osmotic challenges: osmoregulation through accumulation and release of compatible solutes in bacteria, In: Storz, G. and Hengge-Aronis, R. (eds) *Bacterial Stress Responses*. ASM Press, Washington, DC, pp 79–97.

Buckel, W. (1980) Analysis of the fermentation pathways of clostridia using double-labelled glutamate. *Archives of Microbiology* 127, 167–169.

Buckel, W. (2001) Unusual enzymes involved in five pathways of glutamate fermentation. *Applied Microbiology and Biotechnology* 57, 263–273.

Buckel, W. and Barker, H.A. (1974) Two pathways of glutamate fermentation by anaerobic bacteria. *Journal of Bacteriology* 117, 1248–1260.

Buckel, W. and Golding, B.T. (1998) Radical species in the catalytic pathways of enzymes from anaerobes. *FEMS Microbiology Reviews* 22, 523–541.

Buckel, W. and Golding, B.T. (2006) Radical enzymes in anaerobes, *Annual Reviews in Microbiology* 60, 27–49.

Buckel, W., and Keese, R. (1995) One-electron redox reactions of CoASH esters in anaerobic bacteria – a mechanistic proposal. *Angewandte Chemie International Edition in English* 34, 1502–1506.

Buckel, W. and Miller, S.L. (1987) Equilibrium constants of several reactions involved in the fermentation of glutamate. *European Journal of Biochemistry* 164, 565–569.

Buckel, W. and Semmler, R. (1983) Purification, characterisation and reconstitution of glutaconyl-CoA decarboxylase, a biotin-dependent sodium pump from anaerobic bacteria. *European Journal of Biochemistry* 136, 427–434.

Buckel, W. and Thauer, R.K. (2013) Energy conservation via electron bifurcating ferredoxin reduction and proton/Na^+ translocating ferredoxin oxidation. *Biochimica et Biophysica Acta* 1827, 94–113.

Buckel, W., Janssen, P., Schuhmann, A., Eikmanns, U., Messner, P., Sleytr, U.B. and Liesack, W. (1994) *Clostridium viride* sp. nov., a strictly anaerobic bacterium using 5-aminovalerate as growth substrate, previously assigned to *Clostridium aminovalericum*. *Archives of Microbiology* 162, 387–394.

Buckel, W., Bröker, G., Bothe, H., Pierik, A.J. and Golding, B.T. (1999) Glutamate mutase and 2-methyleneglutarate mutase. In: Banerjee, R. (ed.) *Chemistry and Biochemistry of B_{12}*. Wiley, New York, pp 757–781.

Buckel, W., Kratky, C. and Golding, B.T. (2005) Stabilisation of methylene radicals by cob(II)alamin in coenzyme B_{12}-dependent mutases. *Chemistry* 12, 352–362.

Buckel, W., Zhang, J., Friedrich, P., Parthasarathy, A., Li, H., Djurdejevic, I., Dobbek, H. and Martins, B.M. (2012) Enzyme catalyzed radical dehydrations of hydroxy acids. *Biochimica et Biophysica Acta* 1824, 1278–1290.

Chen, J.S. and Mortenson, L.E. (1974) Purification and properties of hydrogenase from *Clostridium pasteurianum* W5. *Biochimica et Biophysica Acta* 371, 283–298.

Chen, D., Walsby, C., Hoffman, B.M. and Frey, P.A. (2003) Coordination and mechanism of reversible cleavage of *S*-adenosylmethionine by the [4Fe-4S] center in lysine 2,3-aminomutase. *Journal of the American Chemical Society* 125, 11788–11789.

Chih, H.-W. and Marsh, E.N.G. (2000) Mechanism of glutamate mutase: identification and kinetic competence of acrylate and glycyl radical as intermediates in the rearrangement of glutamate to methylaspartate. *Journal of the American Chemical Society* 122, 10732–10733.

Chirpich, T.P., Zappia, V., Costilow, R.N. and Barker, H.A. (1970) Lysine 2,3-aminomutase. Purification and properties of a pyridoxal phosphate and *S*-adenosylmethionine-activated enzyme. *The Journal of Biological Chemistry* 245, 1778–1789.

Chowdhury, N.P., Mowafy, A.M., Demmer, J.K., Upadhyay, V., Kölzer, S., Jayamani, E., Kahnt, J., Hornung, M., Demmer, U., Ermler, U. and Buckel, W. (2014) Studies on the mechanism of electron bifurcation catalyzed by electron transferring flavoprotein (Etf) and butyryl-CoA dehydrogenase (Bcd) of *Acidaminococcus fermentans*. *The Journal of Biological Chemistry* 289, 5145–5157.

Chowdhury, N.P., Kahnt, J. and Buckel, W. (2015) Reduction of ferredoxin or oxygen by flavin-based electron bifurcation in *Megasphaera elsdenii*. *The FEBS Journal* 282, 3149–3160.

Chyan, Y.J., Poeggeler, B., Omar, R.A., Chain, D.G., Frangione, B., Ghiso, J. and Pappolla, M.A. (1999) Potent neuroprotective properties against the Alzheimer beta-amyloid by an endogenous melatonin-related indole structure, indole-3-propionic acid. *The Journal of Biological Chemistry* 274, 21937–21942.

Cone, J.E., Del Rio, R.M., Davis, J.N. and Stadtman, T.C. (1976) Chemical characterization of the selenoprotein component of clostridial glycine reductase: identification of selenocysteine as the organoselenium moiety. *Proceedings of the National Academy of Sciences of the United States of America* 73, 2659–2663.

Dickert, S., Pierik, A.J., Linder, D. and Buckel, W. (2000) The involvement of coenzyme A esters in the dehydration of (*R*)-phenyllactate to (*E*)-cinnamate by *Clostridium sporogenes. European Journal of Biochemistry* 267, 3874–3884.

Dietrichs, D., Meyer, M., Rieth, M. and Andreesen, J.R. (1991) Interaction of selenoprotein PA and the thioredoxin system, components of the NADPH-dependent reduction of glycine in *Eubacterium acidaminophilum* and *Clostridium litorale. Journal of Bacteriology* 173, 5983–5991.

Dineen, S.S., McBride, S.M. and Sonenshein, A.L. (2010) Integration of metabolism and virulence by *Clostridium difficile* CodY. *Journal of Bacteriology* 192, 5350–5362.

Djurdjevic, I., Zelder, O. and Buckel, W. (2010) Method for the production of glutaconate. International patent application WO/2010/081885.

Djurdjevic, I., Zelder, O. and Buckel, W. (2011) Production of glutaconic acid in a recombinant *Escherichia coli* strain. *Applied and Environmental Microbiology* 77, 320–322.

Dürre, P., Andersch, W. and Andreesen, J.R. (1981) Isolation and characterization of an adenine-utilizing, anaerobic sporeformer, Clostridium purinolyticum sp. nov. *International Journal of Systematic and Evolutionary Microbiology* 31, 184–194.

Dyer, J.K. and Costilow, R.N. (1968) Fermentation of ornithine by *Clostridium sticklandii. Journal of Bacteriology* 96, 1617–1622.

Eikmanns, U. and Buckel, W. (1991a) Crystalline green 5-hydroxyvaleryl-CoA dehydratase from *Clostridium aminovalericum. European Journal of Biochemistry* 197, 661–668.

Eikmanns, U. and Buckel, W. (1991b) A green 2,4-pentadienoyl-CoA reductase from *Clostridium aminovalericum. European Journal of Biochemistry* 198, 263–266.

Fedotcheva, N.I., Kazakov, R.E., Kondrashova, M.N. and Beloborodova, N.V. (2008) Toxic effects of microbial phenolic acids on the functions of mitochondria. *Toxicology Letters* 180, 182–188.

Fendrich C., Hippe, H. and Gottschalk, G. (1990) *Clostridium halophilum* sp. nov., and *Clostridium litorale* sp. nov., an obligate halophilic and a marine species degrading betaine in Stickland reactions. *Archives of Microbiology* 154, 127–132.

Frey, P.A., Reed, G.H., Moss, M.L., Petrovich, R.M., Ballinger, M.D., Lieder, D., Wu, W., Chang, C.H., Bandarian, V., Ruzicka, F.J. *et al.* (1998) The role of *S*-adenosylmethionine as a poor man's adenosylcobalamin in the reaction of lysine 2,3-aminomutase. In: Kräutler, B., Arigoni, D. and Golding, B.T. (eds) *Vitamin B_{12} and B_{12}-proteins.* Wiley-VCH, Weinheim, Germany, pp 435–445.

Friedrich, P., Darley, D.J., Golding, B.T. and Buckel, W. (2008) The complete stereochemistry of the enzymatic dehydration of 4-hydroxybutyryl coenzyme A to crotonyl coenzyme A. *Angewandte Chemie International Edition in English* 47, 3254–3257.

Gerhardt, A., Cinkaya, I., Linder, D., Huisman, G. and Buckel, W. (2000) Fermentation of 4-aminobutyrate by *Clostridium aminobutyricum*: cloning of two genes involved in the formation and dehydration of 4-hydroxybutyryl-CoA. *Archives of Microbiology* 174, 189–199.

Golding, B.T., Anderson, R.J., Ashwell, S., Edwards, C.H., Garnett, I., Kroll, F. and Buckel, W. (1998) A mechanistic overview of B_{12}-dependent processes. In: Kräutler, B., Arigoni, D., and Golding, B.T. (eds) *Vitamin B_{12} and B_{12}-proteins.* Wiley-VCH, Weinheim, Germany, pp 201–216.

Härtel, U. and Buckel, W. (1996) Sodium ion-dependent hydrogen production in *Acidaminococcus fermentans. Archives of Microbiology* 166, 350–356.

Hasan, S.M. and Hall, J.B. (1975) The physiological function of nitrate reduction in *Clostridium perfringens. Journal of General Microbiology* 87, 120–128.

Hasan, M. and Hall, J.B. (1977) Dissimilatory nitrate reduction in *Clostridium tertium. Zeitschrift für Allgemeine Mikrobiologie* 17, 501–506.

Herrmann, T.G.E. (2008) Enzymes of two clostridial amino-acid fermentation pathways. PhD thesis, Philipps-Universität, Marburg, Germany.

Herrmann, G., Jayamani, E., Mai, G. and Buckel, W. (2008) Energy conservation via electron-transferring flavoprotein in anaerobic bacteria. *Journal of Bacteriology* 190, 784–791.

Hess, V., Schuchmann, K. and Müller, V. (2013) The ferredoxin:NAD+ oxidoreductase (Rnf) from the acetogen *Acetobacterium woodii* requires Na^+ and is reversibly coupled to the membrane potential. *The Journal of Biological Chemistry* 288, 31496–31502.

Hofmeister, A.E.M., Grabowski, R., Linder, D. and Buckel, W. (1993) L-Serine and L-threonine dehydratase from *Clostridium propionicum*, two enzymes with different prosthetic groups. *European Journal of Biochemistry* 215, 341–349.

Imkamp, F., Biegel, E., Jayamani, E., Buckel, W. and Müller, V. (2007) Dissection of the caffeate respiratory chain in the acetogen *Acetobacterium woodii*: identification of an Rnf-type NADH dehydrogenase as a potential coupling site. *Journal of Bacteriology* 189, 8145–8153.

Ishimoto, M., Umeyama, M. and Chiba, S. (1974) Alteration of fermentation products from butyrate to acetate by nitrate reduction in *Clostridium perfringens. Zeitschrift für Allgemeine Mikrobiologie* 14, 115–121.

Jackson, S., Calos, M., Myers, A. and Self, W.T. (2006) Analysis of proline reduction in the nosocomial pathogen *Clostridium difficile. Journal of Bacteriology* 188, 8487–8495.

Jackson-Rosario, S., Cowart, D., Myers, A., Tarrien, R., Levine, R.L., Scott, R.A and Self, W.T. (2009) Auranofin disrupts selenium metabolism in *Clostridium difficile* by forming a stable Au-Se adduct. *Journal of Biological Inorganic Chemistry* 14, 507–519.

Kabisch, U., Gräntzdörffer, A., Schierhorn, A., Rücknagel, K.P., Andreesen, J.R. and Pich, A. (1999) Identification of D-proline reductase from *Clostridium sticklandii* as a selenoenzyme and indications for a catalytically active pyruvoyl group derived from a cysteine residue by cleavage of a proprotein. *The Journal of Biological Chemistry* 274, 8445–8454.

Kandasamy, V., Vaidyanathan, H., Djurdjevic, I., Jayamani, E., Ramachandran, K.B., Buckel, W., Jayaraman, G. and Ramalingam, S. (2013) Engineering *Escherichia coli* with acrylate pathway genes for propionic acid synthesis and its impact on mixed-acid fermentation. *Applied Genetics and Molecular Biotechnology* 97, 1191–1200.

Kaneko, M. and Ishimoto, M. (1977) Effect of nitrate reduction on metabolic products and growth of *Propionibacterium acidi-propionici. Zeitschrift für Allgemeine Mikrobiologie* 17, 211–220.

Karlsson, S., Burman L.G. and Akerlund, T. (1999) Suppression of toxin production in *Clostridium difficile* VPI 10463 by amino acids. *Microbiology* 145, 1683–1693.

Kato, Y. and Asano, Y. (1997) 3-Methylaspartate ammonia-lyase as a marker enzyme of the mesaconate pathway for (*S*)-glutamate fermentation in *Enterobacteriaceae*. *Archives of Microbiology* 168, 457–463.

Kim, J., Hetzel, M., Boiangiu, C.D. and Buckel, W. (2004) Dehydration of (*R*)-2-hydroxyacyl-CoA to enoyl-CoA in the fermentation of α-amino acids by anaerobic bacteria. *FEMS Microbiology Reviews* 28, 455–468.

Kim, J., Darley, D.J., Buckel, W. and Pierik, A.J. (2008) An allylic ketyl radical intermediate in clostridial amino-acid fermentation. *Nature* 452, 239–243.

Klees, A.-G., Linder, D. and Buckel, W. (1992) 2-Hydroxyglutaryl-CoA dehydratase from *Fusobacterium nucleatum* (subsp. *nucleatum*): an iron–sulfur flavoprotein. *Archives of Microbiology* 158, 294–301.

Knauer, S.H., Buckel, W. and Dobbek, H. (2011) Structural basis for reductive radical formation and electron recycling in (*R*)-2-hydroxyisocaproyl-CoA dehydratase. *Journal of the American Chemical Society* 133, 4342–4347.

Kölzer, S. (2008) Aufreiningung und Charakterisierung des Butyryl-CoA Dehydrogenase/ETF Komplexe aus *Clostridium tetanomorphum*. Diploma thesis, Philipps-Universität, Marburg, Germany.

Kreimeyer, A., Perret, A., Lechaplais, C., Vallenet, D., Médigue, C., Salanoubat, M. and Weissenbach, J. (2007) Identification of the last unknown genes in the fermentation pathway of lysine. *The Journal of Biological Chemistry* 282, 7191–7197.

Kreis, W. and Hession, C. (1973) Isolation and purification of L-methionine-α-deamino-γ-mercaptomethane-lyase (L-methioninase) from *Clostridium sporogenes*. *Cancer Research* 33, 1862–1865.

Lancaster, C.R., Kröger, A., Auer, M. and Michel, H. (1999) Structure of fumarate reductase from *Wolinella succinogenes* at 2.2 Å resolution. *Nature* 402, 377–385.

Lewis, D. and Elsden, S.R. (1955) The fermentation of L-threonine, L-serine, L-cysteine and acrylic acid by a Gram-negative coccus. *Biochemistry* 60, 683–692.

Li, F., Hinderberger, J., Seedorf, H., Zhang, J., Buckel, W. and Thauer, R.K. (2008) Coupled ferredoxin and crotonyl coenzyme A (CoA) reduction with NADH catalyzed by the butyryl-CoA dehydrogenase/Etf complex from *Clostridium kluyveri. Journal of Bacteriology* 190, 843–850.

Li, H. (2014) (*R*)-Indolelactyl-CoA dehydratase, the key enzyme of tryptophan reduction to indolepropionate in *Clostridium sporogenes*. PhD thesis, Philipps Universität, Marburg, Germany.

Macieira, S., Zhang, J., Velarde, M., Buckel, W. and Messerschmidt, A. (2009) Crystal structure of 4-hydroxybutyrate CoA-transferase from *Clostridium aminobutyricum*, *Biological Chemistry* 390, 1251–1263.

Matthies, C., Evers, S., Ludwig W. and Schink, B. (2000) *Anaerovorax odorimutans* gen. nov., sp. nov., a putrescine-fermenting, strictly anaerobic bacterium. *International Journal of Systematic and Evolutionary Microbiology* 50, 1591–1594.

Meister, A., Sober, H.A. and Tice, S.V. (1951) Enzymatic decarboxylation of aspartic acid to α-alanine. *The Journal of Biological Chemistry* 189, 577–590.
Mitchell, P. (1975) The protonmotive Q cycle: a general formulation. *FEBS Letters* 59, 137–139.
Mishra, S. and Imlay, J.A. (2012) Why do bacteria use so many enzymes to scavenge hydrogen peroxide? *Archives of Biochemistry and Biophysics* 525, 145–160.
Mishra, S. and Imlay, J.A. (2013) An anaerobic bacterium, *Bacteroides thetaiotaomicron*, uses a consortium of enzymes to scavenge hydrogen peroxide. *Molecular Microbiology* 90, 1356–1371.
Müh, U., Cinkaya, I., Albracht, S.P. and Buckel, W. (1996) 4-Hydroxybutyryl-CoA dehydratase from *Clostridium aminobutyricum*: characterization of FAD and iron–sulfur clusters involved in an overall non-redox reaction. *Biochemistry* 35, 11710–11718.
Muth, W.L. and Costilow, R.N. (1974) Ornithine cyclase (deaminating). *The Journal of Biological Chemistry* 249(23), 7463–7467.
Nanninga, H.J., Drent, W.J. and Gottschal, J.C. (1987) Fermentation of glutamate by *Selenomonas acidaminophila* sp. nov. *Archives of Microbiology* 147, 152–157.
Örlygsson, J., Krooneman, J., Collins, M.D., Pascual, C. and Gottschal, J.C. (1996) *Clostridium acetireducens* sp. nov., a novel amino acid-oxidizing, acetate-reducing anaerobic bacterium. *International Journal of Systematic Bacteriology* 46, 454–459.
Parthasarathy, A., Buckel, W. and Smith, D.M. (2010) On the thermodynamic equilibrium between (*R*)-2-hydroxyacyl-CoA and 2-enoyl-CoA. *The FEBS Journal* 277, 1738–1746.
Parthasarathy, A., Pierik, A.J., Kahnt, J., Zelder, O. and Buckel, W., (2011) Substrate specificity of 2-hydroxyglutaryl-CoA dehydratase from *Clostridium symbiosum*: toward a bio-based production of adipic acid. *Biochemistry* 50, 3540–3550.
Parthasarathy, A., Kahnt, J., Chowdhury, N.P. and Buckel, W. (2013) Phenylalanine catabolism in *Archaeoglobus fulgidus* VC-16. *Archives of Microbiology* 195, 781–797.
Ramezani, M., Resmer, K.L. and White, R.L. (2011) Glutamate racemization and catabolism in *Fusobacterium varium. The FEBS Journal* 278, 2540–2551.
Reitzer, R., Gruber, K., Jogl, G., Wagner, U.G., Bothe, H., Buckel, W. and Kratky, C. (1999) Glutamate mutase from *Clostridium cochlearium*: the structure of a coenzyme B_{12}-dependent enzyme provides new mechanistic insights. *Structure with Folding and Design* [now *Structure*] 7, 891–902.
Roeder, J. and Schink, B. (2009) Syntrophic degradation of cadaverine by a defined methanogenic coculture. *Applied and Environmental Microbiology* 75, 4821–4828.
Roon, R.J. and Barker, H.A. (1972) Fermentation of agmatine in *Streptococcus faecalis*: occurrence of putrescine transcarbamoylase. *Journal of Bacteriology* 109, 44–50.
Roymoulik, I., Moon, N., Dunham, W.R., Ballou D.P. and Marsh, E.N.G. (2000) Rearrangement of L-2-hydroxyglutarate to L-*threo*-3-methylmalate catalyzed by adenosylcobalamin-dependent glutamate mutase. *Biochemistry* 39, 10340–10346.
Ruzicka, F.J. and Frey, P.A. (2007) Glutamate 2,3-aminomutase: a new member of the radical SAM superfamily of enzymes. *Biochimica et Biophysica Acta* 1774, 286–296.
Scherf, U. and Buckel, W. (1991) Purification and properties of 4-hydroxybutyrate coenzyme A transferase from *Clostridium aminobutyricum. Applied and Environmental Microbiology* 57, 2699–2702.
Scherf, U. and Buckel, W. (1993) Purification and properties of an iron–sulfur and FAD-containing 4-hydroxybutyryl-CoA dehydratase/vinylacetyl-CoA delta 3-delta 2-isomerase from *Clostridium aminobutyricum. European Journal of Biochemistry* 215, 421–429.
Scherf, U., Söhling, B., Gottschalk, G., Linder, D. and Buckel, W. (1994) Succinate-ethanol fermentation in *Clostridium kluyveri*: purification and characterisation of 4-hydroxybutyryl-CoA dehydratase/vinylacetyl-CoA delta 3-delta 2-isomerase. *Archives of Microbiology* 161, 239–245.
Schmehl, M., Jahn, A., Meyer zu Vilsendorf, A., Hennecke, S., Masepohl, B., Schuppler, M., Marxer, M., Oelze, J. and Klipp, W. (1993) Identification of a new class of nitrogen fixation genes in *Rhodobacter capsulatus*: a putative membrane complex involved in electron transport to nitrogenase. *Molecular and General Genetics: MGG* 241, 602–615.
Scott, R., Näser, U., Friedrich, P., Selmer, T., Buckel, W. and Golding, B.T. (2004) Stereochemistry of hydrogen removal from the 'unactivated' C-3 position of 4-hydroxybutyryl-CoA catalysed by 4-hydroxybutyryl-CoA dehydratase. *Chemical Communications* 1210–1211.
Sebaihia, M., Peck, M.W., Minton, N.P., Thomson, N.R., Holden, M.T., Mitchell, W.J., Carter, A.T., Bentley, S.D., Mason, D.R., Crossman, L. *et al.* (2007) Genome sequence of a proteolytic (Group I) *Clostridium botulinum* strain Hall A and comparative analysis of the clostridial genomes. *Genome Research* 17, 1082–1092.

Smith, D.M., Buckel, W. and Zipse, H. (2003) Deprotonation of enoxy radicals: theoretical validation of a 50-year-old mechanistic proposal. *Angewandte Chemie International Edition in English* 42, 1867–1870.

Stadtman, T.C. (1973) Lysine metabolism by clostridia, In: Meister A. (ed.) *Advances in Enzymology and Related Areas of Molecular Biology*. Wiley, New York, pp. 413–448.

Stadtman, T.C. and Renz, P. (1968) Anaerobic degradation of lysine. V. Some properties of the cobamide conezyme-dependent β-lysine mutase of *Clostridium sticklandii. Archives of Biochemistry and Biophysics* 125, 226–239.

Stadtman, T.C. and White, F.H. (1954) Tracer studies on ornithine, lysine, and formate metabolism in an amino acid fermenting *Clostridium. Journal of Bacteriology* 67, 651–657.

Steuber, J., Vohl, G., Casutt, M.S., Vorburger, T., Diederichs, K. and Fritz, G. (2014) Structure of the *V. cholerae* Na^+-pumping NADH:quinone oxidoreductase. *Nature* 516, 62–67.

Stickland, L.H. (1935) XXXV. Studies in the metabolism of the strict anaerobes (genus *Clostridium*) II. The reduction of proline. *Biochemical Journal* 29, 288–290.

Takahashi, N., Sato, T. and Yamada, T. (2000) Metabolic pathways for cytotoxic end product formation from glutamate- and aspartate-containing peptides by *Porphyromonas gingivalis. Journal of Bacteriology* 182, 4704–4710.

Thamer, W., Cirpus, I., Hans, M., Pierik, A.J., Selmer, T., Bill, E., Linder, D. and Buckel, W. (2003) A two [4Fe-4S]-cluster-containing ferredoxin as an alternative electron donor for 2-hydroxyglutaryl-CoA dehydratase from *Acidaminococcus fermentans. Archives of Microbiology* 179, 197–204.

Thauer, R.K., Jungermann, K. and Decker, K. (1977) Energy conservation in chemotrophic anaerobic bacteria. *Bacteriological Reviews* 41, 100–180.

Tremblay, P.L., Zhang, T., Dar, S.A., Leang, C. and Lovley, D.R. (2012) The Rnf complex of *Clostridium ljungdahlii* is a proton-translocating ferredoxin:NAD^+ oxidoreductase essential for autotrophic growth. *mBio* 4, e00406–e00412.

Uyeda, K. and Rabinowitz, J.C. (1971) Pyruvate-ferredoxin oxidoreductase 3. Purification and properties of the enzyme. *The Journal of Biological Chemistry* 246, 3111–3119.

Wachsman, J.T. and Barker, H.A. (1954) The accumulation of formamide during the fermentation of histidine by *Clostridium tetanomorphum. Journal of Bacteriology* 69, 83–88.

Wickremasinghe, R.L. and Fry, B.A. (1954) The formation of urocanic acid and glutamic acid in the fermentation of histidine by *Clostridium tetanomorphum. Biochemistry* 58, 268–278.

Weinberg, M.V., Jenney, F.E. Jr, Cui, X. and Adams, M.W. (2004) Rubrerythrin from the hyperthermophilic archaeon *Pyrococcus furiosus* is a rubredoxin-dependent, iron-containing peroxidase. *Journal of Bacteriology* 186, 7888–7895.

Willadsen, P. and Buckel, W. (1990) Assay of 4-hydroxybutyryl-CoA dehydratase from *Clostridium aminobutyricum. FEMS Microbiology Letters* 58, 187–191.

Winkler, M.R. and Williams, V.R. (1967) New substrates for β-methylaspartase. *Biochimica et Biophysica Acta, Enzymology* 146, 287–289.

Zhang, J., Friedrich, P., Pierik, A.J., Martins, B.M. and Buckel, W. (2015) Substrate-induced radical formation in 4-hydroxybutyryl coenzyme A dehydratase from *Clostridium aminobutyricum. Applied and Environmental Microbiology* 81, 1071–1084.

Zhang, L., Shi, Z.Y., Wu, Q. and Chen, G.Q. (2009) Microbial production of 4-hydroxybutyrate, poly-4-hydroxybutyrate, and poly(3-hydroxybutyrate-co-4-hydroxybutyrate) by recombinant microorganisms. *Applied Microbiology and Biotechnology* 84, 909–916.

30 Utilization of *N*-Methylated Amino Acids by Bacteria

M.J. Wargo*

Department of Microbiology and Molecular Genetics, University of Vermont College of Medicine, Burlington, Vermont, USA

30.1 Abstract

N-methylated amino acids are sources of bacterial nutrition and contribute to multiple aspects of bacterial biology, including osmoprotection. This chapter will summarize current knowledge of the sources, functions and catabolic pathways for bacterial utilization of the major *N*-methylated amino acids: glycine betaine, proline betaine and histidine betaine, and their noteworthy metabolites, including dimethylglycine, sarcosine and ergothioneine. Much of our knowledge on the metabolism of these molecules comes from work on *Proteobacteria* found in the environment.

30.2 Introduction

Most of the research on *N*-methylated amino acids comes from studies on methylated forms of glycine, proline and histidine. This chapter will focus on these three compounds, but it also briefly covers *N*-methylated tryptophans and tyrosine betaine.

Glycine betaine (GB) and proline betaine (Pro-B) are the betaine (quaternary amine and carboxylic acid-containing molecules) forms of glycine and proline, respectively (Fig. 30.1). Both are potent osmoprotectants for many organisms and both can be used as sole carbon, nitrogen, and energy sources by diverse bacteria. The pathways for aerobic catabolism of GB and Pro-B are remarkably similar, and evidence for bacterial enzymes with overlapping specificity for GB and Pro-B suggests evolutionary interactions between these two metabolites in bacterial biology. However, GB and Pro-B do differ in their sources. GB is synthesized by plants, animals, most fungi, and some bacteria. Conversely, Pro-B is only known to be synthesized in plants and eukaryotic algae, making it a potential nutritional mediator specifying chemical interactions between bacteria and photosynthetic eukaryotes (Goldmann *et al.*, 1991).

The biological roles of histidine betaine (His-B), the *N*-methylated forms of tryptophan and tyrosine betaine (Tyr-B) are less well studied in bacteria. His-B is the precursor for the synthesis of ergothioneine, a molecule with some specific antioxidant and anti-radical properties. *N*-methylated tryptophan can be catabolized by bacteria, but along with Tyr-B, the biological relevance of these compounds for bacteria in the environment is not well understood.

*E-mail address: matthew.wargo@med.uvm.edu

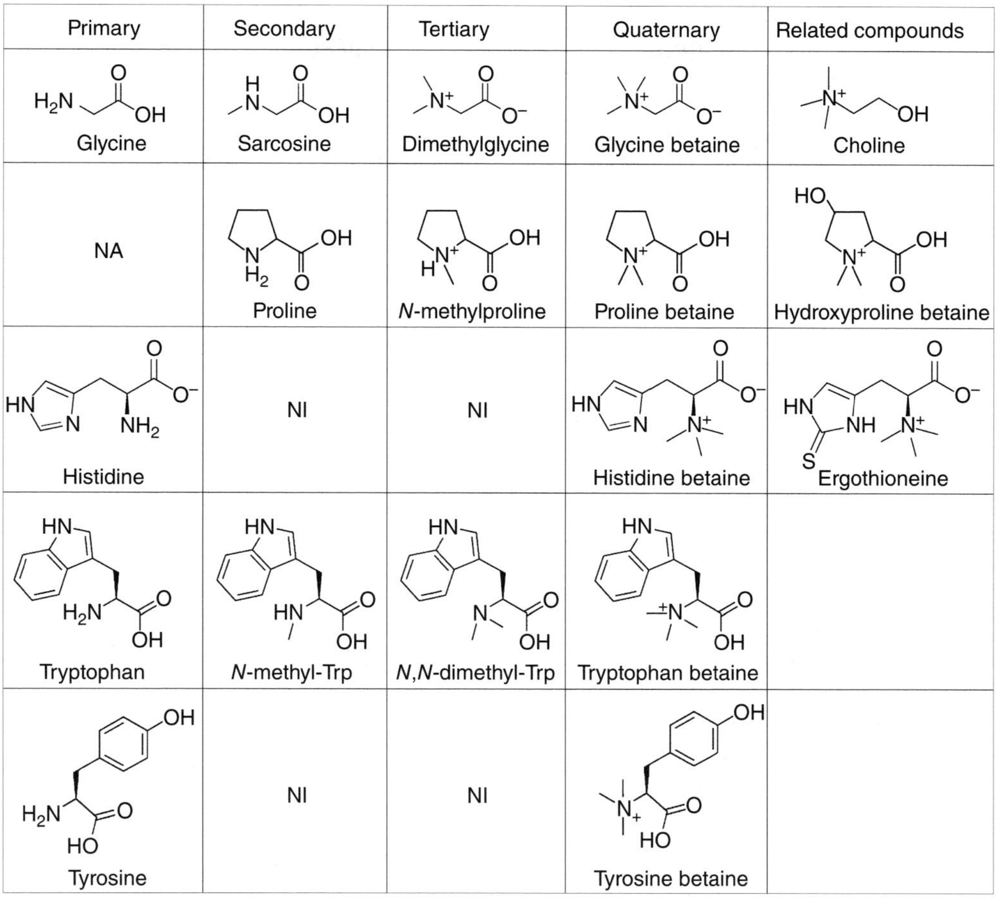

Fig. 30.1. Compounds discussed in the chapter arranged by source amino acid and degree of amine substitution (primary, secondary, tertiary, quaternary). Abbreviations: NA, not applicable; NI, not identified in bacteria; Trp, tryptophan.

30.3 Glycine Betaine and Its Metabolites

The *N*-methylated forms of glycine include sarcosine (*N*-methylglycine), dimethylglycine (*N,N′*-dimethylglycine), and glycine betaine (*N,N′,N″*-trimethylglycine, or GB) (see Fig. 30.1). These compounds are generated as intermediates in the synthesis of GB and/or in the catabolism of choline. They play important roles as osmoprotectants and stress protectants, and can be metabolized by diverse bacteria as sole sources of carbon and nitrogen. This section describes these pathways and their known roles in bacterial osmoprotection, virulence and symbiosis.

30.3.1 Sources

Given its utility as an osmoprotectant for nearly all forms of life (Yancey *et al.*, 1982) and its role as a nutrient source in some bacterial species (Anthoni *et al.*, 1991), it is not surprising that GB is scarce as a freely soluble resource in most environments (Kiene, 1998; Warren, 2014). However, it is present in a number of systems, including bacterial mats in hypersaline ponds, where it can account for ~20% of the soluble nitrogen (King, 1988). Other rich environmental sources include the marsh grass *Spartina* (Cavalieri and Huang, 1981), the tissues and cells of marine invertebrates (Beers, 1967; Pierce, 1982), and earthworms (Liebeke and Bundy, 2013).

GB is also found at ~50 μM in human blood and urine (Lever *et al.*, 1994). In most of these cases, the GB that is available for bacterial use is excreted by other cells or is released during cell death; such release may be particularly important in the *Spartina*-dominated coastal marsh ecosystems (Cavalieri and Huang, 1981).

GB can be generated by the bacterial metabolism of choline and carnitine (Aurich *et al.*, 1967; Kortstee, 1970), as described below; both of these compounds can be derived either from other bacteria or from eukaryotes. Choline is the predominant polar headgroup on the outer leaflet of eukaryotic plasma membranes (Zachowski, 1993; van Meer *et al.*, 2008), while carnitine is abundant in cells that metabolize fat, as it is part of the shuttle for transporting fatty acids for β-oxidation (Bremer, 1983).

Sarcosine is a metabolite of GB (Bater and Venables, 1977), and creatine and glyphosate can also be metabolized to sarcosine (Nimmo-Smith and Appleyard, 1956; Kishore and Jacob, 1987). Creatine is found in many animal tissues and fluids (Wyss and Kaddurah-Daouk, 2000). Glyphosate, a widely used herbicide, is applied to agricultural fields and is rapidly metabolized in the soil (Sprankle *et al.*, 1975; Rueppel *et al.*, 1977). One of the metabolic pathways for glyphosate metabolism yields sarcosine (Kishore and Jacob, 1987).

30.3.2 Roles in pathogenesis, symbiosis and osmoprotection

The primary role for GB, and to a lesser extent for dimethylglycine too, is osmoprotection (Yancey *et al.*, 1982). GB protects human cells from osmotic stress in the kidneys (Chambers and Kunin, 1987), protects marine invertebrates from salt stress (Pierce, 1982), and can be used by nearly all bacteria as a protective molecule against osmotic stress and temperature stress (Csonka and Hanson, 1991; Wood, 2011). There are two modes of bacterial interaction with GB: (i) solely as an osmoprotectant and protectant against stress; and (ii) as a dual function molecule acting as both an osmoprotectant and nutrient source. The former relationship is typified by the *Enterobacteriaceae* (Le Rudulier *et al.*, 1984; Perroud and Le Rudulier, 1985; Landfald and Strøm, 1986); the latter relationship is demonstrated in numerous bacteria found in the environment, and especially within the *Proteobacteria* (Boncompagni *et al.*, 1999; Wargo, 2013a), though this observation may be due more to bias in the bacteria studied than an informative statement of the distribution of this activity.

In addition to its role as an osmoprotectant, GB is important for bacterial pathogenesis during some infections and for interactions with eukaryotic hosts in the context of commensalism and mutualism. For pathogenic interactions, the acquisition of GB and/or conversion of host-derived choline to GB is important for the survival of *Pseudomonas aeruginosa* in the mammalian lung (Wargo, 2013b), of *Escherichia coli* in urine (Kunin *et al.*, 1992) and of *Listeria monocytogenes* intracellularly (Sleator *et al.*, 2003). For commensal and mutualistic interactions, the presence of GB plays a role in the *Rhizobium*/legume symbiosis (Djordjevic *et al.*, 2003) and in biofilm formation on crustaceans for pathogenic *Vibrio* sp. (Kapfhammer *et al.*, 2005). The genes for GB catabolism are also present in bacteria intimately associated with eukaryotes, but not in free-living members of the same clade. Examples include the *Verminephrobacter* (Gram-negative *Betaproteobacteria*)/earthworm (*Eisenia foetida*) symbiosis (Pinel *et al.*, 2008) and the symbiosis between deep-sea gutless marine polychaetes of the genus *Osedax* and intracellular endosymbionts belonging to the order *Oceanospirillales* (*Gammaproteobacteria*), both demonstrated using BLASTP (Protein Basic Local Alignment Search Tool) and STRING database analyses (Szklarczyk *et al.*, 2015).

For some of these associations, GB is likely to be important as an osmoprotectant, but GB and its metabolite dimethylglycine also induce the production of a choline-specific phospholipase C in a number of bacteria (Ivanov and Kostadinova, 1993; Lisa *et al.*, 1994; Stonehouse *et al.*, 2002; Korbsrisate *et al.*, 2007), and this phospholipase is a potent virulence factor in *P. aeruginosa* (Ostroff *et al.*, 1989; Rahme *et al.*, 2000; Wargo *et al.*, 2011). For *P. aeruginosa* infection in the lung, GB-dependent induction of the phospholipase C appears to account for most of the bacterial GB requirement (Wargo, 2013b). The induction of *Vibrio* biofilm formation (Kapfhammer *et al.*, 2005) can also probably be thought of as more of a signalling role for GB rather than a role solely for osmoprotection.

30.3.3 Biosynthesis

GB is often acquired by bacteria from their environment (reviewed in Wargo, 2013a), but the free form has an extremely short half-life in marine ecosystems (Kiene, 1998), and the compound can also be synthesized via three pathways: (i) the oxidation of endogenous or exogenous choline; (ii) the metabolism of L-carnitine; or (iii) the sequential reductive methylation of glycine. These pathways are shown in Fig 30.2A.

The best characterized route of GB synthesis occurs via oxidation of the hydroxyl carbon in choline to a carboxylic acid, passing through an aldehyde intermediate, i.e. choline oxidation to betaine aldehyde, followed by betaine aldehyde oxidation to GB. In many bacteria, this pathway is operated by two separate enzymes, a choline oxidase (BetA in *E. coli*) and a betaine aldehyde dehydrogenase (BetB in *E. coli*) (Landfald and Strøm, 1986). However, in some bacteria, such as *Arthrobacter*, a single subunit enzyme catalyses both oxidation steps (Ikuta *et al.*, 1977).

An alternate route of GB synthesis occurs via the aerobic metabolism of L-carnitine (Aurich *et al.*, 1967). In this pathway, the hydroxylated carbon-3 of carnitine is oxidized to form dehydrocarnitine and the carbon-3 is subsequently attacked by free coenzyme A (CoA) to generate glycine betainyl-CoA and acetyl-CoA; this route is catalysed by enzymes homologous to *P. aeruginosa* CdhC (Bastard *et al.*, 2014). Glycine betainyl-CoA is then oxidized to regenerate free CoA and GB (Wargo and Hogan, 2009). For a comprehensive review on carnitine metabolism, see Meadows and Wargo (2015).

The final route of GB synthesis is through successive methylation of glycine to GB (see Fig. 30.2A). This pathway has been described in marine and halophilic cyanobacteria, including *Aphanothece halophytica* (Waditee *et al.*, 2003), *Synechococcus* (Lu *et al.*, 2006) and *Prochlorococcus* (Klähn *et al.*, 2010). In these organisms, the first two methylation steps are carried out by an *S*-adenosylmethionine (SAM)-utilizing enzyme, glycine/sarcosine-*N*-methyltransferase, to generate dimethylglycine, which is methylated by dimethylglycine-*N*-methyltransferase, which also utilizes SAM as a methyl donor.

30.3.4 Metabolism

Glycine betaine catabolism can proceed through four distinct pathways: (i) using only oxygen-dependent demethylases; (ii) using an *S*-adenosylhomocystine-dependent GB methyltransferase that generates SAM, followed by oxygen-dependent

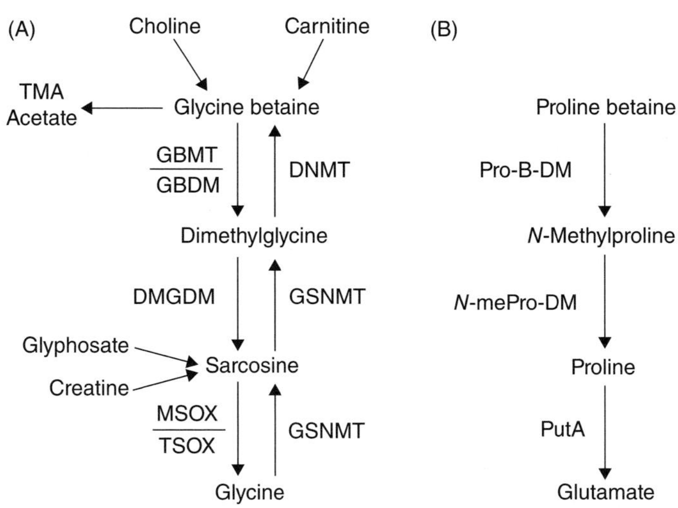

Fig. 30.2. Schematic representation of the bacterial metabolism of (A) glycine betaine and (B) proline betaine. Abbreviations: DMGDM, dimethylglycine demethylase; GBDM, glycine betaine demethylase; GBMT, glycine betaine methyltransferase; MSOX, monomeric sarcosine oxidase; *N*-mePro-DM, *N*-methylproline demethylase; Pro-B-DM, proline betaine demethylase; PutA, proline dehydrogenase; TMA, trimethylamine; TSOX, tetrameric sarcosine oxidase.

demethylases; (iii) using a cobalamin-dependent methyltransferase; or (iv) by deamination. These pathways are summarized in Fig. 30.2A.

In many Gram-negative bacteria, GB is demethylated by a Rieske family oxygenase activity to yield dimethylglycine and formaldehyde, with the two-subunit demethylase encoded by *gbcAB* (for glycine betaine catabolism A and B) *in P. aeruginosa* (Wargo *et al.*, 2008). The demethylation of dimethylglycine is catalysed by a heterodimeric flavin-linked oxidoreductase (Wargo *et al.*, 2008; Fitzsimmons *et al.*, 2011) whose enzymology is best studied in *Arthrobacter* spp. (Meskys *et al.*, 2001). The demethylation of sarcosine is conducted by one of two enzyme types, monomeric or heterotetrameric. Monomeric sarcosine oxidases produce glycine, hydrogen peroxide and formaldehyde via sarcosine demethylation, and are best characterized in *Arthrobacter* (Wagner and Schuman Jorns, 1997). The heterotetrameric sarcosine oxidases can assimilate the *N*-methyl group of sarcosine into the one-carbon metabolite pool using a 5,10-methylene-tetrahydrofolate intermediate (Wagner and Schuman-Jorns, 1997; Zeller *et al.*, 1989). The enzymology of the heterotetrameric sarcosine oxidase has been best characterized in *Corynebacterium* and *Stenotrophomonas* (Hasan-Abdallah *et al.*, 2005; Moriguchi *et al.*, 2010), and via genetic and functional studies in *P. aeruginosa* (Diab *et al.*, 2006; Wargo *et al.*, 2008; Willsey and Wargo, 2015).

In the *Rhizobiaceae*, the demethylation of GB has been proposed to occur via a GB methyl transferase (GBMT) mechanism (Smith *et al.*, 1988; Barra *et al.*, 2006). While the *Rhizobiaceae* have GBMT enzyme activity that correlates with growth on GB, they also have well-supported GbcA and GbcB homologues (Wargo, 2013a), whose contribution to GB catabolism has not been tested in these bacteria. After the initial demethylation, the remainder of this pathway is similar to that described above for other aerobic bacteria.

There is a newly described GB methyltransferase in *Desulfitobacterium hafniense* that functions in a similar manner to a trimethylamine methyltransferase and uses GB and cobalamin to generate dimethylglycine and methylcobalamin (Ticak *et al.*, 2014). In *D. hafniense*, there is an adjacently encoded methylcobalamin:tetrahydrofolate methyltransferase that could lead to the metabolic incorporation of the liberated methyl group in a manner similar to that derived from sarcosine by bacteria using the tetrameric sarcosine oxidase (Ticak *et al.*, 2014).

Under anaerobic conditions, GB can be metabolized to trimethylamine and acetate in *Desulfuromonas* (Heijthuijsen and Hansen, 1989), *Clostridium sporogenes* (Naumann *et al.*, 1983) and *Eubacterium aminoacidaminophilum* (Hormann and Andreesen, 1989). This pathway may be particularly important in hypersaline environments and may provide the primary source of the trimethylamine odour from hypersaline sediment (Oren, 1990). The enzyme responsible for GB deamination has not yet been identified, but a good candidate would be a member of the CntAB group of heterodimeric Rieske-family (oxygenase/reductase) enzymes, which were recently shown to enable the deamination of L-carnitine to trimethylamine and malic semialdehyde (Zhu *et al.*, 2014). The CntAB enzymes are homologous to GbcAB and are members of a large family of enzymes that can split C-N bonds in quaternary amine compounds (Wargo, 2013a; Zhu *et al.*, 2014).

GB catabolism is regulated by salinity such that it is inhibited by increasing salt concentrations (Smith *et al.*, 1998; Diab *et al.*, 2006; Li *et al.*, 2013). This type of regulation probably ensures that the osmoprotectant is retained during times of salt stress (Wargo, 2013a).

30.4 *N*-Methylated Prolines

The *N*-methylated forms of L-proline are *N*-methylproline (also known as hygrinic acid), and *N,N'*-dimethylproline (also known as proline betaine – Pro-B, or stachydrine) (Fig. 30.1). Similar to the *N*-methylated forms of glycine described above, *N*-methylated prolines are important osmoprotectants and can be metabolized by some bacteria as sole sources of carbon and nitrogen. This section describes the known Pro-B catabolic pathway and the demonstrated and suspected roles of *N*-methylprolines in symbiosis and osmoprotection.

30.4.1 Sources

The best information on the role of Pro-B in the environment comes from studies in legumes and their associated *Rhizobiaceae* bacteria, which can transport and catabolize Pro-B (Goldmann *et al.*, 1991; Phillips *et al.*, 1998; Alloing *et al.*, 2006).

Pro-B was initially characterized from lucerne (*Medicago sativa*) hay (Steenbock, 1918), and is present in and on the seeds of many leguminous species. In lucerne, ~16.4 nmol of Pro-B per seed can be collected using a water wash (Phillips *et al.*, 1992). This quantity would lead to ~550 μM Pro-B if evenly diffused into a 1 mm shell around the seed; this is a surprisingly high concentration of freely available nutrient source released by the plant. Pro-B has been hypothesized to function as a nutritional mediator that enables the plant to sculpt the rhizosphere microbial community, even prior to germination (Goldmann *et al.*, 1991). In addition to occurring in legumes such as lucerne, Pro-B can also be found at relatively high concentrations of 30–140 μmol g^{-1} in fresh tissue (where μmol g^{-1} is roughly equivalent to mM) in the leaves of members of the *Lamiaceae* (Wyn Jones and Storey, 1981).

Lucerne can metabolize its abundant Pro-B to *N*-methylproline and subsequently to proline both during and after germination (Le Rudulier *et al.*, 1995). This metabolic transformation by the plant means that in germinating legumes there is most likely a pulse of *N*-methylproline generated during Pro-B metabolism after seed germination (Phillips *et al.*, 1998).

30.4.2 Roles in symbiosis and osmoprotection

N-methylproline and Pro-B both play roles in the association of *Sinorhizobium meliloti* with *Medicago*. In *S. meliloti*, Pro-B induces expression of the *nod* genes (Phillips *et al.*, 1992), which are critical for the establishment of the root nodule nitrogen-fixing symbiosis between these two species. Mutants that failed to catabolize Pro-B were disabled for root colonization and nodule formation (Goldmann *et al.*, 1994). The presence of the Pro-B demethylase adjacent to the *fix* and *nod* genes on the symbiotic megaplasmid, pSym, strongly suggests that this is an important catabolic process in establishment of the symbiosis (Goldmann *et al.*, 1991). *S. meliloti* with mutations in the *N*-methylproline demethylase are moderately defective in competition for root attachment in the presence of wild-type cells (Phillips *et al.*, 1998). These researchers also postulated that the temporal sequence of compound availability – initially high Pro-B concentration followed by a pulse of *N*-methyproline, and finally proline – may be an important developmental sequence to promote progression of the symbiosis and sculpting of the rhizosphere community.

Pro-B is a potent osmoprotectant that can be transported and utilized by diverse bacteria for this purpose (Le Rudulier *et al.*, 1984; Bashir *et al.*, 2014; Kumar *et al.*, 2014; also reviewed in Wood, 2011). In a manner similar to GB, the *Enterobacteriaceae* import Pro-B only during salt stress and quickly export it upon removal of the stressor (Le Rudulier *et al.*, 1984; Lamark *et al.*, 1992). Conversely, and also similar to GB, many environmental bacteria can import Pro-B under any saline condition and have the capacity to catabolize Pro-B (Gloux and Le Rudulier, 1989; Burnet *et al.*, 2000), thus preventing excretion of a nutrient source that may be valuable, particularly during competition for limiting nitrogen in the soil.

30.4.3 Biosynthesis

To my knowledge, *de novo* syntheses of *N*-methylproline and/or Pro-B have not been reported from bacteria. Pro-B is synthesized in the plant through the *N*-methylation of proline via a SAM-dependent pathway (Robertson and Marion, 1960), but based on literature searches, the enzymes responsible for this methylation have not been identified.

30.4.4 Catabolism

The catabolism of Pro-B was initially described for *S. meliloti* in which the gene *stc2*, located on pSym, encodes the Pro-B demethylase, Pro-B-DM, (Goldmann *et al.*, 1991; Burnet *et al.*, 2000), *stcD* encodes the *N*-methylproline demethylase, *N*-mePro-DM (Phillips *et al.*, 1998) and, finally, proline catabolism occurs via the *put* (proline dehydrogenase) system (Fig. 30.2B). Stc2 is a Rieske family oxygenase that catalyses the oxidative demethylation of Pro-B to *N*-methylproline (Burnet *et al.*, 2000). The catalytic mechanism of the enzyme was partially elucidated by crystallizing Stc2 in complex with Pro-B, and Pro-B shown to contribute directly to the coordination of the mononuclear non-haem iron in the active site (Daughtry *et al.*, 2012).

The Rieske family of oxidative demethylases are fairly promiscuous for any compound that can dock in the substrate pocket, which primarily recognizes a single functional group in target molecules, as best described for the ring hydroxylating versions of these proteins (reviewed in Nam *et al.*, 2001). Therefore, it may not be surprising that Burnet *et al.* (2000) showed that *stc2* was also required for the catabolism of GB derived from carnitine, thus providing strong evidence that GB metabolism proceeds through the GbcA-like Stc2 enzyme in *S. meliloti* and not via a GB methyltransferase as discussed in Section 30.3.4. The StcD *N*-methylproline demethylase is more akin to the dimethylglycine and sarcosine demethylases, being an FMN-linked oxidoreductase/demethylase. This early work established the basics of the pathway for Pro-B catabolism, but the components of the entire pathway were not fully categorized and directly tested.

More recently, Kumar *et al.* (2014) demonstrated a full pathway for Pro-B catabolism in *Paracoccus denitrificans* and *Rhodobacter sphaeroides*. This group of researchers used small-molecule docking predictions and bioinformatics analyses of gene synteny to predict the likely path for Pro-B catabolism in the study organisms and followed the predictions up with analyses of gene expression, assessment of metabolic blockages in targeted mutants, complementation analysis and some enzymology. The pathway so elucidated agrees well with that compiled from previous separate studies, and also demonstrates that this pathway is not limited to the *Rhizobiaceae*. The enzymes involved in the two demethylations of Pro-B (HpbB-HpbC, and HpbA) turned out to be identical to those previously identified for the *N*-demethylations of hydroxylproline betaine (hence the Hpb names) (Zhao *et al.*, 2013).

In *S. meliloti*, *P. denitrificans* and *R. sphaeroides*, Pro-B catabolism is regulated by salinity such that increasing salt concentrations inhibit Pro-B catabolism (Gloux and Le Rudulier, 1989; Kumar *et al.*, 2014). This is a homologous mechanism to salt regulation by GB catabolism in *P. aeruginosa* and other bacteria (Diab *et al.*, 2006; Fitzsimmons *et al.*, 2012; further review in Wargo, 2013a); each of these studies proposed that such regulation ensures that the osmoprotectant is retained during times of salt stress.

30.5 Histidine Betaine and Its Metabolites

The *N*-methylated forms of histidine are histidine betaine (His-B, also called hercynine) and the His-B derivative, ergothioneine (ERG). Most of the available research concerns ERG, so this will be the primary focus of the current section. Unlike the other compounds discussed in this chapter, His-B and ERG appear to be synthesized only by bacteria and fungi, but they accumulate to high levels in many plants and animals.

30.5.1 Sources

Some bacteria, particularly members of the *Actinobacteria* and *Cyanobacteria*, can synthesize His-B and ERG (Genghof and van Damme, 1964; Genghof, 1970), and *Mycobacterium smegmatis* has been shown to secrete ERG into the environment (Sao Emani *et al.*, 2013). Fungi also produce His-B and ERG to relatively high levels (Genghof, 1970). Furthermore, ERG is accumulated to high levels in a variety of plants and animals in concentrations of up to 11 mM (Ey *et al.*, 2007). In animals, ERG is particularly enriched in cells of the liver and erythrocytes, and is actively transported by the OCTN1 organic cation integral membrane transporter (Gründemann *et al.*, 2005), where it can function as an anti-teratogen and anti-inflammatory compound (Mayumi *et al.*, 1982; Hartman, 1990; Rahman *et al.*, 2003). In erythrocytes, ERG may function to inhibit methaemoglobin formation (Gründemann *et al.*, 2005). The OCTN1 transporter also mediates the uptake of Pro-B, but this is rapidly excreted and therefore does not accumulate (Gründemann *et al.*, 2005; Kato *et al.*, 2010). Liebeke and Bundy (2013) also detected His-B in earthworms of various species, though they did not determine whether the source of these compounds was endogenous or dietary. Based on the general lack of this pathway in other animals, the likely source is from the soil diet of the worms. It should also be noted that the NMR detection method used in this study may not have distinguished between His-B and ERG, as the ^{1}H and ^{13}C signals of the quaternary methyl groups were used for identification.

30.5.2 Role in bacterial biology

Apart from its ability to be synthesized or degraded, the biological roles of His-B and ERG in bacteria have not been well studied. Some evidence points to an accessory role for ERG as an antioxidant in bacteria, but the effects are relatively small (Hartman, 1990; Ta *et al.*, 2011; Sao Emani *et al.*, 2013). However, the abundance of His-B and ERG in some taxa and the broad distribution of synthetic and catabolic enzymes suggest that these compounds may be more important than is currently appreciated.

30.5.3 Biosynthesis

Bacterial His-B synthesis has been best studied in mycobacteria. His-B is synthesized by the SAM-dependent methylase EgtD from histidine (Fig. 30.3) (Seebeck, 2010). To generate ERG, His-B reacts with glutamylcysteine and oxygen in a reaction catalysed by the EgtB enzyme. Glutamate is removed from the intermediate by EgtC, and the resultant intermediate is processed by the pyridoxal phosphate (PLP)-requiring enzyme EgtE to give ERG, pyruvate and ammonia. Seebeck (2010) was not able to directly demonstrate EgtE activity from purified protein, and Khonde and Jardine (2015) were likewise limited to measuring activity in cell lysates. Song *et al.* (2015) succeeded in purifying the *M. smegmatis* EgtE and conducted a full enzymatic characterization, which showed that the removal of the cysteinyl moiety from the ERG precursor proceeds through a sulfenic acid intermediate.

EgtD and EgtB are evolutionarily conserved in organisms that are known or likely to synthesize ERG, whereas the full *egtABCDE* operon is only conserved in the *Actinobacteria* (Jones *et al.*, 2014). As many *Cyanobacteria* can synthesize ERG (Jones *et al.*, 2014), and other enzymes can promote the downstream steps (Seebeck, 2010), it is probable that it is the trimethylation of histidine and the initial glutamylcysteinyl modification that are the critical specific steps in ERG biosynthesis in bacteria. Likely orthologues of EgtD and EgtB are also present in many *Burkholderia* spp. (Jones *et al.*, 2014).

30.5.4 Catabolism

A number of bacteria can utilize ERG as a nitrogen source or express proteins that can metabolize ERG to thioluroconic acid and trimethylamine by deamination (Fig. 30.3) (Yanasugondha and Appleman, 1957; Kelly and Appleman, 1961; Wolff, 1962). The first ergothionase was identified by Baran *et al.* (2013) in *Shewanella oneidensis* using a metabolomic profiling method. The *SO3057* gene was found to encode a histidine/phenylalanine ammonia lyase family member (CD00332) that when expressed and purified could deaminate ERG and His-B. Therefore, this enzyme should also be sufficient for direct His-B metabolism, although this has not been examined *in vivo* by genetic analysis. Murumatsu *et al.* (2013) also cloned an ergothionase from *Burkholderia*, and this was a member of the same enzyme family as that found for *S. oneidensis*. The *Burkholderia* study also demonstrated that ergothionase was a homotetramer, with each subunit ~55 kDa. The presence of an ergothionase in some *Burkholderia* that also have the genes for His-B and ERG synthesis means that these bacteria can be a *de novo* source and a sink for these two compounds.

30.6 *N*-Methylated Tryptophan

The *N*-methylated forms of tryptophan are *N*-methyl tryptophan (abrine) and tryptophan

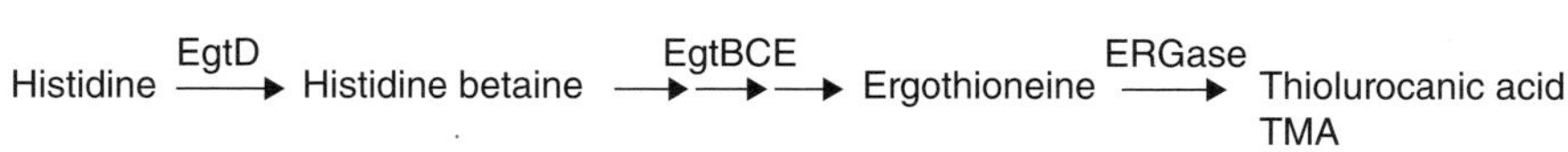

Fig. 30.3. Schematic representation of histidine betaine (His-B) and ergothioneine (ERG) metabolism in bacteria. The protein names for ergothioneine synthesis are from mycobacteria. Abbreviations: EgtBCE, the enzymes synthesizing ERG from His-B (His-B reacts with glutamylcysteine and oxygen catalysed by EgtB; glutamate is removed from the intermediate by EgtC; the resultant intermediate is processed by EgtE to give ERG, pyruvate, and ammonia); EgtD, *S*-adenosylmethionine (SAM)-dependent methylase; ERGase, ergothionase; TMA, trimethylamine.

betaine (Trp-B, also known as hypaphorine) (see Fig. 30.1). *N*-methyl tryptophan is best known as a metabolite made by the legume *Abrus precatorius* as a precursor for the synthesis of the toxin abrin, which is similar in action and chemistry to ricin (Dodge *et al.*, 2015). The first report of the bacterial degradation of *N*-methyl tryptophan came from members of the *Enterobacteriaceae* (Wang, 1957), and it was later shown that *E. coli* encodes an enzyme, SolA, that is similar to the monomeric sarcosine oxidases and highly active towards *N*-methyl tryptophan (Koyama and Ohmori, 1996; Wagner *et al.*, 1999; Khanna and Schuman Jorns, 2001). There is one report of bacterial synthesis of *N*-methyl tryptophan and *N,N*-dimethyl tryptophan in an *Arthrobacter* species when it was fed tryptophan (Forni *et al.*, 1992). The biological consequences of *N*-methyl tryptophan catabolism or synthesis in bacteria are not known.

Trp-B is secreted by ectomycorrhizal fungi (Ditengou *et al.*, 2000; Martin *et al.*, 2001; Wiemken and Boller, 2002) and is present in the latex-producing members of the *Moraceae* plant family (Shrestha and Bisset, 1991). In the ectomycorrhizal symbiosis, Trp-B can alter root hair elongation and main root development (Wiemkin and Boller, 2002). While these two instances are interesting, a search of the literature revealed no mention of specific bacterial utilization of Trp-B.

30.7 *N*-Methylated Tyrosine

Tyrosine betaine (Tyr-B) and *N*-methyl tyrosine have been detected in the lichen *Lobaria laetevirens* (Bernard *et al.*, 1981), which is an interesting three-member symbiosis between a fungus, a eukaryotic alga and a cyanobacterium that may also contain members of the *Rhizobiales* (Erlacher *et al.*, 2015). Tyr-B has also been found in the latex of trees in the *Moraceae* family (Shrestha and Bisset, 1991) and as part of the chemical defence secretions in the leaf beetle *Oreina gloriosa* (Eggenberger and Rowell-Rahier, 1993). A more recent report by Carollo *et al.* (2010) identified a tyrosine-containing molecule produced by a fungus that is conjugated to a glycine betainyl moiety on the nitrogen, generating an amide compound. These authors did not follow 60 years of nomenclature precedence in naming this compound, as they incorrectly identified it as tryptophan betaine. Nomenclature issues aside, this novel compound is a new addition to microbially produced betaines and may represent an intersection between tyrosine metabolism and carnitine catabolism in which a glycine betainyl-CoA intermediate is produced (Bastard *et al.*, 2014; Meadows and Wargo, 2015). Despite these reports of the presence of Tyr-B in eukaryotic secretions and a symbiotic interaction, a biological role for Tyr-B in bacteria is not currently known.

30.8 Conclusions

Recent years have seen substantial progress on *N*-methylated amino acid metabolism, identifying and characterizing enzymes involved in GB, Pro-B and ERG catabolism. These discoveries provide the toolkit to experimentally test the function of the compounds in a broader group of bacteria and under wider environmental conditions. In particular, the roles of GB and Pro-B catabolism in soil or in *Cyanobacteria*-dominated ecosystems remain unknown, and the roles of His-B and ERG in bacterial associations with eukaryotes are woefully understudied. Finally, we know almost nothing about Trp-B and Tyr-B metabolism in bacteria and much less about their biological functions.

References

Alloing, G., Travers, I., Sagot, B., Le Rudulier, D. and Dupont, L. (2006) Proline betaine uptake in *Sinorhizobium meliloti*: characterization of Prb, an Opp-like ABC transporter regulated by both proline betaine and salinity stress. *Journal of Bacteriology* 188, 6308–6317.

Anthoni, U., Christophersen, C., Hougaard, L. and Nielsen, P.H. (1991) Quaternary ammonium compounds in the biosphere – an example of a versatile adaptive strategy. *Comparative Biochemistry and Physiology Part B: Comparative Biochemistry* 99, 1–18.

Aurich, H., Kleber, H.P. and Schopp, W.D. (1967) An inducible carnitine dehydrogenase from *Pseudomonas aeruginosa*. *Biochimica et Biophysica Acta* 139, 505–507.

Baran, R., Bowen, B.P., Price, M.N., Arkin, A.P., Deutschbauer, A.M. and Northen, T.R. (2013) Metabolic footprinting of mutant libraries to map metabolite utilization to genotype. *ACS Chemical Biology* 8, 189–199.

Barra, L., Fontenelle, C., Ermel, G., Traulwetter, A., Walker, G.C. and Blanco, C. (2006) Interrelations between glycine betaine catabolism and methionine biosynthesis in *Sinorhizobium meliloti* strain 102F34. *Journal of Bacteriology* 188, 7195–7204.

Bashir, A., Hoffman, T., Kempf, B., Xie, X., Smits, S.H.J. and Bremer, E. (2014) Plant-derived compatible solutes proline betaine and betonicine confer enhanced osmotic and temperature stress tolerance to *Bacillus subtilis*. *Microbiology* 160, 2283–2294.

Bastard, K., Smith, A.A.T., Vergne-Vaxelaire, C., Perret, A., Zaparucha, A., De Melo-Minardi, R., Mariage, A., Boutard, M., Debard, A., Lechaplais, C. *et al.* (2014) Revealing the hidden functional diversity of an enzyme family. *Nature Chemical Biology* 10, 42–49.

Bater, A.J. and Venables, W.A. (1977) The characterisation of inducible dehydrogenases specific for the oxidation of D-alanine, allohydroxy-D-proline, choline and sarcosine as peripheral membrane proteins in *Pseudomonas aeruginosa*. *Biochimica et Biophysica Acta* 468, 209–226.

Beers, J.R. (1967) The species distribution of some naturally-occurring quaternary ammonium compounds. *Comparative Biochemistry and Physiology* 21, 11–21.

Bernard, T., Goas, G., Hamelin, J. and Joucla, M. (1981) Characterization of DOPA betaine, tyrosine betaine and *N*-dimethyltyrosine from *Lobaria laetevirens*. *Phytochemistry* 20, 2325–2326.

Boncompagni, E., Osteras, M., Poggi, M.C. and Le Rudulier, D. (1999) Occurrence of choline and glycine betaine uptake and metabolism in the family *Rhizobiaceae* and their roles in osmoprotection. *Applied and Environmental Microbiology* 65, 2072–2077.

Bremer, J. (1983) Carnitine – metabolism and functions. *Physiology Reviews* 63, 1420–1480.

Burnet, M.W., Goldmann, A., Message, B., Drong, R., El Amrani, A., Loreau, J. and Tepfer, D. (2000) The stachydrine catabolism region in *Sinorhizobium meliloti* encodes a multi-enzyme complex similar to the xenobiotic degrading systems in other bacteria. *Gene* 244, 151–161.

Carollo, C.A., Calil, A.L.A., Schiave, L.A., Guaratini, T., Roberts, D.W., Lopes, N.P. and Braga, G.U.L. (2010) Fungal tyrosine betaine, a novel secondary metabolite from conidia of entomopathogenic *Metarhizium* spp. fungi. *Fungal Biology* 114, 473–480.

Cavalieri, A.J. and Huang, A.H.C. (1981) Accumulation of proline and glycine betaine in *Spartina alterniflora* Loisel in response to NaCl and nitrogen in the marsh. *Oecologia* 49, 224–228.

Chambers, S.T. and Kunin, C.M. (1987) Isolation of glycine betaine and proline betaine from human urine. *Journal of Clinical Investigation* 79, 731–737.

Csonka, L.N. and Hanson, A.D. (1991) Prokaryotic osmoregulation: genetics and physiology. *Annual Reviews of Microbiology* 45, 569–606.

Daughtry, K.D., Xiao, Y., Stoner-Ma, D., Cho, E., Orville, A.M., Liu, P. and Allen, K.N. (2012) Quaternary ammonium oxidative demethylation: X-ray crystallographic, resonance Raman, and UV-visible spectroscopic analysis of a Rieske-type demethylase. *Journal of the American Chemical Society* 134, 2823–2834.

Diab, F., Bernard, T., Bazire, A., Haras, D., Blanco, D. and Jebbar, M. (2006) Succinate-mediated catabolite repression control on the production of glycine betaine catabolic enzymes in *Pseudomonas aeruginosa* PAO1 under low and elevated salinities. *Microbiology* 152, 1395–1406.

Ditengou, F.A., Béguiristain, T. and Lapeyrie, F. (2000) Root hair elongation is inhibited by hypaphorine, the indole alkaloid from the ectomycorrhizal fungus *Pisolithus tinctorius*, and restored by indole-3-acetic acid. *Planta* 211, 722–728.

Djordjevic, M.A., Chen, H.C., Natera, S., van Noorden, G., Menzel, C., Taylor, S., Renard, C., Geiger, O., the Sinorhizobium DNA Sequencing Consortium and Weiller, G.F. (2003) A global analysis of protein expression profiles in *Sinorhizobium meliloti*: discovery of new genes for nodule occupancy and stress adaptation. *Molecular Plant–Microbe Interactions* 16, 508–524.

Dodge, A.G., Carrasquillo, K., Rivera, L., Xu, L., Wackett, L.P. and Sadowsky, M.J. (2015) Rapid method using two microbial enzymes for detection of L-abrine in food as a marker for the toxic protein abrin. *Applied and Environmental Microbiology* 81, 1610–1615.

Eggenberger, F. and Rowell-Rahier, M. (1993) Physiological sources of variation in chemical defenses of *Oreina gloriosa* (Coleoptera: Chrysomelidae). *Journal of Chemical Ecology* 19, 395–410.

Erlacher, A., Cernava, T., Cardinale, M., Soh, J., Sensen, C.W., Grube, M. and Berg, G. (2015) *Rhizobiales* as functional and endosymbiontic members in the lichen symbiosis of *Lobaria pulmonaria* L. *Frontiers in Microbiology* 6, 1–9.

Ey, J., Schömig, E. and Taubert, D. (2007) Dietary sources and antioxidant effects of ergothioneine. *Journal of Agricultural and Food Chemistry* 55, 6466–6474.
Fitzsimmons, L.F., Flemer, S., Wurthmann, A.S., Deker, P.B., Sarkar, I.N. and Wargo, M.J. (2011) Small-molecule inhibition of choline catabolism in *Pseudomonas aeruginosa* and other aerobic choline-catabolizing bacteria. *Applied and Environmental Microbiology* 77, 4383–4389.
Fitzsimmons, L.F., Hampel, K.J. and Wargo, M.J. (2012) Cellular choline and glycine betaine pools impact osmoprotection and phospholipase C production in *Pseudomonas aeruginosa*. *Journal of Bacteriology* 194, 4718–4726.
Forni, C., Riov, J., Caiola, M.G. and Tel-Or, E. (1992) Indole-3-acetic acid (IAA) production by *Arthrobacter* species isolated from *Azolla*. *Journal of General Microbiology* 138, 377–381.
Genghof, D.S. (1970) Biosynthesis of ergothioneine and hercynine by fungi and *Actinomycetales*. *Journal of Bacteriology* 103, 475–478.
Genghof, D.S. and van Damme, O. (1964) Biosynthesis of ergothioneine and hercynine by mycobacteria. *Journal of Bacteriology* 87, 852–862.
Gloux, K. and Le Rudulier, D. (1989) Transport and catabolism of proline betaine in salt-stressed *Rhizobium meliloti*. *Archives of Microbiology* 151, 143–148.
Goffredi, S.K., Yi, H., Zhang, Q., Klann, J.E., Struve, I.A., Vrijenhoek, R.C. and Brown, C.T. (2014) Genomic versatility and functional variation between two dominant heterotrophic symbionts of deep-sea *Osedax* worms. *The ISME Journal* 8, 908–924.
Goldmann, A., Boivin, C., Fleury, V., Message, B., Lecoeur, L., Maille, M. and Tepfer, D. (1991) Betaine use by rhizosphere bacteria: genes essential for trigonelline, stachydrine, and carnitine catabolism in *Rhizobium meliloti* are located on pSym in the symbiotic region. *Molecular Plant–Microbe Interactions* 4, 571–578.
Goldmann, A., Lecoeur, L., Message, B., Delarue, M., Schoonejans, E. and Tepfer, D. (1994) Symbiotic plasmid genes essential to the catabolism of proline betaine, or stachydrine, are also required for efficient nodulation by *Rhizobium meliloti*. *FEMS Microbiology Letters* 115, 305–311.
Gründemann, D., Harlfinger, S., Golz, S., Geerts, A., Lazar, A., Berkels, R., Jung, N., Rubbert, A. and Schömig, E. (2005) Discovery of the ergothioneine transporter. *Proceedings of the National Academy of Sciences of the United States of America* 102, 5256–5261.
Hartman, P.E. (1990) Ergothioneine as antioxidant. *Methods in Enzymology* 186, 310–318.
Hassan-Abdallah, A., Zhao, G., Eschenbrenner, M., Chen, Z.W., Mathews, F.S. and Jorns, M.S. (2005) Cloning, expression and crystallization of heterotetrameric sarcosine oxidase from *Pseudomonas maltophilia*. *Protein Expression and Purification* 43, 33–43.
Heijthuijsen, J.H.F.G. and Hansen, T.A. (1989) Betaine fermentation and oxidation by marine *Desulfuromonas* strains. *Applied and Environmental Microbiology* 55, 965–969.
Hormann, K. and Andreesen, J.R. (1989) Reductive cleavage of sarcosine and betaine by *Eubacterium acidaminophilum* via enzyme systems different from glycine reductase. *Archives of Microbiology* 153, 50–59.
Ikuta, S., Imamura, S., Misaki, H. and Horiuti, Y. (1977) Purification and characterization of choline oxidase from *Arthrobacter globiformis*. *Journal of Biochemistry* 82, 1741–1749.
Ivanov, A. and Kostadinova, S. (1993) Phosphatidylcholine-specific phospholipase C from *Achromobacter xylosoxidans*. *Acta Microbiologica Bulgaria* 29, 3–8.
Jones, G.W., Doyle, S. and Fitzpatrick, D.A. (2014) The evolutionary history of the genes involved in the biosynthesis of the antioxidant ergothioneine. *Gene* 549, 161–170.
Kapfhammer, D., Karatan, E., Pflughoeft, K.J. and Watnick, P.I. (2005) Role for glycine betaine transport in *Vibrio cholerae* osmoadaptation and biofilm formation within microbial communities. *Applied and Environmental Microbiology* 71, 3840–3847.
Kato, Y., Kubo, Y., Iwata, D., Kato, S., Sudo, T., Sugiura, T., Kagaya, T., Wakayama, T., Hirayama, A., Sugimoto, M. *et al.* (2010) Gene knockout and metabolome analysis of carnitine/organic cation transporter OCTN1. *Pharmaceutical Research* 27, 832–840.
Kelly, B. and Appleman, M.D. (1961) Degradation of ergothioneine by cell-free extracts of *Alcaligenes faecalis*. *Journal of Bacteriology* 81, 715–720.
Khanna, P. and Schuman Jorns, M. (2001) Characterization of the FAD-containing *N*-methyltryptophan oxidase from *Escherichia coli*. *Biochemistry* 40, 1441–1450.
Khonde, P.L. and Jardine, A. (2015) Improved synthesis of the super antioxidant, ergothioneine, and its biosynthesis pathway intermediates. *Organic and Biomolecular Chemistry* 13, 1415–1419.
Kiene, R.P. (1998) Uptake of choline and its conversion to glycine betaine by bacteria in estuarine waters. *Applied and Environmental Microbiology* 64, 1045–1051.

King, G.M. (1988) Methanogenesis from methylated amines in a hypersaline algal mat. *Applied and Environmental Microbiology* 54, 130–136.

Kishore, G.M. and Jacob, G.S. (1987) Degradation of glyphosate by *Pseudomonas* sp. PG2982 via a sarcosine intermediate. *The Journal of Biological Chemistry* 262, 12164–12168.

Klähn, S., Steglich, C., Hess, W.R. and Hagemann, M. (2010) Glucosylglycerate: a secondary compatible solute common to marine cyanobacteria from nitrogen-poor environments. *Environmental Microbiology* 12, 83–94.

Korbsrisate, S., Tomaras, A.P., Damnin, S., Ckumdee, J., Srinon, V., Lenwehasatit, I., Vasil, M.L. and Suparak, S. (2007) Characterization of two distinct phospholipase C enzymes from *Burkholderia pseudomallei*. *Microbiology* 153, 1907–1915.

Kortstee, G.J. (1970) The aerobic decomposition of choline by microorganisms. I. The ability of aerobic organisms, particularly coryneform bacteria, to utilize choline as the sole carbon and nitrogen source. *Archiv für Mikrobiologie* 71, 235–244.

Koyama, Y. and Ohmori, H. (1996) Nucleotide sequence of the *Escherichia coli solA* gene encoding a sarcosine oxidase-like protein and characterization of its product. *Gene* 181, 179–183.

Kumar, R., Zhao, S., Vetting, M.W., Wood, B.M., Sakai, A., Cho, K., Solbiati, J., Almo, S.C., Sweedler, J.V., Jacobson, M.P. *et al.* (2014) Prediction and biochemical demonstration of a catabolic pathway for the osmoprotectant proline betaine. *mBio* 5(1): e00933-13.

Kunin, C.M., Hua, T.H., Van Arsdale White, L. and Villarejo, M. (1992) Growth of *Escherichia coli* in human urine: role of salt tolerance and accumulation of glycine betaine. *Journal of Infectious Diseases* 166, 1311–1315.

Lamark, T., Styrvold, O.B. and Strøm, A.R. (1992) Efflux of choline and glycine betaine from osmoregulating cells of *Escherichia coli*. *FEMS Microbiology Letters* 75, 149–154.

Landfald, B. and Strøm, A.R. (1986) Choline-glycine betaine pathway confers a high level of osmotic tolerance in *Escherichia coli*. *Journal of Bacteriology* 165, 849–855.

Le Rudulier, D., Bernard, T., Goas, G. and Hamelin, J. (1984) Osmoregulation in *Klebsiella pneumoniae*: enhancement of anaerobic growth and nitrogen fixation under stress by proline betaine, gamma-butyrobetaine, and other related compounds. *Canadian Journal of Microbiology* 30, 299–305.

Le Rudulier, D., Gloux, D.K., Poggi, M.C. and Noel, J.P. (1995) Accumulation, biosynthesis and fate of proline betaine in nodulated alfalfa plants (*Medicago sativa* L.) subjected to salinity stress. In: Leigh, R.A. and Blake-Kalff, M.M.A.M. (eds) *Proceedings of the Second STRESSNET Conference, Salsomaggiore, Italy, 21–23 September 1995*. European Commission Directorate General VI, Brussels, pp. 257–262.

Lever, M., Sizeland, P.C., Bason, L.M., Hayman, C.M. and Chambers, S.T. (1994) Glycine betaine and proline betaine in human blood and urine. *Biochimica et Biophysica Acta* 1200, 259–264.

Li, S., Yu, X. and Beattie, G.A. (2013) Glycine betaine catabolism contributes to *Pseudomonas syringae* tolerance to hyperosmotic stress by relieving betaine-mediated suppression of compatible solute synthesis. *Journal of Bacteriology* 195, 2415–2423.

Liebeke, M. and Bundy, J.G. (2013) Biohemical diversity of betaines in earthworms. *Biochemical and Biophysical Research Communications* 430, 1306–1311.

Lisa, T.A., Lucchesi, G.I. and Domenech, C.E. (1994) Pathogenicity of *Pseudomonas aeruginosa* and its relationship to the choline metabolism through the action of cholinesterase, acip phosphatase, and phospholipase C. *Current Microbiology* 29, 193–199.

Lu, W.D., Chi, Z.M. and Su, C.D. (2006) Identification of glycine betaine as compatible solute in *Synechococcus* sp. WH8102 and characterization of its *N*-methyltransferase genes involved in betaine synthesis. *Archives of Microbiology* 186, 495–506.

Martin, F., Duplessis, S., Diengou, F., Lagrange, H., Voiblet, C. and Lapeyrie, F. (2001) Developmental cross talking in the ectomycorrhizal symbiosis: signals and communication genes. *New Phytologist* 151, 145–154.

Mayumi, T., Okamoto, K., Yoshida, K., Kawai, Y., Kawano, H., Hama, T. and Tanaka, K. (1982) Studies on ergothioneine. VIII. Preventive effects of ergothioneine on cadmium-induced teratogenesis. *Chemical and Pharmaceutical Bulletin* 30, 2141–2146.

Meadows, J.A. and Wargo, M.J. (2015) Carnitine in bacterial physiology and metabolism. *Microbiology* 161, 1161–1174.

Meskys, R., Harris, R.J., Casaite, V., Basran, J. and Scrutton, N.S. (2001) Organization of the genes involved in dimethylglycine and sarcosine degradation in *Arthrobacter* spp.: implications for glycine betaine catabolism. *European Journal of Biochemistry* 268, 3390–3398.

Moriguchi, T., Ida, K., Hikima, T., Ueno, G., Yamamoto, M. and Suzuki, H. (2010) Channeling and conformational changes in the heterotetrameric sarcosine oxidase from *Corynebacterium* sp. U-96. *Journal of Biochemistry* 148, 491–505.

Muramatsu, H., Matsuo, H., Okada, N., Ueda, M., Yamamoto, H., Kato, S. and Nagata, S. (2013) Characterization of ergothionase from *Burkholderia* sp. HME13 and its application to enzymatic quantification of ergothioneine. *Applied Microbiology and Biotechnology* 97, 5389–5400.

Nam, J.-W., Hideaki, N., Yoshida, T., Habe, H., Yamane, H. and Omori, T. (2001) New classification system for oxygenase components involved in ring-hydroxylating oxygenations. *Bioscience, Biotechnology, and Biochemistry* 65, 254–263.

Naumann, E., Hippe, H. and Gottschalk, G. (1983) Betaine: new oxidant in the Stickland reaction and methanogenesis from betaine and L-alanine by a *Clostridium sporogenes–Methanosarcina barkeri* coculture. *Applied and Environmental Microbiology* 45, 474–483.

Nimmo-Smith, R.H. and Appleyard, G. (1956) Studies with a pseudomonad able to grow with creatine as main source of carbon and nitrogen. *Journal of General Microbiology* 14, 336–350.

Oren, A. (1990) Formation and breakdown of glycine betaine and trimethylamine in hypersaline environments. *Antonie van Leeuwenhoek* 58, 291–298.

Ostroff, R.M., Wretlind, B. and Vasil, M.L. (1989) Mutations in the hemolytic-phospholipase C operon result in decreased virulence of *Pseudomonas aeruginosa* PAO1 grown under phosphate-limiting conditions. *Infection and Immunity* 57, 1369–1373.

Perroud, B. and Le Rudulier, D. (1985) Glycine betaine transport in *Escherichia coli*: osmotic modulation. *Journal of Bacteriology* 161, 393–401.

Phillips, D.A., Joseph, C.M. and Maxwell, C.A. (1992) Trigonelline and stachydrine released from alfalfa seeds activate NodD2 protein in *Rhizobium meliloti*. *Plant Physiology* 99, 1526–1531.

Phillips, D.A., Sande, E.S., Vriezen, J.A.C., de Bruijn, F.J., Le Rudulier, D. and Joseph, C.M. (1998) A new genetic locus in *Sinorhizobium meliloti* is involved in stachydrine utilization. *Applied and Environmental Microbiology* 64, 3954–3960.

Pierce, S.K. (1982) Invertebrate cell volume control mechanisms: a coordinated use of intracellular amino acids and inorganic ions as osmolyte solute. *Biological Bulletin* 163, 405–419.

Pinel, N., Davidson, S.K. and Stahl, D.A. (2008) *Verminephrobacter eiseniae* gen. nov., sp. nov., a nephridial symbiont of the earthworm *Eisenia foetida* (Savigny). *International Journal of Systematic and Evolutionary Microbiology* 58, 2147–2157.

Rahman, I., Gilmour, P.S., Jimenez, L.A., Biswas, S.K., Antonicelli, F. and Aruoma, O.I. (2003) Ergothioneine inhibits oxidative stress- and TNF-α-induced NF-κB activation and interleukin-8 release in alveolar epithelial cells. *Biochemical and Biophysical Research Communications* 302, 860–864.

Rahme, L.G., Ausubel, F.M., Cao, H., Drenkard, E., Goumnerov, B.C., Lau, G.W., Mahajan-Miklos, S., Plotnikova, J., Tan, M.W., Tsongalis, J. *et al.* (2000) Plants and animals share functionally common bacterial virulence factors. *Proceedings of the National Academy of Sciences of the United States of America* 97, 8815–8821.

Robertson, A.V. and Marion, L. (1960) The biogenesis of alkaloids: XXV. The role of hygric acid in the biogenesis of stachydrine. *Canadian Journal of Chemistry* 38, 396–398.

Rueppel, M.L., Brightwell, B.B., Schaefer, J. and Marvel, J.T. (1977) Metabolism and degradation of glyphosate in soil and water. *Journal of Agricultural and Food Chemistry* 25, 517–528.

Sao Emani, C., Williams, M.J., Wild, I.J., Hiten, N.F., Viljoen, A.J., Pietersen, R.-D.D., van Helden, P.D. and Baker, B. (2013) Ergothioneine is a secreted antioxidant in *Mycobacterium smegmatis*. *Antimicrobial Agents and Chemotherapy* 57, 3202–3207.

Seebeck, F.P. (2010) *In vitro* reconstitution of mycobacterial ergothioneine biosynthesis. *JACS Communications* 132, 6632–6633.

Shrestha, T. and Bisset, N.G. (1991) Quaternary nitrogen compounds from South American *Moraceae*. *Phytochemistry* 30, 3285–3287.

Sleator, R.D., Francis, G.A., O'Bierne, D., Gahan, C.G.M. and Hill, C. (2003) Betaine and carnitine uptake systems in *Listeria monocytogenes* affect growth and survival in foods and during infection. *Journal of Applied Microbiology* 95, 839–846.

Smith, L.T., Pocard, J.A., Bernard, T. and Le Rudulier, D. (1988) Osmotic control of glycine betaine biosynthesis and degradation in *Rhizobium meliloti*. *Journal of Bacteriology* 170, 3142–4149.

Song, H., Hu, W., Naowarojna, N., Her, A.S., Wang, S., Desai, R., Chen, X. and Liu, P. (2015) Mechanistic studies of a novel C-S lyase in ergothioneine biosynthesis: the involvement of a sulfenic acid intermediate. *Scientific Reports* 5: 11870.

Sprankle, P., Meggitt, W.F. and Penner, D. (1975) Adsorption, mobility, and microbial degradation of glyphosate in the soil. *Weed Science* 23, 229–234.

Steenbock, H. (1918) Isolation and identification of stachydrin from alfalfa hay. *The Journal of Biological Chemistry* 35, 1–13.

Stonehouse, M.J., Cota-Gomez, A., Parker, S.K., Martin, W.E., Hankin, J.A., Murphy, R.C., Chen, W., Lim, K.B., Kackett, M., Vasil, A.I. and Vasil, M.L. (2002) A novel class of microbial phosphocholine-specific phospholipases C. *Molecular Microbiology* 46, 661–676.

Szklarczyk, D., Franeschini, A., Wyder, S., Forslund, K., Heller, D., Huerta-Cepas, J., Simonovic, M., Roth, A., Santos, A., Tsafou, K.P. *et al.* (2015) STRING v10: protein–protein interaction networks, integrated over the tree of life. *Nucleic Acids Research* 43, D447–D452.

Ta, P., Buchmeier, N., Newton, G.L., Rawat, M. and Fahey, R.C. (2011) Organic hydroperoxide resistance protein and ergothioneine compensate for loss of mycothiol in *Mycobacterium smegmatis* mutants. *Journal of Bacteriology* 193, 1981–1990.

Ticak, T., Kountz, D.J., Girosky, K.E., Krzycki, J.A. and Ferguson, D.J. (2014) A nonpyrrolysine member of the widely distributed trimethylamone methyltransferase family is a glycine betaine methyltransferase. *Proceedings of the National Academy of Sciences of the United States of America* 11(43) E4668–E4676.

van Meer, G., Voelker, D.R. and Feigenson, G.W. (2008) Membrane lipids: where they are and how they behave. *Nature Reviews Molecular Cell Biology* 9, 112–124.

Waditee, R., Tanaka, Y., Aoki, K., Hibino, T., Jikuya, H., Takano, J., Takabe, T. and Takabe, T. (2003) Isolation and functional characterization of *N*-methyltransferases that catalyze betaine synthesis from glycine in a halotolerant photosynthetic organism *Aphanothece halophytica*. *The Journal of Biological Chemistry* 278, 4932–4942.

Wagner, M.A. and Schuman Jorns, M. (1997) Folate utilization by monomeric versus heterotetrameric sarcosine oxidases. *Archives of Biochemistry and Biophysics* 342, 176–181.

Wagner, M.A., Khanna, P. and Schuman Jorns, M. (1999) Structure of the flavocoenzyme of two homologous amine axidases: monomeric sarcosine oxidase and *N*-methyltryptophan oxidase. *Biochemistry* 38, 5588–5595.

Wang, S.-P (1957) Demethylase of bacteria. *Journal of the Formosan Medical Association* 56, 363–369.

Wargo, M.J. (2013a) Homeostasis and catabolism of choline and glycine betaine: lessons from *Pseudomonas aeruginosa*. *Applied and Environmental Microbiology* 79, 2112–2120.

Wargo, M.J. (2013b) Choline catabolism to glycine betaine contributes to *Pseudomonas aeruginosa* survival during murine lung infection. *PLoS ONE* 8(2): e56850.

Wargo, M.J. and Hogan, D.A. (2009) Identification of genes required for *Pseudomonas aeruginosa* carnitine catabolism. *Microbiology* 155, 2411–2419.

Wargo, M.J., Szwergold, B.S. and Hogan, D.A. (2008) Identification of two gene clusters and a transcriptional regulator required for *Pseudomonas aeruginosa* glycine betaine catabolism. *Journal of Bacteriology* 190, 2690–2699.

Wargo, M.J., Gross, M.J., Rajamani, S., Allard, J.L., Lundblad, L.K., Allen, G.B., Vasil, M.L., Leclair, L.W. and Hogan, D.A. (2011) Hemolytic phospholipase C inhibition protects lung function during *Pseudomonas aeruginosa* infection. *American Journal of Respiratory and Critical Care Medicine* 184, 345–354.

Warren, C.R. (2014) Response of osmolytes in soil to drying and rewetting. *Soil Biology and Biochemistry* 70, 22–32.

Wiemken, V. and Boller, T. (2002) Ectomycorrhiza: gene expression, metabolism and the wood-wide web. *Current Opinion in Plant Biology* 5, 1–7.

Willsey, G.G. and Wargo, M.J. (2015) Sarcosine catabolism in *Pseudomonas aeruginosa* is transcriptionally regulated by SouR. *Journal of Bacteriology* 198, 301–310.

Wolff, J.B. (1962) Ergothionase from *Escherichia coli*. *The Journal of Biological Chemistry* 237, 874–881.

Wood, J.M. (2011) Bacterial osmoregulation: a paradigm for the study of cellular homeostasis. *Annual Reviews of Microbiology* 65, 215–238.

Wyn Jones, R.G. and Storey, R. (1981) Betaines. In: Paleg, L.G. and Aspinall, D. (eds) *Physiology and Biochemistry of Drought Resistance in Plants*. Academic Press, New York, pp. 171–204.

Wyss, M. and Kaddurah-Daouk, R. (2000) Creatine and creatinine metabolism. *Physiology Reviews* 80, 1107–1213.

Yanasugondha, D. and Appleman, M.D. (1957) Degradation of ergothioneine by *Alcaligenes faecalis*. *Journal of Bacteriology* 74, 381–385.

Yancey, P.H., Clark, M.E., Hand, S.C., Bowlus, R.D. and Somero, G.N. (1982) Living with water stress: evolution of osmolyte systems. *Science* 217, 1214–1222.

Zachowski, A. (1993) Phospholipids in animal eukaryotic membranes: transverse asymmetry and movement. *Biochemical Journal* 294, 1–14.

Zeller, H.D., Hille, R. and Jorns, M.S. (1989) Bacterial sarcosine oxidase: identification of novel substrates and a biradical reaction intermediate. *Biochemistry* 28, 5145–5154.

Zhao, S., Kumar, R., Sakai, A., Vetting, M.W., Wood, B.M., Brown, S., Bonanno, J.B., Hillerich, B.S., Seidel, R.D., Babbitt, P.C. *et al.* (2013) Discovery of new enzymes and metabolic pathways by using structure and genome context. *Nature* 502, 698–702.

Zhu, Y., Jameson, E., Crosatti, M., Schäfer, H., Rajakumar, K., Bugg, T.D.H. and Chen, Y. (2014) Carnitine metabolism to trimethylamine by an unusual Rieske-type oxygenase from human microbiota. *Proceedings of the National Academy of Sciences of the United States of America* 111, 4268–4273.

31 Biofilm Formation: Amino Acid Biomarkers in *Candida Albicans*

Y. Cao* and Z. Liao
School of Pharmacy, Second Military Medical University, Shanghai, China

31.1 Abstract

Candida albicans biofilms are multicellular communities formed by individual yeast cells which are extremely resistant to a wide spectrum of antifungal drugs. Based on studies carried out *in vitro*, *C. albicans* biofilm formation can be partitioned into four basic stages: the adherence stage; the initiation stage; the mature stage; and the dispersal stage. Metabonomic studies have shown that the levels of amino acids or metabolites related to amino acids varies in the different stages of biofilm formation. Biosynthesis of the sulfur-containing amino acids methionine (Met) and cysteine (Cys) mediated by the gene *ECM17* was found to be important to *C. albicans* biofilm formation by affecting cell adhesion. Straight after adherence, ornithine, putrescine, tyrosine and nitrogen supplies were determined as regulators of germ tube growth in the initiation phase. When the biofilm is mature, a series of amino acids or amino acid-related metabolites upregulate and contribute to the production of extracellular matrix, and during the maturation phase, metabolic pathways for amino acids are vital in maintaining the stabilization of the *C. albicans* biofilm. Additionally, quorum sensing molecules that might be associated with the metabolism of tyrosine have been found to be crucial for the dispersal of the cells of *C. albicans* biofilms. Consistently, the genes involved in related metabolites have been found to be dissimilarly expressed in a phase-dependent manner.

Changed levels of pivotal intermediates of tricarboxylic acid (TCA) cycle metabolites were also observed during *C. albicans* biofilm growth, indicating the inhibition of mitochondrial respiration. This inhibition might be attributed to the accumulation of amino acids or amino acid-associated metabolites. Previous research on amino acid-related metabolites has indicated that the polyamine biosynthesis pathway might be a favourable target to investigate for enhancing the anti-biofilm activity of widely employed antifungal drugs. However, more screenings are required of specific amino acids that could be used as biomarkers in order to improve the efficacy of antifungal drugs towards *C. albicans* biofilms.

31.2 Introduction

C. albicans resides on mucosal surfaces and in the gastrointestinal and genitourinary tracts, and is a commensal of humans. The increased prevalence of immunosuppressive and antibiotic therapy, as well as infection by human immunodeficiency virus (HIV), have made it a leading fungal pathogen and the causative agent of a wide spectrum

*Corresponding author. E-mail: caoyingying608@163.com

of diseases (Nett *et al.*, 2009; Finkel and Mitchell, 2011). Infections caused by *C. albicans* remain the dominant nosocomial bloodstream infections, especially in Europe and the USA (Falagas *et al.*, 2010). In addition, the emergence of resistance to antifungal agents makes the infections caused by *C. albicans* an increasingly critical clinical problem. The formation of biofilms, which are especially resistant to commonly employed antifungal agents, is an adaptation used by *C. albicans* to survive in a changing environment (Murillo *et al.*, 2005). Through the years, the wide used of implanted medical devices, such as various types of catheter, has also been shown to facilitate biofilm formation by *C. albicans* (Nicastri *et al.*, 2001).

A biofilm is a multicellular community in which the cells are embedded in a polymeric extracellular matrix (ECM). In terms of studies carried out *in vitro*, biofilm formation can be roughly divided into four developmental phases: the adherence phase, in which the fungal cells adhere to substrate; the initiation phase, which involves the proliferation of yeast cells throughout the surface and their growth into filamentous forms; the maturation phase, during which the extracellular matrix accumulates and the communities acquire a high level of drug resistance; and the dispersal phase, in which yeast-form cells are released from the biofilm to seed new sites (Murillo *et al.*, 2005; Nett *et al.*, 2009).

Although natural biofilm formation *in vivo* is more complicated, the *in vitro* model provides a useful framework to study the mechanism of biofilm formation. Compared with planktonic specimens, the transcriptional levels of various *C. albicans* genes related to amino acids are strikingly different during biofilm formation (see Table 31.1).

Table 31.1. List of genes related to amino acids differentially expressed in mature *C. albicans* biofilm (24 h old) compared with planktonic cells. Changes given as trend compared with planktonic cells (upregulated or downregulated) and times difference (fold, which may be positive or negative).

Gene	Trend	Fold
AGP2 (hydrogen:amino acid symporter)	↑	4.90
CPA1 (carbamoyl-phosphate synthase activity)	↑	4.48
CAN1 (basic amino acid transporter activity)	↑	4.38
GDH3 (glutamate dehydrogenase activity)	↑	3.41
ARG1 (argininosuccinate synthase activity)	↑	3.32
CPA2 (carbamoyl-phosphate synthase activity)	↑	3.19
GNP1 (amino acid permease activity)	↑	2.91
ARG3 (ornithine carbamoyltransferase activity)	↑	2.6
DAO2 (amino acid permease activity)	↑	2.37
GAP6 (amino acid permease activity)	↑	2.36
DIP5 (amino acid permease activity)	↑	2.2
ALT1 (alanine transaminase)	↑	1.82
MET1 (uroporphyrin-III C-methyltransferase)	↑	1.78
HOM3 (aspartate kinase)	↑	1.75
orf19.6959 (aspartate kinase)	↑	1.74
STR3 (cystathionine β-lyase activity)	↑	1.64
CAT2 (carnitine *O*-acetyltransferase activity)	↑	1.61
GLT1 (glutamate synthase activity)	↓	−1.53846
GCN2 (protein kinase activity)	↓	−1.5625
SER2 (phosphoserine phosphatase activity)	↓	−1.5873
YHR020W (proline-tRNA ligase activity)	↓	−1.6129
IPF1943.2 (cysteine synthase activity)	↓	−1.75439
SER1 (phosphoserine transaminase activity)	↓	−1.75439
GAD1 (glutamate decarboxylase activity)	↓	−1.81818
MIS11 (formate-tetrahydrofolate ligase activity)	↓	−1.81818
MMF1 (isoleucine biosynthesis)	↓	−1.85185
SHM2 (glycine hydroxymethyltransferase activity)	↓	−1.92308
GLY1 (threonine aldolase activity)	↓	−2.04082
CAR2 (ornithine-oxo-acid transaminase activity)	↓	−2.32558
LYS9 (saccharopine dehydrogenase activity)	↓	−1.69492

Concomitantly and consistently with these changes in the transcription of various genes, there are also marked differences in the metabolites found in planktonic *C. albicans* cells and in the biofilm; these involve metabolites related to the TCA cycle, lipid synthesis, glycolysis and oxidative stress, as well as those related to amino acid metabolism (Yeater *et al.*, 2007; Nobile *et al.*, 2012; Zhu *et al.*, 2013). By developing a metabonomic method using gas chromatography–mass spectrometry (GC/MS), it has been observed that various amino acids, including alanine, aspartate, asparagine, threonine, valine and glycine were upregulated during almost the whole process of *C. albicans* biofilm development compared with planktonic cells (Table 31.2) (Zhu *et al.*, 2013).

In further experiments, a total of 22 key amino acids were screened as potential biomarkers in biofilms by means of an ion-pairing chromatography-porous graphitic carbon column coupled with a time-of-flight mass spectrometry (IP-PGC-TOF/MS) system (Table 31.3) (Chen *et al.*, 2014). As amino acids are essential for the normal physiological function of *C. albicans* in biofilm formation, attention to the metabolism of amino acids in the development of *C. albicans* biofilms would be conducive to guiding the development of therapies for *C. albicans* biofilm-related infections.

31.3 The Adherence Phase

During the establishment of infection, the adherence of the pathogens to host cells is considered to be an essential step. Previous studies have shown that *C. albicans* is able to adhere to a large variety of cells, as well as to medical implants, which property is critical for the development of *C. albicans* biofilms. In addition, it has been shown that the drug efflux proteins Cdr1 and Mdr1, both of which play an important role in drug resistance, are upregulated following the adherence of *C. albicans* to a glass slide (Mateus *et al.*, 2004). Multiple mechanisms have been proposed to unmask how *C. albicans* adheres and then forms a biofilm *in vivo* and *in vitro*; in these, the molecular interactions related to adherence could be classified into non-specific (including electrostatic or hydrophobic) interactions and specific ligand–receptor interactions (as in some fungal proteins, such as the mannoprotein adhesins). In addition to these proposals, global gene expression analysis has been undertaken to investigate the transcript profiles of early biofilm *in vitro*, and this has identified some pathways involved in the metabolism or regulation of amino acids (Yeater *et al.*, 2007). Some transcription factors have also been identified that clarify cell–substrate adherence in *C. albicans* (Marchais *et al.*, 2005; Finkel *et al.*, 2012);

Table 31.2. List of *C. albicans* metabolites involved in amino acid metabolism showing differential production during biofilm formation (at 6, 12, 24 and 48 h old) compared with planktonic cells. Changes given as trend compared with planktonic cells (upregulated or downregulated) and times difference (fold, which may be positive or negative).

	6 h		12 h		24 h		48 h	
Metabolite	Trend	Fold	12 h	Fold	24 h	Fold	48 h	Fold
Alanine			↑	3.28	↑	12.51	↑	1.03
Aspartate			↑	12.35			↑	2.53
Asparagine			↑	2.54	↑	9.15	↑	5.95
N-Acetyl-glutamate							↑	3.88
γ-Aminobutyrate	↑	7.15	↑	1.46	↑	21.88		
Glutamate	↑	1393.41	↑	1.14	↑	2.19	↓	−1.36
Glycine			↑	2.97	↑	28.71	↑	3.88
Isoleucine	↑	12.01	↑	4.55	↑	227.65	↑	39.25
Phenylalanine			↑	12.52				
Pyroglutamate	↑	5.62	↑	2.54			↑	2.2
Sarcosine	↑	9.81			↑	107.85		
Serine			↑	2.95	↑	4.97	↑	3.09
Threonine			↑	3.59	↑	3.77	↑	2.75
Valine			↑	8.26			↑	4.97

Table 31.3. List of amino acids differentially produced in mature *Candida albicans* biofilm (24 h old) compared with planktonic cells as analysed by IP-PGC-TOF/MS.[a] Changes given as trend compared with planktonic cells (upregulated or downregulated) and times difference (fold, which may be positive or negative).

Metabolite	Trend	Fold
Alanine	↑	1.05
Arginine	↓	−1.09
Asparagine	↑	1.32
Aspartic acid	↑	1.05
Citrulline	↑	1.45
Glutamic acid	↑	1.27
Glutamine	↑	1.16
Glycine	↑	1.46
Histidine	↑	1.32
Hydroxyproline	↑	1.04
Isoleucine	↑	1.64
Leucine	↑	1.58
Lysine	↑	2.11
Methionine	↑	1.21
Ornithine	↓	−2.02
Phenylalanine	↑	2.18
Proline	↓	−1.18
Serine	↑	1.31
Threonine	↑	1.31
Tryptophan	↑	2.00
Tyrosine	↑	1.86
Valine	↑	1.80

[a]IP-PGC-TOF/MS, ion-pairing chromatography in a porous graphitic carbon column coupled with a time-of-flight mass spectrometry.

among these is the gene *MET4*, which may play a role in sulfur amino acid metabolism has been indicated to be a cell surface target of adherence regulators. The expression of genes associated with amino acids during adherence also indicated the important role of pathways related to amino acid biosynthesis or metabolism in the adherence stage of biofilm formation in *C. albicans*.

Murillo *et al.* (2005) compared gene expression patterns between non-adherent and adherent cells within 30 min of adherence to an abiotic surface by genome-wide transcription profiling, which identified striking differences in gene expression patterns. In particular, the data indicated that genes related to the methionine (Met)/cysteine (Cys) biosynthetic pathway were upregulated during the early adherence period and that glutathione production might also participate in biofilm formation. Based on the work of Murillo *et al.* (2005), Li *et al.* (2013) studied the essential role of Met and Cys in biofilm formation. These are sulfur-containing amino acids that can transform one to the other (see Chapter 14) to meet the requirements of cells. They are associated with the formation of disulfide bonds, which are important for maintaining the physiological functions of many proteins. The gene *ECM17*, which is thought to encode key enzymes that control the generation of hydrogen sulfide (H_2S, the precursor of sulfur-containing amino acids like Met and Cys), was essential for the adhesion and filamentation of *C. albicans* cells, as well as for biofilm formation. Li *et al.* (2013) found that the deletion of *ECM17* in *C. albicans* results in a functional defect to synthesize Met and Cys. Besides this, the *ecm17Δ/Δ* mutant was defective in adherence, hyphal growth and biofilm formation, which could be related to changes in the expression of adhesion/filamentation genes, or the cAMP–PKA signalling pathway. The results of Li *et al.* (2013) confirm that the Met/Cys biosynthesis pathway mediated by *ECM17* is essential for *C. albicans* biofilm formation through its effects on *C. albicans* cell adhesion.

The regulation of a novel mating factor-response pathway is also reported to be involved in cell adherence. In recent years, *C. albicans* has been found to be able to mate. However, in addition to secretion of the proper mating pheromones, the mating response also requires the switching of cells from the mating-incompetent 'white' (W) cell type to the mating-competent 'opaque' (O) cell type (Bennett *et al.*, 2003; Zordan *et al.*, 2007). A comparison of the genome-wide expression profiles of the W and O cell types identified a remarkable difference between them in their patterns of expression of the genes associated with amino acid metabolism. For example, the genes for CDG1 (cysteine dioxygenase), CHA1 (serine/threonine dehydratase), and CHA2 (serine/threonine dehydratase) are all preferentially expressed in O cells, while the amino acid permeases encoded by the CAN3, GAP1 and AGP2 genes are expressed more in W cells (Lan *et al.*, 2002). Remarkably, *N*-acetylglucosamine, an amino acid-related metabolite as well as a component of bacterial and fungal cell walls and of the monomeric unit of chitin, was shown to activate the cyclic AMP (cAMP) signalling pathway and then induce W–O cell switching, followed by altered morphological characteristics and gene expression in homozygous mating type forms of *C. albicans* (Tong *et al.*, 2014). The more specific quantification of the amino acids that vary between W and O cells requires further investigation.

31.4 The Initiation Phase

Once the single cells of *C. albicans* adhere to the substratum, they proliferate across the surface and form microcolonies which then form a confluent but trabeculated monolayer. Meanwhile, the blastospores develop elongated projections that grow into pseudohyphae or extensive hyphae. Compared with the yeast form of *C. albicans*, the mycelial form is more conducive to fungal adherence and invasion of host tissues. Hyphae are required for stable biofilm formation, and *C. albicans* strains that were incapable of hyphal formation produced only the basal yeast layer under conditions of hyphal induction (Baillie and Douglas, 1999). In the initiation phase, the upregulation of genes related to amino acid metabolism has been identified by DNA microarray analysis, and the upregulation of some of these genes overlaps with hyphal growth. Consistently, the levels of various amino acids, including alanine, aspartate, asparagine, threonine, valine and glycine, were abundant beginning within this phase (Nett *et al.*, 2009; Zhu *et al.*, 2013).

As already mentioned, the emergence of hyphae is a hallmark of the initiation phase, so it is unsurprising to find that a variety of amino acids or amino acid biosynthetic pathways are involved in *C. albicans* mycelial growth, and that this then affects biofilm formation. Polyamines such as putrescine and spermidine are polybasic molecules that are essential for a variety of biological processes, including cell growth, cell death, and gene regulation and translation (Igarashi and Kashiwagi, 2000). In *C. albicans*, putrescine is derived from ornithine via a pathway that is activated by ornithine decarboxylase (ODC); it is then further processed into the more abundant polyamines spermidine and spermine. It has been reported that ODC inhibitors, such as 1,4-diaminobutanone, inhibit the yeast-to-hyphal transition in *C. albicans*. During germ tube emission, ODC activity is transiently increased, but the deletion of the gene *ODC* results in low polyamine levels and defective filaments when the cells are exposed to known inducers of filamentous growth, while the capacity to switch from the yeast to the filamentous form is restored by an increased level of polyamine concentration (Herrero *et al.*, 1999).

These results demonstrate that polyamine levels regulate the switch of the yeast-form cells to a filamentous pattern in *C. albicans*. In agreement with that, the consumption of ornithine for putrescine biosynthesis mediated by ODC function was found to be vital for biofilm formation, and the ODC-negative strain (a mutant with a low level of ornithine) was only able to form biofilm that was mainly composed of pseudohyphae and yeast cells. In contrast, natural biofilms exhibit a significantly upregulated level of putrescine, and the addition of putrescine to the mutant restored its ability to form normal biofilms (Chen *et al.*, 2014).

As indicated previously, the benefit of the Met/Cys biosynthesis pathway for biofilm formation is not only due to the enhancement of adherence but also to the enhancement of filament formation (Li *et al.*, 2013). Several other pathways involved in amino acid metabolism might be also vital for hyphal growth, which can then influence biofilm formation. Quorum sensing (QS) is a cell density-dependent mechanism of microbial communication that can control several behaviours in bacteria, such as virulence traits and biofilm formation. Previous studies have shown that various alcohols, by-products of amino acid metabolism, can function in the alteration of cellular morphology (Martins *et al.*, 2007). Several QS molecules, the most well studied of which in *C. albicans* are *E,E*-farnesol and tyrosol, were observed to accumulate in mature biofilms.

Farnesol is an inhibitor of hyphal formation and, if exogenous farnesol is added early enough, it can inhibit biofilm formation during the time of adherence (Alem *et al.*, 2006). Tyrosol is an aromatic alcohol derived from tyrosine, and its synthesis requires an intact aromatic amino acid biosynthetic pathway (Chen *et al.*, 2004). It has been reported to play an opposite role to that of farnesol, indeed, that tyrosol could promote hyphal formation and partially reverse biofilm inhibition by exogenous farnesol. In addition, a sharp increase in tyrosol concentration was detected in the intermediate stages of biofilm formation, which indicated its vital role in hyphal formation during the early and intermediate stages of biofilm development (Alem *et al.*, 2006). Interestingly, an increased level of tyrosine was observed in mature biofilm cells but not in the intermediate stage (Chen *et al.*, 2014), which might be explained by the normal transformation of tyrosine to tyrosol in the intermediate phase of biofilm formation, while this transformation is inhibited in the mature phase. Many other aromatic alcohols are produced by *C. albicans* in addition to tyrosol; these include phenylethanol and tryptophol, which have been reported to inhibit both hyphal formation and biofilm development (Chen and Fink, 2006). Whether these molecules are involved in amino acid metabolism in *C. albicans*, as well as their detailed enzymology, remain to be worked out.

Nitrogen availability, which has profound significance in fungal biology, is one of the key factors that initiates morphological changes. The GATA transcription factors, a conserved family of zinc finger domain-containing transcriptional regulators that are characterized by their ability to bind to the DNA sequence GATA, ensure the efficient utilization of available nitrogen sources by fungi. *gat1Δ* mutants, which cannot encode GATA transcription factors, are defective in utilizing isoleucine, tyrosine and tryptophan as nitrogen sources, which implies that this compromise might also influence the morphology transformation. However, Limjindaporn *et al.* (2003) found that filamentation was not prevented in *gat1Δ* mutants, although they were attenuated in virulence in a murine model of disseminated candidiasis. In another study, Land *et al.* (1975) found that the utilization of some amino acids, particularly proline, arginine and alanine, promoted the conversion of *C. albicans* to the hyphal form. It is, therefore, clear that further analyses are required to clarify the role of amino acids in hyphal formation and biofilm formation.

31.5 The Maturation and Dispersal Phase

As *C. albicans* cells grow with the accumulation of the ECM, mature *Candida* biofilms form as a framework composed of a basal layer of yeast cells with subsequent layering of filamentous morphotypes embedding within the extensive matrix. There is a large increase in biomass in mature biofilms, which is partly due to the accumulation of large amounts of carbohydrate-rich ECM. Previous *in vitro* studies have shown that the components of the matrix include carbohydrate, protein, hexosamine, phosphorus and uronic acid (Al-Fattani and Douglas, 2006). The acquisition of high-level resistance to antifungals in biofilms may be due to the significantly enhanced production of ECM (Al-Fattani and Douglas, 2006).

Nine amino acids, including alanine, asparagine, γ-aminobutyrate, glutamate, isoleucine, glycine, sarcosine, serine and threonine, were all identified as significant metabolites in planktonic and biofilm cells cultivated for 24 h (see Table 31.2), and all of them were upregulated in biofilm cells (Zhu *et al.*, 2013). The initiation of the synthesis of the ECM was found to be in the same phase as the beginning of amino acid upregulation. Thus, we can hypothesize that the abundance of amino acids in biofilms might be favourable for the formation of ECM. Another major extracellular carbohydrate constituent with increased production by biofilm cells in comparison with planktonic cells has been found to be monosaccharides, such as β-1,3 glucan (Nett *et al.*, 2007). Galactose contributes 3% of the dry weight of the extracellular polymeric material of *C. albicans* biofilm and, consistently, metabolomic experiments have confirmed that three monosaccharides (glucose, galactose and mannose) are upregulated in biofilms; the upregulation of galactose production was during the whole course of biofilm development, while glucose and mannose upregulation were in intermediate and mature biofilms. Specific proteins associated with the biofilm cell surface and extracellular DNA were also revealed to be conducive to the structure and stability of a mature biofilm.

The study by Chen *et al.* (2014), in which key amino acids, as well as several amino acid-related metabolites, were extracted and analysed as potential biomarkers in mature (24 h old) *C. albicans* biofilms by means of IP-PGC-TOF/MS has already been outlined in Section 31.1, and some of the results have been presented in Table 31.3. Several potential amino acid and amino acid-related biomarkers were characterized, screened and shown to be present at high levels in the mature *C. albicans* biofilms, including glutamine, citrulline, leucine, lysine, methionine, histidine, saccharopine, acetyllysine, tyrosine and tryptophan. In agreement with these findings, previous microarray data has shown that several genes related to amino acid biosynthesis, such as *MET3* and *HOM6* (methionine biosynthesis), *HIS1* (histidine biosynthesis), *ARO8* (aromatic amino acid biosynthesis), *LEU1* (3-isopropylmalate dehydratase), *TRP4* (tryptophan biosynthesis) and *BAT22* (branched chain amino acid aminotransferase) were overexpressed in intermediate or mature phases of biofilms (Yeater *et al.*, 2007; Nett *et al.*, 2009).

In the Chen *et al.* (2014) study just described, several amino acid-related metabolites incorporating betaine, putrescine, *N*-acetylputrescine, acetyllysine and saccharopine were also identified by means of untargeted profiling. Saccharopine (an intermediate of the amino adipic acid pathway of lysine biosynthesis) and acetyllysine (a degradation product of lysine) are both lysine-related metabolites, and were both upregulated in biofilm cells, with lysine itself increased by 2.11-fold (see Table 31.3). Activation of the lysine biosynthesis/degradation pathway was also observed in biofilm cells. These results are in good agreement with previous reports by García-Sánchez *et al.* (2004) that Gcn4p, a regulator of amino acid metabolism which was shown to be required for normal biofilm growth and that highly regulates lysine biosynthesis, was overrepresented in gene expression patterns. In addition, methionine and threonine, both of which are amino acids involved in sulfur amino acid biosynthesis and salvage, were more abundant in biofilm cells than in planktonic *C. albicans* cells. Aspartate can be metabolized as homoserine, which will then be fed into the biosynthesis of methionine (García-Sánchez *et al.*, 2004). Although the level of aspartate was not found to be increased in biofilms compared with planktonic cells, the amounts of asparagine and aspartic acid, which are derivatives of aspartate, were both significantly enhanced. The presence of metabolites associated with the sulfur amino acids further confirms the key role of the sulfur amino acid biosynthetic pathway in *C. albicans* biofilms.

In mature biofilms, ornithine was found to be downregulated in biofilm cells as determined by both targeted and untargeted profiling approaches using IP-PGC-TOF/MS (Chen *et al.*, 2014). In contrast, putrescine, the polyamine that is produced from ornithine by ornithine decarboxylase (ODC), and *N*-acetylputrescine, the metabolite produced from putrescine in a reaction catalysed by diamine acetyltransferase, were overexpressed. As has been previously noted, the polyamine biosynthetic pathway mediated by ODC is essential in the yeast–hyphal transition and subsequent biofilm formation, so it is unsurprising that the pathway may also be vital in maintaining the stabilization of *C. albicans* biofilms.

The *C. albicans* biofilm life cycle begins with the planktonic yeast cells adhering to a substrate and then proliferating, which ultimately results in a highly structured mature biofilm comprising complex intertwining layers of yeast, pseudohyphae and hyphae embedded in extracellular matrix. After the mature biofilm has formed, yeast cells will be released from the biofilm under specific growing conditions, thought to be biofilm dispersion/detachment or dissolution, and will seed new surfaces. Across the biofilm life cycle of both fungi and bacteria, dispersal is thought to be the least understood and perhaps the most complicated stage (Uppuluri *et al.*, 2010a). Enhanced adherence, filamentation, biofilm formation and increased pathogenicity in a murine model of haematogenously disseminated candidiasis are particular phenotypic properties that dispersed cells of *C. albicans* display that are distinct from their planktonic counterparts. Previous observations have suggested that the transition from yeast to hyphae that occurs during biofilm initiation may be thrown into reverse for dispersal (Uppuluri *et al.*, 2010b). In the early and intermediate phase of biofilm formation, tyrosol activity exceeds that of farnesol, but this is reversed after 24 h when the biofilm is mature. Consistently, the later stage also shows a significant increase in farnesol production, and farnesol was found to accumulate in the supernatants of mature biofilms (Martins *et al.*, 2007). These observations suggest that farnesol may play a critical role in the later stages of the biofilm cycle; it may stimulate the release of yeast cells from the biofilm and so permit the dispersal of the organism to colonize new surfaces, a process that has already been confirmed in bacteria (Alem *et al.*, 2006).

31.6 Amino Acids and the Tricarboxylic Acid (TCA) Cycle

In biofilm cells, respiratory activity and reactive oxygen species (ROS) were observed to be reduced, implying inhibition of the TCA cycle (Seneviratne *et al.*, 2008; Bonhomme *et al.*, 2011). The TCA cycle, which occurs in the mitochondrial matrix in eukaryotic cells, is a key component of the metabolic pathway for generating energy. The pivotal intermediates of TCA cycle metabolites, such as succinate, fumarate, citrate and malate are decreased in mature biofilms compared with planktonic cells. As the primary source of endogenous ROS is the leakage of electrons from the mitochondrial respiratory chain, downregulation of TCA-related metabolites may explain the reduced ROS level in mature biofilms. Conversely, many amino acids have been found to be upregulated in biofilms, and it is well known that amino acids can be metabolized into TCA intermediates and enter this cycle; reduced TCA cycle activity would, therefore, affect the consumption of the corresponding nutrients (Zhu *et al.*, 2013). Thus, the upregulation of amino acids in biofilms implies a reduced consumption of amino acids as a result of the inhibition of TCA cycle activity.

5-Oxoproline, glycine, glutamate, and 4-aminobutyrate, the amino acids involved in the biosynthesis of glutathione, were also found to be differentially regulated between biofilm and planktonic cells. In mature biofilm, glycine remained upregulated, 5-oxoproline and 4-aminobutyrate fluctuating, and glutamate downregulated. Cysteine, glycine, and glutamate are precursor amino acids that synthesize glutathione; γ-glutamylamine is then produced through transpeptidation of glutathione, and can, in turn, be converted to 5-oxoproline by the enzyme γ-glutamylcyclotransferase. The 5-oxoproline will subsequently be converted to glutamate. In cells, glutathione is a key molecule that confers protection against oxidative damage through the detoxification of ROS and other naturally occurring deleterious compounds. As the reduced ROS level in mature biofilms may be due to decreased TCA cycle activity, which then inhibits mitochondrial respiration, the upregulation of amino acids that could increase the synthesis of intracellular glutathione may contribute to the low ROS activity in early and intermediate biofilms compared with planktonic cells (Zhu *et al.*, 2013). Farnesol, a QS molecule in *C. albicans*, has been reported to conjugate with glutathione. The resulting conjugate could serve as a substrate for ATP-dependent ABC transporters and would be extruded from cells, resulting in total depletion of glutathione, oxidative stress and, ultimately, cell death (Zhu *et al.*, 2011).

31.7 Drug Therapies and *C. albicans* Biofilms

C. albicans biofilms are highly recalcitrant to antifungal therapy. This could be due to the expression of multidrug resistance pumps, decreased expression of drug targets or of biosynthetic genes, leading to a diminished (for example) ergosterol content (Mukherjee *et al.*, 2003), or the presence of the exopolymer matrix (Baillie and Douglas, 2000). Antifungal-tolerant persister cells were found to be another key cause responsible for the high resistance of biofilms to antifungal agents (LaFleur *et al.*, 2006). The experiments indicated that the attachment phase was essential for persister formation. As the adherence phase is one of the vital steps in the *C. albicans* biofilm cycle, and involves many amino acids or amino acid biosynthetic pathways such as the Met/Cys biosynthesis pathway (Li *et al.*, 2013), the disturbance of these pathways might be a good strategy for reversing resistance to antifungal agents.

The significance of ODC activity in the dimorphic transition and biofilm formation of *C. albicans* has been clarified (Martinez *et al.*, 1990; Chen *et al.*, 2014), and ODC function shown to be vital for ornithine consumption and putrescine biosynthesis during biofilm formation. Thus, pathways related to ODC (Chapter 8) could also be favourable drug targets. Biofilm formation was consistently found to be inhibited in an ODC-negative (*spe1Δ*) mutant of *C. albicans* constructed by the deletion of the ODC gene, *SPE1*, in which the biofilm was mainly composed of pseudohyphae and yeast cells. However, the addition of putrescine to the medium was able to compensate for the deletion, and the appearance of the biofilm of the *spe1Δ* mutant was apparently converted to its wild type. These results indicate that *SPE1*, or ODC itself, especially through ODC inhibitors, could be designed for target clinical applications (Chen *et al.*, 2014). ODC is the first and rate-limiting enzyme in polyamine biosynthesis, and polyamine biosynthesis inhibitors, such as 1,4-diamino-2-butanone (DAB) and α-difluoromethylornithine (DFMO), have been found to be competitive inhibitors of ODC (Martinez *et al.*, 1990; Cao *et al.*, 2013). Recently, it was found that the anti-biofilm activities of several commonly used antifungal agents, including amphotericin B (AmB), caspofungin, nystatin and miconazole, could be enhanced by polyamine biosynthesis inhibitors; here, the enhancement of the anti-biofilm activity of AmB by polyamine biosynthesis inhibitors required the caspase pathway mediated by *CaMCA1*, which encodes metacaspase (Liao *et al.*, 2015). Therefore, targeting the polyamine biosynthetic pathway could be a favourable strategy for enhancing the anti-biofilm activities of antifungal agents.

Metabonomic studies have demonstrated that the levels of metabolites associated with amino acid metabolism, including *N,N*-dimethylglycine (Chapter 30), leucine, valine, glycine, threonine, pyroglutamate, aspartate, glutaminate, ornithine and lysine were changed after treatment with AmB compared with the untreated group (Cao *et al.*, 2013). Also, secretory aspartyl proteases, an important virulence factor of *C. albicans* which aids adhesion, invasion and tissue destruction (Ramage *et al.*, 2012), was found to be overproduced in *C. albicans* biofilms exposed to subinhibitory concentrations of fluconazole (Mores *et al.*, 2011). Interestingly, the QS molecule homoserine lactone, which is secreted by *Pseudomonas aeruginosa*, can mimic the action of farnesol and inhibit *C. albicans* filamentation, thereby preventing biofilm formation (Hogan *et al.*, 2004). Whether any of these observations could contribute to the antifungal susceptibility of *C. albicans* biofilms through targeting the biosynthesis of related amino acids or other metabolic pathways remains to be investigated.

31.8 Conclusions

The metabolic circuit controlling *C. albicans* biofilm formation is complicated and requires further elucidation. Many studies have been undertaken to define the genes that govern biofilm formation, unravel the mechanisms that account for biofilm resistance to antifungal agents and investigate credible ways to improve the effectiveness of antifungal agents. In recent years, various methods, especially metabonomics, have been developed to analyse the metabolites that are involved in *C. albicans* biofilm formation.

Amino acids are biologically important organic compounds in cells, which can be used to synthesize proteins and other biomolecules or oxidized to urea and carbon dioxide as a source of energy. Time course analysis of *C. albicans* metabolites during biofilm development has demonstrated that the changed levels of amino acids that are seen vary among specific phases of biofilm formation. Amino acids and amino acid-related metabolites have also been screened as potential biomarkers of biofilm formation. The observations that have been made are consistent with the results of transcriptional analysis, which indicates that the genes associated with amino acid biosynthesis are also differentially expressed during *C. albicans* biofilm formation. Based on previous work, polyamine biosynthetic pathways regulated by ornithine decarboxylase were found to be a promising target for enhancing the anti-biofilm activities of commonly employed antifungal agents.

However, although existing methods for finding biomarkers during biofilm formation have been well established and widely used, there are still many puzzles to be worked out, some of which are attributed to the decomposition and rearrangement of amino acids and to the precision of the instruments used. Moreover, *C. albicans* biofilm formation *in vivo* is more intricate than it is *in vitro*; for example, while biofilm formation *in vitro* can be viewed as a series of sequential steps, these steps may occur concurrently *in vivo*. Most importantly, the studies that have focused on variations in metabolites during *C. albicans* biofilm formation are insufficient. There has now an urgent need to resolve these issues because of the impact of microbial biofilms both in the food industry and on human health.

References

Alem, M.A., Oteef, M.D., Flowers, T.H. and Douglas, L.J. (2006) Production of tyrosol by *Candida albicans* biofilms and its role in quorum sensing and biofilm development. *Eukaryotic Cell* 5, 1770–1779.

Al-Fattani, M.A. and Douglas, L.J. (2006) Biofilm matrix of *Candida albicans* and *Candida tropicalis*: chemical composition and role in drug resistance. *Journal of Medical Microbiology* 55, 999–1008.

Baillie, G.S. and Douglas, L.J. (1999) Role of dimorphism in the development of *Candida albicans* biofilms. *Journal of Medical Microbiology* 48, 671–679.

Baillie, G.S. and Douglas, L.J. (2000) Matrix polymers of *Candida* biofilms and their possible role in biofilm resistance to antifungal agents. *Journal of Antimicrobial Chemotherapy* 46, 397–403.

Bennett, R.J., Uhl, M.A., Miller, M.G. and Johnson, A.D. (2003) Identification and characterization of a *Candida albicans* mating pheromone. *Molecular and Cellular Biology* 23, 8189–8201.

Bonhomme, J., Chauvel, M., Goyard, S., Roux, P., Rossignol, T. and d'Enfert, C. (2011) Contribution of the glycolytic flux and hypoxia adaptation to efficient biofilm formation by *Candida albicans*. *Molecular Microbiology* 80, 995–1013.

Cao, Y., Zhu, Z., Chen, X., Yao, X., Zhao, L., Wang, H., Yan, L., Wu, H., Chai, Y. and Jiang, Y. (2013) Effect of amphotericin B on the metabolic profiles of *Candida albicans*. *Journal of Proteome Research* 12, 2921–2932.

Chen, H. and Fink, G.R. (2006) Feedback control of morphogenesis in fungi by aromatic alcohols. *Genes and Development* 20, 1150–1161.

Chen, H., Fujita, M., Feng, Q., Clardy, J. and Fink, G.R. (2004) Tyrosol is a quorum-sensing molecule in *Candida albicans*. *Proceedings of the National Academy of Sciences of the United States of America* 101, 5048–5052.

Chen, X., Wu, H., Cao, Y., Yao, X., Zhao, L., Wang, T., Yang, Y., Lv, D., Chai, Y., Cao, Y. and Zhu, Z. (2014) Ion-pairing chromatography on a porous graphitic carbon column coupled with time-of-flight mass spectrometry for targeted and untargeted profiling of amino acid biomarkers involved in *Candida albicans* biofilm formation. *Molecular BioSystems* 10, 74–85.

Falagas, M.E., Roussos, N. and Vardakas, K.Z. (2010) Relative frequency of albicans and the various non-albicans *Candida* spp among candidemia isolates from inpatients in various parts of the world: a systematic review. *International Journal of Infectious Diseases* 14, e954–e966.

Finkel, J.S. and Mitchell, A.P. (2011) Genetic control of *Candida albicans* biofilm development. *Nature Reviews Microbiology* 9, 109-18.

Finkel, J.S., Xu, W., Huang, D., Hill, E.M., Desai, J.V., Woolford, C.A., Nett, J.E., Taff, H., Norice, C.T., Andes, D.R. *et al.* (2012) Portrait of *Candida albicans* adherence regulators. *PLoS Pathogens* 8(2): e1002525.

García-Sánchez, S., Aubert, S., Iraqui, I., Janbon, G., Ghigo, J. M. and d'Enfert, C. (2004) *Candida albicans* biofilms: a developmental state associated with specific and stable gene expression patterns. *Eukaryotic Cell* 3, 536–545.

Herrero, A.B., López, M.C., García, S., Schmidt, A., Spaltmann, F., Ruiz-Herrera, J. and Dominguez, A. (1999) Control of filament formation in *Candida albicans* by polyamine levels. *Infection and Immunity* 67, 4870–4878.

Hogan, D.A., Vik, A. and Kolter, R. (2004) A *Pseudomonas aeruginosa* quorum-sensing molecule influences *Candida albicans* morphology. *Molecular Microbiology* 54, 1212–1223.

Igarashi, K. and Kashiwagi, K. (2000) Polyamines: mysterious modulators of cellular functions. *Biochemical and Biophysical Research Communications* 271, 559–564.

LaFleur, M.D., Kumamoto, C.A. and Lewis, K. (2006) *Candida albicans* biofilms produce antifungal-tolerant persister cells. *Antimicrobial Agents and Chemotherapy* 50, 3839–3846.

Lan, C.Y., Newport, G., Murillo, L.A., Jones, T., Scherer, S., Davis, R.W. and Agabian, N. (2002) Metabolic specialization associated with phenotypic switching in *Candida albicans*. *Proceedings of the National Academy of Sciences of the United States of America* 99, 14907–14912.

Land, G.A., McDonald, W.C., Stjernholm, R.L. and Friedman, T.L. (1975) Factors affecting filamentation in *Candida albicans*: relationship of the uptake and distribution of proline to morphogenesis. *Infection and Immunity* 11, 1014–1023.

Li, D.D., Wang, Y., Dai, B.D., Li, X.X., Zhao, L.X., Cao, Y.B., Yan, L. and Jiang, Y.Y. (2013) ECM17-dependent methionine/cysteine biosynthesis contributes to biofilm formation in *Candida albicans*. *Fungal Genetics and Biology* 51, 50–59.

Liao, Z., ZhangGuan, X., Zhu, Z., Yao, X., Yang, Y., Jiang, Y. and Cao, Y. (2015) Enhancement of the antibiofilm activity of amphotericin B by polyamine biosynthesis inhibitors. *International Journal of Antimicrobial Agents* 46, 45–52.

Limjindaporn, T., Khalaf, R.A. and Fonzi, W.A. (2003) Nitrogen metabolism and virulence of *Candida albicans* require the GATA-type transcriptional activator encoded by GAT1. *Molecular Microbiology* 50, 993–1004.

Marchais, V., Kempf, M., Licznar, P., Lefrancois, C., Bouchara, J.P., Robert, R. and Cottin, J. (2005) DNA array analysis of *Candida albicans* gene expression in response to adherence to polystyrene. *FEMS Microbiology Letters* 245, 25–32.

Martinez, J.P., Lopez-Ribot, J.L., Gil, M.L., Sentandreu, R. and Ruiz-Herrera, J. (1990) Inhibition of the dimorphic transition of *Candida albicans* by the ornithine decarboxylase inhibitor 1,4-diaminobutanone: alterations in the glycoprotein composition of the cell wall. *Journal of General Microbiology* 136, 1937–1943.

Martins, M., Henriques, M., Azeredo, J., Rocha, S.M., Coimbra, M.A. and Oliveira, R. (2007) Morphogenesis control in *Candida albicans* and *Candida dubliniensis* through signaling molecules produced by planktonic and biofilm cells. *Eukaryotic Cell* 6, 2429–2436.

Mateus, C., Crow, S.A. Jr and Ahearn, D.G. (2004) Adherence of *Candida albicans* to silicone induces immediate enhanced tolerance to fluconazole. *Antimicrobial Agents and Chemotherapy* 48, 3358–3366.

Mores, A.U., Souza, R.D., Cavalca, L., de Paula e Carvalho, A., Gursky, L.C., Rosa, R.T., Samaranayake, L.P. and Rosa, E.A. (2011) Enhancement of secretory aspartyl protease production in biofilms of *Candida albicans* exposed to sub-inhibitory concentrations of fluconazole. *Mycoses* 54, 195–201.

Mukherjee, P.K., Chandra, J., Kuhn, D.M. and Ghannoum, M.A. (2003) Mechanism of fluconazole resistance in *Candida albicans* biofilms: phase-specific role of efflux pumps and membrane sterols. *Infection and Immunity* 71, 4333–4340.

Murillo, L.A., Newport, G., Lan, C.Y., Habelitz, S., Dungan, J. and Agabian, N.M. (2005) Genome-wide transcription profiling of the early phase of biofilm formation by *Candida albicans*. *Eukaryotic Cell* 4, 1562–1573.

Nett, J., Lincoln, L., Marchillo, K., Massey, R., Holoyda, K., Hoff, B., VanHandel, M. and Andes, D. (2007) Putative role of β-1,3 glucans in *Candida albicans* biofilm resistance. *Antimicrobial Agents and Chemotherapy* 51, 510–520.

Nett, J.E., Lepak, A.J., Marchillo, K. and Andes, D.R. (2009) Time course global gene expression analysis of an *in vivo Candid*a biofilm. *The Journal of Infectious Diseases* 200, 307–313.

Nicastri, E., Petrosillo, N., Viale, P. and Ippolito, G. (2001) Catheter-related bloodstream infections in HIV-infected patients. *Annals of the New York Academy of Sciences* 946, 274–290.

Nobile, C.J., Fox, E.P., Nett, J.E., Sorrells, T.R., Mitrovich, Q.M., Hernday, A.D., Tuch, B.B., Andes, D.R. and Johnson, A.D. (2012) A recently evolved transcriptional network controls biofilm development in *Candida albicans*. *Cell* 148, 126–138.

Ramage, G., Coco, B., Sherry, L., Bagg, J. and Lappin, D.F. (2012) *In vitro Candida albicans* biofilm induced proteinase activity and SAP8 expression correlates with *in vivo* denture stomatitis severity. *Mycopathologia* 174, 11–19.

Seneviratne, C.J., Wang, Y., Jin, L., Abiko, Y. and Samaranayake, L.P. (2008) *Candida albicans* biofilm formation is associated with increased anti-oxidative capacities. *Proteomics* 8, 2936–2947.

Tong, Y., Cao, C., Xie, J., Ni, J., Guan, G., Tao, L., Zhang, L. and Huang, G. (2014) *N*-acetylglucosamine-induced white-to-opaque switching in *Candida albicans* is independent of the Wor2 transcription factor. *Fungal Genetics and Biology* 62, 71–77.

Uppuluri, P., Chaturvedi, A.K., Srinivasan, A., Banerjee, M., Ramasubramaniam, A.K., Kohler, J.R., Kadosh, D. and Lopez-Ribot, J.L. (2010a) Dispersion as an important step in the *Candida albicans* biofilm developmental cycle. *PLoS Pathogens* 6(3): e1000828.

Uppuluri, P., Pierce, C.G., Thomas, D.P., Bubeck, S.S., Saville, S.P. and Lopez-Ribot, J.L. (2010b) The transcriptional regulator Nrg1p controls *Candida albicans* biofilm formation and dispersion. *Eukaryotic Cell* 9, 1531–1537.

Yeater, K.M., Chandra, J., Cheng, G., Mukherjee, P.K., Zhao, X., Rodriguez-Zas, S.L., Kwast, K.E., Ghannoum, M.A. and Hoyer, L.L. (2007) Temporal analysis of *Candida albicans* gene expression during biofilm development. *Microbiology* 153, 2373–2385.

Zhu, J., Krom, B., Sanglard, D., Intapa, C., Dawson, C.C., Peters, B.M., Shirtliff, M.E. and Jabra-Rizk, M.A. (2011) Farnesol-induced apoptosis in *Candida albicans* is mediated by Cdr1-p extrusion and depletion of intracellular glutathione. *PloS One* 6(12): e28830.

Zhu, Z., Wang, H., Shang, Q., Jiang, Y., Cao, Y. and Chai, Y. (2013) Time course analysis of *Candida albicans* metabolites during biofilm development. *Journal of Proteome Research* 12, 2375–2385.

Zordan, R.E., Miller, M.G., Galgoczy, D.J., Tuch, B.B. and Johnson, A.D. (2007) Interlocking transcriptional feedback loops control white-opaque switching in *Candida albicans*. *PLoS Biology* 5(10): e256.

32 Recent Advances Underpinning Innovative Strategies for the Future

J.P.F. D'Mello*
Formerly of SAC, University of Edinburgh Campus, Edinburgh, UK

32.1 Abstract

It is instructive to evaluate recent evidence as presented in the preceding chapters of this volume. It is immediately clear that significant advances have been reported on the mainstream biochemistry of all the major families of amino acids in a diverse range of microorganisms. Emerging developments include enhanced understanding and manipulation of gene expression, the evaluation of enzyme structure–function relationships, assessment of the relative importance of competing biochemical pathways and elucidation of mechanisms of transcriptional and post-translational regulation. There has also been considerable understanding gained of the peripheral pathways of amino acid metabolism, particularly in certain bacteria. In addition, there is a perceptible shift in focus over recent years towards unlocking practical value for the potential resolution of intractable issues in food quality and safety, pathogen characterization, host–pathogen interactions and drug resistance. This change in emphasis is based on the premise that amino acid metabolism is a fundamental determinant of microbial viability, growth, pathogenicity and virulence.

The chapters in this volume reflect the significant advances that have emerged over recent decades on the amino acid characterization of diverse species of microorganisms. In this respect, considerable data on the metabolic pathways in *Lactobacillus*, *Brevibacterium linens* and *Geotrichum candidum* have enhanced perceptions relating to the quality of dairy products. Additionally, amino acid metabolism has been investigated in enteric pathogens including *Salmonella typhimurium*, *Campylobacter jejuni*, *Listeria monocytogenes* and *Escherichia coli*, although much more work is still required to unlock practical value. A number of prominent disease pathogens such as *Mycobacterium tuberculosis*, *Chlamydia trachomatis*, *Candida albicans*, trypanosomatids and helminths have been characterized in greater detail within the context of amino acid metabolism, and there is now an enhanced basis for promising developments, particularly in the elucidation of host–microbe interactions and metabolic drug design.

Regarding specific amino acids, the evidence for serine metabolism is compelling, extending to important enteropathogens such as *S. typhimurium*, *S. enterica* and *C. jejuni*. However, the metabolism of other amino acids in these microorganisms has only been confirmed for the inhibition of glutamine synthetase by serine, alanine and glycine in *S. typhimurium*. Similar characterization in *L. monocytogenes* and mycotoxin-producing fungi remains to be done. The characterization of *Pesudomonas syringae* has been established for

*E-mail address: jpfdmello@hotmail.co.uk

glycine betaine utilization, L-amino acid ligase and the inhibition of ornithine/arginine biosynthesis, while in *P. aeruginosa*, the transcriptional response to phenylalanine and tyrosine has been reported. The biochemical characterization of *M. tuberculosis* is significantly more comprehensive, including glutamine synthetase export, serine metabolism, structural determination of the glycine cleavage system component H, sulfur assimilation, acetohydroxyacid synthase inhibitors, branched-chain amino acid (BCAA) aminotransferase activity, the role of the *ilvB1* (interleukin 1 beta) gene and shikimate pathway enzymes and tryptophan biosynthesis enzymes as drug targets.

Recent developments have facilitated the proposal of a model for host–pathogen interactions in humans relating to tryptophan metabolism after infection with *Chlamydia trachomatis*. In response to this infection, tryptophan catabolism is induced by mechanisms within the host, thereby depriving the bacterium of an essential amino acid. However, in uropathogenic *E. coli*, the induction of tryptophan degradation may cause innate responses to epithelial infection to be compromised. Thus, the regulation of tryptophan biosynthesis and indole turnover is increasingly perceived as a site of interaction between pathogen and host.

Features of pathogenicity and virulence in disease microorganisms are frequently reflected in, or linked with amino acid transport and metabolism. Indeed, the concept of 'nutritional virulence' has recently been enunciated and may well be validated on the basis of host–pathogen interactions in amino acid uptake and utilization. For example, an amino acid ATP-binding cassette transporter has been proposed as an important pathogenesis factor in *C. jejuni*, while the modulation of phagosomal pH, requiring *Stp2p*, a regulator of amino acid transport, appears to be an effective virulence adaptation in *C. albicans*. This latter pathogen utilizes imported amino acids to promote the neutralization of the phagosomal pH, hyphal morphogenesis and escape from macrophages. It has been suggested that virulent strains of *Francisella tularensis* rely exclusively on dedicated uptake systems to satisfy their BCAA requirements. Several bacterial pathogens also rely on export systems for virulence. Thus, mycobacteria export significant quantities of glutamine synthetase into their extracellular environment to promote pathogenicity. In terms of specific amino acids, the upregulation of serine utilization has been associated with a hypervirulent strain of *C. jejuni*, whereas an intact proline catabolism pathway is required for the homeostasis of reactive oxygen species (ROS) and virulence of the fungal pathogen *Cryptococcus neoformans*. From a study on proline-dependent regulation of Stickland metabolism, it is maintained that a detailed understanding of amino acid metabolism may provide important insights into *Clostridium difficile* pathogenesis. BCAA transport and metabolism exert profound effects on virulence in diverse microbial pathogens, including *Streptococcus pneumoniae*, *S. mutans*, *M. tuberculosis*, *Actinobacillus pleuropneumoniae* and potentially other pulmonary pathogens. Other evidence has implicated BCAA transaminase activity in the synthesis of HC-toxin and the pathogenicity of *Cochliobolus carbonum* in maize, while *fgilv5* (the gene for a homologue of the *Saccharomyces cerevisiae* keto-acid reductoisomerase) is required for BCAA biosynthesis and full virulence in the phytopathogenic fungus *Fusarium graminearum*. Furthermore, branched-chain keto acid dehydrogenase, which is involved in the catabolism of BCAAs, is required for the full virulence of *Toxoplasma gondii* and *Plasmodium berghei*. Hence, the microbial metabolism of BCAAs is emerging as a crucial factor for consideration in future work.

The quest for novel pharmaceuticals continues with the increasing development of drug-resistant strains of well-known bacterial pathogens, such as *M. tuberculosis* and methicillin-resistant *Staphylococcus aureus* (MRSA). The transport and metabolism of amino acids in microbes, particularly involving pathways that are absent in the host, provide potential targets for drug discovery. Examples of metabolic targets in *M. tuberculosis* include the export of glutamine synthetase, glutamate racemase, dihydrodipicolinate reductase and acetohydroxyacid synthase, which are essential for the *de novo* synthesis of BCAAs in microbes, inactivation of the *ilvB1* gene, the shikimate pathway and tryptophan biosynthesis. Other potential targets include glutamate dehydrogenase (GDH) in *Bacillus subtilis* and D-amino acids and haemin conjugates in *S. aureus*. In the case of *Plasmodium falciparum* there is limited evidence that isoleucine and methionine transport and utilization as well as lysine acetylation may provide potential targets for antimalarial drug development. In *Leishmania infantum*, tyrosine aminotransferase has been advanced as a new potential drug target.

Associated with the foregoing and other issues is the question of interactions at the microniche level involving microbial adaptations in amino acid metabolism in response to local changes in reactive oxygen status, pH and anoxic/hypoxic and nitrosative stress, as might prevail in biofilms and in the mammalian/avian microbiome. Amino acid metabolism and exchanges in specialized ecosystems such as the root nodules of leguminous plants and the rumen continue to attract attention. Advances are emerging in relation to the overall dynamics and legume–*Rhizobium* interdependence in amino acid metabolism, for example as affecting polyamine dynamics.

In summary, the elucidation of enzyme structure–function relationships and the mechanisms in the regulation of amino acid metabolism have been a distinctive feature of recent developments, although the evidence is fragmentary, encompassing a wide range of amino acids and microbial species. Consequently, a more focused and integrated approach is now essential in order to ensure delivery of practical results, particularly in relation to food-borne and disease pathogens. The relationships between virulence and host immune/defence reactions are still in the initial phases of development. In this respect, it is necessary to resolve the complexities of microbial–amino acid interactions in a wide spectrum of ecological niches, including, among others, biofilms, the human microbiome, the root nodules of leguminous plants and the rhizosphere. The elucidation of host–pathogen interactions may well be compounded by amino acid auxotrophy or biosynthetic competence (autotrophy) in the specific microorganism under investigation. In addition, the effects of physical factors such as hypoxia, acid resistance and thermophilic adaptations need to be addressed. Insight into these questions should provide the basis of innovative applications in food surveillance and in the design of novel metabolic antimicrobial compounds for the control of drug-resistant pathogens.

Thus, it may be concluded that much has been achieved, but an even more concerted effort is required if the questions concerning food quality and safety, host–pathogen interactions, the design of metabolic drugs, biotechnology applications and environmental bioremediation are to be resolved. In addressing these issues, it is suggested that studies in microbial amino acid metabolism should remain at the centre of future attempts to deliver practical value. These objectives may well be fulfilled if particular attention is focused on the metabolism of arginine, serine, methionine, BCAAs and tryptophan in human pathogens.

32.2 Overview

The impetus for further research in the microbial metabolism of amino acids is destined to continue as we seek solutions to a diverse range of intractable issues, particularly within the disciplines of food quality and safety, plant pathology and human disease, and drug-resistant strains of bacteria. The unprecedented volume of existing published data has not diminished interest; rather, it has increased the scope for future work by revealing important gaps in knowledge. Key aspects have emerged in recent years to justify continuing investment in research and development on this topic.

This volume covers a wide and rapidly evolving field of research, in terms of both the range of amino acids and the diversity of microorganisms under investigation (Table 32.1). It is, therefore, instructive to review salient themes that have emerged over the past two or three decades. The need for a multidisciplinary approach in the future is as imperative as it has been in the recent past. The role of molecular biologists and biochemists will be crucial in our efforts to resolve outstanding issues. A robust programme of research has to be drawn up within a long-term strategy in order ensure that diminishing resources are used more effectively.

The following review is based primarily on the conclusions of the chapters in *The Handbook of Microbial Metabolism of Amino Acids*. However, these findings are integrated within a more comprehensive perspective in order to highlight opportunities and constraints to future developments. Nevertheless, even within a volume of over 500 pages, some editorial selection has been necessary.

Although this volume has been designed along conventional lines, with chapters on the metabolism of individual amino acids, it is important that emerging developments are perceived and

Table 32.1. Enzymes and pathways of amino acid metabolism in microorganisms of major relevance in food quality and safety and as human pathogens: a summary based on the evidence presented in the foregoing chapters of this volume.

Enzyme(s)/pathway(s) of amino acid metabolism	Microorganisms of major interest
Glutamate dehydrogenase	*Bacillus cereus*, *B. subtilis*, *Escherichia coli*, *Psychrobacter* sp., *Ruminococcus flavefaciens*, *Pseudomonas aeruginosa*, *Saccharomyces cerevisiae*, *Agaricus bisporus* and *Neurospora crassa*
Glutamate decarboxylase	*E. coli*, *Brucella microti*, *Clostridium perfringens*, *Lactobacillus brevis*, *L. paracasei*, *L. plantarum*, *Lactococcus lactis* and *Thermococcus kodakaraensis*
GABA (γ-aminobutyric acid) shunt	*S. cerevisiae*
Dihydrodipicolinate synthase	*E. coli*
Dihydrodipicolinate reductase	*Mycobacterium tuberculosis*
Diaminopimelate epimerase	*E. coli*
meso-Diaminopimelate dehydrogenase	*Ureibacillus thermosphaericus*
L,L-Diaminopimelate aminotransferase	*Chlamydomonas reinhardtii*
Arginine deiminase	*P. aeruginosa*, *Bacillus licheniformis*, *Enterococcus faecalis*, *Burkholderia pseudomallei* and *Lactobacillus sakei*
Arginase in microbial pathogenesis in the lungs	*Streptococcus pneumoniae*
Polyamine biosynthesis: role of arginine and methionine	*Trypanosoma* sp.
Ornithine and lysine decarboxylation	*Lactobacillus saerimneri*, enterobacteria and *Staphylococcus epidermidis*
Nitric oxide synthase	*S. cerevisiae*
Hydroxyproline metabolism	*P. aeruginosa*, *P. putida* and *Thermococcus litoralis*
Serine–glycine interconversion	*S. cerevisiae*
Threonine degradation	*T. litoralis*, *T. kodakarensis*, *Pyrococcus horikoshii* and *Thermotoga maritima*
Methionine biosynthesis	*E. coli*, *Corynebacterium glutamicum*, *B. subtilis* and *Staphylococcus aureus*
Regulation of sulfur amino acid metabolism	*N. crassa*, and *Aspergillus nidulans*
O-Acetylserine sulfhydrylase	*Trichomonas vaginalis*, *M. tuberculosis* and *Salmonella enterica*
Valine production	*C. glutamicum*
Leucine metabolism	Lactic acid bacteria
Degradation of phenolic amino acids	*Ferroglobus placidus*
Anthranilate synthase:	
Homodimer	*M. tuberculosis*
Heterotetramer	*Serratia marcescens*
Anthranilate phosphoribosyltransferase	*Sulfolobus sulfataricus*
Tryptophan biosynthesis: drug targets	*M. tuberculosis*
Kynurenine as a key metabolite in tryptophan catabolism	*Pseudomonas fluorescens*, *P. aeruginosa*, *N. crassa*, *Aspergillus niger*, *Penicillium roqueforti* and *S. cerevisiae*
Histidine degradation	*Pseudomonas* sp., *B. subtilis*, *Klebsiella pneumoniae* and enteric bacteria
Histidine phosphatase superfamily: post-translational activity	*M. tuberculosis* and *Francisella tularensis*
D-Amino acid biosynthesis	*E.coli*, *B. subtilis*, lactic acid bacteria, *Cochliobolus carbonum* and many other microorganisms
Rhizobial amino acid metabolism and polyamine biosynthesis	*Sinorhizobium meliloti*

Continued

Table 32.1. Continued.

Enzyme(s)/pathway(s) of amino acid metabolism	Microorganisms of major interest
Amino acid biosynthesis in endosymbiosis	*Strigomonas culicis*
Amino acid metabolism in helminths	*Haemonchus contortus*, *Teladorsagia circumcincta* and *Nippostrongylus brasiliensis*
Anaerobic metabolism of amino acids	*Clostridium sticklandii*, *C. perfringens* and *C. difficile*
Utilization of *N*-methylated amino acids	*P. aeruginosa*, *Listeria monocytogenes*, *S. meliloti* and *M. smegmatis*
Biofilm amino acid metabolism	*Candida albicans*

presented in terms of their relevance to practical issues in food microbiology and pathogen characterization. There are also ecological and environmental implications that should be considered in the light of recent evidence. This chapter is, therefore, and attempt to present an overarching view of the diverse issues that are now emerging in microbial amino acid metabolism.

32.3 Unlocking Practical Value

Research for strategic purposes is undeniably desirable in order to elicit potential benefits for future applications. In this respect, it is salutary to note that differences in drug resistance can be correlated with mutation rates in bacterial pathogens (Ford *et al.*, 2013), demonstrating the importance of constantly updating data about the biochemical genetics of microorganisms. Furthermore, knowledge of the fundamental biochemistry of microorganisms in general, and of bacteria in particular, will need to be significantly enhanced if efforts at environmental bioremediation are to succeed. At the same time, however, it is important to extract practical value from existing data on the microbial metabolism of amino acids.

Regarding food quality, the relationship between microbial metabolism of amino acids and the biogenesis of cheese flavour compounds has been well established. Particular emphasis has been given to the role of catabolic pathways in the generation of a variety of aroma compounds (Yvon and Rijnen, 2001). Initial studies focused on the biosynthesis of flavour components by the reaction of amino acids with carbonyl compounds (Griffith and Hammond, 1989), by the addition of α-ketoglutarate (Yvon *et al.*, 1998) or free amino acids (Wallace and Fox, 1998), through interactions with storage temperature (Pinho *et al.*, 2001) or between *Lactococcus lactis* and non-starter lactobacilli (Kieronczyk *et al.*, 2003). However, even by this stage, the role of catabolic enzymes, particularly aminotransferases, was beginning to emerge from established research groups (Yvon *et al.*, 1997). Up-to-date data will be reviewed at appropriate points in this chapter.

Probiotics are designed to confer nutritional and other physiological benefits on livestock and humans. It is considered important that probiotic bacteria should contain genes imparting acid stress resistance to ensure survival in the gastrointestinal tract. It has been suggested that amino acid decarboxylation serves to maintain intracellular pH under these conditions. Ferreira *et al.* (2015) implicated the role of the ornithine decarboxylase (ODC) gene in probiotic *Lactobacillus delbrueckii* in support of this hypothesis. It should be noted that ODC is a key enzyme in polyamine biosynthesis, thereby imparting additional benefits. Similarly, glutamate decarboxylase in probiotic lactobacilli may also assist in acid tolerance owing to its production of γ-aminobutyric acid (GABA).

There is growing interest in promoting certain products as functional foods. The production of GABA in foods fermented with *Lactobacillus brevis* is considered to provide benefits in neurological disorders and hypertension. Yields of GABA may be enhanced by co-culture with *Bifidobacterium bifidum* (Kim *et al.*, 2014) or *Streptococcus thermophilus* (Wu *et al.*, 2015). GABA-rich yogurt fermented with *S. salivarius* appears to confer protection in diabetes (Chen *et al.*, 2016).

In 2003, concern was expressed over continuing and widespread issues in food safety across a range of organic and inorganic contaminants (D'Mello, 2003b). Current evidence serves as a vindication of earlier predictions, particularly with regard to microbial contaminants. Food-associated morbidity remains an intractable public health matter, with outbreaks documented in detail for a wide range of microorganisms in diverse environmental conditions and foods. Thus, the incidence of *Salmonella* in a hospital (Quick *et al.*, 2015) and in poultry production (Lofstrom *et al.*, 2015) have been investigated. Similarly, a multi-province outbreak of listeriosis in Canada has been linked to contaminated processed meat consumed predominantly in institutional settings (Currie *et al.*, 2015), whereas a multistate episode of listeriosis in the USA was attributed to imported cheese (Heiman *et al.*, 2015). The emergence of campylobacteriosis as an issue in global epidemiology has been emphasized (Kaakoush *et al.*, 2015), and its association with undercooked chicken continues to attract attention in food safety research and surveillance (Scott *et al.*, 2015). Meanwhile, other studies relating to *C. difficile* (Lund and Peck, 2015) and *E. coli* (Gaulin *et al.*, 2015) provide evidence of additional routes and sources of contamination.

Fungal infection of plant products is also a continuing and widespread occurrence both prior to harvest and during storage (D'Mello, 2003a). Adverse effects arise in animals and humans from mycotoxins produced by certain plant pathogenic and storage fungi, including *Fusarium*, *Penicillium* and *Aspergillus* species. Relatively low levels of mycotoxins may also occur in blue-veined cheeses (Fontaine *et al.*, 2015).

The transport and metabolism of amino acids exert profound effects on virulence and other characteristics of activity in diverse microbial pathogens, including *S. pneumoniae*, *M. tuberculosis* (Parish and Stoker, 2002; Abendroth *et al.*, 2010), *S. mutans*, *A. pleuropneumoniae* and potentially other pulmonary pathogens. The uptake and enzymes of amino acid metabolism in the opportunistic pathogen *C. albicans* are also under investigation, as discussed later in this chapter (Brega *et al.*, 2004; Suliman *et al.*, 2007). In addition, there is increasing evidence of the role of specific enzymes of amino acid metabolism in parasite infectivity, as in the case of *Leishmania amazonensis* (da Silva *et al.*, 2012b). There is also considerable interest in the characterization of amino acid metabolism in *T. gondii*, *Plasmodium* species (Cobbold *et al.*, 2011; Oppenheim *et al.*, 2014) and *Trypanosoma cruzi* (Manchola *et al.*, 2015).

Host–microbe dynamics in amino acid metabolism have long been a major subject of research across a wide context from symbiosis to parasitology. Thus, it is well known that host-derived free amino acids (as well as peptides and proteins) are utilized to support the proliferation and growth of microbial symbionts (Graf and Ruby, 1998). According to Popp *et al.* (2015), *S. enterica*, a facultative intracellular pathogen, is isolated within a membrane-bound vacuole in host cells. Despite this compartmentation, *Salmonella* is able to efficiently acquire cell-derived amino acids for its own proliferation by redirecting host cell vesicular transport mechanisms.

The development of novel drugs continues with the increasing incidence of drug-resistant strains of important bacterial pathogens, such as *M. tuberculosis* and MRSA. In several cases, amino acid biosynthetic pathways present in microbes, but absent in the host, provide the fundamental validation of such investigations, in conjunction with examination of the crystal structure of key enzymes (Gokulan *et al.*, 2003; Lee *et al.*, 2006; Gupta *et al.*, 2009) and binding sites (Nordquist *et al.*, 2008). As described below, comprehensive strategies for novel therapeutics directed against amino acid metabolism in *M. tuberculosis* have recently emerged.

Legume plants are effective hosts for symbiotic bacteria, enabling the fixation of atmospheric nitrogen and the synthesis and exchange of amino acids within this system. The reduction in the need for nitrogenous fertilizers confers benefits for environmental protection. Metabolism in the rumen ecosystem, comprising specialist anaerobic bacteria, protozoa and fungi, is another prime example of symbiosis, enabling the host animal to derive essential amino acids, energy and other nutrients from basic, often low-grade, feedstocks. As reviewed below, despite decades of research, several issues of amino acid biosynthesis and exchange remain under scrutiny, both in microecological and global terms (Day *et al.*, 2001; Lodwig *et al.*, 2003; Chen *et al.*, 2012; Chungopast *et al.*, 2014).

32.4 Glutamate

All living organisms readily synthesize and utilize glutamate in diverse metabolic reactions. Although glutamate is a major constituent of proteins, its ubiquitous distribution as a free molecule and its signal transduction roles place it in a unique position in the metabolism of amino acids (Chapters 1–3). Its central position in transamination reactions is well established, but its metabolism by glutamate decarboxylase (GAD), GDH and glutamine synthetase should not be underestimated. Indeed, glutamate has been portrayed as an amino acid of 'particular distinction' (Young and Ajami, 2000). Such a statement, however, can become a distraction, serving merely to diminish the roles of other amino acids such as serine, arginine, BCAAs and tryptophan. Nevertheless, emerging evidence indicating a putative role for glutamate metabolism in bacterial responses towards acid and other stresses (Feehily and Karatzas, 2012) should be considered. In particular, the GAD pathway is implicated in intracellular pH homeostasis in several genera of bacteria. Linkage to and regulation of the GABA shunt system (Chapter 3) may also be of relevance here. In addition, Ramond *et al.* (2014) suggested that glutamate utilization provides a coupled mechanism for oxidative stress defence and the tricarboxylic acid (TCA) cycle in the phagosomal escape of *Francisella* and the virulence of this intracellular pathogen. The role of the GAD system continues to attract attention as part of the mechanism of acid tolerance in bacteria with potential for use in biotechnology and bioremediation (Liu *et al.*, 2015). Indeed, Damiano *et al.* (2015) extended interest in GAD-dependent acid resistance to the survival of *Brucella microti* under extremely low pH conditions.

P. falciparum contains three genes potentially encoding GDH; however, the protein encoded by *gdha* is dispensable and not a target during erythrocytic development (Storm *et al.*, 2011). The characterization of glutamine synthetase in terms of serine, glycine and alanine feedback inhibition in *S. typhimurium* has been published (Liaw *et al.*, 1993). In bacteria, the essentiality of glutamate racemase is widely acknowledged and is the subject of continuing research in the quest for potential drug targets specifically directed against *M. tuberculosis* (Morayya *et al.*, 2015).

32.5 Arginine

The universal position of arginine in living organisms remains undisputed, being characterized by its association, directly or ultimately, with the biosynthesis of urea, polyamines and nitric oxide (Chapters 5–9). Arginine has also been implicated in a regulatory role in the utilization of amino acids by intestinal bacteria (Dai *et al.*, 2012).

Arginase is the pivotal enzyme in both biosynthetic and catabolic pathways, but emerging evidence emphasizes significant roles in determining the activity and control of human parasites. For example, in *L. amazonensis*, arginase compartmentation in the glycosome appears to be important for infectivity of the parasite (da Silva *et al.*, 2012b). Aspects of crystal structure and molecular dynamics of arginase in *L. mexicana* have been investigated with the aim of identifying potential sites for drug design, either directly (Mendez-Cuesta *et al.*, 2012) or via the inhibition of polyamine biosynthesis in parasitic infections (D'Antonio *et al.*, 2013). In this respect, specific compounds have now been identified. Thus, the leishmanicidal activity of the flavonols, quercetin and quercitrin has been attributed to the targeting of arginase in *L. amazonensis* (da Silva *et al.*, 2012a). Dietary flavonoids may also be effective: for example, Manjolin *et al.* (2013) concluded that fisetin, luteolin and associated derivatives inhibit arginase and may provide protection in *Leishmania* infection.

In publishing evidence for the crystal structure of arginase in *P. falciparum*, Dowling *et al.* (2010) implied that a boronic acid inhibitor might serve as a starting point for the development of antimalarial therapeutics. The crystal structure of *Schistosoma mansoni* arginase has also been examined, revealing prospects for drug development for the treatment of this parasite (Hai *et al.*, 2014).

32.6 Serine

In microorganisms, serine is significant both as a free amino acid and for its post-translational activity. Serine is a primary participant in carbon metabolism in its interrelationship with glycine (Chapter 11) and is also the precursor of cysteine.

Changes in the pathways of energy generation and serine utilization may contribute to virulence, as has been suggested recently (Wu *et al.*, 2013), but there is extensive historical evidence of the preferential role of serine in microorganisms. The glycine–serine interconversion and the glycine cleavage enzyme system (Salcedo *et al.*, 2005; Abendroth *et al.*, 2010) are distinctive pathways common to all living organisms and are deeply embedded in the biochemical history of one-carbon metabolism (Stauffer and Brenchley, 1974; Stauffer *et al.*, 1989). Enzyme characterization and structural aspects of the glycine-cleavage complex in *Thermus thermophilus* (Nakai *et al.*, 2003) and *C. albicans* (McNeil *et al.*, 2000) have been undertaken. Cysteine biosynthesis in bacteria proceeds via a two-step process and continues to be an issue of importance due to the potential for the identification of specific drug targets (Qiu *et al.*, 2013; Campanini *et al.*, 2015).

The molecular characterization of serine metabolism in food-associated and pathogenic microorganisms is now well under way, as summarized in Table 32.2. This summary, set out in chronological order for each species, illustrates the progressive advancement of research from pathways and enzyme characterization (Becker *et al.*, 1969) towards increasing emphasis on regulation and the determination of factors affecting colonization and virulence. Identifying the genes encoding key enzymes has been a major feature underlying recent developments. The use of mutant strains has enabled the elucidation of key elements of serine metabolism, including gene expression (Hulanicka and Kredich, 1976). Increasing emphasis is now emerging for 'multi-omics' approaches to virulence (Wu *et al.*, 2013; Saint-Marc *et al.*, 2015).

32.7 Methionine

Research at the gene level in microorganisms reveals novel aspects in methionine biosynthesis (Chapter 13). Thus, work with *Agrobacterium tumefaciens* indicates the enzyme encoded by the *metA* gene is functionally a transacetylase rather than a transsuccinylase and that the gene itself is under regulation by a heat shock transcriptional activator (Roten *et al.*, 2013). In *F. graminearum*, cystathionine γ-synthase, encoded by *FgMETB*, appears to be essential for methionine biosynthesis and for the response of this fungal pathogen to demethylation inhibitors (Fu *et al.*, 2013). The rice blast fungus, *Magnaporthe oryzae*, is unable to assimilate sufficient methionine from plant tissues and relies on *de novo* synthesis of this amino acid in order to fulfil its sulfur amino acid requirements for infection (Saint-Macary *et al.*, 2015).

The microbial catabolism of methionine proceeds along established pathways. In *B. subtilis*, the conversion to cysteine occurs efficiently and the role of three enzymatic proteins in the regulation of this process has been reported (Hullo *et al.*, 2007). In *Helicobacter pylori*, cysteine is derived from products of an incomplete activated methyl cycle in a process of reverse transsulfuration (Doherty *et al.*, 2010).

Of particular relevance here is the degradation of methionine to volatile sulfur compounds (VSCs), which are considered to be essential for flavour development in cheese. The identification and functional analysis of the gene encoding methionine-γ-lyase in *B. linens* have implicated this enzyme as exerting a key role in the formation of VSCs (Amarita *et al.*, 2004). However, interaction with the yeast *G. candidum* may influence this process. Thus, Arfi *et al.* (2006) observed that methionine degradation occurred via a single step in *B. linens*, whereas a two-step pathway involving 4-methylthio-2-oxobutyric acid and 4-methylthio-2-hydroxybutyric acid intermediates was employed in *G. candidum*. It was suggested that as this yeast develops earlier than *B. linens* during cheese ripening, these compounds could also be used as precursors for VSCs by *B. linens*. In lactic acid bacteria, methanethiol has been cited as a precursor of VSCs that contribute to cheese flavour (Martinez-Cuesta *et al.*, 2013).

32.8 Branched-chain Amino Acids

The BCAAs are associated with profound metabolic and regulatory roles in all living organisms, eliciting manifestations of complex interactions particularly within the group (D'Mello, 2015) but also with other amino acids. Evidence concerning BCAA metabolism in microbes continues to reflect a wide range of interests from fundamental to applied aspects, including food quality

Table 32.2. Serine utilization and metabolism in diverse microorganisms. This summary, set out in chronological order for each species, illustrates the progressive advancement of research from pathways and enzyme characterization to increasing emphasis on regulation and the determination of factors affecting colonization and virulence. Identifying genes encoding key enzymes has been a major feature underlying recent developments. The use of mutant strains has enabled the elucidation of key elements of serine metabolism. Increasing emphasis is now advocated for multi-omics approaches to virulence.

Microorganism	Aspects and significance of serine metabolism	Reference
Salmonella typhimurium	Biosynthesis of serine; comparison with *Escherichia coli*	Umbarger *et al*. (1963)
S. typhimurium	Purification and characterization of *O*-acetylserine sulfhydrylase A	Becker *et al*. (1969)
S. typhimurium	Regulation of cysteine biosynthesis: effects of varying sulfur sources and *O*-acetylserine on gene expression	Kredich (1971)
S. typhimurium	Involvement of serine transhydroxymethylase in serine and glycine interconversions	Stauffer and Brenchley (1974)
S. typhimurium	Regulation of serine transhydroxymethylase activity	Stauffer *et al*. (1974)
S. typhimurium	Inhibition by 1,2,4-triazole mediated via effect on cysteine biosynthesis from serine	Kredich *et al*. (1975)
S. typhimurium	Expression of the gene coding for serine transacetylase	Hulanicka and Kredich (1976)
S. typhimurium	Influence of methionine biosynthesis on serine transhydroxymethylase regulation in *S. typhimurium* LT2	Stauffer and Brenchley (1977)
S. typhimurium	Cysteine synthetase: aggregation, kinetic behaviour and effect of modifiers	Cook and Wedding (1978)
S. typhimurium	Regulation of *O*-acetylserine sulfhydrylase by cysteine	Hulanicka *et al*. (1979)
S. typhimurium	CysB protein (a regulator of the cysteine regulon) interaction with *O*-acetylserine	Miller and Kredich (1987)
S. typhimurium	Serine utilization as a potential competition mechanism between *Salmonella* and a caecal bacterium	Ha *et al*. (1994)
S. typhimurium	Conformational dynamics of *O*-acetylserine sulfhydrylase; functional characteristics of two isoenzymes	Chattopadhyay *et al*. (2007)
S. typhimurium	D-serine deaminase: substrate specificity and catalysis	Bharath *et al*. (2011)
S. enterica	Inactivated serine hydroxymethyltransferase in mutants linked to decreased coenzyme A levels	Flynn *et al*. (2013)
S. enterica	Endogenous synthesis of 2-aminoacrylate contributes to cysteine sensitivity; previous studies focused on the dehydration of serine as a source of 2-aminoacrylate	Ernst *et al*. (2014)
Campylobacter jejuni	Complete sequence of the glyA gene encoding serine hydroxymethytransferase	Chan and Bingham (1991)
C. jejuni	Serine catabolism via an oxygen-labile serine dehydratase is essential for colonization of the avian gut	Velayudhan *et al*. (2004)
C. jejuni	Amino acid-dependent growth, preferentially for serine, aspartate, glutamate and proline	Guccione *et al*. (2008)
C. jejuni	Serine utilization depends on serine transporter and serine ammonia-lyase. Identifying substrates that fuel infection and growth of the bacterium	Hofreuter *et al*. (2012)
C. jejuni	Multi-omics approach: pathways of energy generation and serine utilization significantly upregulated in hypervirulent clone, but proline, glutamate, aspartate and lactate utilization significantly downregulated	Wu *et al*. (2013)
C. jejuni	Preferential utilization of serine relative to aspartate, asparagine and glutamine	Wagley *et al*. (2014)
C. coli 49941	Serine (and asparagine) allowed greater respiration of the bacterium at 37°C than at 42°C; temperature may trigger colonization or expression of virulence factors	Line *et al*. (2011)

Continued

Table 32.2. Continued.

Microorganism	Aspects and significance of serine metabolism	Reference
Streptococcus faecalis	Characterization of serine dehydratase activity	Farias *et al.* (1988)
Mycobacterium tuberculosis	Identification and characterization of serine acetyltransferase encoded by the Rv2335 gene	Qiu *et al.* (2013)
Lactobacillus casei	Factors influencing deamination of serine	Kristoffersen and Nelson (1955)
L. casei	*CysK* encodes a protein with *O*-acetylserine sulfhydrylase and cysteine desulfurization activity	Bogicevic *et al.* (2012)
L. plantarum	Serine metabolism by *L. plantarum* adjunct	Liu *et al.* (2002)
L. plantarum	Serine metabolism	Liu *et al.* (2003)
Saccharomyces cerevisiae	A novel feedforward/feedback mechanism in response to serine availability	Montefusco *et al.* (2012)
S. cerevisiae	Serine hydroxymethyltransferase: a key player linking purine, folate and methionine metabolism	Saint-Marc *et al.* (2015)
Aspergillus nidulans	Mutant lacking serine transacetylase: evidence for two pathways of cysteine biosynthesis	Pieniazek *et al.* (1974)
A. nidulans	*CysA* gene encodes a novel type of serine *O*-acetyltransferase	Grynberg *et al.* (2000)
Penicillium chrysogenum	Identification and purification of *O*-acetylserine sulfhydrylase	Ostergaard *et al.* (1998)
Fusarium graminearum	Serine *O*-acetyltransferase is important but not essential for cysteine and methionine synthesis; existence of alternative pathway	Fu *et al.* (2014)

and safety, specific pathogens, drug design (discussed later) and ecological interactions.

At the fundamental level, and also reflecting interactions with other amino acids, Ludwig *et al.* (2002) concluded that a *B. subtilis* mutant *ccpA* with insufficient expression of the *ilv-leu* operon, which encodes enzymes for the synthesis of BCAAs, needs BCAAs and methionine for optimal growth. Tuite *et al.* (2005) demonstrated that homocysteine toxicity in *E. coli* is caused by perturbation of BCAA biosynthesis. In *Streptomyces coelicolor*, a common transcriptional activator has been linked with both methionine and leucine biosynthesis (Kim *et al.*, 2012). A clear relationship between BCAAs and sulfur amino acids has also been noted for glucosinolate biosynthesis in plants (D'Mello, 2015).

The particular relevance of bacterial metabolism of BCAAs in the manufacture of dairy and cured meat products is worthy of consideration (Chapter 17). For example, the characterization of two aminotransferases in *L. lactis* indicated a broad spectrum of activity in the catabolism of BCAAs and methionine (Engels *et al.*, 2000). In addition, the expression in *L. lactis* of functional genes related to amino acid catabolism and the formation of cheese aroma is modulated by BCAAs (García-Cayuela *et al.*, 2012). The association of BCAA aminotransferases in *Lactobacillus sakei* with aroma formation is considered to occur via catabolism to methyl-branched volatiles, but the further conversion of α-keto acids to corresponding aldehydes, alcohols and carboxylic acids should also be considered (Freiding *et al.*, 2012). More specifically, the role of leucine catabolism in *Brevibacterium* species and the provision of precursors for the production of *S*-methyl thioesters has been examined by Sourabié *et al.* (2012). Products included *S*-methyl thioisovalerate, *S*-methyl thioacetate or *S*-methylthioisobutyrate.

The association of *S. mutans* with dental decay is generally acknowledged and the ability of the bacterium to withstand organic acids from carbohydrate metabolism constitutes a major virulence factor in the induction of carious damage. The results of Santiago *et al.* (2012) imply that at least part of this acid tolerance in *S. mutans* may be attributed to the expression of the BCAA aminotransferase encoded by the *ilvE* gene. Furthermore, Santiago *et al.* (2013) implicated two proteins in the regulation of the BCAA aminotransferase linked to acid adaptation of the bacterium.

It is generally accepted that BCAA transport and availability are important for pathogenicity and virulence in disease-causing microorganisms. For example, Basavanna *et al.* (2009) concluded that functional BCAA transporters were required for the full virulence of *S. pneumoniae*. BCAA status is also important for the survival and virulence of *A. pleuropneumoniae*, and potentially for other pulmonary pathogens as well (Subashchandrabose *et al.*, 2009). However, during commensalism, the acquisition of BCAAs appears not to be a determining factor for *C. jejuni*, as concluded by Ribardo and Hendrixson (2011), following an analysis of the LIV (leucine, isoleucine and valine) transport system in this bacterium.

The pathogenic fungus *F. graminearum* causes head (ear) blight in cereal plants, often accompanied by mycotoxin contamination of the harvested grain. It is now proposed that the *fgilv5* gene required for BCAA biosynthesis also contributes to virulence in this pathogen (Liu *et al.*, 2014).

The regulatory role of BCAAs is clearly seen in *Enterobacter cloacae*, a soil bacterium producing the rhizosphere signalling molecule, indole-3-acetic acid (IAA; Parsons *et al.*, 2015). It was observed that IAA was influenced by both BCAAs and aromatic amino acids, and a broader role was implied beyond plant interaction.

32.9 Aromatic Amino Acids

The microbial metabolism of phenylalanine, tyrosine and tryptophan (Chapters 18–21) has assumed enhanced significance in recent years due to impacts on food quality, human pathology and the synthesis of bioactive compounds. In general, the catabolism of aromatic amino acids exerts a significant role in the formation of undesirable floral aroma in Cheddar-type cheeses. Inhibition of the initial reaction involving aromatic aminotransferase by gene inactivation prevents the expression of floral aroma without affecting the development of beneficial volatile flavour compounds from BCAAs and methionine (Rijnen *et al.*, 1999).

In pathology, characterization of the *P. aeruginosa* transcriptional response to phenylalanine and tyrosine in relation to cystic fibrosis has been reported (Palmer *et al.*, 2010). Another example includes the influence of the tryptophan-indole-cytokine axis on human genital *C. trachomatis* infection, which exemplifies one of several host–pathogen interactions to be discussed later in this chapter (Aiyar *et al.*, 2014). Naz *et al.* (2013) concluded that the metabolism of aromatic amino acids in *G. candidum* results in the synthesis of antimicrobial products, including phenyllactate, hydroxyphenyllactate, phenylethyl alcohol and indoleacetate. Phenyllactate is attributed with broad and effective inhibitory activity towards both bacteria and fungi. The aromatic amino acids are also precursors of a diverse array of metabolites that may serve as scaffolds for drug design and development (Alkhalaf and Ryan, 2015).

32.9.1 Phenylalanine and tyrosine

Lee and Richard (1984) identified 42 microorganisms in cheese, including yeasts and bacteria. In yeasts, phenylalanine was metabolized via phenylpyruvate to phenethyl alcohol, whereas in strains of *Moraxella* spp. and *Staphylococcus saprophyticus*, the product named was phenylacetaldehyde. Groot and de Bont (1998) reported that the conversion of phenylalanine to benzaldehyde in *Lactobacillus plantarum* was mediated initially by an aminotransferase reaction. The resulting phenylpyruvic acid was then utilized for benzaldehyde biosynthesis. In continuing studies on the theme of off-flavours in cheese, Gummalla and Broadbent (2001) used *L. casei* and *L. helveticus* under simulated Cheddar cheese ripening conditions, specifically investigating the effects of the phenolic amino acids. These were catabolized by successive, reversible transamination and dehydrogenation reactions, the major products being phenyllactate, phenylacetate and benzoic acid from phenylalanine and *p*-hydroxyphenyllactate and *p*-hydroxyphenylacetate. The formation of some of these products by non-enzymatic processes was also considered.

In contrast, bacterial formation of leucine- and phenylalanine-derived volatile compounds in cured meat model systems led Moller *et al.* (1998) to recommend that *Moraxella phenylpyruvica* be considered as a potential starter for ensuring superior meat flavour. The products of catabolism included 3-methylbutanal and 3-methylbutanol from leucine and benzacetaldehyde from phenylalanine.

32.9.2 Tryptophan

In the cheese adjuncts, *L. casei* and *L. helveticus*, tryptophan may be converted to indole-3-lactate via successive transamination and dehydrogenation reactions (Gummalla and Broadbent, 1999). Furthermore, the action of decarboxylases in cheese-associated bacteria has been correlated with the production of biologically active amines, including tryptamine (Curtin and McSweeney, 2003). It is now acknowledged the IAA is one of the major products of tryptophan catabolism in microorganisms involving multiple mechanisms (Mujahid *et al.*, 2011).

In *E. coli*, the enzyme tryptophanase TnaA hydrolyses exogenous tryptophan to indole, pyruvate and ammonia. Indole acts as signalling molecule in the mammalian gut and may facilitate the protection of *E. coli* against several antibiotics (Li and Young, 2013). The indolepyruvate pathway in the host and the microbiome may also be influenced by perturbations in human genetic disorders of tyrosine metabolism and concomitant drug treatment (Gertsman *et al.*, 2015).

Consistent with observations in higher organisms, the microbial metabolism of tryptophan is associated with the biosynthesis of a wide range of bioactive compounds. Alkhalaf and Ryan (2015) attribute this diversity to the property of tryptophan to undergo multiple enzymatic and spontaneous reactions. Several of these molecules are considered to be potential scaffolds for drug discovery.

Catabolic pathways of tryptophan metabolism (Chapter 21) provide opportunities for the development of other novel compounds. To this end, Zummo *et al.* (2012) identified three genes coding for tryptophan-2,3-dioxygenase (TDO), kynureninase and kynurenine formamidase in *S. coelicolor* which could be inactivated to increase antibiotic production. Akiyama *et al.* (2014) identified the thermophilic bacterium *Laceyella sacchari* as an underexploited biomedical resource, based on studies with the synthesis of bioactive tryptophan metabolites.

32.10 Secondary Metabolism

The synthesis of secondary compounds is a significant feature of microbial metabolism, with amino acids serving as direct precursors or as sources of smaller molecules. For example, isoleucine-derived coronamic acid functions as an intermediate in the biosynthesis of coronatine, a chlorosis-inducing phytotoxin produced by *P. syringae* (Ullrich and Bender, 1994). Secondary compounds may also cause perturbations in amino acid metabolism. Tabtoxin, for example, originates in the lysine biosynthetic pathway in certain *Pseudomonas* species (Liu and Shaw, 1997), but tabtoxin in *P. tabaci* has been associated with the downregulation of glutamine synthetase in the host, *Nicotiana tabacum* (Turner, 1981). In addition, mango toxin, a novel antimetabolite produced by a strain of *P. syringae* isolated from the mango tree, inhibits ornithine/arginine biosynthesis by targeting ornithine *N*-acetyltransferase (Arrebola *et al.*, 2003). With reference to marine organisms and mycosporine-like amino acids, Carreto and Carignan (2011) commented on relevant secondary metabolites and ecological implications. Secondary compounds produced by bacteria are used as antibiotics, anticancer drugs and other pharmaceutical applications. Most of these metabolites are polyketides or non-ribosomal peptides (Craney *et al.*, 2012). There is also potential for the use of L-amino acid ligase in *P. syringae* producing tabtoxin for the enzymatic synthesis of several functional peptides with physiological activity in humans (Arai *et al.*, 2013).

Secondary metabolism in fungi is also associated with the synthesis of a wide range of bioactive compounds. Those that are detrimental to plants are divided into two groups: host-specific and non-specific toxins. Host-specific toxins (HSTs) are essentially pathogen effectors, inducing toxicity and disease exclusively in specific genotypes of host plants (Friesen *et al.*, 2008). Species of *Cochliobolus*, *Alternaria* and *Pyrenophora* are pre-eminent examples of fungal pathogens with HST activity. Thus, the cyclic tetrapeptide HC-toxin is required for the pathogenicity of *C. carbonum* in maize. A pathotype of *A. alternata* that causes black spot of Japanese pear produces the HST designated as AK-toxin, while another pathogenic strain infecting tomato produces the AAL-toxin.

Fungal metabolites that exert adverse effects in humans and other vertebrates are termed mycotoxins (D'Mello, 2003a). The principal fungi associated with mycotoxins include *Claviceps purpurea*, producing ergot alkaloids and three *Aspergillus* species frequently implicated in the synthesis of aflatoxins, cyclopiazonic acid and ochratoxins. A number of *Penicillium* species are regularly linked with ochratoxins, citrinin, patulin and citreoviridin. For instance, two distinct fungal species are responsible for the synthesis of the ochratoxins. A number of *Fusarium* species synthesize a diverse range of toxins, including several trichothecenes, zearalenone, fumonisins and fusaric acid. Secondary compounds produced by *A. alternata* include tenuazonic acid, alternariol and altenuene. Cereal grains, nuts and certain fruits are regularly found to contain mycotoxins and there is continuing surveillance to monitor levels that may be harmful to human and animal health.

Substrate utilization and biosynthetic pathways in fungi are as diverse as the chemistry and toxicology of the associated secondary metabolites (D'Mello, 2003a). However, in several instances there are direct or indirect connections to amino acids as structural components, precursors or donors of active groups or via enzymes. It has long been known that aspartate (together with acetate) is incorporated into fusaric acid (Desaty *et al.*, 1968). Similarly, glycine is incorporated directly into AAL-toxin, while methionine provides methyl groups (Caldas *et al.*, 1998). Additionally, phenylalanine is an integral component of the ochratoxins in which the amino acid is linked to an isocoumarin molecule (Varga *et al.*, 2001; Gallo *et al.*, 2012). In the case of the ergot alkaloids, tryptophan provides the structure of the tetracyclic ergoline ring (Gerhards *et al.*, 2014). The biosynthesis of fumonisins involves the addition of a more complex range of active groups derived from methionine, glutamate and alanine to a hydrocarbon backbone (Blackwell *et al.*, 2013). In *C. carbonum*, a putative BCAA transaminase gene has been implicated in HC-toxin biosynthesis (Cheng *et al.*, 1999).

32.11 Osmoprotection

The following excerpt on proline, modified to remove the references that were cited, appeared in the review of D'Mello (2015) relating to amino acid metabolism in higher plants. As a derivative of glutamate, proline is ubiquitous in living organisms and is now emerging as a key multifunctional amino acid in primary metabolism and in the response of microbes and plants to environmental stress. Proline is associated with specific attributes that include the regulation of cellular homeostasis, acting as a signal transduction molecule that modulates mitochondrial function and with a role in gene expression, particularly under adverse conditions. Additionally, the physical properties of proline that arise from its cyclic structure which restricts conformational flexibility, as well as its activity as a compatible solute in the protection of cellular components, have been emphasized. It has been suggested that efforts to manipulate proline metabolism might assist in unlocking more potential to enhance stress tolerance in plants.

The role of proline and other compatible solutes in bacteria is of considerable importance, reflecting both fundamental and ecological issues (Chapter 30). Hoffmann *et al.* (2012), in their work with *B. subtilis*, reiterated the role of proline as a defence molecule against high osmolarity and for the maintenance of optimal hydration of the cytoplasm and turgor of the cell. The functions of the transporter OpuE in the scavenging and recapture of proline were also elaborated. Zaprasis *et al.* (2014) identified GABA permease (GabP) as another proline transporter in *B. subtilis* and furthermore examined the uptake of amino acids and the metabolic conversions to proline associated with osmoprotection in this bacterium (Zaprasis *et al.*, 2015). The dynamics supplement previous observations of osmoprotection in *B. subtilis* through the import and proteolysis of proline-containing peptides (Zaprasis *et al.*, 2013).

Concurrently, Bashir *et al.* (2014) reported that dimethylglycine provided salt as well as temperature stress protection to *B. subtilis*, while in *P. syringae*, glycine betaine catabolism exerts tolerance to hyperosmotic stress by relieving betaine-induced depression of compatible solute synthesis (Li *et al.*, 2013).

At this juncture, it is salutary to emphasize that certain osmoprotectants are endowed with other attributes in microorganisms. Thus, proline in particular suppresses apoptosis in *Colletotrichum trifolii*, a fungal pathogen of lucerne. Proline acts as an antioxidant and inhibitor of programmed cell death (Chen and Dickman, 2005). Subsequently, Lee *et al.* (2013) showed that ROS homeostasis and virulence in the fungal pathogen *C. neoformans* require an intact proline catabolism pathway. Notwithstanding these comments, it is pertinent to enquire on the extent and mechanisms involved in the relationship between osmoregulation and mycotoxin production in different field and storage fungi (Duran *et al.*, 2010).

32.12 Host–microbe Interactions

The issue of host–microbe interactions is relevant here due to the inevitable impact of microbes on the amino acid metabolism of the host and the concomitant response to colonization or infection. It is important to state that virtually all aspects of current understanding are based on speculative models, and that limited experimental evidence is only now emerging in studies with bacterial and fungal pathogens. Nevertheless, it is instructive to consider the relative merits and limitations of recent evidence.

Host–microbe interactions may be mutually beneficial as in the case of symbiotic relationships in leguminous plants and in the ruminant animal, although in the latter contrasting effects have been reported (see below). Alternatively, the effects may be antagonistic, as in host–pathogen relationships, with the final outcome being determined by microbial auxotrophy and the success of defence or immune mechanisms. In both cases, there are, invariably, profound effects on the metabolism of amino acids.

32.12.1 Symbiosis

The legume–*Rhizobium* association in root nodules remains the pre-eminent example of symbiosis, conferring benefits and exerting demands to both constituents in this axis (Chapter 26). In terms of overall dynamics, Day *et al.* (2001) elucidated further the complexities of amino acid transport mechanisms across symbiotic membranes in the nodules of N-fixing legumes. In addition, Lodwig *et al.* (2003) indicated that amino acid exchange drives N fixation in the legume–*Rhizobium* interaction, stating that this interdependence prevents the symbiosis being dominated by the plant. In terms of intermediary metabolism, specific aspects are now emerging. Thus, the biosynthesis of BCAAs is considered to be essential for efficient symbioses between beta-rhizobia and the tropical legume, *Mimosa pudica* (Chen *et al.*, 2012). Furthermore, at the enzyme level, Chungopast *et al.* (2014) observed that glutamine synthetase I deficiency in *Mesorhizobium loti* differentially affected nodule development and activity in *Lotus japonicus*.

Metabolism in the rumen ecosystem, which comprises specialist anaerobic bacteria, protozoa and fungi, is another prime example of symbiosis, enabling the host animal to derive essential amino acids, energy and other nutrients from basic, often low-grade, feedstocks. The rumen microbes depend, to an appreciable extent, on *de novo* synthesis of amino acids from ammonia and a variety of carbon precursors produced during the fermentation of carbohydrates. Rumen microbial degradation of intact feed proteins and free amino acids is equally extensive. Details of the exact nature of the microorganisms involved and the individual anabolic and catabolic pathways for each amino acid are now widely available (see Atasoglu and Wallace, 2003).

32.12.1.1 Microbial metabolism of non-protein amino acids: contrasting observations

The diverse array of unconjugated non-protein amino acids occurring in living organisms has been particularly well characterized in higher plants (see D'Mello, 2015). Mimosine and *S*-methylcysteine sulfoxide (SMCO) have been selected here to highlight contrasting aspects of mutualism. Mimosine contributes significantly to the free amino acid content of the leaves and seeds of the ubiquitous tropical forage legume, *Leucaena leucocephala*, whereas SMCO occurs in leafy and root *Brassica* crops, also in a readily available form.

The microbial breakdown of mimosine is now well established, particularly in symbiotic relationships with legumes and certain species of animals. In *Rhizobium* species (strain TAL 1145), it has been suggested that the *midK*-encoded pyruvate carboxylase is required for the efficient conversion of mimosine to 3-hydroxy-4(1H)-pyridone (3,4-DHP; see D'Mello, 2015). Adverse effects of mimosine in affected animals include loss of wool, teratogenic abnormalities, organ damage and mortality. The metabolic fate of mimosine in ruminants (for example cattle, sheep and goats) is largely determined by the rate and extent of its degradation by the rumen microorganisms. During this breakdown, 3,4-DHP is produced. However, this metabolite may also arise in the leaf of *Leucaena* itself following post-harvest enzymatic action. Ruminants consuming this legume macerate the leaves and add alkaline saliva, thus providing optimum conditions for the autolysis of mimosine to the extent that a substantial proportion of the non-protein amino acid may be converted to 3,4-DHP before even reaching the rumen. Additional quantities of 3,4-DHP arise during the rumen fermentation of mimosine and another isomer, 2,3-DHP, may also be produced. Some rumen microbes are capable of degrading the two forms of DHP to as yet unidentified compounds. Despite these transformations, considerable quantities of mimosine and 3,4-DHP may escape degradation to emerge in the faeces. Furthermore, conjugated forms of 3,4-DHP may also appear in the faeces and urine, while the parent amino acid may itself undergo decarboxylation within the tissues of the ruminant to yield mimosinamine, which is then excreted in the urine. All of these transformations are subject to geographical differences in the microbial ecology of ruminants. Moreover, the ability to degrade DHP may decline or disappear entirely on removing these animals from *Leucaena* pasture (D'Mello, 2015).

In the plant, SMCO co-occurs with its metabolite, methyl methane thiosulfinate. The deleterious effects of SMCO in ruminant animals fed mainly or solely on brassica forage are well recognized, and include haemolytic anaemia, loss of appetite, reduced milk yield, organ damage and death (see D'Mello, 2015). Adverse effects are precipitated after the synthesis of a reactive derivative (dimethyl disulfide) during fermentation of SMCO by rumen microorganisms. In this particular instance, the symbiotic rumen microorganisms enhance reactivity instead of protecting the host animal, as in the case of mimosine.

32.12.1.2 The host–pathogen axis

A simple hypothesis for host–pathogen interactions in humans is presented in Fig. 32.1, which shows tryptophan metabolism after infection with *C. trachomatis* (Aiyar *et al.*, 2014). In response to this infection, the primary cytokine interferon gamma (IFNγ) is synthesized and released in host tissues, inducing the catabolism of tryptophan and thereby depriving the bacterium of an essential amino acid. *In vitro* observations suggest that *C. trachomatis* adopts a quiescent mode and that reactivation of the bacterium could be induced by restoration of tryptophan supply. However, in uropathogenic *E. coli*, the induction of tryptophan degradation may cause innate responses to epithelial infection to be compromised (Loughman and Hunstead, 2012). Thus, the regulation of tryptophan biosynthesis and indole flux is widely perceived as an interface between pathogen and host, and as a putative target for drug development (Andersen, 2012). Other examples include infections attributed to *M. tuberculosis* (Zhang *et al.*, 2013) and to *Leishmania* parasites (Makala, 2012). It is important to note that the antimicrobial or immunoregulatory role of tryptophan biosynthesis and catabolism may be pathogen specific (Divanovic *et al.*, 2011).

It has been suggested by Subashchandrabose *et al.* (2009) that in mammalian lungs, BCAA availability may be limited, but that *A. pleuropneumoniae* responds to this restricted supply by regulating the expression of genes required for survival and virulence. Such as strategy would involve increased synthesis of BCAAs by the pathogen.

The opportunistic pathogen *P. aeruginosa* is associated with a wide range of infections, particularly in immunocompromised patients. It has been demonstrated that alanine is the preferred source of carbon and that two genes are induced for its catabolism during *in vivo* proliferation in mammalian tissue and during *in vitro* growth in the lung mucus of individuals with cystic fibrosis (Boulette *et al.*, 2009).

The transport and metabolism of BCAAs may also modulate, or be affected by, host–pathogen interactions. It has been suggested that virulent strains of *F. tularensis* rely exclusively on dedicated uptake systems to satisfy their BCAA requirements (Gesbert *et al.*, 2015).

C. albicans, one of the most important fungal pathogens of humans, is characterized by certain features of amino acid metabolism relating to morphogenesis and virulence. In particular, *C. albicans* utilizes an amino acid sensor, Csy1p, to seek and coordinate its nutritional requirements and thereby determine its morphological state (Brega *et al.*, 2004). The pathogen also requires Stp2p, a

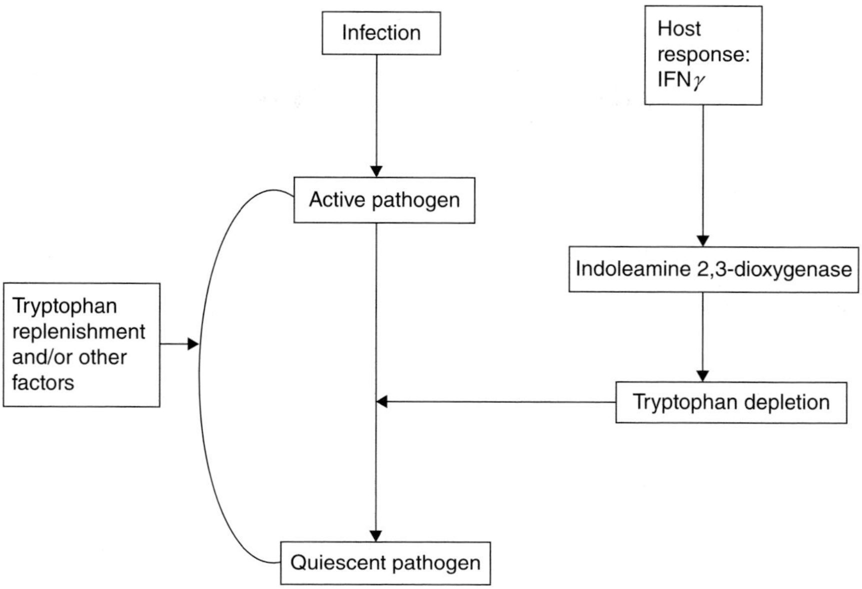

Fig. 32.1. Tryptophan dynamics in human pathogenesis associated with microbial infections (based on the hypothesis advanced by Aiyar *et al.*, 2014, in studies with *Chlamydia trachomatis*). Regulation of tryptophan biosynthesis and indole flux is widely perceived as an interface between the pathogen and host and as a putative target for drug development. Examples include infections attributed to *Mycobacterium tuberculosis* (Zhang *et al.*, 2013), uropathogenic *Escherichia coli* (Loughman and Hunstead, 2012) and *Leishmania* parasites (Makala, 2012). The antimicrobial or immunoregulatory role of tryptophan biosynthesis and catabolism may be pathogen specific (Divanovic *et al.*, 2011). Abbreviation: IFNγ, interferon gamma.

regulator of amino acid transport, to promote the neutralization of phagosomal pH, hyphal morphogenesis and escape from macrophages (Vylkova and Lorenz, 2014). This sequence is perceived as an important virulence adaptation in *C. albicans*. The anti-*Candida* efficacy of cispentacin appears to be dependent upon active transport into the cell by amino acid permeases (Capobianco *et al.*, 1993). The intervention of multidrug transporters has also been invoked to explain azole antifungal resistance in *C. albicans* isolated from AIDS patients (Sanglard *et al.*, 1995).

In trypanosomes, BCAAs are essential nutrients and important substrates for energy production. *T. cruzi*, the causal pathogen of Chagas disease, alternates between mammalian and invertebrate animals, securing its supply of BCAAs by uptake from the extracellular medium in host tissues (Manchola *et al.*, 2015). Thus, the depletion of BCAAs in the host is an inevitable consequence of trypanosome infection, but the disease provokes a marked immune response, possibly linked with host arginine status (Bronzon da Costa *et al.*, 2014). The issue of parasite-induced arginine metabolism in the host has also been raised by Stadelmann *et al.* (2013) in studies with *Giardia intestinalis*.

Food-borne bacterial pathogens rely on the host for a supply of key nutrients. Such a dependence is demonstrated by the utilization of host-derived cysteine-containing peptides to overcome the restricted sulfur metabolism of *C. jejuni* (Vorwerk *et al.*, 2014).

A general model for microbial stress metabolism in higher plants has been advanced (Fig. 32.2). It is readily admitted that this scheme is too simplistic to accommodate the complex chain of molecular events that eventually result in the expression of defence responses or incidence of disease. Nevertheless, the scheme shown in Fig. 32.2 should be perceived as an instructive initial model that may be developed further as the full scale of biochemical interactions are established for particular pathogens.

There is an emerging consensus that the primary stress responses in plants are broadly analogous to those observed in animals in terms of passive and active immunity, and the mediation

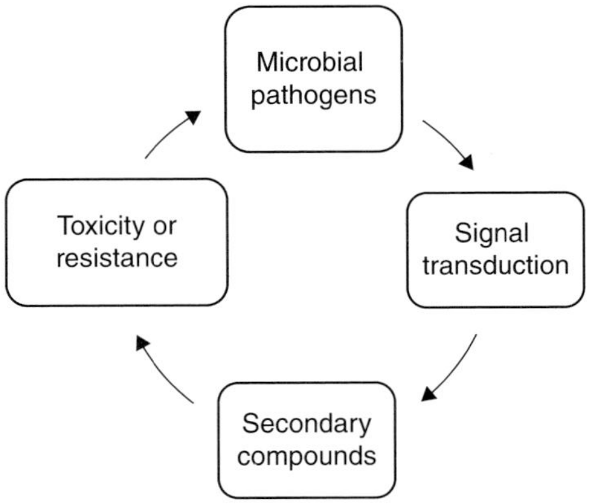

Fig. 32.2. Host–parasite interactions: a general model for biotic stress dynamics in higher plants, as exemplified by responses to microbial pathogens. Adapted from D'Mello, 2015. The relevance to amino acid metabolism in higher plants and the dynamics of host–parasite relationships, as portrayed here, are based on hypotheses and evidence presented in D'Mello (2015). In essence, certain amino acids, including methionine, phenylalanine and tryptophan, serve as precursors of, respectively, ethylene (ET), salicylic acid (SA) and indoleacetic acid (IAA), as well as parent molecules for the biosynthesis of a diverse array of secondary compounds implicated in plant defences. Nitration of tyrosine residues in certain proteins is central to peroxynitrite signalling following pathogen infection. In addition, isoleucine activates jasmonic acid (JA) and, separately, leucine-rich repeat receptor kinases and leucine zipper transcription factors are involved in signal transduction. There is evidence of phytopathogen adaptation to secondary defence compounds, indicative of a dynamic flux in host–parasite relationships.

of signalling and other regulatory molecules. In addition, local and systemic defence processes are also recognized in both plants and animals. Furthermore, the question of host–parasite relationships is a common issue relating in particular to pathogen adaptation to defence compounds produced *in vivo* or administered exogenously. The biochemical impact of stress in plants, as outlined in Fig. 32.2, is generally assumed to commence with signal transduction from injured or infected tissues. Signal molecules generated there are translocated via the phloem to undamaged tissues where they are assumed to be detected. In response, a diverse array of low molecular weight secondary compounds and proteins are synthesized and released to augment pre-existing constitutive defences against phytopathogens. The net effects include intoxication or resistance to the biotic challenge, depending upon the source of injury or infection.

Amino acids may act directly as signalling molecules or may yield other compounds with similar functions. Additionally, amino acids may act indirectly by facilitating the functional capacity of other signalling compounds such as salicylic acid (SA), brassinosteroids and jasmonic acid (JA). In response to signals, defence molecules may be deployed to reduce or neutralize the physiological impact of the pathogen. A wide range of secondary and polymeric compounds are attributed with defence functions including, among many others, amino acid-derived polyphenolics, serotonin, phytoalexins, indole glucosinolates and GABA, as well as specific proteins with catalytic or even inhibitory activities.

The induction of defence compounds is a complex process involving the interaction of a diverse range of signal transduction networks. Thus, camalexin deployment in response to fungal infection requires ROS and SA signal transduction, but glutathione status may also be an important component of the pathway. Targets of this signalling cascade include the tryptophan and camalexin biosynthetic genes which are transcriptionally upregulated in affected tissues. Although particular pathogens are effectively inhibited by camalexin, other fungi are resistant due to active degradation of this indole compound (see D'Mello, 2015).

Specific changes in host plant metabolism may, in turn, affect relationships with pathogens. Stuttmann *et al.* (2011), for example, demonstrated that the perturbation of *Arabidopsis* amino acid metabolism caused incompatability with the adapted biotrophic pathogen *Hyaloperonospora arabidopsidis*, thereby influencing its capacity to cause disease.

32.13 Antimicrobial Chemotherapy: Potential Targets in Pathways of Amino Acid Metabolism

Pathways and enzymes of amino acid metabolism in pathogens provide the potential for the development of novel drugs and vaccines for clinical use (Zohar *et al.*, 2003; Nordquist *et al.*,

2008; Wang *et al.*, 2013; see also Chapters 4, 5, 7, 13, 15 and 20). In several cases, biosynthetic pathways present in microbes, but absent in the host, form the basis of such investigations in conjunction with examination of the crystal structure of key enzymes (Gokulan *et al.*, 2003; Lee *et al.*, 2006; Gupta *et al.*, 2009) and binding sites (Nordquist *et al.*, 2008). The production of mutants with inactivated genes is an integral part of this strategy (Parish *et al.*, 1999). In a study of the role of amino acid permeases in *C. neoformans*, Fernandes *et al.* (2015) concluded that the tryptophan biosynthetic pathway is a promising pharmacological target. One of the target enzymes in the cysteine biosynthetic pathway is *O*-acetylserine sulfhydrylase (Chapter 15). Mazumder and Gourinath (2016) have reviewed a structure-based approach for the design of potential inhibitors of this enzyme that may be suitable for a variety of diseases in humans.

Due to the worldwide health issues presented by tuberculosis and the development of drug resistance, much attention is currently being given to the identification and characterization of drug targets and attenuated strains of the causal organism, *M. tuberculosis*. The evidence presented in Table 32.3 demonstrates the diverse aspects of amino acid metabolism currently under investigation with this pathogen. On the one hand, extracellular protein, as exemplified by the export of glutamine synthetase, is perceived as a potential target for novel drug design (Harth and Horwitz, 1999). On the other, studies have focused on various sites in biosynthetic pathways, particularly those that do not overlap with endogenous enzyme systems in the host. Thus, acetohydroxyacid synthase (AHAS) is essential for the *de novo* synthesis of BCAAs in microbes involving a pathway that does not exist in mammalian host tissues. Consequently, AHAS is perceived as a valid target for the development of drugs for the treatment of tuberculosis (Table 32.3). The crystal structure of key biosynthetic enzymes is also of relevance in drug development (Gokulan *et al.*, 2003; Koon *et al.*, 2004). The X-ray structure determination of the glycine cleavage system protein H of *M. tuberculosis* may also have implications for pharmaceutical applications (Abendroth *et al.*, 2010). Furthermore, Giffin *et al.* (2012) identified an *ald* gene encoding both alanine dehydrogenase and the putative glycine dehydrogenase, thereby adding another potential target in the molecular profile of *M. tuberculosis*. In addition, and of particular note, is the unique role of racemases in the essential synthesis of peptidoglycans in bacteria, which provides yet another target for drug development (Morayya *et al.*, 2015).

Working with *S. typhimurium*, LaRossa *et al.* (1987) reported the toxic accumulation of α-ketobutyrate caused by the inhibition of acetolactate synthase, which is involved in the biosynthesis of BCAAs. It is conceivable that this enzyme may be an appropriate target for drug development aimed at a wide range of bacterial pathogens due to its absence in mammalian tissues.

The quest for novel drugs continues with the increasing development of antibiotic-resistant strains of well-known bacterial pathogens, such as MRSA. The development of daptomycin analogues by tryptophan replacement with other amino acids has been advanced by He *et al.* (2012). Okorochenkov *et al.* (2013) reported that arginine-containing haemin conjugates showed high antibacterial activity *in vitro* towards Gram-positive bacteria, including methicillin- and vancomycin-resistant strains. Meanwhile, Ahn *et al.* (2013) developed a histidine-derived lipo-amino acid for the synthesis of antimicrobial peptidomimetics with potent efficacy, salt resistance and protease stability. Another strategy is based on the use of non-protein or D-amino acids to enhance the activity of antimicrobial agents (Knerr *et al.*, 2012; Sanchez *et al.*, 2014).

Novel antifungal β-amino acids have been developed and tested *in vitro* for efficacy against *C. albicans*, with β-aa21 (BAY 10-8888/PLD-118) emerging as the product with the most favourable activity–tolerability profile for potential oral treatment of yeast infections (Mittendorf *et al.*, 2003). In *C. albicans*, the gene for cobalamin-independent methionine synthase has been designated as essential and proposed as a 'plausible' target for the design of antifungal agents (Suliman *et al.*, 2007). There is continuing interest in exploiting inhibitors of amino acid biosynthesis as antifungal agents (Jastrzebowska and Gabriel, 2015). Here, the selection of an appropriate enzyme may be a determining factor. Thus, Fu *et al.* (2014) concluded that serine *O*-acetyltransferase (SAT) is important but not essential for cysteine and methionine synthesis in *F. graminearum* due to the existence of an alternative pathway. It might be presumed,

Table 32.3. Potential targets in pathways of amino acid metabolism for development of pharmaceuticals. The evidence below is presented to exemplify the diverse aspects of amino acid metabolism in *Mycobacterium tuberculosis* that may provide a basis for drug and vaccine design.

Pathways/enzymes/ genes in *M. tuberculosis*	Observations	Reference
General	Production of mutants in amino acid biosynthesis genes as a technique for identifying potential drug targets for understanding the mechanisms underlying infection and establishment of disease	Parish *et al.* (1999)
	In silico analysis of pathways of pathogen and host indicated potential drug targets in amino acid, lipid, carbohydrate and nucleotide metabolism	Anishetty *et al.* (2005)
D-Alanine ligase	A target of the anti-tuberculosis drug D-cycloserine	Bruning *et al.* (2011)
Glutamine synthetase	Evaluation of the amino acid binding site for drug discovery	Nordquist *et al.* (2008)
Glutamine synthetase (GS)	An inhibitor of exported GS selectively blocks the growth of pathogenic mycobacteria in axenic culture and in human monocytes: extracellular protein as a potential drug target	Harth and Horwitz (1999)
Glutamate racemase (GR)	Essential role in peptidoglycan synthesis suggests GR should be considered a potential drug target	Morayya *et al.* (2015)
Dihydrodipicolinate reductase (DDR)	Inhibitors of DDR, a key enzyme of the diaminopimelate pathway, reveal an underlying mechanism for drug development	Paiva *et al.* (2001)
Aspartate-β-semialdehyde dehydrogenase (ASD)	ASD is one of the key enzymes in diaminopimelate biosynthesis and is a validated drug target	Shafiani *et al.* (2005)
Diaminopimelate decarboxylase (DPDC)	Crystal structure of DPDC, an essential enzyme in bacterial lysine biosynthesis; a potential drug target as DPDC is absent in host tissues	Gokulan *et al.* (2003)
Nitric oxide synthase (NOS)	Inhibitors of NOS confirm the importance of reactive N species in resistance to *M. tuberculosis*	Chan *et al.* (1995)
Selenomethionine (SeMet)	Crystal structure of bacterioferritin: incorporation of SeMet causes cleavage and demetallation of haem; structural basis for drug design	Gupta *et al.* (2009)
Sulfur metabolism	Microbial S metabolic pathways largely absent in humans and therefore represent unique drug targets	Bhave *et al.* (2007)
Acetohydroxyacid synthase (AHAS)	AHAS from *M. avium* and its inhibition by sulfonylureas and imidazolinones	Zohar *et al.* (2003)
	Biochemical and transcriptional analysis of AHAS isoforms identified as experimental targets for drug development	Singh *et al.* (2011)
	Novel AHAS inhibitors identified as active agents by virtual screening and bioassay	Wang *et al.* (2013)
α-Isopropylmalate synthase (LeuA)	Crystal structure of LeuA, a key enzyme in leucine biosynthesis and a potential target for design of novel drugs	Koon *et al.* (2004)
Interleukin 1 beta (*ilvB1*) gene	Inactivation of the *ilvB1* gene in *M. tuberculosis* leads to branched-chain amino acid auxotrophy and attenuation of virulence; potential for development of vaccine	Awasthy *et al.* (2009)

Continued

Table 32.3. Continued.

Pathways/enzymes/ genes in *M. tuberculosis*	Observations	Reference
Branched-chain amino acid (BCAA) biosynthesis and aminotransferase	Inhibitors of BCAA biosynthesis as potential anti-tuberculosis agents	Grandoni *et al.* (1998)
	BCAA aminotransferase inhibited by selected aminooxy compounds which are also effective in preventing cell growth *in vitro*; these compounds may provide the basis for the synthesis of inhibitors with enhanced efficacy and reduced toxicity	Venos *et al.* (2004)
Aromatic amino acid biosynthesis	Gene disruption proposed as a method for generating attenuated vaccine	Parish and Stoker (2002)
Shikimate pathway	Mycobacterial shikimate pathway enzymes as targets for drug design	Basso and Santos (2007)
Anthranilate phosphoribosyl-transferase (Trp D)	Trp D is essential for lung colonization by *M. tuberculosis*: implications for drug development	Lee *et al.* (2006)

therefore, that SAT is unlikely to be a suitable target for the development of antifungal agents.

The intraerythrocytic malaria parasite (*Plasmodium falciparum*) is auxotrophic for most amino acids, relying on degradation of its host haemoglobin for its nutrient requirements (Babbitt *et al.*, 2012). However, its isoleucine needs are derived extracellularly owing to deficiency of this amino acid in haemoglobin. In addition, an exogenous source of methionine may also be important, at least in some parasite strains (Cobbold *et al.*, 2011). It is, therefore, plausible that isoleucine and methionine transport and utilization may provide potential targets for the development of antimalarial drugs (Istvan *et al.* 2011). Additionally, Trenholme *et al.* (2014) suggested that lysine acetylation in sexual stage malaria parasites may be a target for antimicrobial pharmaceuticals.

In *L. infantum*, tyrosine aminotransferase has been advanced as a new potential drug target (Moreno *et al.*, 2014). Tyrosine, a non-essential amino acid for humans, is readily synthesized from phenylalanine, minimizing any risk in adopting this approach for further investigations.

32.13.1 Drug/fungicide resistance

The question of drug/fungicide-resistance in pathogenic microorganisms continues to dominate the attitudes of all of those concerned with the treatment of disease, whether in humans or in crop plants. The observations of Gale and Rodwell (1948) in a paper on the amino acid metabolism of penicillin-resistant staphylococci remains a motivating factor to stimulate more research. The conclusion that GDH affects resistance to cell wall antibiotics in *B. subtilis* provides some insight into the relevance of amino acid metabolism in this context (Lee *et al.*, 2012). The regulatory roles of D-amino acids in bacteria may lead to the development of alternative therapeutics (Cava *et al.*, 2011). Tapas *et al.* (2013) provide another approach in their work on difluoromethyl ornithine-resistant ornithine decarboxylase in *Entamoeba histolytica*, which is implicated in amoebiasis, amoebic liver abscess and amoebic colitis in humans.

32.14 Summary of Outcomes

Even the most perfunctory perusal of this volume will indicate the remarkable achievements in discerning amino acid metabolism across a wide range of microorganisms and in greater depth than ever before. Recent advances also relate to both fundamental and applied aspects, with ramifications in particular micro-ecosystems. It is now instructive to summarize the salient features in an integrative perspective.

Although it is important that emerging developments are perceived and presented in terms of relevance to practical issues in food microbiology and pathogen characterization, much of the evidence reviewed in this volume underlines the case for more fundamental research.

32.14.1 Food quality

With regard to aroma development in dairy foods, Tanous *et al.* (2006) suggested that amino acid utilization, at least in *L. lactis* may be limited by the availability of α-ketoglutarate. It was concluded that glutamate dehydrogenase activity can be transmitted naturally to the microbe to rectify any substrate deficiency.

It is consistently reported that the microbial catabolism of BCAAs both contributes to and regulates the biosynthesis of volatiles and other components associated with desirable characteristics in cheese. However, tyrosine and phenylalanine are regularly linked with the formation of off-flavours. It would be over-simplistic to regard recent findings employing simulated cheese ripening conditions as conclusive due to other interactions with microorganisms and the complexity of matrix constituents. Considering that Lee and Richard (1984) isolated 42 species and strains of microorganisms from cheese, there are formidable challenges ahead in determining taste, flavour and aroma interactions, but the difficulties have not deterred investigators (Niimi *et al.*, 2016).

32.14.2 Food safety

Elucidation of the metabolism of amino acids in food-borne pathogens has been a major feature of recent developments, and with ample justification. Food-associated morbidity remains an intractable public health issue, with outbreaks documented in detail for a wide range of microorganisms in diverse environmental conditions and primary foods. The metabolism of serine in *S. typhimurium* and *C. jejuni* has been investigated in appreciable detail (Table 32.2), whereas similar characterization in other microorganisms such as *L. monocytogenes* and mycotoxin-producing fungi remains to be determined. Furthermore, studies are required to investigate the metabolism of BCAAs and aromatic amino acids in food-borne pathogens. It is consistently reported that BCAA transport and metabolism exert profound effects on virulence in diverse microbial pathogens. The characterization of tryptophan in these microorganisms may also be worthwhile in view of pathogen-specific relationships.

32.14.3 Enzyme and biochemical characterization of microorganisms

The use of metabolic criteria to characterize microorganisms remains a major feature of recent developments. The evidence presented in this volume demonstrates the potential for characterizing diverse microorganisms in terms of their amino acids and the associated enzymes of metabolism.

Arginase is the pivotal enzyme in both biosynthetic and catabolic pathways, but emerging evidence concerning its crystal structure and molecular dynamics emphasizes that it may have significant roles in determining the activity and control of mammalian parasites, including *L. amazonensis*, *P. falciparum* and *S. mansoni*.

With regard to serine characterization of food-related microorganisms, Table 32.2 summarizes the progressive advancement of research from pathways and enzyme characterization to increasing emphasis on regulation and the determination of factors affecting colonization and virulence. Identifying genes encoding key enzymes has been a major feature underlying recent developments. The use of mutant strains has enabled the elucidation of key elements of serine metabolism.

Identification and functional analysis of the gene encoding methionine-γ-lyase in *B. linens* have implicated this enzyme as exerting a key role in the formation of volatile sulfur compounds in cheese flavour development. However, interaction with the yeast *G. candidum* may influence this process.

At a fundamental level, and also reflecting interactions with other amino acids (see below), a *B. subtilis* mutant, *ccpA*, which has insufficient expression of the *ilv-leu* operon, needs BCAAs and methionine for optimal growth. The characterization of two aminotransferases in *L. lactis* indicated a broad spectrum of activity in the catabolism of BCAAs and methionine. Other results imply that at least part of the acid tolerance in *S. mutans* may be attributed to the expression of the BCAA aminotransferase encoded by the *ilvE* gene.

In short, the enzyme and biochemical characterization of a wide range of food-borne and disease pathogens in terms of amino acid metabolism has emerged as a defining feature of recent research. The evidence for serine metabolism is comprehensive, extending to important enteropathogens such as *S. typhimurium*, *S. enterica* and *C. jejuni* (Table 32.2). The metabolism of other amino acids in these microorganisms remains to be established in any significant detail, although information has been published on the inhibition of glutamine synthetase by serine, alanine and glycine in *S. typhimurium*. The biochemistry of *P. syringae* has been characterized in limited terms, including glycine betaine utilization, L-amino acid ligase and the inhibition of ornithine/arginine biosynthesis, while in *P. aeruginosa*, the transcriptional response to phenylalanine and tyrosine has been reported. In contrast, the biochemical characterization of *M. tuberculosis* is more comprehensive, and includes glutamine synthetase export, serine metabolism, structural determination of the glycine cleavage system component H, sulfur assimilation, acetohydroxyacid synthase inhibitors, BCAA aminotransferase activity, the role of the *ilvB1* gene and shikimate pathway enzymes as drug targets. The identification of the *ald* gene as encoding both alanine dehydrogenase and the putative glycine dehydrogenase adds yet another important dimension to the molecular profile of *M. tuberculosis*. The biochemical characterization of *P. falciparum* has been expressed in terms of arginase activity, but the role of the glycine cleavage complex in pathogenesis is an emerging concept; in addition, three genes encoding potential glutamate dehydrogenases have been identified.

32.14.4 Metabolic pathways

Studies with *Lactobacillus* species, *B. linens*, *G. candidum* and other microorganisms are relevant not merely for evaluating quality of dairy products but also for supplementing the mainstream pathways in the biochemistry of amino acids as presented in standard texts. Thus, methionine degradation occurs via a single step in *B. linens*, whereas a two-step pathway involving 4-methylthio-2-oxobutyric acid and 4-methylthio-2-hydroxybutyric acid intermediates is employed in *G. candidum*. The association of BCAA aminotransferases with aroma formation in *L. sakei* is considered to occur via catabolism to methyl-branched volatiles, but the further conversion of α-keto acids to the corresponding aldehydes, alcohols and carboxylic acids should also be considered: more specifically, the role of leucine catabolism in *Brevibacterium* species and the provision of precursors for the production of *S*-methyl thioesters, including *S*-methyl thioisovalerate, *S*-methyl thioacetate and *S*-methylthioisobutyrate. In *L. casei* and *L. helveticus*, phenolic amino acids are catabolized by successive, reversible transamination and dehydrogenation reactions, the major products being phenyllactate, phenylacetate, benzoic acid, ρ-hydroxyphenyllactate and ρ-hydroxyphenylacetate. In *L. casei* and *L. helveticus*, tryptophan may be converted to indole-3-lactate via successive transamination and dehydrogenation reactions. Furthermore, the action of decarboxylases in cheese-associated bacteria has been correlated with the production of biologically active amines, including tryptamine. It is now acknowledged that indole-3-acetate is one of the major products of tryptophan catabolism in microorganisms involving multiple mechanisms. In *C. purpurea*, which is associated with the synthesis of the ergot alkaloids, tryptophan provides the structure of the tetracyclic ergoline ring.

32.14.5 Antagonisms and synergism

There are numerous and diverse examples of amino acid interactions at the whole-organism level and particularly so in the molecular mechanisms of cellular metabolism. Conventional perceptions commonly relate to deleterious interactions between structurally related amino acids such as lysine and arginine, as well as mutual antagonisms among BCAAs. However, synergistic relationships between unrelated families of amino acids are now emerging as important aspects of metabolism in specialized niche microorganisms, including foodborne and disease pathogens. The Stickland reaction remains the classical model of synergism in anaerobic microorganisms (Chapter 29), but there are other significant developments worth noting. For example, Engels *et al.* (2000) characterized two aminotransferases occurring in *L. lactis* that are involved in the catabolism of both methionine and BCAAs. In *M. tuberculosis*, BCAA aminotransferase catalyses the synthesis of methionine from ketomethiobutyrate (Venos *et al.*, (2004). In addition, Kim *et al.* (2012) implicated a common transcriptional activator in methionine and leucine biosynthesis in *S. coelicolor*. In the light of other data linking homocysteine toxicity in *E. coli* with the perturbation of BCCA biosynthesis (Tuite *et al.*, 2005), there is now convincing evidence of a firm metabolic association between sulfur amino acids and BCAAs.

32.14.6 Amino acid metabolism and disease

A model for host–pathogen interactions in humans is presented in Fig. 32.1 that relates to tryptophan metabolism after infection with *C. trachomatis*. In response to this infection, tryptophan catabolism is induced by mechanisms within the host, thereby depriving the bacterium of an essential amino acid. However, in uropathogenic *E. coli*, the induction of tryptophan degradation may cause innate responses to epithelial infection to be compromised. Thus, the regulation of tryptophan biosynthesis and indole turnover is increasingly perceived as a site of interaction between the pathogen and its host. The transport and uptake of amino acids in general and BCAAs in particular are regularly implicated in the pathogenicity of *C. albicans*, *F. tularensis* and *T. cruzi*. A general model for fungal-induced stress metabolism in higher plants has been advanced (Fig. 32.2). In response to fungal attack, plants synthesize signalling molecules which, in turn, promote the deployment of defence compounds, thereby conferring resistance to the pathogen. It is to be noted here that a wide range of signalling and defence compounds in the host plant are derived from amino acids.

32.14.7 Pathogenicity and virulence

Issues of pathogenicity and virulence in disease microorganisms are frequently linked, directly or indirectly, with amino acid transport, enzymes and metabolism. For example, the amino acid ATP-binding cassette transporter has been proposed as an important pathogenesis factor in *C. jejuni* (Lin *et al.*, 2009), while the modulation of phagosomal pH, requiring *Stp2p*, a regulator of amino acid transport, appears to be an effective virulence adaptation in *C. albicans*. This pathogen utilizes imported amino acids to promote neutralization of the phagosomal pH, hyphal morphogenesis and escape from macrophages (Vylkova and Lorenz, 2014). It has been suggested that virulent strains of *F. tularensis* rely exclusively on dedicated uptake systems to satisfy their BCAA requirements (Gesbert *et al.*, 2015). Several bacterial pathogens also rely on export systems for virulence. For example, mycobacteria export significant quantities of glutamine synthetase into their extracellular environment to promote pathogenicity (Harth and Horwitz, 1999).

In terms of specific amino acids, Wu *et al.* (2013), employing a multi-omics approach, concluded that serine utilization was significantly upregulated in a hypervirulent strain of *C. jejuni*. Lee *et al.* (2013) indicated that an intact proline catabolism pathway was required for ROS homeostasis and virulence of the fungal pathogen *C. neoformans*. In a study on proline-dependent regulation of Stickland metabolism, Bouillaut *et al.* (2013) maintained that a detailed understanding of amino acid metabolism may provide important insights into the pathogenesis of *C. difficile*.

BCAA transport and metabolism exert profound effects on virulence in diverse microbial pathogens. For example, Basavanna *et al.* (2009) concluded that functional BCAA transporters were required for full virulence of *S. pneumoniae*. Furthermore, Awasthy *et al.* (2009) demonstrated that inactivation of the *ilvB1* gene in *M. tuberculosis* leads to BCAA auxotrophy and attenuation of virulence in mice. Meanwhile, Santiago *et al.* (2012) indicated that the BCAA aminotransferase encoded by *ilvE* is involved in acid tolerance in *S. mutans*, implying that this represented a major virulence factor in the formation of carious lesions. BCAA status is also important for the survival and virulence of *A. pleuropneumoniae* and, potentially, in other pulmonary pathogens (Subashchandrabose *et al.*, 2009). In addition, Cheng *et al.* (1999) implicated BCAA transaminase activity in the synthesis of HC-toxin and the pathogenicity of *C. carbonum* in maize. Liu *et al.* (2014) concluded that *fgilv5* is required for BCAA biosynthesis and full virulence in the phytopathogenic fungus *F. graminearum*. Also, the branched-chain keto acid dehydrogenase involved in the catabolism of BCAAs is required for the full virulence of *T. gondii* and *P. berghei* (Oppenheim *et al.*, 2014).

In its reduced form, sulfur is utilized for cysteine and methionine biosynthesis in microorganisms. Bhave *et al.* (2007) suggested that the acquisition and metabolism of sulfur is essential for mycobacterial virulence and survival.

Lee *et al.* (2006) proposed that anthranilate phosphoribosyltransferase (Trp D) is a metabolic enzyme that is essential for lung colonization by *M. tuberculosis*, while Li and Young (2013) implicated indole as a contributing factor in virulence.

An intriguing mechanism has been uncovered by Cui *et al.* (2013), who demonstrated that the *P. syringae* type III effector promotes pathogen virulence in *Arabidopsis* by stimulating auxin/indole protein turnover in the plant. Host–microbe relationships in tryptophan metabolism may be represented by the model presented in Fig. 32.1, but the implications for virulence may be pathogen specific (Divanovic *et al.*, 2011).

32.14.8 Drug targets

The transport, metabolism and post-translational activity of amino acids provide potential targets for the development of novel antimicrobial agents and vaccines as alternatives to conventional antibiotics in circumstances where resistance is a persistent problem or where existing prophylactics are ineffective. The quest for novel drugs continues with the increasing development of drug-resistant strains of well-known bacterial pathogens, such as *M. tuberculosis* and MRSA. In several cases, biosynthetic pathways present in microbes, but absent in the host, form the basis of such investigations. Studies on the crystal structure of key enzymes and binding sites provide additional data for the development of pharmaceuticals. The evidence presented in Table 32.3 demonstrates the diverse aspects of amino acid metabolism currently under investigation in *M. tuberculosis*. Examples of metabolic targets currently under investigation include the export of glutamine synthetase, glutamate racemase, dihydrodipicolinate reductase and acetohydroxyacid synthase, which are essential for the *de novo* synthesis of BCAAs in microbes, inactivation of the *ilvB1* gene, tryptophan biosynthesis (Chapter 20) and the shikimate pathway. The conclusion that GDH affects resistance to cell wall antibiotics in *B. subtilis* provides further insight into the relevance of amino acid metabolism in the context of drug development. With *S. aureus*, the activities of D-amino acids and haemin conjugates are currently underway in separate reviews. There may be potential in exploiting inhibitors of amino acid biosynthesis as antifungal agents, but the selection of an appropriate enzyme may be a determining factor due to alternative pathways of metabolism. In the case of *P. falciparum* there is limited evidence that isoleucine and methionine transport and utilization, as well as lysine acetylation, may provide potential targets for antimalarial drug development. In *L. infantum*, tyrosine aminotransferase has been advanced as a new potential drug target.

32.15 Outlook

The preceding chapters attest to the significant advances that have emerged over several decades in virtually all disciplines covered in this edition of the *The Handbook of Microbial Metabolism of Amino Acids*. It is abundantly clear that vibrant groups of scientists are actively pursuing a diverse set of objectives. However, work is still in progress, with several issues requiring immediate attention and in much greater depth. Each chapter in this volume provides a review of future priorities relating to the fundamental biochemistry of microorganisms, food quality and safety, human and animal health, plant pathology, drug design and ecology.

32.15.1 Constraints

The spectre of austerity is likely to dominate strategic planning for many years to come and will inevitably determine the outlook for the research programmes that have been described in *The Handbook of Microbial Metabolism of Amino Acids*. Worldwide rationalization and erosion of the research base continues unabated to the dismay of both recently qualified and highly motivated and established scientists. In the European Union member countries, for example, there appears to be a lack of strategic planning for the biological sciences, with just coincidental support for amino acid research. Several of the topics covered in this volume may well be perceived as 'near-market' and, therefore, not eligible for state funding. The question arises as to whether corporate support will emerge to compensate for the shortfall in finances.

Efforts have been thwarted by exaggerated and undeliverable claims of a number of modellers who maintain that empiricism is outdated, recommending instead a new order based on simulation and predictive methodologies designed to yield sustainable ideas and advances in the future. It has been suggested that complex biochemical pathways and interactions can only be accommodated and developed within dynamic and flexible simulation models. The potential of *in silico* methodologies in the identification of drug targets in amino acid metabolism is undeniable. Nevertheless, even within this scenario, validation through assessment of clinical efficacy is still an essential component in the design and use of novel pharmaceutical products. It is also argued that the identification of research priorities is more effectively established via logic generated within simulation models. However, it is abundantly clear that virtually all the advances reviewed in this volume have been the result of deliberate and meticulous use of empirical techniques, particularly in the development of mutants, the analysis of gene–enzyme interactions and the application of structure-based methodologies for the design of metabolic drugs. It is now increasingly clear that a combination of approaches will be necessary in the future and that opinion should not be polarized by academic debates over methodology.

32.15.2 Opportunities

It is likely that future directions will be determined by contingency rather than by forward planning, in reaction to external factors such as food poisoning episodes, chronic illnesses at the regional level, disease pandemics and biosecurity risks. Geopolitical issues arising from recent disease outbreaks indicate the need for a revised approach in order to minimize the impact of severe health and safety issues.

It is regrettable to observe that Research Council support for amino acid investigations in Europe is meagre and unlikely to generate the critical mass that is required and justified for current needs. However, as microbes evolve, so must our strategies and technologies if the continuing food safety and health issues are to be addressed in an orderly manner rather than as part of a random crisis-management response. Food-associated morbidity remains an intractable public health matter, with outbreaks documented in detail for a wide range of microorganisms, including *S. typhimurium*, *L. monocytogenes*, and *C. jejuni*, while other studies relating to *C. difficile* and *E. coli* provide evidence of additional routes and sources of contamination. Limited characterization of amino acid metabolism in these and other disease pathogens such as *M. tuberculosis* have been conducted, but there is considerable need for more research. Future practical value can only be unlocked by continuing research at the fundamental level by employing an array of techniques in molecular biology and biochemistry. These studies should be aimed at unravelling the impact of amino acid metabolism in the multifaceted aspects of pathogenicity and virulence. Further objectives should include a thorough molecular analysis of host–pathogen interactions and the development of effective treatments and vaccines, given the increasing evidence of drug-resistant strains of microbial pathogens.

References

Abendroth, J., McCormick, M.S., Edwards, T.E., Staker, B., Loewen, R., Gifford, M., Rifkin, J., Mayer, C., Guo, W., Zhang, Y. *et al.* (2010) X-ray structure determination of the glycine cleavage system protein H of *Mycobacterium tuberculosis* using an inverse Compton synchrotron X-ray source. *Journal of Structural and Functional Genomics* 11, 91–100.

Ahn, M., Murugan, R.N., Jacob, B., Hyun, J.K., Cheong, C., Hwang, E., Park, H.N., Seo, J.H., Srinivasrao, G., Lee, K.S. *et al.* (2013) Discovery of novel histidine-derived lipo-amino acids: applied in the synthesis of ultra-short antimicrobial peptidomimetics having potent antimicrobial activity, salt resistance and protease stability. *European Journal of Medicinal Chemistry* 68, 10–18.

Aiyar, A., Quayle, A.J., Buckner, L.R., Sherchand, S.P., Chang, T.L., Zea, A.H., Martin, D.H. and Belland, R.J. (2014) Influence of the tryptophan–indole-IFNγaxis on human genital *Chlamydia trachomatis* infection: role of vaginal infection. *Frontiers in Cellular and Infection Microbiology* 4:72.

Akiyama, H., Oku, N., Kasai, H., Shizuri, Y., Matsumoto, S. and Igarashi, Y. (2014) Metabolites from thermophilic bacteria I: *N*-propionylanthranilic acid, a co-metabolite of the bacillamide class antibiotics and tryptophan metabolites with herbicidal activity from *Laceyella sacchari*. *The Journal of Antibiotics* 67, 795–798.

Alkhalaf, L.M. and Ryan, K.S. (2015) Biosynthetic manipulation of tryptophan in bacteria: pathways and mechanisms. *Chemistry and Biology* 22, 317–328.

Amarita, F., Yvon, M., Nardi, M., Delettre, J. and Bonnarme, P. (2004) Identification and functional analyses of the gene encoding methionine-γ-lyase in *Brevibacterium linens*. *Applied and Environmental Microbiology* 70, 7348–7354.

Andersen, M.H. (2012) The specific targeting of immune regulation: T-cell responses against indoleamine 2,3-dioxygenase. *Cancer Immunology, Immunotherapy* 6, 1289–1297.

Anishetty, S., Pulimi, M. and Pennathur, G. (2005) Potential drug targets in *Mycobacterium tuberculosis* through metabolic pathway analysis. *Computational Biology and Chemistry* 29, 363–378.

Arai, T., Arimura, Y., Ishikura, S. and Kino, K. (2013) L-Amino acid ligase from *Pseudomonas syringae* producing tabtoxin can be used for enzymatic synthesis of various functional peptides. *Applied and Environmental Microbiology* 79, 5023–5029.

Arfi, K., Landaud, S. and Bonnarme, P. (2006) Evidence for distinct L-methionine catabolic pathways in the yeast *Geotrichum candidum* and the bacterium *Brevibacterium linens*. *Applied and Environmental Microbiology* 72, 2155–2162.

Arrebola, E., Cazorla, F.M., Durán, V.E, Rivera, E., Olea, F., Codina, J.C., Pérez-García, A. and de Vicente, A. (2003) Mangotoxin: a novel antimetabolite toxin produced by *Pseudomonas syringae* inhibiting ornithine/arginine biosynthesis. *Physiological and Molecular Plant Pathology* 63, 117–127.

Atasoglu, C. and Wallace, R.J. (2003) In: D'Mello, J.P.F. (ed.) *Amino Acids in Animal Nutrition*. CAB International, Wallingford, pp.265–290.

Awasthy, D., Gaonkar, S., Shandil, R.K., Yadav, R., Bharath, S., Marcel, N., Subbulakshmi, V. and Sharma, U. (2009) Inactivation of the *ilvB1* gene in *Mycobacterium tuberculosis* leads to branched-chain amino acid auxotrophy and attenuation of virulence in mice. *Microbiology* 155, 2978–2987.

Babbitt, S.E., Altenhofen, L., Cobbold, S.A., Istvan, E.S., Fennell, C., Doerig, C., Llinás, M. and Goldberg, D.E. (2012) *Plasmodium falciparum* responds to amino acid starvation by entering into a hibernatory state. *Proceedings of the National Academy of Sciences of the United States of America* 109, E3278–E3287.

Bashir, A., Hoffmann, T., Smits, S.H.J. and Bremer, E. (2014) Dimethylglycine provides salt and temperature stress protection to *Bacillus subtilis*. *Applied and Environmental Microbiology* 80, 2773–2785.

Basavanna, S., Khandavilli, S., Yuste, J., Cohen, J.M., Hosie, A.H.F., Webb, A.J., Thomas, G.H. and Brown, J.S. (2009) Screening of *Streptococcus pneumoniae* ABC transporter mutants demonstrates that LivJHMGF, a branched-chain ABC transporter, is necessary for disease pathogenesis. *Infection and Immunity* 77, 3412–3423.

Basso, D.R.G. and Santos, D.S. (2007) Mycobacterial shikimate pathway enzymes as target for drug design. *Current Drug Targets* 8, 423–435.

Becker, M.A., Kredich, N.M. and Tomkins, G.M. (1969) The purification and characterization of *O*-acetylserine sulfhydrylase A from *Salmonella typhimurium*. *The Journal of Biological Chemistry* 244, 2418–2427.

Bharath, S.R., Bisht, S., Savithri, H.S. and Murthy, M.R.N. (2011) Crystal structures of open and closed forms of D-serine deaminase from *Salmonella typhimurium* – implications on substrate specificity and catalysis. *FEBS Journal* 278, 2879–2891.

Bhave, D.P., Muse, W.B. III and Carroll, K.S. (2007) Drug targets in mycobacterial sulfur metabolism. *Infectious Disorders – Drug Targets* 7, 140–158.

Blackwell, B.A., Edwards, O.E., Fuchier, A., ApSimon, J.W. and Miller, J.D. (2013) NMR structural studies of fumonisin B_1 and related compounds from *Fusarium moniliforme*. *Advances in Experimental Medicine and Biology* 392, 75–92.

Bogicevic, B., Berthoud, H., Portmann, R., Meile, L. and Immler, S. (2012) *CysK* from *Lactobacillus casei* encodes a protein with *O*-acetylserine sulfhydrylase and cysteine desulfurization activity. *Applied Microbiology and Biotechnology* 94, 1209–1220.

Bouillaut, L., Self, W.T. and Sonensheim, A.L. (2013) Proline-dependent regulation of *Clostridium difficile* Stickland metabolism. *Journal of Bacteriology* 195, 844–854.

Boulette, M.L., Bayaham, P.J., Jorth, P.A., Kukavica-Ibrulj, I., Longoria, A., Barrera, K., Levesque, R.C. and Whiteley, M. (2009) Characterization of alanine catabolism in *Pseudomonas aeruginosa* and its importance for proliferation *in vivo*. *Journal of Bacteriology* 191, 6329–6334.

Brega, E., Zufferey, R. and Mamoun, C.B. (2004) *Candida albicans* Csy1p is a nutrient sensor important for activation of amino acid uptake and hyphal morphogenesis. *Eukaryotic Cell* 3, 135–143.

Bronzon da Costa, C.M., Barbosa de Freitas, M.R., Brazão, V., Domingues dos Santosa, C., Angel Sala, M., Clovis do Prado Júnior, J. and Carraro Abrahão, A.A. (2014) Does L-arginine availability during early pregnancy alter the response of *Trypanosoma cruzi* infected and pregnant Wistar rats? *Experimental Parasitology* 142, 59–66.

Bruning, J.B., Murillo, A.C., Chacon, O., Barletta, R.G. and Sacchettini, J. (2011) Structure of *Mycobacterium tuberculosis* D-alanine: D-alanine ligase, a target of the antituberculosis drug D-cycloserine. *Antimicrobial Agents and Chemotherapy* 55, 291–301.

Caldas, E.D., Sadilkova, K., Ward, B.L., Jones, A.G., Winter, C. and Gilchrist, D.G. (1998) Biosynthetic studies of fumonisin B_1 and AAL toxins. *Journal of Agricultural and Food Chemistry* 46, 4734–4743.

Campanini, B., Benoni, R., Bettati, S., Beck, C.M., Hayes, C.S. and Mozzarelli, A. (2015) Moonlighting *O*-acetylserine sulfhydrylase: new functions for an old protein. *Biochimica et Biophysica – Proteins and Proteomics* 1854, 1184–1193.

Capobianco, J.O., Zakula, D., Coen, M.L. and Goldman, R.C. (1993) Anti-*Candida* activity of cispentacin: the active transport by amino acid permeases and possible mechanisms of action. *Biochemical and Biophysical Research Communications* 190, 1037–1044.

Carreto, J.I. and Carignan, M.O. (2011) Mycosporine-like amino acids: relevant secondary metabolites, chemical and ecological aspects. *Marine Drugs* 9, 387–446.

Cava, F., Lam, H., de Pedro, M.A. and Waldor, M.K. (2011) Emerging knowledge of regulatory roles of D-amino acids in bacteria. *Cellular and Molecular Life Sciences* 68, 817–831.

Chan, V.L. and Bingham, H.L. (1991) Complete sequence of the *Campylobacter jejuni glyA* gene encoding serine hydroxymethyltransferase. *Gene* 101, 51–58.

Chan, J., Tanka, K., Carroll, D., Flynn, J. and Bloom, B.R. (1995) Effects of NO synthase inhibitors on murine infection with *Mycobacterium tuberculosis*. *Infection and Immunity* 63, 736–740.

Chattopadhyay, A., Meier, M., Burkhard, P., Speroni, F., Campanini, B., Bettati, S., Mozzarelli, A., Rabeh, W.M., Li, L. and Cook, P.F. (2007) Structure, mechanism and conformational dynamics of *O*-acetylserine sulfhydrylase from *Salmonella typhimurium*: comparison of A and B isozymes. *Biochemistry* 46, 8315–8330.

Chen, C. and Dickman, M.B. (2005) Proline suppresses apoptosis in the fungal pathogen *Colletotrichum trifolii*. *Proceedings of the National Academy of Sciences of the United States of America* 102, 3175–3176.

Chen, L., Zhao, H., Zhang, C., Lu, Y., Zhu, X. and Lu, Z. (2016) γ-Aminobutyric acid-rich yogurt fermented by *Streptococcus salivarius* subsp. Thermophiles fmb5 appears to have anti-diabetic effect on streptozotocin-induced diabetic mice. *Journal of Functional Foods* 20, 267–275.

Chen, W.-M., Prell, J., James, E.K., Sheu, D.-S. and Sheu, S.-Y. (2012) Biosynthesis of branched-chain amino acids is essential for effective symbioses between *betarhizobia* and *Mimosa pudica*. *Microbiology* 158, 1758–1766.

Cheng, Y.-Q., Ahn, J.-H. and Walton, J.D. (1999) A putative branched-chain amino acid transaminase gene required for HC-toxin biosynthesis and pathogenicity in *Cochliobolus carbonum*. *Microbiology* 145, 3539–3546.

Chungopast, S., Thapanapongworakul, P., Matsuura, H., Van Dao, T., Asahi, T., Tada, K. and Nomura, M. (2014) Glutamine synthetase I deficiency in *Mesorhizobium loti* differentially affects nodule development and activity in *Lotus japonicus*. *Journal of Plant Physiology* 171, 104–108.

Cobbold, S.A., Martin, R.E. and Kirk, K. (2011) Methionine transport in the malaria parasite, *Plasmodium falciparum*. *International Journal for Parasitology* 41, 125–135.

Cook, P.F. and Wedding, R.T. (1978) Cysteine synthetase from *Salmonella typhimurium* LT-2. Aggregation, kinetic behaviour and effect of modifiers. *The Journal of Biological Chemistry* 253, 7874–7879.

Craney, A., Ozimok, C., Capretta, A. and Nodwell, J.R. (2012) Chemical perturbation of secondary metabolism demonstrates important links to primary metabolism. *Chemistry and Biology* 19, 1020–1027.

Cui, F., Wu, S., Sun, W., Coaker, G., Kunkel, B., He, P. and Shan, L. (2013) The *Pseudomonas syringae* type III effector AvrRpt2 promotes pathogen virulence via stimulating *Arabidopsis* auxin/indole acetic acid protein turnover. *Plant Physiology* 162, 1018–1029.

Currie, A., Farber, J.M., Nadon, C., Sharma, D., Whitfield, Y., Gaulin, C., Galanis, E., Bekal, S., Flint, J., Tschetter, L. *et al.* (2015) Multi-province listeriosis outbreak linked to contaminated deli meat consumed primarily in institutional settings, Canada 2008. *Foodborne Pathogens and Disease* 12, 645–652.

Curtin, A.C. and McSweeney, P.L.H. (2003) Catabolism of aromatic amino acids in cheese-related bacteria: aminotransferase and decarboxylase activities. *Journal of Dairy Research* 70, 249–252.

Dai, Z.-L., Li,X.-L., Xi, P.-B., Zhang, J., Wu, G. and Zhu, W.-Y. (2012) Regulatory role for L-arginine in the utilization of amino acids by pig small-intestinal bacteria. *Amino Acids* 43, 233–244.

Damiano, M.A., Bastianelli, D., Al Dahouk, S., Köhler, S., Cloeckaert, A., De Biase, D. and Occhialini, A. (2015) Glutamate decarboxylase-dependent acid resistance in *Brucella* spp.: distribution and contribution to fitness under extremely acidic conditions. *Applied and Environmental Microbiology* 81, 578–586.

D'Antonio, E.L., Ullman, B., Roberts, S., Dixit, U.G., Wilson, M.E., Hai, Y. and Christianson, D.W. (2013) Crystal structure of arginase from *Leishmania mexicana* and implications for the inhibition of polyamine biosynthesis in parasitic infections. *Archives of Biochemistry and Biophysics* 535, 163–176.

da Silva, E.R., do Carmo Maquiaveli, C. and Magalhaes, P.P. (2012a) The leishmanicidal flavonols quercetin and quercitrin target *Leishmania* (*Leishmania*) *amazonensis* arginase. *Experimental Parasitology* 130, 183–188.

da Silva, M.F.L., Zampieri, R.A., Muxel, S.M., Beverley, S.M. and Floeter-Winter, L.M. (2012b) *Leishmania amazonensis* arginase compartmentalization in the glycosome is important for parasite infectivity. *PLoS ONE* 7(3): e34022.

Day, D.A., Poole, P.S., Tyerman, S.D. and Rosendahl, L. (2001) Ammonia and amino acid transport across symbiotic membranes in nitrogen-fixing legume nodules. *Cellular and Molecular Life Sciences* 58, 61–71.

Desaty, D., McInnes, A.G., Smith, D.G. and Vining, L.C. (1968) Use of ^{13}C in biosynthetic studies: incorporation of isotopically labelled acetate and aspartate into fusaric acid. *Canadian Journal of Biochemistry* 46, 1293–1300.

Divanovic, S., Sawtell, N.M., Trompette, A., Warning, J.I., Dias, A., Cooper, A.M., Yap, G.S., Arditi, M., Shimada, K., DuHadaway, J.D. *et al.* (2011) Opposing biological functions of tryptophan catabolizing enzymes during intracellular infection. *Journal of Infectious Diseases* 205, 152–161.

D'Mello, J.P.F. (2003a) Mycotoxins in cereal grains, nuts and other plant products. In: D'Mello, J.P.F. (ed.) *Food Safety: Contaminants and Toxins*. CAB International, Wallingford, pp. 65–90.

D'Mello, J.P.F. (2003b) Widespread and continuing concerns over food safety. In: D'Mello, J.P.F. (ed.) *Food Safety: Contaminants and Toxins*. CAB International, Wallingford, pp. 409–437.

D'Mello, J.P.F. (2015) Delivering innovative solutions and paradigms for a changing environment. In: D'Mello, J.P.F. (ed.) *Amino Acids in Higher Plants*. CAB International, Wallingford, pp. 538–583.

Doherty, N.C., Shen, F., Halliday, N.M., Barrett, D.R., Hardie, K.R., Winzer, K. and Atherton, J.C. (2010) In *Helicobacter pylori*, LuxS is a key enzyme in cysteine provision through a reverse transsulfuration pathway. *Journal of Bacteriology* 192, 1184–1192.

Dowling, D.P., Ilies, M., Olszewski, K.L., Portugal, S., Mota, M.M., Llinás, M. and Christianson, D.W. (2010) Crystal structure of arginase from *Plasmodium falciparum* and implications for L-arginine depletion in malarial infection. *Biochemistry* 49, 5600–5608.

Duran, R., Cary, J.W. and Calvo, A.M. (2010) Role of the osmotic stress regulatory pathway in morphogenesis and secondary metabolism in filamentous fungi. *Toxins* 2, 367–381.

Engels, W.J.M., Alting, A.C., Arntz, M.M.T.G., Gruppen, H., Voragen, A.G.J., Smit, G. and Visser, S. (2000) Partial purification and characterization of two aminotransferases from *Lactococcus lactis* subsp. *Cremoris* B78 involved in the catabolism of methionine and branched-chain amino acids. *International Dairy Journal* 10, 443–452.

Ernst, D.C., Lambrecht, J.A., Schomer, R.A. and Downs, D.M. (2014) Endogenous synthesis of 2-aminoacrylate contributes to cysteine sensitivity in *Salmonella enterica*. *Journal of Bacteriology* 196, 3335–3342.

Farias, M.E., de Ruiz Holgado, A.A.P. and Oliver, G. (1988) Characterization of L-serine dehydratase activity from *Streptococcus faecalis*. *Le Lait* 68, 177–187.

Feehily, C. and Karatzas, K.A.G. (2012) Role of glutamate metabolism in bacterial responses towards acid and other stresses. *Journal of Applied Microbiology* 114, 11–24.

Fernandes, J.D.S., Martho, K., Tofik, V. and Pascon, R.C. (2015) The role of amino acid permeases and tryptophan biosynthesis in *Cryptococcus neoformans* survival. *PLoS ONE* 10(7): e0132369.

Ferreira, A.B., de Oliveira, M.N.V., Freitas, F.S., Paiva, A.D., Alfenas-Zerbini, P., da Silva D.F., de Queiroz, M.V., Borges, A.C. and de Moraes, C.A. (2015) Involvement of the ornithine decarboxylase gene in acid stress response in probiotic *Lactobacillus delbrueckii* UFV H2b20. *Beneficial Microbes* 6, 719–725.

Flynn, J.M., Christopherson, M.R. and Downs, D.M. (2013) Decreased coenzyme A levels of *ridA* mutant strains of *Salmonella enterica* result from inactivated serine hydroxymethyltransferase. *Molecular Microbiology* 89, 751–759.

Fontaine, K., Passeró, E., Vallone, L., Hymery, N., Coton, M., Jany, J.-L., Mounier, J. and Coton, E. (2015) Occurrence of roquefortine C, mycophenolic acid and aflatoxin M_1, mycotoxins in blue-veined cheeses. *Food Control* 47, 634–640.

Ford, C.B., Shah, R.R., Maeda, M.K., Gagneux, S., Murray, M.B., Cohen, T., Johnston, J.C., Gardy, J., Lipsitch, M. and Fortune, S.M. (2013) *Mycobacterium tuberculosis* mutation rate estimates from different lineages predict substantial differences in the emergence of drug-resistant tuberculosis. *Nature Genetics* 45, 784–790.

Freiding, S., Ehrmann, M.A. and Vogel, R.F. (2012) Comparison of different ilvE aminotransferases in *Lactobacillus sakei* and investigation of their contribution to aroma formation from branched-chain amino acids. *Food Microbiology* 29, 205–214.

Friesen, T.L., Faris, J.D., Solomon, P.S. and Oliver, R.P. (2008) Host-specific toxins: effectors of necrotrophic pathogenicity. *Cellular Microbiology* 10, 1421–1428.

Fu, J., Wu, J., Jiang, J., Wang, Z. and Ma, Z. (2013) Cystathionine gamma-synthase is essential for methionine biosynthesis in *Fusarium graminearum*. *Fungal Biology* 117, 13–21.

Fu, J., Zhang, X., Chen, X., Yin, Y. and Ma, Z. (2014) Serine *O*-acetyltransferase is important but not essential for cysteine-methionine synthesis in *Fusarium graminearum*. *World Journal of Microbiology and Biotechnology* 30, 1219–1228.

Gale, E.F. and Rodwell, A.W. (1948) Amino acid metabolism of penicillin-resistant staphylococci. *Journal of Bacteriology* 55, 161–167.

Gallo, A., Bruno, K.S., Solfrizzo, M., Perrone, G., Mulè, G., Visconti, A. and Baker, S.E. (2012) New insight into the ochratoxin A biosynthetic pathway through deletion of a nonribosomal peptide synthetase gene in *Aspergillus carbonarius*. *Applied and Environmental Microbiology* 78, 8208–8218.

García-Cayuela, T., Gómez de Cadiñanos, L.P., Peláez, C. and Requena, T. (2012) Expression in *Lactococcus lactis* of functional genes related to amino acid catabolism and cheese aroma formation is influenced by branched-chain amino acids. *International Journal of Food Microbiology* 159, 207–213.

Gaulin, C., Ramsay, D., Catford, A. and Bekal, S. (2015) *Escherichia coli* O157:H7 outbreak associated with the consumption of beef and veal in the Province of Quebec, Canada in 2013. *Foodborne Pathogens in Disease* 12, 612–618.

Gerhards, N., Neubauer, L., Tudzyncki, P. and Li, S.M. (2014) Biosynthetic pathways of ergot alkaloids. *Toxins* 6, 3281–3295.

Gertsman, I., Gangoiti, J.A., Nyhan, W.L. and Barshop, B.A. (2015) Perturbations of tyrosine metabolism promote the indolepyruvate pathway via tryptophan in host and microbiome. *Molecular Genetics and Metabolism* 114, 431–437.

Gesbert, G., Ramond, E., Tros, F., Dairou, J., Frapy, E., Barel, M. and Charbit, A. (2015) Importance of branched-chain amino acid utilization in *Francisella* intracellular adaptation. *Infection and Immunity* 83, 173–183.

Giffin, M.M., Modesti, L., Raab, R.W., Wayne, L.G. and Sohaskey, C.D. (2012) *ald* of *Mycobacterium tuberculosis* encodes both the alanine dehydrogenase and its putative glycine dehydrogenase. *Journal of Bacteriology* 194, 1045–1054.

Gokulan, K., Rupp, R., Pavelka, M.S. Jr, Jacobs, W.R. Jr and Sacchettini, J.C. (2003) Crystal structure of *Mycobacterium tuberculosis* diaminopimelate decarboxylase, an essential enzyme in bacterial lysine biosynthesis. *The Journal of Biological Chemistry* 278, 18588–18596.

Graf, J. and Ruby, E.G. (1998) Host-derived amino acids support the proliferation of symbiotic bacteria. *Proceedings of the National Academy of Sciences of the United States of America* 95, 1818–1822.

Grandoni, J.A., Marta, P.T. and Schloss, J.V. (1998) Inhibitors of branched-chain amino acid biosynthesis as potential antituberculosis agents. *Journal of Antimicrobial Chemotherapy* 42, 475–482.

Griffith, R. and Hammond, E.G. (1989) Generation of Swiss cheese flavour components by the reaction of amino acids with carbonyl compounds. *Journal of Dairy Science* 72, 604–613.

Groot, M.N.N. and de Bont, J.A.M. (1998) Conversion of phenylalanine to benzaldehyde initiated by an aminotransferase in *Lactobacillus plantarum*. *Applied and Environmental Microbiology* 64, 3009–3013.

Grynberg, M., Topczewski, J., Godzik, A. and Paszewski, A. (2000) The *Aspergillus nidulans CysA* gene encodes a novel type of serine *O*-acetyltransferase which is homologous to *O*-acetyltransferases. *Microbiology* 146, 2695–2703.

Guccione, E., Leon-Kempis Mdel, R., Pearson, B.M., Hitchin, E., Mulholland, F., van Diemen, P.M., Stevens, M.P. and Kelly, D.J. (2008) Amino acid-dependent growth of *Campylobacter jejuni*: key roles for aspartase (AspA) under microaerobic and oxygen-limited conditions and identification of AspB (Cj0762), essential for growth on glutamate. *Molecular Microbiology* 69, 77–93.

Gummalla, S. and Broadbent, J.R. (1999) Tryptophan catabolism by *Lactobacillus casei* and *Lactobacillus helveticus* cheese flavour adjuncts. *Journal of Dairy Science* 82, 2070–2077.

Gummalla, S. and Broadbent, J.R. (2001) Tyrosine and phenylalanine catabolism by *Lactobacillus* cheese flavour adjuncts. *Journal of Dairy Science* 84, 1011–1019.

Gupta, V., Gupta, R.K., Khare, G., Salunke, D.M. and Tyagi, A.K. (2009) Crystal structure of Bfr A from *Mycobacterium tuberculosis*: incorporation of selenomethionine results in cleavage and demetallation of haem. *PLoS ONE* 4 (11): e8028.

Ha, S.D., Ricke, S.C., Nisbet, D.J. and Corrier, D.E. (1994) Serine utilization as a potential competition mechanism between *Salmonella* and a chicken cecal bacterium. *Journal of Food Protection* 12, 1048–1121.

Hai, Y., Edwards, J.E., Van Zandt, M.C., Hoffmann, K.F. and Christianson, D.W. (2014) Crystal structure of *Schistosoma mansoni* arginase, a potential drug target for the treatment of schistosomiasis. *Biochemistry* 53, 4671–4684.

Harth, G. and Horwitz, M.A. (1999) An inhibitor of exported *Mycobacterium tuberculosis* glutamine synthetase selectively blocks the growth of pathogenic mycobacteria in axenic culture and in human monocytes: extracellular protein as potential drug targets. *The Journal of Experimental Medicine* 189, 1425–1436.

He, Y., Li, J., Yin, N., Herradura, P.S., Martel, L., Zhang, Y., Pearson, A.L., Kulkarni, V., Mascio, C., Howland, K. *et al.* (2012) Reduced pulmonary surfactant interaction of daptomycin analogs via tryptophan replacement with alternative amino acids. *Bioorganic & Medicinal Chemistry Letters* 22, 6248–6251.

Heiman, K.E., Garalde, V.B., Gronostaj, M., Jackson, K.A., Beam, S., Joseph, L. Saupe, A., Ricotta, E., Waechter, H., Wellman, A. *et al.* (2015) Multistate outbreak of listeriosis caused by imported cheese and evidence of cross-contamination of other cheeses, USA 2012. *Epidemiology and Infection* First view, DOI: http://dx.doi.org/10.1017/S095026881500117X.

Hoffmann, T., von Blohn, C., Stanek, A., Moses, S., Barzantny, H. and Bremer, E. (2012) Synthesis, release and recapture of compatible solute proline by osmotically-stressed *Bacillus subtilis* cells. *Applied and Environmental Microbiology* 78, 5753–5762.

Hofreuter, D., Mohr, J., Wensel, O., Rademacher, S., Schreiber, K., Schomburg, D., Gao, B. and Galan, J.E. (2012) Contribution of amino acid catabolism to the tissue specific persistence of *Campylobacter jejuni* in a murine colonization model. *PLOS ONE* 7(11): e50699.

Hulanicka, M.D. and Kredich, N.M. (1976) A mutation affecting the expression of the gene coding for serine transacetylase in *Salmonella typhimurium*. *Molecular and General Genetics* 148, 143–148.

Hulanicka, M.D., Hallquist, S.G. and Kredich, N.M. (1979) Regulation of *O*-acetylserine sulfhydrylase B by cysteine in *Salmonella typhimurium*. *Journal of Bacteriology* 140, 141–146.

Hullo, M.-F., Auger, S., Soutourina, O., Barzu, O., Yvon, M., Danchin, A. and Martin-Verstraete, I. (2007) Conversion of methionine to cysteine in *Bacillus subtilis* and its regulation. *Journal of Bacteriology* 189, 187–197.

Istvan, E.S., Dharia, N.V., Bopp, S.E., Gluzman, I., Winzeler, E.A. and Goldberg, D.E. (2011) Validation of isoleucine utilization targets in *Plasmodium falciparum*. *Proceedings of the National Academy of Sciences of the United States of America* 108, 1627–1632.

Jastrzebowska, K. and Gabriel, I. (2015) Inhibitors of amino acid biosynthesis as antifungal agents. *Amino Acids* 47, 227–249.

Kaakoush, N.O., Castaño-Rodríguez, N., Mitchell, H.M. and Man, S.M. (2015) Global epidemiology of *Campylobacter* infection. *Clinical Microbiology Reviews* 28, 687–720.

Kieronczyk, A., Skeie, S., Langsrud, T. and Yvon, M. (2003) Cooperation between *Lactococcus lactis* and nonstarter lactobacilli in the formation of cheese aroma from amino acids. *Applied and Environmental Microbiology* 69, 734–739.

Kim, J.A., Park, M.S., Kang, S.A. and Ji, G.E. (2014) Production of γ-aminobutyric acid during fermentation of *Gastrodia elata* B1 by co-culture of *Lactobacillus brevis* GABA 100 with *Bifidobacterium bifidum* BG N4. *Food Science and Biotechnology* 23, 459–466.

Kim, S.H., Lee, B.R., Kim, J.N. and Kim, B.G. (2012) NdgR, a common transcriptional activator for methionine and leucine biosynthesis in *Streptomyces coelicolor*. *Journal of Bacteriology* 194, 6837–6846.

Knerr, P.J., Oman, T.J., Garcia De Gonzalo, C.V., Lupoli, T.J., Walker, S. and van der Donk, W.A. (2012) Non-proteinogenic amino acids in lacticin 481 analogues result in more potent inhibition of peptidoglycan transglycosylation. *ACS Chemical Biology* 7, 1791–1795.

Koon, N., Squire, C.J. and Baker, E.N. (2004) Crystal structure of LeuA from *Mycobacterium tuberculosis*, a key enzyme in leucine biosynthesis. *Proceedings of the National Academy of Sciences of the United States of America* 101, 8295–8300.

Kredich, N.M. (1971) Regulation of L-cysteine biosynthesis in *Salmonella typhimurium* 1. Effects of growth on varying sulphur sources and *O*-acetyl-L-serine on gene expression. *The Journal of Biological Chemistry* 246, 3474–3484.

Kredich, N.M., Foote, L.J. and Hulanicka, M.D. (1975) Studies on the mechanism of inhibition of *Salmonella typhimurium* by 1,2,4-triazole. *The Journal of Biological Chemistry* 250, 7324–7331.

Kristoffersen, T. and Nelson, F.E. (1955) Degradation of amino acids by *Lactobacillus casei* and some factors influencing deamination of serine. *Applied Microbiology* 3, 268–269.

LaRossa, R.A., Van Dyk, T.K. and Smulski, D.R. (1987) Toxic accumulation of α-ketobutyrate caused by inhibition of the branched-chain amino acid biosynthetic enzyme acetolactate synthase in *Salmonella typhimurium*. *Journal of Bacteriology* 169, 1372–1378.

Lee, C.W. and Richard, J. (1984) Catabolism of L-phenylalanine by some microorganisms of cheese origin. *Journal of Dairy Research* 51, 461–469.

Lee, C.E., Goodfellow, C., Javid-Majd, F., Baker, E.N. and Lott, J.S. (2006) The crystal structure of Trp D, a metabolic enzyme essential for lung colonization by *Mycobacterium tuberculosis* in complex with its substrate phosphoribosylpyrophosphate. *Journal of Molecular Biology* 355, 784–797.

Lee, I.R., Lui, E.Y.L., Chow, E.W.L., Arras, S.D.M., Morrow, C.A. and Fraser, J.A. (2013) Reactive oxygen species homeostasis and virulence of the fungal pathogen *Cryptococcus neoformans* require an intact proline catabolism pathway. *Genetics* 194, 421–433.

Lee, Y.H., Kingston, A.W. and Helmann, J.D. (2012) Glutamate dehydrogenase affects resistance to cell wall antibiotics in *Bacillus subtilis*. *Journal of Bacteriology* 194, 993–1001.

Li, G. and Young, K.D. (2013) Indole production by the tryptophanase TnaA in *Escherichia coli* is determined by the amount of exogenous tryptophan. *Microbiology* 159, 402–410.

Li, S., Yu, X. and Beattie, G.A. (2013) Glycine betaine catabolism contributes to *Pseudomonas syringae* tolerance to hyperosmotic stress by relieving betaine-mediated suppression of compatible solute synthesis. *Journal of Bacteriology* 195, 2415–2423.

Liaw, S.H., Pan, C. and Eisenberg, D. (1993) Feedback inhibition of fully unadenylylated glutamine synthetase from *Salmonella typhimurium* by glycine, alanine and serine. *Proceedings of the National Academy of Sciences of the United States of America* 90, 4996–5000.

Lin, A.E., Krastel, K., Thompson, S.A. and Gaynor, E.C. (2009) Atypical roles for *Campylobacter jejuni* amino acid ATP binding cassette transporter components *PaqP* and *PaqQ* in bacterial stress tolerance and pathogen–host cell dynamics. *Infection and Immunity* 77, 4912–4924.

Line, J., Hiett, K., Guard, J and Seal, B. (2011) Temperature affects sole carbon utilization patterns of *Campylobacter coli* 49941. *Current Microbiology* 62, 821–825.

Liu, L. and Shaw, P.D. (1997) Characterization of *dapB*, a gene required by *Pseudomonas syringae* pv. *tabaci* BR2.024 for lysine and tabtoxinine-beta-lactam biosynthesis. *Journal of Bacteriology* 179, 507–513.

Liu, S.Q., Holland, R. and Crow, V.L. (2002) Serine metabolism by a *Lactobacillus plantarum* adjunct. *Australian Journal of Dairy Technology* 57, 177–185.

Liu, S.Q., Holland, R., McJarrow, P. and Crow, V.L. (2003) Serine metabolism in *Lactobacillus plantarum*. *International Journal of Food Microbiology* 89, 265–273.

Liu, X., Wang, J., Xu, J. and Shi, J. (2014) *fgilv5* is required for branched-chain biosynthesis and full virulence in *Fusarium graminearum*. *Microbiology* 160, 692–702.

Liu, Y., Tang, H., Lin, Z. and Xu, P. (2015) Mechanisms of acid tolerance in bacteria and prospects in biotechnology and bioremediation. *Biotechnology Advances* 33, 1484–1492.

Lodwig, E.M., Hosie, A.H.F., Bourdès, A., Findlay, K., Allaway, D., Karunakaran, R., Downie, J.A. and Poole, P.S. (2003) Amino acid cycling drives nitrogen fixation in the legume–*Rhizobium* symbiosis. *Nature* 422, 722–726.

Lofstrom, C., Hintzmann, A.-S., Sorensen, G. and Baggesen, D.L. (2015) Outbreak of *Salmonella enterica* serovar Typhimurium phage type DT 41 in Danish poultry production. *Veterinary Microbiology* 178, 167–172.

Loughman, J.A. and Hunstead, D.A. (2012) Induction of indoleamine 2,3-dioxygenase by uropathogenic bacteria attenuates innate responses to epithelial infection. *The Journal of Infectious Diseases* 205, 1830–1839.

Ludwig, H., Meinken, C., Matin, A. and Stulke, J. (2002) Insufficient expression of the *ilv-leu* operon encoding enzymes of branched-chain amino acid biosynthesis limits growth of *Bacillus subtilis ccpA* mutant. *Journal of Bacteriology* 184, 5174–5178.

Lund, B.M. and Peck, M.W. (2015) A possible route for foodborne transmission of *Clostridium difficile*. *Foodborne Pathogens in Disease*. March 2015, 177–182.

Makala, L.H.C. (2012) The role of indoleamine 2,3-dioxygenase in regulating host immunity to *Leishmania* infection. *Journal of Biomedical Science* 19:5.

Manchola, N.C., Rapado, L.N., Barison, M.J. and Silber, A.M. (2015) Biochemical characterization of branched-chain amino acid uptake in *Trypanosoma cruzi*. *Journal of Eukaryotic Microbiology* 60, 299–308.

Manjolin, L.C., dos Reis, M.B., do Carmo Maquiaveli, C., Santos-Filho, O.A. and da Silva, E.R. (2013) Dietary flavonoids fisetin, luteolin and their derived compounds inhibit arginase, a central enzyme in *Leishmania (Leishmania) amazonensis* infection. *Food Chemistry* 14, 2253–2262.

Martinez-Cuesta, M.D.C., Pelaez, C. and Requena, T. (2013) Methionine metabolism: major pathways and enzymes involved and strategies for control and diversification of volatile sulfur compounds in cheese. *Critical Reviews in Food Science and Nutrition* 53, 366–385.

Mazumder, M. and Gourinath, S. (2016) Structure-based design of inhibitors of the crucial cysteine biosynthetic pathway enzyme *O*-acetyl serine sulfhydrylase. *Current Topics in Medicinal Chemistry* 16. DOI: 10.2174/1568026615666150825142422.

McNeil, J.B., Flynn, J., Tsao, N., Monschau, N., Stahmann, K., Haynes, R.H., McIntosh, E.M. and Pearlman, R.E. (2000) Glycine metabolism in *Candida albicans*: characterization of the serine hydroxymethyltransferase (*SHM1*, *SHM2*) and threonine aldolase (*GLY1*) genes. *Yeast* 16, 167–175.

Mendez-Cuesta, C.C., Mendez-Lucio, O. and Castillo, R. (2012) Homology modelling, docking and molecular dynamics of the *Leishmania mexicana* arginase: a description of the catalytic site useful for drug design. *Journal of Molecular Graphics and Modelling* 38, 50–59.

Miller, B.E. and Kredich, N.M. (1987) Purification of the cysB protein from *Salmonella typhimurium*. *The Journal of Biological Chemistry* 262, 6006–6009.

Mittendorf, J., Kunisch, F., Matzke, M., Militzer, H.-C., Schmidt, A. and Schonfeld, W. (2003) Novel antifungal *β*-amino acids: synthesis and activity against *Candida albicans*. *Bioorganic and Medicinal Chemistry Letters* 13, 433–436.

Moller, J.K.S., Hinrichsen, L.L. and Andersen, H.J. (1998) Formation of amino acid (L-leucine, L-phenylalanine) derived volatile flavour compounds by *Moraxella phenylpyruvica* and *Staphylococcus xylosus* in cured meat model systems. *International Journal of Food Microbiology* 42, 101–117.

Montefusco, D.J., Newcomb, B., Gandy, J.L. and Brice, S.E. (2012) Sphingoid bases and the serine catabolic enzyme CHA1 define a novel feedforward/feedback mechanism in the response to serine availability. *The Journal of Biological Chemistry* 287, 9280–9289.

Morayya, S., Awasthy, D., Yadav, R., Ambady, A. and Sharma, U. (2015) Revisiting the essentiality of glutamate racemase in *Mycobacterium tuberculosis*. *Gene* 555, 269–276.

Moreno, M.A., Alonso, P.J., Abramov, A. and Larraga, V. (2014) Tyrosine aminotransferase from *Leishmania infantum*: a new drug target candidate. *International Journal for Parasitology: Drugs and Drug Resistance* 4, 347–354.

Mujahid, M., Sasihala, C. and Ramana, C.V. (2011) Production of indole-3-acetic acid and related indole derivatives from L-tryptophan by *Rubrivivax benzoatilyticus* JA2. *Applied Microbiology and Biotechnology* 89, 1001–1008.

Nakai, T., Ishijama, J., Masui, R., Kuramitsu, S. and Kamiya, N. (2003) Structure of *Thermus thermophilus* HB8 H-protein of the glycine-cleavage system, resolved by a six-dimensional molecular replacement method. *Acta Crystallographica Section D, Biological Crystallography* 59, 1610–1618.

Naz, S., Gueguen-Minerbe, M., Cretenet, M. and Vernoux, J.P. (2013) Aromatic amino acids as precursors of antimicrobial metabolites in *Geotrichum candidum*. *FEMS Microbiology Letters* 344, 39–47.

Niimi, J., Overington, A.R., Silcock, P., Bremer, P.J. and Delahunty, C.M. (2016) Cross-modal taste and aroma interactions: cheese flavour perception and changes in flavour character in multicomponent mixtures. *Food Quality and Preference* 48, 70–80.

Nordquist, A., Nilsson, M.T., Röttger, S., Odell, L.R., Krajewski, W.W., Evalena Andersson, C., Larhed, M., Mowbray, S.L. and Karlén, A. (2008) Evaluation of the amino acid binding site of *Mycobacterium tuberculosis* glutamine synthetase for drug discovery. *Bioorganic and Medicinal Chemistry* 16, 5501–5513.

Okorochenkov, S.A., Zheltukhina, G.A., Mirchink, E.P., Isakova, E.B., Feofanov, A.V. and Nebolsin, V.E. (2013) Synthesis, anti-MRSA and anti-VRE activity of hemin conjugates with amino acids and branched peptides. *Chemical Biology and Drug Design* 82, 410–417.

Oppenheim, R.D., Creek, D.J., Macrae, J.I., Modrzynska, M., Pino, P., Limenitakis, J., Polonais, V., Seeber, F., Barrett, M.P., Billker, O. *et al.* (2014) BCKDH: the missing link in apicomplexan mitochondrial metabolism is required for full virulence of *Toxoplasma gondii* and *Plasmodium berghi*. *PLoS Pathogens* 10(7): e1004263.

Ostergaard, S., Theilgaard, H.A. and Nielsen, J. (1998) Identification and purification of *O*-acetyl-L-serine sulphhydrylase in *Penicillium chrysogenum*. *Applied Microbiology and Biotechnology* 50, 663–668.

Paiva, A.M., Vanderwall, D.E., Blanchard, J.S., Kozarich, J.W., Williamson, D.E. and Kelly, T.M. (2001) Inhibitors of dihydrodipicolinate reductase, a key enzyme of the diaminopimelate pathway of *Mycobacterium tuberculosis*. *Biochimica et Biophysica Acta (BBA) Protein Structure and Molecular Enzymology* 1545, 67–77.

Palmer, G.C., Palmer, K.L., Jorth, P.A. and Whiteley, M. (2010) Characterization of the *Pseudomonas aeruginosa* transcriptional response to phenylalanine and tyrosine. *Journal of Bacteriology* 192, 2722–2728.

Parish, T. and Stoker, N.G. (2002) The common aromatic amino acid biosynthesis pathway is essential in *Mycobacterium tuberculosis*. *Microbiology* 148, 3069–3077.

Parish, T., Gordhan, B.G., McAdam, R.A., Duncan, K., Mizrahi, V. and Stoker, N.G. (1999) Production of mutants in amino acid biosynthesis genes of *Mycobacterium tuberculosis* by homologous recombination. *Microbiology* 145, 3497–3503.

Parsons, C.V., Harris, D.M.M. and Patten, C.L. (2015) Regulation of indole-3-acetic acid biosynthesis by branched-chain amino acids in *Enterobacter cloacae* UW5. *FEMS Microbiology Letters* 362(18):fnv153. DOI: http://dx.doi.org/10.1093/femsle/fnv153.

Pieniazek, N.J., Balbin, E. and Stepien, P.P. (1974) An *Aspergillus nidulans* mutant lacking serine transacetylase: evidence for two pathways of cysteine biosynthesis. *Molecular and General Genetics* 132, 363–366.

Pinho, O., Ferreira, I.M., Mendes, E., Oliveira, B.M. and Ferreira, M. (2001) Effect of temperature on evolution of free amino acid and biogenic amine contents during storage of Azeitao cheese. *Food Chemistry* 75, 287–291.

Popp J., Noster, J., Busch, K., Kehl, A., zur Hellen, G. and Hensel, M. (2015) Role of host cell-derived amino acids in nutrition of intracellular *Salmonella enterica*. *Infection and Immunity* 83, 4466–4475.

Qiu, J., Wang, D., Ma, Y., Jiang, T. and Xin, Y. (2013) Identification and characterization of serine acetyltransferase encoded by *Mycobacterium tuberculosis* Rv 2335 gene. *International Journal of Molecular Medicine* 31, 1229–1233.

Quick, J., Ashton, P., Calus, S., Chatt, C., Gossain, S., Hawker, J., Nair, S., Neal, K., Nye, K., Peters, T. *et al.* (2015) Rapid draft sequencing and real-time nanopore sequencing in a hospital outbreak of *Salmonella*. *Genome Biology* 16, 114.

Ramond, E., Gesbert, G., Rigard, M., Dairou, J., Dupuis, M., Dubail, I., Meibom, K., Henry, T., Barel, M. and Charbit, A. (2014) Glutamate utilization couples oxidative stress defence and the tricarboxylic acid cycle in *Francisella* phagosomal escape. *PloS Pathogens* 10(1): e1003893.

Ribardo, D.A. and Hendrixson, D.R. (2011) Analysis of the LIV system of *Campylobacter jejuni* reveals alternative roles for *LivJ* and *LivK* in commensalism beyond branched-chain amino acid transport. *Journal of Bacteriology* 193, 6233–6243.

Rijnen, L., Delacroix-Buchet, A., Demaizières, D., Le Quéré, L.L., Gripon, J.C. and Yvon, M. (1999) Inactivation of lactococcal aromatic aminotransferase prevents the formation of floral aroma compounds from aromatic amino acids in semi-hard cheese. *International Dairy Journal* 9, 877–885.

Roten, O., Biran, D. and Ron, E.Z. (2013) Methionine biosynthesis in *Agrobacterium tumefaciens*: study of the first enzyme. *Research in Microbiology* 164, 12–16.

Saint-Macary, M.E., Barbisan, C., Gagey, M.J., Frelin, O., Beffa, R., Lebrun, M.H. and Droux, M. (2015) Methionine biosynthesis is essential for infection in the rice blast fungus *Magnaporthe oryzae*. *PLoS ONE* 10(4): e011110.

Saint-Marc, C., Hurlimann, H.C., Daignan-Fornier, B. and Pinson, B. (2015) Serine hydroxymethyltransferase: a key player connecting purine, folate and methionine metabolism in *Saccharomyces cerevisiae*. *Current Genetics* 61, 633–640.

Salcedo, E., Sims, P.F.G. and Hyde, J.E. (2005) A glycine-cleavage complex as part of the folate one-carbon metabolism of *Plasmodium falciparum*. *Trends in Parasitology* 21, 406–411.

Sanchez, C.J. Jr, Akers, K.S., Romano, D.R., Woodbury, R.L., Hardy, S.K., Murray, C.K. and Wenke, J.C. (2014) D-Amino acids enhance the activity of antimicrobials against biofilms of clinical wound isolates of *Staphylococcus aureus* and *Pseudomonas aeruginosa*. *Antimicrobial Agents and Chemotherapy* 58, 4353–4361.

Sanglard, D., Kuchler, K., Ischer, F., Pagani, J.L., Monod, M. and Bille, J. (1995) Mechanism of resistance to azole antifungal agents in *Candida albicans* isolates from AIDS patients involve specific multidrug transporters. *Antimicrobial Agents and Chemotherapy* 39, 2378–2386.

Santiago, B., MacGilvray, M., Faustoferri, R.C. and Quivey, R.G. (2012) The branched-chain amino acid aminotransferase encoded by *ilvE* is involved in acid tolerance in *Streptococcus mutans*. *Journal of Bacteriology* 194, 2010–2019.

Santiago, B., Marek, M., Faustoferri, R.C. and Quivey, R.G. (2013) The *Streptococcus mutans* aminotransferase encoded by *ilvE* is regulated by CodY and CcpA. *Journal of Bacteriology* 195, 3552–3562.

Scott, M.K., Geissler, A., Poissant, T., De Bess, E., Melius, B., Eckmann, K., Salehi, E. and Cieslak, P.R. (2015) Notes from the field: campylobacteriosis outbreak associated with consuming undercooked chicken liver pate – Ohio and Oregon, December 2013–January 2014. *Morbidity and Mortality Weekly Report* 64(14): 399.

Shafiani, S., Sharma, P., Vohra, R.M. and Tewari, R. (2005) Cloning and characterization of aspartate-β-semialdehyde dehydrogenase from *Mycobacterium tuberculosis* H37Rv. *Journal of Applied Microbiology* 98, 832–839.

Singh, V., Chandra, D., Srivastava, B.S. and Srivastava, R. (2011) Biochemical and transcriptional analysis of acetohydroxyacid synthase isoforms in *Mycobacterium tuberculosis* identifies these enzymes as potential targets for drug development. *Microbiology* 157, 29–37.

Sourabié, A.M., Spinnier, H.E., Bourdat-Deschamps, M., Tallon, R., Landaud, S. and Bonnarme, P. (2012) *S*-Methyl thioesters are produced from fatty acids and branched-chain amino acids by brevibacteria: focus on L-leucine catabolic pathway and identification of acyl-CoA intermediates. *Applied Microbiology and Biotechnology* 93, 1673–1683.

Stadelmann, B., Hanevik, K., Andersson, M.K., Bruserud, O., Svärd, S.G. (2013) The role of arginine and arginine-metabolizing enzymes during *Giardia*–host cell interactions *in vitro*. *BMC Microbiology* 13:256.

Stauffer, G.V. and Brenchley, J.E. (1974) Evidence for the involvement of serine transhydroxymethylase in serine and glycine interconversions in *Salmonella typhimurium*. *Genetics* 77, 185–198.

Stauffer, G.V. and Brenchley, J.E. (1977) Influence of methionine biosynthesis on serine transhydroxymethylase regulation in *Salmonella typhimurium* LT2. *Journal of Bacteriology* 129, 740–749.

Stauffer, G.V., Baker, C.A. and Brenchley, J.E. (1974) Regulation of serine transhydroxymethylase activity in *Salmonella typhimurium*. *Journal of Bacteriology* 120, 1017–1025.

Stauffer, G.V., Stauffer, L.T. and Plamann, M.D. (1989) The *Salmonella typhimurium* glycine cleavage enzyme system. *Molecular and General Genetics* 220, 154–156.

Storm, J., Perner, J., Aparicio, I., Patzewitz, E.M., Olszewski, K., Llinas, M., Engel, P.C. and Müller S. (2011) *Plasmodium falciparum* glutamate dehydrogenase a is dispensable and not a drug target during erythrocytic development. *Malaria Journal* 10:193.

Stuttmann, J., Hubberten, H.-M., Rietz, S., Kaur, J., Muskett, P., Guerois, R., Bednarek, P., Hoefgen, R. and Parker, J.E. (2011) Perturbation of *Arabidopsis* amino acid metabolism causes incompatability with the adapted biotrophic pathogen *Hyaloperonospora arabidopsidis*. *The Plant Cell* 23, 2788–2803.

Subashchandrabose, S., LeVeque, R.M., Wagner, T.K., Kirkwood, R.N., Kiupel, M. and Mulks, M.H. (2009) Branched-chain amino acids are required for the survival and virulence of *Actinobacillus pleuropneumoniae* in swine. *Infection and Immunity* 77, 4925–4933.

Suliman, M.S., Appling, D.R. and Robertus, J.D. (2007) The gene for cobalamin-independent methionine synthase is essential in *Candida albicans*: a potential antifungal target. *Archives of Biochemistry and Biophysics* 467, 218–226.

Tanous, C., Chambellon, E., Le Bars, D., Delespaul, G. and Yvon, M. (2006) Glutamate dehydrogenase activity can be transmitted naturally to *Lactococcus lactis* strains to stimulate amino acid conversion to aroma compounds. *Applied and Environmental Microbiology* 72, 1402–1409.

Tapas, P.S., Tomar, P.K.S. and Madhubala, R. (2013) Structural insight into DFMO-resistant ODC from *Entamoeba histolytica*: an inkling to adaptive evolution. *PLoS ONE* 8(1): e53397.

Trenholme, K., Marek, L., Duffy, S., Pradel, G., Fisher, G., Hansen, F.K., Skinner-Adams, T.S., Butterworth, A., Ngwa, C.J., Moecking, J. *et al.* (2014) Lysine acetylation in sexual stage malaria parasites is a target for antimalarial small molecules. *Antimicrobial Agents and Chemotherapy* 58, 3666–3678.

Tuite, N.L., Fraser, K.R., and O'Byrne, C.P. (2005) Homocysteine toxicity in *Escherichia coli* is caused by a perturbation of branched-chain amino acid biosynthesis. *Journal of Bacteriology* 187, 4362–4371.

Turner, J.G. (1981) Tabtoxin produced by *Pseudomonas tabaci* decreases *Nicotiana tabacum* glutamine synthetase *in vivo* and causes accumulation of ammonia. *Physiological Plant Pathology* 19, 57–61.

Ullrich, M. and Bender, C.L. (1994) The biosynthetic gene cluster for coronamic acid, an ethylcyclopropyl amino acid, contains genes homologous to amino acid-activating enzymes and thioesterases. *Journal of Bacteriology* 176, 7574–7586.

Umbarger, H.E., Umbarger, M.A. and Siu, P.M.L. (1963) Biosynthesis of serine in *Escherichia coli* and *Salmonella typhimurium*. *Journal of Bacteriology* 85, 1431–1439.

Varga, J., Rigo, K., Teren, J. and Mesterhazy, A. (2001) Recent advances in ochratoxin research II. Biosynthesis, mode of action and control of ochratoxin. *Cereal Research Communications* 29, 93–100.

Velayudhan, J., Jones, M.A., Barrow, P.A. and Kelly, D.J. (2004) L-serine catabolism via an oxygen-labile-L-serine dehydratase is essential for colonization of the avian gut by *Campylobacter jejuni*. *Infection and Immunity* 72, 260–268.

Venos, E.S., Knodel, M.H., Radford, C.L. and Berger, B.J. (2004) Branched-chain amino acid aminotransferase and methionine formation in *Mycobacterium tuberculosis*. *BMC Microbiology* 4:39.

Vorwerk, H., Mohr, J., Huber, C., Wensel, O., Schmidt-Hohagen, K., Gripp, E., Josenhans, C., Schomburg, D., Eisenreich, W. and Hofreuter, D. (2014) Utilization of host-derived cysteine-containing peptides overcomes the restricted sulphur metabolism of *Campylobacter jejuni*. *Molecular Microbiology* 93, 1224–1245.

Vylkova, S. and Lorenz, M.C. (2014) Modulation of phagosomal pH by *Candida albicans* promotes hyphal morphogenesis and requires Stp 2p, a regulator of amino acid transport. *PLoS Pathogens* 10(3), e 1003995.

Wagley, S., Newcombe, J., Laing, E., Yusuf, E., Sambles, C.M., Studholme, D.J., La Ragione, R.M., Titball, R.W. and Champion, O.L. (2014) Differences in carbon source utilization distinguish *Campylobacter jejuni* from *Campylobacter coli*. *BMC Microbiology* 14:262.

Wallace, J.M. and Fox, P.F. (1998) Effect of adding free amino acids to Cheddar cheese curd on flavour development. *Developments in Food Science* 40, 559–572.

Wang, D.I., Zhu, X., Cui, C., Dong, M., Jiang, H., Li, Z., Liu, Z., Zhu, W. and Wang, J.-G. (2013) Discovery of novel acetohydroxyacid synthase inhibitors as active agents against *Mycobacterium tuberculosis* by virtual screening and bioassay. *Journal of Chemical Information and Modeling* 53, 343–353.

Wu, Q., Law, Y.-S. and Shah, N.P. (2015) Dairy *Streptococcus thermophilus* improves cell viability of *Lactobacillus brevis* NPS-QW-145 and its γ-aminobutyric acid biosynthesis ability in milk. *Scientific Reports* 5: 12885.

Wu, Z., Sahin, O., Shen, Z., Liu, P., Miller, W.G. and Zhang, Q. (2013) Multi-omics approaches to deciphering a hypervirulent strain of *Campylobacter jejuni*. *Genome Biology and Evolution* 5, 2217–2230.

Young, V.R. and Ajami, A.M. (2000) Glutamate: an amino acid of particular distinction. *Journal of Nutrition* 130, 892S–900S.

Yvon, M. and Rijnen, L. (2001) Cheese flavour formation by amino acid catabolism. *International Dairy Journal* 11, 185–201.

Yvon, M., Thirouin, S., Rijnen, L., Fromentier, D. and Gripon, J.C. (1997) An aminotransferase from *Lactococcus lactis* initiates conversion of amino acids to cheese flavour compounds. *Applied and Environmental Microbiology* 63, 414–419.

Yvon, M., Berthelot, S. and Gripon, J.G. (1998) Adding α-ketoglutarate to semi-hard cheese curd highly enhances the conversion of amino acids to aroma compounds. *International Dairy Journal* 8, 889–898.

Zaprasis, A., Brill, J., Thüring, M., Wünsche, G., Heun, M., Barzantny, H., Hoffmann, T. and Bremer (2013) Osmoprotection of *Bacillus subtilis* through the import and proteolysis of proline-containing peptides. *Applied and Environmental Microbiology* 79, 576–587.

Zaprasis, A., Hoffmann, T., Stannek, L., Gunka, K., Commichau, F.M. and Bremer, E. (2014) The γ-aminobutyrate permease GabP serves as the third proline transporter of *Bacillus subtilis*. *Journal of Bacteriology* 196, 515–526.

Zaprasis, A., Bleisteiner, M., Kerres, A., Hoffmann, T. and Bremer, E. (2015) Uptake of amino acids and their metabolic conversion into the compatible solute proline confers osmoprotection to *Bacillus subtilis*. *Applied and Environmental Microbiology* 81, 250–259.

Zhang, Y.J., Reddy, M.C., Ioerger, T.R., Rothchild, A.C., Dartois, V., Schuster, B.M., Trauner, A., Wallis, D., Galaviz, S., Huttenhower, C. *et al.* (2013) Tryptophan biosynthesis protects mycobacteria from CD4 T-cell-mediated killing. *Cell* 155, 1296–1308.

Zohar, Y., Einav, M., Chipman, D.M. and Barak, L. (2003) Acetohydroxyacid synthase from *Mycobacterium avium* and its inhibition by sulfonylureas and imidazolinones. *Biochimica et Biophysica Acta (BBA) Proteins and Proteomics* 1649, 97–105.

Zummo, F.P., Marineo, S., Pace, A., Civiletti, F., Giardina, A. and Puglin, A.M. (2012) Tryptophan catabolism via kynurenine production in *Streptomyces coelicolor*: identification of three genes coding for the enzymes of tryptophan to anthranilate pathway. *Applied Microbiology and Biotechnology* 94, 719–728.

Index

Page numbers in **bold** type refer to figures and tables.